Programming the Finite Element Method

Third Edition

Programming the Finite Element Method

Third Edition

I. M. Smith
University of Manchester, UK

D. V. Griffiths
Colorado School of Mines, USA

JOHN WILEY & SONS

Chichester · New York · Weinheim · Brisbane · Singapore · Toronto

National 01243 779777
International (+44) 1243 779777
e-mail (for orders and customer service enquiries): cs-books@wiley.co.uk
Visit our Home Page on http://www.wiley.co.uk
or http://www.wiley.com

Third edition reprinted April 1998

Front cover illustration: Finite element model of excavations for Large Hadron Collider facility at CERN, Switzerland. Courtesy Golder Associates.

Other Wiley Editorial Offices

John Wiley & Sons, Inc., 605 Third Avenue,
New York, NY 10158-0012, USA

Wiley-VCH Verlag GmbH, Pappelallee 3,
D-69469 Weinheim, Germany

Jacaranda Wiley Ltd, 33 Park Road, Milton,
Queensland 4064, Australia

John Wiley & Sons (Canada) Ltd, 22 Worcester Road,
Rexdale, Ontario M9W 1L1, Canada

John Wiley & Sons (Asia) Pte Ltd, 2 Clementi Loop #02-01,
Jin Xing Distripark, Singapore 129809

Library of Congress Cataloging-in-Publication Data

Smith, I. M.
Programming the finite element method/I. M. Smith and D. V. Griffiths.—3rd ed.
p. cm.
Includes bibliographical references and index.
ISBN 0-471-96542-1 (cloth).—ISBN 0-471-96543-X (pbk.)
1. Finite element method—Computer programs. 2. Subroutines (Computer programs) 3. Computer-aided engineering. I. Griffiths, D. V. II. Title.
TA347.F5S64 1997
620′.001′51535—DC21 96-54905
CIP

British Library Cataloguing in Publication Data

A catalogue record for this book is available from the British Library

ISBN 0-471-96542-1 0-471-96543-X (pbk).

Typeset in 10/12pt Times from the author's disks by Thomson Press (India) Limited, New Delhi
Printed and bound in Great Britain by Biddles Ltd, Guildford and King's Lynn.
This book is printed on acid-free paper responsibly manufactured from sustainable forestation, for which at least two trees are planted for each one used for paper production.

Contents

Preface to Third Edition

Building on the success of the first two editions, significant modifications and improvements have been made. Foremost amongst these are the complete rewriting of all subroutine libraries and example programs in Fortran 90, and the inclusion of "mesh-free" or "element-by-element" methods in all chapters to supplement the more traditional "mesh-dependent" or "global assembly" methods. Examples of "analytical" rather than numerical element integration are given.

Examples for students to solve are included at the end of most chapters, making the book more suitable for use as a teaching text.

Chapter 1 is devoted to a reasonably lengthy exposition of modern programming considerations and of the principal features of the new standard, Fortran 90, which are useful in finite element programming.

In Chapter 3, iterative techniques, in particular the preconditioned conjugate gradient method, are introduced for the solution of the linear systems of algebraic equations which arise from the analysis of problems of static equilibrium. It is shown how these facilitate an "element-by-element" approach to the the solution of very large problems within the memory limitations of relatively small computers. Similar techniques carry over to eigenvalue and transient problems and it is shown that Fortran 90 is a particularly suitable programming vehicle.

In Chapter 5, issues of vectorisation and parallelisation of programs are addressed for the first time, and analytical element integration is provided as an alternative to conventional numerical quadrature.

The book is intended for general use in a wide variety of finite element applications but in Chapter 6 the treatment of plasticity is aimed especially at applications in geomechanics. Because of this, construction processes involving building up and excavating are added for the first time. Also, in Chapter 9, nonlinear coupled problems of interest in geomechanics are new additions.

Although pre-processing (mesh generation) and post-processing (results display) programs are not included, examples of post-processing illustrate the power of these techniques. In practice, displays involving for example animation will be essential, particularly for large problems. To assist with pre- and post-processing, general programs are included for the first time. These enable data to be supplied by user-provided mesh-generators rather than by the simple "geometry" subroutines which are intended more for explanatory and teaching purposes.

COMPUTER PROGRAMS

All software (libraries and programs) described in this book are available on ftp at ftp://golden.eng.man.ac.uk/pub/fe and on the web at http://www.eng.man.ac.uk/geotech/software.

1 Preliminaries: Computer Strategies

1.0 INTRODUCTION

Many textbooks exist which describe the principles of the finite element method of analysis and the wide scope of its applications to the solution of practical engineering problems. Usually, little attention is devoted to the construction of the computer programs by which the numerical results are actually produced. It is presumed that readers have access to pre-written programs (perhaps to rather complicated "packages") or can write their own. However, the gulf between understanding in principle what to do, and actually doing it, can still be large for those without years of experience in this field.

The present book bridges this gulf. Its intention is to help readers assemble their own computer programs to solve particular engineering problems by using a "building block" strategy specifically designed for computations via the finite element technique. At the heart of what will be described is not a "program" or a set of programs but rather a collection (library) of procedures or subroutines which perform certain functions analogous to the standard functions (`SIN`, `SQRT`, `ABS`, etc.) provided in permanent library form in all useful scientific computer languages. Because of the matrix structure of finite element formulations, most of the building block routines are concerned with manipulation of matrices.

The building blocks are then assembled in different patterns to make test programs for solving a variety of problems in engineering and science. The intention is that one of these test programs then serves as a platform from which new applications programs are developed by interested users.

The aim of the present book is to teach the reader to write intelligible programs and to use them. Super-efficiency has not been paramount in program construction. However, all building block routines (numbering over 100) and all test programs (numbering over 50) have been verified on a wide range of computers.

The chosen programming language is the latest dialect of FORTRAN, called Fortran 90. Later in this chapter, a fairly full description of the features of Fortran 90 which influence the programming of the finite element method will be given. At present, all that need be said is that Fortran 90 represents a very radical improvement compared with the previous standard, FORTRAN 77 (which was used in earlier editions of this book), and that Fortran remains, overwhelmingly, the most popular language for writing large engineering and scientific programs.

1.1 HARDWARE

In principle, any computing machine capable of compiling and running Fortran programs can execute the finite element analyses described in this book. In practice, hardware will range from personal computers for more modest analyses and teaching purposes to "super" computers for very large (especially nonlinear three-dimensional (3-d)) analyses. It is a powerful feature of the programming strategy proposed that the *same* software will run on all machine ranges. The special features of vector and parallel processors are described in a separate section (1.3).

The user's choice of hardware is a matter of accessibility and of cost. For example, in very rough terms, a personal computer costing £1.5K ($2.5K) has a processing speed about 10 times slower than that of a "workstation" costing £15K ($25K). Thus a job taking five minutes on the "workstation" takes one hour on the PC. Which hardware is "better" clearly depends on individual circumstances. The main advice that can be tendered is against using hardware that is too weak for the task; that is, the user is advised not to operate at the extremes of the hardware's capability. If this is done turn-round times become too long to be of value in any design cycle.

1.2 MEMORY MANAGEMENT

In the programs in this book it will be assumed that sufficient main random access memory is available for the storage of data and the execution of programs. However, the arrays processed in finite element calculations might be of size, say, 100 000 by 1000. Thus a computer would have to have a main memory of 1×10^8 words to hold this information, and while some such computers exist, they are still comparatively rare. A much more typical memory size is still of the order of 1×10^7 words.

One strategy to get round this problem is for the programmer to write "out of memory" routines which arrange for the processing of chunks of arrays in memory and the transfer of the appropriate chunks to and from back-up storage.

Alternatively, store management is removed from the user's control and given to the system hardware and software. The programmer sees only a single level of memory of very large capacity and information is moved from secondary memory to main memory and out again by the supervisor or executive program which schedules the flow of work through the machine. This concept, namely of a very large "virtual" memory, was first introduced on the ICL ATLAS in 1961, and is now very common.

Clearly it is necessary for the system to be able to translate the virtual address of variables into a real address in memory. This translation usually involves a complicated bit-pattern matching called "paging". The virtual store is split into segments or pages of fixed or variable size referenced by page tables, and the supervisor program tries to "learn" from the way in which the user accesses data in order to manage the store in a predictive way. However, memory management can *never* be totally removed from the user's control. It must always be assumed that the programmer is acting in a reasonably logical manner, accessing array elements in sequence (by rows or columns as organised by the compiler and the language). If the user accesses a virtual memory of 10^8 words in a random fashion the paging requests will ensure that very little execution of the program can take place (see e.g. Willé, 1995).

In the immediate future, "large" finite element analyses, say involving more than 100 000 unknowns, are likely to be processed by the vector and parallel processing hardware described in the next section. When using such hardware there is usually a considerable time penalty if the programmer interrupts the flow of the computation to perform out-of-memory transfers or if automatic paging occurs. Therefore, in Chapter 3 of this book, special strategies are described whereby large analyses can still be processed "in memory".

1.3 VECTOR AND PARALLEL PROCESSORS

Early digital computers performed calculations "serially", that is, if a thousand operations were to be carried out, the second could not be initiated until the first had been completed and so on. When operations are being carried out on arrays of numbers, however, it is perfectly possible to imagine that computations in which the result of an operation on two array elements has no effect on an operation on another two array elements, can be carried out simultaneously. The hardware feature by means of which this is realised in a computer is called a "pipeline" and in general all modern computers use this feature to a greater or lesser degree. Computers which consist of specialised hardware for pipelining are called "vector" computers. The "pipelines" are of limited length and so for operations to be carried out simultaneously it must be arranged that the relevant operands are actually in the pipeline at the right time. Furthermore the condition that one operation does not depend on another must be respected. These two requirements (amongst others) mean that some care must be taken in writing programs so that best use is made of the vector processing capacity of many machines. It is moreover an interesting side-effect that programs well structured for vector machines will tend to run better on any machine because information tends to be in the right place at the right time (in a special cache memory for example) and modern so-called "scalar" computers tend to contain some vector-type hardware. In this book, beginning at Chapter 5, programs which "vectorise" well will be illustrated.

True vector hardware tends to be expensive and an alternative means of achieving simultaneous processing of parts of arrays of data is called "parallel" computation. In this concept (of which there are many variants) there are several physically distinct processors (a few expensive ones or a lot of cheaper ones for example). Programs and/or data can reside on different processors which have to communicate with one another.

There are two foreseeable ways in which this communication can be organised (rather like memory management which was described earlier). Either the programmer takes control of the communication process, using a programming feature called "message passing", or it is done automatically, without user control. The second strategy is of course appealing and has led to the development of "High Performance Fortran" or HPF (see e.g. Koelbel *et al.*, 1995) which has been designed as an extension to Fortran 90. "Directives", which are treated as comments by non-HPF compilers, are inserted into the Fortran 90 programs and allow data to be mapped on to parallel processors together with the specification of the operations on such data which can be carried out in parallel. The attractive feature of this strategy is that programs are "portable", that is they can be easily transferred from computer to computer. One would also anticipate that manufacturers could produce compilers which made best use of their specific type of hardware. At the time of writing, the first implementations of HPF are just being reported.

The alternative strategy of message passing under programmer control can be realised in an environment such as MPI ("message passing interface"). This is also intended to preserve portability between machines but it does involve very significant additional programming and certainly destroys the compact appearance of the non-parallel programs which make up the bulk of the present book.

An example of a finite element program, with MPI additions, is given in Chapter 5. The hope is that HPF will ultimately be successful but, in the mean time, there is a regrettable lack of standardisation and many parallel machine manufacturers provide their own specific message-passing environment.

1.4 SOFTWARE

Since all computers have different hardware (instruction formats, vector capability, etc.) and different store management strategies, programs which would make the most effective use of these varying facilities would of course differ in structure from machine to machine. However, for excellent reasons of program portability and programmer training, engineering computations on all machines are usually programmed in "high-level" languages which are intended to be machine-independent.

The high-level language is translated into the machine order code by a program called a "compiler".

FORTRAN is by far the most widely used language for programming engineering and scientific calculations and in this section the principal features of the latest standard, called Fortran 90, will be described with particular reference to features of the language which are useful in finite element computations.

Shown in Figure 1.1 is a typical simple program written in Fortran 90 (Smith, 1995). It concerns an opinion poll survey and serves to illustrate the basic structure of the language for those used to its predecessor, FORTRAN 77, or to other languages.

It can be seen that programs are written in "free source" form. That is, statements can be arranged on the page or screen at the user's discretion. Other features to note are:

- Upper and lower case characters may be mixed at will.
- Multiple statements can be placed on one line, separated by ";".

```
PROGRAM GALLUP_POLL
! TO CONDUCT A GALLUP POLL SURVEY
IMPLICIT NONE
INTEGER : :  SAMPLE, I, COUNT, THIS_TIME, LAST_TIME, TOT_REP, TOT_ &
             MAV, TOT_DEM, TOT_OTHER, REP_TO_MAV, DEM_TO_MAV, &
             CHANGED_MIND
READ*, SAMPLE
COUNT=0; TOT_REP=0; TOT_MAV=0; TOT_DEM=0; TOT_OTHER=0; REP_ &
       TO_MAV=0
DEM_TO_MAV=0; CHANGED_MIND=0
OPEN (10, FILE='GALLUP.DAT')
DO I = 1, SAMPLE
  COUNT=COUNT+1
  READ (10, '(I3,I2)', ADVANCE='NO') THIS_TIME, LAST_TIME
  VOTES: SELECT CASE (THIS_TIME)
    CASE (1); TOT_REP=TOT_REP+1
    CASE (3); TOT_MAV=TOT_MAV+1
     IF (LAST_TIME/=3) THEN
       CHANGED_MIND=CHANGED_MIND+1
       IF (LAST_TIME==1)REP_TO_MAV=REP_TO_MAV+1
       IF (LAST_TIME==2)DEM_TO_MAV=DEM_TO_MAV+1
     END IF
    CASE (2); TOT_DEM=TOT_DEM+1
    CASE DEFAULT; TOT_OTHER=TOT_OTHER+1
END SELECT VOTES
END DO
  PRINT*,  'PERCENT REPUBLICAN IS', REAL (TOT_REP)/REAL (COUNT)*100.0
  PRINT*,  'PERCENT MAVERICK IS', REAL (TOT_MAV)/REAL(COUNT)*100.0
  PRINT*,  'PERCENT DEMOCRAT IS', REAL (TOT_MAV)/REAL(COUNT)*100.0
  PRINT*,  'PERCENT OTHERS IS', REAL (TOT_OTHER)/REAL(COUNT)*100.0
  PRINT*,  'PERCENT CHANGING REP TO MAV IS', REAL (REP_TO_MAV)/&
            REAL (CHANGED_MIND)*100.0
  PRINT*,  'PERCENT CHANGING DEM TO MAV IS', REAL (DEM_TO_MAV)/&
           REAL(CHANGED_MIND)*100.0
END PROGRAM GALLUP_POLL
```

Figure 1.1 A typical program written in Fortran 90

- Long lines can be extended by "&" at the end of the line, and ideally another "&" at the start of the continuation line(s).
- Comments, placed after "!", are ignored.
- Long names (up to 31 characters, including the underscore) allow meaningful identifiers.
- The "`IMPLICIT NONE`" statement forces the declaration of all variable and constant names. This is a great help in debugging programs.
- Declarations involve the "::" double colon convention.
- There are no labelled statements.

1.4.1 ARITHMETIC

Finite element processing is computationally intensive (see e.g. Chapters 6 and 10) and a reasonably safe numerical precision to aim for is that provided by a 64-bit machine word length. Fortran 90 contains some useful intrinsic procedures for determining, and changing, processor precision. For example the statement

```
IWP=SELECTED_REAL_KIND(P=15)
```

would return an integer `IWP` which is the "`KIND`" of variables on a particular processor which is necessary to achieve 15 decimal place precision. If the processor cannot achieve this order of accuracy, `IWP` would be returned as negative.

Having established the necessary value of `IWP`, Fortran 90 declarations of `REAL` quantities then take the form

```
REAL(KIND=IWP)::A,B,C
```

and assignments the form

```
A=1.0_IWP; B=2.0_IWP; C=3.0_IWP
```

and so on.

To maintain conciseness, the programs listed in this book make the assumption that the default `KIND` value is adequate and so the `_IWP` appendage is not necessary. Readers are warned that the listed results were obtained on a 64-bit processor and that deviations from these results will be obtained on processors with shorter word lengths. Indeed, in some sensitive calculations, less than 64-bit computation may lead to seriously erroneous results or to a failure to arrive at any result at all.

1.4.2 CONDITIONS

There are two basic structures for conditional statements in Fortran 90. The first corresponds to the classical `IF ... THEN ... ELSE` structure found in most high-level languages. It takes the form

```
NAME_OF_CLAUSE:IF (logical expression) THEN
                                  .     first block
                                  .     of statements
                              ELSE
                                  .     second block
                                  .     of statements
END IF NAME_OF_CLAUSE
```

For example

```
CHANGE_SIGN: IF(A/=B) THEN
                                A=-A
                            ELSE
                                B=-B
END IF CHANGE_SIGN
```

The name of the conditional statement, NAME_OF_CLAUSE or CHANGE_ SIGN in the above examples, is optional and can be left out.

The second conditional structure is illustrated in the program in Figure 1.1 and involves the "SELECT CASE" construct. If choices are to be made in particularly simple circumstances, for example, an INTEGER, LOGICAL or CHARACTER scalar has a given value, then the form

```
SELECT_CASE_NAME: SELECT CASE (VARIABLE or EXPRESSION)
                        CASE (SELECTOR)
                     first block of statements
                        CASE (SELECTOR)
                     second block of statements
                              .
                              .
                              .
                         CASE DEFAULT
                     default block of statements
END SELECT_CASE_NAME
```

can be used. This replaces the ugly "computed go to" construct in FORTRAN 77.

1.4.3 LOOPS

There are two constructs in Fortran 90 for repeating blocks of instructions. In the first, the block is repeated a fixed number of times, for example

```
FIXED_ITERATIONS: DO I = 1,N
                                block of statements
END DO FIXED_ITERATIONS
```

In the second, the loop is left or continued depending on the result of some condition. For example

```
EXIT_TYPE: DO
                  block of statements
                    IF (conditional statement) EXIT
                  block of statements
END DO EXIT_TYPE
```

or

```
CYCLE_TYPE: DO
                  block of statements
                    IF (conditional statement) CYCLE
                  block of statements
END DO CYCLE_TYPE
```

The first variant transfers control out of the loop to the first statement after END DO. The second variant transfers control to the beginning of the loop, skipping the remaining statements between CYCLE and END DO.

In the above examples, as was the case for conditions, the naming of the loops is optional. In the programs in this book, loops and conditions of major significance tend to be named and simpler ones not.

1.4.4 ARRAY FEATURES

1.4.4.1 Dynamic arrays

Fortran 90 has remedied perhaps the greatest deficiency of earlier FORTRANs for large-scale array computations such as occur in finite element analysis in that it allows "dynamic" declaration of arrays. That is, array sizes do not have to be specified at program compilation time but can be "allocated" after some data has been read into the program, or some intermediate results computed. A simple illustration follows:

```
PROGRAM DYNAMIC
   ! just to illustrate dynamic array allocation
    IMPLICIT NONE
   ! declare variable space for two- dimensional array A
    REAL, ALLOCATABLE :: A (:,:)
   ! now read in the bounds for A
    READ*,M,N
   ! allocate actual space for A
    ALLOCATE (A(M,N))
    READ*, A
    PRINT*, 2.*SQRT (A) +3.
    DEALLOCATE (A)! A no longer needed
END PROGRAM DYNAMIC
```

This simple program also illustrates some other very useful features of the new standard. "Whole array" operations are permissible, so that the whole of an array is read in, or the square root of all its elements computed, by a single statement. The efficiency with which these features are implemented by practical compilers is variable.

1.4.4.2 "Broadcasting"

A feature called "broadcasting" enables operations on whole arrays by scalars such as 2. or 3. in the above example. These scalars are said to be "broadcast" to all the elements of the array so that what will be printed out are the square roots of all the elements of the array having been multiplied by 2.0 and added to 3.0.

1.4.4.3 Constructors

Array elements can be assigned values in the normal way but Fortran 90 also permits the "construction" of 1-d arrays, or vectors, such as the following:

$$\mathtt{v} = (/1.0, 2.0, 3.0, 4.0, 5.0/)$$

which is equivalent to

$$\mathtt{v}(1) = 1.0;\ \mathtt{v}(2) = 2.0;\ \mathtt{v}(3) = 3.0;\ \mathtt{v}(4) = 4.0;\ \mathtt{v}(5) = 5.0$$

Array constructors can themselves be arrays, for example

$$\mathtt{w} = (/\mathtt{v},\mathtt{v}/)$$

would have the obvious result for the 10 numbers in `w`.

1.4.4.4 Vector subscripts

Integer vectors can be used to define subscripts of arrays, and this is very useful in the "gather" and "scatter" operations involved in finite element (and other numerical) methods. Figure 1.2 shows a portion of a finite element mesh of eight-node quadrilaterals with its nodes numbered "globally" at least up to 107 in the example shown. When "local" calculations have to be done involving individual elements, for example to determine element strains or fluxes, a local index vector could hold the node numbers of each element, that is

82	76	71	72	73	77	84	83	for element 65
93	87	82	83	84	88	95	94	for element 73

and so on. This index or "steering" vector could be called STEERING_VECTOR. When a LOCAL vector has to be gathered from a GLOBAL one,

```
LOCAL = GLOBAL (STEERING_VECTOR)
```

is valid, and for scattering

```
GLOBAL (STEERING_VECTOR) = LOCAL
```

In this example LOCAL and STEERING_VECTOR would be eight-long vectors whereas GLOBAL could have a length of thousands or millions.

1.4.4.5 Array sections

Parts of arrays or "subarrays" can be referenced by giving an integer range for one or more of their subscripts. If the range is missing for any subscript, the

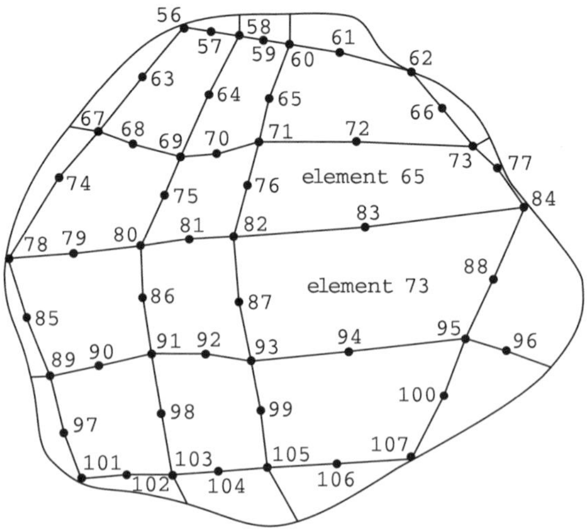

Figure 1.2 Portion of a finite element mesh with node and element numbers

whole extent of that dimension is implied. Thus if A and B are 2-d arrays, A(:,1:3) and B(11:13,:) refer to all the terms in the first three columns of A, and all the terms in rows 11 to 13 of B respectively. If array sections "conform", that is, have the right number of rows and columns, they can be manipulated just like "whole" arrays.

1.4.4.6 Whole array manipulations

It is worth emphasising that the array computation features in Fortran 90 remove the need for several subroutines which were essential in FORTRAN 77 and used in earlier editions of this book. For example, if A,B and C conform and S is a scalar, the following are valid:

```
A=B+C   ! no need for a matrix add, involving DO loops
A=B-C   ! no need for a matrix subtract
A=S*B   ! no need for a matrix-scalar multiply by s
A=B/S   ! no need for a matrix-scalar divide by s
A=0.    ! no need for a matrix null
```

However, although A=B*C has a meaning for conforming arrays A, B and C, its consequence is the computation of the element-by-element products of B and C and is not to be confused with the matrix multiply described in the next subsection.

1.4.4.7 Intrinsic procedures for arrays

To supplement whole array arithmetic operations, Fortran 90 provides a few intrinsic procedures (functions) which are very useful in finite element work. These can be grouped conveniently into those involving array computations, and those involving array inspection. The array computation functions are

```
FUNCTION MATMUL (A,B)            ! returns matrix product of a and b
FUNCTION DOT_PRODUCT (V1,V2)     ! returns dot product of v1 and v2
FUNCTION TRANSPOSE (A)           ! returns transpose of a.
```

All three are heavily used in the programs in this book and replace the user-written subroutines which had to be provided in previous FORTRAN 77 editions.

The array inspection functions include

```
FUNCTION MAXVAL (A)   ! returns the element of an array a of
                      ! maximum value (not absolute maximum)
FUNCTION MINVAL (A)   ! returns the element of an array a of
                      ! minimum value (not absolute minimum)
```

```
FUNCTION MAXLOC (A)       ! returns the location of the maximum element
                          ! of array a
FUNCTION MINLOC (A)       ! returns the location of the minimum element
                          ! of array a
FUNCTION PRODUCT (A)      ! returns the product of all the elements of a
FUNCTION SUM (A)          ! returns the sum of all the elements of a
FUNCTION LBOUND (A,1)     ! returns the first lower bound of a, etc.
FUNCTION UBOUND (A,1)     ! returns the first upper bound of a, etc.
```

The first six of these procedures allow an optional argument called a "masking" argument. For example the statement

```
ASUM = SUM (COLUMN, MASK = COLUMN >=.0)
```

will result in ASUM containing the sum of the positive elements of array COLUMN.

Useful procedures whose only argument is a "mask" are

```
ALL (MASK=COLUMN >= .0)     ! true if all elements of column are positive
ANY (MASK=COLUMN >= .0)     ! true if any elements of column are positive
COUNT (MASK=COLUMN <= .0)   ! number of elements of column
                              which are negative
```

For multidimensional arrays, operations such as SUM can be carried out on a particular dimension of the array. When a mask is used, the dimension argument must be specified even if the array is 1-d. Referring to Figure 1.2, the "half-bandwidth" of an assembled system of equation coefficients for such a finite element mesh could be found from the element freedom steering vectors, G, by the statement

```
BANDWIDTH = MAXVAL (G,1,G>0) - MINVAL (G,1,G>0)
```

allowing for the possibility of zero entries in G.

1.4.5 ADDITIONAL FORTRAN 90 FEATURES

The programs in this book are written in a style of Fortran 90 not too far removed from that of FORTRAN 77.

The example in Figure 1.3 shows gains in conciseness from whole array operations, array intrinsic functions and dynamic arrays, but no complete revolution in programming style has been implemented.

Fortran 90 contains features such as derived data types, pointers, operator overloading and user-defined operators which programmers used to another style might implement to bring about a more radical revision of FORTRAN 77.

Fortran 90

```
KM = 0.0
GAUSS_POINTS : DO I=1,NGP; DO J=1, NGP
  CALL FORMLN (DER,FUN,SAMP,I,J)
  CALL TWOBYTWO (MATMUL (DER, COORD), JAC1, DET)
  DERIV = MATMUL (JAC1, DER)
  BEE = 0.0; CALL FORMB (BEE, DERIV)
KM = KM + MATMUL (MATMUL (TRANSPOSE (BEE), DEE),BEE) &
     *DET*SAMP (I, 2) *SAMP (J, 2)
END DO; END DO GAUSS_POINTS
```

FORTRAN 77

```
      CALL NULL (KM, IKM, IDOF, IDOF)
      DO 20 I=1, NGP
      DO 20 J=1, NGP
      CALL FORMLN (DER, IDER, FUN, SAMP, ISAMP, I, J)
      CALL MATMUL (DER, IDER, COORD, ICOORD, JAC, IJAC, IT, NOD, IT)
      CALL TWOBYTWO (JAC, IJAC, JAC1, IJAC1, DET)
      CALL MATMUL (JAC1, IJAC1, DER, IDER, DERIV, IDERIV, IT, IT, NOD)
      CALL NULL (BEE, IBEE, IH, IDOF)
      CALL FORMB (BEE, IBEE, DERIV, IDERIV, NOD)
      CALL MATMUL (DEE, IDEE, BEE, IBEE, DBEE, IDBEE, IH, IH, IDOF)
      CALL MATRAN (BT, IBT, BEE, IBEE, IH, IDOF)
      CALL MATMUL (BT, IBT, DBEE, IDBEE, BTDB, IBTDB, IDOF, IH, IDOF)
      QUOT = DET*SAMP(I, 2)*SAMP(J, 2)
      CALL MSMULT (BTDB, IBTDB, QUOT, IDOF, IDOF)
20    CALL MATADD (KM, IKM, BTDB, IBTDB, IDOF, IDOF)
```

Figure 1.3 Comparison of a portion of a finite element program in Fortran 90 with FORTRAN 77

This is a matter of taste. One completely new feature of Fortran 90 which has been implemented in the programs which follow is the idea of a "module".

A module is a program unit separate from the main program unit in the way that subroutines and functions are. However, in its simplest form, it may contain no executable statements at all and just be a list or collection of declarations or data which is globally accessible to the program unit which invokes it by a `USE` statement. Its main use later in the book will be to contain either a collection of subroutines and functions which constitute a "library" or to contain the "interfaces" between such a library and a program which uses it.

1.4.6 SUBPROGRAM LIBRARIES

It was stated in the introduction to this chapter that what will be presented in Chapter 4 onwards is not a monolithic program but rather a collection of test programs which all access a common subroutine library which contains about 100 subroutines and functions. In the simplest implementation of Fortran 90 the library routines could simply be appended to the main program after a `CONTAINS` statement as follows:

```
PROGRAM TEST_ONE
      –
      –
      –
      –
CONTAINS
  SUBROUTINE ONE (P1,P2,P3)
      –
      –
      –
  END SUBROUTINE ONE
  SUBROUTINE TWO ( )
      etc.
      –
      –
END PROGRAM TEST_ONE
```

This would be tedious because a sub-library would really be required for each test program, containing only the needed subroutines. Secondly, compilation of the library routines with each test program compilation is wasteful.

What is required, therefore, is for the whole subroutine library to be precompiled and for the test programs to link only to the parts of the library which are needed.

The designers of Fortran 90 seem to have intended this to be done in the following way. The subroutines would be placed in a file:

```
SUBROUTINE ONE ( )
      –
      –
      –
SUBROUTINE NINETY_NINE ( )
      –
      –
      –
END SUBROUTINE NINETY_NINE
```

and compiled.

A module would constitute the interface between library and calling program. It would take the form

```
MODULE INTERFACES
  INTERFACE
    SUBROUTINE ONE ( )
      (PARAMETER DECLARATIONS)
    END SUBROUTINE ONE
           -
           -
           -
    SUBROUTINE NINETY_NINE (     )
      (PARAMETER DECLARATIONS)
    END SUBROUTINE NINETY_NINE
  END INTERFACE
END MODULE INTERFACES
```

Thus the interface module would contain only the subroutine "headers", i.e. the subroutine's name, argument list, and declaration of argument types. This is deemed to be safe because the compiler can check the number and type of arguments in each call (one of the greatest sources of error in FORTRAN 77).

The libraries would be interfaced by a statement "`USE INTERFACES`" at the beginning of each test program. For example

```
PROGRAM TEST_PROGRAM_1
  USE INTERFACES
      -
      -
      -
END PROGRAM TEST_PROGRAM_1
```

However, it is still quite tedious to keep updating two files when making changes to a library (the library and the interface module). Users with straightforward Fortran 90 libraries may well prefer to omit the interface stage altogether and just create a module containing the subroutines themselves. These would then be accessed by "`USE LIBRARY_ROUTINES`" in the example shown below. This still allows the compiler to check the numbers and types of subroutine arguments when the test programs are compiled.

For example

```
MODULE LIBRARY_ROUTINES
  CONTAINS
    SUBROUTINE ONE (  )
        -
        -
    END SUBROUTINE ONE
        -
        -
    SUBROUTINE NINETY_NINE ( )
        -
        -
    END SUBROUTINE NINETY_NINE
END MODULE LIBRARY_ROUTINES
```

and then

```
PROGRAM TEST_PROGRAM_2
  USE LIBRARY_ROUTINES
        -
        -
        -
END PROGRAM_TEST_PROGRAM_2
```

1.5 STRUCTURED PROGRAMMING

The finite element programs which will be described are strongly "structured" in the sense of Dijkstra (1976). The main feature exhibited by our programs will be seen to be a *nested* structure and we will use representations called "structure charts" (Lindsey, 1977) rather than flow charts to describe their actions. The main features of these charts are

1. The block:

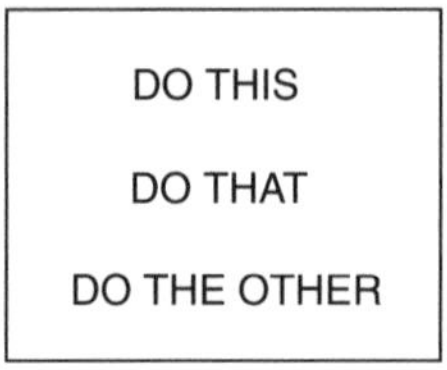

This will be used for the outermost level of each structure chart. Within a block, the indicated actions are to be performed sequentially.

2. The choice:

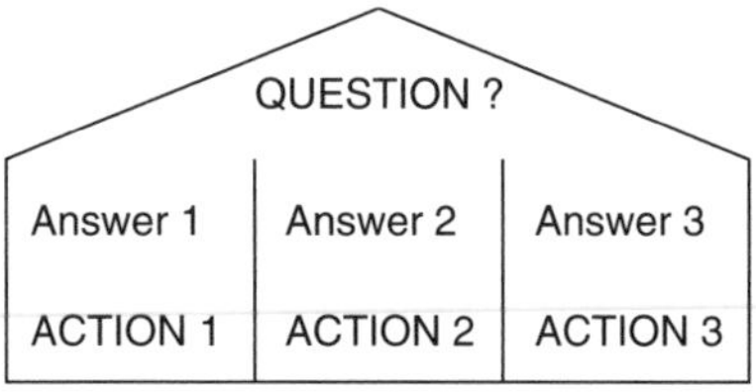

This corresponds to the `IF` ... `THEN` ... `ELSE IF` ... `THEN` ... `END IF` or `SELECT CASE` types of construct.

3. The loop: This comes in various forms, but we shall usually be concerned with "DO" loops, either for a fixed number of repetitions or "for ever" (so-called because of the danger of the loop never being completed).

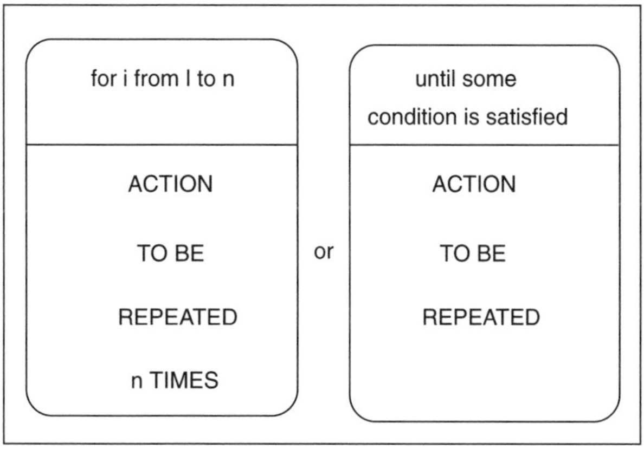

In particular, the structure chart notation discourages the use of GOTO statements. Using this notation, a matrix multiplication program would be represented as follows:

The nested nature of a typical program can be seen quite clearly.

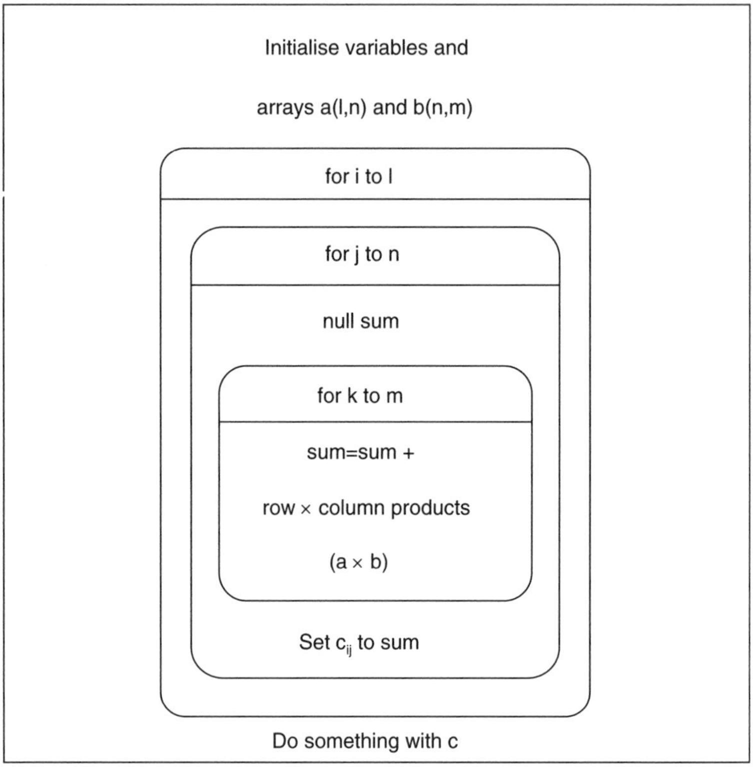

1.6 CONCLUSIONS

Computers on which finite element computations can be done vary widely in their size and architecture. Because of its entrenched position FORTRAN is the language in which computer programs for engineering applications had best be written in order to assure maximum readership and portability. Using Fortran 90, a library of subroutines can be created which is held in compiled form on backing store and accessed by programs in just the way that a manufacturer's permanent library is.

Using this philosophy, a library of over 100 subroutines has been assembled, together with some 50 example programs which access the library. These programs are written in a reasonably "structured" style, and discs containing library and programs will be supplied to readers on request. Versions are at present available for all the common machine ranges and Fortran 90 compilers.

The structure of the remainder of the book is as follows. Chapter 2 shows how the differential equations governing the behaviour of solids and fluids are semi-discretised in space using finite elements.

Chapter 3 describes the sub-program library and the basic techniques by which main programs are constructed to solve the equations listed in Chapter 2. Two basic solution strategies are described, one involving element matrix assembly to form global matrices, which can be used for small- to medium-sized problems, and the other using "element-by-element" matrix techniques to avoid assembly and therefore permit the solution of very large problems.

The remaining Chapters 4 to 11 are concerned with applications, partly in the authors' field of geomechanics. However, the methods and programs described are equally applicable in many other fields of engineering and science such as structural mechanics, fluid dynamics, electromagnetics and so on. Chapter 4 leads off with static analysis of skeletal structures. Chapter 5 deals with static analysis of linear solids, while Chapter 6 discusses extensions to deal with material nonlinearity. Programs dealing with the common geotechnical process of construction (element addition during the analysis) and excavation (element removal during the analysis) are given. Chapter 7 is concerned with problems of fluid flow in the steady state while transient states with inclusion of transport phenomena (diffusion with advection) are treated in Chapter 8. In Chapter 9, coupling between solid and fluid phases is treated, with applications to consolidation processes. A second type of "coupling" which is treated involves the Navier–Stokes equations. Chapter 10 contains programs for the solution of steady-state vibration problems, involving the determination of natural modes, by various methods. Integration of the equations of motion in time is described in Chapter 11.

In every applications chapter, test programs are listed and described, together with specimen input and output.

At the conclusion of most chapters, additional examples, with solutions, are given to enable the book to be used as a teaching text.

1.7 REFERENCES

Dijkstra, E. W. (1976) *A Discipline of Programming*. Prentice-Hall, Englewood Cliffs, New Jersey.

Koelbel, C. H., Loveman, D. B., Schreiber, R. S., Steele, G. L. and Zosel, M. E. (1995) *The High Performance Fortran Handbook*. MIT Press, Cambridge Mass.

Lindsey, C. H. (1977) Structure charts: a structured alternative to flow charts. *SIGPLAN Notices*, **12**(11), 36.

Smith, I. M. (1995) *Programming in Fortran 90*. Wiley, Chichester.

Willé, D. R. (1995) *Advanced Scientific Fortran*. Wiley, Chichester.

2 Spatial Discretisation by Finite Elements

2.0 INTRODUCTION

The finite element method is a technique for solving partial differential equations by first discretising these equations in their space dimensions. The discretisation is carried out locally over small regions of simple but arbitrary shape (the finite elements). This results in matrix equations relating the input at specified points in the elements (the nodes) to the output at these same points. In order to solve equations over large regions, the matrix equations for the smaller sub-regions can be summed node by node, resulting in global matrix equations, or "element-by-element" techniques can be employed to avoid creating (large) global matrices. The method is already described in many texts, for example Zienkiewicz (1991), Strang and Fix (1973), Cook (1989), and Rao (1989), but the principles will briefly be described in this chapter in order to establish a notation and to set the scene for the later descriptions of programming techniques.

2.1 ROD ELEMENT

Figure 2.1(a) shows the simplest solid element, namely an elastic rod, with end nodes 1 and 2. The element has length L while u denotes the longitudinal displacements of points on the rod which is subjected to axial loading only.

If P is the axial force in the rod at a particular section and F is an applied body force with units of force per unit length then

$$P = \sigma A = EA\varepsilon = EA\frac{\partial u}{\partial x} \tag{2.1}$$

and for equilibrium from Figure 2.1(b),

$$\frac{\partial P}{\partial x} + F = 0 \tag{2.2}$$

Hence the differential equation to be solved is

$$EA\frac{\partial^2 u}{\partial x^2} + F = 0 \tag{2.3}$$

(Although this is an ordinary differential equation, future equations, for example (2.13), preserve the same notation.)

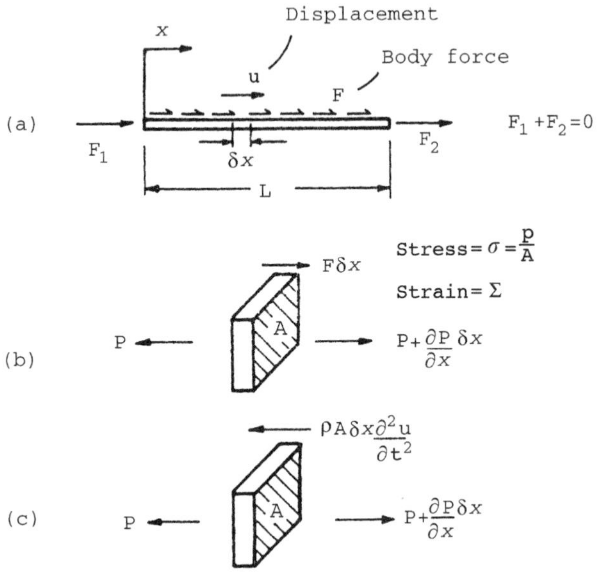

Figure 2.1 Equilibrium of a rod element

In the finite element technique, the continuous variable u is approximated in terms of its nodal values, u_1 and u_2, through simple functions of the space variable called "shape functions". That is

$$u \simeq N_1 u_1 + N_2 u_2$$

$$u \simeq [N_1 \quad N_2]\begin{Bmatrix} u_1 \\ u_2 \end{Bmatrix} = \mathbf{Nu} \tag{2.4}$$

In simple examples the approximate equality in (2.4) could be made exact by setting

$$N_1 = \left(1 - \frac{x}{L}\right), \qquad N_2 = \frac{x}{L} \tag{2.5}$$

if the true variation of u is linear. If the true variation in u is higher order, as will often be the case, greater accuracy could be achieved by introducing higher-order shape functions or by including more linear subdivisions.

When (2.4) is substituted in (2.3) we have

$$EA\frac{\partial^2}{\partial x^2}[N_1 \quad N_2]\begin{Bmatrix} u_1 \\ u_2 \end{Bmatrix} + F = R \tag{2.6}$$

where R is a measure of the error in the approximation and is called the residual. The partial differential equation has thus been replaced by an equation in the discretised space variables u_1 and u_2. The problem now reduces to one of finding "good" values for u_1 and u_2 by minimising the residual R.

Many methods could be used to achieve this, and Griffiths and Smith (1991) discuss collocation, subdomain, Galerkin and least squares techniques. Of these, Galerkin's method, e.g. Finlayson (1972), is the most widely used in finite element work. The method consists of multiplying or "weighting" the residual in equation (2.6) by each shape function in turn, integrating over the element and equating to zero. Thus

$$\int_0^L \begin{Bmatrix} N_1 \\ N_2 \end{Bmatrix} EA \frac{\partial^2}{\partial x^2} [N_1 \quad N_2] \begin{Bmatrix} u_1 \\ u_2 \end{Bmatrix} \mathrm{d}x + \int_0^L \begin{Bmatrix} N_1 \\ N_2 \end{Bmatrix} F \,\mathrm{d}x = 0 \tag{2.7}$$

Note that in the present example in which the shape functions are linear, double differentiation of these functions would cause them to vanish and yet we know the correct shape may not be of higher order than linear. This difficulty is resolved by applying Green's theorem (integration by parts) to yield typically

$$\int N_i \frac{\partial^2 N_j}{\partial x^2} \,\mathrm{d}x = -\int \frac{\partial N_i}{\partial x}\frac{\partial N_j}{\partial x} \,\mathrm{d}x + \text{boundary terms which we usually ignore} \tag{2.8}$$

Hence, assuming EA and F are not functions of x, (2.7) becomes

$$EA \int_0^L \begin{Bmatrix} \dfrac{\partial N_1}{\partial x}\dfrac{\partial N_1}{\partial x} & \dfrac{\partial N_1}{\partial x}\dfrac{\partial N_2}{\partial x} \\ \dfrac{\partial N_2}{\partial x}\dfrac{\partial N_1}{\partial x} & \dfrac{\partial N_2}{\partial x}\dfrac{\partial N_2}{\partial x} \end{Bmatrix} \mathrm{d}x \begin{Bmatrix} u_1 \\ u_2 \end{Bmatrix} - F \int_0^L \begin{Bmatrix} N_1 \\ N_2 \end{Bmatrix} \mathrm{d}x = 0 \tag{2.9}$$

On evaluation of the integrals,

$$EA \begin{bmatrix} \dfrac{1}{L} & -\dfrac{1}{L} \\ -\dfrac{1}{L} & \dfrac{1}{L} \end{bmatrix} \begin{Bmatrix} u_1 \\ u_2 \end{Bmatrix} - F \begin{Bmatrix} \dfrac{L}{2} \\ \dfrac{L}{2} \end{Bmatrix} = \begin{Bmatrix} 0 \\ 0 \end{Bmatrix} \tag{2.10}$$

The above case is for a uniformly distributed force F acting along the rod. For the case in Figure 2.1(a) where the loading is applied only at the nodes we have

$$\frac{EA}{L} \begin{bmatrix} 1 & -1 \\ -1 & 1 \end{bmatrix} \begin{Bmatrix} u_1 \\ u_2 \end{Bmatrix} = \begin{Bmatrix} f_1 \\ f_2 \end{Bmatrix} \tag{2.11}$$

which represents the rod element stiffness relationship. In matrix notation we may write

$$\mathbf{KMu} = \mathbf{f} \tag{2.12}$$

where **KM** is called the "element stiffness matrix".

2.2 ROD INERTIA MATRIX

Consider now the case of an unrestrained rod in longitudinal motion. Figure 2.1(c) shows the equilibrium of a segment in which the body force is now given by Newton's law as mass times acceleration. If the mass per unit volume is ρ, the differential equation becomes

$$EA\frac{\partial^2 u}{\partial x^2} - \rho A\frac{\partial^2 u}{\partial t^2} = 0 \tag{2.13}$$

On discretising u in space by finite elements as before, the first term in (2.13) clearly leads again to **KM**. The second term takes the form

$$-\int_0^L \begin{Bmatrix} N_1 \\ N_2 \end{Bmatrix} \rho A \frac{\partial^2}{\partial t^2}[N_1 \quad N_2] \begin{Bmatrix} u_1 \\ u_2 \end{Bmatrix} \mathrm{d}x \tag{2.14}$$

and assuming that ρA is not a function of x,

$$-\rho A \int_0^L \begin{bmatrix} N_1N_1 & N_1N_2 \\ N_2N_1 & N_2N_2 \end{bmatrix} \mathrm{d}x \frac{\partial^2}{\partial t^2} \begin{Bmatrix} u_1 \\ u_2 \end{Bmatrix} \tag{2.15}$$

Evaluation of integrals yields

$$-\rho AL \begin{bmatrix} \frac{1}{3} & \frac{1}{6} \\ \frac{1}{6} & \frac{1}{3} \end{bmatrix} \frac{\partial^2}{\partial t^2} \begin{Bmatrix} u_1 \\ u_2 \end{Bmatrix} \tag{2.16}$$

or in matrix notation

$$-\mathbf{MM}\frac{\partial^2 \mathbf{u}}{\partial t^2}$$

where **MM** is the "element mass matrix". Thus the full matrix statement of equation (2.13) is

$$\mathbf{KMu} + \mathbf{MM}\frac{\mathrm{d}^2\mathbf{u}}{\mathrm{d}t^2} = \mathbf{0} \tag{2.17}$$

which is a set of ordinary differential equations.

Note that **MM** as formed in this manner is the "consistent" mass matrix and differs from the "lumped" equivalent which would lead to $1/2\rho AL$ terms on the diagonal with zeros off-diagonal.

2.3 THE EIGENVALUE EQUATION

Equation (2.17) is sometimes integrated directly (Chapter 11) but is also the starting point for derivation of the eigenvalues of the stiffness matrix of an element or mesh of elements.

Suppose the elastic rod element is undergoing free harmonic motion. Then all nodal displacements will be harmonic, of the form

$$\mathbf{u} = \mathbf{a}\sin(\omega t + \psi) \tag{2.18}$$

where $\mathbf{a}$ are amplitudes of the motion, ω its frequencies and ψ its phase shifts. When (2.18) is substituted in (2.17) the equation

$$\mathbf{KMa} - \omega^2 \mathbf{MMa} = \mathbf{0} \tag{2.19}$$

is obtained, which can easily be rearranged as a standard eigenvalue equation. Chapter 10 describes solution of equations of this type.

2.4 BEAM ELEMENT

As a second 1-d solid element, consider the slender beam in Figure 2.2. The end nodes 1 and 2 are subjected to shear forces and moments which result in translations and rotations. Each node, therefore, has two "degrees of freedom".

The element shown in Figure 2.2 has length L, flexural rigidity EI and carries a uniform transverse load of q per unit length acting in the downwards (negative) direction. The well-known equilibrium equation for this system is given by

$$EI\frac{\partial^4 w}{\partial x^4} = q \tag{2.20}$$

Again the continuous variable, w in this case, is approximated in terms of discrete nodal values, but we introduce the idea that not only w itself but also its derivatives can be used in the approximation. Thus we choose to write

$$w \simeq [N_1 \quad N_2 \quad N_3 \quad N_4]\begin{Bmatrix} w_1 \\ \theta_1 \\ w_2 \\ \theta_2 \end{Bmatrix} \tag{2.21}$$

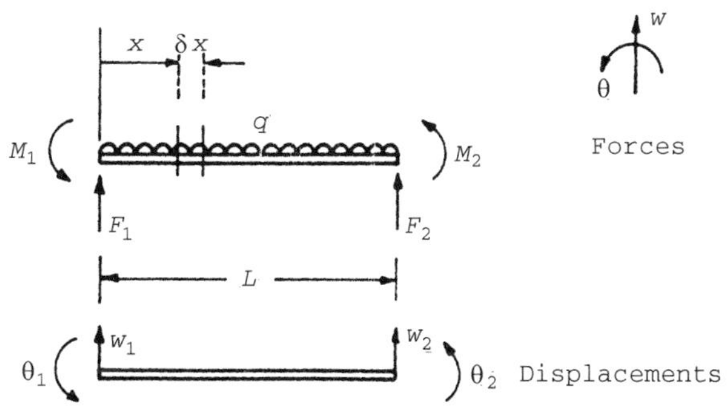

Figure 2.2 Slender beam element

where $\theta_1 = \partial w/\partial x$ at node 1 and so on. In this case, equation (2.21) can often be made exact by choosing the cubic shape functions:

$$\begin{aligned} N_1 &= \frac{1}{L^3}(L^3 - 3Lx^2 + 2x^3) \\ N_2 &= \frac{1}{L^2}(L^2x - 2Lx^2 + x^3) \\ N_3 &= \frac{1}{L^3}(3Lx^2 - 2x^3) \\ N_4 &= \frac{1}{L^2}(x^3 - Lx^2) \end{aligned} \tag{2.22}$$

Note that the shape functions have the property that they, or their derivatives in this case, equal one at a specific node and zero at all others.

Substitution in (2.20) and application of Galerkin's method leads to the four element equations:

$$\int_0^L \begin{Bmatrix} N_1 \\ N_2 \\ N_3 \\ N_4 \end{Bmatrix} EI \frac{\partial^4}{\partial x^4} [N_1 \quad N_2 \quad N_3 \quad N_4] \begin{Bmatrix} w_1 \\ \theta_1 \\ w_2 \\ \theta_2 \end{Bmatrix} \mathrm{d}x = \int_0^L \begin{Bmatrix} N_1 \\ N_2 \\ N_3 \\ N_4 \end{Bmatrix} q\, \mathrm{d}x \tag{2.23}$$

Again Green's theorem is used to avoid differentiating four times; for example

$$\int N_i \frac{\partial^4 N_j}{\partial x^4}\, \mathrm{d}x \cong -\int \frac{\partial N_i}{\partial x}\frac{\partial^3 N_j}{\partial x^3}\, \mathrm{d}x \cong \int \frac{\partial^2 N_i}{\partial x^2}\frac{\partial^2 N_j}{\partial x^2}\, \mathrm{d}x + \text{neglected terms} \tag{2.24}$$

Hence assuming that EI and q are not functions of x, (2.23) become

$$EI \int_0^L \frac{\partial^2 N_i}{\partial x^2}\frac{\partial^2 N_j}{\partial x^2}\, \mathrm{d}x \begin{Bmatrix} w_1 \\ \theta_1 \\ w_2 \\ \theta_2 \end{Bmatrix} = q \int_0^L \begin{Bmatrix} N_1 \\ N_2 \\ N_3 \\ N_4 \end{Bmatrix} \mathrm{d}x \tag{2.25}$$

where $i, j = 1, 2, 3, 4$.

Evaluation of the integrals gives

$$EI \begin{bmatrix} \frac{12}{L^3} & \frac{6}{L^2} & -\frac{12}{L^3} & \frac{6}{L^2} \\ & \frac{4}{L} & -\frac{6}{L^2} & \frac{2}{L} \\ & \text{symmetrical} & \frac{12}{L^3} & -\frac{6}{L^2} \\ & & & \frac{4}{L} \end{bmatrix} \begin{Bmatrix} w_1 \\ \theta_1 \\ w_2 \\ \theta_2 \end{Bmatrix} = q \begin{Bmatrix} \frac{L}{2} \\ \frac{L^2}{12} \\ \frac{L}{2} \\ -\frac{L^2}{12} \end{Bmatrix} \tag{2.26a}$$

which recovers the standard "slope-deflection" equations for beam elements.

The above case is for a uniformly distributed load applied to the beam. For the case where loading is applied only at the nodes we have

$$EI\begin{bmatrix} \frac{12}{L^3} & \frac{6}{L^2} & -\frac{12}{L^3} & \frac{6}{L^2} \\ & \frac{4}{L} & -\frac{6}{L^2} & \frac{2}{L} \\ & & \frac{12}{L^3} & -\frac{6}{L^2} \\ \text{symmetrical} & & & \frac{4}{L} \end{bmatrix}\begin{Bmatrix} w_1 \\ \theta_1 \\ w_2 \\ \theta_2 \end{Bmatrix}=\begin{Bmatrix} f_1 \\ m_1 \\ f_2 \\ m_2 \end{Bmatrix} \tag{2.26b}$$

which represents the beam element stiffness relationship.

Hence, in matrix notation we again have

$$\mathbf{KMw} = \mathbf{f} \tag{2.27}$$

Beam–column elements, in which axial and bending effects are combined from (2.26b) and (2.11), are described further in Chapter 4.

2.5 BEAM INERTIA MATRIX

If the element in Figure 2.2 were vibrating transversely it would be subjected to an additional restoring force $-\rho A(\partial^2 w/\partial t^2)$. The inertia or mass matrix, by analogy with (2.15), is just

$$-\rho A\int_0^L \begin{bmatrix} N_1N_1 & N_1N_2 & N_1N_3 & N_1N_4 \\ N_2N_1 & N_2N_2 & N_2N_3 & N_2N_4 \\ N_3N_1 & N_3N_2 & N_3N_3 & N_3N_4 \\ N_4N_1 & N_4N_2 & N_4N_3 & N_4N_4 \end{bmatrix} \mathrm{d}x \frac{\partial^2}{\partial t^2}\begin{Bmatrix} w_1 \\ \theta_1 \\ w_2 \\ \theta_2 \end{Bmatrix} \tag{2.28}$$

Evaluation of the integrals yields

$$-\frac{\rho AL}{420}\begin{bmatrix} 156 & 22L & 54 & -13L \\ & 4L^2 & 13L & -3L^2 \\ & & 156 & -22L \\ \text{symmetrical} & & & 4L^2 \end{bmatrix}\frac{\partial^2}{\partial t^2}\begin{Bmatrix} w_1 \\ \theta_1 \\ w_2 \\ \theta_2 \end{Bmatrix} \tag{2.29}$$

In this instance, the approximation of the consistent mass terms by lumped ones can lead to large errors in the prediction of beam frequencies as shown by Leckie and Lindberg (1963). Combination of (2.29) and (2.16) is required in the dynamic analysis of framed structures and this is described further in Chapter 10.

2.6 BEAM WITH AN AXIAL FORCE

If the beam element in Figure 2.2 is subjected to an additional axial force P (Figure 2.3), a simple modification to (2.20) results in the differential equation

$$EI\frac{\partial^4 w}{\partial x^4} \pm P\frac{\partial^2 w}{\partial x^2} = q \tag{2.30}$$

where the positive sign corresponds to a compressive axial load and vice versa.

Finite element discretisation and application of Galerkin's method lead to an additional matrix associated with the axial force contribution

$$\mp P\int_0^L \frac{\partial N_i}{\partial x}\frac{\partial N_j}{\partial x}\,\mathrm{d}x\begin{Bmatrix} w_1 \\ \theta_1 \\ w_2 \\ \theta_2 \end{Bmatrix} \tag{2.31}$$

where $i, j = 1, 2, 3, 4$.

Evaluation of these integrals yields for compressive P

$$+\frac{P}{30}\begin{bmatrix} \frac{36}{L} & 3 & -\frac{36}{L} & 3 \\ & 4L & -3 & -L \\ & & \frac{36}{L} & -3 \\ \text{symmetrical} & & & 4L \end{bmatrix}\begin{Bmatrix} w_1 \\ \theta_1 \\ w_2 \\ \theta_2 \end{Bmatrix} \tag{2.32}$$

If this matrix is designated by **KA** the equilibrium equation becomes

$$(\mathbf{KM} - \mathbf{KA})\mathbf{w} = \mathbf{f} \tag{2.33}$$

Buckling of a member can be investigated by solving the eigenvalue problem where $\mathbf{f} = \mathbf{0}$, by increasing the compressive force on the element (corresponding to **KA** in 2.33) until large deformations result or in simple cases by determinant search. Equations (2.32) and (2.33) represent an approximation of the approach to modifying the element stiffness involving stability functions (Lundquist and Kroll, 1944; Horne and Merchant, 1965). The accuracy of the approximation

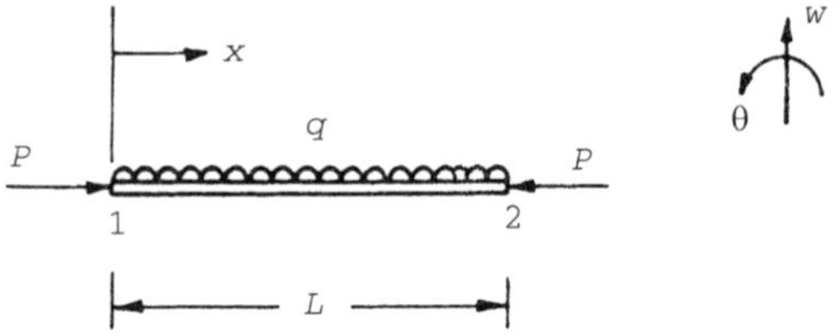

Figure 2.3 Beam with axial force

depends on the value of P/P_{E} for each member, where P_{E} is the Euler load. Over the range $-1 \leq P/P_{\mathrm{E}} \leq 1$ the approximation introduces errors no greater than 7% (Livesley, 1975). For larger positive values of P/P_{E}, however, equation (2.33) can become inaccurate unless more element subdivisions are used. Program 4.6 describes analyses of this kind for the stability of plane frames.

2.7 BEAM ON ELASTIC FOUNDATION

In Figure 2.4 a continuous elastic support has been placed beneath the basic element. If this support has stiffness k (force/ length^2) then clearly the transverse load is resisted by an extra force $+kw$ leading to the differential equation

$$EI\frac{\partial^4 w}{\partial x^4} + kw = q \tag{2.34}$$

(although this is again an ordinary differential equation "partial" notation is retained).

By comparison with the inertia restoring force $-\rho A(\partial^2 w/\partial t^2)$ it will be apparent that application of the Galerkin process to (2.34) will result in a foundation stiffness matrix that is identical to the consistent mass matrix apart from the multiple $+kL$ in place of $-\rho AL$. An example of a consistent finite element solution to (2.34) is given by Program 4.3. A "lumped mass" approach to this problem is also possible by simply adding the appropriate spring stiffness to the diagonal terms of the beam stiffness matrix (see e.g. Griffiths, 1989).

2.8 GENERAL REMARKS ON THE DISCRETISATION PROCESS

Enough examples have now been described for a general pattern to emerge of how terms in a differential equation appear in matrix form after discretisation. Table 2.1 gives a summary, N being the shape functions.

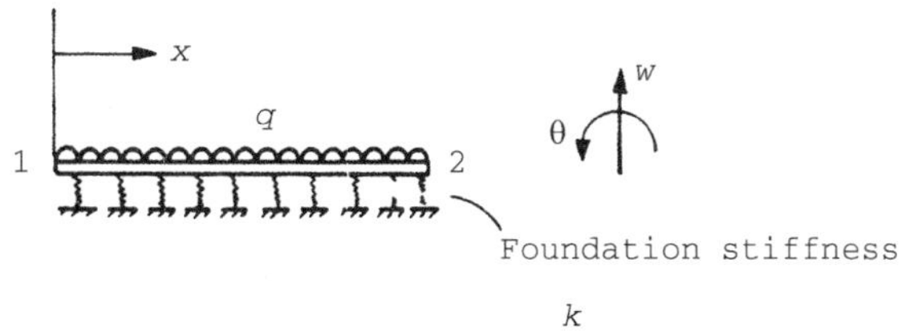

Figure 2.4 Beam on continuous elastic foundation

Table 2.1 Semi-discretisation of partial differential

Term appearing in differential equation	Typical term in finite element matrix equation	Symmetry?
u	$\int N_i N_j \, dx$	Yes
$\dfrac{\partial u}{\partial x}$	$\int N_i \dfrac{\partial N_j}{\partial x} \, dx$	No
$\dfrac{\partial^2 u}{\partial x^2}$	$-\int \dfrac{\partial N_i}{\partial x} \dfrac{\partial N_j}{\partial x} \, dx$	Yes
$\dfrac{\partial^4 u}{\partial x^4}$	$\int \dfrac{\partial^2 N_i}{\partial x^2} \dfrac{\partial^2 N_j}{\partial x^2} \, dx$	Yes

In fact, first-order terms such as $\partial u/\partial x$ have not yet arisen. They are unique in Table 2.1 in leading to matrix equations which are not symmetrical, as indeed would be the case for any odd order of derivative. We shall return to terms of this type in due course, in relation to advection in fluid flow.

2.9 ALTERNATIVE DERIVATION OF ELEMENT STIFFNESS

Instead of working from the governing differential equation, element properties can often be derived by an alternative method based on a consideration of energy. For example, the strain energy stored due to bending of a very small length δx of the elastic beam element in Figure 2.2 is

$$\delta U = \frac{1}{2} \frac{M^2}{EI} \delta x \tag{2.35}$$

where M is the "bending moment" and by conservation of energy this must be equal to the work done by the external loads q; thus

$$\delta W = \tfrac{1}{2} q w \delta x \tag{2.36}$$

To proceed, we discretise the displacements in terms of the element shape functions (2.21 and 2.22) to give

$$w = N_1 w_1 + N_2 \theta_1 + N_3 w_2 + N_4 \theta_2 = \mathbf{Nw} \tag{2.37}$$

The bending moment M is related to w through the "moment–curvature" expression

$$M = -EI \frac{\partial^2 w}{\partial x^2}$$

or

$$M = \mathbf{DA}w \tag{2.38}$$

where $\mathbf{D}$ is the material property EI and $\mathbf{A}$ is the operator $-\partial^2/\partial x^2$. Writing (2.35) in the form

$$\delta U = \frac{1}{2}\left(-\frac{\partial^2 w}{\partial x^2}\right) M\,\delta x \tag{2.39}$$

we have

$$\delta U = \frac{1}{2}(\mathbf{A}w)^{\mathrm{T}} M\,\delta x \tag{2.40}$$

Using (2.37) and (2.38) this becomes

$$\begin{aligned}\delta U &= \frac{1}{2}(\mathbf{ANw})^{\mathrm{T}}\mathbf{DANw}\,\delta x \\ &= \frac{1}{2}\mathbf{w}^{\mathrm{T}}(\mathbf{AN})^{\mathrm{T}}\mathbf{DANw}\,\delta x\end{aligned} \tag{2.41}$$

The total strain energy of the element is thus

$$U = \frac{1}{2}\int_0^L \mathbf{w}^{\mathrm{T}}(\mathbf{AN})^{\mathrm{T}}\mathbf{DANw}\,\mathrm{d}x \tag{2.42}$$

The matrix $\mathbf{AN}$ is usually written as $\mathbf{B}$, and since $\mathbf{w}$ are nodal values and therefore constants

$$U = \frac{1}{2}\mathbf{w}^{\mathrm{T}}\int_0^L \mathbf{B}^{\mathrm{T}}\mathbf{DB}\,\mathrm{d}x\,\mathbf{w} \tag{2.43}$$

Similar operations on (2.36) lead to the total external work done and hence the stored potential energy of the beam is given by

$$\begin{aligned}\Pi &= U - W \\ &= \frac{1}{2}\mathbf{w}^{\mathrm{T}}\int_0^L \mathbf{B}^{\mathrm{T}}\mathbf{DB}\,\mathrm{d}x\,\mathbf{w} - \frac{1}{2}\mathbf{w}^{\mathrm{T}}q\int_0^L \mathbf{N}^{\mathrm{T}}\mathrm{d}x\end{aligned} \tag{2.44}$$

A state of stable equilibrium is achieved when π is a minimum with respect to all $\mathbf{w}$. That is

$$\frac{\partial\Pi}{\partial\mathbf{w}^{\mathrm{T}}} = \int_0^L \mathbf{B}^{\mathrm{T}}\mathbf{DB}\,\mathrm{d}x\,\mathbf{w} - q\int_0^L \mathbf{N}^{\mathrm{T}}\,\mathrm{d}x = 0 \tag{2.45}$$

or

$$\int_0^L \mathbf{B}^{\mathrm{T}}\mathbf{DB}\,\mathrm{d}x\,\mathbf{w} = q\int_0^L \mathbf{N}^{\mathrm{T}}\,\mathrm{d}x \tag{2.46}$$

which is simply another way of writing (2.25).

Thus we see from (2.27) that the elastic element stiffness matrix **KM** can be written in the form

$$\mathbf{KM} = \int_0^L \mathbf{B}^{\mathrm{T}}\mathbf{D}\mathbf{B}\,\mathrm{d}x \tag{2.47}$$

which will prove to be a useful general matrix form for expressing stiffnesses of all elastic solid elements. The computer programs for analysis of solids developed in the next chapter use this notation and method of stiffness formation.

The "energy" formulation described above is clearly valid only for "conservative" systems and Galerkin's method is more generally applicable.

2.10 TWO-DIMENSIONAL ELEMENTS: PLANE STRESS AND STRAIN

The elements so far described have not been true finite elements because they have been used to solve differential equations in one space variable only. Thus the real problem involving two or three space variables has been replaced by a hypothetical, equivalent 1-d problem before solution. The elements we have considered can be joined together at points (the nodes) and complete continuity (compatibility) and equilibrium achieved. In this way we can sometimes obtain exact solutions to our hypothetical problems and these solutions will be unaffected by the number of elements chosen to represent uniform line segments.

This situation changes radically when problems in two or three space dimensions are analysed. For example, consider the plane shear wall with openings shown in Figure 2.5(a). The wall has been subdivided into rectangular elements of sides a and b of which Figure 2.5(b) is typical. These elements have four corner nodes so that when the idealised wall is assembled, the elements will only be attached at these points.

If the wall can be considered to be of unit thickness and in a state of plane stress, see Timoshenko and Goodier (1951), the equations to be solved are the following:

1. Equilibrium

$$\begin{aligned}\frac{\partial \sigma_x}{\partial x} + \frac{\partial \tau_{xy}}{\partial y} + F_x = 0\\ \frac{\partial \tau_{xy}}{\partial x} + \frac{\partial \sigma_y}{\partial y} + F_y = 0\end{aligned} \tag{2.48}$$

where σ_x, σ_y and τ_{xy} are the only zon-zero stress components and F_x, F_y are body forces, per unit volume.

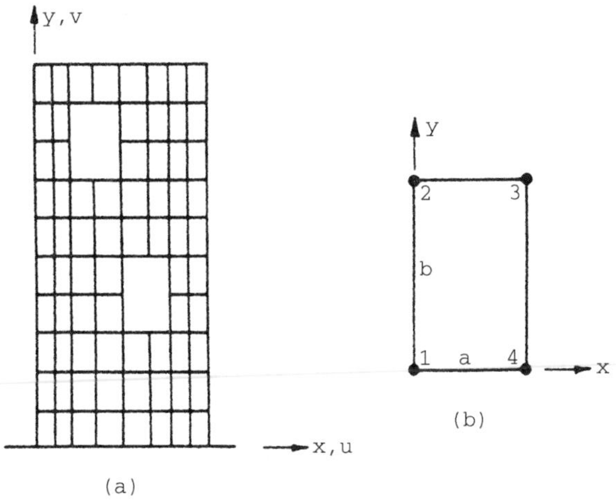

Figure 2.5 (a) Shear wall with openings. (b) Typical rectangular element with four corner nodes

2. Constitutive (plane stress)

$$\begin{Bmatrix} \sigma_x \\ \sigma_y \\ \tau_{xy} \end{Bmatrix} = \frac{E}{1-\nu^2}\begin{bmatrix} 1 & \nu & 0 \\ \nu & 1 & 0 \\ 0 & 0 & \dfrac{1-\nu}{2} \end{bmatrix}\begin{Bmatrix} \varepsilon_x \\ \varepsilon_y \\ \gamma_{xy} \end{Bmatrix} \tag{2.49}$$

where E is Young's modulus, ν is Poisson's ratio and $\varepsilon_x, \varepsilon_y$ and γ_{xy} are the independent small strain components.

3. Strain-displacement

$$\begin{Bmatrix} \varepsilon_x \\ \varepsilon_y \\ \gamma_{xy} \end{Bmatrix} = \begin{bmatrix} \dfrac{\partial}{\partial x} & 0 \\ 0 & \dfrac{\partial}{\partial y} \\ \dfrac{\partial}{\partial y} & \dfrac{\partial}{\partial x} \end{bmatrix}\begin{Bmatrix} u \\ v \end{Bmatrix} \tag{2.50}$$

where u, v are the components of displacement in the x, y directions.

Using the notation of the previous section with $\mathbf{A}$ as the strain–displacement operator and $\mathbf{D}$ as the constitutive stress–strain matrix, these three equations become

$$\begin{aligned} \mathbf{A}^{\mathrm{T}}\boldsymbol{\sigma} &= -\mathbf{f} \\ \boldsymbol{\sigma} &= \mathbf{D}\boldsymbol{\varepsilon} \\ \boldsymbol{\varepsilon} &= \mathbf{A}\mathbf{e} \end{aligned} \tag{2.51}$$

where

$$\mathbf{e} = \begin{Bmatrix} u \\ v \end{Bmatrix}, \qquad \mathbf{A} = \begin{Bmatrix} \dfrac{\partial}{\partial x} & 0 \\ 0 & \dfrac{\partial}{\partial y} \\ \dfrac{\partial}{\partial y} & \dfrac{\partial}{\partial x} \end{Bmatrix}, \qquad \mathbf{D} = \frac{E}{1-\nu^2}\begin{bmatrix} 1 & \nu & 0 \\ \nu & 1 & 0 \\ 0 & 0 & \dfrac{1-\nu}{2} \end{bmatrix} \tag{2.52}$$

We shall only be concerned in this book with "displacement" formulations in which $\boldsymbol{\sigma}$ and $\boldsymbol{\varepsilon}$ are eliminated from (2.51) as follows:

$$\begin{aligned} \mathbf{A}^{\mathrm{T}}\boldsymbol{\sigma} &= -\mathbf{f} \\ \mathbf{A}^{\mathrm{T}}\mathbf{D}\boldsymbol{\varepsilon} &= -\mathbf{f} \\ \mathbf{A}^{\mathrm{T}}\mathbf{D}\mathbf{A}\mathbf{e} &= -\mathbf{f} \end{aligned} \tag{2.53}$$

Writing out (2.53) in full we have

$$\frac{E}{1-\nu^2}\begin{Bmatrix} \dfrac{\partial^2 u}{\partial x^2} + \dfrac{1-\nu}{2}\dfrac{\partial^2 u}{\partial y^2} + \nu\dfrac{\partial^2 v}{\partial x\partial y} + \dfrac{1-\nu}{2}\dfrac{\partial^2 v}{\partial y\partial x} \\ \dfrac{\nu\partial^2 u}{\partial y\partial x} + \dfrac{1-\nu}{2}\dfrac{\partial^2 u}{\partial x\partial y} + \dfrac{1-\nu}{2}\dfrac{\partial^2 v}{\partial x^2} + \dfrac{\partial^2 v}{\partial y^2} \end{Bmatrix} = \begin{Bmatrix} -F_x \\ -F_y \end{Bmatrix} \tag{2.54}$$

which is a pair of simultaneous partial differential equations in the continuous space variables u and v.

As usual these can be solved by discretising over each element using shape functions (here we assume the same functions in the x and y directions)

$$u = [N_1 \quad N_2 \quad N_3 \quad N_4]\begin{Bmatrix} u_1 \\ u_2 \\ u_3 \\ u_4 \end{Bmatrix} = \mathbf{N}\mathbf{u} \tag{2.55}$$

and

$$v = [N_1 \quad N_2 \quad N_3 \quad N_4]\begin{Bmatrix} v_1 \\ v_2 \\ v_3 \\ v_4 \end{Bmatrix} = \mathbf{N}\mathbf{v} \tag{2.56}$$

where in the case of the rectangular element shown in Figure 2.5(b) the N_i functions were first derived by Taig (1961) to be

$$\begin{aligned} N_1 &= \left(1 - \frac{x}{a}\right)\left(1 - \frac{y}{b}\right) \\ N_2 &= \left(1 - \frac{x}{a}\right)\frac{y}{b} \\ N_3 &= \frac{x}{a}\frac{y}{b} \\ N_4 &= \frac{x}{a}\left(1 - \frac{y}{b}\right) \end{aligned} \tag{2.57}$$

These result in linear variations in strain across the element which is sometimes called the "linear strain rectangle".

Discretisation and application of Galerkin's method (Szabo and Lee, 1969), using Table 2.1, lead to the stiffness equations for a typical element:

$$\frac{E}{1-\nu^2}\int_0^a\int_0^b \begin{bmatrix} \left(\frac{\partial N_i}{\partial x}\frac{\partial N_j}{\partial x} + \frac{1-\nu}{2}\frac{\partial N_i}{\partial y}\frac{\partial N_j}{\partial y}\right) & \left(\nu\frac{\partial N_i}{\partial x}\frac{\partial N_j}{\partial y} + \frac{1-\nu}{2}\frac{\partial N_i}{\partial y}\frac{\partial N_j}{\partial x}\right) \\ \left(\nu\frac{\partial N_i}{\partial y}\frac{\partial N_j}{\partial x} + \frac{1-\nu}{2}\frac{\partial N_i}{\partial x}\frac{\partial N_j}{\partial y}\right) & \left(\frac{\partial N_i}{\partial y}\frac{\partial N_j}{\partial y} + \frac{1-\nu}{2}\frac{\partial N_i}{\partial x}\frac{\partial N_j}{\partial x}\right) \end{bmatrix}$$

$$\mathrm{d}y\,\mathrm{d}x\begin{Bmatrix}\mathbf{u}\\ \mathbf{v}\end{Bmatrix} = \begin{Bmatrix}\mathbf{f}_x\\ \mathbf{f}_y\end{Bmatrix} \tag{2.58}$$

where $i, j = 1, 2, 3, 4$ or

$$\mathbf{KMr} = \mathbf{f}$$

Evaluation of the first term in the plane stress stiffness matrix yields

$$KM_{1,1} = \frac{E}{1-\nu^2}\left(\frac{b}{3a} + \frac{1-\nu}{2}\frac{a}{3b}\right) \tag{2.59}$$

and so on.

Note that the size of the element does not appear in this expression, only the ratio a/b or b/a which is called the "aspect ratio" of the element.

In the course of this evaluation, integration by parts now involves integrals of the type

$$\iint N_i \frac{\partial^2 N_j}{\partial x^2}\,\mathrm{d}x\,\mathrm{d}y = -\iint \frac{\partial N_i}{\partial x}\frac{\partial N_j}{\partial x}\,\mathrm{d}x\,\mathrm{d}y + \oint N_i \frac{\partial N_j}{\partial x} l\,\mathrm{d}S \tag{2.60}$$

where l is the direction cosine of the normal to boundary S and we assume that the contour integral in (2.60) is zero between elements. This assumption is generally reasonable but extra care is needed at mesh boundaries. Only if the elements become vanishingly small can our solution be the correct one (an infinite number of elements) except in trivial cases. Physically, in a displacement method, it is usual to satisfy compatibility everywhere in a mesh but to satisfy equilibrium only at the nodes. It is also possible to violate compatibility, but none of the elements described in this book does.

2.11 ENERGY APPROACH

As was done in the case of the elastic beam element, the principle of minimum potential energy can be used to provide an alternative derivation of (2.58) for elastic plane elements. The element strain energy per unit thickness is

$$\begin{aligned} U &= \iint \tfrac{1}{2}\boldsymbol{\sigma}^{\mathrm{T}}\boldsymbol{\varepsilon}\,\mathrm{d}\omega\,\mathrm{d} \\ &= \frac{1}{2}\mathbf{r}^{\mathrm{T}}\iint (\mathbf{AS}^{\mathrm{T}}\mathbf{D}(\mathbf{AS})\,\mathrm{d}x\,\mathrm{d}y\,\mathbf{r} \\ &= \frac{1}{2}\mathbf{r}^{\mathrm{T}}\iint \mathbf{B}^{\mathrm{T}}\mathbf{DB}\,\mathrm{d}x\,\mathrm{d}y\,\mathbf{r} \end{aligned} \tag{2.61}$$

where **A** and **D** are defined in (2.52) and

$$\mathbf{S} = \begin{bmatrix} N_1 & N_2 & N_3 & N_4 & 0 & 0 & 0 & 0 \\ 0 & 0 & 0 & 0 & N_1 & N_2 & N_3 & N_4 \end{bmatrix} \tag{2.62}$$

Thus we have again for this element

$$\mathbf{KM} = \iint \mathbf{B}^{\mathrm{T}}\mathbf{DB}\,\mathrm{d}x\,\mathrm{d}y \tag{2.63}$$

which is the form in which it will be computed in Chapter 3.

Exactly the same expression holds in the case of plane strain, but the elastic **D** matrix becomes (Timoshenko and Goodier, 1951), for unit thickness,

$$\mathbf{D} = \frac{E(1-\nu)}{(1+\nu)(1-2\nu)}\begin{bmatrix} 1 & \dfrac{\nu}{1-\nu} & 0 \\ \dfrac{\nu}{1-\nu} & 1 & 0 \\ 0 & 0 & \dfrac{1-2\nu}{2(1-\nu)} \end{bmatrix} \tag{2.64}$$

2.12 PLANE ELEMENT INERTIA MATRIX

When inertia is significant (2.54) are supplemented by forces $-\rho(\partial^2 u/\partial t^2)$ and $-\rho(\partial^2 v/\partial t^2)$ respectively, where ρ is the mass of the element per unit volume.

For an element of unit thickness this leads, in exactly the same way as in (2.14), to the element mass matrix which has terms given by

$$MM_{ij} = \rho \int\int N_i N_j \, dx \, dy \tag{2.65}$$

and hence to an eigenvalue equation the same as (2.19).

2.13 AXISYMMETRIC STRESS AND STRAIN

Solids of revolution subjected to axisymmetric loading possess only two independent components of displacement and can be analysed as if they were 2-d. For example, Figure 2.6(a) shows a thick tube subjected to radial pressure p and axial pressure q. Only a typical radial cross-section need be analysed and is subdivided into rectangular elements in the figure. The cylindrical coordinate system, Figure 2.6(b), is the most convenient and when it is used the element stiffness equation equivalent to (2.58) is

$$\mathbf{KM} = \int\int\int \mathbf{B}^{\mathrm{T}}\mathbf{DB} r \, dr \, d\theta \, dz \tag{2.66}$$

which, when integrated over one radian, becomes

$$\mathbf{KM} = \int\int \mathbf{B}^{\mathrm{T}}\mathbf{DB} r \, dr \, dz$$

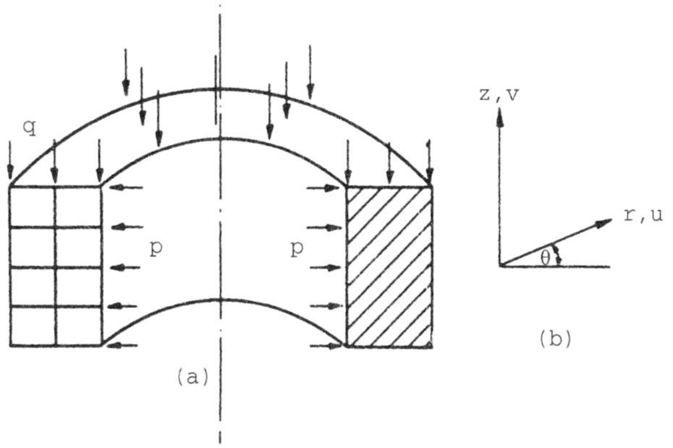

Figure 2.6 Cylinder under axial and radial pressure

where the strain–displacement relations are now (Timoshenko and Goodier, 1951)

$$\begin{Bmatrix} \varepsilon_r \\ \varepsilon_z \\ \gamma_{rz} \\ \varepsilon_\theta \end{Bmatrix} = \begin{bmatrix} \dfrac{\partial}{\partial r} & 0 \\ 0 & \dfrac{\partial}{\partial z} \\ \dfrac{\partial}{\partial z} & \dfrac{\partial}{\partial r} \\ \dfrac{1}{r} & 0 \end{bmatrix} \begin{Bmatrix} u \\ v \end{Bmatrix} \tag{2.67}$$

or

$$\boldsymbol{\varepsilon} = \mathbf{Ae}$$

and as usual $\mathbf{B} = \mathbf{AS}$, where for elements of rectangular cross-section $\mathbf{S}$ could again be defined by (2.57) and (2.62). The stress–strain matrix must be redefined as

$$\mathbf{D} = \frac{E(1-\nu)}{(1+\nu)(1-2\nu)} \begin{bmatrix} 1 & \dfrac{\nu}{1-\nu} & 0 & \dfrac{\nu}{1-\nu} \\ \dfrac{\nu}{1-\nu} & 1 & 0 & \dfrac{\nu}{1-\nu} \\ 0 & 0 & \dfrac{1-2\nu}{2(1-\nu)} & 0 \\ \dfrac{\nu}{1-\nu} & \dfrac{\nu}{1-\nu} & 0 & 1 \end{bmatrix} \tag{2.68}$$

2.14 THREE-DIMENSIONAL STRESS AND STRAIN

When (2.48), (2.49) and (2.50) are extended to the three space-displacement variables u, v, w, three simultaneous partial differential equations equivalent to (2.54) result. Discretisation proceeds as usual, and again the familiar element stiffness properties are derived as

$$\mathbf{KM} = \iiint \mathbf{B}^{\mathrm{T}}\mathbf{DB}\,\mathrm{d}x\,\mathrm{d}y\,\mathrm{d}z \tag{2.69}$$

where the full strain-displacement relations are (Timoshenko and Goodier, 1951)

$$\begin{Bmatrix} \varepsilon_x \\ \varepsilon_y \\ \varepsilon_z \\ \gamma_{xy} \\ \gamma_{yz} \\ \gamma_{zx} \end{Bmatrix} = \begin{bmatrix} \frac{\partial}{\partial x} & 0 & 0 \\ 0 & \frac{\partial}{\partial y} & 0 \\ 0 & 0 & \frac{\partial}{\partial z} \\ \frac{\partial}{\partial y} & \frac{\partial}{\partial x} & 0 \\ 0 & \frac{\partial}{\partial z} & \frac{\partial}{\partial y} \\ \frac{\partial}{\partial z} & 0 & \frac{\partial}{\partial x} \end{bmatrix} \begin{Bmatrix} u \\ v \\ w \end{Bmatrix} \tag{2.70}$$

or

$$\boldsymbol{\varepsilon} = \mathbf{Ae}$$

and as before $\mathbf{B} = \mathbf{AS}$. For example, the rectangular brick-shaped element shown in Figure 2.7, which has eight corner nodes, would have shape functions of the form

$$N_1 = \left(1 - \frac{x}{a}\right)\left(1 - \frac{y}{b}\right)\left(1 - \frac{z}{c}\right) \tag{2.71}$$

and so on. The full $\mathbf{S}$ matrix would be of the form

$$\mathbf{S} = \begin{bmatrix} \mathbf{N}_u & \mathbf{0} & \mathbf{0} \\ \mathbf{0} & \mathbf{N}_v & \mathbf{0} \\ \mathbf{0} & \mathbf{0} & \mathbf{N}_w \end{bmatrix} \tag{2.72}$$

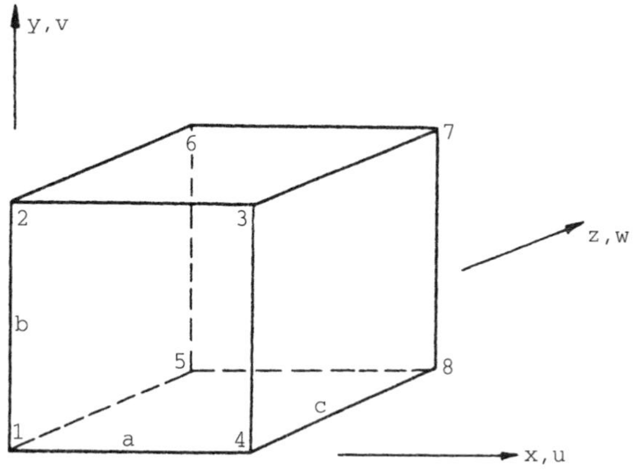

Figure 2.7 Rectangular brick element

where, assuming the same shape functions in x, y and z,

$$\mathbf{N}_u = \mathbf{N}_v = \mathbf{N}_w = [N_1 \quad N_2 \quad N_3 \quad N_4 \quad N_5 \quad N_6 \quad N_7 \quad N_8] \tag{2.73}$$

leading to the assembly of **B**. In the programs, **S** is in fact arranged with the u, v, w contributions from node 1 first, then node 2 and so on. The stress–strain matrix is in this case

$$\mathbf{D} = \frac{E(1-\nu)}{(1+\nu)(1-2\nu)} \begin{bmatrix} 1 & \frac{\nu}{1-\nu} & \frac{\nu}{1-\nu} & 0 & 0 & 0 \\ \frac{\nu}{1-\nu} & 1 & \frac{\nu}{1-\nu} & 0 & 0 & 0 \\ \frac{\nu}{1-\nu} & \frac{\nu}{1-\nu} & 1 & 0 & 0 & 0 \\ 0 & 0 & 0 & \frac{1-2\nu}{2(1-\nu)} & 0 & 0 \\ 0 & 0 & 0 & 0 & \frac{1-2\nu}{2(1-\nu)} & 0 \\ 0 & 0 & 0 & 0 & 0 & \frac{1-2\nu}{2(1-\nu)} \end{bmatrix} \tag{2.74}$$

2.15 SUMMARY OF ELEMENT EQUATIONS FOR SOLIDS

The preceding sections have demonstrated the essential similarity of all problems in linear elastic solid mechanics when formulated in terms of finite elements. The statement of element properties is to be found in two expressions, namely the element stiffness matrix

$$\mathbf{KM} = \int \mathbf{B}^{\mathrm{T}}\mathbf{DB}\,\mathrm{d(element)} \tag{2.75}$$

and the element mass matrix

$$\mathbf{MM} = \rho \int \mathbf{N}^{\mathrm{T}}\mathbf{N}\,\mathrm{d(element)} \tag{2.76}$$

These expressions then appear in the three main classes of problem which concern us in engineering practice, namely:

1. Static equilibrium problems $\mathbf{KMr} = \mathbf{f}$ (2.77)
2. Eigenvalue problems $(\mathbf{KM} - \omega^2\,\mathbf{MM})\mathbf{a} = \mathbf{0}$ (2.78)
3. Propagation problems $\mathbf{KMr} + \mathbf{MM}\dfrac{\mathrm{d}^2\mathbf{r}}{\mathrm{d}t^2} = \mathbf{f}(t)$ (2.79)

Equations (2.77) are simultaneous equations which can be solved for known forces **f** to give equilibrium displacements **r**. Equations (2.78) may be solved by various techniques (iteration, QR algorithm, etc.; see Bathe (1996) or Jennings and Mckeown (1992) to yield mode shapes **a** and natural frequencies ω of elastic systems, while equations (2.79) can be solved by advancing step by step in time from a known initial condition.

Later chapters in the book describe programs that enable the user to solve practical engineering problems which are governed by these three basic equations. Additional features, such as treatment of nonlinearity, damping, etc., will be dealt with in these chapters as they arise.

2.16 FLOW OF FLUIDS: NAVIER–STOKES EQUATIONS

We shall be concerned only with the equations governing the motion of viscous, incompressible fluids. These equations are widely developed elsewhere, e.g. Schlichting (1960). Preserving an analogy with previous sections on 2-d solids, u and v now become velocities in the x and y directions respectively and ρ is the mass density as before. Also as before, F_x and F_y are body forces in the appropriate directions.

Conservation of mass leads to

$$\frac{\partial \rho}{\partial t} + \frac{\partial}{\partial x}(\rho u) + \frac{\partial}{\partial y}(\rho v) = 0 \tag{2.80}$$

but due to incompressibility this may be reduced to

$$\frac{\partial u}{\partial x} + \frac{\partial v}{\partial y} = 0 \tag{2.81}$$

Conservation of momentum leads to

$$\begin{aligned} \rho\left(\frac{\partial u}{\partial t} + u\frac{\partial u}{\partial x} + v\frac{\partial u}{\partial y}\right) &= F_x + \left(\frac{\partial \sigma_x}{\partial x} + \frac{\partial \tau_{xy}}{\partial y}\right) \\ \rho\left(\frac{\partial v}{\partial t} + u\frac{\partial v}{\partial x} + v\frac{\partial v}{\partial y}\right) &= F_y + \left(\frac{\partial \sigma_y}{\partial y} + \frac{\partial \tau_{xy}}{\partial x}\right) \end{aligned} \tag{2.82}$$

where σ_x, σ_y and τ_{xy} are stress components as previously defined for solids.

Introducing the simplest constitutive parameters μ (the molecular viscosity), λ (taken to be $-2/3\mu$) and p (the fluid pressure) the following form of the stress equations is reached:

$$\begin{aligned} \sigma_x &= -p + \lambda\left(\frac{\partial u}{\partial x} + \frac{\partial v}{\partial y}\right) + 2\mu\frac{\partial u}{\partial x} \\ \sigma_y &= -p + \lambda\left(\frac{\partial u}{\partial x} + \frac{\partial v}{\partial y}\right) + 2\mu\frac{\partial v}{\partial y} \\ \tau_{xy} &= \tau_{yx} = \mu\left(\frac{\partial u}{\partial y} + \frac{\partial v}{\partial x}\right) \end{aligned} \tag{2.83}$$

Combining (2.81) to (2.83) a form of the "Navier–Stokes" equations can be written:

$$\frac{\partial u}{\partial t}+u\frac{\partial u}{\partial x}+v\frac{\partial u}{\partial y}=\frac{1}{\rho}F_x-\frac{1}{\rho}\frac{\partial p}{\partial x}+\frac{1}{3}\frac{\mu}{\rho}\frac{\partial}{\partial x}\left(\frac{\partial u}{\partial x}+\frac{\partial v}{\partial y}\right)+\frac{\mu}{\rho}\left(\frac{\partial^2 u}{\partial x^2}+\frac{\partial^2 u}{\partial y^2}\right)$$
$$\frac{\partial v}{\partial t}+u\frac{\partial v}{\partial x}+v\frac{\partial v}{\partial y}=\frac{1}{\rho}F_y-\frac{1}{\rho}\frac{\partial p}{\partial y}+\frac{1}{3}\frac{\mu}{\rho}\frac{\partial}{\partial y}\left(\frac{\partial u}{\partial x}+\frac{\partial v}{\partial y}\right)+\frac{\mu}{\rho}\left(\frac{\partial^2 v}{\partial x^2}+\frac{\partial^2 v}{\partial y^2}\right)$$
$$(2.84)$$

On introduction of the incompressibility condition, these can be further simplified to

$$\begin{aligned}\frac{\partial u}{\partial t}+u\frac{\partial u}{\partial x}+v\frac{\partial u}{\partial y}&=\frac{1}{\rho}F_x-\frac{1}{\rho}\frac{\partial p}{\partial x}+\frac{\mu}{\rho}\left(\frac{\partial^2 u}{\partial x^2}+\frac{\partial^2 u}{\partial y^2}\right)\\ \frac{\partial v}{\partial t}+u\frac{\partial v}{\partial x}+v\frac{\partial v}{\partial y}&=\frac{1}{\rho}F_y-\frac{1}{\rho}\frac{\partial p}{\partial y}+\frac{\mu}{\rho}\left(\frac{\partial^2 v}{\partial x^2}+\frac{\partial^2 v}{\partial y^2}\right)\end{aligned}\qquad(2.85)$$

For steady-state conditions the terms $\partial u/\partial t$ and $\partial v/\partial t$ can be dropped resulting in "coupled" equations in the "primitive" variables, u, v and p. The equations are also, in contrast to those of solid elasticity, nonlinear due to the presence of products like $u(\partial u/\partial x)$.

Ignoring body forces for the present, the steady-state equations to be solved are

$$\begin{aligned}u\frac{\partial u}{\partial x}+v\frac{\partial u}{\partial y}+\frac{1}{\rho}\frac{\partial p}{\partial x}-\frac{\mu}{\rho}\left(\frac{\partial^2 u}{\partial x^2}+\frac{\partial^2 u}{\partial y^2}\right)&=0\\ u\frac{\partial v}{\partial x}+v\frac{\partial v}{\partial y}+\frac{1}{\rho}\frac{\partial p}{\partial y}-\frac{\mu}{\rho}\left(\frac{\partial^2 v}{\partial x^2}+\frac{\partial^2 v}{\partial y^2}\right)&=0\end{aligned}\qquad(2.86)$$

Proceeding as before, and for the moment assuming the same shape functions are applied to all variables, $u=\mathbf{Nu}$, $v=\mathbf{Nv}$, $p=\mathbf{Np}$ we have for a single element, treating the terms u and v as constants $\bar{u}=\mathbf{Nu}_0$ and $\bar{v}=\mathbf{Nv}_0$ for the purposes of integration:

$$\begin{aligned}\bar{u}\frac{\partial\mathbf{N}}{\partial x}\mathbf{u}+\bar{v}\frac{\partial\mathbf{N}}{\partial y}\mathbf{u}+\frac{1}{\rho}\frac{\partial\mathbf{N}}{\partial x}\mathbf{p}-\frac{\mu}{\rho}\frac{\partial^2\mathbf{N}}{\partial x^2}\mathbf{u}-\frac{\mu}{\rho}\frac{\partial^2\mathbf{N}}{\partial y^2}\mathbf{u}&=\mathbf{0}\\ \bar{u}\frac{\partial\mathbf{N}}{\partial x}\mathbf{v}+\bar{v}\frac{\partial\mathbf{N}}{\partial y}\mathbf{v}+\frac{1}{\rho}\frac{\partial\mathbf{N}}{\partial y}\mathbf{p}-\frac{\mu}{\rho}\frac{\partial^2\mathbf{N}}{\partial x^2}\mathbf{v}-\frac{\mu}{\rho}\frac{\partial^2\mathbf{N}}{\partial y^2}\mathbf{v}&=\mathbf{0}\end{aligned}\qquad(2.87)$$

Multiplying by the weighting functions and integrating as usual yields

$$\begin{aligned}&\iint\mathbf{N}\bar{u}\frac{\partial\mathbf{N}}{\partial x}\mathbf{u}\,\mathrm{d}x\,\mathrm{d}y+\iint\mathbf{N}\bar{v}\frac{\partial\mathbf{N}}{\partial y}\mathbf{u}\,\mathrm{d}x\,\mathrm{d}y+\frac{1}{\rho}\iint\mathbf{N}\frac{\partial\mathbf{N}}{\partial x}\mathbf{p}\,\mathrm{d}x\,\mathrm{d}y\\ &\quad-\frac{\mu}{\rho}\iint\mathbf{N}\frac{\partial^2\mathbf{N}}{\partial x^2}\mathbf{u}\,\mathrm{d}x\,\mathrm{d}y-\frac{\mu}{\rho}\iint\mathbf{N}\frac{\partial^2\mathbf{N}}{\partial y^2}\mathbf{u}\,\mathrm{d}x\,\mathrm{d}y=\mathbf{0}\end{aligned}\qquad(2.88)$$

and

$$\iint \mathbf{N}\bar{u}\frac{\partial \mathbf{N}}{\partial x}\mathbf{v}\,dx\,dy + \iint \mathbf{N}\bar{v}\frac{\partial \mathbf{N}}{\partial y}\mathbf{v}\,dx\,dy + \frac{1}{\rho}\iint \mathbf{N}\frac{\partial \mathbf{N}}{\partial y}\mathbf{p}\,dx\,dy - \frac{\mu}{\rho}\iint \mathbf{N}\frac{\partial^2 \mathbf{N}}{\partial x^2}\mathbf{v}\,dx\,dy - \frac{\mu}{\rho}\iint \mathbf{N}\frac{\partial^2 \mathbf{N}}{\partial y^2}\mathbf{v}\,dx\,dy = \mathbf{0}$$

Integrating products by parts where necessary and neglecting resulting contour integrals gives

$$\begin{aligned}
&\iint \mathbf{N}\bar{u}\frac{\partial \mathbf{N}}{\partial x}\,dx\,dy\,\mathbf{u} + \iint \mathbf{N}\bar{v}\frac{\partial \mathbf{N}}{\partial y}\,dx\,dy\,\mathbf{u} + \frac{1}{\rho}\iint \mathbf{N}\frac{\partial \mathbf{N}}{\partial x}\,dx\,dy\,\mathbf{p} \\
&\qquad + \frac{\mu}{\rho}\iint \frac{\partial \mathbf{N}}{\partial x}\frac{\partial \mathbf{N}}{\partial x}\,dx\,dy\,\mathbf{u} + \frac{\mu}{\rho}\iint \frac{\partial \mathbf{N}}{\partial y}\frac{\partial \mathbf{N}}{\partial y}\,dx\,dy\,\mathbf{u} = \mathbf{0} \\
&\iint \mathbf{N}\bar{u}\frac{\partial \mathbf{N}}{\partial x}\,dx\,dy\,\mathbf{v} + \iint \mathbf{N}\bar{v}\frac{\partial \mathbf{N}}{\partial y}\,dx\,dy\,\mathbf{v} + \frac{1}{\rho}\iint \mathbf{N}\frac{\partial \mathbf{N}}{\partial y}\,dx\,dy\,\mathbf{p} \\
&\qquad + \frac{\mu}{\rho}\iint \frac{\partial \mathbf{N}}{\partial x}\frac{\partial \mathbf{N}}{\partial x}\,dx\,dy\,\mathbf{v} + \frac{\mu}{\rho}\iint \frac{\partial \mathbf{N}}{\partial y}\frac{\partial \mathbf{N}}{\partial y}\,dx\,dy\,\mathbf{v} = \mathbf{0}
\end{aligned} \tag{2.89}$$

The set of equations is completed by the continuity condition

$$\iint \mathbf{N}\left(\frac{\partial \mathbf{N}}{\partial x}\mathbf{u} + \frac{\partial \mathbf{N}}{\partial y}\mathbf{v}\right)dx\,dy = \mathbf{0} \tag{2.90}$$

Collecting terms in **u**, **p** and **v** respectively leads to an equilibrium equation (Taylor and Hughes, 1981):

$$\begin{bmatrix} \mathbf{C}_{11} & \mathbf{C}_{12} & \mathbf{C}_{13} \\ \mathbf{C}_{21} & \mathbf{C}_{22} & \mathbf{C}_{23} \\ \mathbf{C}_{31} & \mathbf{C}_{32} & \mathbf{C}_{33} \end{bmatrix} \begin{Bmatrix} \mathbf{u} \\ \mathbf{p} \\ \mathbf{v} \end{Bmatrix} = \begin{Bmatrix} \mathbf{0} \\ \mathbf{0} \\ \mathbf{0} \end{Bmatrix} \tag{2.91}$$

where

$$\begin{aligned}
\mathbf{C}_{11} &= \iint \left(\mathbf{N}\bar{u}\frac{\partial \mathbf{N}}{\partial x} + \mathbf{N}\bar{v}\frac{\partial \mathbf{N}}{\partial y} + \frac{\mu}{\rho}\frac{\partial \mathbf{N}}{\partial x}\frac{\partial \mathbf{N}}{\partial x} + \frac{\mu}{\rho}\frac{\partial \mathbf{N}}{\partial y}\frac{\partial \mathbf{N}}{\partial y}\right)dx\,dy \\
\mathbf{C}_{12} &= \iint \frac{1}{\rho}\mathbf{N}\frac{\partial \mathbf{N}}{\partial x}\,dx\,dy \\
\mathbf{C}_{13} &= \mathbf{0} \\
\mathbf{C}_{21} &= \iint \mathbf{N}\frac{\partial \mathbf{N}}{\partial x}\,dx\,dy \\
\mathbf{C}_{22} &= \mathbf{0} \\
\mathbf{C}_{23} &= \iint \mathbf{N}\frac{\partial \mathbf{N}}{\partial y}\,dx\,dy \\
\mathbf{C}_{31} &= \mathbf{0} \\
\mathbf{C}_{32} &= \iint \frac{1}{\rho}\mathbf{N}\frac{\partial \mathbf{N}}{\partial y}\,dx\,dy \\
\mathbf{C}_{33} &= \mathbf{C}_{11}
\end{aligned} \tag{2.92}$$

Referring to Table 2.1 we now have many terms of the type $N_i(\partial N_j/\partial x)$ which imply unsymmetrical structures for $\mathbf{C}_{ij}$. Thus special solution algorithms will be necessary. Computational details are left until Chapter 9.

2.17 SIMPLIFIED FLOW EQUATIONS

In many practical instances it may not be necessary to solve the complete coupled system described in the previous section. The pressure p can be eliminated from (2.85) and if vorticity ω is defined as

$$\omega = \frac{\partial u}{\partial y} - \frac{\partial v}{\partial x} \tag{2.93}$$

this results in a single equation:

$$\frac{\partial \omega}{\partial t} + u\frac{\partial \omega}{\partial x} + v\frac{\partial \omega}{\partial y} = \frac{\mu}{\rho}\left(\frac{\partial^2 \omega}{\partial x^2} + \frac{\partial^2 \omega}{\partial y^2}\right) \tag{2.94}$$

Defining a stream function ψ such that

$$\begin{aligned} u &= \frac{\partial \psi}{\partial y} \\ v &= -\frac{\partial \psi}{\partial y} \end{aligned} \tag{2.95}$$

an alternative coupled system involving ψ and ω can be devised, given here for steady-state conditions:

$$\begin{aligned} \frac{\partial^2 \psi}{\partial x^2} + \frac{\partial^2 \psi}{\partial y^2} &= \omega \\ \frac{\mu}{\rho}\left(\frac{\partial^2 \omega}{\partial x^2} + \frac{\partial^2 \omega}{\partial y^2}\right) &= \frac{\partial \psi}{\partial y}\frac{\partial \omega}{\partial x} - \frac{\partial \psi}{\partial x}\frac{\partial \omega}{\partial y} \end{aligned} \tag{2.96}$$

This clearly has the advantage that only two unknowns are involved rather than the previous three. However, the solution of (2.96) is still a relatively complicated process and flow problems are sometimes solved via equation (2.94) alone, assuming that u and v can be approximated by some independent means or measured. In this form, equation (2.94) is an example of the "diffusion–convection" equation, the second–order space derivatives corresponding to a "diffusion" process and the first order ones to a "convection" process. The equation arises in various areas of engineering, for example sediment transport and pollutant disposal (Smith, 1976, 1979).

If there is no convection, the resulting equation is of the type

$$\frac{\partial \omega}{\partial t} = \frac{\mu}{\rho}\left(\frac{\partial^2 \omega}{\partial x^2} + \frac{\partial^2 \omega}{\partial y^2}\right) \tag{2.97}$$

which is the "heat conduction" or 'diffusion' equation well known in many areas of engineering.

A final simplification is a reduction to steady-state conditions, in which case

$$\frac{\partial^2 \omega}{\partial x^2} + \frac{\partial^2 \omega}{\partial y^2} = 0 \tag{2.98}$$

leaving the familiar "Laplace" equation. In the following sections, finite element formulations of these simplified flow equations are described, in order of increasing complexity.

2.18 SIMPLIFIED FLUID FLOW: STEADY STATE

The form of Laplace's equation (2.98) which arises in geomechanics, for example concerning groundwater flow in an aquifer (Muskat, 1937), is

$$k_x \frac{\partial^2 \phi}{\partial x^2} + k_y \frac{\partial^2 \phi}{\partial y^2} = 0 \tag{2.99}$$

where ϕ is the fluid potential and k_x, k_y are permeabilities in the x and y directions. The finite element discretisation process reduces the differential equation to a set of equilibrium type simultaneous equations of the form

$$\mathbf{KP}\boldsymbol{\phi} = \mathbf{q} \tag{2.100}$$

where **KP** is the symmetrical "conductivity matrix" and **q** is a vector of net nodal inflows/outflows.

Reference to Table 2.1 shows that typical terms in the matrix **KP** are of the form

$$\int\int \left(k_x \frac{\partial N_i}{\partial x}\frac{\partial N_j}{\partial x} + k_y \frac{\partial N_i}{\partial y}\frac{\partial N_j}{\partial y} \right) \mathrm{d}x\,\mathrm{d}y \tag{2.101}$$

With the usual finite element discretisation

$$\phi = \mathbf{N}\boldsymbol{\phi} \tag{2.102}$$

a convenient way of expressing the matrix **KP** in (2.100) is

$$\mathbf{KP} = \int\int \mathbf{T}^{\mathrm{T}}\mathbf{KT}\,\mathrm{d}x\,\mathrm{d}y \tag{2.103}$$

where the property matrix **K** is analogous to the stress–strain matrix **D** in solid mechanics; thus

$$\mathbf{K} = \begin{bmatrix} k_x & 0 \\ 0 & k_y \end{bmatrix} \tag{2.104}$$

(assuming that the axes of the permeability tensor coincide with x and y). The **T** matrix is similar to the **B** matrix of solid mechanics and is given by

$$\mathbf{T} = \begin{bmatrix} \frac{\partial N_1}{\partial x} & \frac{\partial N_2}{\partial x} & \frac{\partial N_3}{\partial x} & \frac{\partial N_4}{\partial x} \\ \frac{\partial N_1}{\partial y} & \frac{\partial N_2}{\partial y} & \frac{\partial N_3}{\partial y} & \frac{\partial N_4}{\partial y} \end{bmatrix} \tag{2.105}$$

The similarity between (2.103) for a fluid and (2.63) for a solid enables the corresponding programs to look similar in spite of the governing differential equations being quite different. This unity of treatment is utilised in describing the programming techniques in Chapter 3.

Finally, it is worth noting that (2.103) can also be arrived at from energy considerations. The equivalent energy statement is that the integral

$$\iint \left[\frac{1}{2} k_x \left(\frac{\partial \phi}{\partial x} \right)^2 + \frac{1}{2} k_y \left(\frac{\partial \phi}{\partial y} \right)^2 \right] \mathrm{d}x \, \mathrm{d}y \tag{2.106}$$

shall be a minimum for all possible $\phi(x, y)$.

Example solutions to steady-state problems described by (2.99) are given in Chapter 7.

2.19 SIMPLIFIED FLUID FLOW: TRANSIENT STATE

Transient conditions must be analysed in many physical situations, for example in the case of Terzaghi "consolidation" in soil mechanics. The governing consolidation diffusion equation for excess pore pressure u_w takes the form

$$c_x \frac{\partial^2 u_w}{\partial x^2} + c_y \frac{\partial^2 u_w}{\partial y^2} = \frac{\partial u_w}{\partial t} \tag{2.107}$$

where c_x, c_y are the coefficients of consolidation. Discretisation of the left-hand side of (2.107) clearly follows that of (2.99) while the time derivative will be associated with a matrix of the "mass matrix" type without the multiple ρ. Hence the discretised system is

$$\mathbf{KPu}_w + \mathbf{PM} \frac{\mathrm{d}\mathbf{u}_w}{\mathrm{d}t} = \mathbf{0} \tag{2.108}$$

This set of first-order, ordinary differential equations can be solved by many methods, the simplest of which discretise the time derivative by finite differences. The algorithms are described in Chapter 3 with example solutions in Chapter 8.

2.20 SIMPLIFIED FLUID FLOW WITH ADVECTION

If pollutants, sediments, tracers, etc., are transported by a laminar flow system they are at the same time translated or "advected" by the flow and diffused within it. The governing differential equation for the 2-d case is (Smith et al, 1973)

$$c_x \frac{\partial^2 \phi}{\partial x^2} + c_y \frac{\partial^2 \phi}{\partial y^2} - u \frac{\partial \phi}{\partial x} - v \frac{\partial \phi}{\partial y} = \frac{\partial \phi}{\partial t} \tag{2.109}$$

where u and v are the fluid velocity components in the x and y directions (compare equation 2.94).

The extra advection terms $-u(\partial\phi/\partial x)$ and $-v(\partial\phi/\partial y)$ compared with (2.107) lead, as shown in Table 2.1, to unsymmetric components of the "stiffness" matrix of the type

$$\iint \left(-uN_i \frac{\partial N_j}{\partial x} - vN_i \frac{\partial N_j}{\partial y} \right) \mathrm{d}x\,\mathrm{d}y \tag{2.110}$$

which must be added to the symmetric, diffusion components given in (2.101). When this has been done, equilibrium equations like (2.100) or transient equations like (2.108) are regained.

Mathematically, equation (2.109) is a differential equation which is not self-adjoint (Berg, 1962), due to the presence of the first-order spatial derivatives. From a finite element point of view, equations which are not self-adjoint will always lead to unsymmetrical stiffness matrices.

A second consequence of non-self-adjoint equations is that there is no energy formulation equivalent to (2.106). It is clearly a benefit of the Galerkin approach that it can be used for all types of equation and is not restricted to self-adjoint systems.

Equation (2.109) can be rendered self-adjoint by using the transformation

$$\phi = h \exp\left(\frac{ux}{2c_x}\right) \exp\left(\frac{vy}{2c_y}\right) \tag{2.111}$$

but this is not recommended unless u and v are small compared with c_x and c_y, as shown by Smith et al (1973).

Equation (2.109) and the use of (2.111) are described in Chapter 8.

2.21 FURTHER COUPLED EQUATIONS: BIOT CONSOLIDATION

Thus far in this chapter, analyses of solids and fluids have been considered separately. However, Biot formulated the theory of coupled solid–fluid interaction which finds application in soil mechanics (Smith and Hobbs,

1976). The soil skeleton is treated as a porous elastic solid and the laminar pore fluid is coupled to the solid by the conditions of compressibility and of continuity. Thus Biot's governing equation is given by

$$\frac{K'}{\gamma_w}\left[k_x\frac{\partial^2 u_w}{\partial x^2}+k_y\frac{\partial^2 u_w}{\partial y^2}+k_z\frac{\partial^2 u_w}{\partial z^2}\right]=\frac{\partial u_w}{\partial t}-\frac{\partial p}{\partial t} \tag{2.112}$$

where K' is the soil bulk modulus and p is the mean total stress.

For 2-d equilibrium in the absence of body forces, the gradient of *effective* stress from (2.48) must be augmented by the gradients of the fluid pressure u_w as follows:

$$\begin{aligned}\frac{\partial \sigma'_x}{\partial x}+\frac{\partial \tau_{xy}}{\partial y}+\frac{\partial u_w}{\partial x}&=0\\ \frac{\partial \tau_{xy}}{\partial x}+\frac{\partial \sigma'_y}{\partial y}+\frac{\partial u_w}{\partial y}&=0\end{aligned} \tag{2.113}$$

where σ'_x etc. are the effective stresses $(\sigma_x - u_w)$.

The constitutive laws are those previously defined for the solid and fluid respectively; hence in plane strain

$$\begin{Bmatrix}\sigma'_x\\ \sigma'_y\\ \tau_{xy}\end{Bmatrix}=\frac{E'(1-\nu')}{(1+\nu')(1-2\nu')}\begin{bmatrix}1 & \dfrac{\nu'}{1-\nu'} & 0\\ \dfrac{\nu'}{1-\nu'} & 1 & 0\\ 0 & 0 & \dfrac{1-2\nu'}{2(1-\nu')}\end{bmatrix}\begin{Bmatrix}\varepsilon_x\\ \varepsilon_y\\ \gamma_{xy}\end{Bmatrix} \tag{2.114}$$

and

$$\begin{Bmatrix}q_x\\ q_y\end{Bmatrix}=\frac{1}{\gamma_w}\begin{bmatrix}k_x & 0\\ 0 & k_y\end{bmatrix}\begin{Bmatrix}\partial u_w/\partial x\\ \partial u_w/\partial y\end{Bmatrix} \tag{2.115}$$

where q_x and q_y are the volumetric flow rates per unit area into and out of the element and γ_w is the unit weight of water. The solid strain–displacement relations are still given by (2.50), and the final condition is that for full saturation and, in this case incompressibility, outflow from an element of soil equals the reduction in volume of the element. Hence

$$\frac{\partial q_x}{\partial x}+\frac{\partial q_y}{\partial y}=-\frac{\mathrm{d}}{\mathrm{d}t}\left(\frac{\partial u}{\partial x}+\frac{\partial v}{\partial y}\right) \tag{2.116}$$

and from (2.115) the third differential equation is given by

$$\frac{k_x}{\gamma_w}\frac{\partial^2 u_w}{\partial x^2}+\frac{k_y}{\gamma_w}\frac{\partial^2 u_w}{\partial y^2}+\frac{\mathrm{d}}{\mathrm{d}t}\left(\frac{\partial u}{\partial x}+\frac{\partial v}{\partial y}\right)=0 \tag{2.117}$$

As usual in a displacement method, σ and ε are eliminated in terms of u and v so that the final coupled variables are u, v and u_w. These are now discretised in the normal way:

$$\begin{aligned} u &= \mathbf{Nu} \\ v &= \mathbf{Nv} \\ u_{\mathrm{w}} &= \mathbf{Nu}_{\mathrm{w}} \end{aligned} \tag{2.118}$$

In practice, it may be preferable to use a higher order of discretisation for u and v compared with u_{w} but, for the present, the same shape functions are assumed to describe all three variables.

When discretisation and the Galerkin process are completed, (2.113) and (2.117) lead to the pair of equilibrium and continuity equations:

$$\begin{aligned} \mathbf{KMr} + \mathbf{Cu}_{\mathrm{w}} &= \mathbf{f} \\ \mathbf{C}^{\mathrm{T}} \frac{\mathrm{d}\mathbf{r}}{\mathrm{d}t} - \mathbf{KPu}_{\mathrm{w}} &= \mathbf{0} \end{aligned} \tag{2.119}$$

where, for a four-noded element,

$$\mathbf{r} = \begin{Bmatrix} u_1 \\ v_1 \\ u_2 \\ v_2 \\ u_3 \\ v_3 \\ u_4 \\ v_4 \end{Bmatrix} \quad \text{and} \quad \mathbf{u}_{\mathrm{w}} = \begin{Bmatrix} u_{\mathrm{w}1} \\ u_{\mathrm{w}2} \\ u_{\mathrm{w}3} \\ u_{\mathrm{w}4} \end{Bmatrix} \tag{2.120}$$

KM and **KP** are the familiar elastic and fluid "stiffness" matrices and **C** is a new rectangular coupling matrix consisting of terms of the form

$$\iint \frac{\partial N_j}{\partial x} N_i \, \mathrm{d}x \, \mathrm{d}y \tag{2.121}$$

f is the external loading vector. After assembly into global matrices, equations (2.119) must be integrated in time by some method such as finite differences and this is described further in Chapter 3. Examples of such solutions in practice are given in Chapter 9.

2.22 CONCLUSIONS

When viewed from a finite element standpoint, all static equilibrium problems, whether involving solids or fluids, take the same form, namely

$$\mathbf{KMr} = \mathbf{f}$$

or

$$\mathbf{KP}\boldsymbol{\phi} = \mathbf{q} \tag{2.122}$$

For simple uncoupled problems the solid **KM** and fluid **KP** matrices have similar symmetrical structures and computer programs to construct them will be similar. However, for coupled problems such as are described by the Navier–Stokes equations, the matrices are unsymmetrical and appropriate alternative software will be necessary.

In the same way, eigenvalue, propagation and transient problems all involve the mass matrix **MM** (or a simple multiple of it, **PM**). Therefore, coding of these different types of solution can be expected to contain sections common to all three problems.

So far, single elements have been considered in the discretisation process, and only the simplest line and rectangular elements have been described. The next chapter is mainly devoted to a description of programming strategy, but before this, the finite element concept is extended to embrace meshes of interlinked elements and elements of general shape.

2.23 REFERENCES

Bathe, K. J. (1996) *Numerical Methods in Finite Element Analysis*, 3rd ed. Prentice-Hall, Englewood Cliffs, New Jersey.

Berg, P. N. (1962) Calculus of variations. In *Handbook of Engineering Mechanics*, Chapter 16, ed. W. Flugge, McGraw-Hill, New York.

Cook, R. D. (1989) *Concepts and Applications of Finite Element Analysis,* 3rd edn. Wiley, New York.

Finlayson, B. A. (1972) *The Method of Weighted Residuals and Variational Principles.* Academic Press, New York.

Griffiths, D. V. (1989) Advantages of consistent over lumped methods for analysis of beams on elastic foundations, *Comm. Appl. Num. Meths.*, **5**, 53–60.

Griffiths, D. V. and Smith, I. M. (1991) *Numerical Methods for Engineers*, Blackwell, Oxford.

Horne, M. R. and Merchant, W. (1965) *The Stability of Frames.* Pergamon Press, Oxford.

Jennings, A. and Mckeown, J. (1992) *Matrix Computation for Engineers and Scientists*, 2nd edn. John Wiley, London.

Leckie, F. A. and Lindberg, G. M. (1963) The effect of lumped parameters on beam frequencies. *The Aeronautical Quarterly*, **14**, 234.

Livesley, R. K. (1975) *Matrix Methods of Structural Analysis*. Pergamon Press, Oxford.

Lundquist, E. E. and Kroll, W. D. (1944) *NACA Report ARR, No. 4824.*

Muskat, M. (1937) *The Flow of Homogeneous Fluids through Porous Media.* McGraw-Hill, New York.

Rao, S. S. (1989) *The Finite Element Method in Engineering,* 2nd edn. Pergamon Press, Oxford.

Schlichting, H. (1960) *Boundary Layer Theory.* McGraw-Hill, New York.

Smith, I. M. (1976) Integration in time of diffusion and diffusion–convection equations. In *Finite Elements in Water Resources*, ed. W. G. Gray, G. F. Pinder and C. A. Brebbia, pp. 1.3–1.20. Pentech Press, Southampton.

Smith, I. M. (1979) The diffusion-convection equation. In *A Survey of Numerical Methods for Partial Differential Equations,* ed. I. Gladwell and R. Wait, pp. 195–211. Oxford University Press.

Smith I. M., Farraday, R. V. and O'Connor, B. A. (1973) Rayleigh–Ritz and Galerkin finite elements for diffusion-convection problems. *Water Resources Research*, **9**(3), 593.

Smith, I. M. and Hobbs, R. (1976) Biot analysis of consolidation beneath embankments. *Géotechnique*, **26**(1), 149.

Strang, G. and Fix, G. J. (1973) *An Analysis of the Finite Element Method.* Prentice-Hall, Englewood Cliffs, New Jersey.

Szabo, B. A. and Lee, G. C. (1969) Derivation of stiffness matrices for problems in plane elasticity by the Galerkin method. *Int. J. Num. Meth. Eng.*, **1**, 301.

Taig, I. C. (1961) Structural analysis by the matrix displacement method. *English Electric Aviation Report No. SO17*, Preston.

Taylor, C. and Hughes, T. G. (1981) *Finite Element Programming of the Navier–Stokes Equation.* Pineridge Press Ltd, Swansea.

Timoshenko, S. P. and Goodier, J. N. (1951) *Theory of Elasticity.* McGraw-Hill, New York.

Zienkiewicz, O. C. (1991) *The Finite Element Method in Engineering Science*, 4th edn. McGraw-Hill, London.

3 Programming Finite Element Computations

3.0 INTRODUCTION

In Chapter 2, the finite element spatial discretisation process was described, whereby partial differential equations can be replaced by matrix equations which take the form of linear and nonlinear algebraic equations, eigenvalue equations or ordinary differential equations in the time variable. The present chapter describes how programs can be constructed in order to formulate and solve these kinds of equations.

Before this, two additional features must be introduced. First, we have so far dealt only with the simplest shapes of elements, namely lines and rectangles. Obviously if differential equations are to be solved over regions of general shape, elements must be allowed to assume general shapes as well. This is accomplished by introducing general triangular, quadrilateral, tetrahedral and hexahedral elements together with the concept of a coordinate system local to the element.

Second, we have so far considered only a single element, whereas useful solutions will normally be obtained by many elements, usually from hundreds to millions in practice, joined together at the nodes. Also, various types of boundary conditions may be prescribed which constrain the solution in some way.

Local coordinate systems, multi-element analyses and incorporation of boundary conditions are all explained in the sections that follow.

3.1 LOCAL COORDINATES FOR QUADRILATERAL ELEMENTS

Figure 3.1 shows two types of plane four-noded quadrilateral elements. The shape functions for the rectangle were shown to be given by equations (2.57), namely $N_1 = (1 - x/a)(1 - y/b)$ and so on. If it is attempted to construct similar shape functions in the "global" coordinates (x, y) for the general quadrilateral, very complex algebraic expressions will result, which are best generated by computer algebra packages (Griffiths, 1994).

Traditionally the approach has been to work in a local coordinate system as shown in Figure 3.2, originally proposed by Taig (1961), and to evaluate resulting integrals numerically. The general point $P(\xi, \eta)$ within the quadrilateral is located at the intersection of two lines which cut opposite sides of the quadrilateral in equal proportions. For reasons associated with

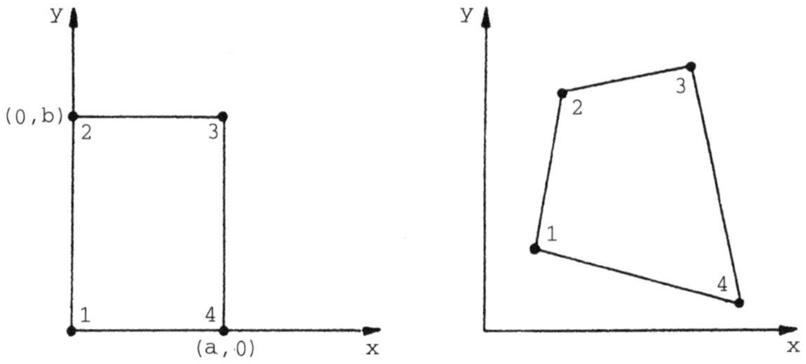

Figure 3.1 (a) Plane rectangular element. (b) Plane general quadrilateral element

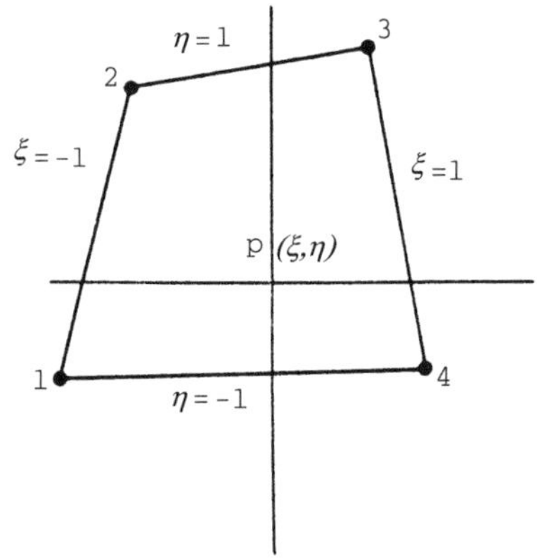

Figure 3.2 Local coordinate system for quadrilateral elements

subsequent numerical integrations it proves to be convenient to "normalise" the coordinates so that side 1 2 has $\xi = -1$, side 3 4 has $\xi = +1$, side 1 4 has $\eta = -1$ and side 2 3 has $\eta = +1$. In this system the intersection of the bisectors of opposite sides of the quadrilateral is the point (0, 0), while the corners 1, 2, 3, 4 are $(-1, -1)$, $(-1, 1)$, $(1, 1)$, $(1, -1)$ respectively.

When this choice is adopted, the shape functions for a four-noded quadrilateral with corner nodes take the simple form

$$\begin{aligned}
N_1 &= 1/4(1-\xi)(1-\eta)\\
N_2 &= 1/4(1-\xi)(1+\eta)\\
N_3 &= 1/4(1+\xi)(1+\eta)\\
N_4 &= 1/4(1+\xi)(1-\eta)
\end{aligned} \tag{3.1}$$

and these can be used to describe the variation of unknowns such as displacement or fluid potential in an element as before.

Under special circumstances the same shape functions can also be used to specify the relation between the global (x, y) and local (ξ, η) coordinate systems. If this is so the element is of a type called "isoparametric" (Ergatoudis et al., 1968; Zienkiewicz et al., 1969); the four-node quadrilateral is an example. The coordinate transformation is therefore

$$\begin{aligned} x &= N_1 x_1 + N_2 x_2 + N_3 x_3 + N_4 x_4 \\ &= \mathbf{N}\mathbf{x} \\ y &= N_1 y_1 + N_2 y_2 + N_3 y_3 + N_4 y_4 \\ &= \mathbf{N}\mathbf{y} \end{aligned} \tag{3.2}$$

where the $\mathbf{N}$ are given by (3.1).

In the previous chapter it was shown that element properties involve not only $\mathbf{N}$ but also their derivatives with respect to the global coordinates (x, y) which appear in matrices such as $\mathbf{B}$ and $\mathbf{T}$. Further, products of these quantities need to be integrated over the element area or volume.

Derivatives are easily converted from one coordinate system to the other by means of the chain rule of partial differentiation, best expressed in matrix form by

$$\begin{Bmatrix} \dfrac{\partial}{\partial \xi} \\ \dfrac{\partial}{\partial \eta} \end{Bmatrix} = \begin{Bmatrix} \dfrac{\partial x}{\partial \xi} & \dfrac{\partial y}{\partial \xi} \\ \dfrac{\partial x}{\partial \eta} & \dfrac{\partial y}{\partial \eta} \end{Bmatrix} \begin{Bmatrix} \dfrac{\partial}{\partial x} \\ \dfrac{\partial}{\partial y} \end{Bmatrix} = \mathbf{J} \begin{Bmatrix} \dfrac{\partial}{\partial x} \\ \dfrac{\partial}{\partial y} \end{Bmatrix} \tag{3.3}$$

or

$$\begin{Bmatrix} \dfrac{\partial}{\partial x} \\ \dfrac{\partial}{\partial y} \end{Bmatrix} = \mathbf{J}^{-1} \begin{Bmatrix} \dfrac{\partial}{\partial \xi} \\ \dfrac{\partial}{\partial \eta} \end{Bmatrix} \tag{3.4}$$

where $\mathbf{J}$ is the Jacobian matrix. The determinant of this matrix, $\det|\mathbf{J}|$, must also be evaluated because it is used in the transformed integrals as follows:

$$\iint \mathrm{d}x\,\mathrm{d}y = \int_{-1}^{1}\int_{-1}^{1} \det|\mathbf{J}|\,\mathrm{d}\xi\,\mathrm{d}\eta \tag{3.5}$$

Under certain circumstances, for example that shown in Figure 3.3(b), the Jacobian becomes indeterminate. When using quadrilateral elements, reflex interior angles should be avoided.

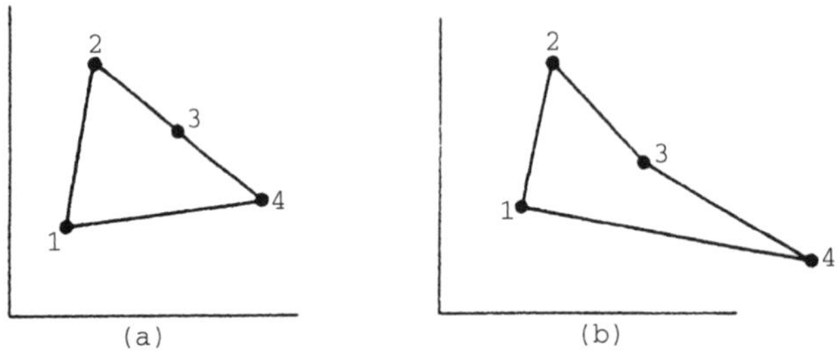

Figure 3.3 (a) Degenerate quadrilateral. (b) Unacceptable quadrilateral

3.2 NUMERICAL INTEGRATION FOR QUADRILATERALS

Although some integrals of this type can be evaluated analytically, this has traditionally been impractical for complicated functions, particularly in the general case when (ξ, η) become curvilinear. In most finite element programs (3.5) are evaluated numerically, using Gauss-Legendre quadrature over quadrilateral regions (Irons, 1966a, b). The quadrature rules in two dimensions are all of the form

$$\begin{aligned}\int_{-1}^{1}\int_{-1}^{1} f(\xi,\eta)\mathrm{d}\xi\,\mathrm{d}\eta &\simeq \sum_{i=1}^{n}\sum_{j=1}^{n} w_i w_j f(\xi_i,\eta_j)\\ &\simeq \sum_{i=1}^{\text{nip}} W_i f(\xi,\eta)_i\end{aligned} \tag{3.6}$$

where $\text{nip} = n^2$ (total number of integrating points), w_i and w_j or W_i are weighting coefficients and ξ_i, η_j are coordinate positions within the element. These values for n equal to 1, 2 and 3 are shown in Table 3.1 and complete tables are available in other sources, e.g. Kopal (1961). The table assumes that the range of integration is ± 1, hence the reason for normalising the local coordinate system in this way.

Table 3.1 Coordinates and weights in Gaussian quadrature formulae

n	nip	$\xi_i \eta_j$	$w_i w_j$
1	1	0	2
2	4	$\pm\frac{1}{\sqrt{3}}$	1
3	9	$\pm\frac{\sqrt{3}}{\sqrt{5}}$	$\frac{5}{9}$
		0	$\frac{8}{9}$

The approximate equality in (3.6) is exact for cubic functions when $n = 2$ and for quintics when $n = 3$. Usually one attempts to perform integrations over finite elements exactly, but in special circumstances (Zienkiewicz et al, 1971) "reduced" integration whereby integrals are evaluated approximately can improve the quality of solutions.

3.3 ANALYTICAL INTEGRATION FOR QUADRILATERALS

Recently, considerable progress has been made in the development of "computer algebra systems" (CAS) such as "REDUCE" and "Maple". These enable algebraic expressions (for example the finite element shape functions) to be manipulated essentially "analytically". Expressions can be differentiated, integrated, factorised and so on, leading to explicit formulations of element matrices avoiding the need for numerical integration. Particularly for 3-d elements, this approach can lead to substantial savings in integration times. A further point is that for new elements (for example a 14-node hexahedron described later in this chapter) the shape functions are so complex algebraically that it is doubtful if they could be isolated at all without the help of computer algebra.

For the four-node quadrilateral element in the context of plane elasticity, the element stiffness matrix has been shown in Chapter 2 to be given by integrals of the form

$$\mathbf{KM} = \int_{\text{vol}} \mathbf{B}^{\mathrm{T}}\mathbf{D}\mathbf{B}\,\mathrm{d}(\text{vol}) \tag{3.7}$$

where **B** and **D** represent the strain–displacement and stress–strain matrices respectively.

If the element is rectangular with its sides parallel to the x and y axes, the term under the integral consists of simple polynomial terms which can be easily integrated in closed form by separation of the variables resulting in compact terms like (2.59). In general, however, quadrilateral elements will lead to very complicated expressions under the integral sign which can only be tackled numerically.

Noting that "two-point" Gaussian quadrature, i.e. nip $= 4$, leads to an exact solution for the stiffness matrix of a four-node quadrilateral, a compromise approach is to evaluate the contribution to stiffness for each of the "Gauss points" algebraically and add them together, thus:

$$\mathbf{KM} = \sum_{i=1}^{2}\sum_{j=1}^{2} w_i w_j (\det \mathbf{J})_{ij}\, \mathbf{B}^{\mathrm{T}}\mathbf{D}\mathbf{B}_{ij} \tag{3.8}$$

where **J** is the Jacobian matrix described previously.

This at first leads to rather long expressions, but a considerable amount of cancelling and simplification is possible (for example the $1/\sqrt{3}$ term that appears in the sampling points of the Gaussian formula disappears in the simplification process). The algebraic expressions can be generated with the help of a CAS (e.g. Maple) and the risk of typographical errors can be virtually eliminated by outputting the results in Fortran format.

The simplified algebraic expressions that form the stiffness matrix of the four-node quadrilateral element have been isolated. These expressions form the basis of subroutine ANALY4 used in Program 5.2 of this book. In the same program, a similar approach has also been used to form the algebraic version of the **B** matrix of the four-node quadrilateral element corresponding to any given local coordinate (ξ, η) resulting in subroutine BEE4.

All stiffness terms are of the form

$$KM_{ij} = \frac{1}{2}\left\{\frac{A_2(E^*s_1 + Gs_2) + f_1(E^*s_3 + Gs_4)}{3A_2^2 - f_1^2} + \frac{A_2(E^*t_1 + Gt_2) + f_2(E^*t_3 + Gt_4)}{3A_2^2 - f_2^2}\right\} \tag{3.9}$$

where E^* contains elastic properties E and ν, G is the shear modulus and

$$\begin{aligned} A_2 &= (x_4 - x_2)(y_3 - y_1) - (x_3 - x_1)(y_4 - y_2) \\ &= \text{twice the area of the element} \end{aligned} \tag{3.10}$$

The functions

$$f_1 = (x_1 + x_3)(y_4 - y_2) - (y_1 + y_3)(x_4 - x_2) - 2(x_2y_4 - x_4y_2) \tag{3.11}$$

$$f_2 = (y_2 + y_4)(x_3 - x_1) - (x_2 + x_4)(y_3 - y_1) - 2(x_3y_1 - x_1y_3) \tag{3.12}$$

while s_1, s_2, s_3, s_4, t_1, t_2, t_3 and t_4 depend on the nodal coordinates, and are given in Griffiths (1994).

The same technique can be applied to other element types and other element matrices (e.g. eight-node quadrilaterals, 3-d elements, mass, conductivity, etc.).

The technique is to be found again in Program 7.1 of the book where the conductivity matrices of four-node quadrilateral elements are computed algebraically using subroutine SEEP4 to solve Laplace's equation.

3.4 LOCAL COORDINATES FOR TRIANGULAR ELEMENTS

Figure 3.4 shows how a general triangular element can be mapped into a right-angled isosceles triangle. This approach is identical to using area coordinates (Zienkiewicz et al, 1971) in which any point within the triangle can be referenced using three local coordinates (L_1, L_2, L_3). Clearly for a plane region, only two independent coordinates are necessary; hence the third coordinate is a function of the other two:

$$L_3 = 1 - L_1 - L_2 \tag{3.13}$$

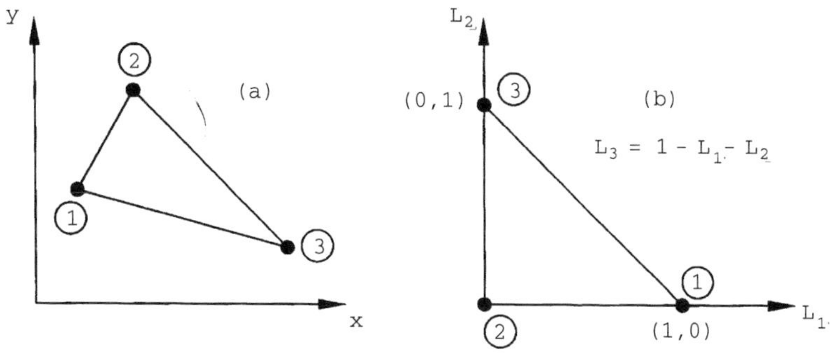

Figure 3.4 (a) General triangular element. (b) Element mapped in local coordinates

However, it often leads to more elegant formulations algebraically if all three coordinates are retained. For example, the shape functions for a three-noded triangular element ("constant strain" triangle) take the form

$$\begin{aligned} N_1 &= L_1 \\ N_2 &= L_2 \\ N_3 &= L_3 \end{aligned} \tag{3.14}$$

and the isoparametric property is that

$$\begin{aligned} x &= N_1x_1 + N_2x_2 + N_3x_3 \\ y &= N_1y_1 + N_2y_2 + N_3y_3 \end{aligned} \tag{3.15}$$

Equations (3.3) and (3.4) from a previous paragraph still apply regarding the Jacobian matrix but equation (3.5) must be modified for triangles to give

$$\iint \mathrm{d}x\,\mathrm{d}y = \int_0^1 \int_0^{1-L_1} \det|\mathbf{J}|\mathrm{d}L_2\mathrm{d}L_1 \tag{3.16}$$

3.5 NUMERICAL INTEGRATION FOR TRIANGLES

Numerical integration over triangular regions is similar to that for quadrilaterals except that the sampling points do not occur in such "convenient" positions.

The quadrature rules take the general form

$$\int_0^1 \int_0^{1-L_1} f(L_1, L_2)\,\mathrm{d}L_2\mathrm{d}L_1 \simeq \sum_{i=1}^{\mathrm{nip}} W_i f(L_1, L_2)_i \tag{3.17}$$

Table 3.2 Coordinates and weights for integration over triangular areas

nip	L_1^i	L_2^i	W_i
1	$\frac{1}{3}$	$\frac{1}{3}$	$\frac{1}{2}$
	$\frac{1}{2}$	$\frac{1}{2}$	1/6
3	$\frac{1}{2}$	0	1/6
	0	$\frac{1}{2}$	1/6

where W_i is the weighting coefficient corresponding to the sampling point $(L_1, L_2)_i$ and nip represents the number of sampling points. Typical values of the weights and sampling points are given in Table 3.2.

As with quadrilaterals, numerical integration can be exact for certain polynomials. For example, in Table 3.2, the one-point rule is exact for integration of first-degree polynomials and the three-point rule is exact for polynomials of second degree. Reduced integration can again be beneficial in some instances.

Computer formulations involving local coordinates, transformations of coordinates and numerical integration are described in subsequent paragraphs.

3.6 MULTI-ELEMENT ASSEMBLIES

Properties of elements in isolation have been shown to be given by matrix equations, for example the equilibrium equation

$$\mathbf{KP}\phi = \mathbf{q} \tag{3.18}$$

describing steady laminar fluid flow. In Figure 3.5 is shown a small mesh containing three quadrilateral elements, all of which have properties defined by (3.18). If an assembly strategy is chosen (but see section 3.7), the next problem

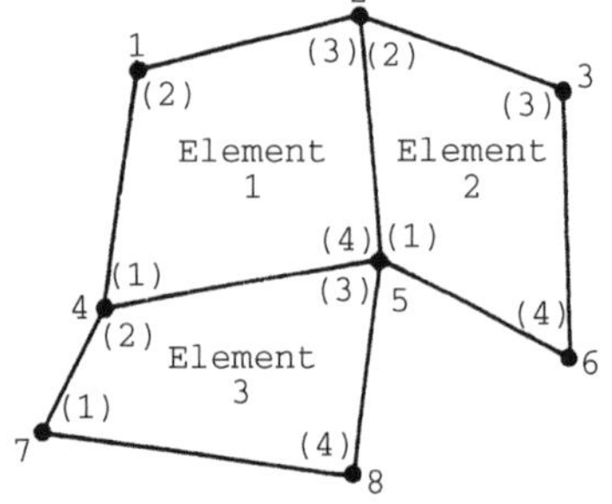

Figure 3.5 Mesh of quadrilateral elements

is to assemble the elements and so derive the properties of the three-element system. Each element possesses node numbers, shown in parentheses, which follow the scheme in Figures 3.1 to 3.3, namely numbering clockwise from the lower left-hand corner. Since there is only one unknown at every node, the fluid potential, each individual element equation can be written

$$\begin{bmatrix} KP_{1,1} & KP_{1,2} & KP_{1,3} & KP_{1,4} \\ KP_{2,1} & KP_{2,2} & KP_{2,3} & KP_{2,4} \\ KP_{3,1} & KP_{3,2} & KP_{3,3} & KP_{3,4} \\ KP_{4,1} & KP_{4,2} & KP_{4,3} & KP_{4,4} \end{bmatrix} \begin{Bmatrix} \phi_1 \\ \phi_2 \\ \phi_3 \\ \phi_4 \end{Bmatrix} = \begin{Bmatrix} q_1 \\ q_2 \\ q_3 \\ q_4 \end{Bmatrix} \tag{3.19}$$

However, in the mesh numbering system, not in parentheses, mesh node 4 corresponds to element node (1) in element 1 and to element node (2) in element 3. The total number of equations for the mesh is 8 and, within this system, term $KP_{1,1}$ from element 1 and term $KP_{2,2}$ from element 3 would be added together and would appear in location 4,4 and so on. The total system matrix for Figure 3.5 is given in Table 3.3, where the superscripts refer to element numbers.

This total system matrix is symmetrical provided its constituent matrices are symmetrical. The matrix also possesses the useful property of "bandedness" which means that the terms are concentrated around the "leading diagonal" which stretches from the upper left to the lower right of the table. In this example, no term in any row can be more than four locations removed from the leading diagonal so the system is said to have a "semi-bandwidth" `NBAND` of 4. This can be obtained by inspection from Figure 3.5 by subtracting the lowest from the highest freedom number in each element. Complicated meshes have variable bandwidths and useful computer programs make use of banding when storing the system matrices.

The importance of efficient mesh numbering is illustrated for a mesh of line elements in Figure 3.6 where the scheme in parentheses has `NBAND` = 13 compared with the other scheme with `NBAND` = 2.

Table 3.3 System stiffness matrix for mesh in Figure 3.5. Superscripts indicate element numbers

$KP^1_{2,2}$	$KP^1_{2,3}$	0	$KP^1_{2,1}$	$KP^1_{2,4}$	0	0	0
$KP^1_{3,2}$	$KP^1_{3,3}+KP^2_{2,2}$	$KP^2_{2,3}$	$KP^1_{3,1}$	$KP^1_{3,4}+KP^2_{2,1}$	$KP^2_{2,4}$	0	0
0	$KP^2_{3,2}$	$KP^2_{3,3}$	0	$KP^2_{3,1}$	$KP^2_{3,4}$	0	0
$KP^1_{1,2}$	$KP^1_{1,3}$	0	$KP^1_{1,1}+KP^3_{2,2}$	$KP^1_{1,4}+KP^3_{2,3}$	0	$KP^3_{2,1}$	$KP^3_{2,4}$
$KP^1_{4,2}$	$KP^1_{4,3}+KP^2_{1,2}$	$KP^2_{1,3}$	$KP^1_{4,1}+KP^3_{3,2}$	$KP^1_{4,4}+KP^2_{1,1}$ $+KP^3_{3,3}$	$KP^2_{1,4}$	$KP^3_{3,1}$	$KP^3_{3,4}$
0	$KP^2_{4,2}$	$KP^2_{4,3}$	0	$KP^2_{4,1}$	$KP^2_{4,4}$	0	0
0	0	0	$KP^3_{1,2}$	$KP^3_{1,3}$	0	$KP^3_{1,1}$	$KP^3_{1,4}$
0	0	0	$KP^3_{4,2}$	$KP^3_{4,3}$	0	$KP^3_{4,1}$	$KP^3_{4,4}$

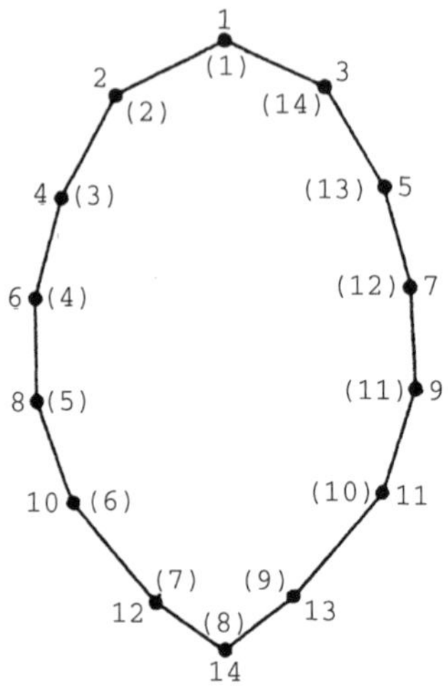

Figure 3.6 Alternative mesh numbering schemes

If system symmetry exists it should also be taken into account. For example, if the system in Table 3.3 is symmetrical, there are only 30 unique components (the leading diagonal terms plus a maximum of four terms to the left or right of the diagonal in each row). Often, with a slight decrease of efficiency, the symmetrical half of a band matrix is stored as a rectangular array with a size equal to the number of system equations times the semi-bandwidth plus 1. In this case zeros are filled into the extra locations in the first few or last few rows depending on whether the lower or upper "half" of the matrix is stored. Special storage schemes, for example "skyline" techniques (Bathe, 1996), are also considered later in the chapter (see Figure 3.19) and are important in cases where bandwidths vary greatly.

Later in this chapter, procedures are described whereby system matrices like that in Table 3.3 can be automatically assembled in band form, with or without the "skyline" option, from the constituent element matrices.

3.7 "ELEMENT-BY-ELEMENT" OR "MESH-FREE" TECHNIQUES

Our purpose is to solve classes of problem summarised for solids for example by equations (2.77) to (2.79), namely

$$\text{Static equilibrium problems} \quad \mathbf{KMr} = \mathbf{f} \tag{3.20}$$

$$\text{Eigenvalue problems} \quad (\mathbf{KM} - \omega^2\mathbf{MM})\mathbf{a} = \mathbf{0} \tag{3.21}$$

$$\text{Propagation problems} \quad \mathbf{KMr} + \mathbf{MM}\frac{\mathrm{d}^2\mathbf{r}}{\mathrm{d}t^2} = \mathbf{f}(t) \tag{3.22}$$

Traditionally, computer programs have been based on the assembly techniques described in the previous section. For static equilibrium problems, all the element **KM**s would be assembled to form a system matrix like that shown in Table 3.3. Then the linear algebraic system (3.20) would be solved, typically by some form of Gaussian elimination. In the previous section it was emphasised that this strategy depends on efficient storage of the system coefficient matrices. In Gaussian elimination processes, "fill-in" means that coefficients in Table 3.3 like the fourth one in the third row, will not remain zero during the elimination process. Therefore, all coefficients contained within a "band" or "skyline" must be stored and manipulated.

As problem sizes grow, this storage requirement becomes a burden, even on a modern computer. For meshes of 3-d elements 100 000 equations are likely to have a semi-bandwidth of the order of 1000. Thus 10^8 "words" of storage, typically 800 Mb, would be required to hold the coefficient matrix. If this space is not available, out-of-memory techniques or "paging" (see Chapter 1) cause a serious deterioration in machine speeds.

For this reason, alternative solution strategies to Gaussian elimination have been sought, and there has been a resurgence of interest in iterative techniques for the solution of large systems like (3.20). Griffiths and Smith (1991) describe a number of algorithms of this type, the most popular being based on the method of "conjugate gradients" (Jennings and McKeown, 1992).

Solution of the linear algebraic system (3.20), which we could write as

$$\mathbf{Ax} = \mathbf{b} \tag{3.23}$$

proceeds by the following steps:

$$\mathbf{p}^0 = \mathbf{r}^0 = \mathbf{b} - \mathbf{Ax}^0 \tag{3.24}$$

where $\mathbf{r}^0$ is the "residual" or error for a first trial $\mathbf{x}^0$, and then k steps of the process

$$\begin{aligned}
\mathbf{u}^k &= \mathbf{Ap}^k \\
\alpha^k &= \frac{(\mathbf{r}^k)^T\mathbf{r}^k}{(\mathbf{p}^k)^T\mathbf{u}^k} \\
\mathbf{x}^{k+1} &= \mathbf{x}^k + \alpha^k\mathbf{p}^k \\
\mathbf{r}^{k+1} &= \mathbf{r}^k - \alpha^k\mathbf{u}^k \\
\beta^k &= \frac{(\mathbf{r}^{k+1})^{\mathrm{T}}\mathbf{r}^{k+1}}{(\mathbf{r}^k)^T\mathbf{r}^k} \\
\mathbf{p}^{k+1} &= \mathbf{r}^{k+1} + \beta^k\mathbf{r}^k
\end{aligned} \tag{3.25}$$

until the difference between $\mathbf{x}^{k+1}$ and $\mathbf{x}^k$ is "sufficiently" small. In the above, **u**,**p** and **r** are vectors of length `NEQ`, the number of equations to be solved, while α and β are scalars.

It can be seen that the algorithm described by equations (3.24) and (3.25) consists of simple vector operations of the type vector = vector ± scalar * vector which are neatly coded in Fortran 90 using whole arrays (Chapter 1), inner products of the type $\mathbf{r}^{\mathrm{T}}\mathbf{r}$ which are computed by the Fortran 90 intrinsic procedure `DOT_PRODUCT`, and a single 2-d array operation $\mathbf{u} = \mathbf{Ap}$ which can be computed by the Fortran 90 intrinsic procedure `MATMUL`.

Vitally, however, if **A** is a system stiffness matrix given for example by Table 3.3, and all that is required is the product **Ap** where **p** is a known vector, this product can be carried out "element-by-element" without ever assembling **A** at all. That is

$$\mathbf{u} = \sum \mathbf{KP}_i \mathbf{p}_i \tag{3.26}$$

where $\mathbf{KP}_i$ is the element stiffness matrix of the ith element and $\mathbf{p}_i$ the appropriate part of **p**, gathered as $[p(7)\ p(4)\ p(5)\ p(8)]^T$ for $i = 3$ in Figure 3.5 and so on.

The storage required by such an algorithm, compared with the 800 Mb discussed earlier for a 3-d system of 100 000 unknowns, would be an order of magnitude less, and would grow linearly with the increase in number of elements or equations rather than as the square. Iterative strategies of this type for the solution of (3.21) and (3.22) as well will be described in due course. In practice "preconditioning" (Griffiths and Smith, 1991) is used to accelerate convergence of the iterative process for solving (3.20).

3.8 INCORPORATION OF BOUNDARY CONDITIONS

Eigenvalues of stiffness matrices of freely floating elements or meshes are sometimes required but normally in eigenvalue problems and always in equilibrium and propagation problems additional boundary information has to be supplied before solutions can be obtained. For example, the system matrix defined in Table 3.3 is singular and this set of equations has no solution.

The simplest type of boundary condition occurs when the dependent variable in the solution is known to be zero at various points in the region (and hence nodes in the finite element mesh). When this occurs, the equation components associated with these nodes are not required in the solution and information is given to the assembly routine which prevents these components from ever being assembled into the final system. Thus only the non-zero nodal values are solved for.

A variation of this condition occurs when the dependent variable has known, but non-zero, values at various locations. Although an elimination procedure could be devised, the way this condition is handled in practice is by adding a "large" number or "penalty" term, say 10^{12}, to the leading diagonal of the "stiffness" matrix in the row in which the prescribed value is required. The term in the same row of the right-hand side vector is then set to the prescribed value multiplied by the augmented "stiffness" coefficient. For example, suppose the value of the fluid head at node 5 in Figure 3.5 is known to be 57.0 units. The unconstrained set of equations (Table 3.3) would be assembled and term (5,5) augmented by adding 10^{12}. In the subsequent solution there would be an equation

$$(K_{5,5} + 10^{12})\phi_5 + \text{small terms} = 57.0 \times (K_{5,5} + 10^{12}) \tag{3.27}$$

which would have the effect of making ϕ_5 equal to 57.0. Clearly this procedure is only successful if indeed "small terms" are small relative to 10^{12}.

This method could also be used to enforce the boundary condition $\phi = 0$ and has some attractions in simplicity of programming.

Boundary conditions can also involve gradients of the unknown in the forms

$$\frac{\partial \phi}{\partial n} = 0 \tag{3.28}$$

$$\frac{\partial \phi}{\partial n} = C_1 \phi \tag{3.29}$$

$$\frac{\partial \phi}{\partial n} = C_2 \tag{3.30}$$

where n is the normal to the boundary and C_1, C_2 are constants.

To be specific, consider a solution of the diffusion–advection equation (2.109) subject to boundary conditions (3.28), (3.29) and (3.30) respectively. When the second-order terms $c_x(\partial^2\phi/\partial x^2)$ and $c_y(\partial^2\phi/\partial y^2)$ are integrated by parts, boundary integrals of the type

$$\int_s c_n \mathbf{N}^{\mathrm{T}} \frac{\partial \phi}{\partial n} l_n \mathrm{d}s \tag{3.31}$$

arise, where s is a length of boundary and l_n the direction cosine of the normal. Clearly the case $\partial\phi/\partial n = 0$ presents no difficulty since the contour integral (3.31) vanishes and this is the default boundary condition obtained at any free surface of a finite element mesh.

However, (3.29) gives rise to an extra integral, which for the boundary element shown in Figure 3.7 is

$$\int_j^k c_y \mathbf{N}^{\mathrm{T}} C_1 \phi l_y \mathrm{d}s \tag{3.32}$$

When ϕ is expanded as $\mathbf{N\phi}$ we get an additional matrix

$$\frac{-C_1 c_y (x_k - x_j)}{6} \begin{bmatrix} 0 & 0 & 0 & 0 \\ 0 & 2 & 1 & 0 \\ 0 & 1 & 2 & 0 \\ 0 & 0 & 0 & 0 \end{bmatrix} \tag{3.33}$$

which must be added to the left-hand side of the element equations.

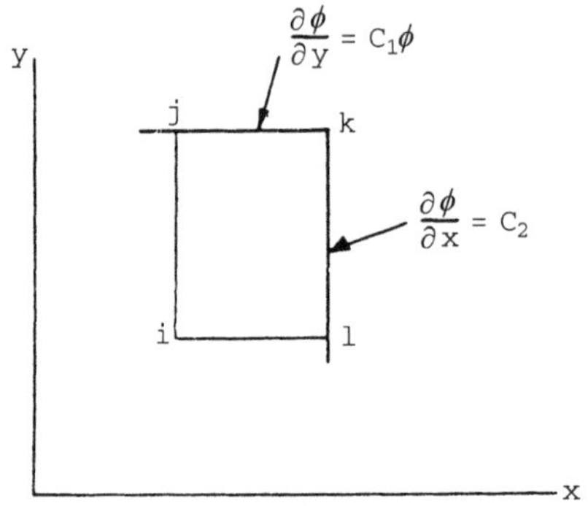

Figure 3.7 Boundary conditions involving non-zero gradients of the unknown

For boundary condition (3.30) the additional term is

$$\int_k^l c_x \mathbf{N}^{\mathrm{T}} C_2 l_x \mathrm{d}s \tag{3.34}$$

which is just a vector

$$\frac{C_2 c_x (y_k - y_l)}{2} \begin{Bmatrix} 0 \\ 0 \\ 1 \\ 1 \end{Bmatrix} \tag{3.35}$$

which would be added to the right-hand side of the element equations. For a further discussion of boundary conditions see Smith (1979).

In summary, boundary conditions of the type $\phi = 0$ or $\partial\phi/\partial n = 0$ are the most common and are easily handled in finite element analyses. The cases $\phi =$ constant or $\partial\phi/\partial n =$ constant$^*\phi$ are somewhat more complicated but can be appropriately treated. Examples of the use of these types of boundary specification are included in the examples chapters.

3.9 PROGRAMMING USING BUILDING BLOCKS

The programs in subsequent chapters are constituted from 100 or so building blocks in the form of Fortran 90 functions and subroutines which perform the tasks of computing and integrating the element matrices, assembling these into system matrices if necessary and carrying out the appropriate equilibrium, eigenvalue or propagation calculations.

It is anticipated that users will elect to pre-compile all of the building blocks and to hold these permanently in a library. The library should then be automatically accessible to the calling programs by means of a simple "`USE`" statement at the beginning of the program (see section 1.4.6).

A summary of the building blocks is listed in Appendix 4 where their actions and input/output parameters are described. A separation has been made into "black box" routines (concerned with some matrix operations), whose mode of action the reader need not necessarily know in detail, and special purpose routines which are the basis of specific finite element computations. These special purpose routines are listed in Appendix 5. The black box routines should be thought of as an addition to the permanent library functions such as `MATMUL` or `DOT_PRODUCT`, and could well be substituted with equivalents from a mathematical subroutine library perhaps tuned to a specific machine.

3.10 BLACK BOX ROUTINES

Readers are reminded of the much improved array-handling facilities of Fortran 90, compared with earlier FORTRANs. Section 1.4.4 summarised such features as whole array operations and intrinsic array procedures which mean that most simple array manipulations can be done using the power of the language itself and do not need to be user-supplied.

In the programs which follow from Chapter 4 onwards, only two simple procedures have been added to those provided as standard in the language. These are

```
DETERMINANT
INVERT
```

the first of which returns the determinant of a 1×1, 2×2 or 3×3 matrix (usually the Jacobian matrix) and the second returns the inverse of a (small) square matrix, again usually the Jacobian matrix $\mathbf{J}$ (see equation (3.3)).

A second batch of subroutines is concerned with the solution of linear algebraic equations. The subroutines have been split into factorisation and forward/backward re-substitution phases:

Factorisation	`BANRED`	`CHOLIN`	`GAUSS_BAND`	`SPARIN`
Backsubstitution	`BACSUB`	`CHOBAC`	`SOLVE_BAND`	`SPABAC`
Method	Gauss	Choleski	Gauss	Choleski
Equation coefficients	Symmetrical band	Symmetrical band	Unsymmetrical full band	Symmetrical skyline

In some programs, it is necessary to isolate the forward and backward substitution operations for which purpose `CHOBAC` is subdivided into

`CHOBK1` (forward substitution)

`CHOBK2` (backward substitution)

Several subroutines are associated with eigenvalue and eigenvector determination, for example

Tridiagonalisation	`BANDRED`
Finds eigenvalues	`BISECT`
Method	Jacobi
Equation coefficients	Symmetrical, banded, upper triangle stored

The first tridiagonalises the matrix and the second extracts all of the eigenvalues. It should be noted that these routines, although robust and accurate, can be inefficient both in storage requirements and in run-time and should not be used for solving very large problems, for which in any case it is unlikely that the full range of eigenmodes would be required. The various vector iteration methods (Bathe, 1996) should be resorted to in such cases.

One of the most effective of these is the Lanczos method, and routines

`LANCZ1`

`LANCZ2`

are used to calculate the eigenvalues and eigenvectors of a matrix.

When the Lanczos procedure is described in more detail (section 3.13) it will be found that, in common with its close relation the conjugate gradient procedure (section 3.7), the method requires a matrix–vector product where the matrix is essentially the global system stiffness matrix, followed by a series of whole-vector operations. To save storage, the matrix–vector product can again be done "element-by-element" and this feature is taken advantage of in Chapter 10.

Although simple matrix-by-vector multiplications can be accomplished by the intrinsic procedure `MATMUL`, advantage is usually taken of banding of global system matrix coefficients whenever possible. To allow for this, four special matrix-by-vector multiplication routines are provided:

Subroutine	`LINMUL`	`BANMUL`	`BANTMUL`	`LINMUL_SKY`
Matrix coefficients	Symmetrical	Symmetrical	Unsymmetrical	Symmetrical
Storage of matrix	Vector	Lower triangle	Full band	Skyline

Further information on when these routines should be used is given in section 3.12.

In a teaching text such as this, elaborate input and output procedures are avoided. It is expected that users may pre-process their input data using independent programs and plot the output graphically.

In order to describe the action of the remaining special purpose subroutines, which are listed in full in Appendix 5, it is necessary first to consider the properties of individual finite elements and then the representation of continua from assemblages of these elements. Static linear problems (including eigenproblems) are considered first. Thereafter modifications to programs to incorporate time dependence are added.

3.11 SPECIAL PURPOSE ROUTINES

The job of these routines is to compute the element matrix coefficients, for example the "stiffness", to integrate these over the element area or volume and finally, if necessary, to assemble the element submatrices into a global system matrix or matrices. The black box routines for equation solution, eigenvalue determination and so on then take over to produce the final results.

3.11.1 ELEMENT MATRIX CALCULATION

In the remainder of this chapter the notation adopted is that used in the subroutine listings in Appendix 5. Wherever possible, mnemonics are used so that local coordinate ξ becomes `XI` in the subroutines and so on.

3.11.1.1 Plane strain (stress) analysis of elastic solids using quadrilateral elements

As an example of element matrix calculation, consider the computation of the element stiffness matrix for plane elasticity given by (2.63):

$$\mathbf{KM} = \int\int \mathbf{B}^{\mathrm{T}}\mathbf{D}\mathbf{B}\,\mathrm{d}x\,\mathrm{d}y \tag{3.36}$$

In program terminology this becomes

$$\texttt{KM} = \int\int \texttt{TRANSPOSE(BEE)*DEE*BEE}\,\mathrm{d}x\,\mathrm{d}y \tag{3.37}$$

and its formation is described by the inner loop of the structure chart in Figure 3.8.

It is assumed for the moment that the element nodal coordinates (x, y) have been calculated and stored in the array `COORD`. For example, for a four-node quadrilateral

$$\texttt{COORD} = \begin{bmatrix} x_1 & y_1 \\ x_2 & y_2 \\ x_3 & y_3 \\ x_4 & y_4 \end{bmatrix} \tag{3.38}$$

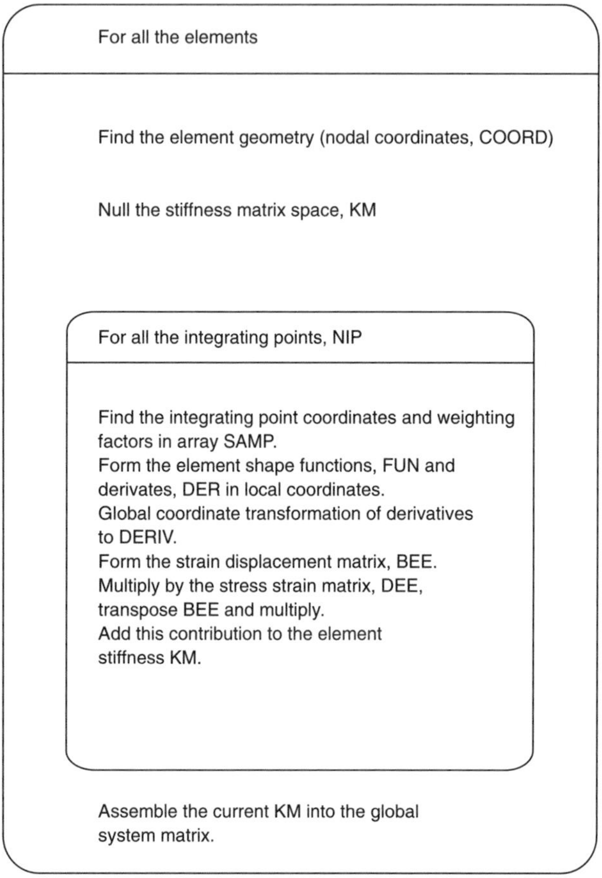

Figure 3.8 Structure chart for element matrix assembly assuming numerical integration

The shape functions **N** are held in array `FUN`, in terms of local coordinates, as specified in (3.1) by

$$\mathtt{FUN} = \begin{Bmatrix} 1/4(1-\mathtt{XI})(1-\mathtt{ETA}) \\ 1/4(1-\mathtt{XI})(1+\mathtt{ETA}) \\ 1/4(1+\mathtt{XI})(1+\mathtt{ETA}) \\ 1/4(1+\mathtt{XI})(1-\mathtt{ETA}) \end{Bmatrix}^{\mathrm{T}} \tag{3.39}$$

The **B** matrix contains derivatives of the shape functions with respect to global coordinates and these are easily computed in the local coordinate system as

$$\text{DER} = \begin{bmatrix} \dfrac{\partial \text{FUN}^{\text{T}}}{\partial \xi} \\ \dfrac{\partial \text{FUN}^{\text{T}}}{\partial \eta} \end{bmatrix}$$

or

$$\text{DER} = \frac{1}{4}\begin{bmatrix} -(1-\text{ETA}) & -(1+\text{ETA}) & (1+\text{ETA}) & (1-\text{ETA}) \\ -(1-\text{XI}) & (1-\text{XI}) & (1+\text{XI}) & -(1+\text{XI}) \end{bmatrix} \tag{3.40}$$

The information in (3.39) and (3.40) for a four-node quadrilateral (NOD $= 4$) is formed by the subroutines

SHAPE_FUN

SHAPE_DER

for the specific Gaussian integration points $(\text{XI, ETA})_{\text{I}}$ held in the array POINTS where I runs from 1 to NIP, the total number of Gauss points specified in each element. Figure 3.9(a) shows the typical layout and ordering for "two-point" Gaussian integration. Since there are two integrating points in each coordinate direction NIP is 4 in this example. In all cases POINTS and their corresponding WEIGHTS (equation (3.6)) are found by the subroutine

SAMPLE

where NIP can take the values 1, 4 and 9 for quadrilaterals.

The derivatives DER must then be converted into their counterparts in the (x, y) coordinate system, DERIV, by means of the Jacobian matrix transformation (3.3) or (3.4). From the isoparametric property,

$$\begin{Bmatrix} x \\ y \end{Bmatrix} = \text{COORD}^{\text{T}} * \text{FUN}^{\text{T}} \tag{3.41}$$

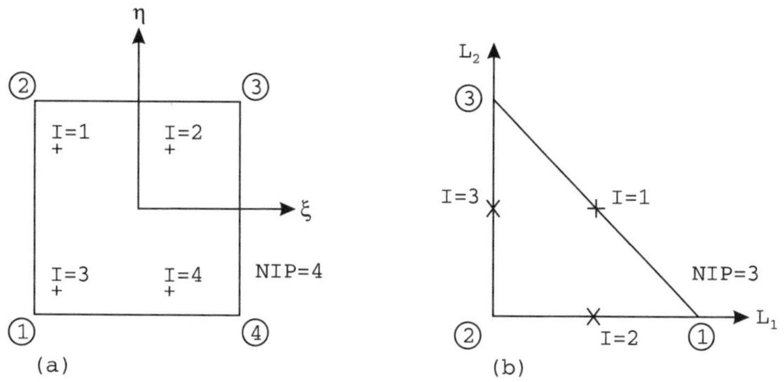

Figure 3.9 Integration schemes for (a) quadrilaterals and (b) triangles

and since the Jacobian matrix is given by

$$\begin{bmatrix} \dfrac{\partial x}{\partial \xi} & \dfrac{\partial y}{\partial \xi} \\ \dfrac{\partial x}{\partial \eta} & \dfrac{\partial y}{\partial \eta} \end{bmatrix} \tag{3.42}$$

it is clear that (3.42) can be obtained by differentiating (3.41) with respect to the local coordinates. In this way

$$\text{JAC} = \text{DER*COORD} \tag{3.43}$$

In order to compute `DERIV` we must invert `JAC` using `INVERT` and finally carry out the multiplication of this inverse by `DER` to give `DERIV`:

$$\text{DERIV} = (\text{INVERSE OF JAC})\text{*DER} \tag{3.44}$$

Thus the sequence of operations

```
CALL SHAPE_DER (DER,POINTS,I);JAC=MATMUL (DER,COORD)
DET=DETERMINANT (JAC);CALL INVERT (JAC)                    (3.45)
DERIV=MATMUL (JAC,DER)
```

where `NDIM` is the number of coordinate dimensions (two in this case), will be found in all programs which require `DERIV`. After these operations have been performed, the derivatives of the element shape functions with respect to the Cartesian coordinates are held in `DERIV` while `DET` is the determinant of the Jacobian matrix, required later for the purposes of numerical integration.

The matrix `BEE` in (3.37) can now be assembled as it consists of components of `DERIV`. This assembly is performed by the subroutine

`BEEMAT`

Thus the strain–displacement relations are

$$\text{EPS} = \text{BEE*ELD} \tag{3.46}$$

where in the case of a four-node quadrilateral

$$\text{ELD} = \{u_1 v_1 u_2 v_2 u_3 v_3 u_4 v_4\}^T \tag{3.47}$$

The variables u and v are simply the nodal displacements in the x and y directions respectively assuming the nodal ordering of Figure 3.5.

The components of the integral of `TRANSPOSE (BEE) * DEE * BEE`, evaluated at the integrating points given by all values of `I`, can now be computed by transposing `BEE` using the Fortran 90 intrinsic function `TRANSPOSE`, and by forming the stress–strain matrix using the subroutine

`DEEMAT`

(for plane strain) with NST, the number of components of stress and strain being three.

In this case

$$\texttt{DEE} = \frac{\texttt{E}(1-\texttt{V})}{(1+\texttt{V})(1-2\texttt{V})}\begin{bmatrix} 1 & \dfrac{\texttt{V}}{1-\texttt{V}} & 0 \\ \dfrac{\texttt{V}}{1-\texttt{V}} & 1 & 0 \\ 0 & 0 & \dfrac{1-2\texttt{V}}{2(1-\texttt{V})} \end{bmatrix} \tag{3.48}$$

and the multiplications

$$\texttt{TRANSPOSE(BEE)*DEE*BEE} \tag{3.49}$$

to give, say, BTDB, can be carried out.

A plane stress analysis would be obtained by simply replacing DEEMAT by FMDSIG.

The integral is evaluated numerically by

$$\texttt{KM} = \sum_{\texttt{I}=1}^{\texttt{NIP}} \texttt{DET}_\texttt{I}\texttt{*WEIGHTS(I)*BTDB}_\texttt{I} \tag{3.50}$$

where WEIGHTS(I) are the numerical integration weighting coefficients. Note that WEIGHTS are the products $w_i w_j$ in equation (3.6).

As soon as the element matrix has been formed from (3.50) it can be assembled into the global system matrix (or matrices) by special subroutines described later in this chapter. Since the assembly process is common to all problems, modifications to the element matrix calculation for different situations will first be described.

In cases where the stiffness matrix KM is obtained analytically, the integration loop in Figure 3.8 is replaced by a single call to the subroutine ANALY4. When strains and stresses are back-calculated (usually at the integrating points) analytic subroutine BEE4 replaces BEEMAT. The two "analytic" routines are valid only for four-node quadrilaterals.

3.11.1.2 Plane strain (stress) analysis of elastic solids using triangular elements

The previous section showed how the stiffness matrix of a typical four-node quadrilateral could be built up. In order to use triangular elements, very few alterations are required. For example, for a six-node triangular element (NOD = 6), the values of the shape functions and their derivatives with respect to local coordinates at a particular location (L_1, L_2, L_3) are still formed by the subroutines

SHAPE_FUN

SHAPE_DER

which deliver the shape functions

$$\mathtt{FUN} = \begin{Bmatrix} (2L_1 - 1)L_1 \\ 4L_1L_2 \\ (2L_2 - 1)L_2 \\ 4L_2L_3 \\ (2L_3 - 1)L_3 \\ 4L_3L_1 \end{Bmatrix}^{\mathrm{T}} \tag{3.51}$$

and their derivatives with respect to L_1 and L_2

$$\mathtt{DER} = \begin{bmatrix} \dfrac{\partial \mathtt{FUN}^{\mathrm{T}}}{\partial L_1} \\ \dfrac{\partial \mathtt{FUN}^{\mathrm{T}}}{\partial L_2} \end{bmatrix} = \begin{bmatrix} (4L_1-1) & 4L_2 & 0 & -4L_2 & -(4L_3-1) & 4(L_3-L_1) \\ 0 & 4L_1 & (4L_2-1) & 4(L_3-L_2) & -(4L_3-1) & -4L_1 \end{bmatrix} \tag{3.52}$$

The nodal numbering and the order in which the integration points are sampled for a typical three-point scheme are shown in Figure 3.9(b).

For integration over triangles, the sampling points in local coordinates (L_1, L_2) are held in the array `POINTS` and the corresponding weighting coefficients in the array `WEIGHTS`. Both of these items are provided by the subroutine

`SAMPLE`

The version of this subroutine described in Appendix 5 allows the total number of integrating points (`NIP`) for triangles to take the values, 1, 3, 6, 7, 12 or 16. The coding should be referred to in order to determine the sequence in which the integrating points are sampled for `NIP` > 3.

Exactly the same sequence of operations (3.45) as was used for quadrilaterals places the required derivatives with respect to (x, y) in `DERIV` and finds the Jacobian determinant `DET`.

Finally, numerical integration is performed by

$$\mathtt{KM} = \sum_{\mathtt{I}=1}^{\mathtt{NIP}} \mathtt{DET}_{\mathtt{I}} * \mathtt{WEIGHTS(I)} * \mathtt{BTDB}_{\mathtt{I}} \tag{3.53}$$

A higher-order triangular element with 15 nodes is also considered in Chapter 5. The shape functions and derivatives for this element are again provided by routines `SHAPE_FUN` and `SHAPE_DER`.

3.11.1.3 Axisymmetric strain of elastic solids

The strain–displacement relations can again be written by (3.46) but in this case `BEE` must be formed by the subroutine

```
BMATAXI
```

where the cylindrical coordinates (r, z) replace their plane strain counterparts (x, y). The stress–strain matrix is still given by an expression similar to (3.48) (see (2.68)) and is returned by subroutine `DEEMAT` with `NST`, the number of stress and strain components now being four.

In this case the integrated element stiffness is (2.66), namely

$$\mathrm{KM} = \iint \mathrm{TRANSPOSE(BEE)} * \mathrm{(DEE)}{*}\mathrm{BEE}{*}r\,\mathrm{d}r\,\mathrm{d}z \tag{3.54}$$

Considering a four-node element, the isoparametric property gives

$$\begin{aligned} r &= \mathrm{SHAPE\_FUN}\begin{Bmatrix} r_1 \\ r_2 \\ r_3 \\ r_4 \end{Bmatrix} \\ &= \sum_{\mathrm{K}=1}^{\mathrm{NOD}} \mathrm{SHAPE\_FUN(K)}{*}\mathrm{COORD(K,1)} \\ &= \mathrm{RADIUS} \end{aligned} \tag{3.55}$$

Hence we have

$$\mathrm{KM} = \sum_{\mathrm{I}=1}^{\mathrm{NIP}} \mathrm{RADIUS_I}{*}\mathrm{DET_I}{*}\mathrm{WEIGHTS(I)}{*}\mathrm{BTDB_I} \tag{3.56}$$

By comparison with (3.50) it may be seen that when evaluated numerically the algorithms for axisymmetric and plane stiffness formation will be essentially the same, despite the fact that they are algebraically quite different. This is very significant from the points of view of programming effort and of program flexibility.

However (3.56) now involves numerical evaluation of integrals involving $1/r$ which do not have simple polynomial representations. Therefore, in contrast to plane problems, it will be impossible to evaluate (3.56) exactly by numerical means, especially as r (i.e. `RADIUS`) approaches zero. Provided integration points do not lie on the $r = 0$ axis, however, reasonable results are usually achieved using a similar order of quadrature to that used in plane analysis. More accurate numerical integration schemes for axisymmetric elements are available (Griffiths, 1991).

3.11.1.4 Plane steady laminar fluid flow

It was shown in (2.103) that a fluid element has a "stiffness" defined by

$$\mathbf{KP} = \iint \mathbf{T}^{\mathrm{T}}\mathbf{K}\mathbf{T}\,\mathrm{d}x\,\mathrm{d}y \tag{3.57}$$

which becomes in program terminology

$$\mathrm{KP} = \iint \mathtt{TRANSPOSE(DERIV)*KAY*DERIV}\,\mathrm{d}x\,\mathrm{d}y \tag{3.58}$$

and the similarity to (3.37) is obvious. The matrix `DERIV` simply contains the derivatives of the element shape functions with respect to (x, y) which were previously needed in the analysis of solids and are formed by the sequence (3.45) while `KAY` contains the permeability properties of the element in the form

$$\mathtt{KAY} = \begin{bmatrix} \mathtt{PERMX} & 0 \\ 0 & \mathtt{PERMY} \end{bmatrix} \tag{3.59}$$

In the computations,

$$\mathtt{TRANSPOSE\ (DERIV)*KAY*DERIV} \tag{3.60}$$

`DTKD`, say, is evaluated by appropriate matrix transposes and multiplications and the final matrix summation for a quadrilateral element is

$$\mathtt{KP} = \sum_{\mathtt{I=1}}^{\mathtt{NIP}} \mathtt{DET_I*WEIGHTS(I)*DTKD_i} \tag{3.61}$$

By comparison with (3.50) it will be seen that these physically very different problems are likely to require similar solution algorithms.

3.11.1.5 Mass matrix formation

The mass matrix was shown in Chapter 2, e.g. (2.65), to take the general form

$$\mathbf{MM} = \rho \iint \mathbf{N}^{\mathrm{T}}\mathbf{N}\,\mathrm{d}x\,\mathrm{d}y \tag{3.62}$$

where $\mathbf{N}$ are just the shape functions. In the case of plane fluid flow, since there is only one degree of freedom per node (`NDOF` = 1), the "mass" matrix is particularly simple in program terminology, namely

$$\mathrm{MM} = \mathrm{RHO} * \iint \mathtt{FUN^T*FUN}\,\mathrm{d}x\,\mathrm{d}y \tag{3.63}$$

By defining the product of the shape functions

$$\mathtt{FUN^T*FUN} \tag{3.64}$$

as FTF, we have

$$\mathtt{MM} = \mathrm{RHO} * \sum_{\mathtt{I}=1}^{\mathtt{NIP}} \mathtt{DET}_{\mathtt{i}} * \mathtt{WEIGHTS(I)} * \mathtt{FTF}_{\mathtt{I}} \tag{3.65}$$

where RHO is the mass density.

In the case of plane stress or strain of solids, because of the arrangement of the displacement vector in (3.47) it is convenient to use a special subroutine

ECMAT

to form the terms of the mass matrix as ECM before integration. Thereafter

$$\begin{aligned} \mathtt{MM} &= \mathrm{RHO} * \iint \mathtt{ECM}\, dx\, dy \\ &= \mathrm{RHO} * \sum_{\mathtt{I}=1}^{\mathtt{NIP}} \mathtt{DET}_{\mathtt{I}} * \mathtt{WEIGHTS(I)} * \mathtt{ECM}_{\mathtt{I}} \end{aligned} \tag{3.66}$$

When "lumped" mass approximations are used **MM** becomes a diagonal matrix. For a four-noded quadrilateral (NOD = 4), for example,

$$\mathtt{MM} = (\mathrm{RHO} * \mathrm{AREA}/\mathrm{NOD}) * \mathbf{I} \tag{3.67}$$

where AREA is the element area and **I** the unit matrix. For higher-order elements, however, all nodes may not receive equal (and indeed intuitively "obvious") weighting (see especially Chapters 10 and 11).

3.11.1.6 Higher order quadrilateral elements

To emphasise the ease with which element types can be interchanged in programs, consider the next member of the isoparametric quadrilateral group, namely the 'quadratic' quadrilateral with mid-side nodes shown in Figure 3.10. The same local coordinate system is retained and the coordinate matrix becomes

$$\mathrm{COORD} = \begin{bmatrix} x_1 & y_1 \\ x_2 & y_2 \\ x_3 & y_3 \\ x_4 & y_4 \\ x_5 & y_5 \\ x_6 & y_6 \\ x_7 & y_7 \\ x_8 & y_8 \end{bmatrix} \tag{3.68}$$

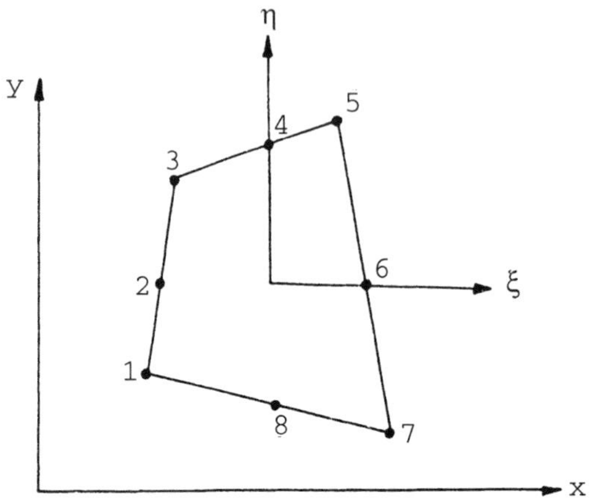

Figure 3.10 General quadratic quadrilateral element

The shape functions are now

$$\text{FUN} = \begin{Bmatrix} \tfrac{1}{4}(1-\text{XI})(1-\text{ETA})(-\text{XI}-\text{ETA}-1) \\ \tfrac{1}{2}(1-\text{XI})(1-\text{ETA}^2) \\ \tfrac{1}{4}(1-\text{XI})(1+\text{ETA})(-\text{XI}+\text{ETA}-1) \\ \tfrac{1}{2}(1-\text{XI}^2)(1+\text{ETA}) \\ \tfrac{1}{4}(1+\text{XI})(1+\text{ETA})(\text{XI}+\text{ETA}-1) \\ \tfrac{1}{2}(1+\text{XI})(1-\text{ETA}^2) \\ \tfrac{1}{4}(1+\text{XI})(1-\text{ETA})(\text{XI}-\text{ETA}-1) \\ \tfrac{1}{2}(1-\text{XI}^2)(1-\text{ETA}) \end{Bmatrix}^T \tag{3.69}$$

formed by SHAPE_FUN. The number of nodes (NOD = 8) serves to identify the appropriate values of FUN.

Their derivatives with respect to local coordinates, DER, are again formed by the subroutine

SHAPE_DER

The sequence of operations described by (3.45) again places the required derivatives with respect to (x, y) in DERIV and finds the Jacobian determinant DET.

Another plane element used in the programs later in this book is the Lagrangian nine-node element. This element uses "complete" polynomial interpolation in each direction, but requires a ninth node at its centre (Figure

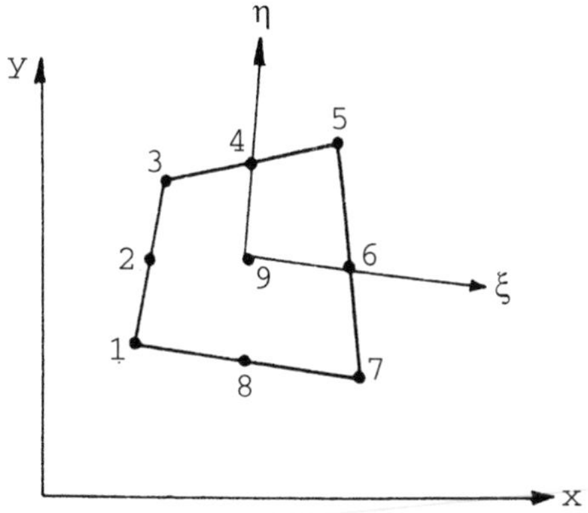

Figure 3.11 The Lagrangian nine-node element

3.11). The shape functions for this element are given by

$$\mathrm{FUN} = \begin{Bmatrix} \tfrac{1}{4}(\mathtt{XI})(\mathtt{XI}-1)(\mathtt{ETA})(\mathtt{ETA}-1) \\ -\tfrac{1}{2}(\mathtt{XI})(\mathtt{XI}-1)(\mathtt{ETA}+1)(\mathtt{ETA}-1) \\ \tfrac{1}{4}(\mathtt{XI})(\mathtt{XI}-1)(\mathtt{ETA})(\mathtt{ETA}+1) \\ -\tfrac{1}{2}(\mathtt{XI}+1)(\mathtt{XI}-1)(\mathtt{ETA})(\mathtt{ETA}+1) \\ \tfrac{1}{4}(\mathtt{XI})(\mathtt{XI}+1)(\mathtt{ETA})(\mathtt{ETA}+1) \\ -\tfrac{1}{2}(\mathtt{XI})(\mathtt{XI}+1)(\mathtt{ETA}+1)(\mathtt{ETA}-1) \\ \tfrac{1}{4}(\mathtt{XI})(\mathtt{XI}+1)(\mathtt{ETA})(\mathtt{ETA}-1) \\ -\tfrac{1}{2}(\mathtt{XI}+1)(\mathtt{XI}-1)(\mathtt{ETA})(\mathtt{ETA}-1) \\ (\mathtt{XI}+1)(\mathtt{XI}-1)(\mathtt{ETA}+1)(\mathtt{ETA}-1) \end{Bmatrix}^{T} \tag{3.70}$$

and are formed together with their derivatives with respect to local coordinates by the subroutines

```
SHAPE_FUN
SHAPE_DER
```

as usual where NOD is now 9. Thus programs using different element types will be identical, although operating on different sizes of arrays.

3.11.1.7 Three-dimensional cuboidal elements

As was the case with changes of plane element types, changes of element dimensions are readily made. For example, the eight-node hexahedral "brick" element in Figure 3.12 is the 3-d extension of the four-noded quadrilateral.

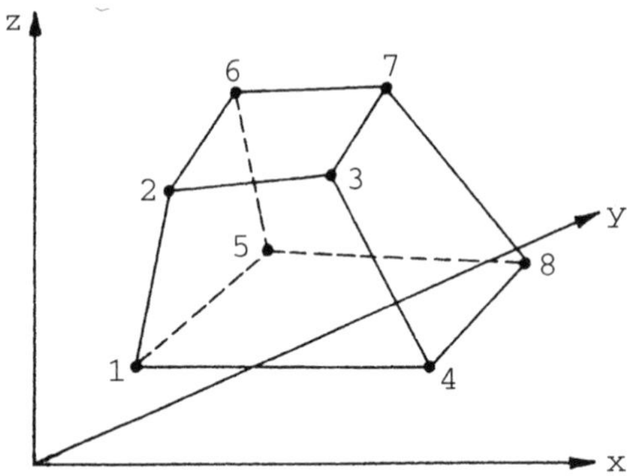

Figure 3.12 General linear brick element

Using the local coordinate system (ξ, η, ζ), the coordinate matrix is

$$\texttt{COORD} = \begin{bmatrix} x_1 & y_1 & z_1 \\ x_2 & y_2 & z_2 \\ x_3 & y_3 & z_3 \\ x_4 & y_4 & z_4 \\ x_5 & y_5 & z_5 \\ x_6 & y_6 & z_6 \\ x_7 & y_7 & z_7 \\ x_8 & y_8 & z_8 \end{bmatrix} \tag{3.71}$$

The shape functions are

$$\texttt{FUN} = \left\{ \begin{array}{l} 1/8(1-\texttt{XI})(1-\texttt{ETA})(1-\texttt{ZETA}) \\ 1/8(1-\texttt{XI})(1-\texttt{ETA})(1+\texttt{ZETA}) \\ 1/8(1+\texttt{XI})(1-\texttt{ETA})(1+\texttt{ZETA}) \\ 1/8(1+\texttt{XI})(1-\texttt{ETA})(1-\texttt{ZETA}) \\ 1/8(1-\texttt{XI})(1+\texttt{ETA})(1-\texttt{ZETA}) \\ 1/8(1-\texttt{XI})(1+\texttt{ETA})(1+\texttt{ZETA}) \\ 1/8(1+\texttt{XI})(1+\texttt{ETA})(1+\texttt{ZETA}) \\ 1/8(1+\texttt{XI})(1+\texttt{ETA})(1-\texttt{ZETA}) \end{array} \right\}^{T} \tag{3.72}$$

which together with their derivatives with respect to local coordinates are as usual formed by the subroutines

SHAPE_FUN

SHAPE_DER

with NDIM $= 3$ and NOD $= 8$.

The sequence of operations described by (3.45) results in DERIV, the required gradients with respect to (x, y, z) and the Jacobian determinant DET.

A higher-order brick element with 20 nodes (Figure 3.13) is also used in programs later in the book. The shape functions and their derivatives for this element are as usual provided by the subroutines

SHAPE_FUN

SHAPE_DER

although they are not quoted here (see the coding in Appendix 5).

For the 3-d elastic solid the element stiffness is given by

$$\text{KM} = \iiint \text{TRANSPOSE (BEE)*DEE*BEE}\, dx\, dy\, dz \tag{3.73}$$

where BEE and DEE must be formed by the subroutines

BEEMAT

DEEMAT

as usual but with NST, the number of components of stress and strain now being six. The final summation is

$$\text{KM} = \sum_{\text{I}=1}^{\text{NIP}} \text{DET}_\text{I} * \text{WEIGHTS(I)} * \text{BTDB}_\text{I} \tag{3.74}$$

where WEIGHTS(I) are the weighting multipliers obtained from SAMPLE.

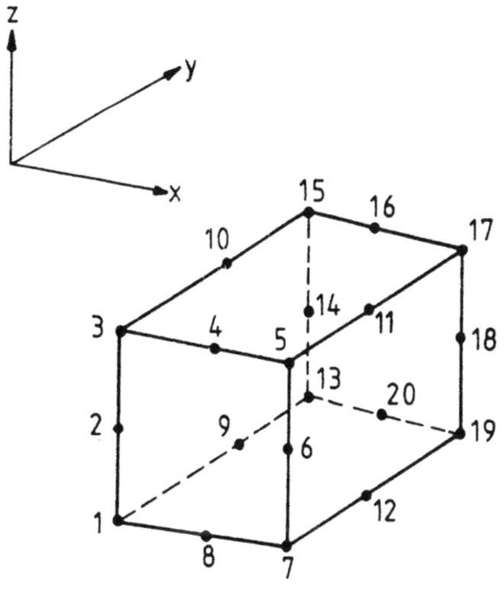

Figure 3.13 A 20-node brick element

Gaussian integration is expensive for this element. Smith and Kidger (1991) show that full $3 \times 3 \times 3$ (NIP=27) point integration is essential if spurious "zero energy" eigenmodes are to be avoided in the element stiffness. Routine SAMPLE allows Irons's (1971) 13,14 and 15 point rules to be used instead.

For 3-d steady laminar fluid flow the element "stiffness" is

$$\begin{aligned} \mathtt{KP} &= \iiint \mathtt{TRANSPOSE\ (DERIV)*KAY*DERIV}\, \mathrm{d}x\, \mathrm{d}y\, \mathrm{d}z \\ &= \iiint \mathtt{DTKD}\, \mathrm{d}x\, \mathrm{d}y\, \mathrm{d}z \quad \text{(say)} \end{aligned} \tag{3.75}$$

where KAY is the principal axes permeability tensor

$$\mathtt{KAY} = \begin{bmatrix} \mathtt{PERMX} & 0 & 0 \\ 0 & \mathtt{PERMY} & 0 \\ 0 & 0 & \mathtt{PERMZ} \end{bmatrix} \tag{3.76}$$

Gaussian integration gives

$$\mathtt{KP} = \sum_{\mathtt{I}=1}^{\mathtt{NIP}} \mathtt{DET}_{\mathtt{I}} * \mathtt{WEIGHTS(I)} * \mathtt{DTKD}_{\mathtt{I}} \tag{3.77}$$

which is similar to (3.74).

The "mass" matrix for potential flow is

$$\begin{aligned} \mathrm{MM} &= \iiint \mathtt{TRANSPOSE\ (FUN)*FUN}\, \mathrm{d}x\, \mathrm{d}y\, \mathrm{d}z \\ &= \iiint \mathtt{FTF}\, \mathrm{d}x\, \mathrm{d}y\, \mathrm{d}z \quad \text{(say)} \end{aligned} \tag{3.78}$$

which is replaced by quadrature as

$$\mathtt{MM} = \sum_{\mathtt{I}=1}^{\mathtt{NIP}} \mathtt{DET}_{\mathtt{I}} * \mathtt{WEIGHTS(I)} * \mathtt{FTF}_{\mathtt{I}} \tag{3.79}$$

In solid mechanics, a 3-d equivalent of ECMAT would be required. This development is left to the reader.

3.11.1.8 A 14-node hexahedral element

The 20-node element described above is rather cumbersome and its stiffness can be expensive to compute in nonlinear analyses. It is essential to employ "full" $3 \times 3 \times 3$ Gaussian (or some equivalent) integration (Smith and Kidger, 1991).

An alternative (Smith and Kidger, 1992) is to use a 14-node element. As shown in Figure 3.14, this has eight corner and six mid-face nodes. These nodes "populate" 3-d space more uniformly than 20 nodes do, since the latter are

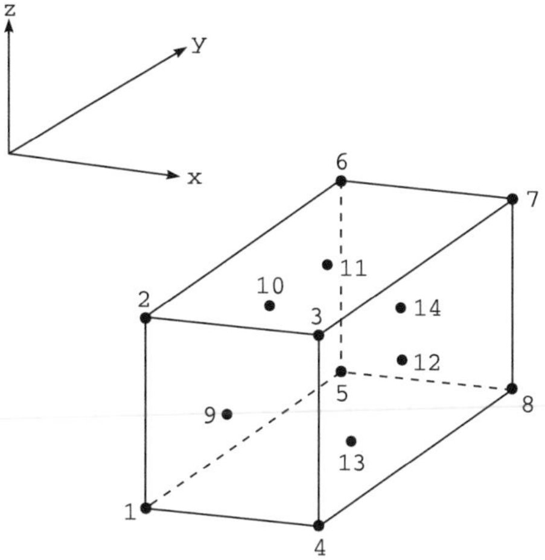

Figure 3.14 A 14-node brick element

concentrated along the mesh lines. However, there is no unique choice of shape functions for a 14-node element. Figure 3.15 shows the "Pascal pyramid" of polynomials in x, y, z (or ξ, η, ζ) and one could experiment with various combinations of terms. Smith and Kidger (1992) tried six permutations which were called "Types 1 to 6". For example Type 1 contained all 10 polynomials down to the second "plane" of the pyramid plus the terms xyz, x^2y, y^2z and z^2x from the third "plane". Type 6 selectively contained terms as far down as the fifth "plane". Computer algebra was essential when deriving the shape functions which are listed below for the Type 6 element.

Type 6. Interval $-1 \leq x, y, z \leq 1$

$$
\begin{aligned}
\text{FUN}(1) &= ((x*y+x*z+2.*x+y*z+2.*y+2.*z+2.)*(x-1.)*(y-1.)*(z-1.))/8.\\
\text{FUN}(2) &= ((x*y-x*z-2.*x+y*z+2.*y-2.*z-2.)*(x-1.)*(y+1.)*(z-1.))/8.\\
\text{FUN}(3) &= ((x*y+x*z+2.*x-y*z-2.*y-2.*z-2.)*(x+1.)*(y-1.)*(z-1.))/8.\\
\text{FUN}(4) &= ((x*y-xz-2.*x-y*z-2.*y+2.*z+2.)*(x+1.)*(y+1.)*(z-1.))/8.\\
\text{FUN}(5) &= -((x*y-x*z+2.*x-y*z+2.*y-2.*z+2.)*(x-1.)*(y-1.)*(z+1.))/8.\\
\text{FUN}(6) &= -((x*y+x*z-2.*x-y*z+2.*y+2.*z-2.)*(x-1.)*(y+1.)*(z+1.))/8.\\
\text{FUN}(7) &= -((x*y-x*z+2.*x+y*z-2.*y+2.*z-2.)*(x+1.)*(y-1.)*(z+1.))/8.\\
\text{FUN}(8) &= -((x*y+x*z-2.*x+y*z-2.*y-2.*z+2.)*(x+1.)*(y+1.)*(z+1.))/8.\\
\text{FUN}(9) &= -((x+1.)*(x-1.)*(y-1.)*(z-1.))/2.\\
\text{FUN}(10) &= ((x+1.)*(x-1.)*(y+1.)*(y-1.)*(z+1.))/2.\\
\text{FUN}(11) &= -((x+1.)*(x-1.)*(y-1.)*(z+1.)*(z-1.))/2.\\
\text{FUN}(12) &= ((x+1.)*(x-1.)*(y+1.)*(z+1.)*(z-1.))/2.\\
\text{FUN}(13) &= -((x-1.)*(y+1.)*(y-1.)*(z+1.)*(z-1.))/2.\\
\text{FUN}(14) &= ((x+1.)*(y+1.)*(y-1.)*(z+1.)*(z-1.))/2.
\end{aligned}
\tag{3.80}
$$

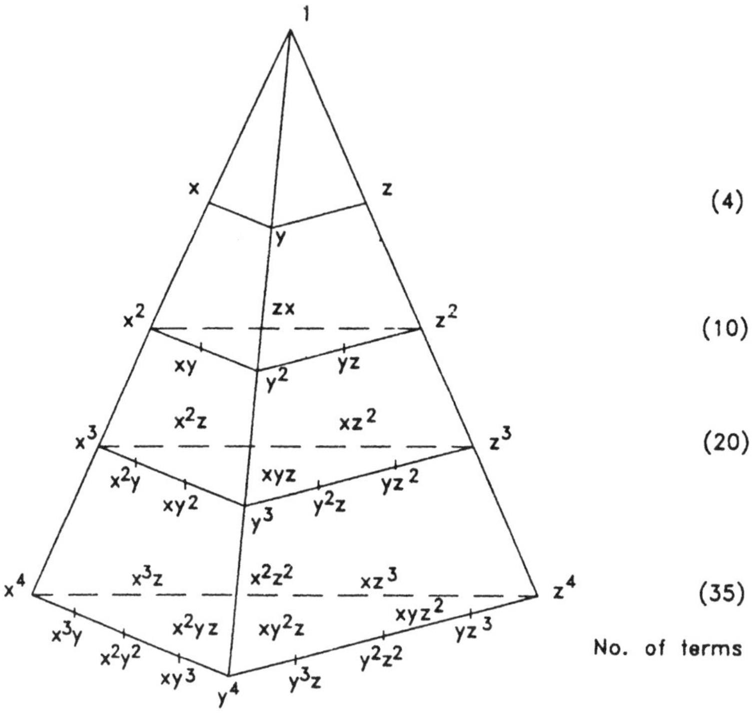

Figure 3.15 The Pascal pyramid

For the values of `DER`, simply obtained from (3.80) by computer algebra, see the coding in Appendix 5.

3.11.1.9 Three-dimensional tetrahedral elements

An alternative element for 3-d analysis is the tetrahedron, the simplest of which has four corner nodes. The local coordinate system makes use of volume coordinates as shown in Figure 3.16. For example, point P can be identified by four volume coordinates (L_1, L_2, L_3, L_4) where

$$
\begin{aligned}
L_1 &= \frac{\text{P432}}{V_\text{T}} \\
L_2 &= \frac{\text{P413}}{V_\text{T}} \\
L_3 &= \frac{\text{P421}}{V_\text{T}} \\
L_4 &= \frac{\text{P123}}{V_\text{T}}
\end{aligned}
\tag{3.81}
$$

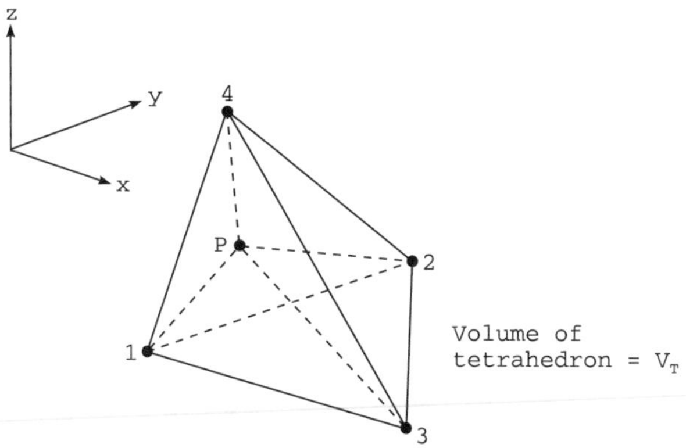

Figure 3.16 A four-node tetrahedron element

As is to be expected, one of these coordinates is redundant due to the identity

$$L_1 + L_2 + L_3 + L_4 = 1 \tag{3.82}$$

The shape functions for the constant strain tetrahedron are

$$\texttt{FUN} = \begin{Bmatrix} L_1 \\ L_2 \\ L_3 \\ L_4 \end{Bmatrix}^{\mathrm{T}} \tag{3.83}$$

and these, together with their derivatives with respect to L_1, L_2 and L_3 are formed by the usual subroutines

SHAPE_FUN

SHAPE_DER

for NDIM $= 3$ and NOD $= 4$.

The sequence of operations (3.45) results in DERIV which is needed to form the element matrices and DET which is used in the numerical integrations.

The integration is performed by

$$\texttt{KM} = \sum_{\texttt{I}=1}^{\texttt{NIP}} \texttt{DET}_{\texttt{I}} * \texttt{WEIGHTS(I)} * \texttt{BTDB}_{\texttt{I}} \tag{3.84}$$

where WEIGHTS(I) is the weighting coefficient corresponding to the particular integrating point.

The addition of mid-side nodes results in the 10-noded tetrahedron which represents the next member of the family. Reference to Figure 3.15 shows that the tetrahedral family of shape functions maps naturally on to the Pascal pyramid (4, 10, 20, 35 etc. nodes). These elements could easily be implemented by the interested reader.

Transient, coupled poro-elastic transient and elastic–plastic analysis all involve manipulations of the few simple element property matrices described above. Before describing such applications, methods of assembling elements and of solving linear equilibrium and eigenvalue problems will first be described.

3.11.2 ASSEMBLY OF ELEMENTS

The seven special purpose subroutines

FORMNF	FORMKV	FORMKU
FKDIAG	FORMKB	FORMTB
FSPARV		

are concerned with assembling the individual element matrices to form the system matrices that approximate the desired continuum if assembly is preferred to an "element-by-element" approach. Allied to these there must be a specification of the geometrical details, in particular the nodal coordinates of each element and the element's place in some overall node numbering scheme.

Large finite element programs contain mesh generation code which is usually of some considerable complexity. Indeed, in much finite element work, the most expensive and time-consuming task is the preparation of the input data using the mesh generation routines. In the present book, this aspect of the computations is essentially ignored and attention is restricted to simple classes of geometry which can be automatically built up by small subroutines. Examples are the simple meshes made up of four-noded rectangles shown in Figure 3.17.

Alternatively, more general purpose programs are presented later in which the element geometries and nodal connectivities are simply read into the analysis program as data, having previously been worked out by an independent mesh generator.

In the present work, a typical program might use plane rectangular elements, so subroutines such as

```
GEOMETRY_4QY
```

(four-node quadrilaterals numbering in the y direction) are provided to generate coordinates and node and freedom numbering.

A full list of "geometry" routines is given in Appendix 3.

With reference to Figure 3.17, the nodes of the mesh are first assigned numbers as economically as possible (i.e. always numbering in the "shorter" direction to minimise the bandwidth). Associated with each node are degrees of freedom (displacements, fluid potentials and so on) which are numbered in the same order as the nodes. However, account is taken at this stage of whether a degree of freedom exists or whether, generally at the boundaries of the region, the freedom is suppressed, in which case that freedom number is assigned the value zero.

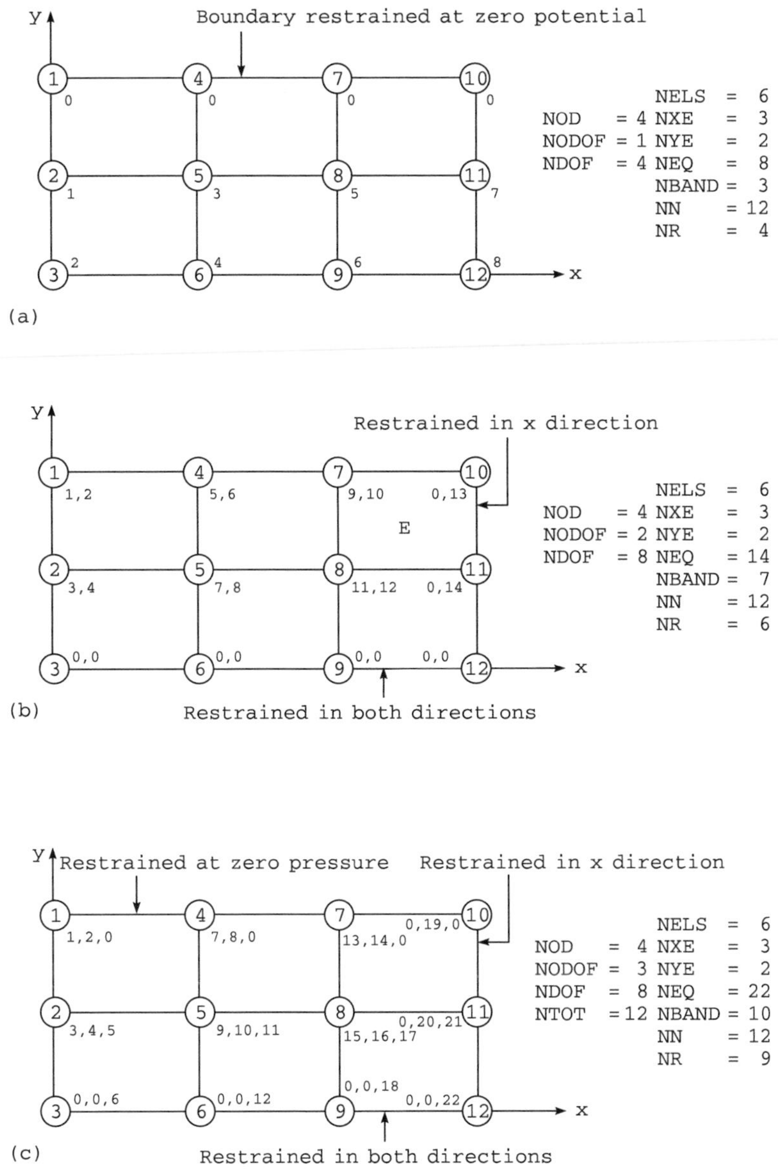

Figure 3.17 Numbering system and data for regular meshes. (a) One degree of freedom per node. (b) Two degrees of freedom per node. (c) Coupled problem with three degrees of freedom per node

The variables in Figure 3.17 have the following meaning:

NXE	elements counting in x direction
NYE	elements counting in y direction
NELS	total number of elements
NEQ	total number of (non-zero) freedoms in problem
NBAND	semi-bandwidth
NN	total number of nodes in problem
NR	number of restrained nodes
NOD	nodes per element
NODOF	freedoms per node
NDOF	freedoms per element
NTOT	total freedoms per element (for coupled problems)

The values of NEQ and NBAND can be calculated by the program.

In scalar potential problems described by these types of element there is one degree of freedom possible per node, the potential ϕ. In plane or axisymmetric strain there are two, namely the u and v components of displacement specified in that order. In coupled solid–fluid problems the order is u, v, u_w (where u_w = excess pressure) and in 3-d displacement problems u, v, w (x, y, z). For Navier–Stokes applications the order u, p, v (where p = pressure) is adopted.

This information about the degrees of freedom present in specific problems is stored in an integer array NF called the "node freedom array", formed by the buiding block subroutine

```
FORMNF
```

The node freedom array NF has NODOF rows, one for each degree of freedom per node, and NN columns, one for each node in the problem analysed. Formation of NF is achieved by specifying, as data to be read in, the number of any node whose freedom is restrained in some way, followed by the digit 0 if the node is restrained in that sense and by the digit 1 if it is not. The appropriate Fortran 90 coding is

```
IF(NR > 0) READ (10,*)(K,NF(:,K),I = 1,NR)
```

followed by

```
CALL FORMNF(NF)
```

For example, to create NF for the problem shown in Figure 3.17(b) the data specified and the resulting NF are listed in Table 3.4.

Building block subroutines of the GEOMETRY_4QY type can then be constructed using NF as input. For example, the rectangular meshes need to be specified by the number of elements in the $x(r)$ and $y(z)$ directions respectively (NXE, NYE), together with their sizes (AA, BB). The subroutine has

Table 3.4 Formation of typical nodal freedom array

Data			Resulting NF array											
3	0	0	1	3	0	5	7	0	9	11	0	0	0	0
6	0	0	2	4	0	6	8	0	10	12	0	13	14	0
9	0	0												
10	0	1												
11	0	1												
12	0	0												

Table 3.5

Building block subroutine	Banding considered?	Symmetry of coefficients	Upper or lower triangle?	Storage scheme
FORMKV	Yes	Yes	Upper	Columns
FORMKB	Yes	Yes	Lower	Rows
FORMKU	Yes	Yes	Upper	Rows
FORMTB	Yes	No	Both	Rows
FSPARV	Yes	Yes	Lower	Rows (skyline)

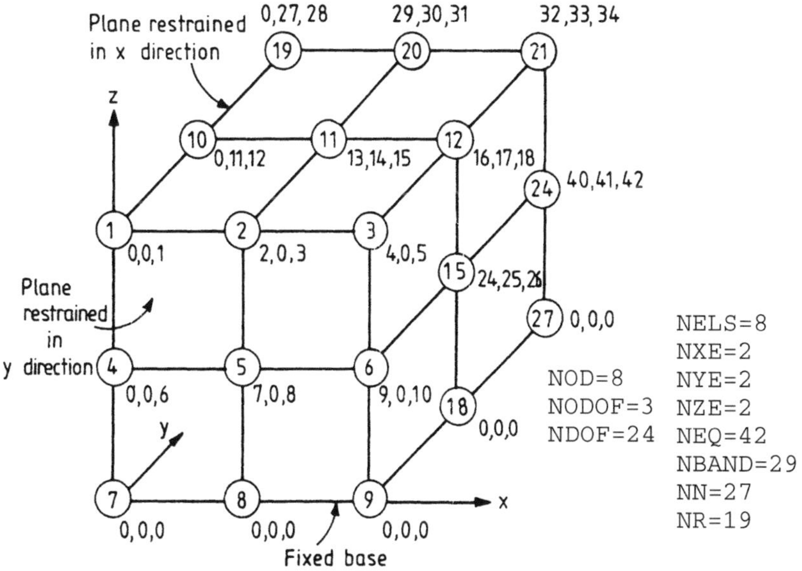

Figure 3.18 A simple 3-d mesh

to work out the nodal coordinates COORD of each element together with a vector G called the "steering vector", which contains the numbers of the degrees of freedom associated with that particular element in accordance with the nodal order 1234 in Figure 3.1.

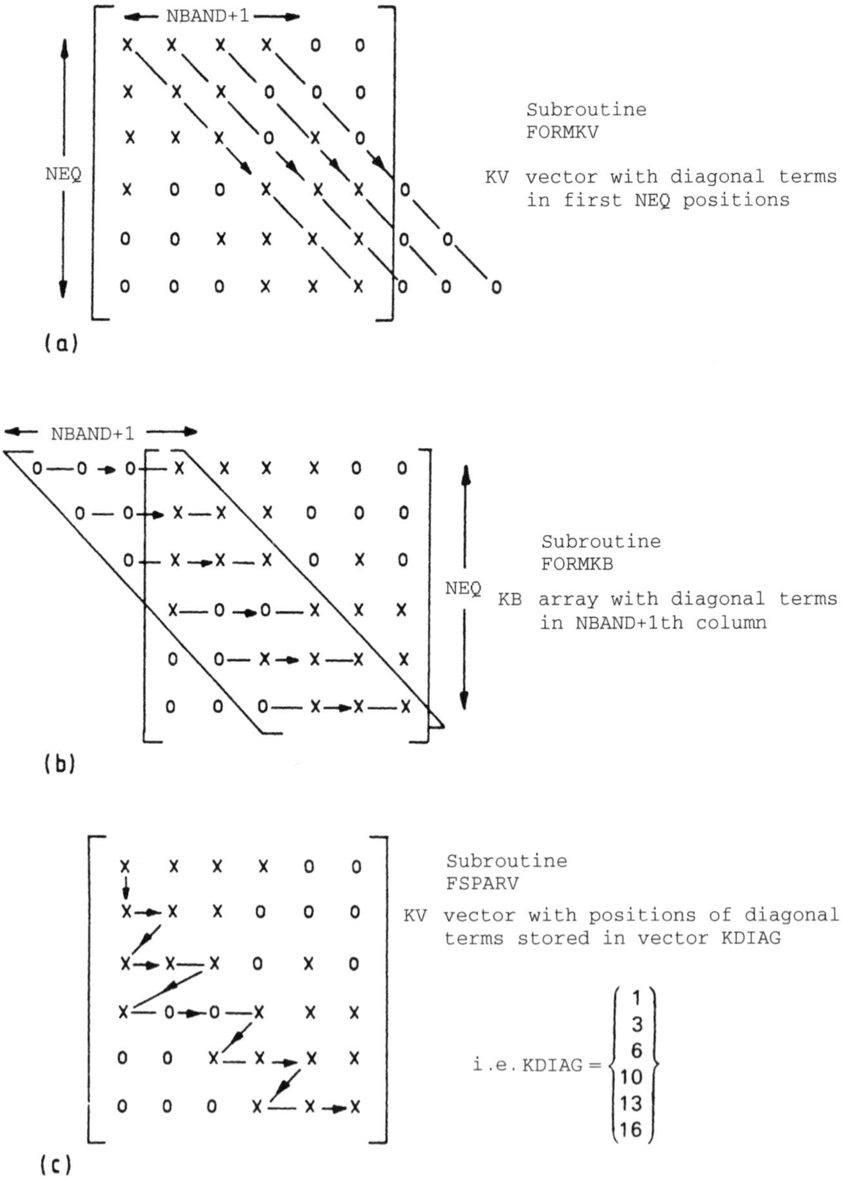

Figure 3.19 Different storate strategies

For example, element E in Figure 3.17(b) has the steering vector

$$\mathtt{G}^{\mathrm{T}} = [11 \quad 12 \quad 9 \quad 10 \quad 0 \quad 13 \quad 0 \quad 14] \tag{3.85}$$

In its turn G can be used to assemble the coefficients of the element property matrices such as **KM**, **KP** and **MM** into the appropriate places in the overall coefficient matrix. This is done according to the following scheme shown in Table 3.5

A simple 3-d mesh is shown in Figure 3.18 and the system coefficients can again be assembled using the same building blocks.

Although the user of these subroutines does not strictly need to know how the storage is carried out, examples are given in Figure 3.19 for the most commonly used assembly routines FORMKV, FORMKB and FSPARV.

Subroutine FORMKV stores the global stiffness matrix as a vector of length NEQ*(NBAND + 1) with the diagonal terms occupying the first NEQ positions (Figure 3.19a).

Subroutine FORMKB stores the global stiffness matrix as a rectangular matrix with NEQ rows and (NBAND + 1) columns. The diagonal terms are held in the (NBAND + 1)th column (Figure 3.19b).

Both these strategies include a number of zeros in the storage due to the variability of the bandwidth and the inclusion of "fictitious" zeros outside the actual array to simplify the subscripting.

Subroutine FSPARV stores only those numbers within the 'skyline' (Figure 3.19c) and can lead to improvements in storage requirements. Information regarding the position of the diagonal terms within the resulting vector must be held in the integer vector KDIAG formed by subroutine FKDIAG. Some examples of this type of storage are given later in the book (e.g. Program 5.10).

3.12 SOLUTION OF EQUILIBRIUM EQUATIONS

If an assembly strategy is chosen, after specification of boundary conditions, a typical equilibrium equation is

$$\mathtt{KV*DISPS} = \mathtt{LOADS} \tag{3.86}$$

in which the coefficients of KV have been formed by FORMKV and LOADS are usually just input as data. The black box routines for equation solution depend of course on the method of coefficient storage according to the scheme shown in Table 3.6.

In fact, to save storage, all of the solution routines overwrite the right-hand side by the solution. That is, on completion of (3.86) the vector LOADS holds the solution DISPS. The storage strategy adopted by the compiler and the hardware, as described in Chapter 1, strongly influences the solution method

Table 3.6

Solution routines	Coefficients formed by	Matrix/vector product
BANRED, BACSUB	FORMKV	LINMUL
CHOLIN, CHOBK1 CHOBAC, CHOBK2	FORMKB	BANMUL
GAUSS_BAND, SOLVE_BAND	FORMTB	BANTML
SPARIN, SPABAC	FSPARV	LINMUL_SKY

that should be adopted for large problems. For example, on a virtual store machine in which Fortran 90 stores by columns, the CHOLIN, CHOBAC options should be avoided, because of the excessive paging requests that might be involved.

For very large systems, assembly would not be used and the iterative preconditioned conjugate gradient (pcg) process described in section 3.7 would be substituted.

3.13 EVALUATION OF EIGENVALUES AND EIGENVECTORS

It is usually assumed that mass matrices can be "lumped", i.e. diagonalised. Before solution, it is necessary to reduce the eigenvalue equation to "standard form"

$$\mathbf{A}\mathbf{x} = \omega^2 \mathbf{x} \tag{3.87}$$

where $\mathbf{A}$ is symmetric.

3.13.1 JACOBI ALGORITHM

The eigenproblem to be solved is

$$\text{KB}*\text{X}=\text{OMEGA}^2*\text{MB}*\text{X} \tag{3.88}$$

where KB and MB are the system stiffness and mass matrices respectively. The stiffness is stored as a banded NEQ * (NBAND+1) matrix having been formed by FORMKU. The mass is either a banded matrix MB with the same structure as KB or a diagonal matrix which can be stored in a vector DIAG.

In order to reduce (3.88) to the required form (3.87) it is necessary to factorise the mass matrix by forming

$$\mathtt{MB} = L * L^{\mathrm{T}} \tag{3.89}$$

While this is essentially a Choleski factorisation, it is particularly simple in the case of a diagonal matrix. The non-zero terms in L are simply the square roots of the terms in `DIAG` and the inverse of L merely consists of the reciprocals of these square roots. Thus

$$\mathtt{KB}*\mathrm{X} = \mathtt{OMEGA}^2 * L * L^{\mathrm{T}} * \mathrm{X} \tag{3.90}$$

which is then reduced to standard form by making the substitution

$$L^{\mathrm{T}} * \mathrm{X} = \mathrm{Z} \tag{3.91}$$

Then

$$L^{-1} * \mathtt{KB} * L^{-\mathrm{T}} * \mathrm{Z} = \mathtt{OMEGA}^2 * \mathrm{Z} \tag{3.92}$$

is of the desired form. Having solved for `Z`, the required eigenvectors `X` are readily recovered using (3.91).

The subroutines

```
BANDRED
BISECT
```

deliver the appropriate eigenvalues.

3.13.2 LANCZOS ALGORITHM

The transformation technique previously described is robust in that it will not fail to find an eigenvalue or to detect multiple roots. However, it is expensive to use on large problems for which, in general, vector iteration methods are preferable. Since the heart of all of these involves a matrix-by-vector product they are ideal for "element-by-element" manipulation, in which case the lumped mass matrix must be used.

Programs described in Chapter 10 use subroutines modified from those described by Parlett and Reid (1981). These subroutines calculate the eigenvalues and eigenvectors of a symmetric matrix, say $\mathbf{A}$, requiring the user only to compute the product and sum $\mathbf{Av} + \mathbf{u}$ for $\mathbf{u}$ and $\mathbf{v}$ provided by the subroutines. These operations can be carried out element-wise as was described for the preconditioned conjugate gradient method in section 3.7.

The Lanczos vector subroutines are

```
LANCZ1
LANCZ2
```

There is a slight additional complexity in that, as in (3.90) we often have the generalised eigenvalue problem to solve. For consistent mass approximations $L*L^T$ in (3.90) is formed by Choleski factorisation using `CHOLIN`. Then on each Lanzos iteration, wherever $\mathbf{u} = \mathbf{Av} + \mathbf{u}$ is called for, we compute

1. $L^T z_1 = v$ (`CHOBK2`)
2. $z_2 = KBz_1$ (`BANMUL`)
3. $Lz_3 = z_2$ (`CHOBK1`)
4. $u = u + z_3$ whole array vector addition

Finally, the true eigenvectors are recovered from the transformed ones using backward substitution (`CHOBK2`) as in (3.91).

For lumped mass approximations, essentially the same operations are performed but the diagonal nature of L means that subroutine calls can be dispensed with.

This algorithm has proved to be reasonably robust and is certainly efficient. However, it is possible for clustered eigenvalues to cause difficulties.

3.14 SOLUTION OF FIRST-ORDER TIME-DEPENDENT PROBLEMS

A typical equation (compare (2.108)) is given by

$$\mathbf{KP}\boldsymbol{\phi} + \mathbf{PM}\frac{\mathrm{d}\boldsymbol{\phi}}{\mathrm{d}t} = \mathbf{q} \tag{3.93}$$

where $\mathbf{q}$ may be a function of time. There are many ways of integrating this set of ordinary differential equations, and modern methods for small numbers of equations would probably be based on variable order, variable time-step methods with error control. However, for large engineering systems more primitive methods are still mainly used, involving linear interpolations and fixed time steps Δt. The basic algorithm is written:

$$\mathbf{KP}[\theta\,\boldsymbol{\phi}_1 + (1-\theta)\boldsymbol{\phi}_0] + \mathbf{PM}\left[\theta\frac{\mathrm{d}\boldsymbol{\phi}_1}{\mathrm{d}t} + (1-\theta)\frac{\mathrm{d}\boldsymbol{\phi}_0}{\mathrm{d}t}\right] = \theta\mathbf{q}_1 + (1-\theta)\mathbf{q}_0 \tag{3.94}$$

leading to the following recurrence relation between time steps "0" and "1":

$$(\mathbf{PM} + \theta\Delta t\mathbf{KP})\boldsymbol{\phi}_1 = [\mathbf{PM} - (1-\theta)\Delta t\mathbf{KP}]\boldsymbol{\phi}_0 + \theta\Delta t\mathbf{q}_1 + (1-\theta)\Delta t\mathbf{q}_0 \tag{3.95}$$

This system is only unconditionally "stable" (i.e. errors will not grow unboundedly) if $\theta \geq 1/2$. Common choices would be $\theta = 1/2$, giving the "Crank-Nicolson" method (assuming for the moment $\mathbf{q} = \mathbf{0}$):

$$\left(\mathbf{PM} + \frac{\Delta t}{2}\mathbf{KP}\right)\boldsymbol{\phi}_1 = \left(\mathbf{PM} - \frac{\Delta t}{2}\mathbf{KP}\right)\boldsymbol{\phi}_0 \tag{3.96}$$

and $\theta = 1$ giving the "fully implicit" method:

$$(\mathbf{PM} + \Delta t\mathbf{KP})\boldsymbol{\phi}_1 = \mathbf{PM}\boldsymbol{\phi}_0 \tag{3.97}$$

If an element assembly strategy is used these element equations are summed into system equations using the usual assembly procedures with **PM**s being assembled into, say, **BP** and **KP**s being assembled into, say, **BK.** The solution of problems will clearly involve a marching process from time 0 to time 1. First a matrix by vector multiplication must be carried out on the right-hand side of the assembled (3.96) or (3.97) using `LINMUL` (assuming `FORMKV` has been used for assembly) or its equivalent, depending on the storage strategy, and second, a set of linear equations must be solved on each time step. If the **KP** and **PM** matrices do not change with time, the factorisation of the left-hand side coefficients, which is a time-consuming operation, need only be performed once, as shown in the structure chart in Figure 3.20.

The "implicit" strategies described above are quite effective for linear problems (constant **KP** and **PM**). However, storage requirements can be considerable, and in nonlinear problems the necessity to refactorise $\mathbf{BP} + \theta\Delta t\mathbf{BK}$ can lead to lengthy calculations.

Storage can be saved by replacing `LINMUL` by an element-by-element matrix-vector multiply and by solving the simultaneous equations iteratively (by pcg

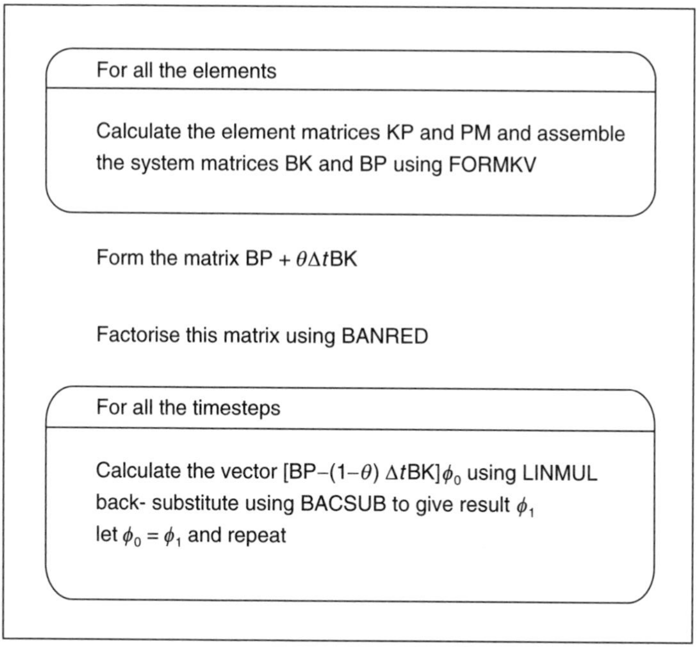
For all the elements
Calculate the element matrices KP and PM and assemble the system matrices BK and BP using FORMKV
Form the matrix BP + $\theta\Delta t$BK
Factorise this matrix using BANRED
For all the timesteps
Calculate the vector [BP−(1−θ) ΔtBK]ϕ_0 using LINMUL
back- substitute using BACSUB to give result ϕ_1
let $\phi_0 = \phi_1$ and repeat

Figure 3.20 Structure chart for first-order time-dependent problems by implicit methods using an assembly strategy

for example). Such a strategy can be attractive for parallel processing and examples are given in Chapter 8.

There is another alternative, widely used for second-order problems (see section 3.17), in which θ is set to zero and the **PM** matrix is "lumped" (see section 2.2). In that case the system to be solved at element level is

$$\mathbf{PM}\boldsymbol{\phi}_1 = (\mathbf{PM} - \Delta t\mathbf{KP})\boldsymbol{\phi}_0 \tag{3.98}$$

where **PM** is a diagonal matrix and hence $\boldsymbol{\phi}_1$ can be obtained "explicitly" from $\boldsymbol{\phi}_0$ without resorting to equation solution. As usual in an element-by-element strategy, appropriate manipulation of element **PM** and **KP** matrices means no large "global" arrays are necessary. The disadvantage is that (3.98) is only stable on condition that Δt is small, and in practice perhaps so small that real times of interest would require an excessive number of steps.

Yet another element-by-element approach which conserves computer storage while preserving the stability properties of implicit methods involves "operator splitting" on an element-by element product basis (Hughes et al, 1983; Smith et al, 1989). Although not necessary for the operation of the method, the simplest algorithms result from "lumping" **PM**. The global equations are, for $\mathbf{q} = \mathbf{0}$,

$$\boldsymbol{\phi}_1 = (\mathbf{BP} + \theta\Delta t\mathbf{BK})^{-1}[\mathbf{BP} - (1-\theta)\Delta t\mathbf{BK}]\boldsymbol{\phi}_0 \tag{3.99}$$

Thus **BP** is the global **PM**:

$$\mathbf{BP} = \sum_{\text{elements}} \mathbf{PM} \tag{3.100}$$

and **BK** is the global **KP**

$$\mathbf{BK} = \sum_{\text{elements}} \mathbf{KP} \tag{3.101}$$

The element-by-element splitting methods are based on binomial theorem expansions of $(\mathbf{BP} + \theta\Delta t\mathbf{BK})^{-1}$ which neglect product terms. When **BP** is lumped (diagonal) the method is particularly straightforward because **BP** can be effectively be replaced by **I**, the unit matrix.

$$\begin{aligned}(\mathbf{I} + \theta\Delta t\mathbf{BK})^{-1} &= (\mathbf{I} + \theta\Delta t\Sigma\mathbf{KP})^{-1} \\ &\cong \prod_{\text{elements}} (\mathbf{I} + \theta\Delta t\mathbf{KP})^{-1}\end{aligned} \tag{3.102}$$

where $\prod$ indicates a product. As was the case with implicit methods, optimal accuracy consistent with stability is achieved for $\theta = 1/2$. It is shown by Hughes et al (1983) that further optimisation is achieved by splitting further to

$$\mathbf{KP} = (\tfrac{1}{2}\mathbf{KP}) + (\tfrac{1}{2}\mathbf{KP}) \tag{3.103}$$

and carrying out the product (3.102) by sweeping twice through the elements. They suggest from first to last and back again, but clearly various choices of sweeps could be employed. It can be shown that as $\Delta t \to 0$, any of these processes converges to the true solution of the global problem (for lumped **PM** of course). A structure chart for the process is shown in Figure 3.21 and examples of all the methods described in this section are described in Chapter 8. Consistent mass versions are described by Gladwell et al (1989).

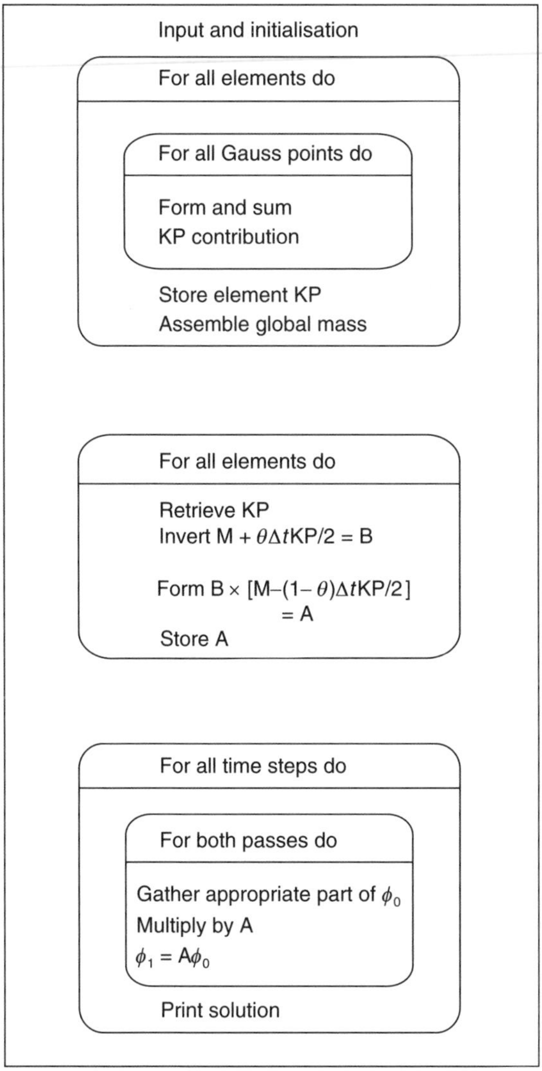

Figure 3.21 Structure chart for the element-by-element product algorithm (two-pass)

3.15 SOLUTION OF COUPLED NAVIER-STOKES PROBLEMS

For steady-state conditions, it was shown in section 2.16 that a nonlinear system of algebraic equations had to be solved, involving, at element level, submatrices C_{11},C_{12}, etc. These element matrices contained velocities $\bar{u}$ and $\bar{v}$ (called UBAR and VBAR in the programs) together with shape functions and their derivatives for the velocity and pressure variables. It was mentioned that it would be possible to use different shape functions for the velocity (vector) quantity and pressure (scalar) quantity and this is what is done in the program in Chapter 9.

The velocity shape functions are designated as FUN and the pressure shape functions as FUNF. Similarly, the velocity derivatives are DERIV and the pressure derivatives DERIVF.

Thus the element integrals which have to be evaluated numerically are (2.92), typically

$$\begin{aligned} \texttt{c11}(=\texttt{c33}) = & \iint \texttt{TRANSPOSE(DERIV)*KAY*DERIV}\, dx\, dy \\ & + \iint \texttt{UBAR*FUN*DERIV(1,:)} dx\, dy \\ & + \iint \texttt{VBAR*FUN*DERIV(2,:)} dx\, dy \end{aligned} \tag{3.104}$$

In this equation, DERIV (1,:) signifies the first row of DERIV and so on in the usual Fortran 90 style. The diagonal terms in KAY represent the reciprocal of the Reynolds number. Note the identity of the first term of c11 with (3.58) for uncoupled flow.

The remaining submatrices are

$$\begin{aligned} \texttt{c12} &= \frac{1}{\texttt{RHO}} \iint \texttt{FUN * DERIVF(1,:)} dx\ dy \\ \texttt{c32} &= \frac{1}{\texttt{RHO}} \iint \texttt{FUN * DERIVF(2,:)} dx\ dy \\ \texttt{c21} &= \iint \texttt{FUNF * DERIV(1,:)} dx\ dy \\ \texttt{c23} &= \iint \texttt{FUNF * DERIV(2,:)} dx\ dy \end{aligned} \tag{3.105}$$

where RHO is the mass density.

The (unsymmetrical) matrix built up from these submatrices is formed by a special subroutine called

FORMUPV

and the global, unsymmetrical, band matrix is assembled using FORMTB. The appropriate equation solution routines are GAUSS_BAND and SOLVE_BAND.

These problems are unique in the present book in that no element-by-element strategy is presented for their solution.

3.16 SOLUTION OF COUPLED TRANSIENT PROBLEMS

The element equations for Biot consolidation were shown in section 2.21 to be given by

$$\begin{aligned} \mathbf{KMr} + \mathbf{Cu}_{\mathrm{w}} &= \mathbf{f} \\ \mathbf{C}^{\mathrm{T}} \frac{\mathrm{d}\mathbf{r}}{\mathrm{d}t} - \mathbf{KPu}_{\mathrm{w}} &= \mathbf{0} \end{aligned} \tag{3.106}$$

where **KM** and **KP** are the now familiar solid and fluid stiffnesses. The matrix **C** is the connection matrix which, if it is assumed that the same linear shape functions govern the variation of porewater pressure and displacements within an element, is given for plane strain by

$$\mathtt{C} = \iint \mathtt{VOL} * \mathtt{FUN}\,\mathrm{d}x\mathrm{d}y \tag{3.107}$$

The matrix product is called `VOLF` and `C` is numerically integrated as

$$\mathtt{C} = \sum_{\mathtt{I}=1}^{\mathtt{NIP}} \mathtt{DET}_{\mathtt{I}} * \mathtt{WEIGHTS(I)} * \mathtt{VOLF}_{\mathtt{I}} \tag{3.108}$$

The vector `VOL` is related to the volumetric strain in an element. In plane strain, for example, the strain–displacement relation is

$$\begin{Bmatrix} \varepsilon_x \\ \varepsilon_y \\ \gamma_{xy} \end{Bmatrix} = \mathtt{BEE} * \mathtt{ELD} \tag{3.109}$$

where `ELD` are the element displacements and

$$\mathtt{BEE} = \begin{bmatrix} \frac{\partial N_1}{\partial x} & 0 & \frac{\partial N_2}{\partial x} & 0 & \frac{\partial N_3}{\partial x} & 0 & \frac{\partial N_4}{\partial x} & 0 \\ 0 & \frac{\partial N_1}{\partial y} & 0 & \frac{\partial N_2}{\partial y} & 0 & \frac{\partial N_3}{\partial y} & 0 & \frac{\partial N_4}{\partial y} \\ \frac{\partial N_1}{\partial y} & \frac{\partial N_1}{\partial x} & \frac{\partial N_2}{\partial y} & \frac{\partial N_2}{\partial x} & \frac{\partial N_3}{\partial y} & \frac{\partial N_3}{\partial x} & \frac{\partial N_4}{\partial y} & \frac{\partial N_4}{\partial x} \end{bmatrix} \tag{3.110}$$

The element volumetric strain can therefore be written

$$\varepsilon_x + \varepsilon_y = \mathtt{TRANSPOSE(VOL)} * \mathtt{ELD} \tag{3.111}$$

where

$$\texttt{TRANSPOSE (VOL)} = \left[\frac{\partial N_1}{\partial x}\ \frac{\partial N_1}{\partial y}\ \frac{\partial N_2}{\partial x}\ \frac{\partial N_2}{\partial y}\ \frac{\partial N_3}{\partial x}\ \frac{\partial N_3}{\partial y}\ \frac{\partial N_4}{\partial x}\ \frac{\partial N_4}{\partial y}\right] \tag{3.112}$$

and this vector is worked out by the subroutine `VOL2D`.

To integrate equations (3.106) with respect to time there are again many methods available, but we consider only the simplest linear interpolation in time using finite differences thus:

$$\begin{aligned} \theta\mathbf{KMr}_1 + \theta\mathbf{Cu}_{w_1} &= (\theta - 1)\mathbf{KMr}_0 + (\theta - 1)\mathbf{Cu}_{w_0} + \mathbf{f} \\ \theta\mathbf{C}^{\mathrm{T}}\mathbf{r}_1 - \theta^2\Delta t\mathbf{KPu}_{w_1} &= \theta\mathbf{C}^{\mathrm{T}}\mathbf{r}_0 - \theta(\theta - 1)\Delta t\mathbf{KPu}_{w_0} \end{aligned} \tag{3.113}$$

where **f** is assumed here to be independent of time and the second equation has been multiplied through by θ to preserve symmetry on the left-hand side of (3.113).

In the Crank–Nicolson type of approximation, θ is made equal to $1/2$ in both of (3.113), leading to the recurrence relation:

$$\begin{bmatrix} \mathbf{KM} & \mathbf{C} \\ \mathbf{C}^{\mathrm{T}} & -\frac{\Delta t}{2}\mathbf{KP} \end{bmatrix} \begin{Bmatrix} \mathbf{r}_1 \\ \mathbf{u}_{w_1} \end{Bmatrix} = \begin{bmatrix} -\mathbf{KM} & -\mathbf{C} \\ \mathbf{C}^{\mathrm{T}} & \frac{\Delta t}{2}\mathbf{KP} \end{bmatrix} \begin{Bmatrix} \mathbf{r}_0 \\ \mathbf{u}_{w_0} \end{Bmatrix} + \begin{Bmatrix} 2\mathbf{f} \\ \mathbf{0} \end{Bmatrix} \tag{3.114}$$

It will be shown in Chapter 9 that this approximation can lead to oscillatory results. The oscillations can be smoothed out either by using the fully implicit version of (3.113) with $\theta = 1$, which leads to the recurrence relation

$$\begin{bmatrix} \mathbf{KM} & \mathbf{C} \\ \mathbf{C}^{\mathrm{T}} & -\Delta t\mathbf{KP} \end{bmatrix} \begin{Bmatrix} \mathbf{r}_1 \\ \mathbf{u}_{w_1} \end{Bmatrix} = \begin{bmatrix} \mathbf{0} & \mathbf{0} \\ \mathbf{C}^{\mathrm{T}} & \mathbf{0} \end{bmatrix} \begin{Bmatrix} \mathbf{r}_0 \\ \mathbf{u}_{w_0} \end{Bmatrix} + \begin{Bmatrix} \mathbf{f} \\ \mathbf{0} \end{Bmatrix} \tag{3.115}$$

or by writing the first of (3.114) with $\theta = 1$ and the second with $\theta = 1/2$. This hybrid method leads to a recurrence relation

$$\begin{bmatrix} \mathbf{KM} & \mathbf{C} \\ \mathbf{C}^{\mathrm{T}} & -\frac{\Delta t}{2}\mathbf{KP} \end{bmatrix} \begin{Bmatrix} \mathbf{r}_1 \\ \mathbf{u}_{w_1} \end{Bmatrix} = \begin{bmatrix} \mathbf{0} & 0 \\ \mathbf{C}^{\mathrm{T}} & \frac{\Delta t}{2}\mathbf{KP} \end{bmatrix} \begin{Bmatrix} \mathbf{r}_0 \\ \mathbf{u}_{w_0} \end{Bmatrix} + \begin{Bmatrix} \mathbf{f} \\ \mathbf{0} \end{Bmatrix} \tag{3.116}$$

Results of calculations using (3.114) and (3.115) are presented in Chapter 9. The algorithms will clearly be of the same form as those described previously for uncoupled equations (3.96, 3.97) and the structure chart of Figure 3.20. A right-hand side matrix-by-vector multiplication is followed by an equation solution for each time step. As before, a saving in computer time can be achieved if **KM**, **C** and **KP** are independent of time, and constant Δt is used, because the left-hand side matrix needs to be factorised only once. Note that the left-hand side matrix is always symmetrical whereas the right hand side is not. Therefore `FORMKV` or `FORMKB` can be used to assemble the left-hand side system equations from (3.114), (3.115) or (3.116) where `FORMTB` must be used to assemble the right-hand side system. Element-by-element summation is most effective in the right-hand side operations in (3.115) and (3.116) due to sparsity, and element-by-element (mesh-free) algorithms can be developed for

(3.114) to (3.116) based for example on p.c.g. equation solution as shown in Chapter 9.

3.16.1 INCREMENTAL FORMULATION

In (3.114) to (3.116) **f** is the total force applied and these equations are appropriate to linear systems. Later in this book we shall be concerned with nonlinear systems in which **KM** and/or **KP** in (3.106) vary with time and **f** has to be applied in (small) increments $\Delta\mathbf{f}$. If $\Delta\mathbf{f}$ is the change in load between successive times, the incremental form of the first of (3.106) is

$$\mathbf{KM}\Delta\mathbf{r} + \mathbf{C}\Delta\mathbf{u}_\mathrm{w} = \Delta\mathbf{f} \tag{3.117}$$

where $\Delta\mathbf{r}$ and $\Delta\mathbf{u}_\mathrm{w}$ are the resulting changes in displacement and excess porewater pressure respectively. Linear interpolation in time using the θ-method yields

$$\Delta\mathbf{r} = \left[(1-\theta)\frac{\mathrm{d}\mathbf{r}_0}{\mathrm{d}t} + \theta\frac{\mathrm{d}\mathbf{r}_1}{\mathrm{d}t}\right]\Delta t \tag{3.118}$$

and the second of (3.106) can be written at the two time levels to give expressions for $\mathrm{d}\mathbf{r}_0/\mathrm{d}t$ and $\mathrm{d}\mathbf{r}_1/\mathrm{d}t$. When this is done

$$\mathbf{C}^\mathrm{T}\Delta\mathbf{r} - \mathbf{KP}(\mathbf{u}_\mathrm{w} + \theta\Delta\mathbf{u}_\mathrm{w})\Delta t = \mathbf{0} \tag{3.119}$$

or

$$\mathbf{C}^\mathrm{T}\Delta\mathbf{r} - \theta\Delta t\mathbf{KP}\Delta\mathbf{u}_\mathrm{w} = \Delta t\mathbf{KP}\mathbf{u}_\mathrm{w} \tag{3.120}$$

and finally

$$\begin{bmatrix} \mathbf{KM} & \mathbf{C} \\ \mathbf{C}^\mathrm{T} & -\theta\Delta t\mathbf{KP} \end{bmatrix} \begin{bmatrix} \Delta\mathbf{r} \\ \Delta\mathbf{u}_\mathrm{w} \end{bmatrix} = \begin{Bmatrix} \Delta\mathbf{f} \\ \Delta t\mathbf{KP}\mathbf{u}_\mathrm{w} \end{Bmatrix} \tag{3.121}$$

is the incremental form of (3.114) to (3.116). (see Hicks, 1995).

3.17 SOLUTION OF SECOND-ORDER TIME-DEPENDENT PROBLEMS

The basic second-order propagation type of equation was derived in Chapter 2 and takes the form of (2.79), namely

$$\mathbf{KMr} + \mathbf{MM}\frac{\mathrm{d}^2\mathbf{r}}{\mathrm{d}t^2} = \mathbf{f}(t) \tag{3.123}$$

where in the context of solid mechanics, **KM** is the element elastic stiffness and **MM** the element mass. In addition to these elastic and inertial forces, solids in motion experience a third type of force whose action is to dissipate energy. For example, the solid may deform so much that plastic strains result, or may be subjected to internal or external friction. Although these phenomena are nonlinear in character and can be treated by the nonlinear analysis techniques given in Chapter 6, it has been common to linearise the dissipative forces, for example by assuming that they are proportional to velocity. This allows (3.122) to be modified to

$$\mathbf{KMr} + \mathbf{CM}\frac{\mathrm{d}\mathbf{r}}{\mathrm{d}t} + \mathbf{MM}\frac{\mathrm{d}^2\mathbf{r}}{\mathrm{d}t^2} = \mathbf{f}(t) \tag{3.123}$$

where **CM** is assumed to be a constant damping matrix. The most generally applicable technique for integrating (3.123) with respect to time is "direct integration" in an analogous way to that previously described for first-order problems. Again a choice between assembly and element-by-element techniques is available.

Although in principle **CM** could be independently measured or assessed, it is common practice to assume that **CM** is taken to be a linear combination of **MM** and **KM**,

$$\mathbf{CM} = \alpha\mathbf{MM} + \beta\mathbf{KM} \tag{3.124}$$

where α and β are scalars, the so-called "Rayleigh" damping coefficients. They can be related to the more usual "damping ratio" γ (Timoshenko et al, 1974) by means of

$$\gamma = \frac{\alpha + \beta\omega^2}{2\omega} \tag{3.125}$$

where ω is the particular (usually fundamental) frequency of vibration.

There are many methods for advancing the solution of (3.123) with respect to time by direct integration, but attention is first focused on two of the simplest popular implicit methods. In both of these, integration is advanced by one time interval Δt, the values of the displacement and its derivatives at one instant in time being sufficient to determine these values at the subsequent instant by means of recurrence relations. Both preserve unconditional stability, and are implemented in Chapter 11 via element assembly, or element-by-element.

3.17.1 NEWMARK OR CRANK-NICOLSON METHOD

If Rayleigh damping is assumed, a class of recurrence relations based on linear interpolation in time can again be constructed, involving the scalar parameter θ which varies between $1/2$ and 1 in the same way as was done for first-order problems.

Writing the differential equation (3.123) at both the "0" and "1" stations, using notation for a single element:

$$\begin{aligned}\mathbf{KMr}_0 + (\alpha\mathbf{MM} + \beta\mathbf{KM})\frac{\mathrm{d}\mathbf{r}_0}{\mathrm{d}t} + \mathbf{MM}\frac{\mathrm{d}^2\mathbf{r}_0}{\mathrm{d}t^2} &= \mathbf{f}_0\\ \mathbf{KMr}_1 + (\alpha\mathbf{MM} + \beta\mathbf{KM})\frac{\mathrm{d}\mathbf{r}_1}{\mathrm{d}t} + \mathbf{MM}\frac{\mathrm{d}^2\mathbf{r}_1}{\mathrm{d}t^2} &= \mathbf{f}_1\end{aligned} \tag{3.126}$$

and assuming linear interpolation in time giving

$$\begin{aligned} \mathbf{r}_1 &= \mathbf{r}_0 + \Delta t\left[(1-\theta)\frac{\mathrm{d}\mathbf{r}_0}{\mathrm{d}t} + \theta\frac{\mathrm{d}\mathbf{r}_1}{\mathrm{d}t}\right] \\ \frac{\mathrm{d}\mathbf{r}_1}{\mathrm{d}t} &= \frac{\mathrm{d}\mathbf{r}_0}{\mathrm{d}t} + \Delta t\left[(1-\theta)\frac{\mathrm{d}^2\mathbf{r}_0}{\mathrm{d}t^2} + \theta\frac{\mathrm{d}^2\mathbf{r}_1}{\mathrm{d}t^2}\right] \end{aligned} \tag{3.127}$$

rearrangement of these equations leads to the following three recurrence relations:

$$\begin{aligned} &\left[\left(\alpha + \frac{1}{\theta\Delta t}\right)\mathbf{MM} + (\beta + \theta\Delta t)\mathbf{KM}\right]\mathbf{r}_1 \\ &\quad = \theta\Delta t\mathbf{f}_1 + (1-\theta)\Delta t\mathbf{f}_0 + \left(\alpha + \frac{1}{\theta\Delta t}\right)\mathbf{MMr}_0 \\ &\quad + \frac{1}{\theta}\mathbf{MM}\frac{\mathrm{d}\mathbf{r}_0}{\mathrm{d}t} + [\beta - (1-\theta)\Delta t]\mathbf{KMr}_0 \end{aligned} \tag{3.128}$$

$$\frac{\mathrm{d}\mathbf{r}_1}{\mathrm{d}t} = \frac{1}{\theta\Delta t}(\mathbf{r}_1 - \mathbf{r}_0) - \frac{1-\theta}{\theta}\frac{\mathrm{d}\mathbf{r}_0}{\mathrm{d}t} \tag{3.129}$$

$$\frac{\mathrm{d}^2\mathbf{r}_1}{\mathrm{d}t^2} = \frac{1}{\theta\Delta t}\left(\frac{\mathrm{d}\mathbf{r}_1}{\mathrm{d}t} - \frac{\mathrm{d}\mathbf{r}_0}{\mathrm{d}t}\right) - \frac{1-\theta}{\theta}\frac{\mathrm{d}^2\mathbf{r}_0}{\mathrm{d}t^2} \tag{3.130}$$

In the special case when $\theta = 1/2$ this method is Newmark's "$\beta = 1/4$" method, which is also the exact equivalent of the Crank–Nicolson method used in first-order problems. There are other variants of the Newmark type but this is the most common.

The principal recurrence relation (3.128) is clearly similar to those which arose in first-order problems, for example (3.96). Although substantially more matrix-by-vector multiplications are involved on the right-hand side, together with matrix and vector additions, the recurrence again consists essentially of an equation solution per time step. Advantage can as usual be taken of a constant left-hand side matrix should this occur, and element-by-element strategies are easily implemented via p.c.g.

3.17.2 WILSON METHOD

This method advances the solution of (3.123) from some known state $\mathbf{r}_0$, $\mathrm{d}\mathbf{r}_0/\mathrm{d}t$, $\mathrm{d}^2\mathbf{r}_0/\mathrm{d}t^2$ to the new solution $\mathbf{r}_1$ $\mathrm{d}\mathbf{r}_1/\mathrm{d}t$, $\mathrm{d}^2\mathbf{r}_1/\mathrm{d}t^2$ an interval Δt later by first linearly extrapolating to a hypothetical solution, say $\mathbf{r}_2, \mathrm{d}\mathbf{r}_2/\mathrm{d}t, \mathrm{d}^2\mathbf{r}_2/\mathrm{d}t^2$ an interval $\delta t = \theta\Delta t$ later where $1.4 \leq \theta \leq 2$.

If Rayleigh damping is again assumed, $\mathbf{r}_2$ is first computed from

$$\left[\left(\frac{6}{\theta^2\Delta t^2}+\frac{3\alpha}{\theta\Delta t}\right)\mathbf{MM}+\left(\frac{3\beta}{\theta\Delta t}+1\right)\mathbf{KM}\right]\mathbf{r}_2$$
$$=\mathbf{f}_2+\mathbf{MM}\left[\left(\frac{6}{\theta^2\Delta t^2}+\frac{3\alpha}{\theta\Delta t}\right)\mathbf{r}_0+\left(\frac{6}{\theta\Delta t}+2\alpha\right)\frac{\mathrm{d}\mathbf{r}_0}{\mathrm{d}t}+\left(2+\frac{\alpha\theta\Delta t}{2}\right)\frac{\mathrm{d}^2\mathbf{r}_0}{\mathrm{d}t^2}\right]$$
$$+\mathbf{KM}\left(\frac{3\beta}{\theta\Delta t}\mathbf{r}_0+2\beta\frac{\mathrm{d}\mathbf{r}_0}{\mathrm{d}t}+\frac{\beta\theta\Delta t}{2}\frac{\mathrm{d}^2\mathbf{r}_0}{\mathrm{d}t^2}\right) \tag{3.131}$$

The acceleration at the hypothetical station can then be computed from

$$\frac{\mathrm{d}^2\mathbf{r}_2}{\mathrm{d}t^2}=\frac{6}{\theta^2\Delta t^2}(\mathbf{r}_2-\mathbf{r}_0)-\frac{6}{\theta\Delta t}\frac{\mathrm{d}\mathbf{r}_0}{\mathrm{d}t}-2\frac{\mathrm{d}^2\mathbf{r}_0}{\mathrm{d}t^2} \tag{3.132}$$

and thus the acceleration at the true station can be interpolated or "averaged" using

$$\frac{\mathrm{d}^2\mathbf{r}_1}{\mathrm{d}t^2}=\frac{\mathrm{d}^2\mathbf{r}_0}{\mathrm{d}t^2}+\frac{1}{\theta}\left(\frac{\mathrm{d}^2\mathbf{r}_2}{\mathrm{d}t^2}-\frac{\mathrm{d}^2\mathbf{r}_0}{\mathrm{d}t^2}\right) \tag{3.133}$$

A Crank–Nicolson equation then gives the desired velocity

$$\frac{\mathrm{d}\mathbf{r}_1}{\mathrm{d}t}=\frac{\mathrm{d}\mathbf{r}_0}{\mathrm{d}t}+\frac{\Delta t}{2}\left(\frac{\mathrm{d}^2\mathbf{r}_0}{\mathrm{d}t^2}+\frac{\mathrm{d}^2\mathbf{r}_1}{\mathrm{d}t^2}\right) \tag{3.134}$$

and finally the displacement

$$\mathbf{r}_1=\mathbf{r}_0+\Delta t\frac{\mathrm{d}\mathbf{r}_0}{\mathrm{d}t}+\frac{\Delta t^2}{3}\frac{\mathrm{d}^2\mathbf{r}_0}{\mathrm{d}t^2}+\frac{\Delta t^2}{6}\frac{\mathrm{d}^2\mathbf{r}_1}{\mathrm{d}t^2} \tag{3.135}$$

Finally, $\mathbf{f}_2$ in (3.131) must be replaced by $\mathbf{f}_0+\theta(\mathbf{f}_1-\mathbf{f}_0)$ to complete the algorithm.

The principal recurrence relation (3.131) is again of the familiar type for all one-step time integration methods.

3.17.3 EXPLICIT METHODS AND OTHER STORAGE-SAVING STRATEGIES

The implicit methods described above are relatively safe to use due to their unconditional stability. However, as was the case for first-order problems, storage demands become considerable for large systems, and so can solution times for nonlinear problems (even although refactorisation of the left-hand side of (3.128) or (3.131) is not usually necessary, the nonlinear effects having been transposed to the right-hand side).

Using a p.c.g. strategy, the implicit equation solution can always be done element-by-element, but the simplest option, which at least minimises storage, is the analogue of (3.98), in which θ is set to zero and the mass matrix lumped. In the resulting explicit algorithm, operations are carried out element-wise and no global system storage is necessary. Of course the drawback is potential loss of stability, so that stable time steps can be very small indeed.

Since stability is governed by the highest natural frequency of the numerical approximation and since such high frequencies are derived from the stiffest elements in the system, it is quite possible to implement hybrid methods in which the very stiff elements are integrated implicitly, but the remainder are integrated explicitly. Equation solution as implied by (3.128), for example, is still necessary, but the half-bandwidth NBAND of the rows in the coefficient matrix associated with freedoms in explicit elements not connected to implicit ones is only one. Thus great savings in storage can be made (Smith, 1984).

Another alternative is to resort to operator splitting, as was done in first-order problems.

In Chapter 11, implicit, explicit and mixed implicit/explicit algorithms are described and listed, with alternative assembly or (EBE element-by-element) solution for the implicit cases. Although product EBE methods have been developed (Wong et al., 1989) they are beyond the scope of the present book.

3.18 CONCLUSIONS

The principles by which finite element computer programs may be constructed from building block subroutines have been outlined in this chapter. In general, local coordinates are used to express the element shape functions, and the element matrices are numerically integrated. However, in the next chapters we include programs in which the element stiffness matrix can be explicitly stated, or obtained by computer algebra.

Element assembly into a global system of equations is done automatically using a nodal numbering system, a "node freedom array" NF and a "steering vector" G. This vector is also used in the "scatter" and "gather" operations inherent in element-by-element strategies which avoid assembly altogether. Simple boundary conditions are taken care of automatically at this stage. The most common global matrices are the system "stiffness" and "mass" matrices.

These matrices are then manipulated to solve three basic types of problem: equilibrium, eigenvalue and propagation. The matrix operations involved are linear equation solution, eigenvalue extraction and linear equation solution with additional matrix-by-vector multiplications and additions.

Extra features will be introduced as they occur in the examples chapters, but these will be found to involve minor adaptations of what has already been described. For example, axisymmetric structures under non-axisymmetric loads will be considered in Chapter 5, while nonlinear solid problems take up Chapter 6. However, these are solved by linearising incrementally, so that a minimum of new methodology is necessary.

3.19 REFERENCES

Bathe, K.J. (1996) *Numerical Methods in Finite Element Analysis*, 3rd edn. Prentice-Hall, Englewood Cliffs, New Jersey.

Ergatoudis, J., Irons, B.M. and Zienkewicz, O.C. (1968) Curved isoparametric quadrilateral elements for finite element analysis. *Int. J. Solids and Structures*, **4**, 31.

Gladwell, I., Smith, I.M., Gilvary, B., Wong, S.W. (1989) A consistent mass EBE algorithm for linear parabolic systems, *Com. App. Num. Meth.*, **5**, 229–235.

Griffiths, D.V. (1991) Numerical integration of moments. *Int. J. Numer Methods Eng.*, 32(1), 129–147.

Griffiths, D.V. (1994) Stiffness matrix of the four-node quadrilateral element in closed form. *Int. J. Num. Meth. Eng.*, **37**(6), 1027–1038.

Griffiths, D.V. and Smith, I.M. (1991) *Numerical Methods for Engineers*. Blackwell, Oxford.

Hicks, M.A. (1995) MONICA — A computer algorithm for solving boundary value problems using the double hardening constitutive law Monot: I Algorithm development. *Int J. Num. Analytical Meth. Geotech.*, **19**, 1–27.

Hughes, T.J.R., Levit, I. and Winget, J. (1983) Element by element implicit algorithms for heat conduction. *ASCE J. Eng. Mech.*, **109**(2), 576–585.

Irons, B.M. (1966a) Numerical integration applied to finite element methods. *Proc. Conference on Use of Digital Computers in Structural Engineering*, University of Newcastle.

Irons, B.M. (1966b) Engineering applications of numerical integration in stiffness method. *J.A.I.A.A.*, **14**, 2035.

Irons, B.M. (1971) Quadrature rules for brick-based finite elements. *Int. J. Num. Meth. Eng.*, **3**, 293–294.

Jennings, A. and McKeown, J. (1992) *Matrix Computation for Engineers and Scientists*. John Wiley, London.

Kopal, A. (1961) *Numerical Analysis*, 2nd edn. Chapman & Hall.

Parlett, B.N. and Reid, J.K. (1981) Tracking the progress of the Lanczos algorithm for large symmetric eigenproblems. *IMA J. Num. Anal.*, **1**, 135–155.

Smith, I.M. (1979) Discrete element analysis of pile instability. *Int. J. Num. Anal. Meth. Geomechanics*, **3**, 205–211.

Smith, I.M. (1984) Adaptability of truly modular software. *Engineering Computations*, **1**(1), March, 25–35.

Smith, I.M. and Kidger, D.J. (1991) "Properties of the 20-node brick". *Int. J. Num. Analytical Meth Geomech.*, **15**(12), 871–891.

Smith, I.M. and Kidger, D.J. (1992) Elastoplastic analysis using the 14-node brick element family. *IJNME*, **35**(6), 1263–1275.

Smith, I.M., Wong, S.W., Gladwell, I. and Gilvary, B. (1989) PCG methods in transient FE analysis, Part I: First order problems. *IJMNE*, **28**(7), 1557–1566.

Taig, I.C. (1961) Structural analysis by the matrix displacement method. *English Electric Aviation Report SO17*. Preston.

Timoshenko, S.P., Young, D.H. and Weaver, W. (1974) *Vibration Problems in Engineering*, 4th edn, John Wiley, New York.

Wong, S.W., Smith, I.M. and Gladwell, I. (1989) PCG methods in transient FE analysis, Part II: Second order problems. *IJNME*, **28**(7), 1567–1576.

Zienkiewicz, O.C., Irons, B.M. Ergatoudis, S., Ahmad, S. and Scott, F.C. (1969) Isoparametric and associated element families for two and three dimensional analysis. In *Proceedings of Course on Finite Element Methods in Stress Analysis*, ed. I Holland and K. Bell, Trondheim Technical University.

Zienkiewicz, O.C., Too, J. and Taylor, R.L. (1971) Reduced integration technique in general analysis of plates and shells. *Int. J. Num. Meth. Eng.*, **3**, 275–290.

4 Static Equilibrium of Structures

4.0 INTRODUCTION

Practical finite element analysis had as its starting point matrix analysis of "structures", by which engineers used to mean assemblages of elastic, line elements. The matrix displacement (stiffness) method is a special case of finite element analysis and, since many engineers still begin their acquaintance with the finite element method in this way, the opening applications chapter of this book is devoted to "structural" analysis.

The first program, Program 4.0, permits the analysis of a rod subjected to combinations of axial loads and displacements at various points along its length. Each 1-d rod element can have a different length and axial stiffness but the element stiffness matrices, being simple functions of these two quantities, are easily formed by a subroutine. Indeed, in nearly all the programs in this chapter, the element stiffness matrices consist of simple explicit expressions which are conveniently provided by subroutines. Program 4.1 is a more general rod element program allowing analyses to be performed for 2-d or 3-d pin-jointed frames.

Program 4.2 permits the analysis of slender beams subjected to combinations of transverse loading and moments. Anticipating more complex elements used later in the book, Program 4.3 uses numerical integration to formulate the beam element stiffness and "mass" matrices in the analysis of beams on elastic foundations. Such procedures will be useful for elements whose matrices are too complex to be conveniently stated explicitly.

When 1-d beam and rod elements are superposed, the result is a "beam–rod" element which is a structural element capable of analysing all conventional structures. Program 4.4 is a more general implementation of this element capable of analysing any framed structure in one, two or three dimensions.

Program 4.5 introduces material nonlinearity in the form of an elastic–perfectly plastic moment/curvature relationship for beams. The program can compute plastic collapse of 1-d, 2-d or 3-d structures when subjected to incrementally changing loads. The nonlinearity is dealt with using an iterative constant stiffness (modified Newton–Raphson) approach. At each iteration the internal loads on the structure are altered rather than the stiffness matrix itself. This approach will be used extensively in the elasto-plastic analyses of solids described in Chapter 6.

Program 4.6 performs elastic stability analysis of 1-d or 2-d framed structures. As loads are incrementally increased on the system, axial forces are monitored, and the element stiffness matrices are modified accordingly. This requires an iterative approach which does necessitate the continual updating of the stiffness matrices. A buckling mode is signalled by a change of sign of the global stiffness matrix determinant, and also by a significant increase in the displacement generated by a small disturbing force.

PROGRAM 4.0: EQUILIBRIUM OF AXIALLY LOADED RODS

The main features of this program are the elastic rod element stiffness matrix (equation 2.11) and the global stiffness matrix assembly described in section 3.6. The structure chart in Figure 4.1 gives the main sequence of operations.

Program 4.0 is illustrated by two examples shown in Figures 4.2 and 4.3. In both cases, the rod is restrained at one end and free at the other. The rod in Figure 4.2 is subjected to a uniformly distributed axial force of 5.0/unit length, and the rod in Figure 4.3 is subjected to a fixed displacement of 0.05 at its tip.

The global node numbering system reads from the left as shown in Figure 4.2 and, at the element level, node one is always to the left. As explained in Chapter 3, the nodal freedom numbering associated with each element, accounting for any restraints, is contained in the "steering" vector `G`. Thus, with reference to Figures 4.3 and 4.4, the steering vector for element 1 would be $[0\ 1]^T$ and for element 2, $[1\ 2]^T$, and so on.

For problems such as this where 1-d elements are strung together in a line, it is a simple matter to automate the generation of the `G` vector for each element. This is done by the library subroutine `GEOMETRY_2L` which picks the correct entries out of the nodal freedom array `NF` (see Appendix 3 for listings of geometry subroutines). Nodal freedom data concerning boundary restraints is read by the main program, and takes the form of the number of restrained nodes, `NR`, followed by the restrained node number and a zero (`NR` times). The default is that nodes are not restrained.

All programs in the book use this approach for defining boundary conditions. In cases later on where nodes have more than one degree of freedom some of those nodal freedoms may be restrained and others not. In these cases, the convention will be to give the node number followed by either ones or zeros (in the correct sequence), where the latter implies a restrained degree of freedom and the former an unrestrained freedom.

Returning to the main program, the meanings of the variable names used in Program 4.0 are now given below:

Scalar integers:

`NELS`	number of elements
`NEQ`	number of degrees of freedom in the mesh
`NN`	number of nodes
`NBAND`	half-bandwidth of mesh
`NR`	number of restrained nodes

```
program p40
!-------------------------------------------------------
! program 4.0 equilibrium of axially loaded 1-d rods
!-------------------------------------------------------
use new_library  ;   use  geometry_lib ;   implicit none
integer::nels,neq,nn,nband,nr,nod=2,nodof=1,ndof=2,iel,i,k,ndim=1,          &
         loaded_nodes,fixed_nodes,np_types
!--------------------------dynamic arrays------------------------------------
real,allocatable::km(:,:),eld(:),kv(:),loads(:),coord(:,:),                 &
                    action(:),g_coord(:,:),value(:),ea(:),ell(:)
integer,allocatable::nf(:,:),g(:),num(:),g_num(:,:),no(:),g_g(:,:),         &
                     node(:),etype(:)
!------------------------input and initialisation------------------------------
open (10 , file = 'p40.dat' , status = 'old' ,    action ='read')
open (11 , file = 'p40.res' , status = 'replace', action='write')
read(10,*)nels,np_types; nn=nels+1
allocate (nf(nodof,nn),km(ndof,ndof),coord(nod,ndim),g_coord(ndim,nn),      &
          eld(ndof),action(ndof),g_num(nod,nels),num(nod),g(ndof),          &
          ea(np_types),ell(nels),g_g(ndof,nels),etype(nels))
read(10,*)ea; etype=1; if(np_types>1)read(10,*)etype ; read(10,*)ell,nr
nf=1; if(nr>0)read(10,*)(k,nf(:,k),i=1,nr); call formnf(nf);neq=maxval(nf)
!---------------loop the elements to find global arrays sizes-----------------
nband=0
elements_1: do iel=1,nels
              call geometry_2l(iel,ell(iel),coord,num); call num_to_g(num,nf,g)
              g_num(:,iel)=num; g_coord(:,num)=transpose(coord)
              g_g(:,iel)=g; if(nband<bandwidth(g))nband=bandwidth(g)
end do elements_1
allocate(kv(neq*(nband+1)),loads(0:neq)); kv=0.0
write(11,'(a)')"Global coordinates"
do k=1,nn; write(11,'(a,i5,a,3e12.4)')                                      &
    "Node    ",k,"    ",g_coord(:,k); end do
write(11,'(a)')"Global node numbers"
do k=1,nels; write(11,'(a,i5,a,27i3)')                                      &
    "Element ",k,"      ",g_num(:,k); end do
write(11,'(2(a,i5),/)')                                                     &
    "There are ",neq,"  equations and the half-bandwidth is ",nband
!---------------------global stiffness matrix assembly-------------------------
elements_2: do iel=1,nels
              call rod_km(km,ea(etype(iel)),ell(iel)); g=g_g(:,iel)
              call formkv(kv,km,g,neq)
end do elements_2
!------------------read loads and/or displacements----------------------------
loads=0.0; read(10,*)loaded_nodes
if(loaded_nodes/=0)read(10,*)(k,loads(nf(:,k)),i=1,loaded_nodes)
read(10,*)fixed_nodes
if(fixed_nodes/=0)then
   allocate(node(fixed_nodes),no(fixed_nodes),value(fixed_nodes))
   read(10,*)(node(i),value(i),i=1,fixed_nodes)
   do i=1,fixed_nodes; no(i)=nf(1,node(i)); end do
   kv(no)=kv(no)+1.e20 ; loads(no)=kv(no)*value
end if
!------------------------equation solution ----------------------------------
call banred(kv,neq); call bacsub(kv,loads)
write(11,'(a)')"The nodal displacements are:"
do k=1,nn; write(11,'(i5,a,3e12.4)')k,"    ",loads(nf(:,k)); end do
!--------------------retrieve element end actions-----------------------------
write(11,'(a)')"The element 'actions' are:"
elements_3: do iel=1,nels
              call rod_km(km,ea(etype(iel)),ell(iel))
              g=g_g(:,iel); eld=loads( g ); action=matmul(km,eld)
              write(11,'(i5,a,2e12.4)')iel,"    ",action
end do elements_3
end program p40
```

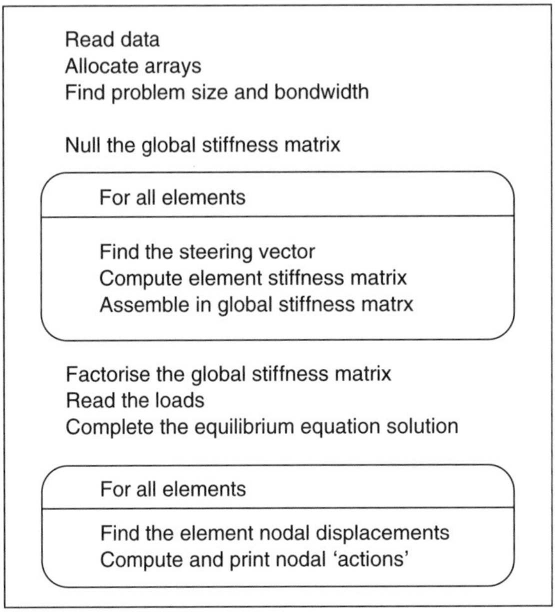
Read data
Allocate arrays
Find problem size and bondwidth
Null the global stiffness matrix
For all elements
Find the steering vector
Compute element stiffness matrix
Assemble in global stiffness matrx
Factorise the global stiffness matrix
Read the loads
Complete the equilibrium equation solution
For all elements
Find the element nodal displacements
Compute and print nodal 'actions'

Figure 4.1 Structure chart for Program 4.0

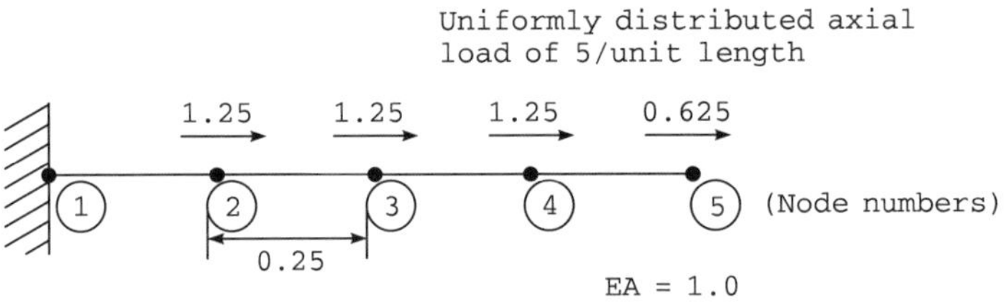
Uniformly distributed axial
load of 5/unit length
1.25
1.25
1.25
0.625
1
2
3
4
5
(Node numbers)
0.25
EA = 1.0

Structure and element data	NELS 4	NP_TYPES 1
	EA 1.0	
	ELL 0.25 0.25 0.25 0.25	
Node freedom data	NR 1	K,NF(:,K),I=1,NR 1 0
Loading data	LOADED_NODES 4	
	K,LOADS(NF(:,K)), I = 1,LOADED_NODES 2 1.25 3 1.25 4 1.25 5 0.625	
	FIXED_NODES 0	

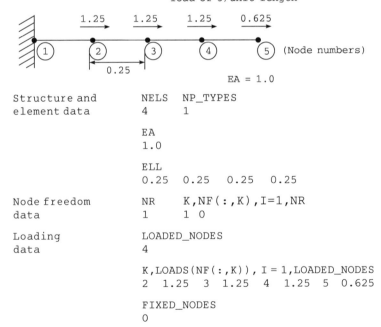

Figure 4.2 Mesh and data for first example with Program 4.0

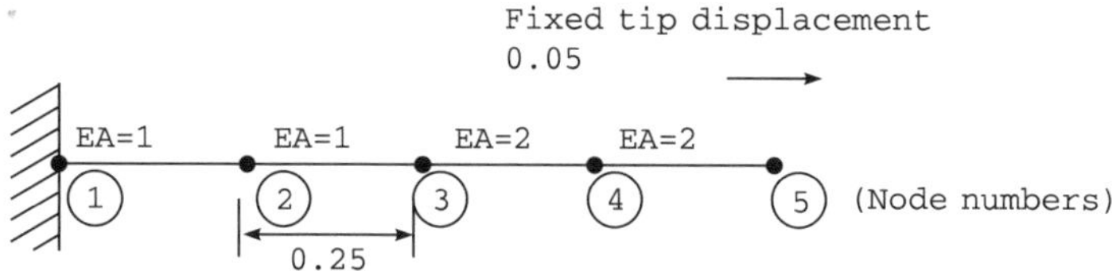


```
Structure and NELS                    NP_TYPES
element data       4                         2

                   EA
                   1.0  2.0

                   ETYPE
                   1  1  2  2

                   ELL
                   0.25  0.25  0.25  0.25

Node freedom       NR      K,NF(:,K),I=1,NR
data               1       1  0

loading            LOADED_NODES
data               0

                   FIXED_NODES
                   1

                   NODE (I),VALUE(I),I=1,FIXED_NODES
                   5  0.05
```

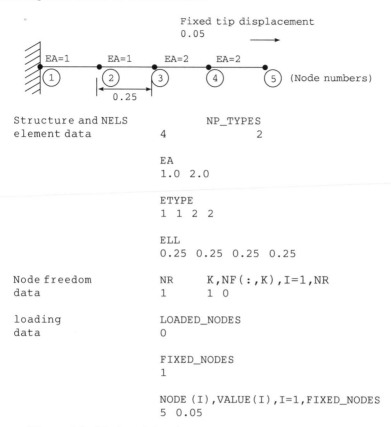

Figure 4.3 Mesh and data for second example with Program 4.0

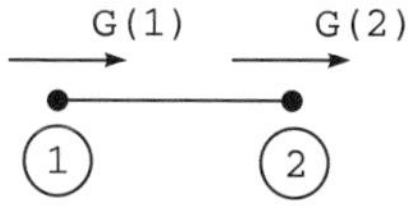

Figure 4.4 Node and freedom numbering for rod elements

NOD	number of nodes per element
NODOF	number of degrees of freedom per node
NDOF	number of degrees of freedom per element
NDIM	number of dimensions
LOADED_NODES	number of loaded nodes
FIXED_NODES	number of fixed displacements
NP_TYPES	number of different property types
IEL,I,K	simple counters

Allocatable real arrays:

KM	element stiffness matrix
ELD	element displacement vector
KV	global stiffness matrix
LOADS	global load (displacement) vector
COORD	element nodal coordinates matrix
ACTION	element nodal action vector
G_COORD	global nodal coordinates matrix
VALUE	fixed displacements vector
EA	element axial stiffnesses vector
ELL	element lengths vector

Allocatable integer arrays:

NF	nodal freedom matrix
G	element steering vector
NUM	element node numbers vector
G_NUM	global element node numbers matrix
NO	fixed displacement freedom numbers vector
G_G	global element steering matrix
NODE	fixed displacement nodes vector
ETYPE	element property type vector

Some quantities are defined by the limitations of the program, for example there can only be two nodes per element (NOD=2) and these are assigned in the declaration lines. In the "input and initialisation" section, data is read relating to the number of elements NELS and the number of property types NP_TYPES. If there is only one property type (NP_TYPES=1) as in the example shown in Figure 4.2, the property is allocated to all elements. If there is more than one property type, as in the example shown in Figure 4.3, the property types are read, followed by an integer vector ETYPE which holds the information relating to which element is assigned to which property group. The boundary condition data labelled "node freedom data", is then read enabling the formation of array NF.

Inside the "global stiffness matrix assembly" section the element stiffnesses and lengths are used by subroutine ROD_KM to compute the element stiffness matrices KM.

The global stiffness matrix (stored as a column vector) is then assembled for all elements in turn by the library subroutine FORMKV once the "steering" vector G has been retrieved. Gaussian elimination on the system equations is divided into a "factorisation" phase performed by library subroutine BANRED and a forward and back-substitution phase performed by library subroutine BACSUB. The "loads" data is then read, and this takes the form of information relating to nodal forces and/or fixed nodal displacements.

In the case of forces, LOADED_NODES is read first, signifying the number of nodes with forces applied. Then for each of these, the node number and the applied force are read. In this rod example there is only one freedom at each node, but in later programs in the chapter where more than one freedom exists at each node, all the "forces" applied at the loaded node must be included in the correct sense (even if some of them are zero).

In the case of displacements, FIXED_NODES is read signifying the number of fixed freedoms in the mesh. Then for each of these, the node number and the value to which the freedom is to be fixed is read. In later programs in the chapter where more than one freedom exists at each node, data must also be read into a vector SENSE, which gives the sequential number of the freedom at the node that is to be fixed. If a particular node has more than one fixed freedom, the node must be entered in the data list for each fixed "sense". If either LOADED_NODES or FIXED_NODES equals zero, no further data relating to that category is required.

In the case shown in Figure 4.2, a rod of uniform stiffness 1.0 is subjected to a uniformly distributed axial load of 5/unit length. The force has been "lumped" at the nodes as was indicated in equation (2.10). In this example, there are no fixed displacements, so FIXED_NODES is read as zero.

In the case shown in Figure 4.3, the rod has a non-uniform stiffness, and since there is more than one property group (NP_TYPES = 2), the element type vector ETYPE must be read indicating in this case that elements 1 and 2 have an axial stiffness of 1.0, and elements 3 and 4 have an axial stiffness of 2.0. There are no loaded nodes so LOADED_NODES is read as zero but a fixed displacement of 0.05 is applied to the tip of the rod (node 5).

Following forward and back-substitution, the nodal displacements (overwritten as LOADS) are computed and printed. A final "post-processing" phase is then entered in which the elements are scanned once more. In this loop, the element nodal displacements (ELD) are retrieved from the global displacements vector and the element stiffness matrices (KM) recomputed. Multiplication of the element nodal displacements by the element stiffness matrix results in the element "actions" vector called ACTION, which holds the internal end reaction forces for each element.

The computed results for both cases are reproduced in Figure 4.5. In the first case, the end deflection at node 5 is given as 2.500 which is the exact solution. Note, however, that the element "actions" indicate that the first element sustains a constant tensile force of 4.375 which is the best this element can do to approximate the true solution of linearly varying load.

In the second case, the nodal displacements indicate the distribution of displacements along the length of the rod up to the end node 5 which has the expected displacement of 0.05. All the elements in the rod are sustaining a tensile load of 0.067 which is correct in this case.

The limitations of the linear shape function assumptions must always be borne in mind.

```
Global coordinates
Node          1       0.0000E+00
Node          2       0.2500E00
Node          3       0.5000E00
Node          4       0.7500E00
Node          5       0.1000E+01
Global node numbers
Element       1          1  2
Element       2          2  3
Element       3          3  4
Element       4          4  5
There are       4   equations and the half-bandwidth is      1

The nodal displacements are:
    1      0.0000E+00
    2      0.1094E+01
    3      0.1875E+01
    4      0.2344E+01
    5      0.2500E+01
The element 'actions' are:
    1     -0.4375E+01   0.4375E+01
    2     -0.3125E+01   0.3125E+01
    3     -0.1875E+01   0.1875E+01
    4     -0.6250E00    0.6250E00                  First example

Global coordinates
Node          1       0.0000E+00
Node          2       0.2500E00
Node          3       0.5000E00
Node          4       0.7500E00
Node          5       0.1000E+01
Global node numbers
Element       1          1  2
Element       2          2  3
Element       3          3  4
Element       4          4  5
There are       4   equations and the half-bandwidth is      1

The nodal displacements are:
    1      0.0000E+00
    2      0.1667E-01
    3      0.3333E-01
    4      0.4167E-01
    5      0.5000E-01
The element 'actions' are:
    1     -0.6667E-01   0.6667E-01
    2     -0.6667E-01   0.6667E-01
    3     -0.6667E-01   0.6667E-01
    4     -0.6667E-01   0.6667E-01                Second example
```

Figure 4.5 Results from examples run with Program 4.0

PROGRAM 4.1: EQUILIBRIUM OF PIN-JOINTED FRAMES

The philosophy used throughout this book is to explore solutions to new problems by making gradual alterations to previously described programs. This program, therefore, is an adaptation of the previous one to allow analysis of rod elements in one, two or three dimensions. Since rod elements can only sustain axial loads, the types of structural systems for which this is applicable are pin-jointed frames or space structures (in 3-d).

The previous program considered rod elements in 1-d joined end to end in which the number of nodes was always one greater than the number of

```
program p41
!---------------------------------------------------------------
! program 4.1 equilibrium of pin-jointed frames using rod elements
!             in 1-, 2- or 3-dimensions
!---------------------------------------------------------------
use  new_library ;  use  geometry_lib  ;     use vlib    ;  implicit none
real::axial
integer::nels,neq,nn,nband,nr,nod=2,nodof,ndof,iel,i,k,ndim,              &
         loaded_nodes,fixed_nodes,np_types
!--------------------------dynamic arrays------------------------------------
real,allocatable::km(:,:),eld(:),kv(:),loads(:),coord(:,:),               &
                  action(:),g_coord(:,:),value(:),ea(:)
integer,allocatable::nf(:,:),g(:),num(:),g_num(:,:),no(:),g_g(:,:),       &
                     node(:),sense(:),etype(:)
!--------------------input and initialisation---------------------------------
open (10 , file = 'p41a.dat' , status = 'old' ,    action ='read')
open (11 , file = 'p41a.res' , status = 'replace', action='write')
read(10,*)nels,nn,ndim,np_types; nodof=ndim; ndof=nod*nodof
allocate(nf(nodof,nn),km(ndof,ndof),coord(nod,ndim),g_coord(ndim,nn),     &
         eld(ndof),action(ndof),g_num(nod,nels),num(nod),g(ndof),         &
         ea(np_types),g_g(ndof,nels),etype(nels))
read(10,*)ea; etype=1; if(np_types>1)read(10,*)etype
read(10,*)g_coord; read(10,*)g_num
read(10,*)nr
nf=1; if(nr>0)read(10,*)(k,nf(:,k),i=1,nr); call formnf(nf); neq=maxval(nf)
!----------------loop the elements to find global array sizes-----------------
nband=0
elements_1: do iel=1,nels
              num=g_num(:,iel) ; call num_to_g ( num , nf , g )
              g_g(:,iel)=g; if(nband<bandwidth(g))nband=bandwidth(g)
end do elements_1
allocate(kv(neq*(nband+1)),loads(0:neq)); kv=0.0
write(11,'(a)')"Global coordinates"
do k=1,nn; write(11,'(a,i5,a,3e12.4)')                                    &
    "Node    ",k,"     ",g_coord(:,k); end do
write(11,'(a)')"Global node numbers"
do k=1,nels; write(11,'(a,i5,a,27i3)')                                    &
    "Element ",k,"      ",g_num(:,k); end do
write(11,'(2(a,i5),/)')                                                   &
    "There are ",neq," equations and the half-bandwidth is ",nband
!--------------------global stiffness matrix assembly-------------------------
elements_2: do iel=1,nels
              num=g_num(:,iel); coord=transpose(g_coord(:,num))
              call pin_jointed(km,ea(etype(iel)),coord); g=g_g(:,iel)
              call formkv(kv,km,g,neq)
end do elements_2
!-------------------read loads and/or displacements---------------------------
loads=0.0; read(10,*)loaded_nodes
if(loaded_nodes/=0)read(10,*)(k,loads(nf(:,k)),i=1,loaded_nodes)
read (10,*)fixed_nodes
if(fixed_nodes/=0)then
   allocate(node(fixed_nodes),no(fixed_nodes),                            &
            sense(fixed_nodes),value(fixed_nodes))
   read(10,*)(node(i),sense(i),value(i),i=1,fixed_nodes)
   do i=1,fixed_nodes; no(i)=nf(sense(i),node(i)); end do
   kv(no)=kv(no)+1.e20; loads(no)=kv(no)*value
end if
!---------------------------equation solution -------------------------------
call banred(kv,neq); call bacsub(kv,loads)
write(11,'(a)')"The nodal displacements are:"
do k=1,nn; write(11,'(i5,a,3e12.4)')k,"    ",loads(nf(:,k)); end do
!-------------------retrieve element end actions------------------------------
write(11,'(a)')"The element 'actions' are:"
elements_3: do iel=1,nels
              num=g_num(:,iel); coord=transpose(g_coord(:,num))
              g=g_g(:,iel); eld=loads(g)
              call pin_jointed(km,ea(etype(iel)),coord);
              action=matmul(km,eld)
              call glob_to_axial(axial,action,coord)
              write(11,'(i5,6e12.4)')iel,action
              write(11,'(a,6e12.4)')"Axial force         ",axial
end do elements_3
end program p41
```

elements. This will no longer be true for general 2-d or 3-d structures, so the number of nodes NN is now included as data together with the dimensionality of the problem NDIM.

The variable names are virtually the same as in the previous program. The real array ELL has been discarded because the rod element lengths are more conveniently calculated from their nodal coordinates, and the real variable AXIAL has been introduced as this will hold the axial force sustained by each member. The other change from Program 4.0 is that the geometry routine

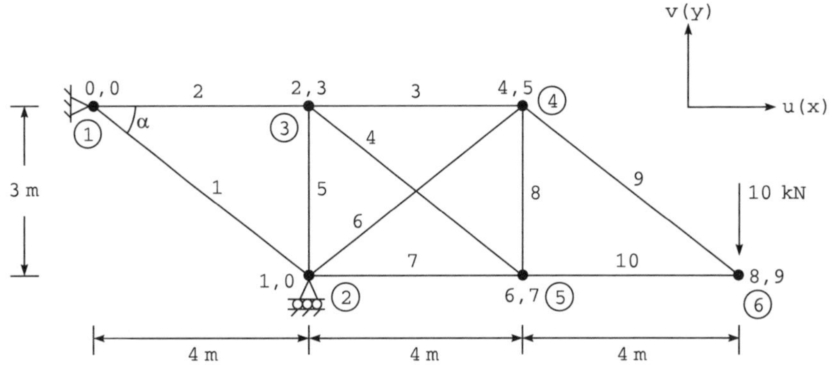

$\alpha = 36.87°$

$EA = 2 \times 10^5$ kN

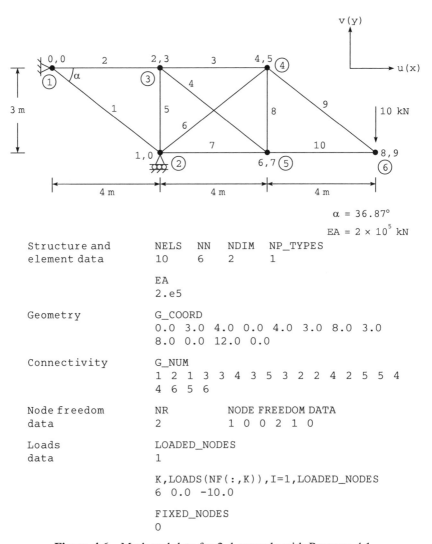

Structure and element data	NELS NN NDIM NP_TYPES 10 6 2 1
	EA 2.e5
Geometry	G_COORD 0.0 3.0 4.0 0.0 4.0 3.0 8.0 3.0 8.0 0.0 12.0 0.0
Connectivity	G_NUM 1 2 1 3 3 4 3 5 3 2 2 4 2 5 5 4 4 6 5 6
Node freedom data	NR NODE FREEDOM DATA 2 1 0 0 2 1 0
Loads data	LOADED_NODES 1
	K,LOADS(NF(:,K)),I=1,LOADED_NODES 6 0.0 -10.0
	FIXED_NODES 0

Figure 4.6 Mesh and data for 2-d example with Program 4.1

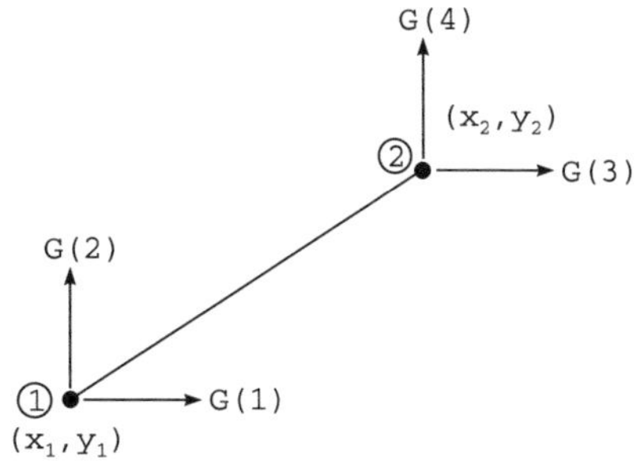

Figure 4.7 Node and freedom numbering for 2-d rod elements

```
Global coordinates
Node         1        0.0000E+00  0.3000E+01
Node         2        0.4000E+01  0.0000E+00
Node         3        0.4000E+01  0.3000E+01
Node         4        0.8000E+01  0.3000E+01
Node         5        0.8000E+01  0.0000E+00
Node         6        0.1200E+02  0.0000E+00
Global node numbers
Element      1          1  2
Element      2          1  3
Element      3          3  4
Element      4          3  5
Element      5          3  2
Element      6          2  4
Element      7          2  5
Element      8          5  4
Element      9          4  6
Element     10          5  6
There are      9  equations and the half-bandwidth is      6

The nodal displacements are:
    1      0.0000E+00  0.0000E+00
    2     -0.1042E-02  0.0000E+00
    3      0.5333E-03 -0.6562E-04
    4      0.9500E-03 -0.3046E-02
    5     -0.1425E-02 -0.2981E-02
    6     -0.1692E-02 -0.7263E-02
The element 'actions' are:
    1  0.2667E+02 -0.2000E+02 -0.2667E+02  0.2000E+02
Axial force           -0.3333E+02
    2 -0.2667E+02  0.0000E+00  0.2667E+02  0.0000E+00
Axial force            0.2667E+02
    3 -0.2083E+02  0.0000E+00  0.2083E+02  0.0000E+00
Axial force            0.2083E+02
    4 -0.5833E+01  0.4375E+01  0.5833E+01 -0.4375E+01
Axial force            0.7292E+01
    5  0.0000E+00 -0.4375E+01  0.0000E+00  0.4375E+01
Axial force           -0.4375E+01
    6  0.7500E+01  0.5625E+01 -0.7500E+01 -0.5625E+01
Axial force           -0.9375E+01
    7  0.1917E+02  0.0000E+00 -0.1917E+02  0.0000E+00
Axial force           -0.1917E+02
    8  0.0000E+00  0.4375E+01  0.0000E+00 -0.4375E+01
Axial force           -0.4375E+01
    9 -0.1333E+02  0.1000E+02  0.1333E+02 -0.1000E+02
Axial force            0.1667E+02
   10  0.1333E+02  0.0000E+00 -0.1333E+02  0.0000E+00
Axial force           -0.1333E+02
```

Figure 4.8 Results from 2-d example with Program 4.1

GEOMETRY_2L has been discarded and replaced by reading data of the global nodal coordinates straight into array G_COORD followed by reading data of the element node numbers into G_NUM. Two new library subroutines are also introduced. Subroutine PIN_JOINTED computes the element stiffness matrix KM, and subroutine GLOB_TO_LOC transforms the element "actions" into axial forces.

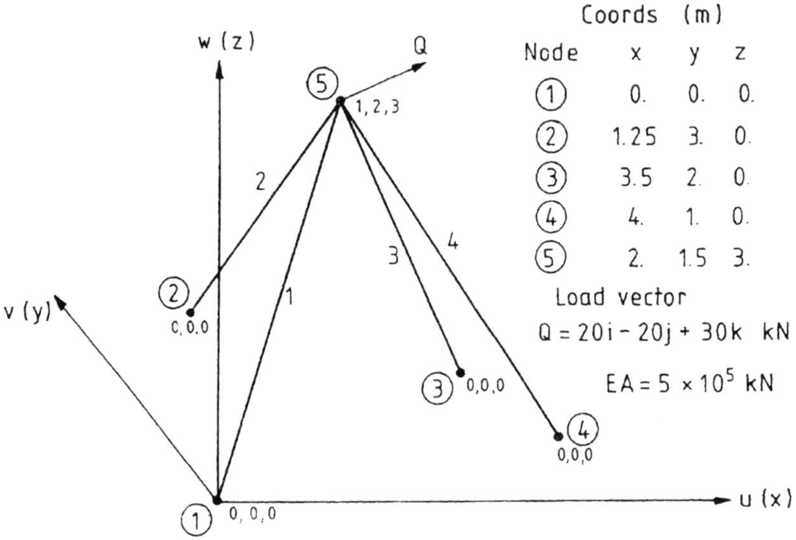

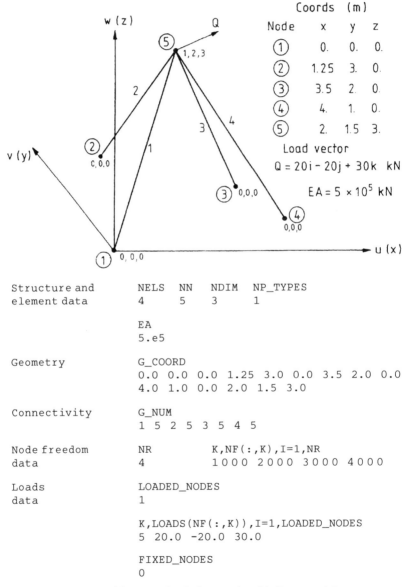

```
Structure and     NELS   NN    NDIM   NP_TYPES
element data      4      5     3      1

                  EA
                  5.e5

Geometry          G_COORD
                  0.0 0.0 0.0 1.25 3.0 0.0 3.5 2.0 0.0
                  4.0 1.0 0.0 2.0 1.5 3.0

Connectivity      G_NUM
                  1 5 2 5 3 5 4 5

Node freedom      NR        K,NF(:,K),I=1,NR
data              4         1 0 0 0  2 0 0 0  3 0 0 0  4 0 0 0

Loads             LOADED_NODES
data              1

                  K,LOADS(NF(:,K)),I=1,LOADED_NODES
                  5 20.0 -20.0 30.0

                  FIXED_NODES
                  0
```

Figure 4.9 Data for 3-d example with Program 4.1

The first example to be solved by Program 4.1 is the 2-d pin-jointed frame shown in Figure 4.6. Each element now has four degrees of freedom with an x and y translation permitted at each node as shown in Figure 4.7.

With respect to the data shown in Figure 4.6, in a small problem such as this, the order in which the nodes are read is immaterial. In larger problems nodal numbering should be made in the most economical order as this will minimise the bandwidth and hence the storage requirements. In very large problems which use an assembly strategy, a bandwidth optimiser would be used. Following the nodal coordinates, the node number at the end of each element is read in. The loaded nodes and fixed nodes data follows the same procedure as described in the previous program. In this example, a single load of -10.0 is applied in the y direction at node 6. Note how the load in the x direction at this node has also to be read in as zero. There are no fixed nodes in this example.

The results given in Figure 4.8 indicate the nodal displacements, followed by the end "actions" and axial force in each element. The results indicate that the displacement under the load is -0.0073 and the axial load in element number 1 is -33.33 (compressive).

The second example to be solved by Program 4.1 is the 3-d pin-jointed frame shown in Figure 4.9. Each element now has six degrees of freedom with an x, y and z translation permitted at each node as shown in Figure 4.10.

The data organisation is virtually the same as in the previous example. The dimensionality is increased to `NDIM=3` and there are correspondingly three coordinates at each node and three freedoms to be defined at each restrained node. The space frame shown in Figure 4.9 represents a pyramid-like structure loaded by a force at its apex with components in the x, y and z directions of 20, -20 and 30 respectively. The computed results shown in Figure 4.11 indicate

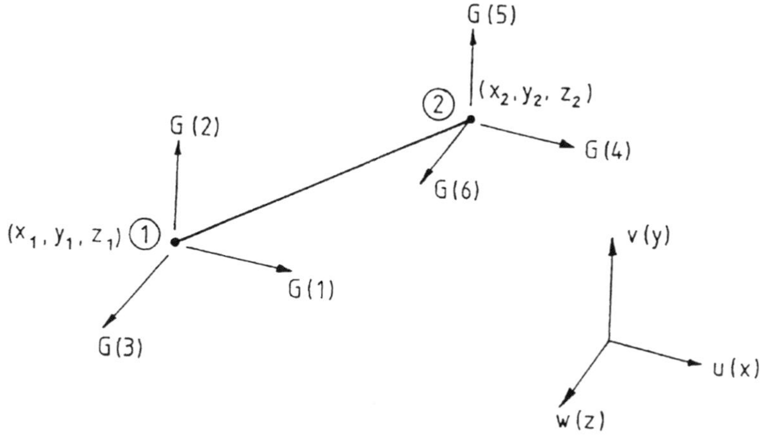

Figure 4.10 Node and freedom numbering for 3-d rod elements

```
Global coordinates
Node         1      0.0000E+00  0.0000E+00  0.0000E+00
Node         2      0.1250E+01  0.3000E+01  0.0000E+00
Node         3      0.3500E+01  0.2000E+01  0.0000E+00
Node         4      0.4000E+01  0.1000E+01  0.0000E+00
Node         5      0.2000E+01  0.1500E+01  0.3000E+01
Global node numbers
Element      1          1  5
Element      2          2  5
Element      3          3  5
Element      4          4  5
There are      3  equations and the half-bandwidth is      2

The nodal displacements are:
    1      0.0000E+00  0.0000E+00  0.0000E+00
    2      0.0000E+00  0.0000E+00  0.0000E+00
    3      0.0000E+00  0.0000E+00  0.0000E+00
    4      0.0000E+00  0.0000E+00  0.0000E+00
    5      0.2490E-03 -0.3991E-03  0.8075E-04
The element 'actions' are:
    1 -0.2378E+01 -0.1783E+01 -0.3567E+01  0.2378E+01  0.1783E+01  0.3567E+01
Axial force           0.4643E+01
    2 -0.9493E+01  0.1899E+02 -0.3797E+02  0.9493E+01 -0.1899E+02  0.3797E+02
Axial force           0.4350E+02
    3  0.1313E+01  0.4375E00 -0.2625E+01 -0.1313E+01 -0.4375E00  0.2625E+01
Axial force           0.2968E+01
    4 -0.9442E+01  0.2360E+01  0.1416E+02  0.9442E+01 -0.2360E+01 -0.1416E+02
Axial force          -0.1718E+02
```

Figure 4.11 Results from 3-d example with Program 4.1

that the corresponding displacement components of the loaded node are 2.49×10^{-4}, -3.99×10^{-4} and 8.07×10^{-5}. The axial force in element number 2 is computed to be 43.5 (tensile).

PROGRAM 4.2: EQUILIBRIUM OF BEAMS

This program has much in common with Program 4.0 for rod elements. A line of beam elements of different stiffnesses and lengths can be analysed for any combination of transverse and/or moment loading. We see the return of library subroutine `GEOMETRY_2L` for the generation of nodal coordinates and "steering" data involving strings of elements, be they rods or beams. The beam element stiffness matrix is then provided by subroutine `BEAM_KM`. The flexural stiffness is read element-by-element into the real array `EI`, as are the element lengths into array `ELL`.

The problem to be solved in this example is shown in Figure 4.12. The non-uniform beam is subjected to a combination of nodal loads and fixed displacements. Node 1 is to be rotated clockwise by 0.001 and node 2 is to be translated vertically downwards by 0.005. In addition, a vertical force of 20 acts between nodes 1 and 2, a uniformly distributed load of 4/unit length acts between nodes 2 and 3, and a linearly decreasing load of 4/unit length to zero acts between nodes 3 and 4.

```
program p42
!------------------------------------------------------------------------
! program 4.2 equilibrium of beams
!------------------------------------------------------------------------
use new_library  ;  use  geometry_lib   ;    implicit none
integer::nels,neq,nn,nband,nr,nod=2,nodof=2,ndof=4,iel,i,k,ndim=1,       &
         loaded_nodes,fixed_nodes,np_types
!----------------------------dynamic arrays-----------------------------------
real,allocatable::km(:,:),eld(:),kv(:),loads(:),coord(:,:),              &
                  action(:),g_coord(:,:),value(:),ei(:),ell(:)
integer,allocatable::nf(:,:),g(:),num(:),g_num(:,:),no(:),g_g(:,:),      &
                     node(:),sense(:),etype(:)
!----------------------input and initialisation--------------------------------
open (10 , file = 'p42.dat' , status = 'old' ,    action ='read')
open (11 , file = 'p42.res' , status = 'replace', action='write')
read (10,*)nels,np_types; nn=nels+1
allocate(nf(nodof,nn),km(ndof,ndof),coord(nod,ndim),g_coord(ndim,nn),     &
         eld(ndof),action(ndof),g_num(nod,nels), num(nod), g(ndof),       &
         ei(np_types),ell(nels),g_g(ndof,nels),etype(nels))
read(10,*)ei; etype=1; if(np_types>1)read(10,*)etype
read(10,*)ell,nr
nf=1; if(nr>0)read(10,*)(k,nf(:,k),i=1,nr); call formnf(nf); neq=maxval(nf)
!----------------loop the elements to find global array sizes------------------
nband=0
elements_1: do iel=1,nels
              call geometry_2l(iel,ell(iel),coord,num);call num_to_g(num,nf,g)
              g_num(:,iel)=num; g_coord(:,num)=transpose(coord)
              g_g(:,iel)=g; if(nband<bandwidth(g))nband=bandwidth(g)
end do elements_1
allocate(kv(neq*(nband+1)),loads(0:neq)); kv=0.0
write(11,'(a)')"Global coordinates"
do k=1,nn; write(11,'(a,i5,a,3e12.4)')                                    &
    "Node    ",k,"    ",g_coord(:,k); end do
write(11,'(a)')"Global node numbers"
do k=1,nels; write(11,'(a,i5,a,27i3)')                                    &
    "Element ",k,"      ",g_num(:,k); end do
write(11,'(2(a,i5),/)')                                                   &
    "There are ",neq," equations and the half-bandwidth is ",nband
!------------------global stiffness matrix assembly---------------------------
elements_2: do iel=1, nels
              call beam_km(km,ei(etype(iel)),ell(iel)); g=g_g(:,iel)
              call formkv(kv,km,g,neq)
end do elements_2
!------------------read loads and/or displacements----------------------------
loads=0.0; read(10,*)loaded_nodes
if(loaded_nodes/=0)read(10,*)(k,loads(nf(:,k)),i=1,loaded_nodes)
read (10,*)fixed_nodes
if(fixed_nodes/=0)then
   allocate(node(fixed_nodes),no(fixed_nodes),                           &
            sense(fixed_nodes),value(fixed_nodes))
   read(10,*)(node(i),sense(i),value(i),i=1,fixed_nodes)
   do i=1,fixed_nodes; no(i)=nf(sense(i),node(i)); end do
   kv(no)=kv(no)+1.e20; loads(no)=kv(no)*value
end if
!------------------------equation solution ---------------------------------
call banred(kv,neq); call bacsub(kv,loads)
write(11,'(a)')"The nodal displacements are:"
do k=1,nn; write(11,'(i5,a,3e12.4)')k,"   ",loads(nf(:,k)); end do
!----------------retrieve element end actions--------------------------------
write(11,'(a)')"The element 'actions' are:"
elements_3: do iel=1,nels
              call beam_km(km,ei(etype(iel)),ell(iel))
              g=g_g(:,iel); eld=loads( g ); action=matmul(km,eld)
              write(11,'(i5,a,6e12.4)')iel,"   ",action
end do elements_3
end program p42
```

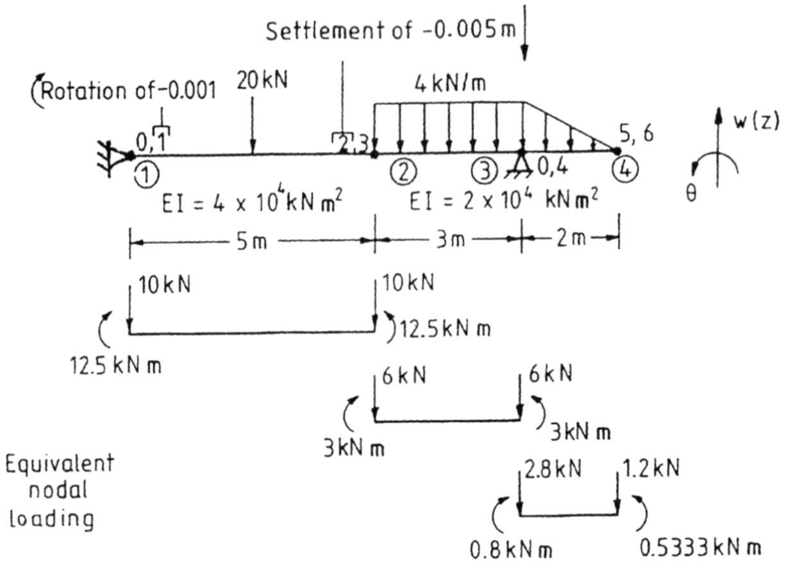

Structure and element data	NELS NP_TYPES 3 2
	EI 4.e4 2.e4
	ETYPE 1 2 2
	ELL 5.0 3.0 2.0
Node freedom data	NR K,NF(:,K),I=1,NR 2 1 0 1 3 0 1
Loads data	LOADED_NODES 3
	K,LOADS(NF(:,K)),I=1,LOADED_NODES 2 0.0 9.5 3 0.0 2.2 4 -1.2 0.53333
	FIXED_NODES 2
	NODE(I),SENSE(I),VALUE(I),I=1,FIXED_NODES 1 2 -0.001 2 1 -0.005

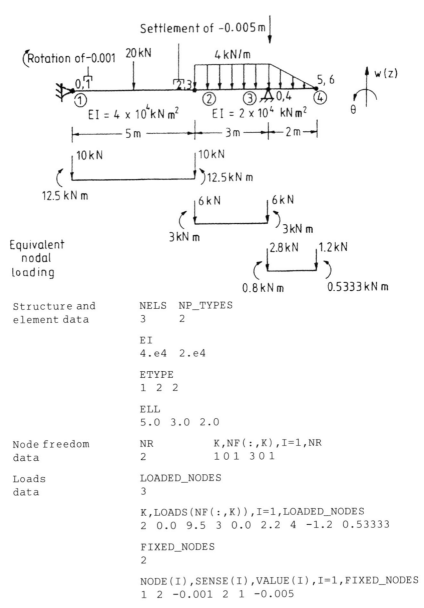

Figure 4.12 Mesh and data for beam example with Program 4.2

At each node, two degrees of freedom are possible, a vertical translation and a rotation in that order. The global node numbering reads from left to right as shown in Figure 4.12, and at the element level, node 1 is always to the left. Each element has four degrees of freedom taken in the order w_1, θ_1, w_2 and θ_2. The

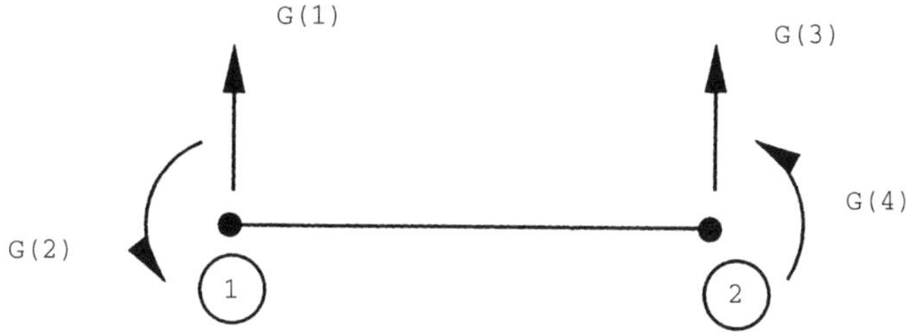

Figure 4.13 Node and freedom numbering for beam elements

nodal freedom numbering associated with each element, accounting for any restraints, is as usual contained in the "steering" vector G. Thus, with reference to Figures 4.12 and 4.13, the steering vector for element 1 would be $[0\ 1\ 2\ 3]^T$ and for element 2, $[2\ 3\ 0\ 4]^T$, and so on. It should be noted that in the "penalty" technique, the freedoms to be "fixed" are assembled into the global stiffness matrix in the usual way prior to augmenting the appropriate diagonal and force terms [see equation. (3.27)].

Nodal loading in the context of beam analysis can take the form of either point loads or moments. In the analysis, loading can only be applied at the nodes, so if forces or moments are required between nodes a set of equivalent nodal loads must be derived for application to the mesh. These equivalent nodal loads can be found by computing the shear forces and moments that would have been generated at each node if the element had been fully restrained at both ends (the fixed end moments and shear forces). The signs of these fixed end values are then reversed and applied to the nodes of the element in the actual problem (see e.g. Przemieniecki, 1968). If loads or moments are required at the nodes themselves, they are applied directly to the node without any further manipulation. If point loads are to be applied to a beam it might be advisable to place a node at that location to avoid the need for the "fixed end" calculation.

In the example of Figure 4.12, all three elements have loading of the "fixed end" type. The appropriate equivalent nodal loads to be applied are shown beneath their respective elements.

When the computed results shown in Figure 4.14 are examined, it will be seen that the fixed freedoms have the expected values. Of the remaining displacements the rotation at node 2, for example, is 0.000205 (anticlockwise).

In order to compute the actual moments and shear forces in the beam, the fixed end shear forces and moments (if applicable) must be added to the corresponding "action" vector printed for each element in the "post-processing" phase. For example, the moments at the nodes in this example for each of the elements are given as follows:

```
Global coordinates
Node        1       0.0000E+00
Node        2       0.5000E+01
Node        3       0.8000E+01
Node        4       0.1000E+02
Global node numbers
Element     1         1  2
Element     2         2  3
Element     3         3  4
There are       6  equations and the half-bandwidth is      2

The nodal displacements are:
    1       0.0000E+00 -0.1000E-02
    2      -0.5000E-02  0.2051E-03
    3       0.0000E+00  0.2410E-02
    4       0.4713E-02  0.2343E-02
The element 'actions' are:
    1       0.1157E+02  0.1928E+02 -0.1157E+02  0.3856E+02
    2      -0.9577E+01 -0.2906E+02  0.9577E+01  0.3333E00
    3       0.1200E+01  0.1867E+01 -0.1200E+01  0.5333E00
```

Figure 4.14 Results from beam example with Program 4.2

Element 1

$$M_1 = 19.28 + 12.50 = 31.78$$

$$M_2 = 38.56 - 12.50 = 26.06$$

Element 2

$$M_1 = -29.06 + 3.00 = -26.06$$

$$M_2 = 0.33 - 3.00 = -2.67$$

Element 3

$$M_1 = 1.87 + 0.80 = 2.67$$

$$M_2 = 0.53 - 0.53 = 0.00$$

Internal equilibrium is maintained, e.g. M_2 of element 1 is equal and opposite to M_1 of element 2, and so on.

PROGRAM 4.3: EQUILIBRIUM OF BEAMS ON ELASTIC FOUNDATIONS

For more complicated elements, or for elements whose properties vary along their length, it is often more convenient to form the element matrices numerically. The most efficient way of doing this is to use Gauss–Legendre quadrature, in which case it is convenient to replace the beam element

```
program p43
!-------------------------------------------------------------------
! program 4.3 beam on an elastic foundation
!             numerically integrated beam and foundation stiffness
!-------------------------------------------------------------------
use new_library  ;  use   geometry_lib   ;   implicit  none
integer::nels,neq,nn,nband,nr,nod=2,nodof=2,ndof=4,iel,i,k,l,ndim=1,        &
         loaded_nodes,fixed_nodes,nip,np_types
real::fs,fs0,fs1,x,samp_pt ; character(len=15) :: element = 'line'
!-----------------------------dynamic arrays---------------------------------
real,allocatable::km(:,:),mm(:,:),eld(:),kv(:),loads(:),coord(:,:),         &
                  action(:),g_coord(:,:),value(:),ftf(:,:),dtd(:,:),        &
                  der2(:),fun(:),mom(:),store_km(:,:,:),points(:,:),        &
                  weights(:),ei(:),ell(:)
integer,allocatable::nf(:,:),g(:),num(:),g_num(:,:),no(:),g_g(:,:),         &
                      node(:),sense(:),etype(:)
!-----------------------input and initialisation-----------------------------
open (10 , file = 'p43.dat' , status = 'old' ,    action ='read')
open (11 , file = 'p43.res' , status = 'replace', action='write')
read(10,*)nels,nip,np_types; nn=nels+1
allocate(nf(nodof,nn),km(ndof,ndof),coord(nod,ndim),g_coord(ndim,nn),       &
         eld(ndof),action(ndof),g_num(nod,nels),num(nod),g(ndof),           &
         g_g(ndof,nels),mm(ndof,ndof),ftf(ndof,ndof),ei(np_types),          &
         ell(nels),dtd(ndof,ndof),store_km(ndof,ndof,nels),der2(ndof),      &
         fun(ndof),mom(nn),points(nip,ndim),weights(nip),etype(nels))
read(10,*)fs0,fs1,ei; etype=1; if(np_types>1)read(10,*)etype
read(10,*)ell,nr
nf=1; if(nr>0)read(10,*)(k,nf(:,k),i=1,nr); call formnf(nf); neq=maxval(nf)
!--------------loop the elements to find global array sizes--------------------
nband=0
elements_1: do iel=1,nels
              call geometry_2l(iel,ell(iel),coord,num);call num_to_g(num,nf,g)
              g_num(:,iel)=num; g_coord(:,num)=transpose(coord)
              g_g(:,iel)=g; if(nband<bandwidth(g))nband=bandwidth(g)
end do elements_1
allocate(kv(neq*(nband+1)),loads(0:neq)); kv=0.0
write(11,'(a)')"Global coordinates"
do k=1,nn; write(11,'(a,i5,a,3e12.4)')                                      &
    "Node    ",k,"     ",g_coord(:,k); end do
write(11,'(a)')"Global node numbers"
do k=1,nels; write(11,'(a,i5,a,27i3)')                                      &
    "Element ",k,"      ",g_num(:,k); end do
write(11,'(2(a,i5),/)')                                                     &
    "There are ",neq," equations and the half-bandwidth is ",nband
!--------numerical integration of beam and foundation stiffness----------------
!----------------global stiffness matrix assembly------------------------------
call sample(element,points,weights)
x=0.0
elements_2: do iel=1,nels
    km=0.0; mm=0.0; g=g_g(:,iel)
    integrating_pts: do i=1,nip
        samp_pt=x+ell(iel)*0.5*(points(i,1)+1.0)
        fs=samp_pt/(ell(iel)*nels)*(fs1-fs0)+fs0
        call fmbeam(der2,fun,points,i,ell(iel))
        do k=1,ndof; do l=1,ndof
          ftf(k,l)=fun(k)*fun(l)*weights(i)*0.5*ell(iel)*fs
          dtd(k,l)=der2(k)*der2(l)*weights(i)*8.0*ei(etype(iel))/(ell(iel)**3)
        end do; end do
        mm=mm+ftf; km=km+dtd
    end do integrating_pts
    km=km+mm; store_km(:,:,iel)=km(:,:); x=x+ell(iel)
    call formkv(kv,km,g,neq)
end do elements_2
!----------------------------read loads----------------------------------------
```

```
loads=0.0; read(10,*)loaded_nodes
if(loaded_nodes/=0)read(10,*)(k,loads(nf(:,k)),i=1,loaded_nodes)
read (10,*)fixed_nodes
if(fixed_nodes /=0)then
  allocate( node(fixed_nodes),no(fixed_nodes),&
  sense(fixed_nodes),value(fixed_nodes))
  read(10,*) (node(i),sense(i),value(i),i=1,fixed_nodes)
  do i=1,fixed_nodes; no(i)=nf(sense(i),node(i)); end do
  kv(no)=kv(no)+1.e20; loads(no)=kv(no)*value
end if
!----------------------------equation solution -------------------------------
call banred(kv,neq); call bacsub(kv,loads)
!----------------------retrieve element end actions--------------------------
elements_3: do iel=1,nels
              km(:,:)=store_km(:,:,iel)
              g=g_g(:,iel); eld=loads( g ); action=matmul(km,eld)
              mom(iel)=action(2)
              if(iel==nels)mom(iel+1)=-action(4)
end do elements_3
write(11,'(a)')"   Node   Displacement Moment "
nodes: do i=1,nn
         write(11,'(i5,a,2e12.4)')i,"    ",loads(2*i-1),mom(i)
end do nodes
end program p43
```

coordinate x in the range [0, L] by the local coordinate ξ in the range [-1,1] using the transformation

$$x = \frac{L}{2}(\xi + 1) \tag{4.1}$$

The cubic beam shape functions from equation (2.22) may thus be written:

$$\begin{aligned} N_1 &= \frac{1}{4}(\xi^3 - 3\xi + 2) \\ N_2 &= \frac{L}{8}(\xi^3 - \xi^2 - \xi + 1) \\ N_3 &= \frac{1}{4}(-\xi^3 + 3\xi + 2) \\ N_4 &= \frac{L}{8}(\xi^3 + \xi^2 - \xi - 1) \end{aligned} \tag{4.2}$$

and these (FUN), together with their second derivatives with respect to ξ (DER2), are formed at the Gauss points by library subroutine FMBEAM.

As discussed in Chapter 2, the stiffness matrix for a beam can be formed from integrals of the type

$$\mathtt{KM}_{k,l} = \mathtt{EI} \int_0^L \frac{d^2 N_k d^2 N_l}{dx^2 dx^2} dx \quad k, l = 1, 2, 3, 4 \tag{4.3}$$

and the foundation stiffness "mass" matrix from integrals of the type

$$\mathtt{MM}_{k,l} = \mathtt{FS} \int_0^L N_k N_l \, dx \quad k, l = 1, 2, 3, 4 \tag{4.4}$$

Converting to local coordinates from equation (4.1) gives

$$\frac{d^2N}{dx^2} = \frac{d^2N}{d\xi^2}\left(\frac{d\xi}{dx}\right)^2 \tag{4.5}$$

and

$$dx = \frac{L}{2}d\xi \tag{4.6}$$

hence equations (4.3) and (4.4) become

$$\mathtt{KM}_{k,l} = \frac{8}{L^3}\mathtt{EI}\int_{-1}^{1}\frac{d^2N_i d^2N_j}{d\xi^2 d\xi^2}d\xi \quad i,\ j = 1,2,3,4 \tag{4.7}$$

and

$$\mathtt{MM}_{k,l} = \int_{-1}^{1}\mathtt{FS}(\xi)N_k N_1\,d\xi \quad k,l = 1,2,3,4 \tag{4.8}$$

Note that in the example to be considered here, the foundation stiffness FS is itself a function of x (or ξ), so it has been moved inside the integral.

In program terminology, the numerical integration summation is of the form

$$\begin{aligned}\mathtt{KM(K,L)} = 8.0 * \mathtt{EI(ETYPE(IEL))/(ELL(IEL)}**3)*\\ \sum_{\mathtt{i=1}}^{\mathtt{nip}}\mathtt{DER2(k)}*\mathtt{DER2(1)}*\mathtt{WEIGHTS(I)}\end{aligned} \tag{4.9}$$

and

$$\mathtt{MM(K,L)} = \mathtt{ELL(IEL)}/2 * \sum_{\mathtt{i=1}}^{\mathtt{nip}}\mathtt{FS_i} * \mathtt{FUN(K)} * \mathtt{FUN(1)} * \mathtt{WEIGHTS(I)} \tag{4.10}$$

where WEIGHTS(I) is the weighting coefficient of the *i*th Gauss point. The weights and sampling points POINTS are provided by the library subroutine SAMPLE for a "LINE" element once the required order of integration NIP has been set. The "mass" matrix of a beam element requires four Gaussian integration points for exact integration. Other variables new to this program include FTF and DTD which hold, respectively, the contribution of each Gauss point to the KM and MM matrices. Array MOM holds the computed moment value at each node, and array STORE_KM stores the element stiffness matrices avoiding the need to recompute them in the post-processing stage.

New real variables include FS, which is the computed foundation stiffness at each Gauss point. The data values FS0 and FS1 represent the foundation stiffnesses at the ends of the whole beam with a linear variation assumed between. The coordinate of the Gauss point is held in SAMP_PT and X keeps track of the *x* coordinate of the left-hand node of each element.

Returning to the main program and following the structure chart of Figure 4.15, it can be seen that an additional loop is required in the stiffness and "mass" matrix formulation, where `I` counts from 1 to `NIP`. The products indicated in equations (4.9) and (4.10) are performed, and the resulting arrays `DTD` and `FTF` are then added into the accumulating matrices `KM` and `MM` repectively. Once all the Gauss point contributions have been accumulated, `KM` and `MM` are added together to give the net stiffness matrix (still called `KM`) for the composite beam/foundation system. The element stiffness matrices are then stored in the 3-d array `STORE_KM`.

The remainder of the program follows a familiar course. Forces are read, and, following equation solution, the global nodal displacements and rotations are

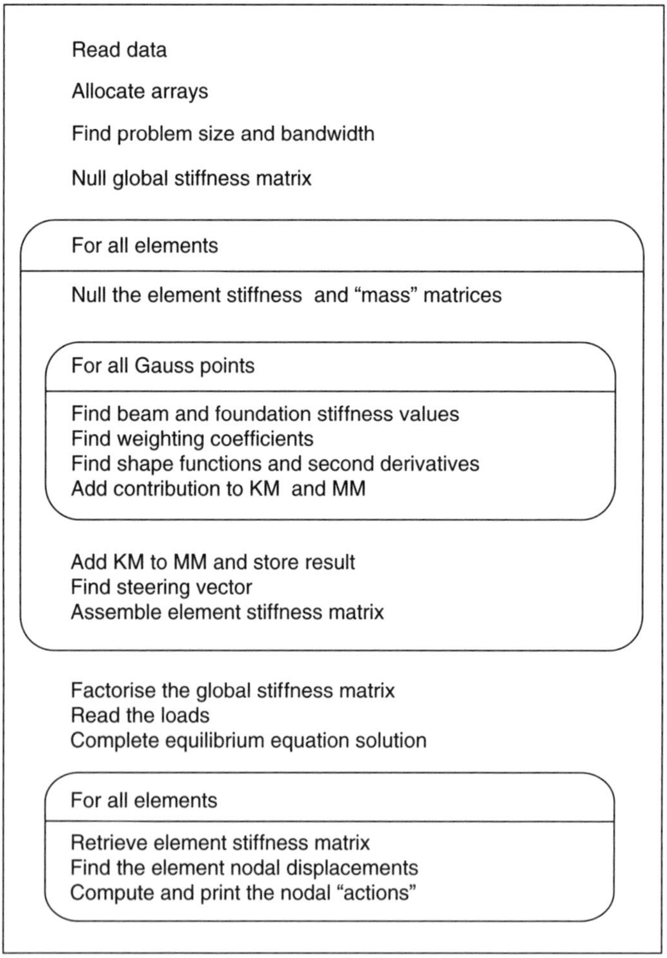
Read data
Allocate arrays
Find problem size and bandwidth
Null global stiffness matrix
For all elements
Null the element stiffness and "mass" matrices
For all Gauss points
Find beam and foundation stiffness values
Find weighting coefficients
Find shape functions and second derivatives
Add contribution to KM and MM
Add KM to MM and store result
Find steering vector
Assemble element stiffness matrix
Factorise the global stiffness matrix
Read the loads
Complete equilibrium equation solution
For all elements
Retrieve element stiffness matrix
Find the element nodal displacements
Compute and print the nodal "actions"

Figure 4.15 Structure chart for Program 4.3

obtained. In the post-processing phase, the element nodal displacements ELD are retrieved, as are the element stiffness matrices KM from STORE_KM. Their product gives the element "actions" which hold the shear forces and moments.

The example given in Figure 4.16 shows a laterally loaded pile made up of five beam elements installed in an elastic soil medium with a modulus varying

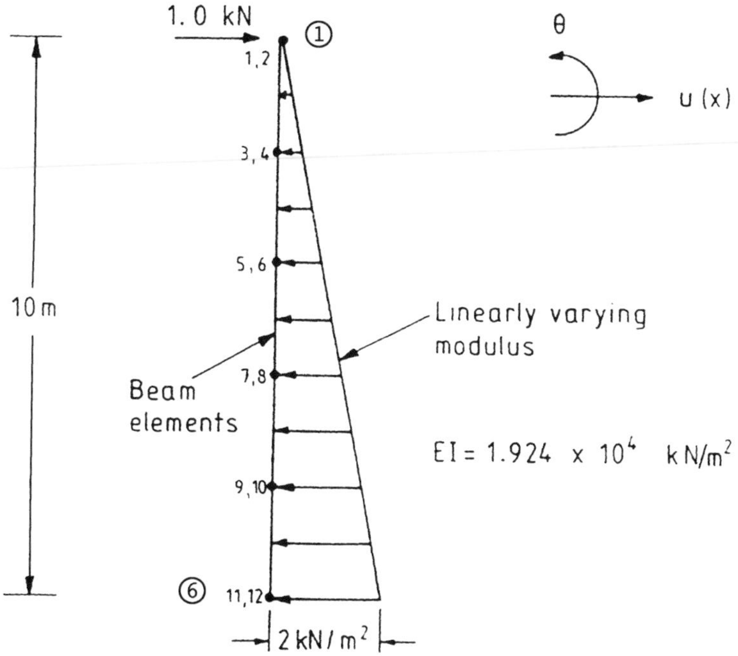

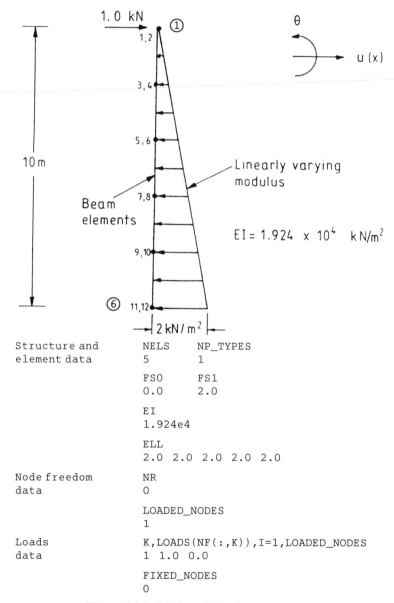

Structure and element data	NELS NP_TYPES 5 1
	FS0 FS1 0.0 2.0
	EI 1.924e4
	ELL 2.0 2.0 2.0 2.0 2.0
Node freedom data	NR 0
	LOADED_NODES 1
Loads data	K,LOADS(NF(:,K)),I=1,LOADED_NODES 1 1.0 0.0
	FIXED_NODES 0

Figure 4.16 Mesh and data for Program 4.3

```
Global coordinates
Node         1       0.0000E+00
Node         2       0.2000E+01
Node         3       0.4000E+01
Node         4       0.6000E+01
Node         5       0.8000E+01
Node         6       0.1000E+02
Global node numbers
Element      1          1  2
Element      2          2  3
Element      3          3  4
Element      4          4  5
Element      5          5  6
There are     12  equations and the half-bandwidth is      3

   Node    Displacement Moment
    1       0.9033E00  -0.2961E-03
    2       0.6618E00  -0.1793E+01
    3       0.4206E00  -0.2596E+01
    4       0.1800E00  -0.2115E+01
    5      -0.6019E-01 -0.8326E00
    6      -0.3002E00   0.2977E-03
```

Figure 4.17 Results from beam/foundation example with Program 4.3

from zero at the ground surface to 2.0 at a depth of 10.0 units. The beam elements have a constant stiffness $EI = 1.924 \times 10^4$. The pile has no restraints placed on the nodes so $NR = 0$. The data FS0 and FS1 represent the foundation stiffness at the top and bottom of the pile respectively. The program automatically interpolates between these values to obtain the foundation stiffness at the Gauss points. A unit horizontal load has been applied to the top of the pile.

The computed results shown in Figure 4.17 show the horizontal translation and moment value at each of the nodes. It is seen that the horizontal translation of node 1 is given as 0.903 which is in close agreement with the analytical solution to this problem from Hetenyi (1946).

PROGRAM 4.4: EQUILIBRIUM OF RIGID-JOINTED FRAMES

The first three programs in this chapter were concerned only with 1-d elements which could either sustain axial loads (rod elements) or transverse loading and moments (beam elements). It is much more common to encounter structures made up of members arbitrarily inclined to one another. Loading of such structures results in displacements due to both axial and bending effects, although the former is often ignored in many approximate methods described in texts on structural analysis. The beam–rod element stiffness matrix used by this program is formed by superposing the beam and rod stiffness matrices described in earlier programs in this chapter. The program described in this section can analyse 2-d or 3-d framed structures, with element stiffness formed by the library subroutine RIGID_JOINTED.

```
program p44
!------------------------------------------------------------------------
! program 4.4 equilibrium of rigid-jointed frames using beam/rod elements
!            in 1-, 2- or 3-dimensions
!------------------------------------------------------------------------
use new_library  ;  use  geometry_lib   ;  use  vlib  ;   implicit none
integer::nels,neq,nn,nband,nr,nod=2,nodof,ndof,iel,i,k,ndim,loaded_nodes, &
         fixed_nodes,nprops,np_types
!---------------------------dynamic arrays------------------------------
real,allocatable::km(:,:),eld(:),kv(:),loads(:),coord(:,:),              &
                  action(:),g_coord(:,:),value(:),prop(:,:),gamma(:)
integer,allocatable::nf(:,:),g(:),num(:),g_num(:,:),no(:),g_g(:,:),     &
                     node(:),sense(:),etype(:)
!------------------------input and initialisation------------------------
open (10 , file = 'p44a.dat' , status = 'old' ,    action ='read')
open (11 , file = 'p44a.res' , status = 'replace', action='write')
read(10,*)nels,nn,ndim,nprops,np_types
select case(ndim)
  case(1); nodof=2; case(2); nodof=3; case(3); nodof=6
  case default; write(11,'(a)')"Wrong number of dimensions input"
end select
ndof=nod*nodof
allocate(nf(nodof,nn),km(ndof,ndof),coord(nod,ndim),g_coord(ndim,nn),     &
         eld(ndof),action(ndof),g_num(nod,nels),num(nod),g(ndof),        &
         gamma(nels),g_g(ndof,nels),prop(nprops,np_types),etype(nels))
read(10,*)prop; etype=1; if(np_types>1)read(10,*)etype
if(ndim == 3)read(10,*)gamma
read(10,*)g_coord; read(10,*)g_num
read(10,*)nr
nf=1; if(nr>0)read(10,*)(k,nf(:,k),i=1,nr); call formnf(nf); neq=maxval(nf)
!---------------loop the elements to find global array sizes------------------
nband=0
elements_1: do iel=1,nels
             num=g_num(:,iel); call num_to_g ( num ,nf  , g  )
             g_g(:,iel)=g; if(nband<bandwidth(g))nband=bandwidth(g)
end do elements_1
allocate(kv(neq*(nband+1)),loads(0:neq)); kv=0.0
write(11,'(a)')"Global coordinates"
do k=1,nn; write(11,'(a,i5,a,3e12.4)')                                    &
    "Node    ",k,"    ",g_coord(:,k); end do
write(11,'(a)')"Global node numbers"
do k=1,nels; write(11,'(a,i5,a,27i3)')                                    &
    "Element ",k,"      ",g_num(:,k); end do
write(11,'(2(a,i5),/)')                                                   &
    "There are ",neq," equations and the half-bandwidth is ",nband
!---------------------global stiffness matrix assembly------------------------
elements_2: do iel=1,nels
             num=g_num(:,iel); coord=transpose(g_coord(:,num))
             call rigid_jointed(km,prop,gamma,etype,iel,coord)
             g=g_g(:,iel)   ;    call formkv(kv,km,g,neq)
end do elements_2
!--------------------read loads and/or displacements--------------------------
loads=0.0; read(10,*)loaded_nodes
if(loaded_nodes/=0)read(10,*)(k,loads(nf(:,k)),i =1,loaded_nodes)
read(10,*)fixed_nodes
if(fixed_nodes/=0)then
  allocate(node(fixed_nodes),no(fixed_nodes),                              &
           sense(fixed_nodes),value(fixed_nodes))
  read(10,*)(node(i),sense(i),value(i),i=1,fixed_nodes)
  do i=1,fixed_nodes; no(i)=nf(sense(i),node(i)); end do
  kv(no)=kv(no)+1.e20; loads(no)=kv(no)*value
end if
!----------------------------equation solution -------------------------------
call banred(kv,neq); call bacsub(kv,loads)
```

```
write(11,'(a)')"The nodal displacements are:"
do k=1,nn; write(11,'(i5,6e12.4)')k,loads(nf(:,k)); end do
!-----------------retrieve element end actions--------------------
write(11,'(a)')"The element 'actions' are:"
elements_3: do iel=1,nels
                num=g_num(:,iel); coord=transpose(g_coord(:,num))
                g=g_g(:,iel); eld=loads(g)
                call rigid_jointed(km,prop,gamma,etype,iel,coord)
                action=matmul(km,eld)
                if(ndim<3)then
                  write(11,'(i5,6e12.4)')iel,action
                else
                  write(11,'(i5,6e12.4)')iel,    action(1: 6:1)
                  write(11,'(a,6e12.4)')"        ",action(7:12:1)
                end if
end do elements_3
end program p44
```

The first example analysed by Program 4.4 is shown in Figure 4.18 and is a 2-d rigid-jointed frame subjected to distributed loads and point loads. In 2-d, the elements have six degrees of freedom as shown in Figure 4.19. At each node there are two translational freedoms in x and y and a rotation. The nodal freedom numbering associated with each element, accounting for any restraints, is as usual contained in the "steering" vector G. Thus, with reference to Figures 4.18 and 4.19, the steering vector for element 1 would be $[0\ 0\ 1\ 2\ 3\ 4]^T$ and for element 4, $[0\ 0\ 0\ 2\ 3\ 4]^T$, and so on. The data organisation is similar to the pin-jointed frame analysis described in Program 4.1. The first line of data provides the number of elements (NELS), the number of nodes (NN), the dimensionality (NDIM), the number of properties NPROPS and the number of property types NP_TYPES.

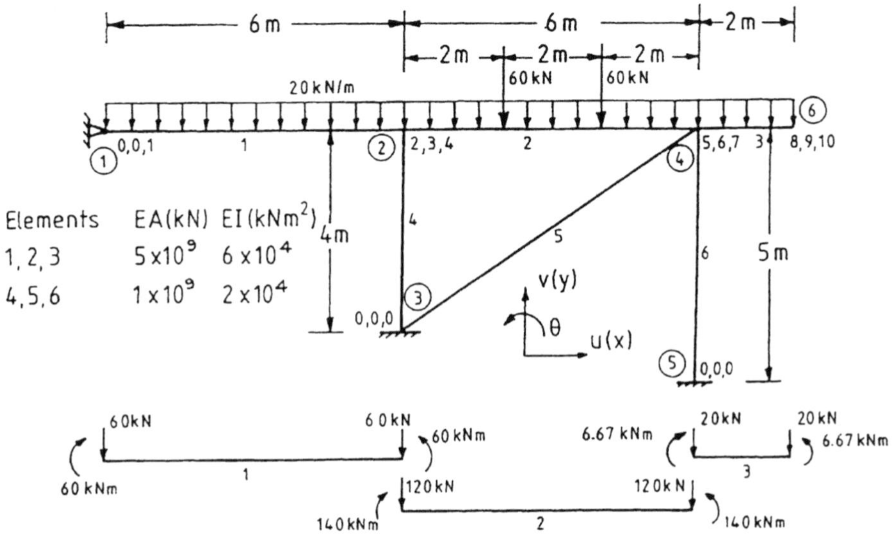

Figure 4.18 *(continued)*

Structure and element data	NELS NN NDIM NPROPS NP_TYPES 6 6 2 2 2
	PROP (EA, EI) 5.e9 6.e4 1.e9 2.e4
	ETYPE 1 1 1 2 2 2
Geometry	G_COORD 0. 0. 6. 0. 6. -4. 12. 0. 12. -5. 14. 0.
Connectivity	G_NUM 1 2 2 4 4 6 3 2 3 4 5 4
Node freedom data	NR K,NF(:,K),I=1,NR 3 1 0 0 1 3 0 0 0 5 0 0 0
Loads data	LOADED_NODES 4
	K,LOADS(NF(:,K)),I=1,LOADED_NODES 1 0.0 0. -60. 2 0.0 -180. -80. 4 0.0 -140. 133.33 6 0.0 -20. 6.67
	FIXED_NODES 0

Figure 4.18 Mesh and data for 2-d example with Program 4.4

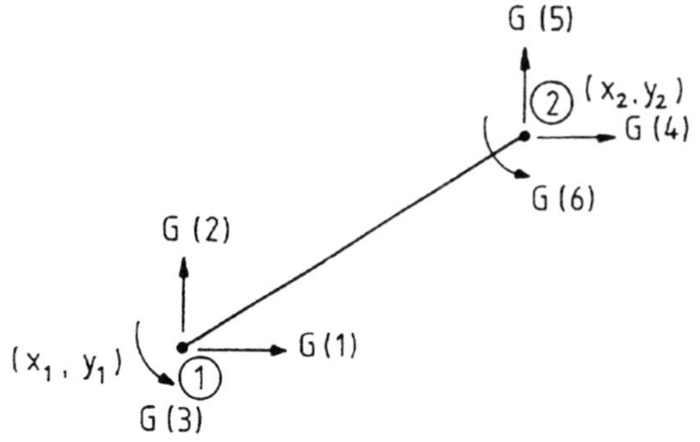

Figure 4.19 Node and freedom numbering for 2-d beam/rod elements

In this program, the number of material properties required depends on the dimensionality so the data now includes input to the integer `NPROPS` which indicates the number of material properties required for each property type. In a 2-d frame problem, there are two material properties required (`EA` and `EI`), so

```
Global coordinates
Node         1        0.0000E+00  0.0000E+00
Node         2        0.6000E+01  0.0000E+00
Node         3        0.6000E+01 -0.4000E+01
Node         4        0.1200E+02  0.0000E+00
Node         5        0.1200E+02 -0.5000E+01
Node         6        0.1400E+02  0.0000E+00
Global node numbers
Element      1          1  2
Element      2          2  4
Element      3          4  6
Element      4          3  2
Element      5          3  4
Element      6          5  4
There are    10  equations and the half-bandwidth is     5

The nodal displacements are:
    1  0.0000E+00  0.0000E+00 -0.1025E-02
    2  0.3645E-07 -0.8319E-06 -0.9497E-03
    3  0.0000E+00  0.0000E+00  0.0000E+00
    4  0.6435E-07 -0.6283E-06  0.1774E-02
    5  0.0000E+00  0.0000E+00  0.0000E+00
    6  0.6435E-07  0.2880E-02  0.1329E-02
The element 'actions' are:
    1 -0.3038E+02 -0.1975E+02 -0.6000E+02  0.3038E+02  0.1975E+02 -0.5849E+02
    2 -0.2325E+02  0.8238E+01 -0.2519E+01  0.2325E+02 -0.8238E+01  0.5195E+02
    3 -0.2161E-05  0.2000E+02  0.3333E+02  0.2161E-05 -0.2000E+02  0.6670E+01
    4  0.7123E+01  0.2080E+03 -0.9497E+01 -0.7123E+01 -0.2080E+03 -0.1899E+02
    5  0.3177E+02  0.2610E+02  0.9839E+01 -0.3177E+02 -0.2610E+02  0.1968E+02
    6 -0.8513E+01  0.1257E+03  0.1419E+02  0.8513E+01 -0.1257E+03  0.2838E+02
```

Figure 4.20 Results from 2-d beam/rod example with Program 4.4

NPROPS = 2. The material property values for each type (NP_TYPES) are then read into the 2-d array PROP.

The material property data is followed by the ETYPE vector (if needed), the global nodal coordinates (G_COORD) and the element node numbering (G_NUM). The loading on the nodes is calculated using the "fixed end" approach described in the section on Program 4.2, and these values are shown for each individual element in Figure 4.18. There are no FIXED_NODES in this example.

The results shown in Figure 4.20 indicate that the rotation at node 1 for example is −0.001025. The ACTION vectors for elements 4, 5 and 6 are final; however, for elements 1, 2 and 3 the fixed end shear forces and moments must be added back, e.g.

Element 1

$$F_{x1} = -30.38 + 0.00 = -30.38$$
$$F_{y1} = -19.75 + 60.00 = 40.25$$
$$M_1 = -60.00 + 60.00 = 0.00$$
$$F_{x2} = 30.38 + 0.00 = 30.38$$
$$F_{y2} = 19.75 + 60.00 = 79.75$$
$$M_2 = -58.49 - 60.00 = -118.49$$

Element 2

$$F_{x1} = -23.25 + 0.00 = -23.25$$
$$F_{y1} = 8.24 + 120.00 = 128.24$$
$$M_1 = -2.52 + 140.00 = 137.48$$
$$F_{x2} = 23.25 + 0.00 = 23.25$$
$$F_{y2} = -8.24 + 120.00 = 111.76$$
$$M_2 = 51.96 - 140.00 = -88.04$$

Element 4

$$F_{x1} = 7.12 + 0.00 = 7.12$$
$$F_{y1} = 207.99 + 0.00 = 207.99$$
$$M_1 = -9.50 + 0.00 = -9.50$$
$$F_{x2} = -7.12 + 0.00 = -7.12$$
$$F_{y2} = -207.99 + 0.00 = -207.99$$
$$M_2 = -18.99 + 0.00 = -18.99$$

Moment equilibrium is established at node 2 where three elements are joined together.

The second example to be analysed is shown in Figure 4.21 and represents a 3-d rigid-jointed frame subjected to a vertical point load of 100. In 3-d, the elements have 12 degrees of freedom as shown in Figure 4.22. At each node there are three translational freedoms in x, y and z and three rotations about each of the global axes. The extension to 3-d is relatively simple, but considerably more care is required in the preparation of data and attention to sign conventions. The data organisation is virtually the same as in the previous example, except `NDIM` is read as data as 3.

In addition to the axial stiffness (`EA`), extra elastic data in 3-d involves the flexural stiffness about the element's local y' and z' axes (`EIY` and `EIZ` respectively) and a torsional stiffness (`GJ`), thus `NPROPS` is 4. The local coordinate x' defines the long axis of the element. The relationship between the global axes and local axes (x', y' and z') must be considered for 3-d space frames because in addition to the six coordinates that define the position of each node of the element in space, a seventh rotational "coordinate" gamma must be read in as data.

For the purposes of data preparation, a "vertical" element is defined as one which lies parallel to the global y axis. For non-vertical elements, the angle γ is defined as the rotation of the element about its local x' axis as shown in Figure 4.23. For "vertical" elements however, γ is defined as the angle between the global z axis and the local z' axis, measured towards the global x axis as shown in Figure 4.24. For "vertical" elements it is essential that the local x' axis points in the same direction as the global y axis.

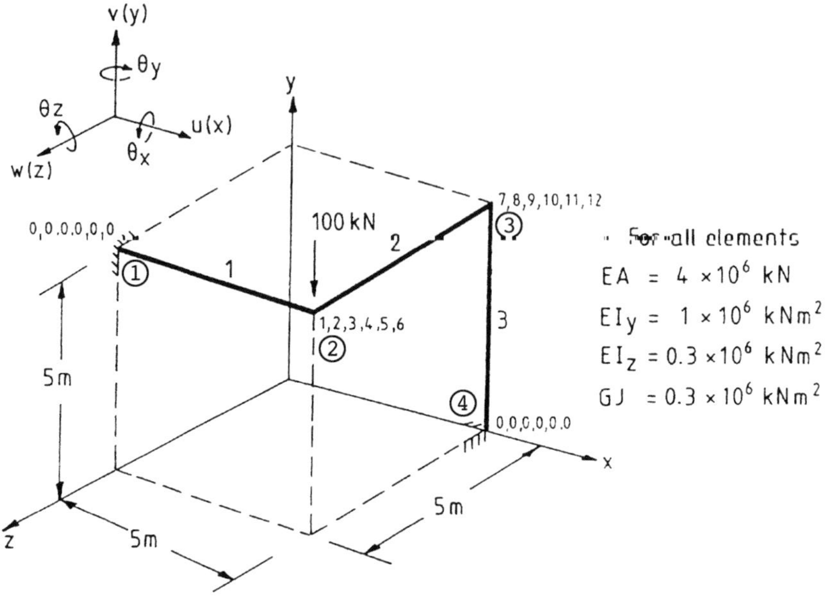

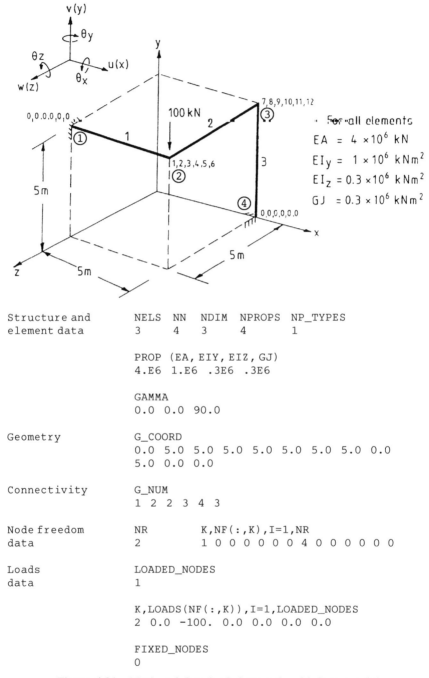

Structure and element data	NELS NN NDIM NPROPS NP_TYPES 3 4 3 4 1
	PROP (EA, EIY, EIZ, GJ) 4.E6 1.E6 .3E6 .3E6
	GAMMA 0.0 0.0 90.0
Geometry	G_COORD 0.0 5.0 5.0 5.0 5.0 5.0 5.0 5.0 0.0 5.0 0.0 0.0
Connectivity	G_NUM 1 2 2 3 4 3
Node freedom data	NR K,NF(:,K),I=1,NR 2 1 0 0 0 0 0 0 4 0 0 0 0 0 0
Loads data	LOADED_NODES 1
	K,LOADS(NF(:,K)),I=1,LOADED_NODES 2 0.0 -100. 0.0 0.0 0.0 0.0
	FIXED_NODES 0

Figure 4.21 Mesh and data for 3-d example with Program 4.4

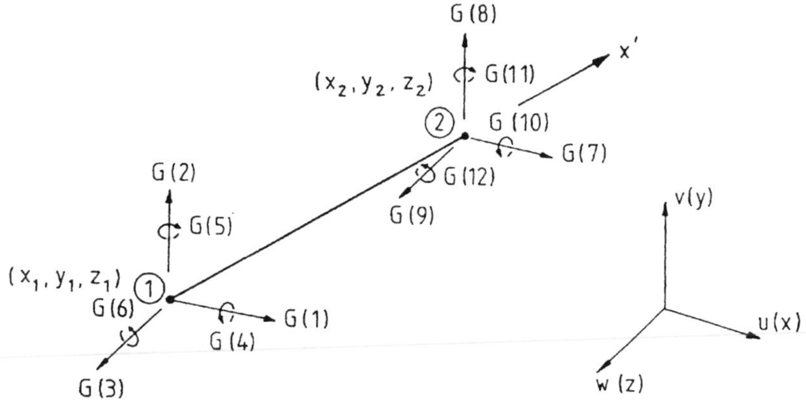

Figure 4.22 Node and freedom numbering for 3-d beam/rod elements

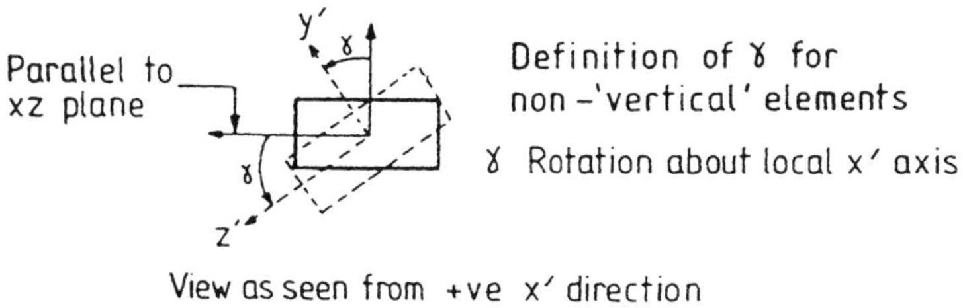

Figure 4.23 Transformation angle for "non-vertical" elements

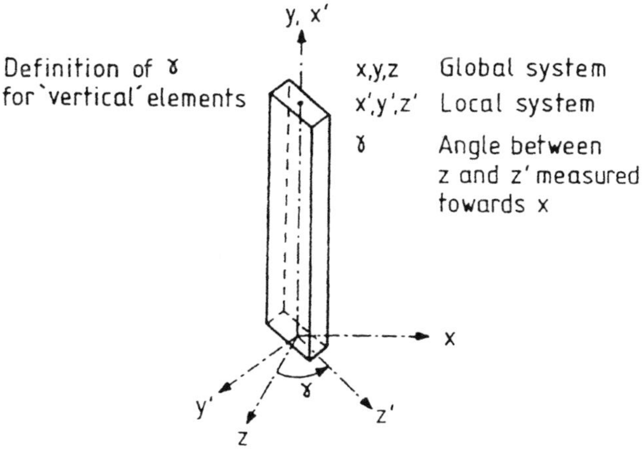

Figure 4.24 Transformation angle for "vertical" elements

Returning to the example problem, it is necessary to establish the local coordinate system for each element as shown in Figure 4.25. As indicated in Figure 4.22, the positive local x' direction is defined by moving from node 1 to node 2, so this is also the order in which the element nodal numbering must be given in the data.

In the example of Figure 4.21, elements 1 and 2 both have their local z' axes parallel to the global xz plane, thus there has been no rotation of these non-vertical elements and γ is set to zero. For vertical element 3, however, γ is set to

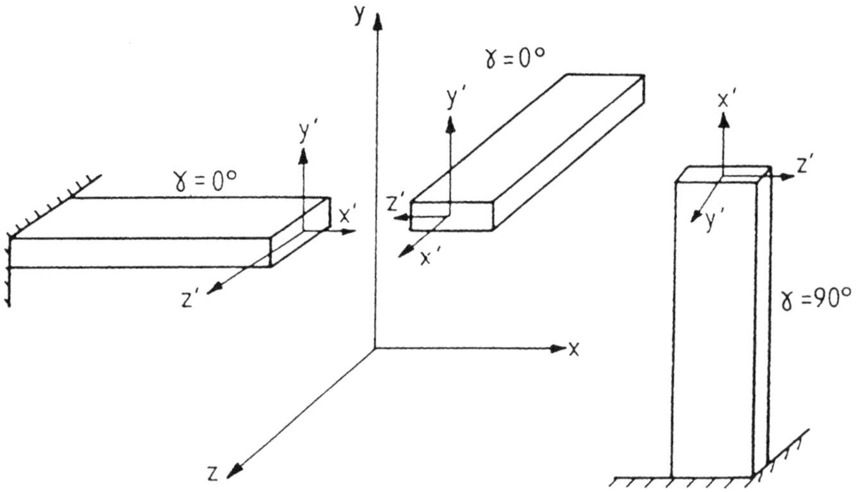

Figure 4.25 Element local coordinate systems

```
Global coordinates
Node         1       0.0000E+00  0.5000E+01  0.5000E+01
Node         2       0.5000E+01  0.5000E+01  0.5000E+01
Node         3       0.5000E+01  0.5000E+01  0.0000E+00
Node         4       0.5000E+01  0.0000E+00  0.0000E+00
Global node numbers
Element      1         1  2
Element      2         2  3
Element      3         4  3
There are    12  equations and the half-bandwidth is     11

The nodal displacements are:
    1  0.0000E+00  0.0000E+00  0.0000E+00  0.0000E+00  0.0000E+00  0.0000E+00
    2 -0.3039E-05 -0.5997E-02  0.8769E-03  0.1129E-02 -0.2360E-03 -0.1514E-02
    3  0.9571E-03 -0.4536E-04  0.9113E-03  0.7470E-03 -0.1582E-03 -0.3727E-03
    4  0.0000E+00  0.0000E+00  0.0000E+00  0.0000E+00  0.0000E+00  0.0000E+00
The element 'actions' are:
    1  0.2431E+01  0.6371E+02 -0.2754E+02 -0.6777E+02  0.1160E+03  0.2501E+03
      -0.2431E+01 -0.6371E+02  0.2754E+02  0.6777E+02  0.2165E+02  0.6846E+02
    2  0.2431E+01 -0.3629E+02 -0.2754E+02 -0.6777E+02 -0.2165E+02 -0.6846E+02
      -0.2431E+01  0.3629E+02  0.2754E+02 -0.1137E+03  0.9490E+01  0.6846E+02
    3 -0.2431E+01  0.3629E+02  0.2754E+02  0.2403E+02  0.9490E+01  0.8062E+02
       0.2431E+01 -0.3629E+02 -0.2754E+02  0.1137E+03 -0.9490E+01 -0.6846E+02
```

Figure 4.26 Results from 3-d beam/rod example with Program 4.4

90°, which is the angle between the global z and local z' axes measured towards the global x axis.

The results of the analysis given in Figure 4.26 indicate that the vertical deflection under the load is -0.005997. As a check on equilibrium under the applied loading of 100.0, the F_{y1} component of the `ACTION` vector at the built-in end (node 1) of each of elements 1 and 3 is 63.71 and 36.29 respectively.

PROGRAM 4.5: PLASTIC ANALYSIS OF RIGID-JOINTED FRAMES

This program is an extension of the preceding program in which a limit is placed on the maximum moment that any member can sustain. As loads on the structure are increased, plastic hinges form progressively and "failure" occurs when a mechanism develops.

This is the first example in the book of a nonlinear analysis in which the moment–curvature behaviour of the members is assumed to be elastic–perfectly plastic as shown in Figure 4.27. To deal with this nonlinearity, an iterative approach is used to find the nodal displacements and element "actions" under a given set of applied loads. Moments in excess of their plastic limits are redistributed to other joints which still have reserves of moment-carrying capacity. Convergence of the iterative process is said to have occurred when, within certain tolerances, moments at the element nodes nowhere exceed their limiting plastic values, and the internal "actions" are in equilibrium with the applied external loads.

The method employed is shown in the structure chart in Figure 4.28. The global stiffness matrix is formed once only with the nonlinearity introduced by iteratively modifying the internal forces on the structure until convergence is achieved. For greater detail of this particular algorithm the reader is referred to Griffiths (1988). Similar procedures are utilised in Chapter 6 in relation to elasto-plastic solids.

The benefits of splitting the solution of the equilibrium equations into two stages, namely factorisation performed once (using subroutine `BANRED`), and a forward and back-substitution performed at each iteration for each new right-hand side (using subroutine `BACSUB`) become clear.

Referring to the program, the additional data `LIMIT`, which refers to the iteration ceiling beyond which the algorithm will stop, and `TOL`, which gives the convergence tolerance, are read. These are followed by material properties which, in addition to the elastic values, must now include the plastic moment values for all members. For 1-d beams or 2-d frames, only one plastic moment (`MP`) is read, thus `NPROPS` is 2 in 1-d and `NPROPS` is 3 in 2-d. For 3-d space frames, however, three plastic moments, (`MPY`, `MPZ` and `MPX`) are read thus `NPROPS` is 7, where `MPY` and `MPZ` represent the limiting bending moments

```
program p45
!-----------------------------------------------------------------------
! program 4.5 elasto-plastic analysis of rigid-jointed frames
! using beam and beam/rod elements in 1-, 2- or 3-dimensions
!-----------------------------------------------------------------------
use new_library  ;  use  geometry_lib  ;   use  vlib   ;   implicit none
integer::nels,neq,nn,nband,nr,nod=2,nodof,ndof,iel,i,k,ndim,loaded_nodes,  &
         incs,limit,iters,iy,nprops,np_types
real::tol,total_load       ;logical::converged
!---------------------------dynamic arrays-------------------------------------
real,allocatable::km(:,:),eld(:),kv(:),loads(:),coord(:,:),action(:),      &
                  g_coord(:,:),gamma(:),prop(:,:),bdylds(:),eldtot(:),     &
                  holdr(:,:),oldsps(:),react(:),val(:,:),dload(:)
integer,allocatable::nf(:,:),g(:),num(:),g_num(:,:),g_g(:,:),no(:),etype(:)
!-----------------------input and initialisation-------------------------------
open (10 , file = 'p45.dat' , status = 'old' ,    action ='read')
open (11 , file = 'p45.res' , status = 'replace', action='write')
read(10,*)nels,nn,ndim,nprops,np_types,limit,tol
select case(ndim)
  case(1); nodof=2; case(2); nodof=3; case(3); nodof=6
  case default; write(11,'(a)')"Wrong number of dimensions input"
end select
ndof=nod*nodof
allocate(nf(nodof,nn),km(ndof,ndof),coord(nod,ndim),g_coord(ndim,nn),      &
         eld(ndof),action(ndof),g_num(nod,nels),num(nod),g(ndof),          &
         gamma(nels),g_g(ndof,nels),holdr(ndof,nels),react(ndof),          &
         prop(nprops,np_types),etype(nels))
read(10,*)prop; etype=1; if(np_types>1) read(10,*)etype
if(ndim==3)read(10,*)gamma
read(10,*)g_coord; read(10,*)g_num
read(10,*)nr
nf=1; if(nr>0)read(10,*)(k,nf(:,k),i=1,nr); call formnf(nf); neq=maxval(nf)
!----------------loop the elements to find global array sizes------------------
nband=0
elements_1: do iel=1,nels
              num=g_num(:,iel) ; call num_to_g (num , nf , g )
              g_g(:,iel)=g; if(nband<bandwidth(g))nband=bandwidth(g)
end do elements_1
allocate(kv(neq*(nband+1)),loads(0:neq),eldtot(0:neq),bdylds(0:neq),       &
           oldsps(0:neq)); kv=0.0; holdr=0.0
write(11,'(a)')"Global coordinates"
do k=1,nn; write(11,'(a,i5,a,3e12.4)')                                      &
    "Node    ",k,"    ",g_coord(:,k); end do
write(11,'(a)')"Global node numbers"
do k=1,nels; write(11,'(a,i5,a,27i3)')                                      &
    "Element ",k,"      ",g_num(:,k); end do
write(11,'(2(a,i5),/)')                                                     &
    "There are ",neq," equations and the half-bandwidth is ",nband
!--------------------global stiffness matrix assembly--------------------------
elements_2: do iel=1,nels
              num=g_num(:,iel); coord=transpose(g_coord(:,num))
              call rigid_jointed(km,prop,gamma,etype,iel,coord); g=g_g(:,iel)
              call formkv(kv,km,g,neq)
end do elements_2
read(10,*)loaded_nodes; allocate(no(loaded_nodes),val(loaded_nodes,nodof))
read(10,*)(no(i),val(i,:),i=1,loaded_nodes)
read(10,*)incs; allocate(dload(incs))  ;   read(10,*)dload
!------------------------equation factorisation--------------------------------
call banred(kv, neq)
!-------------------------load increment loop----------------------------------
total_load=0.0
load_increment: do iy=1,incs
   total_load=total_load+dload(iy)
   oldsps=0.0; iters=0
```

```
   iterations: do
     iters=iters+1; loads=0.0
     do i=1,loaded_nodes
        loads(nf(:,no(i)))=dload(iy)*val(i,:)
     end do
     loads=loads+bdylds; bdylds=0.0
!-----------------------forward/back-substitution----------------------------
     call bacsub(kv, loads)
!---------------------------check convergence--------------------------------
     call checon(loads,oldsps,tol,converged)
!--------------------inspect moments in all elements-------------------------
     elements_3: do iel=1,nels
        num=g_num(:,iel); coord=transpose(g_coord(:,num))
        g=g_g(:,iel); eld=loads(g)
        call rigid_jointed(km,prop,gamma,etype,iel,coord)
        action=matmul(km,eld); react=0.0
!------------if plastic moments exceeded generate correction vector----------
        if(limit/=1)then
           call hinge(coord,holdr,action,react,prop,iel,etype,gamma)
           bdylds(g)=bdylds(g)-react; bdylds(0)=0.0
        end if
!---------------at convergence update element reactions----------------------
      if(iters==limit.or.converged)holdr(:,iel)=holdr(:,iel)+react(:)+action(:)
     end do elements_3
     if(iters==limit .or. converged)exit iterations
   end do iterations
   eldtot=loads+eldtot
   write(11,'(a,e12.4)')"Load factor   ",total_load
   write(11,'(a)')"The nodal displacements are:"
   do i=1,loaded_nodes
     write(11,'(i5,6e12.4)')no(i),eldtot(nf(:,no(i)))
   end do
   write(11,'(a,i5,a,/)')"Converged in  ",iters,"  iterations"
   if(iters==limit .and. limit/=1)exit load_increment
end do load_increment
end program p45
```

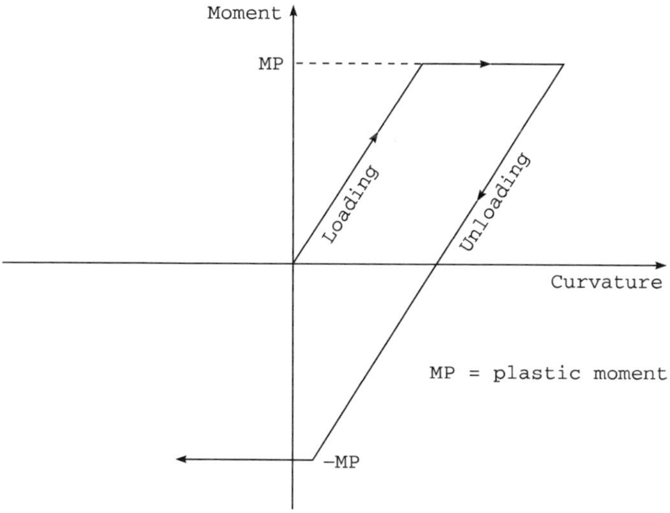

Figure 4.27 Elastic-perfectly plastic moment-curvature relationship

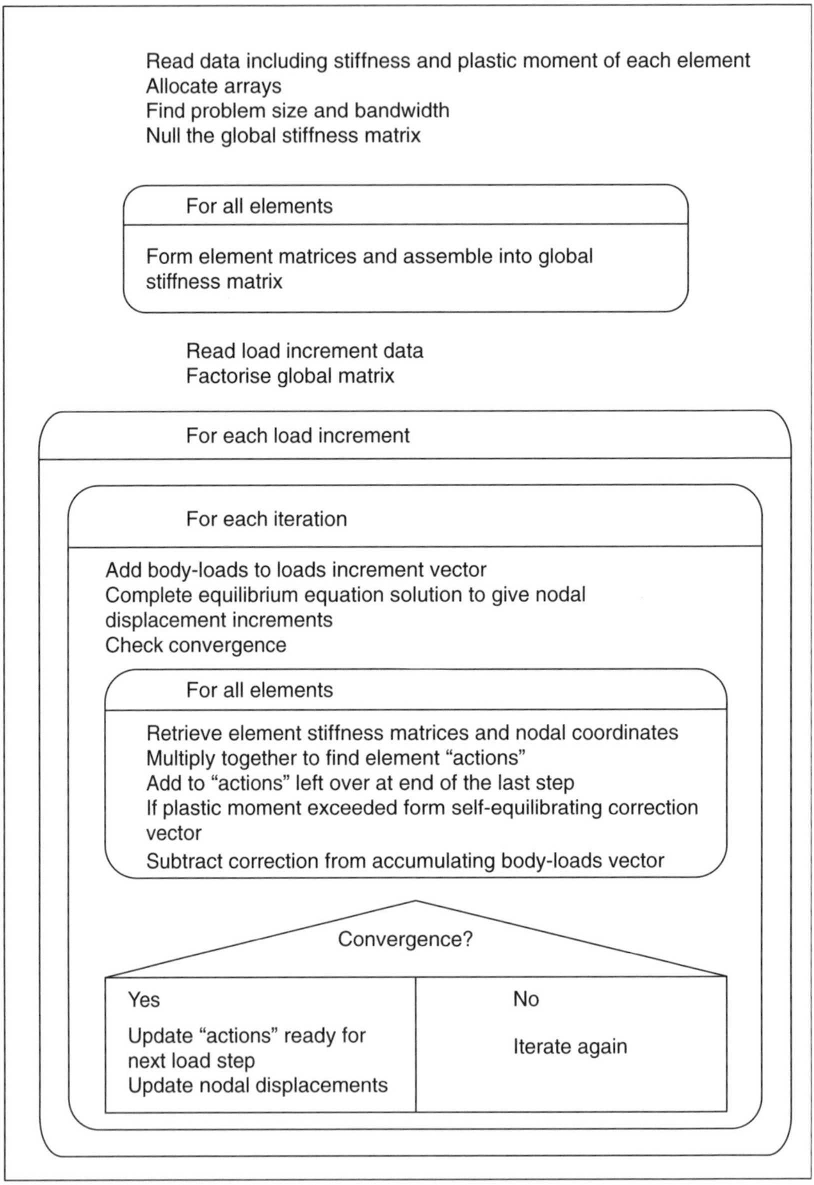

Figure 4.28 Structure chart for Program 4.5

about the major and minor local axes of the member, and MPX represents the limiting torsional moment about the long axis of the member.

Loads are applied in INCS increments to the nodes and the magnitude of each increment is read into the vector DLOAD. The loading remains "proportional" as is customary in plastic hinge analysis, so the relative magnitudes of the nodal loads are read into (NO(I),VAL(I,:),I=1, LOADED_NODES) where NO refers to the node number and VAL(I,:) refers to the loading at each of the freedoms at that node.

Additional arrays are required as part of the algorithm, and their names and functions are listed as follows:

Dynamic real arrays:

OLDSPS	nodal displacements from previous iteration
BDYLDS	added to external LOADS vector to redistribute moments
ELDTOT	keeps a running total of nodal displacements
HOLDR	keeps a running total of element "actions" at convergence
REACT	element self-equilibrating "correction" vector
DLOAD	load increment values
MPX	limiting torsion value
MPY	limiting bending moment about element y'-axis
MPZ	limiting bending moment about element z'-axis
VAL	nodal proportional loading ratios

Dynamic integer arrays:

NO	loaded node numbers

Following assembly of the global stiffness matrix, the program enters the load increment loop. For each iteration ITERS, the external load increments are added to the redistributive internal loads vector BDYLDS. The solution of the equilibrium equations is completed using subroutine BACSUB and the resulting nodal displacement increments compared with their values at the previous iteration using subroutine CHECON. This subroutine observes the relative change in displacement increments from one iteration to the next. If the change is less than TOL then the logical variable CONVERGED is set to .TRUE. and convergence has occurred.

At each iteration, each element is inspected and its ACTION vector computed from the product of nodal "displacements" and the element stiffness matrix. The subroutine HINGE adds the ACTION vector to the values already existing from the previous load step (held in HOLDR) and checks both nodes to see if the plastic moment value has been exceeded. If yield has occurred, the self-equilibrating vector REACT is formed. In Figure 4.29(a), a typical 2-d element is shown in which a particular load increment has pushed the moment value at both nodes over the plastic limit. The correction vector applies a moment to each node equal to the amount of overshoot of the plastic moment values. To preserve equilibrium, a couple is required as shown in the local coordinate system in Figure 4.29(b). Finally as shown in Figure 4.29(c), the couple is transformed into global coordinate directions before being accumulated into the

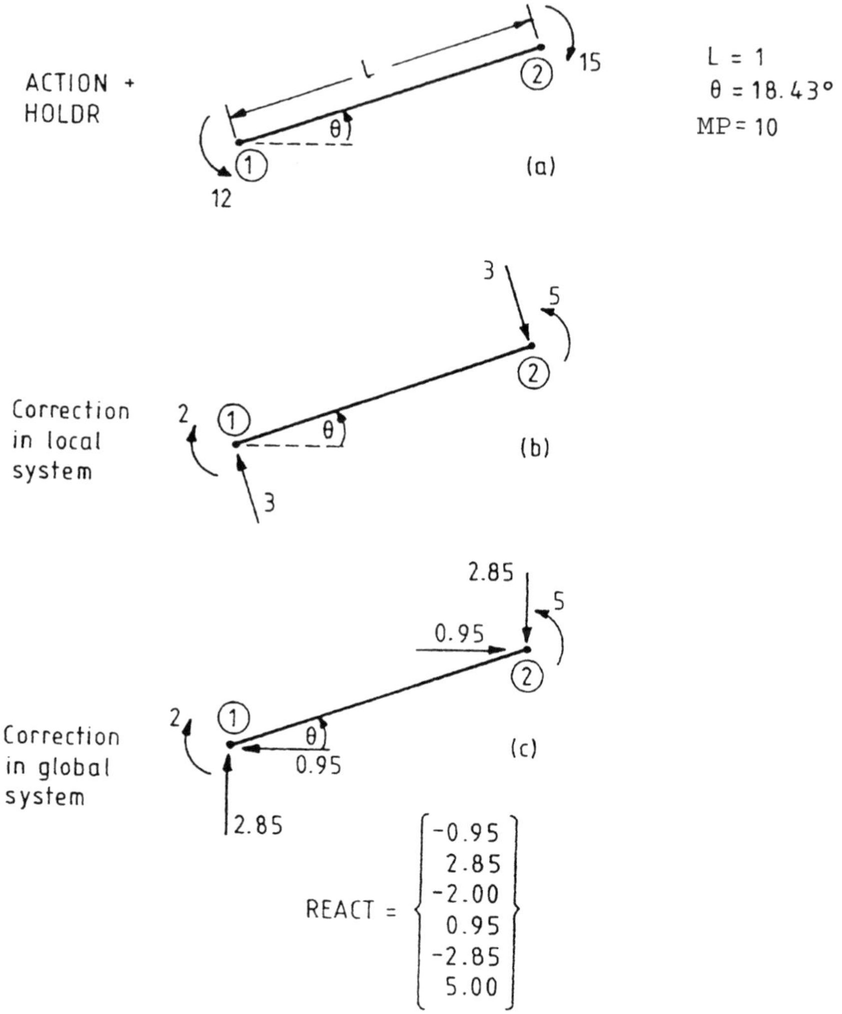

Figure 4.29 Correction vector for "yielding" elements

BDYLDS vector for each element in turn. Only those elements that have moments in excess of the plastic limits will contribute to BDYLDS.

If, at any load step, the algorithm fails to converge within the prescribed iteration ceiling limit then "collapse" of the structure is indicated because the algorithm is unable to find equilibrium without violating the plastic moment values.

A 2-d example shown in Figure 4.30 concerns a plane frame taken from Horne (1971) and represents a two-bay portal frame subjected to proportional loading. After each increment, the output shown in Figure 4.31 gives the

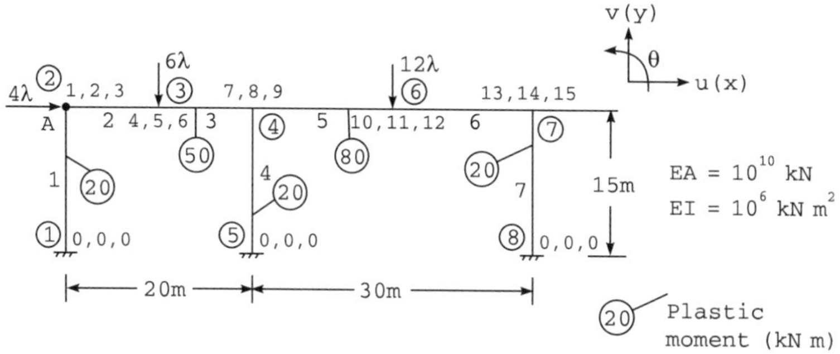

Structure and element data	NELS NN NDIM NPROPS NP_TYPES LIMIT TOL 7 8 2 3 3 200 0.0001
	PROP (EA, EI, MP) 1.e6 1.e6 20. 1.e6 1.e6 50. 1.e6 1.e6 80.
	ETYPE 1 2 2 1 3 3 1
Coordinates	G_COORD 0. 0. 0. 15. 10. 15. 20. 15. 20. 0. 35. 15. 50. 15. 50. 0.
Connectivity	G_NUM 1 2 2 3 3 4 5 4 4 6 6 7 8 7
Node freedom data	NR K,NF(:,K),I=1,NR 3 1 0 0 0 5 0 0 0 8 0 0 0
Loads data	LOADED_NODES 3
	NO(I),VAL(I,:),I=1,LOADED_NODES 2 4.0 0.0 0.0 3 0.0 -6.0 0.0 6 0.0 -12.0 0.0
	INCS 8
	DLOAD 0.5 0.3 0.2 0.2 0.1 0.05 0.02 0.01

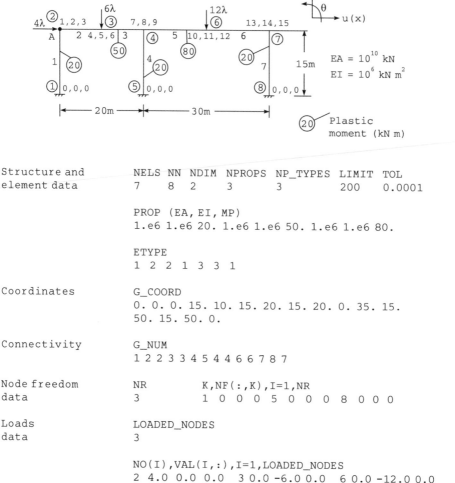

Figure 4.30 Mesh and data for 2-d example with Program 4.5

loading factor λ (held in DLOAD) together with the loaded node displacements and the iteration count. In Figure 4.32, the horizontal movement of point A is plotted against λ and indicates close agreement with the plastic hinge analysis value at failure of $\lambda_f = 1.375$.

```
Global coordinates
Node        1       0.0000E+00  0.0000E+00
Node        2       0.0000E+00  0.1500E+02
Node        3       0.1000E+02  0.1500E+02
Node        4       0.2000E+02  0.1500E+02
Node        5       0.2000E+02  0.0000E+00
Node        6       0.3500E+02  0.1500E+02
Node        7       0.5000E+02  0.1500E+02
Node        8       0.5000E+02  0.0000E+00
Global node numbers
Element     1         1  2
Element     2         2  3
Element     3         3  4
Element     4         5  4
Element     5         4  6
Element     6         6  7
Element     7         8  7
There are    15   equations and the half-bandwidth is      5

Load factor      0.5000E+00
The nodal displacements are:
    2  0.2567E-03 -0.9975E-05 -0.2491E-04
    3  0.2392E-03 -0.1382E-03  0.1035E-04
    6  0.1920E-03 -0.1245E-02 -0.1748E-05
Converged in       2  iterations
...........................................................

Load factor      0.1380E+01
The nodal displacements are:
    2  0.3322E-02 -0.9415E-05 -0.1227E-03
    3  0.3293E-02 -0.1246E-03  0.1101E-03
    6  0.3224E-02 -0.7024E-02 -0.5434E-04
Converged in     200  iterations
```

Figure 4.31 Results from 2-d frame example with Program 4.5

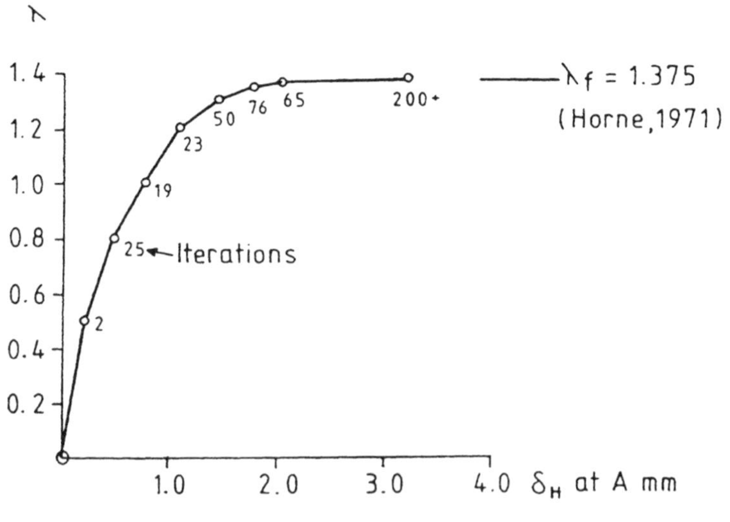

Figure 4.32 Load displacement behaviour from Program 4.5

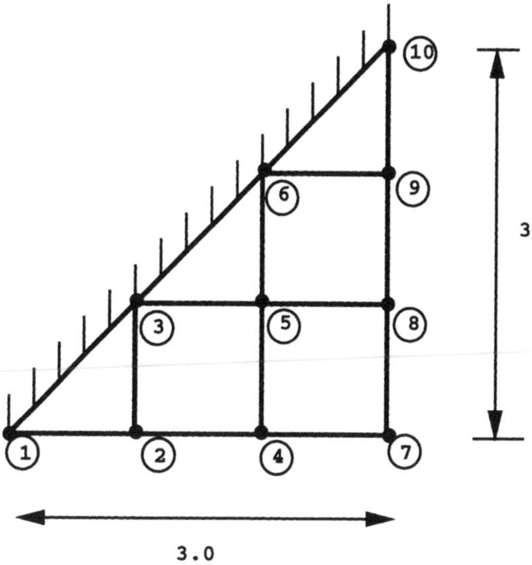

Structure and element data	NELS NN NDIM NPROPS NP_TYPES LIMIT TOL 12 10 3 7 1 200 0.00001
	PROP (EA, EIY, EIZ, GJ, MPY, MPZ, MPX) 1. 1.e4 1.e4 1. 1. 1. 1.e8
	GAMMA 0. 0. 0. 0. 0. 0. 0. 0. 0. 0. 0. 0.
Geometry	G_COORD 0. 0. 0. 1. 0. 0. 1. 1. 0. 2. 0. 0. 2. 1. 0. 2. 2. 0. 3. 0. 0. 3. 1. 0. 3. 2. 0. 3. 3. 0.
Connectivity	G_NUM 1 2 2 3 2 4 3 5 4 5 5 6 4 7 5 8 6 9 7 8 8 9 9 10
Node freedom data	NR K,NF(:,K),I=1,NR 4 1 0 0 0 0 0 0 3 0 0 0 0 0 0 6 0 0 0 0 0 0 10 0 0 0 0 0 0
Loads data	LOADED_NODES 1
	NO(I),VAL(I,:),I=1,LOADED_NODES 7 0.0 0.0 1.0 0.0 0.0 0.0
	INCS DLOAD 4 1.0 0.5 0.05 0.05

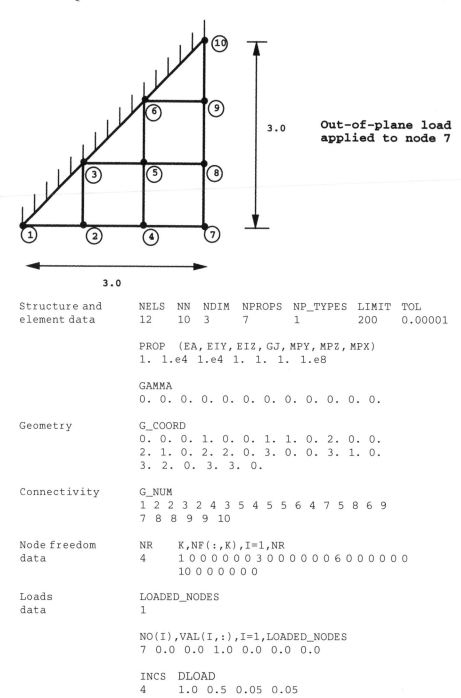

Figure 4.33 Mesh and data for 3-d example with Program 4.5

```
Global coordinates
Node          1       0.0000E+00  0.0000E+00  0.0000E+00
Node          2       0.1000E+01  0.0000E+00  0.0000E+00
Node          3       0.1000E+01  0.1000E+01  0.0000E+00
Node          4       0.2000E+01  0.0000E+00  0.0000E+00
Node          5       0.2000E+01  0.1000E+01  0.0000E+00
Node          6       0.2000E+01  0.2000E+01  0.0000E+00
Node          7       0.3000E+01  0.0000E+00  0.0000E+00
Node          8       0.3000E+01  0.1000E+01  0.0000E+00
Node          9       0.3000E+01  0.2000E+01  0.0000E+00
Node         10       0.3000E+01  0.3000E+01  0.0000E+00
Global node numbers
Element       1          1  2
Element       2          2  3
Element       3          2  4
Element       4          3  5
Element       5          4  5
Element       6          5  6
Element       7          4  7
Element       8          5  8
Element       9          6  9
Element      10          7  8
Element      11          8  9
Element      12          9 10
There are     36  equations and the half-bandwidth is     17

Load factor      0.1000E+01
The nodal displacements are:
     7  0.0000E+00  0.0000E+00  0.2118E-03 -0.1269E-03 -0.1269E-03  0.0000E+00
Converged in       3  iterations

Load factor      0.1500E+01
The nodal displacements are:
     7  0.0000E+00  0.0000E+00  0.3248E-03 -0.1936E-03 -0.1936E-03  0.0000E+00
Converged in      18  iterations

Load factor      0.1550E+01
The nodal displacements are:
     7  0.0000E+00  0.0000E+00  0.1530E-02 -0.8002E-03 -0.8002E-03  0.0000E+00
Converged in     200  iterations
```

Figure 4.34 Results from 3-d grid example with Program 4.5

The 3-d example shown in Figure 4.33 is of a horizontal triangular grid, rigidly supported along one side and subjected to a point transverse load at its apex. All members are given the same stiffness and plastic moment. The output shown in Figure 4.34 indicates failure occurs close to a load of 1.55.

PROGRAM 4.6: STABILITY ANALYSIS OF FRAMES

This program enables buckling analysis to be performed for 1-d and 2-d rigid-jointed frames. This is achieved by modifying the element stiffness matrices to account for axial loading using the "geometric" matrix as described in Chapter 2 [equation. (2.32)]. The simplest example of buckling occurs in a pinned beam or strut, which reaches its first buckling mode when the compressive axial loading equals the "Euler load" of $\pi^2 EI/L^2$. In a similar way, frames made up of several members connected at their joints can become unstable when the

```
 program p46
!-------------------------------------------------------------------------
! program 4.6 stability analysis of rigid-jointed frames using
!            2-d beam/rod elements
!-------------------------------------------------------------------------
use new_library  ;  use  geometry_lib  ;  use vlib   ;implicit none
integer::nels,neq,nn,nband,nr,nod=2,nodof=3,ndof=6,iel,i,k,ndim=2,          &
        loaded_nodes,iy,iters,limit,ksc,incs,nprops,np_types
real::total_load,tol,det  ;  logical::converged
!------------------------dynamic arrays-----------------------------------
real,allocatable::km(:,:),kp(:,:),eld(:),kv(:),loads(:),coord(:,:),         &
                  action(:),g_coord(:,:),val(:,:),prop(:,:),                &
                  axif(:),axip(:),eldtot(:),dload(:),gamma(:),              &
                  oldsps(:),kcop(:),disps(:),local(:)
integer,allocatable::nf(:,:),g(:),num(:),g_num(:,:),no(:),g_g(:,:),etype(:)
!----------------------input and initialisation---------------------------
open (10 , file = 'p46.dat' , status = 'old' ,    action ='read')
open (11 , file = 'p46.res' , status = 'replace', action='write')
read(10,*)nels,nn,nprops,np_types,limit,tol
allocate(nf(nodof,nn),km(ndof,ndof),coord(nod,ndim),g_coord(ndim,nn),       &
         eld(ndof),action(ndof),g_num(nod,nels),num(nod),g(ndof),          &
         g_g(ndof,nels),prop(nprops,np_types),                             &
         axif(nels),axip(nels),local(ndof),kp(ndof,ndof),etype(nels))
read(10,*)prop; etype=1; if(np_types>1)read(10,*)etype
read(10,*)g_coord; read(10,*)g_num
read(10,*)nr
nf = 1; if(nr>0)read(10,*)(k,nf(:,k),i=1,nr); call formnf(nf); neq=maxval(nf)
!---------------loop the elements to find global array sizes-------------------
nband=0
elements_1: do iel=1,nels
             num=g_num(:,iel)  ;  call num_to_g (num , nf , g)
             g_g(:,iel)=g; if(nband<bandwidth(g))nband=bandwidth(g)
end do elements_1
allocate(kv(neq*(nband+1)),kcop(neq*(nband+1)),loads(0:neq),disps(0:neq),  &
        eldtot(0:neq),oldsps(0:neq))
write(11,'(a)')"Global coordinates"
do k=1,nn; write(11,'(a,i5,a,3e12.4)')                                     &
    "Node    ",k,"    ",g_coord(:,k); end do
write(11,'(a)')"Global node numbers"
do k=1,nels; write(11,'(a,i5,a,27i3)')                                     &
    "Element ",k,"     ",g_num(:,k); end do
write(11,'(2(a,i5),/)')                                                    &
    "There are ",neq," equations and the half-bandwidth is ",nband
axif=0.0; axip=0.0; eldtot=0.0
read(10,*)loaded_nodes; allocate(no(loaded_nodes),val(loaded_nodes,nodof))
read(10,*)(no(i),val(i,:),i=1,loaded_nodes)
read(10,*)incs; allocate(dload(incs)) ;    read(10,*)dload
!---------------------load increment loop---------------------------------------
total_load=0.0
load_increments: do iy=1,incs
   total_load=total_load+dload(iy)
   loads=0.0
   do i = 1,loaded_nodes; loads(nf(:,no(i)))=dload(iy)*val(i,:); end do
   oldsps=0.0; iters=0
   iterations: do
      iters=iters+1; kv = 0.0
!-------------------global stiffness matrix assembly---------------------------
      elements_2 : do iel = 1, nels
         num=g_num(:,iel)  ; coord=transpose(g_coord(:,num))
         call rigid_jointed(km,prop,gamma,etype,iel,coord)
         call beam_kp(kp,coord,axif(iel))
         km=km+kp; g=g_g(:,iel)
         call formkv(kv,km,g,neq)
      end do elements_2
```

```
      kcop=kv; call kvdet(kcop,neq,nband,det,ksc); disps=loads
!-------------------------equation solution -----------------------------------
      call banred(kv, neq); call bacsub(kv, disps)
!---------------------------check convergence----------------------------------
      call checon(disps,oldsps,tol,converged)
      elements_3: do iel=1,nels
         num=g_num(:,iel); coord=transpose(g_coord(:,num))
         g=g_g(:,iel); eld=disps(g)
         call rigid_jointed(km,prop,gamma,etype,iel,coord)
         call beam_kp(kp,coord,axif(iel))
         km=km+kp   ;              action=matmul(km,eld)
         call glob_to_loc(local,action,gamma(iel),coord)
         axif(iel)=axip(iel)+local(4)
      end do elements_3
      if(iters==limit .or. converged)exit iterations
   end do iterations
!-------------at convergence update displacements and axial forces-------------
   axip=axif; eldtot=eldtot+disps
   write(11,'(a,e12.4)')"Load factor    ",total_load
   write(11,'(a)')"The nodal displacements are:"
   do i=1,loaded_nodes
     write(11,'(i5,6e12.4)')no(i),eldtot(nf(:,no(i)))
   end do
   write(11,'(a,e12.4)')"The determinant is  ",det
   write(11,'(a,i5,a,/)')"Converged in  ",iters,"  iterations"
   if(iters == limit)exit
end do load_increments
end program p46
```

loading reaches critical levels. The onset of instability can be observed in various ways. The easiest approach for small structures is to monitor the determinant of the global stiffness matrix as the axial loads on the members are increased. Instability corresponds to a singular system in which the determinant is zero. In practice this is indicated by a change in sign of the computed determinant. A combination of loading has been reached which results in a state of neutral equilibrium and a non-unique displacement field. In the context of eigenvalue analysis, the "shape" of the structure at the point of instability is given by the eigenvector.

An alternative approach is to include a small disturbing force in the direction of the anticipated mode of buckling. At first the small force will cause virtually no displacement, but as the instability is approached, the displacement under the disturbing force will increase significantly.

Referring to the program, the following new real arrays are allocated:

AXIF	holds axial forces during iterations
AXIP	holds axial forces at convergence
KCOP	holds a copy of the global stiffness matrix stored as a vector

The modified element stiffness matrices are obtained by subtracting (assuming a compressive axial force) the "geometric" matrix KP which is formed by library subroutine BEAM_KP from the regular element stiffness matrix KM formed by subroutine RIGID_JOINTED. In order to compute the axial force the element "actions" which are in the global directions are converted in local coordinates for each element using library subroutine GLOB_TO_LOC.

As in the previous program for plastic analysis, proportional loading at the nodes is assumed. Within each load step iterations are performed until equilibrium is attained. In contrast to the previous program, however, the global

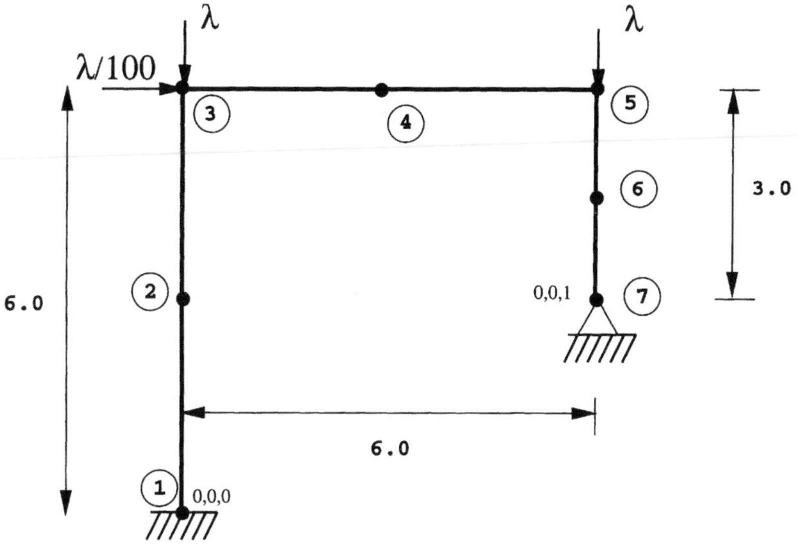

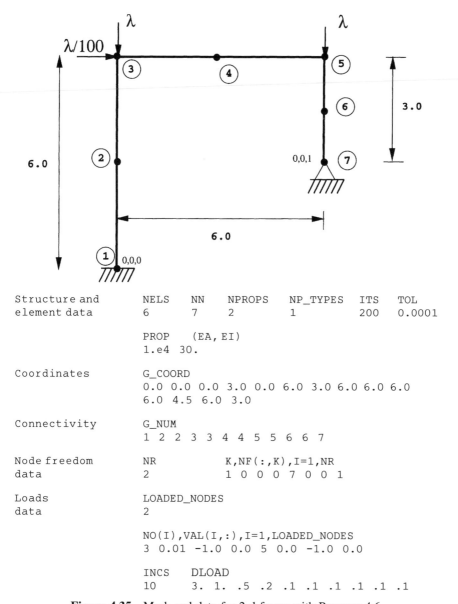

Structure and element data	NELS 6	NN 7	NPROPS 2	NP_TYPES 1	ITS 200	TOL 0.0001

```
PROP    (EA, EI)
1.e4  30.
```

Coordinates
```
G_COORD
0.0 0.0 0.0 3.0 0.0 6.0 3.0 6.0 6.0 6.0
6.0 4.5 6.0 3.0
```

Connectivity
```
G_NUM
1 2 2 3 3 4 4 5 5 6 6 7
```

Node freedom data
```
NR          K,NF(:,K),I=1,NR
2           1 0 0 0 7 0 0 1
```

Loads data
```
LOADED_NODES
2

NO(I),VAL(I,:),I=1,LOADED_NODES
3 0.01 -1.0 0.0 5 0.0 -1.0 0.0

INCS   DLOAD
10     3. 1. .5 .2 .1 .1 .1 .1 .1 .1
```

Figure 4.35 Mesh and data for 2-d frame with Program 4.6

stiffness matrix is modified at each iteration, thus coding describing its assembly will be found inside the iteration loop. At convergence following any load step, the determinant of the global matrix is computed using subroutine KVDET.

For every new load step, the initial element stiffness matrices are computed using the axial forces left over from the previous load step. The resulting displacement increments following solution of the global equilibrium equations enable the axial forces and hence the stiffness matrices to be updated.

```
Global coordinates
Node          1        0.0000E+00   0.0000E+00
Node          2        0.0000E+00   0.3000E+01
Node          3        0.0000E+00   0.6000E+01
Node          4        0.3000E+01   0.6000E+01
Node          5        0.6000E+01   0.6000E+01
Node          6        0.6000E+01   0.4500E+01
Node          7        0.6000E+01   0.3000E+01
Global node numbers
Element       1            1  2
Element       2            2  3
Element       3            3  4
Element       4            4  5
Element       5            5  6
Element       6            6  7
There are     16   equations and the half-bandwidth is      5

Load factor      0.3000E+01
The nodal displacements are:
    3  0.2520E-01 -0.1780E-02 -0.1860E-02
    5  0.2520E-01 -0.9099E-03 -0.4433E-02
The determinant is      0.5532E+38
Converged in        4  iterations

Load factor      0.4000E+01
The nodal displacements are:
    3  0.4118E-01 -0.2368E-02 -0.3062E-02
    5  0.4117E-01 -0.1216E-02 -0.7232E-02
The determinant is      0.2624E+38
Converged in        4  iterations

Load factor      0.4500E+01
The nodal displacements are:
    3  0.5578E-01 -0.2657E-02 -0.4172E-02
    5  0.5577E-01 -0.1372E-02 -0.9785E-02
The determinant is      0.1362E+38
Converged in        4  iterations
...........................................................

Load factor      0.5000E+01
The nodal displacements are:
    3  0.9811E-01 -0.2925E-02 -0.7422E-02
    5  0.9811E-01 -0.1538E-02 -0.1717E-01
The determinant is      0.2002E+37
Converged in        6  iterations

Load factor      0.5100E+01
The nodal displacements are:
    3  0.8921E-01 -0.2992E-02 -0.6722E-02
    5  0.8920E-01 -0.1564E-02 -0.1564E-01
The determinant is     -0.4361E+37
Converged in      200  iterations
```

Figure 4.36 Results from frame example with Program 4.6

The 2-d example shown in Figure 4.35 represents a single-bay frame with one pinned support and one fully fixed support. The critical value of the equal forces (λ) applied to the top of each column to cause instability is to be found. A small disturbing force is also applied in the "sway" direction to the top of the tallest column. From the results given in Figure 4.36, the iteration ceiling of 100 iterations and the change in sign of the determinant both occur at an axial load value of about $\lambda = 5.1$.

It is recommended that when using the "geometric" matrix approach for buckling problems, a minimum of two elements is used for each member.

4.1 CONCLUDING REMARKS

It has been shown how sample programs can be built up from the library of subroutines described in Chapter 3. A central feature of the programs has been their brevity. A typical main program has around 100 lines and is well suited for compilation on a small computer. In subsequent chapters, programs of greater complexity are introduced but the central theme of conciseness is adhered to.

4.2 EXAMPLES

1. Derive the stiffness matrix of a three-noded element (one at each end and one in the middle) suitable for tackling the 1-d rod equation

$$EA\frac{\mathrm{d}^2u}{\mathrm{d}x^2} + F = 0$$

   What advantage might this element have over the two-node element used in Program 4.0?
2. Derive the mass matrix of a three-noded 1-d rod element (one node at each end and one in the middle) of unit length, cross-sectional area and density, given the following shape functions:

$$N_1 = 2(x^2 - 1.5x + 0.5)$$
$$N_2 = -4(x^2 - x)$$
$$N_3 = 2(x^2 - 0.5x)$$

3. A simply supported beam ($L = 1, EI = 1$) supports a unit point transverse load ($Q = 1$) at its mid-span. The beam is also subjected to a compressive axial force P which will reduce the bending stiffness of the beam. Using two ordinary beam elements of equal length, assemble the global matrix equations for this system but do not attempt to solve them. Take full

account of symmetries in the expected deformed shape of the beam to reduce the number of equations.

4. A cantilever $(L = 1, EI = 1)$ rests on an elastic foundation of stiffness $k = 10$. A transverse point load $P = 1$ is applied at the cantilever tip. Using a single finite element, estimate the transverse deflection under the load.
5. The governing equation for axial displacements u of a 1-d rod embedded in an elastic medium is

$$EA\frac{\mathrm{d}^2u}{\mathrm{d}x^2} + ku = F$$

where EA is the axial stiffness of the rod, k is the stiffness of the surrounding elastic material and F is a uniformly distributed axial load.

Use a two element discretisation to estimate the axial displacement at the mid point and tip of the rod shown in Figure 4.37.
6. A propped cantilever is subjected to the loads and displacements indicated in Figure 4.38. Using two finite elements estimate the moment at the centre of the beam.
7. Given the differential equation

$$100\frac{\mathrm{d}^2u}{\mathrm{d}x^2} + 5 = 0$$

with the boundary conditions $u(0) = 0.05$ and $u(1) = 0$ estimate $u(0.25)$, $u(0.5)$ and $u(0.75)$ using a four element distretisation in the range $0 \leq x \leq 1$.

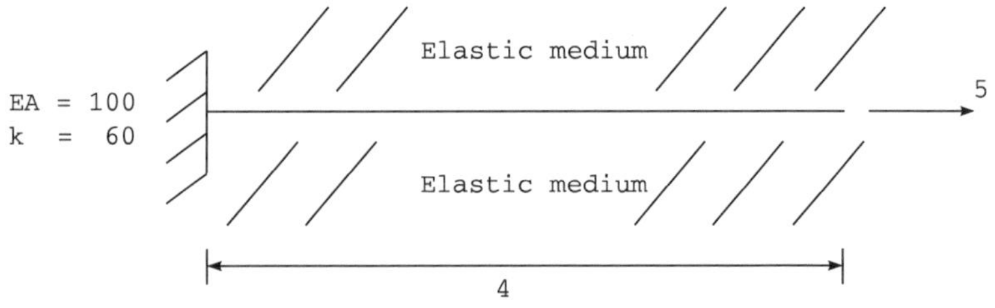

Figure 4.37

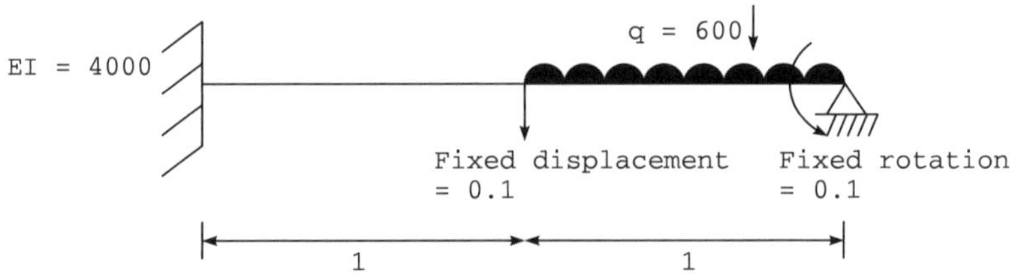

Figure 4.38

8. Use beam elements to form the global stiffness matrix for the cantilever shown in Figure 4.39. and hence compute the translation and rotation at the tip and the rotation at the support.
9. Using two beam–rod finite elements, calculate the deflections and rotations of point A in the frame of Figure 4.40.
10. A cantilever shown in Figure 4.41 is subjected to the loads and displacements indicated. Assemble the global stiffness relationship and solve for the unknown displacements and rotations.
11. A single railway track resting on a ballast subgrade can be approximated as a beam of length L of stiffness EI resting on an elastic foundation

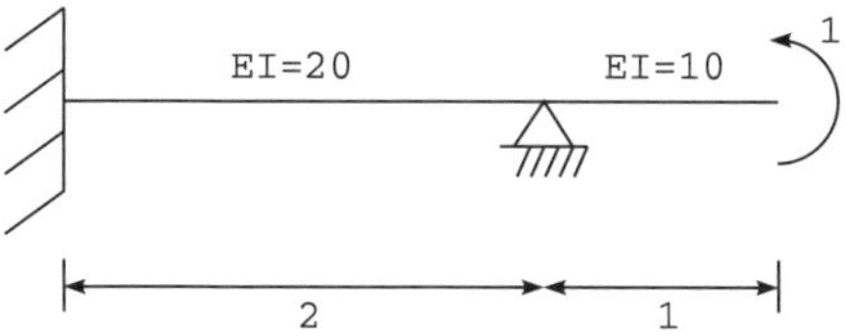

Figure 4.39

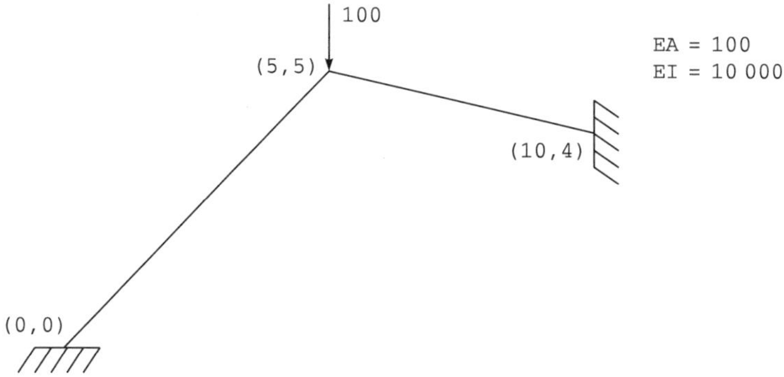

Figure 4.40

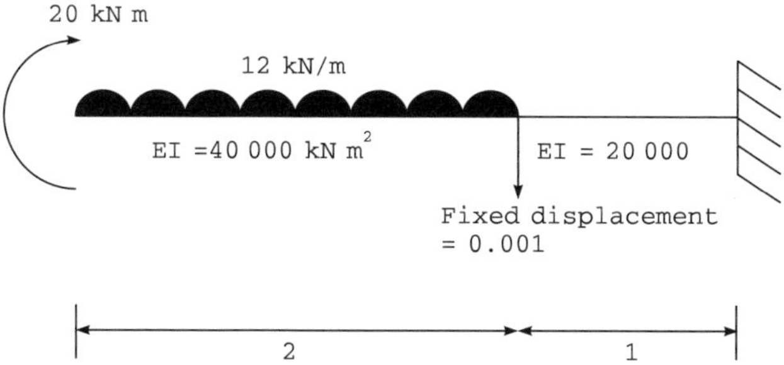

Figure 4.41

of stiffness k. If a single concentrated load P acts on a rail between the ties which can be assumed to be rigid supports, use a single finite element to estimate the relationship between EI, k, P and L so that the rail deflection can be limited to 5 units. (*Hint*: Consider the cases of both simply supported and fully clamped end conditions since reality will lie somewhere in between. In the fully clamped case you will need to consider just half the problem and account for symmetry.)

12. A simply supported beam element of length L, stiffness EI, resting on an elastic foundation of stiffness k supports a uniformly distributed load of q. Estimate the end rotations using a single finite element.
13. A laterally loaded pile is to be modelled as a beam on an elastic foundation system as shown in Figure 4.42. Use a single beam element to estimate the lateral deflection of the pile cap under a unit load. (*Hint*: You may assume the base of the pile is clamped.)
14. Use a single beam element to compute the lowest buckling load of the following cases (you may assume the flexural stiffness and length both equal unity):
    (a) A pin-ended column.
    (b) A column clamped at one end and free at the other. In this case also estimate the ratio of tip rotation to translation when the column buckles.
    (c) A column clamped at both ends.
    (d) A column clamped at both ends with a support preventing deflection at the mid-point.
15. Buckling of a slender beam of stiffness EI resting on a uniform elastic foundation of stiffness k is governed by the equation:

$$EI\frac{\partial^4 w}{\partial x^4} + P\frac{\partial^2 w}{\partial x^2} + kw = 0$$

where w is the transverse deflection of the beam and P the axial load.

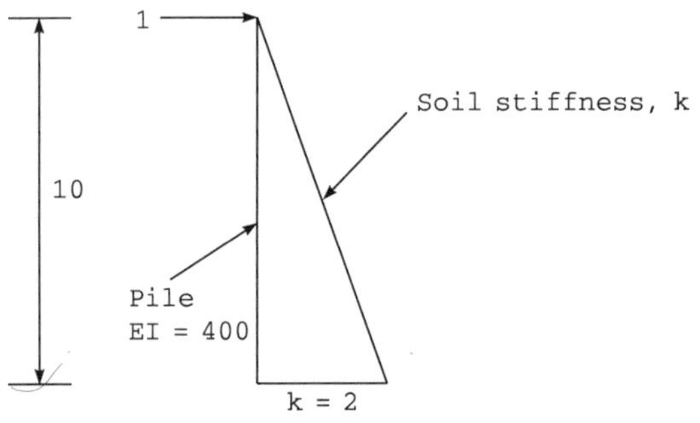

Figure 4.36

Using a single finite element, compute two buckling loads for such a beam of length L, simply supported at its ends. Show that depending on the relationship among k, EI and L, either of these two loads could be the critical one:

$$\text{if } \theta_1 = -\theta_2, \qquad P_{\text{crit}} = \frac{12EI}{L^2} + \frac{kL^2}{10}$$

$$\text{if } \theta_1 = \theta_2 \qquad P_{\text{crit}} = \frac{60EI}{L^2} + \frac{kL^2}{42}$$

and that the transition between these two conditions occurs when

$$k = \frac{630EI}{L^4}$$

4.3 REFERENCES

Griffiths, D.V. (1988) An iterative method for plastic analysis of frames. *Comput. Struct.*, **30**(6), 1347–1354.

Hetenyi, M. (1946) *Beams on Elastic Foundations*. University of Michigan Press, Ann Arbor.

Horne, M.R. (1971) *Plastic Theory of Structures*. Nelson, London.

Przemieniecki, J.S. (1968) *Theory of Matrix Structural Analysis*. McGraw-Hill, New York.

5 Static Equilibrium of Linear Elastic Solids

5.0 INTRODUCTION

This chapter describes 12 programs which can be used to solve equilibrium problems in small strain solid elasticity. The programs differ only slightly from each other and, following the method adopted in Chapter 4, the first is described in some detail with changes gradually introduced into the later programs. The first program deals with a plane stress analysis using constant-strain triangular elements. Later programs incorporate plane strain, axisymmetric strain, non-axisymmetric strain of axisymmetric bodies and 3-d strain conditions. Various types of 2-d and 3-d finite elements are also considered. The majority of examples in this chapter consider problems involving a regular (usually rectangular or cuboidal) geometry. This has been done to simplify the presentation and minimise the volume of data required. The simple geometries enable the nodal coordinates and numbers to be generated automatically. This is done by a series of geometry routines (`GEOMETRY_3TX`, `GEOMETRY_15TYV`, `GEOMETRY_4QY`, etc.) which depend on the element type and numbering system adopted (see Appendix 3). A general program capable of analysing geometrically more complex problems is also included, in which case the user must replace the simple geometry subroutines by some other means of generating nodal coordinates and numbers. A pre-processing mesh generator program affording graphical checks will usually be essential here.

The next stage in all programs is to determine the element "steering vectors" `G`. These can be found from `NUM` and `NF` using the library subroutine `NUM_TO_G`.

In most programs, numerical integration of the element stiffness matrices is employed, but an example using an "analytically" integrated stiffness is also given. For 3-d problems (which will usually be large), the mesh-free p.c.g. technique is provided as well as the conventional one in which stiffnesses are assembled. Also for 3-d problems, vectorisation and parallelisation issues are illustrated by typical programs.

Post-processing of results using graphical software is essential for analyses of any complexity. Plots of deformed meshes and contours of strains, stresses, etc. enable the overall properties of the solution to be taken in at a glance. Machine dependence of such programs, however, has precluded their use in the present book, although typical plots are reproduced to demonstrate the power of this technique.

PROGRAM 5.0: PLANE STRESS ANALYSIS USING THREE-NODE TRIANGLES

```
 program p50
!-----------------------------------------------------------------------
!      program 5.0 plane stress of an elastic
!      solid using uniform 3-node triangular
!      elements numbered in the x direction
!-----------------------------------------------------------------------
 use new_library   ;  use  geometry_lib  ;        implicit none
 integer::nels,nce,neq,nband,nn,nr,nip,nodof=2,nod=3,nst=3,ndof,         &
          loaded_nodes,i,k,iel,ndim=2
 real:: e,v,det,aa,bb    ; character(len=15):: element = 'triangle'
!--------------------- dynamic arrays----------------------------
 real    ,allocatable :: kv(:),loads(:),points(:,:),dee(:,:),coord(:,:),   &
                         jac(:,:), der(:,:),deriv(:,:), weights(:),       &
                         bee(:,:),km(:,:),eld(:),sigma(:),g_coord(:,:)
integer, allocatable :: nf(:,:), g(:) , num(:)  , g_num(:,:) , g_g(:,:)

!----------------input and initialisation---------------------
  open (10,file='p50.dat',status='old',    action='read')
  open (11,file='p50.res',status='replace',action='write')

  read (10,*) nels,nce,nn,nip,aa,bb,e,v
  ndof=nod*nodof
  allocate ( nf(nodof,nn), points(nip,ndim),g(ndof), g_coord(ndim,nn),      &
            dee(nst,nst),coord(nod,ndim),jac(ndim,ndim),weights(nip),       &
            der(ndim,nod), deriv(ndim,nod), bee(nst,ndof), km(ndof,ndof),   &
            eld(ndof),sigma(nst),num(nod),g_num(nod,nels),g_g(ndof,nels))

    nf=1; read(10,*) nr;if(nr>0)read(10,*)(k,nf(:,k),i=1,nr)
    call formnf (nf);neq=maxval(nf)
    nband = 0
! this is a plane stress analysis
  dee=.0; dee(1,1)=e/(1.-v*v);dee(2,2)=dee(1,1);dee(3,3)=.5*e/(1.+v)
  dee(1,2)=v*dee(1,1);dee(2,1)=dee(1,2); call sample(element,points,weights)
!--------loop the elements to find bandwidth and neq-------------------
  elements_1: do iel = 1 , nels
              call geometry_3tx(iel,nce,aa,bb,coord,num);call num_to_g(num,nf,g)
              g_num(:,iel)=num;g_coord(:,num)=transpose(coord);g_g(:,iel) = g
              if(nband<bandwidth(g))nband=bandwidth(g)
  end do elements_1
    write(11,'(a)') "Global coordinates "
    do k=1,nn;write(11,'(a,i5,a,2e12.4)')"Node",k,"       ",g_coord(:,k);end do
    write(11,'(a)') "Global node numbers "
    do k = 1 , nels; write(11,'(a,i5,a,3i5)')                                &
                              "Element ",k,"        ",g_num(:,k); end do
    write(11,'(2(a,i5))')                                                    &
             "There are ",neq,"  equations and the half-bandwidth is ", nband
             allocate(kv(neq*(nband+1)),loads(0:neq)); kv=.0
!------- element stiffness integration and assembly--------------------

 elements_2: do iel = 1 , nels
             num = g_num(:, iel);     g = g_g( : , iel )
             coord = transpose(g_coord(:, num)) ;       km=0.0
          gauss_pts_1: do i = 1 , nip
               call shape_der(der,points,i) ; jac = matmul(der,coord)
               det = determinant(jac); call invert(jac)
               deriv = matmul(jac,der) ; call beemat (bee,deriv)
             km = km + matmul(matmul(transpose(bee),dee),bee) *det* weights(i)
          end do gauss_pts_1
   call formkv (kv,km,g,neq)
 end do elements_2

 loads=.0; read(10,*)loaded_nodes,(k,loads(nf(:,k)),i=1,loaded_nodes)
!------------------------equation solution------------------------------
    call banred(kv,neq) ;call bacsub(kv,loads)
```

```
   write(11,'(a)') "The nodal displacements Are :"
   write(11,'(a)')  "Node          Displacement"
   do k=1,nn; write(11,'(i5,a,2e12.4)')   k,"    ",loads(nf(:,k)); end do
!-------------------recover stresses at centroidal gauss-point------------
   nip = 1; deallocate(points,weights);allocate(points(nip,ndim),weights(nip))
   call sample ( element , points , weights)
       write(11,'(a)') "The central point stresses are :"
 elements_3:do iel = 1 , nels
        write(11,'(a,i5)') "Element No.  ",iel
   num = g_num(: , iel);     coord =transpose( g_coord(: ,num) )
   g = g_g( : ,iel )     ;      eld=loads(g)
   gauss_pts_2: do i = 1 , nip
      call shape_der (der,points,i); jac= matmul(der,coord)
      call invert(jac) ;     deriv= matmul(jac,der)
      call beemat(bee,deriv);  sigma = matmul (dee,matmul(bee,eld))
      write(11,'(a,i5)') "Point  ",i   ;  write(11,'(3e12.4)') sigma
   end do gauss_pts_2
 end do elements_3
end program p50
```

The structure chart in Figure 5.1 illustrates the sequence of calculations for this program. In fact the same chart is essentially valid for all programs in this chapter which use an assembly strategy. The three-node (constant-strain) triangle is not a very good element and is not used much in practice, except when meshes are automatically adapted to improve accuracy (e.g. Hicks and Mar, 1996). In view of its simplicity, however, the first program in this chapter is devoted to it.

As shown in the structure chart, the element stiffness matrices are formed numerically following the procedures described in Chapter 3, equations (3.17), (3.36) and (3.37). For such a simple element only one integrating point is required at each element's centroid.

Figure 5.2 shows a square block of elastic material of unit side length and unit thickness subjected to an equivalent vertical stress of $1\,\mathrm{kN/m^2}$. The boundary conditions imply that two planes of symmetry exist and that only one-quarter of the problem is being considered. The freedom numbers at each node represent possible displacements in the x and y directions respectively. Although this information about freedoms is included for completeness, the programs organise the information by nodes and the user need not be aware of freedom numbers at all. Figure 5.3 shows the nodal numbering system adopted for this example and although it does not matter in which direction nodes are numbered for a case such as this, the most efficient numbering system for general rectangular shapes will count in the direction with the least nodes. The particular geometry assumed in this case identifies elements by counters `IP` and `IQ` running from 1 to `NCE` and `NXE` respectively. The simple geometry generated by routine `GEOMETRY_3TX` assumes that all elements are right-angled, congruent and formed by diagonal lines drawn from the bottom left-hand corner to the top right-hand corner of rectangles. The rectangles are assumed to be of constant size with width `AA` (x direction) and depth `BB` (y direction). Figure 5.4 shows the order of node and freedom numbering at the element level. Node (1) can be any corner, but subsequent corners and freedoms must follow in a clockwise sense. Thus, the top left element (`IP=1, IQ=1`) in Figure 5.2 has a steering vector

$$\mathrm{G} = [0\ 1\ 2\ 3\ 0\ 6]^{\mathrm{T}}$$

and its neighbour (`IP=1, IQ=2`) has a steering vector

$$\mathrm{G} = [7\ 8\ 0\ 6\ 2\ 3]^{\mathrm{T}}$$

It is expected that, where necessary, users will replace the simple geometry routine `GEOMETRY_3TX` by more sophisticated versions. It need only be ensured that the coordinates and node numbers are generated consistently.

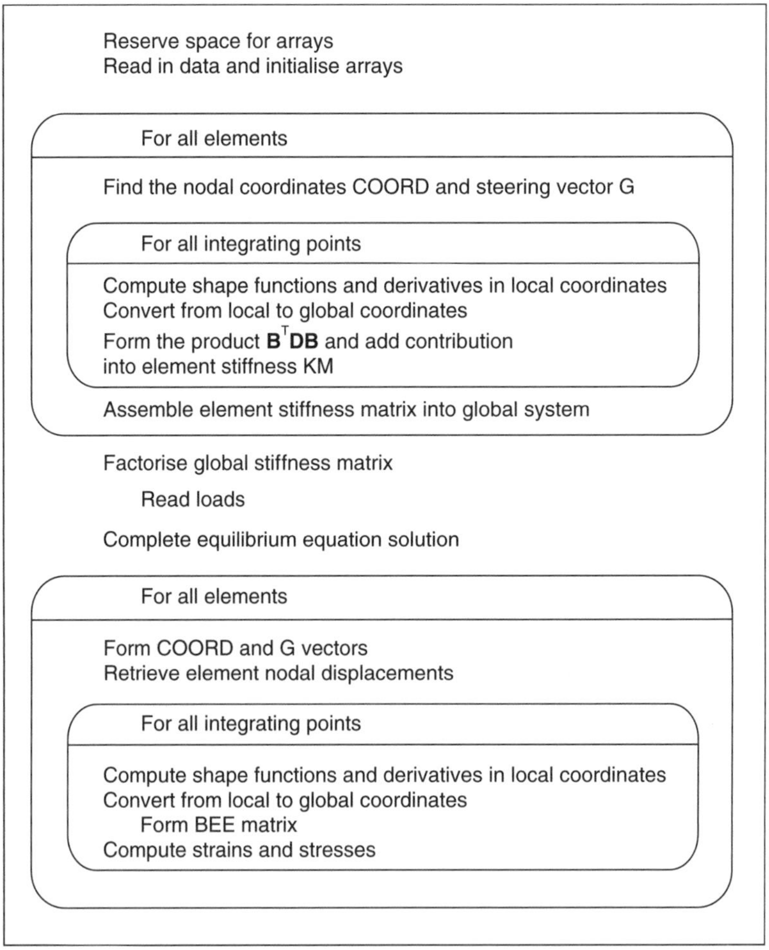

Figure 5.1 Structure chart for all Chapter 5 programs involving assembly

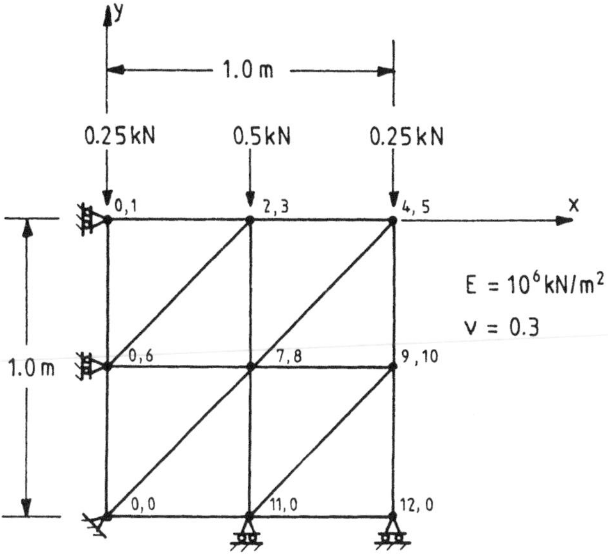


```
Structure and     NELS   NCE   NN     NIP
element data      8      2     9      1

                  AA     BB    E      V
                  .5     .5    1.E6   .3

Node freedom      NR
data              5
                  K,NF(:,K),I=1,NR
                  1 0 1 4 0 1 7 0 0 8 1 0 9 1 0

Loads             LOADED_NODES
data              3
                  (K,LOADS(NF(:,K)),I=1,LOADED_NODES)
                  1 .0 -.25 2 .0 -.5 3 .0 -.25
```

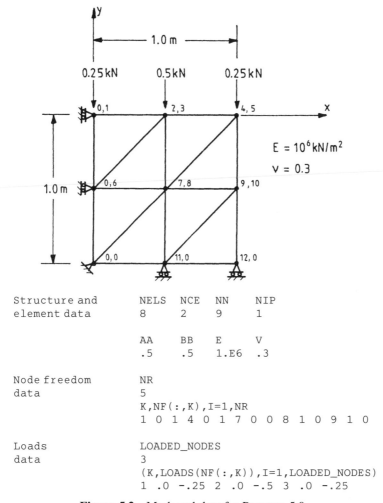

Figure 5.2 Mesh and data for Program 5.0

Returning to the main program, with the exception of simple integer counters `I,K` and `IEL`, the meanings of the variable names with reference to the mesh of Figure 5.2 are given as follows:
Scalar integers:

`NELS`	total number of elements (8)
`NCE`	number of element columns in x direction (2)
`NEQ`	number of degrees of freedom in the mesh (12)
`NBAND`	half-bandwidth of mesh (6)
`NN`	number of nodes in the mesh (9)

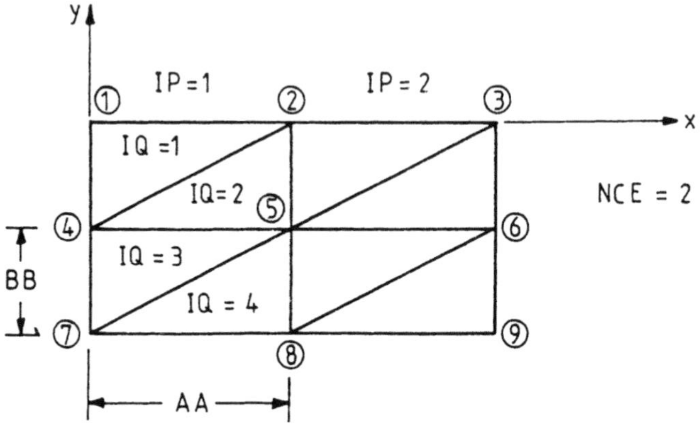

Figure 5.3 Global node and element numbering for mesh of three-node triangle

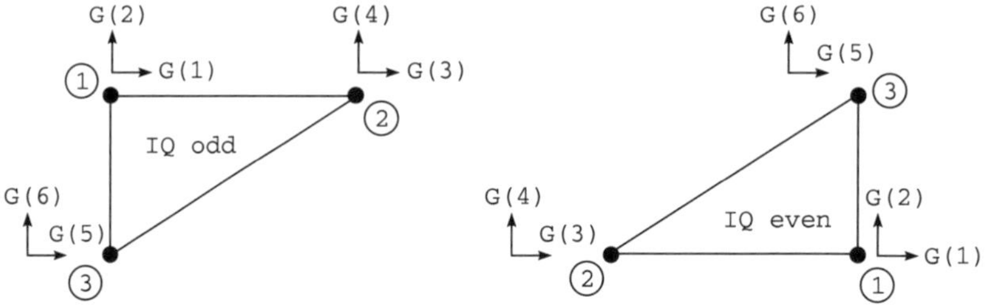

Figure 5.4 Local node and freedom numbering for three-node triangle

`NR`	number of restrained nodes in the mesh (5)
`NIP`	number of integrating points (1)
`LOADED_NODES`	number of loaded nodes (3)
`NDOF`	number of degrees of freedom per element (6)
`NST`	size of stress–strain matrix (3)
`NOD`	number of nodes per element (3)
`NODOF`	number of freedoms per node (2)
`NDIM`	dimensions of problem (2)

Scalar reals:

`E`	Young's modulus (10^6)
`V`	Poisson's ratio (0.3)
`DET`	determinant of Jacobian matrix
`AA`	dimension of elements in x direction (0.5)
`BB`	dimension of elements in y direction (0.5)

Space is reserved for allocatable arrays:

Reals:

DEE	stress–strain matrix
POINTS	quadrature sampling points
COORD	element nodal coordinates
JAC	Jacobian matrix
DER	derivatives of shape functions in local coordinates
DERIV	derivatives of shape function in global coordinates
BEE	strain–displacement matrix
KM	element stiffness matrix
ELD	element displacement vector
SIGMA	element stress vector
FUN	element shape functions in local coordinates
G_COORD	global version of COORD
KV	global stiffness matrix
LOADS	global load (displacement) vector
WEIGHTS	quadrature weights

Integers:

G	element steering vector
NUM	the node number vector
G_NUM	global version of NUM
G_G	global version of G
NF	nodal freedom array

After declaration of arrays whose dimensions are now known the program enters the "input and initialisation" stage. Data concerning the mesh and its properties is now presented together with the nodal freedom data as given in Figure 5.2. This enables the total number of equations in the problem, NEQ, to be calculated. A call to the library subroutine SAMPLE forms the quadrature sampling points and weights.

Since the vast majority of 2-d problems in geomechanics involve plane strain (see Program 5.1 onwards) special coding is inserted for the plane stress DEE matrix formation.

Next the elements can be looped to store "global" arrays containing the element node numbers (G_NUM), the element nodal coordinates (G_COORD) and the element steering vectors (G_G). At the same time the half-bandwidth of the global stiffness matrix, NBAND, can be calculated.

In larger problems, a bandwidth optimiser (e.g. Cuthill and McKee, 1969) will improve efficiency by reordering the nodes.

In the next section of the program, labelled "element stiffness integration and assembly", the elements are inspected one by one. The nodal coordinates COORD and the steering vector G for each element are retrieved.

After the stiffness matrix KM has been nulled, the integration loop is entered. The local coordinates of each integrating point (only one in this case) are extracted from POINTS, and the derivatives of the shape functions with respect to those coordinates (DER) are provided for the three-node element by the library subroutine SHAPE_DER. The conversion of these derivatives to the

```
Global coordinates
Node    1          0.0000E+00  0.0000E+00
Node    2          0.5000E00   0.0000E+00
Node    3          0.1000E+01  0.0000E+00
Node    4          0.0000E+00 -0.5000E00
Node    5          0.5000E00  -0.5000E00
Node    6          0.1000E+01 -0.5000E00
Node    7          0.0000E+00 -0.1000E+01
Node    8          0.5000E00  -0.1000E+01
Node    9          0.1000E+01 -0.1000E+01
Global node numbers
Element     1          1    2    4
Element     2          5    4    2
Element     3          2    3    5
Element     4          6    5    3
Element     5          4    5    7
Element     6          8    7    5
Element     7          5    6    8
Element     8          9    8    6
There are     12  equations and the half-bandwidth is      6
The nodal displacements are :
Node          Displacement
    1      0.0000E+00 -0.1000E-05
    2      0.1500E-06 -0.1000E-05
    3      0.3000E-06 -0.1000E-05
    4      0.0000E+00 -0.5000E-06
    5      0.1500E-06 -0.5000E-06
    6      0.3000E-06 -0.5000E-06
    7      0.0000E+00  0.0000E+00
    8      0.1500E-06  0.0000E+00
    9      0.3000E-06  0.0000E+00
The central point stresses are :
Element No.        1
Point       1
 -0.1223E-06 -0.1000E+01  0.0000E+00
Element No.        2
Point       1
 -0.8483E-07 -0.1000E+01 -0.4373E-07
Element No.        3
Point       1
 -0.8483E-07 -0.1000E+01  0.4373E-07
Element No.        4
Point       1
 -0.3095E-08 -0.1000E+01  0.0000E+00
Element No.        5
Point       1
 -0.4735E-07 -0.1000E+01 -0.4373E-07
Element No.        6
Point       1
  0.4582E-08 -0.1000E+01 -0.3279E-07
Element No.        7
Point       1
  0.3438E-07 -0.1000E+01  0.1093E-07
Element No.        8
Point       1
  0.4206E-07 -0.1000E+01 -0.2186E-07
```

Figure 5.5 Results from Program 5.0

global system (DERIV) requires a sequence of subroutine calls described by equations (3.45). The BEE matrix is then formed by the subroutine BEEMAT.

The next statement calculates the product $\mathbf{B}^{\mathrm{T}}\mathbf{DB}$. The contribution from each integration point is scaled by the weighting factor from WEIGHTS and added into the element stiffness matrix KM. Eventually, the completed KM is assembled into the global stiffness KV using the library subroutine FORMKV which was used extensively in Chapter 4.

When all element stiffnesses have been assembled, the program enters the "equation solution" stage. The global matrix KV is factorised by the subroutine BANRED and the required loads vector is read as data. The final stage of the equation solution is performed by the subroutine BACSUB and the resulting displacement vector (still called LOADS) is printed.

If required, the stresses and strains within the elements can now be computed. Clearly, these could be found anywhere in the elements, but it is convenient and often more accurate to employ the integrating points that were used in the stiffness formulation. In this example only one integrating point (the "centroid") was employed for each element, so it is at these locations that stresses and strains will be calculated. Each element is scanned once more and its nodal displacement (ELD) retrieved from the global displacement vector. The BEE matrix for each integrating point is recalculated and the product of BEE and ELD yields the strains from equation (3.46). Multiplication by the stress–strain matrix DEE gives the stresses SIGMA which are printed.

The computed results for the example shown in Figure 5.2 are given in Figure 5.5. For this simple case the results are seen to be "exact". The vertical displacements under the loads (nodes 1, 2 and 3) all equal 10^{-6} and the Poisson's ratio effect has caused horizontal movement at nodes 3, 6 and 9 equal to 0.3×10^{-6} m. The stress components, printed in the order $\sigma_x, \sigma_y, \tau_{xy}$ are constant and give $\sigma_y = 1\,\mathrm{kN/m^2}$ with $\sigma_x = \tau_{xy} = 0$.

This simple element often does not perform well for problems involving less uniform loading distributions, so the remaining sections of this chapter are devoted to higher-order elements. The program structure, however, will remain essentially unchanged.

PROGRAM 5.1: PLANE STRAIN ANALYSIS USING 15-NODE TRIANGLES

The shape function library subroutines SHAPE_FUN and SHAPE_DER include 6-node triangular elements but the next of the triangular element family to be considered here is the 15-noded "cubic strain" triangle. Plane strain conditions are introduced at this stage, and this is achieved by the library subroutine DEEMAT. Although the triangles are still formed by drawing diagonals of rectangles, a variation is introduced in that the rectangles can vary in size. Thus,

```
program p51
!-----------------------------------------------------------------------
!      program 5.1 plane strain of a rectangular elastic
!      solid using variable-sized 15-node triangular
!      elements numbered in the y direction
!-----------------------------------------------------------------------
 use new_library  ;  use geometry_lib    ;    implicit none
 integer::nels,nce,nye,neq,nband,nn,nr,nip,nodof=2,nod=15,nst=3,ndof,       &
          loaded_nodes,i,k,iel,ndim=2 , nde
 real:: e,v,det    ;      character(len=15):: element = 'triangle'
!--------------------- dynamic arrays-----------------------------------
 real    ,allocatable :: kb(:,:),loads(:),points(:,:),dee(:,:),coord(:,:),   &
                         jac(:,:), der(:,:),deriv(:,:), weights(:),          &
                         bee(:,:),km(:,:),eld(:),sigma(:),g_coord(:,:),      &
                         width(:), depth(:)
 integer, allocatable :: nf(:,:), g(:) , num(:)  , g_num(:,:)  , g_g(: , :)
!----------------input and initialisation--------------------------------
  open (10,file='p51.dat',status=    'old',action='read')
  open (11,file='p51.res',status='replace',action='write')
  read (10,*) nels,nce,nye,nn,nip,e,v
  ndof=nod*nodof     ;  nde = (nye + 2) / 2
  allocate ( nf(nodof,nn), points(nip,ndim),g(ndof), g_coord(ndim,nn),       &
            dee(nst,nst),coord(nod,ndim),jac(ndim,ndim),weights(nip),        &
            der(ndim,nod), deriv(ndim,nod), bee(nst,ndof), km(ndof,ndof),    &
            eld(ndof),sigma(nst),num(nod),g_num(nod,nels),width(nce+1),      &
            depth(nde),g_g(ndof , nels))
  read(10,*) width , depth
  nf=1; read(10,*) nr ; if(nr>0)read(10,*)(k,nf(:,k),i=1,nr)
        call formnf (nf);neq=maxval(nf)
  call deemat (dee,e,v); call sample(element,points,weights)
!----------------loop the elements to find bandwidth and neq-------------------
 nband = 0
  elements_1: do iel = 1 , nels
              call geometry_15tyv(iel,nye,width,depth,coord,num)
              call num_to_g(num,nf,g);  g_num(:,iel) = num
              g_coord(:,num)=transpose( coord );g_g(:,iel)=g
              if(nband<bandwidth(g))nband=bandwidth(g)
  end do elements_1
    write(11,'(a)') "Global coordinates "
    do k=1,nn;write(11,'(a,i5,a,2e12.4)')"Node",k,"       ",g_coord(:,k);end do
    write(11,'(a)') "Global node numbers "
    do k = 1 , nels; write(11,'(a,i5,a,15i3)')
                              "Element ",k,"       ",g_num(:,k); end do
    write(11,'(2(a,i5))')
             "There are ",neq,"  equations and the half-bandwidth is", nband
             allocate(kb(neq,nband+1),loads(0:neq)); kb=.0
!------------- element stiffness integration and assembly--------------------
 elements_2: do iel = 1 , nels
             num = g_num(:,iel);    g = g_g( : , iel )
             coord = transpose(g_coord(: ,num))  ;        km=0.0
          gauss_points_1: do i = 1 , nip
               call shape_der(der,points,i) ; jac = matmul(der,coord)
               det = determinant(jac); call invert(jac)
               deriv = matmul(jac,der) ; call beemat (bee,deriv)
             km = km + matmul(matmul(transpose(bee),dee),bee) *det* weights(i)
          end do gauss_points_1
   call formkb (kb,km,g)
 end do elements_2
 loads=.0; read(10,*)loaded_nodes,(k,loads(nf(:,k)),i=1,loaded_nodes)
!------------------------equation solution-------------------------------
    call cholin(kb) ;call chobac(kb,loads)
    write(11,'(a)') "The nodal displacements are :"
    write(11,'(a)') "Node          Displacement"
    do k=1,nn; write(11,'(i5,a,2e12.4)') k,"   ",loads(nf(:,k)); end do
```

```
!-------------------recover stresses at centroidal point------------------
  nip = 1; deallocate(points,weights);allocate(points(nip,ndim),weights(nip))
    call sample( element ,points , weights )
        write(11,'(a)') "The central  point stresses are :"
 elements_3:do iel = 1 , nels
          write(11,'(a,i5)') "Element no.  ",iel
      num = g_num(:, iel); coord = transpose(g_coord(:,num))
      g = g_g( : ,iel); eld=loads(g)
    gauss_points_2 : do i = 1 , nip
       call shape_der (der,points,i); jac= matmul(der,coord)
       call invert(jac) ;    deriv= matmul(jac,der)
       call beemat(bee,deriv) ; sigma = matmul (dee,matmul(bee,eld))
       write(11,'(a,i5)') "Point  ",i   ;  write(11,'(3e12.4)') sigma
    end do gauss_points_2
 end do elements_3
end program p51
```

instead of constants AA and BB, the widths and depths are read into the arrays WIDTH and DEPTH.

Nodal coordinates COORD and the nodal vector NUM are generated by the "variable" geometry subroutine GEOMETRY_15TYV which assumes nodes are numbered in the "depth"(y) direction. The number of "columns" of elements, is NCE, and there are NYE elements in total in the y direction. The shape function derivatives at each integrating point are provided for this element by the library subroutine SHAPE_DER as always. To form the stiffness matrix for this element exactly in plane strain, NIP is set to 12.

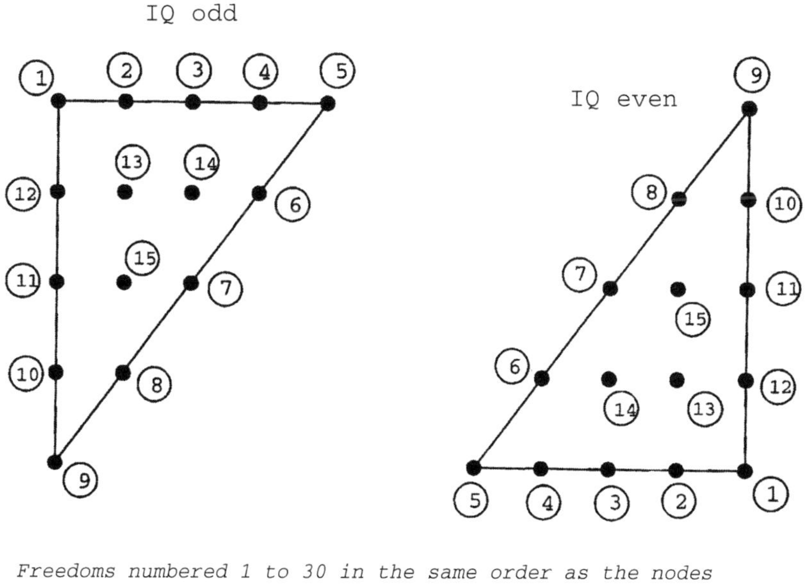

Figure 5.6 Local node numbering for 15-node triangle

The node numbering at the element level follows the pattern of the lower-order triangular element described previously and is given in Figure 5.6. The program is readily derived from Program 5.0 with the difference that Choleski's method is used for equation solution (`CHOLIN` and `CHOBAC` replacing `BANRED` and `BACSUB`). This necessitates the replacement of `FORMKV` for forming the global stiffness by `FORMKB`.

The example and data in Figure 5.7 show half of a flexible footing resting on a uniform elastic layer. The nodal loads imply a uniform stress of 1 kN/m^2 (see

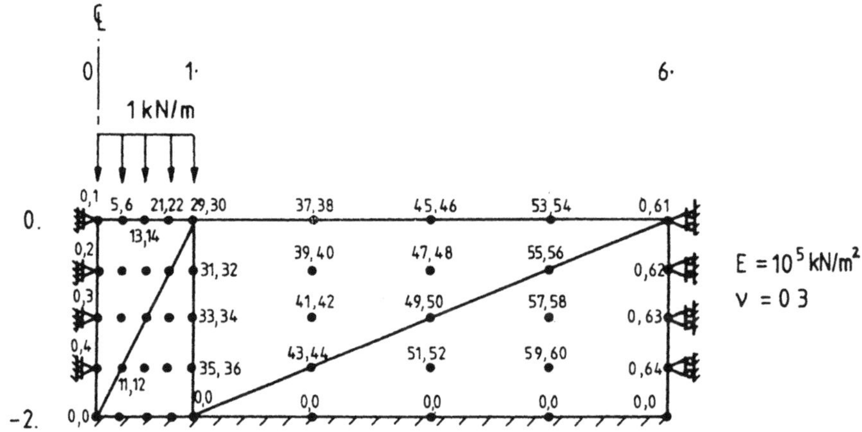

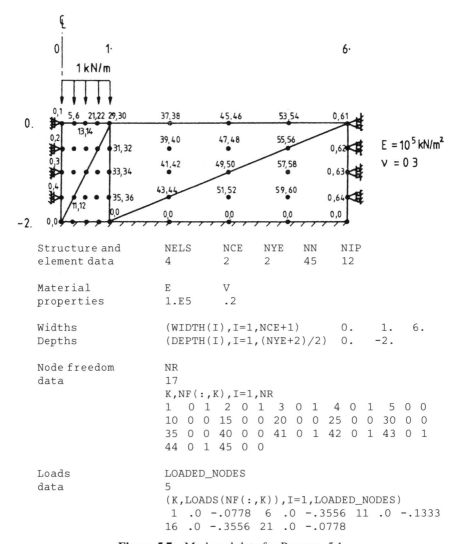

```
Structure and        NELS      NCE     NYE     NN     NIP
element data         4         2       2       45     12

Material             E         V
properties           1.E5      .2

Widths               (WIDTH(I),I=1,NCE+1)        0.     1.     6.
Depths               (DEPTH(I),I=1,(NYE+2)/2)    0.    -2.

Node freedom         NR
data                 17
                     K,NF(:,K),I=1,NR
                     1  0 1   2 0 1   3 0 1   4 0 1   5 0 0
                     10 0 0  15 0 0  20 0 0  25 0 0  30 0 0
                     35 0 0  40 0 0  41 0 1  42 0 1  43 0 1
                     44 0 1  45 0 0

Loads                LOADED_NODES
data                 5
                     (K,LOADS(NF(:,K)),I=1,LOADED_NODES)
                      1 .0 -.0778   6 .0 -.3556  11 .0 -.1333
                     16 .0 -.3556  21 .0 -.0778
```

Figure 5.7 Mesh and data for Program 5.1

Appendix 1). It is seen from the data that due to the number of nodes associated with each element, the bandwidths can become quite large. The high order of the interpolating polynomials, however, suggests that less elements would be required for a typical boundary value problem.

```
Global coordinates
Node    1          0.0000E+00  0.0000E+00
Node    2          0.0000E+00 -0.5000E00
Node    3          0.0000E+00 -0.1000E+01
Node    4          0.0000E+00 -0.1500E+01
Node    5          0.0000E+00 -0.2000E+01
Node    6          0.2500E00   0.0000E+00
Node    7          0.2500E00  -0.5000E00
Node    8          0.2500E00  -0.1000E+01
Node    9          0.2500E00  -0.1500E+01
Node   10          0.2500E00  -0.2000E+01
.......................................................

Node   40          0.4750E+01 -0.2000E+01
Node   41          0.6000E+01  0.0000E+00
Node   42          0.6000E+01 -0.5000E00
Node   43          0.6000E+01 -0.1000E+01
Node   44          0.6000E+01 -0.1500E+01
Node   45          0.6000E+01 -0.2000E+01
Global node numbers
Element     1           1  6 11 16 21 17 13  9  5  4  3  2  7 12  8
Element     2          25 20 15 10  5  9 13 17 21 22 23 24 19 14 18
Element     3          21 26 31 36 41 37 33 29 25 24 23 22 27 32 28
Element     4          45 40 35 30 25 29 33 37 41 42 43 44 39 34 38
There are     64  equations and the half-bandwidth is    32
The nodal displacements are :
Node          Displacement
   1      0.0000E+00 -0.1591E-04
   2      0.0000E+00 -0.1158E-04
   3      0.0000E+00 -0.7226E-05
   4      0.0000E+00 -0.3354E-05
   5      0.0000E+00  0.0000E+00
   6     -0.9321E-06 -0.1559E-04
   7      0.1493E-06 -0.1128E-04
   8      0.4540E-06 -0.7019E-05
   9      0.3347E-06 -0.3255E-05
  10      0.0000E+00  0.0000E+00
..........................................................

  40      0.0000E+00  0.0000E+00
  41      0.0000E+00  0.4022E-07
  42      0.0000E+00 -0.3260E-07
  43      0.0000E+00 -0.3931E-07
  44      0.0000E+00 -0.1906E-07
  45      0.0000E+00  0.0000E+00
The central  point stresses are :
Element no.        1
Point       1
 -0.8302E-01 -0.9098E00  0.7671E-01
Element no.        2
Point       1
 -0.4434E-01 -0.6555E00  0.1123E00
Element no.        3
Point       1
 -0.2042E-01  0.3240E-01 -0.1323E-01
Element no.        4
Point       1
 -0.7382E-02  0.1345E-01 -0.3256E-02
```

Figure 5.8 Results from Program 5.1

The computed results for this example, given in Figure 5.8, indicate a centreline displacement of -1.591×10^{-5} m. This is in good agreement with the closed form solution of -1.53×10^{-5} m given by Poulos and Davis (1974).

PROGRAM 5.2: PLANE STRAIN ANALYSIS USING FOUR-NODE QUADRILATERALS

The next element considered is the four-node 'linear strain' quadrilateral.

In this program, the replacement of numerically integrated stiffness matrices by "analytically" integrated equivalents is demonstrated (Griffiths, 1994). A further change is that "skyline" storage is used for the global stiffness matrix necessitating library subroutine `FSPARV` for global stiffness formation, and `SPARIN` and `SPABAC` for equation solution.

Comparing Program 5.2 with Program 5.0 we see an extra array `KDIAG` which holds the addresses of the skyline storage leading diagonal terms (see Figure 3.19).

The geometry subroutine `GEOMETRY_4QY` counts in the y direction and assumes that all elements are rectangular and equal in size. Figure 5.9 shows a typical mesh of elements together with the global node and element numbering system. Figure 5.10 indicates the sequence of node and freedom numbering at the element level and also the order in which Gauss points are sampled.

Library subroutine `ANALY4` returns the "analytically" computed stiffness matrix for each element and `BEE4` the "analytically" computed `BEE` matrix.

In the particular listing provided here, the stresses are computed at the four Gauss points.

The example of Figure 5.11 shows the mesh and data for a strip load bearing on a uniform elastic layer. Because of symmetry, only half of the layer need be analysed and the width has been arbitrarily terminated at a roller boundary at three times the load width. The computed results in Figure 5.12 give the vertical displacement at nodes 1 and 2 to be 8.601×10^{-6} and 3.771×10^{-6} respectively. Comparison with closed form or other numerical solutions will show that, with such a coarse mesh of these elements, the results can be quite inaccurate. Stresses have been printed at the Gauss points of the elements. The vertical stresses in the element immediately beneath the load are 1.058 kN/m^2 at Gauss point 1, and 0.641 kN/m^2 at Gauss point 2. Thus, locally, the crude solution indicates some decay between the Gauss point stresses and the applied stress. Such discretisation errors are inevitable in finite element work, and it is the user's responsibility to experiment with mesh designs to help discover whether the numerical solution is adequate.

```
program p52
!-----------------------------------------------------------------------
!      program 5.2 plane strain of an elastic solid using uniform 4-node
!      quadrilateral elements numbered in the y direction.
!      Analytical forms of km and bee matrices
!-----------------------------------------------------------------------
use new_library  ;  use  geometry_lib ;  use vlib  ;  implicit none
integer::nels,nye,neq,nn,nr,nip,nodof=2,nod=4,nst=3,ndof,loaded_nodes,    &
         i,k,iel,ndim=2
real:: e,v,det,aa,bb      ;   character(len=15) :: element='quadrilateral'
!---------------------------- dynamic arrays----------------------------------
real  , allocatable  :: kv(:),loads(:),points(:,:),dee(:,:),coord(:,:),   &
                        jac(:,:),der(:,:),deriv(:,:),weights(:),          &
                        bee(:,:),km(:,:),eld(:),sigma(:),g_coord(:,:)
integer, allocatable :: nf(:,:),g(:),num(:),g_num(:,:),g_g(:,:),kdiag(:)
!-------------------------input and initialisation----------------------------
  open (10,file='p52.dat',status=    'old',action='read')
  open (11,file='p52.res',status='replace',action='write')
 read (10,*) nels,nye,nn,nip,aa,bb,e,v
 ndof=nod*nodof
 allocate (nf(nodof,nn),points(nip,ndim),g(ndof),g_coord(ndim,nn),         &
          dee(nst,nst),coord(nod,ndim),jac(ndim,ndim),g_g(ndof,nels),     &
          weights(nip),der(ndim,nod),deriv(ndim,nod),bee(nst,ndof),       &
          km(ndof,ndof),eld(ndof),sigma(nst),num(nod),g_num(nod,nels))
 nf=1; read(10,*) nr ; if(nr>0)read(10,*)(k,nf(:,k),i=1,nr)
 call formnf(nf); neq=maxval(nf)    ;  allocate (kdiag(neq))
 call deemat(dee,e,v)    ;  call sample(element,points,weights)
!-------- loop the elements to set up global geometry and kdiag -------------
 kdiag=0
 elements_1   : do iel =1,nels
                call geometry_4qy(iel,nye,aa,bb,coord,num)
                call num_to_g( num , nf, g);  g_num(:,iel)=num
                g_coord(:,num)=transpose(coord)
                g_g(:,iel)=g    ;    call fkdiag(kdiag,g)
 end do elements_1
   write(11,'(a)') "Global coordinates "
   do k=1,nn;write(11,'(a,i5,a,2e12.4)')"Node",k,"        ",g_coord(:,k);end do
   write(11,'(a)') "Global node numbers "
   do k = 1 , nels; write(11,'(a,i5,a,4i5)')                                  &
                          "Element ",k,"        ",g_num(:,k); end do
 kdiag(1)=1; do i=2,neq; kdiag(i)=kdiag(i)+kdiag(i-1); end do
 write(11,'(2(a,i5))')                                                      &
         "There are",neq," equations and the skyline storage is ",kdiag(neq)
 allocate(kv(kdiag(neq)),loads(0:neq)); kv=0.0
!---------------- element stiffness integration and assembly-------------------
elements_2: do iel = 1 , nels
            num= g_num(: ,iel); coord =transpose(g_coord(:,num));g=g_g(:,iel)
            call analy4(km,coord,e,v) ;  call fsparv (kv,km,g,kdiag)
end do elements_2
loads=0.0 ; read (10,*) loaded_nodes,(k,loads(nf(:,k)), i =1,loaded_nodes)
!-------------------------equation solution-------------------------------
   call sparin(kv,kdiag) ;call spabac(kv,loads,kdiag)
   write(11,'(a)') "The nodal displacements Are :"
   write(11,'(a)') "Node          Displacement"
   do k=1,nn; write(11,'(i5,a,2e12.4)') k,"  ",loads(nf(:,k)); end do
!------------------recover stresses at element Gauss-points-----------------
elements_3:do iel = 1 , nels
            num = g_num(:,iel);  coord =transpose( g_coord(: ,num))
            g = g_g(:,iel)     ;    eld=loads(g)
           write(11,'(a,i5,a)')                                               &
                     "The Gauss Point stresses for element",iel,"  are :"
          integrating_pts_2: do i = 1 , nip
              call bee4(coord,points,i,det,bee)
              sigma = matmul (dee,matmul(bee,eld))
              write(11,'(a,i5)') "Point",i    ;   write(11,'(3e12.4)') sigma
          end do integrating_pts_2
end do elements_3
end  program p52
```

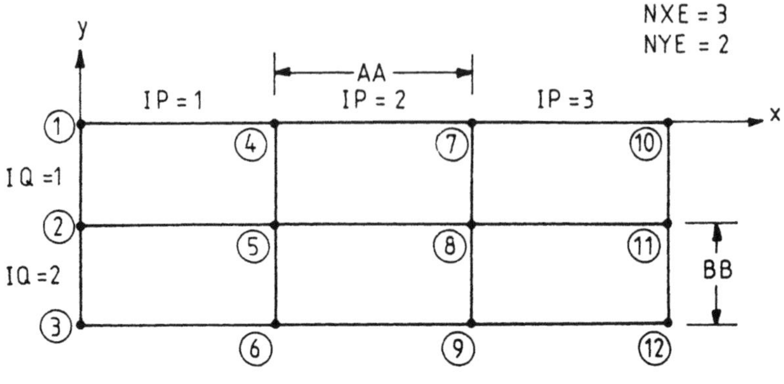

Figure 5.9 Global node and element numbering for mesh of four-node quadrilaterals

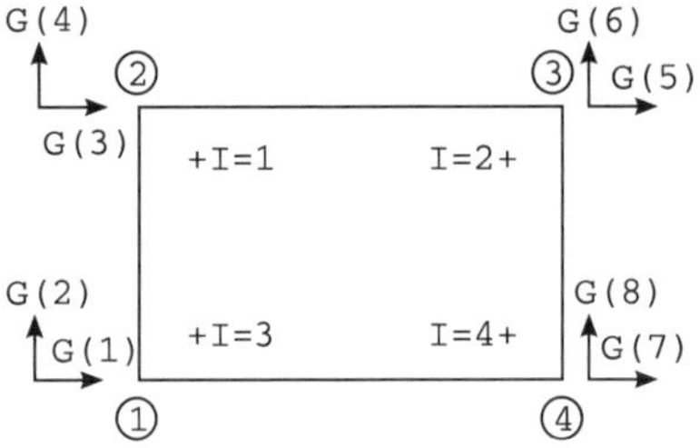

Figure 5.10 Local node, freedom and Gauss point numbering for four-node quadrilateral

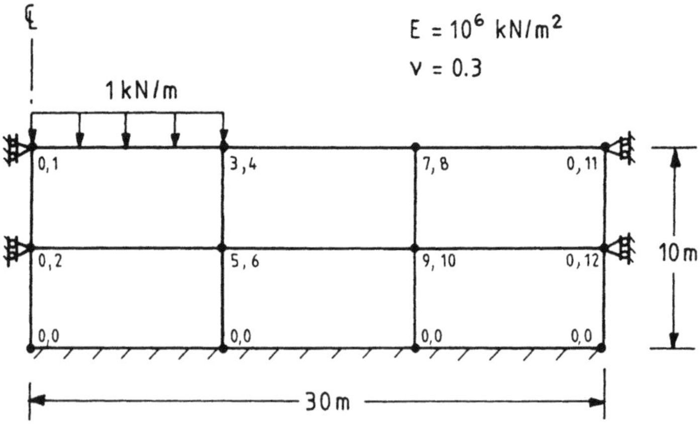

Structure and element data	NELS	NYE	NN	NIP
	6	2	12	4
	AA	BB	E	V
	10.	5.	1.E6	.3
Node freedom data	NR			
	8			
	K,NF(:,K),I=1,NR			

Figure 5.11 *(continued)*

```
                  1 0 1  2 0 1  3 0 0 6 0 0 9 0 0
                  10 0 1 11 0 1 12 0 0

Loads             LOADED_NODES
data              2
                  (K,LOADS(NF(:,K)),I=1,LOADED_NODES)
                  1 .0 -5. 4 .0 -5.
```

Figure 5.11 Mesh and data for Program 5.2

```
Global coordinates
Node    1         0.0000E+00  0.0000E+00
Node    2         0.0000E+00 -0.5000E+01
Node    3         0.0000E+00 -0.1000E+02
Node    4         0.1000E+02  0.0000E+00
Node    5         0.1000E+02 -0.5000E+01
Node    6         0.1000E+02 -0.1000E+02
Node    7         0.2000E+02  0.0000E+00
Node    8         0.2000E+02 -0.5000E+01
Node    9         0.2000E+02 -0.1000E+02
Node   10         0.3000E+02  0.0000E+00
Node   11         0.3000E+02 -0.5000E+01
Node   12         0.3000E+02 -0.1000E+02
Global node numbers
Element     1            2    1    4    5
Element     2            3    2    5    6
Element     3            5    4    7    8
Element     4            6    5    8    9
Element     5            8    7   10   11
Element     6            9    8   11   12
There are   12  equations and the skyline storage is     58
The nodal displacements are :
Node        Displacement
    1      0.0000E+00 -0.8601E-05
    2      0.0000E+00 -0.4056E-05
    3      0.0000E+00  0.0000E+00
    4     -0.5472E-07 -0.3771E-05
    5      0.1225E-05 -0.1910E-05
    6      0.0000E+00  0.0000E+00
    7      0.1792E-06  0.5860E-06
    8      0.6280E-06  0.1707E-06
    9      0.0000E+00  0.0000E+00
   10      0.0000E+00  0.1129E-06
   11      0.0000E+00  0.1066E-06
   12      0.0000E+00  0.0000E+00
The Gauss Point stresses for element    1   are :
Point    1
 -0.4299E00 -0.1058E+01  0.1431E00
Point    2
 -0.2511E00 -0.6411E00  0.8629E-01
Point    3
 -0.3304E00 -0.1016E+01  0.8352E-01
Point    4
 -0.1516E00 -0.5985E00  0.2667E-01
...........................................................

The Gauss Point stresses for element    6   are :
Point    1
 -0.4855E-01  0.1373E-01  0.3616E-01
Point    2
 -0.5281E-01  0.3775E-02  0.8267E-02
Point    3
  0.2640E-03  0.3465E-01  0.3758E-01
Point    4
 -0.4001E-02  0.2469E-01  0.9688E-02
```

Figure 5.12 Results from Program 5.2

PROGRAM 5.3: PLANE STRAIN ANALYSIS USING 8-NODE QUADRILATERALS

```
 program p53
!-------------------------------------------------------------------------
!      program 5.3 plane strain of an elastic solid using uniform
!      8-node quadrilateral elements numbered in the x direction
!-------------------------------------------------------------------------
 use new_library   ;  use geometry_lib   ;   implicit none
 integer::nels,nxe,neq,nband,nn,nr,nip,nodof=2,nod=8,nst=3,ndof,loaded_nodes,&
          i,k,iel,ndim=2
 real::aa,bb,e,v,det ;  character(len=15) :: element = 'quadrilateral'
!--------------------------- dynamic arrays-------------------------------------
 real    ,allocatable :: kb(:,:),loads(:),points(:,:),dee(:,:),coord(:,:),    &
                         jac(:,:), der(:,:),deriv(:,:),weights(:),             &
                         bee(:,:),km(:,:),eld(:),sigma(:),g_coord(:,:)
 integer, allocatable :: nf(:,:), g(:) , num(:)  , g_num(:,:) , g_g(:,:)
!-------------------------input and initialisation-----------------------------
  open (10,file='p53.dat',status=   'old',action='read')
  open (11,file='p53.res',status='replace',action='write')
  read (10,*) nels,nxe,nn,nip,aa,bb,e,v     ;    ndof=nod*nodof
  allocate ( nf(nodof,nn), points(nip,ndim),g(ndof), g_coord(ndim,nn),        &
           dee(nst,nst),coord(nod,ndim),jac(ndim,ndim),weights(nip),          &
           der(ndim,nod), deriv(ndim,nod), bee(nst,ndof), km(ndof,ndof),      &
           eld(ndof),sigma(nst),num(nod),g_num(nod,nels),g_g(ndof,nels))
  nf=1; read(10,*) nr ;if(nr>0)read(10,*)(k,nf(:,k),i=1,nr)
  call formnf(nf);neq=maxval(nf)       ;  nband = 0
  call deemat (dee,e,v); call sample(element,points,weights)
!----------------loop the elements to find bandwidth and neq-------------------
  elements_1: do iel = 1 , nels
              call geometry_8qx(iel,nxe,aa,bb,coord,num); g_num(:,iel) = num
              call num_to_g(num,nf,g); g_coord(:,num)=transpose(coord)
              g_g(:,iel)=g  ;  if(nband<bandwidth(g))nband=bandwidth(g)
  end do elements_1
    write(11,'(a)') "Global coordinates "
    do k=1,nn;write(11,'(a,i5,a,2e12.4)')"Node",k,"      ",g_coord(:,k);end do
    write(11,'(a)') "Global node numbers "
    do k = 1 , nels; write(11,'(a,i5,a,8i5)')                                  &
                             "Element ",k,"       ",g_num(:,k); end do
    write(11,'(2(a,i5))')                                                      &
            "There are ",neq ," equations and the half-bandwidth is", nband
            allocate(kb(neq,nband+1),loads(0:neq)); kb=.0
!--------------- element stiffness integration and assembly--------------------
 elements_2: do iel = 1 , nels
            num = g_num(: , iel);  g = g_g( : , iel)
            coord = transpose(g_coord(:,num)) ; km=0.0
        gauss_pts_1: do i = 1 , nip
              call shape_der (der,points,i) ; jac = matmul(der,coord)
              det = determinant(jac); call invert(jac)
              deriv = matmul(jac,der) ; call beemat (bee,deriv)
            km = km + matmul(matmul(transpose(bee),dee),bee) *det* weights(i)
        end do gauss_pts_1
   call formkb (kb,km,g)
 end do elements_2
 loads=.0; read(10,*)loaded_nodes,(k,loads(nf(:,k)),i=1,loaded_nodes)
!----------------------------equation solution--------------------------------
    call cholin(kb) ;call chobac(kb,loads)
    write(11,'(a)') "The nodal displacements are :"
    write(11,'(a)') "Node          Displacement"
    do k=1,nn; write(11,'(i5,a,2e12.4)') k,"   ",loads(nf(:,k)); end do
!------------------------recover stresses at centroids ------------------------
   i = 1 ; points = .0
       write(11,'(a)') "The centroidal stresses are :"
 elements_3:do iel = 1 , nels
        write(11,'(a,i5)') "Element No.  ",iel
    num = g_num(: , iel); g = g_g(: , iel)
    coord = transpose(g_coord(:,num)); eld=loads(g)
```

```
        call shape_der (der,points,i); jac= matmul(der,coord)
        call invert(jac) ;    deriv= matmul(jac,der)
        call beemat(bee,deriv) ;  sigma = matmul (dee,matmul(bee,eld))
        write(11,'(a,i5)') "Point  ",i   ;  write(11,'(3e12.4)') sigma
  end do elements_3
 end program p53
```

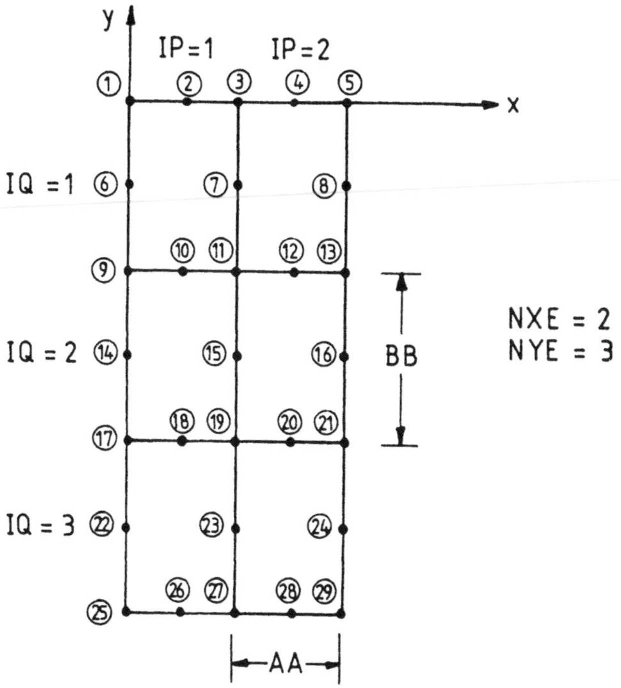

Figure 5.13 Global node and element numbering for mesh of eight-node quadrilaterals

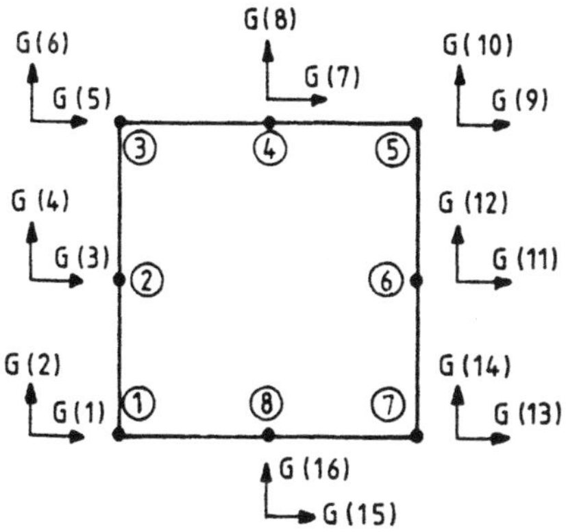

Figure 5.14 Local node and freedom numbering for eight-node quadrilateral

This program illustrates the use of a higher-order element, namely the eight-noded quadrilateral. The element stiffness is numerically integrated and the program is almost identical to Program 5.2. `GEOMETRY_8QX` has replaced `GEOMETRY_4QY` to generate the arrays `COORD` and `NUM`. As the name of the

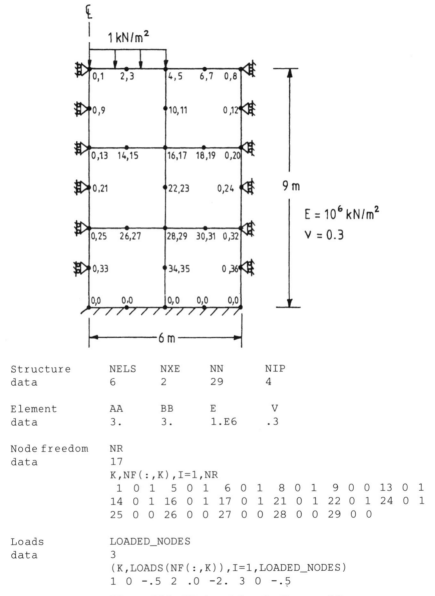

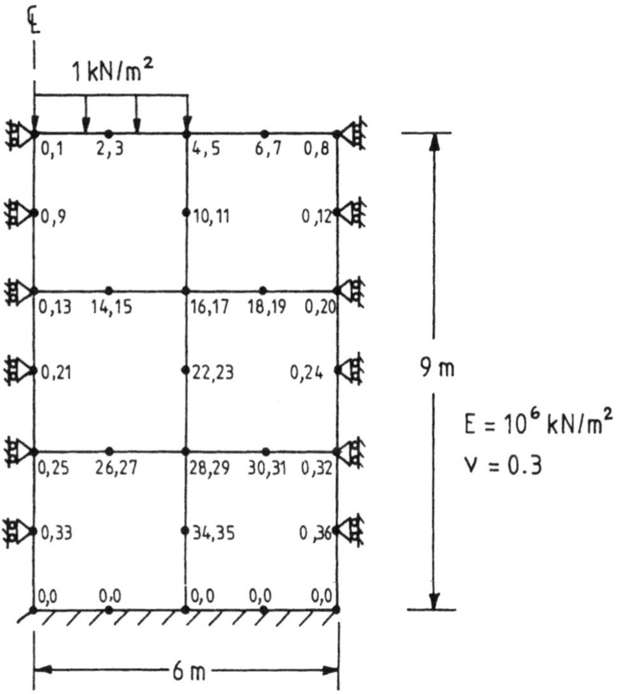


```
Structure      NELS    NXE     NN      NIP
data           6       2       29      4

Element        AA      BB      E        V
data           3.      3.      1.E6    .3

Node freedom   NR
data           17
               K,NF(:,K),I=1,NR
                1 0 1   5 0 1   6 0 1   8 0 1   9 0 0  13 0 1
               14 0 1  16 0 1  17 0 1  21 0 1  22 0 1  24 0 1
               25 0 0  26 0 0  27 0 0  28 0 0  29 0 0

Loads          LOADED_NODES
data           3
               (K,LOADS(NF(:,K)),I=1,LOADED_NODES)
               1 0 -.5 2 .0 -2. 3 0 -.5
```

Figure 5.15 Mesh and data for Program 5.3

```
Global coordinates
Node   1          0.0000E+00  0.0000E+00
Node   2          0.1500E+01  0.0000E+00
Node   3          0.3000E+01  0.0000E+00
Node   4          0.4500E+01  0.0000E+00
Node   5          0.6000E+01  0.0000E+00
Node   6          0.0000E+00 -0.1500E+01
Node   7          0.3000E+01 -0.1500E+01
Node   8          0.6000E+01 -0.1500E+01
Node   9          0.0000E+00 -0.3000E+01
Node  10          0.1500E+01 -0.3000E+01
.............................................................

Node  25          0.0000E+00 -0.9000E+01
Node  26          0.1500E+01 -0.9000E+01
Node  27          0.3000E+01 -0.9000E+01
Node  28          0.4500E+01 -0.9000E+01
Node  29          0.6000E+01 -0.9000E+01
Global node numbers
Element    1           9    6    1    2    3    7   11   10
Element    2          11    7    3    4    5    8   13   12
Element    3          17   14    9   10   11   15   19   18
Element    4          19   15   11   12   13   16   21   20
Element    5          25   22   17   18   19   23   27   26
Element    6          27   23   19   20   21   24   29   28
There are    36  equations and the half-bandwidth is   16
The nodal displacements are :
Node       Displacement
    1     0.0000E+00 -0.5311E-05
    2    -0.4211E-06 -0.5041E-05
    3    -0.7222E-06 -0.3343E-05
    4    -0.4211E-06 -0.1644E-05
    5     0.0000E+00 -0.1375E-05
    6     0.0000E+00 -0.4288E-05
    7     0.3774E-06 -0.2786E-05
    8     0.0000E+00 -0.1283E-05
    9     0.0000E+00 -0.3243E-05
   10     0.2708E-06 -0.2873E-05
................................................................

   25     0.0000E+00  0.0000E+00
   26     0.0000E+00  0.0000E+00
   27     0.0000E+00  0.0000E+00
   28     0.0000E+00  0.0000E+00
   29     0.0000E+00  0.0000E+00
The centroidal stresses are :
Element No.       1
Point       1
 -0.2476E00 -0.9003E00  0.1040E00
Element No.       2
Point       1
 -0.1810E00 -0.9973E-01  0.1040E00
Element No.       3
Point       1
 -0.1683E00 -0.6489E00  0.8714E-01
Element No.       4
Point       1
 -0.2602E00 -0.3511E00  0.8714E-01
Element No.       5
Point       1
 -0.1994E00 -0.5612E00  0.2888E-01
Element No.       6
Point       1
 -0.2292E00 -0.4388E00  0.2888E-01
```

Figure 5.16 Results from Program 5.3

geometry routine suggests, nodes are assumed to be numbered in the x direction for maximum storage efficiency, as shown in Figure 5.13. The local node and freedom numbering for this element (Figure 5.14) indicates that node 1 occurs at a corner and the rest follow in a clockwise sense. The elements are of equal size. The Gauss point numbering follows the pattern indicated by Figure 5.10. The general eight-node quadrilateral element stiffness matrix contains fourth-order polynomial terms and thus requires `NIP` to be 9 for 'exact' integration. It is often the case, however, that the use of "reduced" integration, by putting `NIP` equal to 4, improves the performance of this element. This is found to be particularly true of the plasticity applications described in Chapter 6.

The simple mesh in Figure 5.15 is to be analysed and the consistent nodal loads (Appendix 1) necessary to reproduce a uniform stress field should be noted in the data. The listing provided here prints the nodal displacements and also the strains and stresses at the element "centroids". The computed results are given in Figure 5.16.

PROGRAM 5.4: AXISYMMETRIC STRAIN ANALYSIS USING 4-NODE QUADRILATERALS

This program is the axisymmetric counterpart of Programs 5.1 and 5.2, which dealt with plane strain conditions. The mesh is irregular in a similar way to that described in Program 5.1 (although still "rectangular") while the elastic properties `E` and `V` are allowed to assume different values in each horizontal layer of elements. These properties are stored in the array `PROP`.

The mesh sizes `NRE` and `NDE` for axisymmetric meshes are the counterparts of `NXE` and `NYE` in plane problems.

Nodal coordinates `COORD` and the nodal vector `NUM` are generated by the "variable" geometry subroutine `GEOMETRY_4QYV`, which assumes that nodes are numbered in the "depth" direction. A library subroutine which appears for the first time for axisymmetric conditions is `BMATAXI` which forms the 4×8 strain–displacement `BEE` matrix (see equation (2.67)).

It should be noted that as four components of strain and stress are computed by this program, `NST` is set to 4, and the appropriate `DEE` matrix (2.68) is returned by `DEEMAT`.

The additional real number:

`RADIUS`, the radial coordinate of each Gauss point,

is required in the computation of the weighting factor which integrates over one radian of the solid.

The mesh and data in Figure 5.17 show the axisymmetric problem to be analysed. The nodal loads imply a uniform stress of 1 kN/m^2 is to be applied to

```
program p54
!-------------------------------------------------------------------------
!      program 5.4 axisymmetric strain of a rectangular section elastic
!      solid using variable 4-node quadrilateral elements numbered in x
!      and variable material properties
!-------------------------------------------------------------------------
 use new_library   ;  use  geometry_lib   ;       implicit none
 integer::nels,nre,nde,neq,nband,nn,nr,nip,nodof=2,nod=4,nst=4,ndof,    &
         i,k,iel,ndim=2,loaded_nodes
 real:: e,v,det ,radius    ;  character(len=15) :: element = 'quadrilateral'
!------------------------- dynamic arrays---------------------------------
 real    ,allocatable :: kv(:),loads(:),points(:,:),dee(:,:),coord(:,:), &
                         fun(:),jac(:,:),der(:,:),deriv(:,:),weights(:), &
                         bee(:,:),km(:,:),eld(:),sigma(:),g_coord(:,:),  &
                         width(:),depth(:) ,prop(:,:)
 integer, allocatable :: nf(:,:), g(:)  , num(:)  , g_num(:,:) , g_g(:,:)
!-------------------------input and initialisation-----------------------------
  open (10,file='p54.dat',status=    'old',action='read')
  open (11,file='p54.res',status='replace',action='write')
  read (10,*) nels,nre,nde,nn,nip        ;    ndof=nod*nodof
  allocate ( nf(nodof,nn), points(nip,ndim),dee(nst,nst), g_coord(ndim,nn),  &
            coord(nod,ndim),fun(nod),jac(ndim,ndim), weights(nip),          &
            g_num(nod,nels),der(ndim,nod),deriv(ndim,nod),bee(nst,ndof),    &
            num(nod),km(ndof,ndof), eld(ndof),  sigma(nst), g(ndof),        &
            width(nre+1),depth(nde+1),prop(2,nels), g_g(ndof,nels))
  read(10,*) width ; read(10,*) depth
  read(10,*)(prop(k,:),k=1,2)
  nf=1; read(10,*) nr ;if(nr>0) read(10,*) (k,nf(:,k),i=1,nr)
      call formnf(nf); neq=maxval(nf)  ; call sample(element,points,weights)
!-------- loop the elements to find nband and set up global arrays ----------
      nband=0
     elements_1   : do iel =1,nels
                 call geometry_4qyv(iel,nde,width,depth,coord,num)
                 call num_to_g ( num , nf , g );  g_num(:,iel)=num
                 g_coord(:,num)=transpose(coord);g_g(:,iel)=g
                 if(nband<bandwidth(g))nband=bandwidth(g)
     end do elements_1
    write(11,'(a)') "Global coordinates "
    do k=1,nn;write(11,'(a,i5,a,2e12.4)')"Node",k,"       ",g_coord(:,k);end do
    write(11,'(a)') "Global node numbers "
    do k = 1 , nels; write(11,'(a,i5,a,4i5)')                                 &
                               "Element ",k,"        ",g_num(:,k); end do
        write(11,'(2(a,i5))')                                                 &
             "There are",neq,"  equations and the half-bandwidth is",nband
      allocate( kv(neq*(nband+1)),loads(0:neq)); kv=0.0
!--------------- element stiffness integration and assembly--------------------
  elements_2: do iel=1,nels
                num=g_num(:,iel)  ; coord =transpose( g_coord(:,num))
                g = g_g(: ,iel)   ;     km=0.0
                e = prop(1 , iel);  v = prop(2 , iel); call deemat(dee,e,v)
                integrating_pts_1:  do i=1,nip
                  call shape_fun(fun,points,i); call shape_der(der,points,i)
                  jac=matmul(der,coord) ; det= determinant(jac)
                  call invert(jac);   deriv = matmul(jac,der)
                  call bmataxi(bee,radius,coord,deriv,fun);det =det*radius
                  km= km+matmul(matmul(transpose(bee),dee),bee)*det*weights(i)
                end do integrating_pts_1
               call formkv (kv,km,g,neq)
  end do elements_2
  loads=0.0 ; read (10,*) loaded_nodes,(k,loads(nf(:,k)), i =1,loaded_nodes)
!------------------------equation solution--------------------------------
    call banred(kv,neq) ;call bacsub(kv,loads)
    write(11,'(a)') "The nodal displacements are:"
    do k=1,nn; write(11,'(i5,a,2e12.4)') k,"   ",loads(nf(:,k)); end do
```

```
!--------------------recover stresses at element centroids-----------------
  nip = 1; deallocate(points,weights); allocate(points(nip,ndim),weights(nip))
      call sample (element , points , weights )
 elements_3:do iel=1,nels
             num = g_num(:,iel) ; coord = transpose(g_coord(:,num))
             g = g_g( : , iel)  ;     eld=loads(g)
             write(11,'(a,i5,a)')                                              &
                       "The centroidal stresses for element",iel,"  are :"
              e = prop(1 ,iel);  v = prop(2 , iel); call deemat(dee,e,v)
              integrating_pts_2: do i = 1 , nip
                 call shape_fun(fun,points,i); call shape_der(der,points,i)
                 jac=matmul(der,coord);call invert(jac); deriv=matmul(jac,der)
                 call bmataxi(bee,radius,coord,deriv,fun)
                 sigma=matmul(dee,matmul(bee,eld))
                 write(11,'(a,i5)') "Point",i   ; write(11,'(4e12.4)') sigma
             end do integrating_pts_2
 end do elements_3
end program p54
```

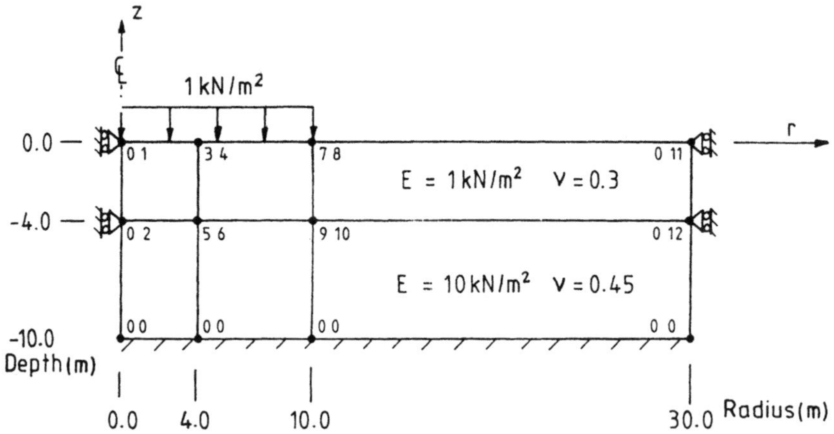

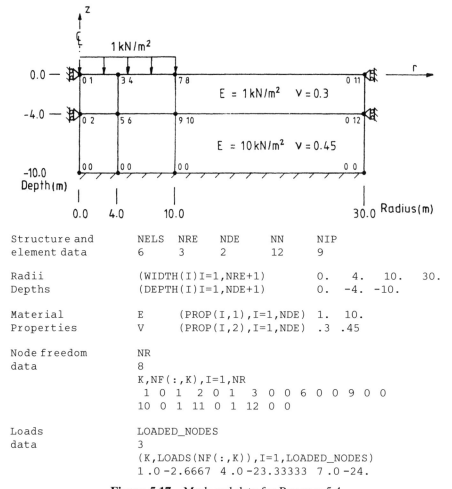

```
Structure and     NELS   NRE    NDE     NN     NIP
element data      6      3      2       12     9

Radii             (WIDTH(I)I=1,NRE+1)          0.    4.   10.   30.
Depths            (DEPTH(I)I=1,NDE+1)          0.   -4.  -10.

Material          E      (PROP(I,1),I=1,NDE)   1.   10.
Properties        V      (PROP(I,2),I=1,NDE)   .3   .45

Node freedom      NR
data              8
                  K,NF(:,K),I=1,NR
                   1  0  1   2  0  1   3  0  0  6  0  0  9  0  0
                  10  0  1  11  0  1  12  0  0

Loads             LOADED_NODES
data              3
                  (K,LOADS(NF(:,K)),I=1,LOADED_NODES)
                  1 .0 -2.6667  4 .0 -23.33333  7 .0 -24.
```

Figure 5.17 Mesh and data for Program 5.4

a circular area (see Appendix 1). The number of integrating points, `NIP`, has been set to 9, but even this is not exact, especially as $r \to 0$. The computed results for this problem, including stresses at the element "centres", are given in Figure 5.18.

```
Global coordinates
Node    1          0.0000E+00  0.0000E+00
Node    2          0.0000E+00 -0.4000E+01
Node    3          0.0000E+00 -0.1000E+02
Node    4          0.4000E+01  0.0000E+00
Node    5          0.4000E+01 -0.4000E+01
Node    6          0.4000E+01 -0.1000E+02
Node    7          0.1000E+02  0.0000E+00
Node    8          0.1000E+02 -0.4000E+01
Node    9          0.1000E+02 -0.1000E+02
Node   10          0.3000E+02  0.0000E+00
Node   11          0.3000E+02 -0.4000E+01
Node   12          0.3000E+02 -0.1000E+02
Global node numbers
Element     1            2    1    4    5
Element     2            3    2    5    6
Element     3            5    4    7    8
Element     4            6    5    8    9
Element     5            8    7   10   11
Element     6            9    8   11   12
There are   12  equations and the half-bandwidth is     7
The nodal displacements are:
    1      0.0000E+00 -0.3176E+01
    2      0.0000E+00 -0.3231E00
    3      0.0000E+00  0.0000E+00
    4      0.1395E00  -0.3991E+01
    5      0.1165E00  -0.2498E00
    6      0.0000E+00  0.0000E+00
    7      0.1704E00  -0.6046E00
    8      0.1330E00  -0.4421E-01
    9      0.0000E+00  0.0000E+00
   10      0.0000E+00  0.2588E00
   11      0.0000E+00  0.3091E-01
   12      0.0000E+00  0.0000E+00
The centroidal stresses for element    1  are :
Point    1
 -0.4140E00 -0.1073E+01 -0.3452E-01 -0.4140E00
The centroidal stresses for element    2  are :
Point    1
 -0.4776E00 -0.9072E00  0.6508E-01 -0.4776E00
The centroidal stresses for element    3  are :
Point    1
 -0.2933E00 -0.7099E00  0.1180E00 -0.2810E00
The centroidal stresses for element    4  are :
Point    1
 -0.4316E00 -0.6101E00  0.1308E00 -0.3796E00
The centroidal stresses for element    5  are :
Point    1
 -0.3200E-01 -0.5814E-01  0.1082E-01 -0.2325E-01
The centroidal stresses for element    6  are :
Point    1
 -0.1090E00 -0.9367E-01  0.4470E-01 -0.7455E-01
```

Figure 5.18 Results from Program 5.4

PROGRAM 5.5: NON-AXISYMMETRIC STRAIN OF AXISYMMETRIC SOLIDS USING EIGHT-NODE QUADRILATERALS

```
 program p55
!-----------------------------------------------------------------------
!     program 5.5 non-axisymmetric strain of an axisymmetric elastic solid
!     using uniform 8-node quadrilateral elements numbered in the x direction
!-----------------------------------------------------------------------
 use new_library   ;   use geometry_lib      ;     implicit none
 integer::nels,nre,nde,neq,nn,nr,nip,nodof=3,nod=8,nst=6,ndof,loaded_nodes, &
         i,k,iel,ndim=2,lth,iflag
 real:: e,v,det,aa,bb,chi,pi,ca,sa,radius
 character (len=15) :: element = 'quadrilateral'
!--------------------------- dynamic arrays---------------------------------
 real   , allocatable :: kv(:),loads(:),points(:,:),dee(:,:),coord(:,:),     &
                         fun(:),jac(:,:),der(:,:),deriv(:,:),weights(:),     &
                         bee(:,:),km(:,:),eld(:),sigma(:),g_coord(:,:),      &
                         value(:)
integer, allocatable :: nf(:,:),g(:),num(:),g_num(:,:),g_g(:,:),kdiag(:),&
                         node(:),no(:),sense(:)
!-----------------------input and initialisation------------------------------
 open (10,file='p55.dat',status=    'old',action='read')
 open (11,file='p55.res',status='replace',action='write')
  read(10,*) nels,nre,nde,nn,nip,aa,bb,e,v
  read(10,*) lth,iflag,chi                    ;     ndof=nod*nodof
  allocate (nf(nodof,nn),points(nip,ndim),g(ndof),g_coord(ndim,nn),          &
            dee(nst,nst),coord(nod,ndim),fun(nod),jac(ndim,ndim),            &
            weights(nip),der(ndim,nod),deriv(ndim,nod),bee(nst,ndof),        &
            km(ndof,ndof),eld(ndof),sigma(nst),num(nod),g_num(nod,nels),     &
            g_g(ndof,nels))
 nf=1; read(10,*) nr ;if(nr>0)read(10,*)(k,nf(:,k),i=1,nr)
       call formnf(nf); neq=maxval(nf) ;  allocate (kdiag(neq))
       call deemat(dee,e,v)    ; call sample( element, points, weights)
  pi=acos(-1.); chi=chi*pi/180.; ca=cos(chi); sa=sin(chi)
!-------- loop the elements to set up global geometry and kdiag -----------
  kdiag=0
  elements_1   : do iel =1,nels
                   call geometry_8qx(iel,nre,aa,bb,coord,num)
                   call num_to_g(num,nf,g);   g_num(:,iel)=num
                   g_coord(:,num)=transpose(coord)
                   g_g( : , iel ) = g        ;    call fkdiag(kdiag,g)
  end do elements_1
    write(11,'(a)') "Global coordinates "
    do k=1,nn;write(11,'(a,i5,a,2e12.4)')"Node",k,"       ",g_coord(:,k);end do
    write(11,'(a)') "Global node numbers "
    do k = 1 , nels; write(11,'(a,i5,a,8i5)')                                &
                              "Element ",k,"        ",g_num(:,k); end do
        kdiag(1)=1; do i=2,neq; kdiag(i)=kdiag(i)+kdiag(i-1); end do
     write(11,'(2(a,i5))')                                                   &
          "There are",neq,"  equations and the skyline storage is",kdiag(neq)
        allocate(kv(kdiag(neq)),loads(0:neq)); kv=0.0
!--------------- element stiffness integration and assembly--------------------
 elements_2: do iel = 1 , nels
               num= g_num(:, iel);  g = g_g( : , iel )
               coord = transpose(g_coord(:,num)) ;  km = .0
          integrating_pts_1: do i = 1 , nip
               call shape_fun(fun,points,i); call shape_der(der,points,i)
               jac = matmul(der,coord); det= determinant(jac)
               call invert(jac); deriv= matmul(jac,der)
               call bmat_nonaxi(bee,radius,coord,deriv,fun,iflag,lth)
               det=det*radius
               km= km+matmul(matmul(transpose(bee),dee),bee)*det*weights(i)
          end do integrating_pts_1
          call fsparv (kv,km,g,kdiag)
 end do elements_2
loads=0.0;read(10,*)loaded_nodes,(k,loads(nf(:,k)),i=1,loaded_nodes)
!-----------------------equation solution-------------------------------
```

```
    call sparin(kv,kdiag) ;call spabac(kv,loads,kdiag)
    write(11,'(a)') "The Nodal Displacements Are :"
    write(11,'(a)') "Node         Displacement"
    do k=1,nn; write(11,'(i5,a,3e12.4)') k,"   ",loads(nf(:,k)); end do
!--------------------recover stresses at  element centroids ----------------
 i = 1; points = .0
 elements_3:do iel = 1 , nels
            num = g_num(:, iel);  coord = transpose(g_coord(:,num))
            g = g_g(: ,iel )  ;    eld=loads(g)
    write(11,'(a,i5,a)') "The centre point stresses for element",iel,"  are :"
        call shape_fun(fun,points,i); call shape_der(der,points,i)
        jac = matmul(der,coord); call invert(jac); deriv= matmul(jac,der)
        call bmat_nonaxi(bee,radius,coord,deriv,fun,iflag,lth)
        bee(1:4,:)=bee(1:4,:)*ca; bee(5:6,:)=bee(5:6,:)*sa
        sigma = matmul(dee,matmul(bee,eld))
       write(11,'(a,i5)') "Point",i    ;    write(11,'(6e12.4)') sigma
 end do elements_3
end program p55
```

This program allows the analysis of axisymmetric solids subjected to non-axisymmetric loads. Variations in displacements, and hence strains and stresses, tangentially are described by Fourier series (e.g. Wilson, 1965; Zienkiewicz, 1988). Although the analysis is genuinely 3-d, with three degrees of freedom at each node, it is only necessary to discretise the problem in a radial plane. The integrals in radial planes are performed using Gaussian quadrature in the usual way. Orthogonality relationships between typical terms in the tangential direction enable the integrals in the third direction to be stated explicitly. The problem therefore takes on the "appearance" of a 2-d analysis with the obvious benefits in terms of storage requirements. The disadvantage of the method over conventional 3-d finite element analysis is that, for complicated loading distributions, several loading harmonic terms may be required and a global stiffness matrix must be stored for each. Several harmonic terms may be required for elasto-plastic analysis (Griffiths, 1986), but for most elastic analyses such as the one described here, one harmonic will often be sufficient.

It is important to realise that the basic stiffness relationships relate amplitudes of load to amplitudes of displacement. Once the amplitudes of a displacement are known, the actual displacement at a particular circumferential location is easily found.

For simplicity, consider only the components of nodal load which are symmetric about the $\theta = 0$ axis of the axisymmetric body. In this case a general loading distribution may be given by

$$\begin{aligned} R &= {}^1\!/_2\bar{R}^0 + \bar{R}^1\cos\theta + \bar{R}^2\cos 2\theta + \dots \\ Z &= {}^1\!/_2\bar{Z}^0 + \bar{Z}^1\cos\theta + \bar{Z}^2\cos 2\theta + \dots \\ T &= \bar{T}^1\sin\theta + \bar{T}^2\sin 2\theta + \dots \end{aligned} \tag{5.1}$$

where R, Z and T represent the load per radian in the radial, depth and tangential directions. The bar terms represent amplitudes of these quantities.

For antisymmetric loading, symmetrical about $\theta = \pi/2$, these expressions become

$$\begin{aligned} R &= \bar{R}^1 \sin\theta + \bar{R}^2 \sin 2\theta + \dots \\ Z &= \bar{Z}^1 \sin\theta + \bar{Z}^2 \sin 2\theta + \dots \\ T &= {}^1\!/_2\,\bar{T}^0 + \bar{T}^1 \cos\theta + \bar{T}^2 \cos 2\theta + \dots \end{aligned} \tag{5.2}$$

Corresponding to these quantities are amplitudes of displacement in the radial, depth and tangential directions.

Since there are now three displacements per node, there are six strains at any point taken in the order

$$\mathtt{EPS}^{\mathrm{T}} = [\varepsilon_r \varepsilon_z \varepsilon_\theta \, \gamma_{rz} \, \gamma_{z\theta} \, \gamma_{\theta r}] \tag{5.3}$$

and six corresponding stresses. The 6×6 stress–strain matrix `DEE` is formed by the subroutine `DEEMAT` as usual. With reference to equation (2.67), the **A** matrix now becomes

$$\mathbf{A} = \begin{bmatrix} \dfrac{\partial}{\partial r} & 0 & 0 \\ 0 & \dfrac{\partial}{\partial z} & 0 \\ \dfrac{1}{r} & 0 & \dfrac{1}{r}\dfrac{\partial}{\partial\theta} \\ \dfrac{\partial}{\partial z} & \dfrac{\partial}{\partial r} & 0 \\ 0 & \dfrac{1}{r}\dfrac{\partial}{\partial\theta} & \dfrac{\partial}{\partial z} \\ \dfrac{1}{r}\dfrac{\partial}{\partial\theta} & 0 & \dfrac{\partial}{\partial r} - \dfrac{1}{r} \end{bmatrix} \tag{5.4}$$

For each harmonic i, the strain–displacement relationship provided by the library subroutine `BMAT_NONAXI` is of the form

$$\mathbf{B}^i = [B_1^i B_2^i B_3^i B_4^i \dots B_j^i \dots B_{\mathrm{NOD}}^i] \tag{5.5}$$

where `NOD` is the number of nodes in an element.

A typical submatrix from the above expression for symmetric loading is given by

$$B_j^i = \begin{bmatrix} \dfrac{\partial N_j}{\partial r}\cos i\theta & 0 & 0 \\ 0 & \dfrac{\partial N_j}{\partial z}\cos i\theta & 0 \\ \dfrac{N_j}{r}\cos i\theta & 0 & \dfrac{iN_j}{r}\cos i\theta \\ \dfrac{\partial N_j}{\partial z}\cos i\theta & \dfrac{\partial N_j}{\partial r}\cos i\theta & 0 \\ 0 & \dfrac{iN_j}{r}\sin i\theta & \dfrac{\partial N_j}{\partial z}\sin i\theta \\ -\dfrac{iN_j}{r}\sin i\theta & 0 & \left(\dfrac{\partial N_j}{\partial r} - \dfrac{N_j}{r}\right)\sin i\theta \end{bmatrix} \tag{5.6}$$

The equivalent expression for antisymmetry is similar to equation (5.6) but with the sine and cosine terms interchanged and the signs of elements (3,3), (5,2) and (6,1) reversed. Additional integer variables required by this subroutine are IFLAG and LTH. The variable IFLAG is set to 1 or -1 for symmetry or antisymmetry respectively, and the variable LTH gives the harmonic on which loads are to be applied.

It should be noted that if LTH = 0 and IFLAG = 1, the analysis reduces to ordinary axisymmetry as described by Program 5.4.

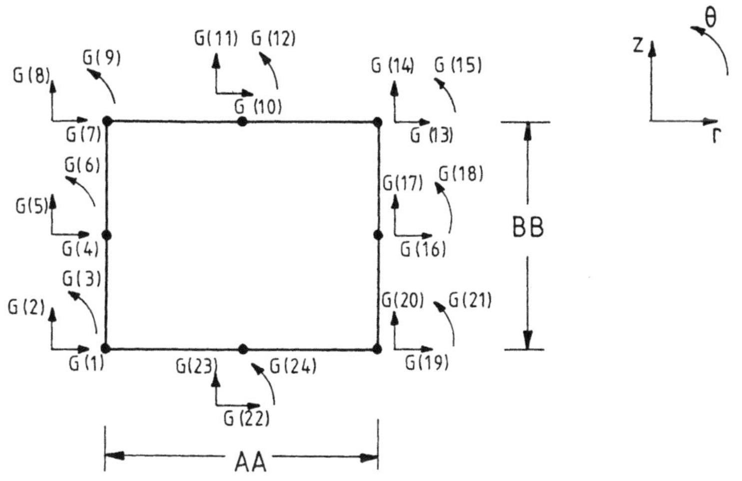

Figure 5.19 Local node and freedom numbering for eight-node quadrilateral (three freedoms per node)

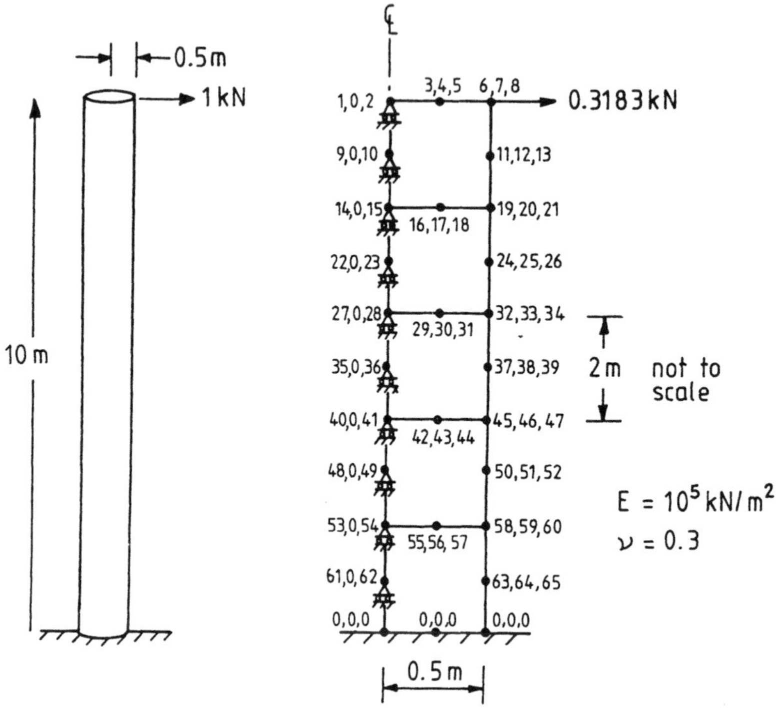

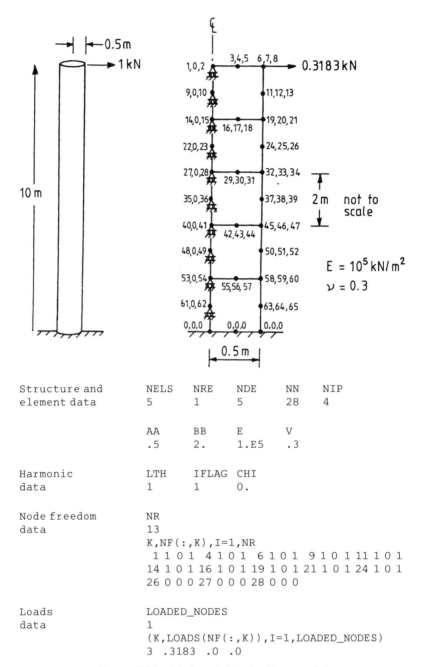

Structure and element data	NELS	NRE	NDE	NN	NIP
	5	1	5	28	4
	AA	BB	E	V	
	.5	2.	1.E5	.3	
Harmonic data	LTH	IFLAG	CHI		
	1	1	0.		

```
Node freedom    NR
data            13
                K,NF(:,K),I=1,NR
                 1 1 0 1  4 1 0 1  6 1 0 1  9 1 0 1 11 1 0 1
                14 1 0 1 16 1 0 1 19 1 0 1 21 1 0 1 24 1 0 1
                26 0 0 0 27 0 0 0 28 0 0 0

Loads           LOADED_NODES
data            1
                (K,LOADS(NF(:,K)),I=1,LOADED_NODES)
                3  .3183  .0  .0
```

Figure 5.20 Mesh and data for Program 5.5

```
Global coordinates
Node    1          0.0000E+00  0.0000E+00
Node    2          0.2500E+00  0.0000E+00
Node    3          0.5000E+00  0.0000E+00
Node    4          0.0000E+00 -0.1000E+01
Node    5          0.5000E+00 -0.1000E+01
Node    6          0.0000E+00 -0.2000E+01
Node    7          0.2500E+00 -0.2000E+01
Node    8          0.5000E+00 -0.2000E+01
.........................................................
Node   22          0.2500E+00 -0.8000E+01
Node   23          0.5000E+00 -0.8000E+01
Node   24          0.0000E+00 -0.9000E+01
Node   25          0.5000E+00 -0.9000E+01
Node   26          0.0000E+00 -0.1000E+02
Node   27          0.2500E+00 -0.1000E+02
Node   28          0.5000E+00 -0.1000E+02
Global node numbers
Element     1           6    4    1    2    3    5    8    7
Element     2          11    9    6    7    8   10   13   12
Element     3          16   14   11   12   13   15   18   17
Element     4          21   19   16   17   18   20   23   22
Element     5          26   24   21   22   23   25   28   27
There are   65  equations and the skyline storage is  871
The Nodal Displacements Are :
Node        Displacement
    1     0.6755E-01  0.0000E+00 -0.6755E-01
    2     0.6755E-01 -0.2528E-02 -0.6755E-01
    3     0.6755E-01 -0.5063E-02 -0.6754E-01
    4     0.5743E-01  0.0000E+00 -0.5743E-01
    5     0.5744E-01 -0.5006E-02 -0.5743E-01
    6     0.4753E-01  0.0000E+00 -0.4752E-01
    7     0.4753E-01 -0.2426E-02 -0.4752E-01
    8     0.4753E-01 -0.4858E-02 -0.4750E-01
    9     0.3801E-01  0.0000E+00 -0.3801E-01
   10     0.3804E-01 -0.4598E-02 -0.3799E-01
   11     0.2914E-01  0.0000E+00 -0.2913E-01
.........................................................
   20     0.8212E-02 -0.2559E-02 -0.8085E-02
   21     0.3764E-02  0.0000E+00 -0.3736E-02
   22     0.3739E-02 -0.9010E-03 -0.3732E-02
   23     0.3754E-02 -0.1809E-02 -0.3687E-02
   24     0.9378E-03  0.0000E+00 -0.9620E-03
   25     0.1052E-02 -0.9219E-03 -0.8806E-03
   26     0.0000E+00  0.0000E+00  0.0000E+00
   27     0.0000E+00  0.0000E+00  0.0000E+00
   28     0.0000E+00  0.0000E+00  0.0000E+00
The centre point stresses for element    1  are :
Point    1
  0.6441E+00 -0.5036E+01 -0.4726E+00  0.8638E-01  0.0000E+00  0.0000E+00
The centre point stresses for element    2  are :
Point    1
  0.2661E+01 -0.1413E+02  0.1144E+01  0.6800E-01  0.0000E+00  0.0000E+00
The centre point stresses for element    3  are :
Point    1
  0.5441E+01 -0.2274E+02  0.3341E+01  0.1484E+00  0.0000E+00  0.0000E+00
The centre point stresses for element    4  are :
Point    1
  0.8040E+01 -0.3164E+02  0.5250E+01  0.3627E+00  0.0000E+00  0.0000E+00
The centre point stresses for element    5  are :
Point    1
  0.1700E+02 -0.3521E+02  0.1580E+02  0.9873E+00  0.0000E+00  0.0000E+00
```

Figure 5.21 Results from Program 5.5

An additional variable input to this program is the angle CHI (in degrees in the range 0–360). This is the angle at which stresses are evaluated and printed. Naturally, stresses could be printed at other locations if required.

The program uses eight-node quadrilateral elements and can be considered a variant of Program 5.3. The geometry routine GEOMETRY_8QX assumes constant size elements with nodes numbered in the x (or radial) direction. Each element has 24 degrees of freedom, as shown in Figure 5.19.

The example shown in Figure 5.20 represents a cylindrical cantilever subjected to a transverse force of 1 kN at its tip. The nature of harmonic loading is such that a radial load amplitude of 1 unit on the first harmonic (LTH = 1) in symmetry (IFLAG = 1) results in a net thrust in the 0° direction of π. Thus the load amplitude applied at the first freedom of node 3 equals $1/\pi$. The nodal freedom data takes account of the fact that there can be no vertical movement along the centreline; hence these freedoms are restrained. The computed displacements in Figure 5.21 give the end deflection of the cantilever to be 6.755×10^{-2} m, compared with the slender beam value of 6.791×10^{-2} m. If the same load amplitude was applied to the second freedom of node 3, it would correspond to a net moment of 0.5 kN m. The computed displacement in this case would be -5.063×10^{-3} m, compared with the slender beam value of -5.093×10^{-3} m.

PROGRAM 5.6: THREE-DIMENSIONAL ANALYSIS USING FOUR-NODED TETRAHEDRA

In cases where many Fourier harmonics are required to define a loading pattern it becomes more efficient to solve fully 3-d problems. The simplest 3-d element is the four-noded tetrahedron. This "constant-strain" element is analogous to the three-noded triangle for plane problems described in Program 5.0 and, like the triangle, is not recommended for practical calculations unless adaptive mesh refinement has been implemented. Due to its simplicity, however, this element is a convenient starting point for 3-d applications, and the program described here can easily be modified to include more sophisticated tetrahedral elements.

Comparing with Program 5.0, few changes are required. The library subroutine SAMPLE provides the integrating points in volume coordinates and weighting coefficients for integration over tetrahedra. It may be noted that this simple element only requires one integrating point as its "centroid". Three-dimensional **B** and **D** matrices are formed by the library subroutines BEEMAT and DEEMAT and the Jacobian matrix is now of dimensions 3 × 3.

For the simple example demonstrated here, the G_COORD and G_NUM arrays are read in as data. These extra READ statements mean that no geometry subroutine is required for this program. The element shape function derivatives are provided by the routine SHAPE_DER as usual. The only other obvious

```
    program p56
!-------------------------------------------------------------------------
!      program 5.6 three dimensional analysis of an elastic
!      solid using 4-node tetrahedral elements
!-------------------------------------------------------------------------
 use new_library      ;  use geometry_lib  ;     implicit none
 integer::nels,neq,nn,nr,nip,nodof=3,nod=4,nst=6,ndof,loaded_nodes,          &
         i,k,iel,ndim=3
 real:: e,v,det    ; character(len=15) :: element = 'tetrahedron'
!-------------------------- dynamic arrays-----------------------------------
 real    ,allocatable :: kv(:),loads(:),points(:,:),dee(:,:),coord(:,:),     &
                         jac(:,:),weights(:), der(:,:), deriv(:,:),bee(:,:), &
                         km(:,:),eld(:),sigma(:),g_coord(:,:)
 integer, allocatable :: nf(:,:), g(:), kdiag(:) ,num(:) ,g_num(:,:),g_g(:,:)
!-------------------------input and initialisation---------------------------
  open (10,file='p56.dat',status=    'old',action='read')
  open (11,file='p56.res',status='replace',action='write')
  read (10,*) nels,nn,nip,e,v          ;     ndof=nod*nodof
  allocate ( nf(nodof,nn), points(nip,ndim),dee(nst,nst),coord(nod,ndim),    &
          jac(ndim,ndim),der(ndim,nod),deriv(ndim,nod),g(ndof),              &
          bee(nst,ndof), km(ndof,ndof),eld(ndof),sigma(nst),g_g(ndof,nels),&
          g_coord(ndim,nn),g_num(nod,nels),weights(nip),num(nod))
  read (10, *) g_coord ; read (10, *)  g_num
  nf=1; read(10,*) nr ;if(nr>0) read(10,*)(k,nf(:,k),i=1,nr)
       call formnf(nf); neq=maxval(nf) ; allocate ( loads(0:neq),kdiag(neq) )
  call deemat (dee,e,v); call sample(element,points,weights)
  kdiag=0
!  ------------- loop the elements to set up  g_g  and  kdiag  ----------------
  elements_1  :  do iel = 1 , nels
     num = g_num(:,iel); call num_to_g(num,nf,g)
     g_g(:,iel) = g ;  call fkdiag(kdiag,g)
  end do elements_1
  kdiag(1)=1; do i=2,neq; kdiag(i)=kdiag(i)+kdiag(i-1); end do
    allocate(kv(kdiag(neq)))  ; kv=0.0
   write(11,'(a)') "Global Coordinates"
   do k=1,nn;write(11,'(a,i5,a,3e12.4)')"Node",k,"       ",g_coord(:,k);end do
   write(11,'(a)') "Global Node Numbers"
   do k=1,nels; write(11,'(a,i5,a,4i5)')                                      &
                       "Element",k,"          ",g_num(:,k); end do
   write(11,'(2(a,i5))')                                                      &
        "There are ",neq,"  equations and the skyline storage is",kdiag(neq)
    loads=0.0 ; read (10,*) loaded_nodes,(k,loads(nf(:,k)),i=1,loaded_nodes)
   write(11,'(a,e12.4)') " The total load is ", sum(loads)
!--------------- element stiffness integration and assembly-------------------
 elements_2: do iel = 1 , nels
             num = g_num(:,iel) ; g = g_g(:,iel)
             coord = transpose(g_coord(:,num)) ;        km=0.0
    gauss_pts_1:  do i =1 , nip
               call shape_der (der,points,i);  jac = matmul(der,coord)
               det = determinant(jac);   call invert (jac)
               deriv = matmul(jac,der); call beemat (bee,deriv)
               km=km+matmul(matmul(transpose(bee),dee),bee)*det*weights(i)
    end do gauss_pts_1
    call fsparv (kv,km,g,kdiag)
 end do elements_2
!---------------------------equation solution---------------------------------
    call sparin(kv,kdiag) ;call spabac(kv,loads,kdiag)
    write(11,'(a)') "The nodal displacements are   :"
    write(11,'(a)') "  Node              Displacement"
    do k=1,nn; write(11,'(i5,a,3e12.4)')k,"      ",loads(nf(:,k)); end do
!----------------------recover stresses at element Gauss-points----------------
 elements_3 : do iel = 1 , nels
                  num = g_num(:,iel); coord = transpose(g_coord( : , num ))
                  g=g_g(:,iel)       ;         eld = loads( g )
```

```
                write(11,'(a,i5,a)')                                          &
                     "The Gauss point stresses for element",iel," are"
  gauss_pts_2: do i = 1,nip
                  call shape_der(der,points,i);jac = matmul(der,coord)
                  call invert(jac);   deriv= matmul(jac,der)
                  call beemat(bee,deriv);sigma = matmul(dee,matmul(bee,eld))
                  write(11,'(a,i5)') "Point  ",i ;write(11,'(6e12.4)')sigma
  end do gauss_pts_2
 end do elements_3
end program p56
```

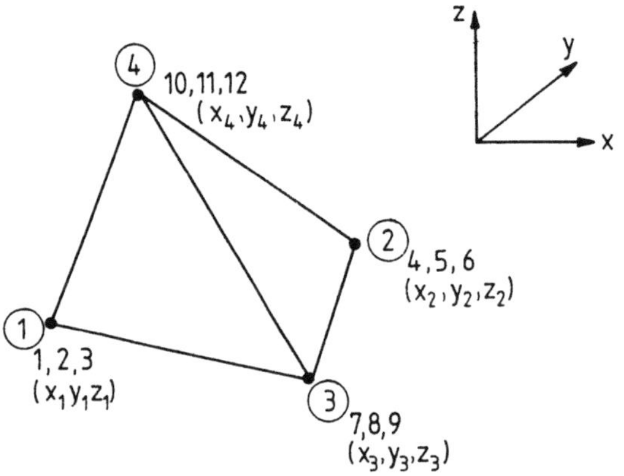

Figure 5.22 Local node and freedom numbering for four-node tetrahedron

difference between this program and Program 5.0 is that the former uses Gaussian elimination to solve the equilibrium equations, whereas this program uses Choleski's method with skyline storage.

The local element node and freedom numbering is indicated in Figure 5.22 and it is important that the nodal coordinates and freedoms are read in a consistent manner. The system adopted here is that nodes (1),(2) and (3) are in a clockwise direction as viewed from node (4).

The example and data given in Figure 5.23 represent a cube made up of six tetrahedra. One corner of the cube is fixed and the three adjacent faces are restrained to move only in their own planes. The four nodal forces applied are equivalent to a uniform vertical compressive stress of 1 kN/m^2.

The computed results given in Figure 5.24 show that the cube compresses uniformly and that the vertical stress σ_z at the centroid of each element is in equilibrium with the applied loads.

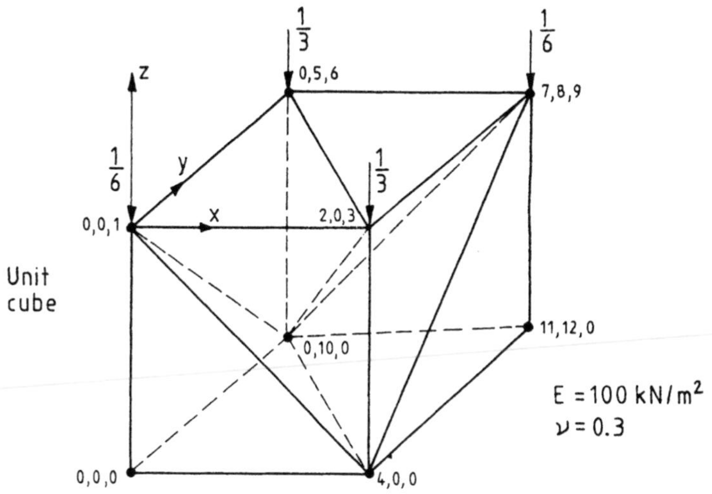

Structure and element data	NELS	NN	NIP	E	V
	6	8	1	100.	.3

```
G_COORD     .0    .0    .0
data        1.    .0    .0
            0.    0.    -1.
            1.    0.    -1.
            .0    1.    .0
            1.    1.    .0
            0.    1.    -1.
            1.    1.    -1.

G_NUM       1  3  4  7
data        1  4  2  7
            1  2  5  7
            6  4  8  7
            6  2  4  7
            6  5  2  7

Node        NR
freedom     7
data        K,NF(:,K),I=1,NR
            1 0 0 1 2 1 0 1 3 0 0 0 4 1 0 0
            5 0 1 1 7 0 1 0 8 1 1 0

Loads       LOADED_NODES
data        4
            (K,LOADS(NF(:,K)),K=1,LOADED_NODES)
            1 .0 .0 -.1667 2 .0 .0 -.333  5 .0 .0 -.333 6 .0 .0 -/1667
```

Figure 5.23 Mesh and data for Program 5.8

```
Global Coordinates
Node    1          0.0000E+00  0.0000E+00  0.0000E+00
Node    2          0.1000E+01  0.0000E+00  0.0000E+00
Node    3          0.0000E+00  0.0000E+00 -0.1000E+01
Node    4          0.1000E+01  0.0000E+00 -0.1000E+01
Node    5          0.0000E+00  0.1000E+01  0.0000E+00
Node    6          0.1000E+01  0.1000E+01  0.0000E+00
Node    7          0.0000E+00  0.1000E+01 -0.1000E+01
Node    8          0.1000E+01  0.1000E+01 -0.1000E+01
Global Node Numbers
Element    1             1    3    4    7
Element    2             1    4    2    7
Element    3             1    2    5    7
Element    4             6    4    8    7
Element    5             6    2    4    7
Element    6             6    5    2    7
There are     12  equations and the skyline storage is    69
 The total load is  -0.1000E+01
The nodal displacements are   :
  Node           Displacement
    1        0.0000E+00  0.0000E+00 -0.1000E-01
    2        0.3000E-02  0.0000E+00 -0.1000E-01
    3        0.0000E+00  0.0000E+00  0.0000E+00
    4        0.3000E-02  0.0000E+00  0.0000E+00
    5        0.0000E+00  0.3000E-02 -0.9999E-02
    6        0.3000E-02  0.3000E-02 -0.1000E-01
    7        0.0000E+00  0.3000E-02  0.0000E+00
    8        0.3000E-02  0.3000E-02  0.0000E+00
The Gauss point stresses for element     1 are
Point       1
 -0.1965E-04 -0.2100E-04 -0.1000E+01  0.0000E+00  0.0000E+00  0.0000E+00
The Gauss point stresses for element     2 are
Point       1
  0.2302E-05  0.1389E-04 -0.1000E+01 -0.6475E-05  0.2974E-04  0.2326E-04
The Gauss point stresses for element     3 are
Point       1
  0.1230E-04  0.2075E-05 -0.9999E+00  0.0000E+00  0.3641E-04  0.2974E-04
The Gauss point stresses for element     4 are
Point       1
 -0.1087E-04 -0.9023E-05 -0.1000E+01  0.1556E-05 -0.7467E-05 -0.9316E-05
The Gauss point stresses for element     5 are
Point       1
  0.6102E-05 -0.1297E-05 -0.1000E+01 -0.8752E-05 -0.2541E-04 -0.3189E-04
The Gauss point stresses for element     6 are
Point       1
  0.9809E-05  0.1536E-04 -0.9999E+00  0.2158E-05 -0.3632E-04 -0.4299E-04
```

Figure 5.24 Results from Program 5.6

PROGRAM 5.7: THREE-DIMENSIONAL ANALYSIS USING 14-NODED BRICK ELEMENTS

The simplest member of the hexahedral or "brick" element family has eight nodes, situated at the corners. The shape functions and their derivatives for this element are included in library subroutines SHAPE_FUN and SHAPE_DER respectively and can be used in programs (see Program 5.9). However the element is quite "stiff" in certain deformation modes and a more commonly available element in commercial programs is the 20-node brick (see Program 5.8). Intermediate 14-node elements have been proposed (Smith and Kidger, 1992). These have the eight corner nodes supplemented by six mid-

```
program p57
!-------------------------------------------------------------------------
! program 5.7 three-dimensional elastic analysis using 14-node brick elements
!-------------------------------------------------------------------------
   use new_library  ;          use   geometry_lib   ;  implicit  none
   integer ::nels,neq,nn,nr,nip,nodof=3,nod=14,nst=6,ndof,fixed_nodes,        &
             iel,i,k,ii,jj,kk,ll,ndim=3
   real     ::e,v,det  ; character(len=15) :: element = 'hexahedron'
!---------------------- dynamic  arrays ------------------------------------
   real     , allocatable :: dee(:,:),points(:,:),weights(:),                 &
                             coord(:,:),jac(:,:),der(:,:),deriv(:,:),        &
                             bee(:,:),km(:,:),eld(:),eps(:),sigma(:),        &
                             kv(:),loads(:),g_coord(:,:), value(:)
   integer  , allocatable ::g(:),nf(:,:),kdiag(:),num(:),g_num(:,:),g_g(:,:), &
                             no(:),sense(:),node(:)
! ---------------------  input and initialisation    ------------------------
   open(10,file='p57.dat',status=    'old',action='read')
   open(11,file='p57.res',status='replace',action='write')
   read(10,*) nels,nn,nip,e,v         ;          ndof = nod * nodof
   allocate( nf(nodof,nn),dee(nst,nst),coord(nod,ndim),num(nod),             &
             jac(ndim,ndim),der(ndim,nod),deriv(ndim,nod),g(ndof),           &
             bee(nst,ndof),km(ndof,ndof),eld(ndof),sigma(nst),eps(nst),      &
             g_g(ndof,nels),g_coord(ndim,nn),g_num(nod,nels))
    nf = 1; read(10,*) nr; if(nr>0) read(10,*)(k,nf(:,k),i=1,nr)
    call formnf(nf); neq = maxval(nf)     ; call deemat(dee,e,v)
   allocate(loads(0:neq), kdiag(neq)) ; loads = .0 ; kdiag = 0
   read(10,*) g_num ; read(10,*) g_coord(:,1:16)
   do i=17,nn
   read(10,*)ii,jj,kk,ll
        g_coord(:,i)=.25*(g_coord(:,ii)+g_coord(:,jj)+g_coord(:,kk)+        &
        g_coord(:,ll))
   end do
   write(11,'(a)') "Global coordinates "
   do k=1,nn;write(11,'(a,i5,a,3e12.4)')"Node",k,"       ",g_coord(:,k);end do
   write(11,'(a)') "Global node numbers "
   do k = 1 , nels; write(11,'(a,i5,a,14i4)')                                  &
                             "Element ",k,"        ",g_num(:,k); end do
! --------  loop the elements to set up global g and find kdiag ---------------
   elements_1 :  do iel = 1 , nels
                   num = g_num(:,iel)  ; call num_to_g (num , nf , g )
                   call fkdiag(kdiag,g);   g_g( : , iel ) = g
   end do elements_1
   kdiag(1)=1; do i=2,neq; kdiag(i)=kdiag(i)+kdiag(i-1); end do
   write(11,'(2(a,i5))')                                                        &
        "There are",neq," equations and the skyline storage is :",kdiag(neq)
   allocate( kv(kdiag(neq))) ;  kv = .0
! -----------------  element stiffness integration and assembly ---------------
  allocate(weights(nip),points(nip,ndim));call sample(element, points, weights)
 elements_2: do iel = 1 , nels
             num = g_num(:,iel) ; g = g_g(:,iel)
             coord = transpose(g_coord(:,num)) ;        km=0.0
   gauss_pts_1:  do i =1 , nip
               call shape_der(der,points,i);  jac = matmul(der,coord)
               det = determinant(jac);   call invert (jac)
               deriv = matmul(jac,der); call beemat (bee,deriv)
               km=km+matmul(matmul(transpose(bee),dee),bee)*det*weights(i)
   end do gauss_pts_1
  call fsparv (kv,km,g,kdiag)
 end do elements_2
 read(10,*) fixed_nodes
 if(fixed_nodes/=0) then
  allocate(no(fixed_nodes),node(fixed_nodes),                                 &
           sense(fixed_nodes),value(fixed_nodes))
  read(10,*)(node(i),sense(i),value(i),i=1,fixed_nodes)
```

```
   do i=1,fixed_nodes; no(i) = nf(sense(i),node(i)); end do
   kv(kdiag(no)) = kv(kdiag(no)) + 1.e20; loads(no)=kv(kdiag(no)) * value
 end if
!------------------------------equation solution------------------------------
    call sparin(kv,kdiag) ;call spabac(kv,loads,kdiag)
    write(11,'(a)') "The nodal displacements are"
    do k=1,nn; write(11,'(i5,a,3e12.4)') k,"    ",loads(nf(:,k)); end do
!------------------recover stresses at element centroids----------------------
 nip = 1; deallocate(points,weights); allocate(points(nip,ndim),weights(nip))
 elements_3 : do iel = 1 , nels
                  num = g_num(:,iel); coord = transpose(g_coord( : , num ))
                  g=g_g(:,iel)       ;          eld = loads( g )
                  write(11,'(a,i5,a)')                                       &
                           "The centroid stresses for element",iel,"  are"
    gauss_pts_2: do i = 1 , nip
                    call shape_der(der,points,i);jac = matmul(der,coord)
                    call invert(jac);    deriv= matmul(jac,der)
                    call beemat(bee,deriv);sigma = matmul(dee,matmul(bee,eld))
                    write(11,'(a,i5)') "Point",i   ; write(11,'(6e12.4)') sigma
   end do gauss_pts_2
 end do elements_3
end program p57
```

face nodes and one of them ("Type 6") is used in the present program (see section 3.11.1.8).

The program uses the skyline storage system introduced in Program 5.2 (3-d problems benefit especially from this method). It also allows nodal freedoms

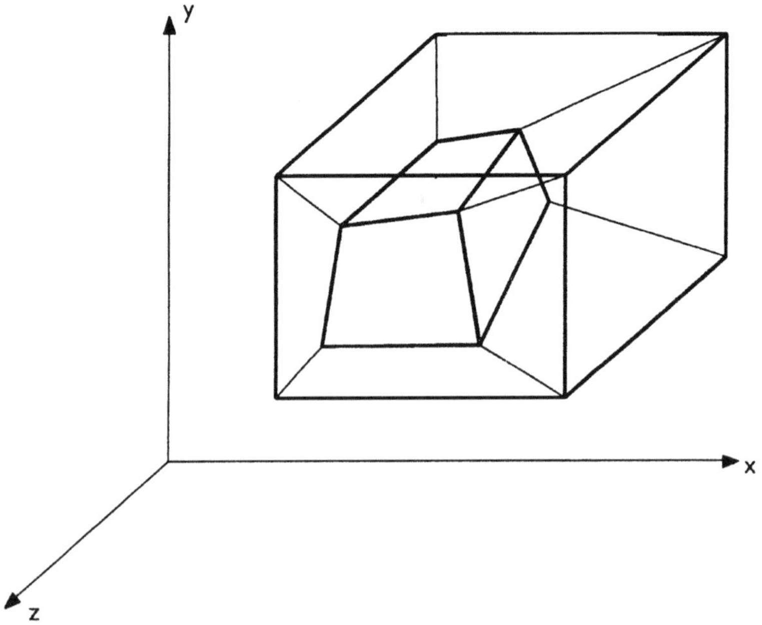

Figure 5.25 Patch test mesh for Program 5.7

to be assigned fixed values by the "penalty" method described in section 3.8. For all `FIXED_NODES` the node number is read in, followed by the "sense" (1 for the x direction, 2 for the y direction and 3 for the z direction) and the fixed value of displacement to be assigned. In the (unusual) case where a node is fixed in more than one sense, it would have to be specified more than once.

The mesh analysed is shown in Figure 5.25. This is a "patch" test mesh suggested by Peano (1987) for testing the admissibility of solid elements. Due to its irregularity, the global node numbers, `G_NUM`, are read directly into the program, followed by the global coordinates, `G_COORD`, of the eight corners of the external regular box and the eight corners of the internal

```
Mesh and          NELS NN   NIP  E    V
element data      7    40   27   1.   .49

Node freedom      NR
data              10
                  K,NF(:,K),I=1,NR
                  1 0 1 1 2 0 1 0  3 0 0 1  4 0 0 0  6 1 1 0
                  7 1 0 1 8 1 0 0 17 0 1 1 20 1 1 0 21 1 0 1
Connectivity      G_NUM
data              12 10 16 14 11 9 15 13 26 25 27 24 23 28
                  4 2 12 10 3 1 11 9 34 33 37 29 17 23
                  10 2 14 6 9 1 13 5 30 32 24 18 29 31
                  16 14 8 6 15 13 7 5 35 36 39 31 28 22
                  4 12 8 16 3 11 7 15 38 40 21 27 37 39
                  4 2 8 6 12 10 16 14 20 26 38 30 34 35
                  11 9 15 13 3 1 7 5 25 19 40 32 33 36
Geometry          G_COORD
data              0. 1. 1.
                  0. 1. 0.
                  0. 0. 1.
                  0. 0. 0.
                  1. 1. 1.
                  1. 1. 0.
                  1. 0. 1.
                  1. 0. 0.
                  .165 .745 .702
                  .272 .750 .230
                  .320 .186 .643
                  .249 .342 .192
                  .788 .693 .644
                  .850 .649 .263
                  .677 .305 .683
                  .826 .288 .288
```

Figure 5.26 (*continued*)

```
                II,JJ,KK,LL
                1 2 3 4
                1 2 5 6
                1 3 5 7
                2 4 6 8
                3 4 7 8
                5 6 7 8
                9  10 11 12
                9  10 13 14
                9  11 13 15
                10 12 14 16
                11 12 15 16
                13 14 15 16
                1 2  9 10
                2 6 10 14
                5 6 13 14
                1 5  9 13
                1 3  9 11
                2 4 10 12
                6 8 14 16
                5 7 13 15
                3 4 11 12
                4 8 12 16
                7 8 15 16
                3 7 11 15
Fixity data     FIXED_NODES
                5
                NODE,SENSE,VALUE
                1  3 .01
                3  3 .01
                5  3 .01
                7  3 .01
                19 3 .01
```

Figure 5.26 Data for Program 5.7

irregular one. The remaining global coordinates are obtained by reading in some nodal data enabling interpolation to be performed. There are five nodes to be fixed on the top face of the exterior box and these are given a vertical displacement of 0.01 (see Figure 5.26). The results are shown in Figure 5.27 where it can be seen that a completely homogeneous strain field with a vertical strain of 0.01 has been produced in the mesh. Thus, the element passes the patch test.

Users could experiment with different orders of integration via `NIP`.

```
Global coordinates
Node    1          0.0000E+00  0.1000E+01  0.1000E+01
Node    2          0.0000E+00  0.1000E+01  0.0000E+00
Node    3          0.0000E+00  0.0000E+00  0.1000E+01
Node    4          0.0000E+00  0.0000E+00  0.0000E+00
Node    5          0.1000E+01  0.1000E+01  0.1000E+01
Node    6          0.1000E+01  0.1000E+01  0.0000E+00
Node    7          0.1000E+01  0.0000E+00  0.1000E+01
.....................................................................

Node   35          0.9190E00  0.4842E00  0.1377E00
Node   36          0.8663E00  0.4995E00  0.8318E00
Node   37          0.1423E00  0.1320E00  0.4588E00
Node   38          0.5187E00  0.1575E00  0.1200E00
Node   39          0.8758E00  0.1482E00  0.4928E00
Node   40          0.4992E00  0.1227E00  0.8315E00
Global node numbers
Element     1          12  10  16  14  11   9  15  13  26  25  27  24  23  28
Element     2           4   2  12  10   3   1  11   9  34  33  37  29  17  23
Element     3          10   2  14   6   9   1  13   5  30  32  24  18  29  31
Element     4          16  14   8   6  15  13   7   5  35  36  39  31  28  22
Element     5           4  12   8  16   3  11   7  15  38  40  21  27  37  39
Element     6           4   2   8   6  12  10  16  14  20  26  38  30  34  35
Element     7          11   9  15  13   3   1   7   5  25  19  40  32  33  36
There are  105  equations and the skyline storage is : 5487
The nodal displacements are
    1       0.0000E+00 -0.4900E-02  0.1000E-01
    2       0.0000E+00 -0.4900E-02  0.0000E+00
    3       0.0000E+00  0.0000E+00  0.1000E-01
    4       0.0000E+00  0.0000E+00  0.0000E+00
    5      -0.4900E-02 -0.4900E-02  0.1000E-01
    6      -0.4900E-02 -0.4900E-02  0.0000E+00
    7      -0.4900E-02  0.0000E+00  0.1000E-01
.................................................................

   35      -0.4503E-02 -0.2373E-02  0.1378E-02
   36      -0.4245E-02 -0.2448E-02  0.8318E-02
   37      -0.6970E-03 -0.6468E-03  0.4588E-02
   38      -0.2542E-02 -0.7718E-03  0.1200E-02
   39      -0.4291E-02 -0.7264E-03  0.4928E-02
   40      -0.2446E-02 -0.6015E-03  0.8315E-02
The centroid stresses for element    1  are
Point    1
 -0.9061E-07 -0.9061E-07  0.1000E-01 -0.2070E-07  0.5048E-07 -0.1765E-07
The centroid stresses for element    2  are
Point    1
  0.1226E-06  0.1822E-06  0.1000E-01 -0.1131E-08  0.5403E-07  0.2040E-07
The centroid stresses for element    3  are
Point    1
 -0.1894E-07 -0.3384E-07  0.1000E-01  0.1599E-07  0.3179E-07  0.1303E-08
The centroid stresses for element    4  are
Point    1
 -0.5127E-07 -0.9597E-07  0.1000E-01 -0.8962E-08  0.3368E-07 -0.2375E-07
The centroid stresses for element    5  are
Point    1
  0.1101E-06  0.1101E-06  0.1000E-01 -0.6355E-08  0.4488E-07 -0.2194E-07
The centroid stresses for element    6  are
Point    1
  0.3345E-07 -0.2615E-07  0.1000E-01 -0.6926E-08  0.8810E-09 -0.5496E-08
The centroid stresses for element    7  are
Point    1
 -0.1881E-06 -0.1136E-06  0.1000E-01 -0.1689E-07  0.4867E-07 -0.2418E-08
```

Figure 5.27 Results from Program 5.7

PROGRAM 5.8: THREE-DIMENSIONAL ANALYSIS USING 20-NODED BRICK ELEMENTS

A more widely used 3-d element, the 20-node hexahedron, is the subject of the next program in this chapter. The element is the 3-d analogue of the eight-noded quadrilateral used in plane problems and described in Program 5.3. Storage requirements rapidly become substantial on even the most powerful computers and later programs in this chapter (5.10–5.11) investigate mesh-free solution methods.

To reduce storage requirements while retaining an assembly technique, the "skyline" storage strategy is adopted.

In programming terms, an additional integer column array `KDIAG` is required as originally introduced in Program 5.2. A preliminary scan of all elements is performed. For each element, the `G` vector is formed and library subroutine `FKDIAG` finds the greatest bandwidth associated with each freedom. This variable bandwidth is accumulated in the vector `KDIAG` as described for a simple case in Figure 3.19. Finally, the total storage requirement for the stiffness matrix is given by `KDIAG(NEQ)`, where `NEQ` is the total number of freedoms. This figure should be considerably less than `NEQ*(NBAND+1)` which is the length of the `KV` vector assuming a constant bandwidth. It may be noted that `NBAND` is no longer required. A saving in storage of the order of 30% is often achieved using this method, and further economies could be found using a bandwidth optimiser.

Due to the storage strategy, appropriate subroutines are required to assemble the global stiffness matrix in this pattern and also to solve the equilibrium equations. Assembly of the global stiffness matrix is performed by the library subroutine `FSPARV` and Choleski factorisation and back-substitution performed by `SPARIN` and `SPABAC` respectively.

Other changes to the program to account for the new element are limited to routine `GEOMETRY_20BXZ` to generate the `COORD` and `NUM` arrays.

The geometry routine assumes that nodes are numbered in xz planes moving in the y direction, as illustrated in Figure 5.28. The node numbering at the element level is given in Figure 5.29.

The example and data of Figure 5.30 are for a simple boundary value problem where an elastic block carries a uniform load over part of its surface. An exact integration scheme has been used (`NIP=27`) in the present analysis. The nodal forces to simulate a uniform stress field involve corner loads which act in the opposite direction to the mid-side loads (Appendix 1).

The computed displacements and centroid stresses are given in Figure 5.31. It can be noted that the length of the vector `KV` was reduced from 7068 to 4388 by the use of the "skyline" storage strategy – an improvement of approximately 38%.

Full Gaussian (27-point) integration is expensive and the reader can experiment with Irons's (1971) approximate rules. Values of `NIP` of 6, 14 and 15 are allowed and in the latter two cases, very similar results to those shown in Figure 5.31 will be found.

```
    program p58
!-------------------------------------------------------------------------
!      program 5.8 three dimensional analysis of an elastic
!      solid using uniform 20-node hexahedral brick elements
!-------------------------------------------------------------------------
 use new_library      ;  use geometry_lib  ;     implicit none
 integer::nels,nxe,nze,neq,nn,nr,nip,nodof=3,nod=20,nst=6,ndof,loaded_nodes, &
        i,k,iel,ndim=3
 real::aa,bb,cc,e,v,det    ;  character(len=15) :: element = 'hexahedron'
!------------------------ dynamic arrays---------------------------------------
 real    ,allocatable :: kv(:),loads(:),points(:,:),dee(:,:),coord(:,:),     &
                         jac(:,:),weights(:), der(:,:), deriv(:,:),bee(:,:), &
                         km(:,:),eld(:),sigma(:),g_coord(:,:)
 integer, allocatable :: nf(:,:), g(:), kdiag(:) ,num(:) ,g_num(:,:),g_g(:,:)
!----------------------input and initialisation-------------------------------
  open (10,file='p58.dat',status=    'old',action='read')
  open (11,file='p58.res',status='replace',action='write')
  read (10,*) nels,nxe,nze,nn,nip,aa,bb,cc,e,v       ;  ndof=nod*nodof
  allocate ( nf(nodof,nn), points(nip,ndim),dee(nst,nst),coord(nod,ndim),     &
            jac(ndim,ndim),der(ndim,nod),deriv(ndim,nod),g(ndof),             &
            bee(nst,ndof), km(ndof,ndof),eld(ndof),sigma(nst),g_g(ndof,nels),&
            g_coord(ndim,nn),g_num(nod,nels),weights(nip),num(nod))
  nf=1; read(10,*) nr ;if(nr>0) read(10,*)(k,nf(:,k),i=1,nr)
        call formnf(nf); neq=maxval(nf)
  allocate  ( loads(0:neq),kdiag(neq)  )
  call deemat (dee,e,v); call sample(element,points,weights)
  kdiag=0
!--------- loop the elements to set up global geometry and kdiag  -------------
  elements_1  :  do iel = 1 , nels
     call geometry_20bxz(iel,nxe,nze,aa,bb,cc,coord,num)
     call num_to_g(num,nf,g);  call fkdiag(kdiag,g);g_num(:,iel)=num
     g_coord(:,num)=transpose(coord); g_g(:,iel)=g
  end do elements_1
  kdiag(1)=1; do i=2,neq; kdiag(i)=kdiag(i)+kdiag(i-1); end do
    allocate(kv(kdiag(neq)))  ; kv=0.0
    write(11,'(a)') "Global coordinates "
    do k=1,nn;write(11,'(a,i5,a,3e12.4)')"Node",k,"       ",g_coord(:,k);end do
    write(11,'(a)') "Global node numbers "
    do k = 1 , nels; write(11,'(a,i5,a,20i3)')                                &
                        "Element ",k,"    ",g_num(:,k); end do
    write(11,'(2(a,i5))')                                                     &
          "There are",neq,"  equations and the skyline storage is",kdiag(neq)
    loads=0.0 ; read (10,*) loaded_nodes,(k,loads(nf(:,k)),i=1,loaded_nodes)
    write(11,'(a,e12.4)') "The total load is ",sum(loads)
!--------------- element stiffness integration and assembly--------------------
 elements_2: do iel = 1 , nels
             num = g_num(:,iel) ; g = g_g(:,iel)
             coord = transpose(g_coord(:,num)) ;        km=0.0
    gauss_pts_1:  do i =1 , nip
               call shape_der (der,points,i);  jac = matmul(der,coord)
               det = determinant(jac) ;   call invert (jac)
               deriv = matmul(jac,der); call beemat (bee,deriv)
               km=km+matmul(matmul(transpose(bee),dee),bee)*det*weights(i)
    end do gauss_pts_1
    call fsparv (kv,km,g,kdiag)
 end do elements_2
!--------------------------equation solution----------------------------------
    call sparin(kv,kdiag) ;call spabac(kv,loads,kdiag)
    write(11,'(a)') "The nodal displacements are"
    do k=1,nn; write(11,'(i5,a,3e12.4)') k,"    ",loads(nf(:,k)); end do
!------------------recover stresses at element centroids----------------------
  nip = 1; deallocate(points,weights); allocate(points(nip,ndim),weights(nip))
 elements_3 : do iel = 1 , nels
                 num = g_num(:,iel); coord = transpose(g_coord( : , num ))
```

```
                g=g_g(:,iel)        ;          eld = loads( g )
                write(11,'(a,i5,a)')                                    &
                       "The centroid stresses for element",iel,"  are"
   gauss_pts_2: do i = 1 , nip
                  call shape_der(der,points,i);jac = matmul(der,coord)
                  call invert(jac);   deriv= matmul(jac,der)
                  call beemat(bee,deriv);sigma = matmul(dee,matmul(bee,eld))
                  write(11,'(a,i5)') "Point",i   ; write(11,'(6e12.4)') sigma
  end do gauss_pts_2
 end do elements_3
end program p58
```

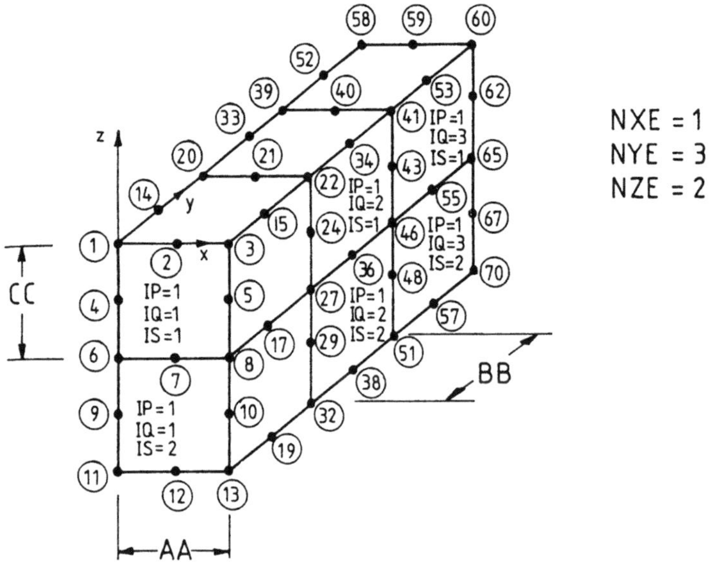

Figure 5.28 Global node and element numbering for 20-node brick

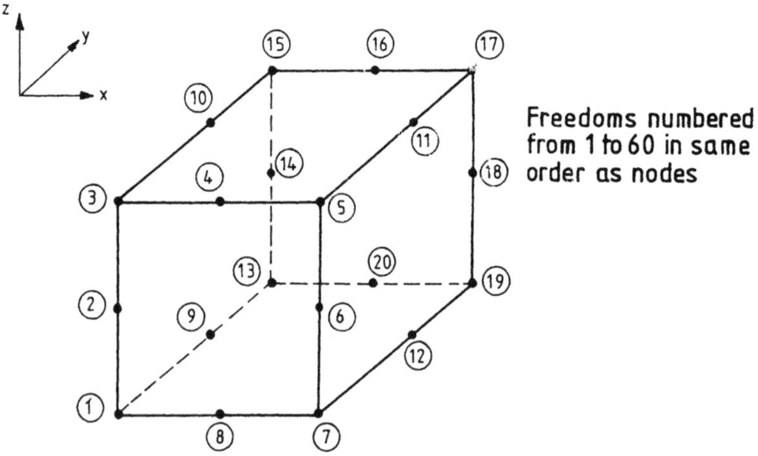

Figure 5.29 Local node numbering for 20-node brick

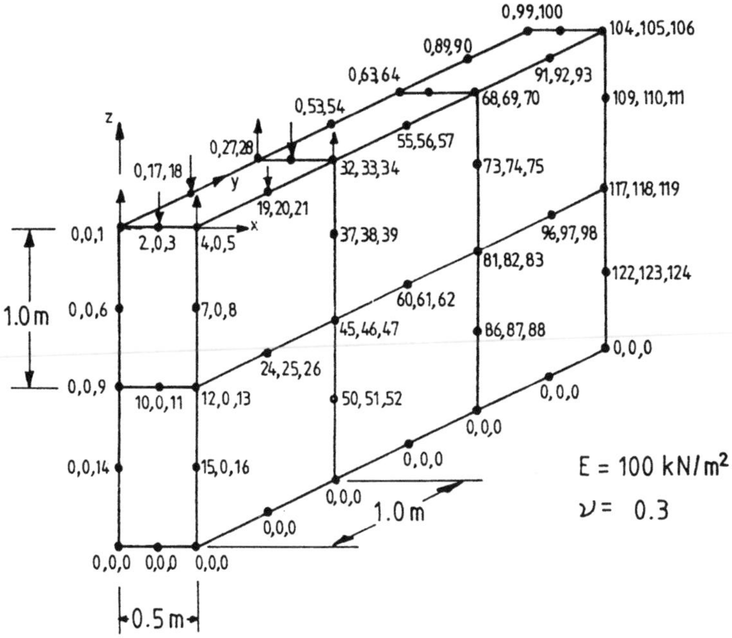


```
Structure and   NELS  NXE  NZE  NN    NIP
element data    6     1    2    70    27

                AA    BB   CC   E     V
                .5    1.   1.   100.  .3

Nodal freedom   NR
data            46
                K,NF(:,K),I=1,NR
                1 0 0 1 2 1 0 1 3 1 0 1 4 0 0 1 5 1 0 1 6 0 0 1
                7 1 0 1 8 1 0 1 9 0 0 1 10 1 0 1 11 0 0 0 12 0 0 0 13 0 0 0 14 0 1 1
                16 0 1 1 18 0 0 0 19 0 0 0 20 0 1 1 23 0 1 1 25 0 1 1 28 0 1 1 30 0 0 0
                31 0 0 0 32 0 0 0 33 0 1 1 35 0 1 1 37 0 0 0 38 0 0 0 39 0 1 1 42 0 1 1
                44 0 1 1 47 0 1 1 49 0 0 0 50 0 0 0 51 0 0 0 52 0 1 1 54 0 1 1 56 0 0 0
                57 0 0 0 58 0 1 1 61 0 1 1 63 0 1 1 66 0 1 1 68 0 0 0 69 0 0 0 70 0 0 0

Loads           LOADED_NODES
data            8
                (K,LOADS(NF(:,K)),I=1,LOADED_NODES)
                1 .0 .0 .0417 2 .0 .0 -.1667 3 .0 .0 .0417 14 .0 .0 -.1667
                15 .0 .0 -.1667 20 .0 .0 .0417 21 .0 .0 -.1667 22 .0
                  .0 .0417
```

Figure 5.30 Mesh and data for Program 5.8

```
Global coordinates
Node    1          0.0000E+00  0.0000E+00  0.0000E+00
Node    2          0.2500E00  0.0000E+00  0.0000E+00
Node    3          0.5000E00  0.0000E+00  0.0000E+00
Node    4          0.0000E+00  0.0000E+00 -0.5000E00
Node    5          0.5000E00  0.0000E+00 -0.5000E00
...............................................................................

Node   67          0.5000E00  0.3000E+01 -0.1500E+01
Node   68          0.0000E+00  0.3000E+01 -0.2000E+01
Node   69          0.2500E00  0.3000E+01 -0.2000E+01
Node   70          0.5000E00  0.3000E+01 -0.2000E+01
Global node numbers
Element       1       6  4  1  2  3  5  8  7 16 14 15 17 25 23 20 21 22 24 27 26
Element       2      11  9  6  7  8 10 13 12 18 16 17 19 30 28 25 26 27 29 32 31
Element       3      25 23 20 21 22 24 27 26 35 33 34 36 44 42 39 40 41 43 46 45
Element       4      30 28 25 26 27 29 32 31 37 35 36 38 49 47 44 45 46 48 51 50
Element       5      44 42 39 40 41 43 46 45 54 52 53 55 63 61 58 59 60 62 65 64
Element       6      49 47 44 45 46 48 51 50 56 54 55 57 68 66 63 64 65 67 70 69
There are  124  equations and the skyline storage is 4388
The total load is  -0.5000E00
The nodal displacements are
    1      0.0000E+00  0.0000E+00 -0.1606E-01
    2      0.1344E-02  0.0000E+00 -0.1616E-01
    3      0.2687E-02  0.0000E+00 -0.1658E-01
    4      0.0000E+00  0.0000E+00 -0.1168E-01
...............................................................................

   67     -0.4145E-05  0.8230E-03 -0.7522E-04
   68      0.0000E+00  0.0000E+00  0.0000E+00
   69      0.0000E+00  0.0000E+00  0.0000E+00
   70      0.0000E+00  0.0000E+00  0.0000E+00
The centroid stresses for element    1  are
Point     1
  0.4610E-02 -0.1648E00 -0.8883E00  0.2472E-02  0.8305E-01  0.2310E-02
The centroid stresses for element    2  are
Point     1
  0.2584E-02 -0.6851E-01 -0.6645E00 -0.6375E-03  0.8298E-01  0.6887E-02
The centroid stresses for element    3  are
Point     1
 -0.6057E-02 -0.9576E-01 -0.1171E00  0.1028E-02  0.8193E-01 -0.2192E-02
The centroid stresses for element    4  are
Point     1
  0.2348E-02 -0.9292E-01 -0.2653E00 -0.1321E-02  0.1252E00  0.3128E-02
The centroid stresses for element    5  are
Point     1
 -0.5789E-03 -0.1013E-01 -0.2536E-02 -0.1240E-02 -0.5978E-02  0.6601E-03
The centroid stresses for element    6  are
Point     1
  0.2176E-02 -0.4532E-01 -0.2886E-01 -0.1753E-03  0.5598E-01  0.3086E-04
```

Figure 5.31 Results from Program 5.8

PROGRAM 5.9: GENERAL PROGRAM FOR ANALYSIS OF ELASTIC SOLIDS IN PLANE STRAIN AND 3-D

Perusal of Programs 5.1, 5.3 and 5.6–5.8 will show that they are essentially identical. The shape function and derivative library subroutines `SHAPE_FUN` and `SHAPE_DER`, the **B** matrix `BEEMAT` and the **D** matrix routine `DEEMAT` can all choose from the data given to the program which element to use for 2-d (plane strain) and 3-d conditions. Program 5.9 utilises this identity of programs to create a single general program. Of course such a program will expect to read the nodal geometry and connectivity details from a file produced by a mesh-

```
program p59
!-------------------------------------------------------------------------
!      program 5.9 general analysis of elastic solids
!-------------------------------------------------------------------------
 use new_library       ;      use geometry_lib    ;    implicit none
 integer::nels,neq,nband,nn,nr,nip,nodof,nod,nst,ndof, &
          i,k,iel,ndim,loaded_nodes,fixed_nodes,nprops,np_types
 real:: det             ;  character (len=15) :: element
!------------------------- dynamic arrays---------------------------------
 real     ,allocatable :: kv(:),loads(:),points(:,:),dee(:,:),coord(:,:),   &
                          jac(:,:),der(:,:),deriv(:,:),weights(:),prop(:,:),&
                          bee(:,:),km(:,:),eld(:),sigma(:),g_coord(:,:),    &
                          value(:)
 integer, allocatable :: nf(:,:), g(:) ,num(:), g_num(:,:) , g_g( :, :),    &
                          no(:),sense(:),node(:) , etype(:)
!-----------------------input and initialisation--------------------------
  open (10,file='p59.dat',status=    'old',action='read')
  open (11,file='p59.res',status='replace',action='write')
  read (10,*) element,nels,nn,nip,nodof,nod,nst,ndim     ;  ndof=nod*nodof
  allocate ( nf(nodof,nn), points(nip,ndim),dee(nst,nst), g_coord(ndim,nn),   &
           coord(nod,ndim),etype(nels),jac(ndim,ndim),weights(nip),num(nod), &
           g_num(nod,nels),der(ndim,nod),deriv(ndim,nod),bee(nst,ndof),      &
           km(ndof,ndof),eld(ndof),sigma(nst),g(ndof),g_g(ndof,nels))
        read(10,*) nprops , np_types
        allocate(prop(nprops,np_types))    ;  read (10,*) prop
        etype=1 ; if(np_types>1) read(10,*) etype
        read(10,*) g_coord ; read(10,*) g_num
       nf=1; read(10,*) nr ; if(nr>0) read(10,*) (k,nf(:,k),i=1,nr)
       call formnf(nf); neq=maxval(nf) ; call sample(element,points,weights)
!------------- loop the elements to find nband and store steering vectors  ----
     nband=0
   elements_1   : do iel =1,nels
                 num=g_num(:,iel) ; call num_to_g(num,nf,g);
                 g_g( : ,iel ) = g; if(nband<bandwidth(g))nband=bandwidth(g)
   end do elements_1
    write(11,'(a)') "Global coordinates "
    do k=1,nn;write(11,'(a,i5,a,3e12.4)')"Node",k,"       ",g_coord(:,k);end do
    write(11,'(a)') "Global node numbers "
    do k = 1 , nels; write(11,'(a,i5,a,27i3)')                                 &
                               "Element ",k,"    ",g_num(:,k); end do
      write(11,'(2(a,i5))')                                                    &
                "There are",neq,"  equations and the half-bandwidth is",nband
   allocate( kv(neq*(nband+1)),loads(0:neq)); kv= .0  ; loads =.0
!--------------- element stiffness integration and assembly--------------------
  elements_2: do iel=1,nels
               call deemat(dee,prop(1,etype(iel)),prop(2,etype(iel)))
               num = g_num(:,iel); coord = transpose(g_coord(: , num))
               g=g_g(:,iel)     ;    km = .0
               integrating_pts_1:  do i=1,nip
                 call shape_der(der,points,i); jac=matmul(der,coord)
                 det= determinant(jac) ; call invert(jac)
                 deriv = matmul(jac,der);call beemat(bee,deriv)
                 km= km+matmul(matmul(transpose(bee),dee),bee)*det*weights(i)
               end do integrating_pts_1
              call formkv (kv,km,g,neq)
  end do elements_2
 read(10,*) loaded_nodes
 if(loaded_nodes/=0)read (10,*)(k,loads(nf(:,k)),i=1,loaded_nodes)
  read(10,*) fixed_nodes
 if(fixed_nodes/=0)then
         allocate(node(fixed_nodes),sense(fixed_nodes),                        &
                  value(fixed_nodes),no(fixed_nodes))
         read(10,*)(node(i),sense(i),value(i),i=1,fixed_nodes)
         do i=1,fixed_nodes; no(i)=nf(sense(i),node(i)); end do
```

```
          kv(no)=kv(no) + 1.e20; loads(no) = kv(no) * value
 end if
!-----------------------------equation solution-------------------------------
    call banred(kv,neq) ;call bacsub(kv,loads)
    write(11,'(a)') "The nodal displacements are:"
    do k=1,nn; write(11,'(i5,a,3e12.4)') k,"    ",loads(nf(:,k)); end do
!---------------------recover stresses at element Gauss-points-----------------
 elements_3:do iel=1,nels
             num = g_num(:,iel) ; coord = transpose(g_coord(:,num))
             g = g_g( : , iel )  ;  eld = loads(g)
             call deemat(dee,prop(1,etype(iel)),prop(2,etype(iel)))
             write(11,'(a,i5,a)')                                              &
                     "The Gauss point stresses for element",iel,"  are :"
             integrating_pts_2: do i=1,nip
                 call shape_der(der,points,i); jac=matmul(der,coord)
                 call invert(jac); deriv=matmul(jac,der)
                 call beemat(bee,deriv); sigma=matmul(dee,matmul(bee,eld))
                 write(11,'(a,i5)') "Point",i  ;  write(11,'(6e12.4)') sigma
             end do integrating_pts_2
 end do elements_3
end program p59
```

generation pre-processor. The element types provided in the library are, for plane strain:

3, 6 and 15-node triangles

4, 8 and 9-node quadrilaterals

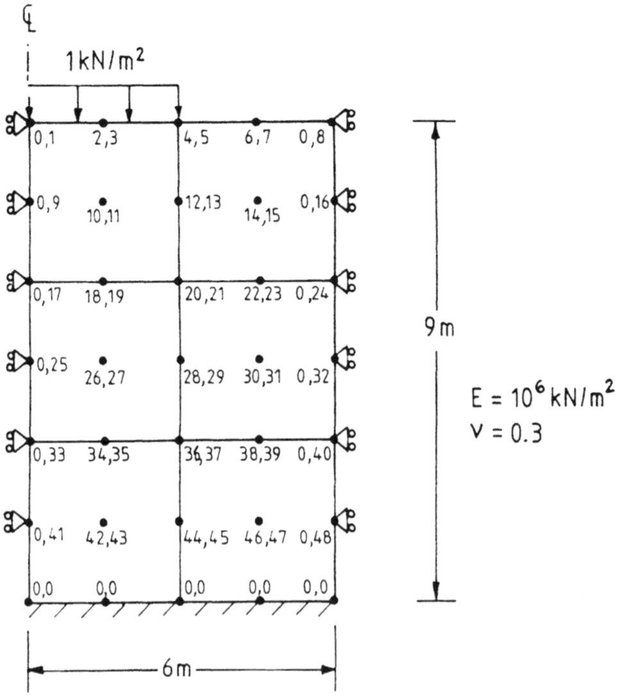

Figure 5.32 Nine-node quadrilateral mesh analysed by Program 5.9

and for 3-d

4-node tetrahedra

8, 14 and 20-node hexahedra (bricks)

An example using nine-node "Lagrangian" elements in plane strain is shown in Figure 5.32 with the local node numbering in Figure 5.33.

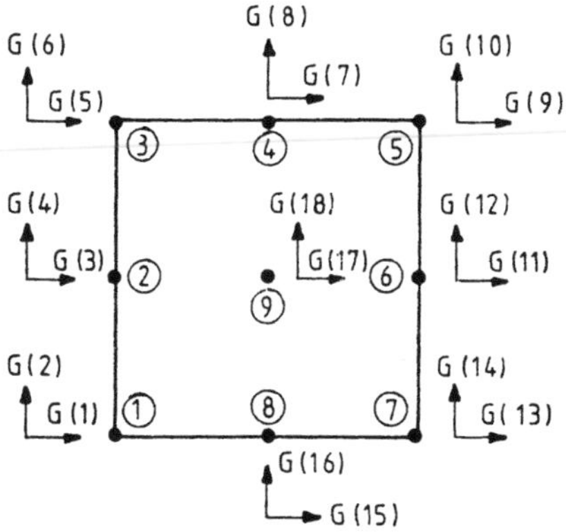

Figure 5.33 Local node numbering for nine-node quadrilateral

```
Mesh and       ELEMENT            NELS  NN  NIP  NODOF  NOD  NST  NDIM
element data   'QUADRILATERAL'    6     35  9    2      9    3    2

Property       NPROPS NP_TYPES
data           2      1
               PROP
               1000000.0  0.3
Geometry       G_COORD
data           0.0000E+00  0.0000E+00
               0.1500E+01  0.0000E+00
               0.3000E+01  0.0000E+00
               0.4500E+01  0.0000E+00
               0.6000E+01  0.0000E+00
               0.0000E+00  -0.1500E+01
               0.1500E+01  -0.1500E+01
               0.3000E+01  -0.1500E+01
               0.4500E+01  -0.1500E+01
               0.6000E+01  -0.1500E+01
               0.0000E+00  -0.3000E+01
               0.1500E+01  -0.3000E+01
               0.3000E+01  -0.3000E+01
               0.4500E+01  -0.3000E+01
               0.6000E+01  -0.3000E+01
```

Figure 5.34 (*continued*)

```
               0.0000E+00  -0.4500E+01
               0.1500E+01  -0.4500E+01
               0.3000E+01  -0.4500E+01
               0.4500E+01  -0.4500E+01
               0.6000E+01  -0.4500E+01
               0.0000E+00  -0.6000E+01
               0.1500E+01  -0.6000E+01
               0.3000E+01  -0.6000E+01
               0.4500E+01  -0.6000E+01
               0.6000E+01  -0.6000E+01
               0.0000E+00  -0.7500E+01
               0.1500E+01  -0.7500E+01
               0.3000E+01  -0.7500E+01
               0.4500E+01  -0.7500E+01
               0.6000E+01  -0.7500E+01
               0.0000E+00  -0.9000E+01
               0.1500E+01  -0.9000E+01
               0.3000E+01  -0.9000E+01
               0.4500E+01  -0.9000E+01
               0.6000E+01  -0.9000E+01
Connectivity   G_NUM
data           11  6  1  2  3  8  13  12  7
               21  16  11  12  13  18  23  22  17
               31  26  21  22  23  28  33  32  27
               13  8  3  4  5  10  15  14  9
               23  18  13  14  15  20  25  24  19
               33  28  23  24  25  30  35  34  29
Node           NR
freedom        17
data
               K,NF(:,K),I=1,NR
               1  0  1  5  0  1  6  0  1  10  0  1  11  0  1  15  0  1
               16  0  1  20  0  1  21  0  1  25  0  1  26  0  1  30  0  1
               31  0  0  32  0  0  33  0  0  34  0  0  35  0  0
Loads          LOADED_NODES
data           3
               (K,LOADS(NF(:,K)),I=1,LOADED_NODES)
               1  0.0  -0.5
               2  0.0  -2.0
               3  0.0  -0.5
               FIXED_NODES
               0
```

Figure 5.34 Input data for Program 5.9

Input data is shown in Figure 5.34 and the resulting output in Figure 5.35. This analysis is deliberately chosen to allow a comparison with a similar problem previously analysed by Program 5.3 using eight-node elements, the results of which were given as Figure 5.16.

```
Global coordinates
Node    1          0.0000E+00  0.0000E+00
Node    2          0.1500E+01  0.0000E+00
.........................................................................

Node   34          0.4500E+01 -0.9000E+01
Node   35          0.6000E+01 -0.9000E+01
Global node numbers
Element     1      11  6  1  2  3  8 13 12  7
Element     2      21 16 11 12 13 18 23 22 17
Element     3      31 26 21 22 23 28 33 32 27
Element     4      13  8  3  4  5 10 15 14  9
Element     5      23 18 13 14 15 20 25 24 19
Element     6      33 28 23 24 25 30 35 34 29
There are   48  equations and the half-bandwidth is   20
The nodal displacements are:
    1     0.0000E+00 -0.5299E-05
    2    -0.4004E-06 -0.4988E-05
.........................................................................

   33     0.0000E+00  0.0000E+00
   34     0.0000E+00  0.0000E+00
   35     0.0000E+00  0.0000E+00
The Gauss point stresses for element    1  are :
Point    1
 -0.6323E00 -0.1035E+01 -0.4108E-01
Point    2
 -0.5703E00 -0.1047E+01  0.5463E-01
Point    3
 -0.3507E00 -0.6910E00  0.1917E00
Point    4
 -0.2434E00 -0.9108E00  0.2695E-01
Point    5
 -0.2597E00 -0.8766E00  0.1085E00
Point    6
 -0.1681E00 -0.5906E00  0.2173E00
Point    7
 -0.1395E00 -0.9083E00  0.5062E-01
Point    8
 -0.1846E00 -0.8070E00  0.1512E00
Point    9
 -0.1714E00 -0.5698E00  0.2648E00
..............................................................................

The Gauss point stresses for element    6  are :
Point    1
 -0.2307E00 -0.4805E00  0.6544E-01
Point    2
 -0.2436E00 -0.4162E00  0.4191E-01
Point    3
 -0.2740E00 -0.3928E00  0.1298E-01
Point    4
 -0.2245E00 -0.4855E00  0.4446E-01
Point    5
 -0.2283E00 -0.4382E00  0.2936E-01
Point    6
 -0.2449E00 -0.4208E00  0.8179E-02
Point    7
 -0.2137E00 -0.4886E00  0.3231E-01
Point    8
 -0.2059E00 -0.4572E00  0.2255E-01
Point    9
 -0.2063E00 -0.4448E00  0.6023E-02
```

Figure 5.35 Results from Program 5.9

PROGRAM 5.10: THREE-DIMENSIONAL ANALYSIS USING 20-NODE BRICK ELEMENTS : MESH-FREE STRATEGY USING PRECONDITIONED CONJUGATE GRADIENTS

```
program p510
!-------------------------------------------------------------------------
!      program 5.10 three dimensional analysis of an elastic
!      solid using 20-node brick elements
!      preconditioned conjugate gradient solver ;  only integrate one element
!      diagonal preconditioner diag_precon
!-------------------------------------------------------------------------
 use new_library  ;   use  geometry_lib    ;      implicit none
 integer::nxe,nze,neq,nn,nr,nip,nodof=3,nod=20,nst=6,ndof,loaded_nodes,      &
         i,k,m,ndim=3,iters,limit,iel,nels
 real::aa,bb,cc,e,v,det,tol,up,alpha,beta
 character(len=15)::element= 'hexahedron';    logical :: converged
!------------------------- dynamic arrays---------------------------------
 real    ,allocatable :: points(:,:),dee(:,:),coord(:,:), weights(:),        &
                         g_coord(:,:), jac(:,:), der(:,:), deriv(:,:),       &
                         bee(:,:), km(:,:),eld(:),eps(:),sigma(:),           &
                         diag_precon(:),p(:),r(:),x(:),xnew(:),              &
                         u(:),pmul(:),utemp(:),d(:)
 integer, allocatable :: nf(:,:), g(:), num(:), g_num(:,:)  , g_g(:,:)
!-------------------------input and initialisation------------------------
  open (10,file='p510.dat',status=    'old',action='read')
  open (11,file='p510.res',status='replace',action='write')
  read (10,*) nels,nxe,nze,nn,nip,aa,bb,cc,e,v,   tol,limit ;  ndof=nod*nodof
  allocate ( nf(nodof,nn), points(nip,ndim),dee(nst,nst),coord(nod,ndim),    &
            jac(ndim,ndim),der(ndim,nod),deriv(ndim,nod),                    &
            bee(nst,ndof),km(ndof,ndof),eld(ndof),eps(nst),sigma(nst),       &
            g(ndof),pmul(ndof),utemp(ndof), g_coord(ndim,nn),                &
            g_num(nod,nels),weights(nip),num(nod),g_g(ndof,nels))
   nf=1; read(10,*) nr ; if(nr>0) read(10,*)(k,nf(:,k),i=1,nr)
         call formnf(nf);neq=maxval(nf)
  allocate(p(0:neq),r(0:neq),x(0:neq),xnew(0:neq),u(0:neq),&
          diag_precon(0:neq),d(0:neq))
            r=0.; p=0.; x=0.; xnew=0.  ; diag_precon=0.
   call deemat(dee,e,v);   call sample(element,points,weights)
!---------- element stiffness integration and build the preconditioner---------
             iel=1
             call geometry_20bxz(iel,nxe,nze,aa,bb,cc,coord,num)
             km=0.0
      gauss_pts_1:  do i=1,nip
                 call shape_der (der,points,i) ; jac = matmul(der,coord)
                 det = determinant(jac); call invert(jac)
                 deriv = matmul(jac,der) ;call beemat (bee,deriv)
                 km=km+matmul(matmul(transpose(bee),dee),bee)*det*weights(i)
      end do gauss_pts_1
     elements_1: do iel = 1,nels
                   call geometry_20bxz(iel,nxe,nze,aa,bb,cc,coord,num)
                   g_num(:,iel) = num; g_coord(:,num) = transpose(coord)
                   call num_to_g ( num, nf, g ) ; g_g(: , iel) = g
                 do m=1,ndof;diag_precon(g(m))=diag_precon(g(m))+km(m,m);end do
     end do elements_1
    write(11,'(a)') "Global coordinates "
    do k=1,nn;write(11,'(a,i5,a,3e12.4)')"Node",k,"        ",g_coord(:,k);end do
    write(11,'(a)') "Global node numbers "
    do k = 1 , nels; write(11,'(a,i5,a,27i3)')                                  &
                                   "Element ",k,"    ",g_num(:,k); end do
      write(11,'(a,i5)') "The number of unknowns is",neq
!--------------------invert the preconditioner and get starting r--------------
        read(10,*) loaded_nodes,(k,r(nf(:,k)),i=1,loaded_nodes)
        write(11,'(a,e12.4)') "The total load is", sum(r)
       diag_precon(1:neq)=1./ diag_precon(1:neq)  ; diag_precon(0) = .0
               d=diag_precon*r  ; p = d
!--------------------preconditioned c. g. iterations---------------------------
      iters = 0
     iterations  :      do
```

```
            iters = iters + 1      ;     u = 0.
      elements_2 : do iel = 1, nels
                   g = g_g( : , iel )  ;    pmul = p(g)
                   utemp = matmul(km,pmul); u(g)  = u(g)+ utemp
      end do elements_2
!-------------------------pcg equation solution-------------------------------
          up=dot_product(r,d); alpha= up/ dot_product(p,u)
          xnew = x + p* alpha ; r=r - u*alpha;  d = diag_precon*r
          beta=dot_product(r,d)/up; p=d+p*beta
          converged = (maxval(abs(xnew-x))/maxval(abs xnew) < tol );  x=xnew
          if(converged .or. iters==limit) exit
     end do iterations
       write(11,'(a,i5)')"The number of iterations to convergence was  ",iters
       write(11,'(a)')   "The nodal displacements are    :"
   do k=1,22; write(11,'(i5,a,3e12.4)') k,"     ",xnew(nf(:,k)); end do
!------------------recover stresses at centroidal gauss-point-----------------
  nip=1;  deallocate(points,weights); allocate(points(nip,ndim),weights(nip))
  elements_3:do iel = 1, nels
               num = g_num(: ,iel)   ; coord =transpose( g_coord(:,num))
               g = g_g( : , iel )  ;      eld=xnew(g)
               write(11,'(a,i5,a)')                                         &
                  "The Centroid point stresses for element",iel,"  are :"
     gauss_pts_2: do i= 1 , nip
       call shape_der(der,points,i); jac= matmul(der,coord)
       call invert (jac);   deriv= matmul(jac,der) ; call beemat(bee,deriv)
       eps   = matmul (bee,eld)  ; sigma = matmul (dee,eps)
       write(11,'(a,i5)') "Point  ",i   ; write(11,'(6e12.4)')  sigma
     end do gauss_pts_2
  end do elements_3
 end program p510
```

Section 3.7 described a mesh-free approach to the solution of linear static equilibrium problems in which the equation solution process could be carried out by the preconditioned conjugate gradient (p.c.g.) technique without ever assembling element matrices into a global (stiffness) matrix. The essential process was described by equations (3.23)–(3.26). Program 5.10 will be used to solve the problem illustrated in Figure 5.30 and previously solved using an assembly technique by Program 5.8. In this very simple problem, all of the elements are the same and so their stiffness matrix, `KM`, needs to be built up by numerical integration once only.

A structure chart is shown in Figure 5.36.

All of the elements are looped in order to build the preconditioning matrix which is simply the inverse of the diagonal terms in what would have been the assembled global stiffness matrix. The preconditioning matrix (vector) is called `DIAG_PRECON`. The loop labelled `ITERATIONS` then carries out the operations demanded by equations (3.25). The matrix–vector multiplication needed in the first of (3.25) is done using (3.26). The steering vector `G` identifies the appropriate components of **p**, which are gathered into `PMUL`. Vector `UTEMP` then holds the part of **u** associated with that particular element and `G` enables `UTEMP` to be scattered out into the appropriate locations in **u**. The section

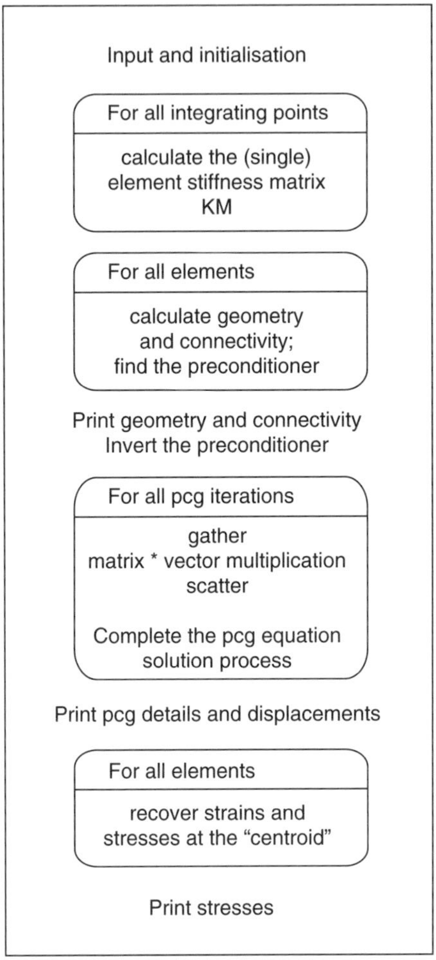

Figure 5.36 Structure chart for mesh-free Program 5.10

labelled "p.c.g. equation solution" then carries out the five vector operations in (3.25). A tolerance `TOL` enables the iterations to be stopped when successive solutions are "close enough" but since `ITERATIONS` is a loop which could carry on "for ever", a maximum number of iterations, `LIMIT`, is specified also. Strains and stresses can then be recovered from the displacements in the usual manner.

```
Structure and  NELS   NXE   NZE  NN   NIP
element data   6      1     2    70   27


               AA     BB    CC   E    V
               .5     1.    1.   100. .3

Iteration      TOL    LIMIT
data           1.E-5  200

Nodal freedom  NR
data           46
               K,NF(:,K),I=1,NR
               1 0 0 1 2 1 0 1 3 1 0 1 4 0 0 1 5 1 0 1 6 0 0 1
               7 1 0 1 8 1 0 1 9 0 0 1 10 1 0 1 11 0 0 0 12 0 0 0 13 0 0 0 14 0 1 1
               16 0 1 1 18 0 0 0 19 0 0 0 20 0 1 1 23 0 1 1 25 0 1 1 28 0 1 1 30 0 0 0
               31 0 0 0 32 0 0 0 33 0 1 1 35 0 1 1 37 0 0 0 38 0 0 0 39 0 1 1 42 0 1 1
               44 0 1 1 47 0 1 1 49 0 0 0 50 0 0 0 51 0 0 0 52 0 1 1 54 0 1 1 56 0 0 0
               57 0 0 0 58 0 1 1 61 0 1 1 63 0 1 1 66 0 1 1 68 0 0 0 69 0 0 0 70 0 0 0


Loads          LOADED_NODES
data           8
               (K,LOADS(NF(:,K)),I=1,LOADED_NODES)
               1 .0 .0 .0417 2 .0 .0 -.1667 3 .0 .0 .0417 14 .0 .0 -.1667
               15 .0 .0 -.1667 20 .0 .0 .0417 21 .0 .0 -.1667 22 .0 .0
                 .0417
```

Figure 5.37 Data for Program 5.10

The data for the program is shown in Figure 5.37. The only change from Figure 5.30 (assembly strategy) is that TOL and LIMIT are included and set to be $1*10^{-5}$ and 200 respectively.

Figure 5.38 shows the results which may be compared with those listed in Figure 5.31. To achieve the specified accuracy of solution took 56 iterations in this case. Since, in perfect arithmetic, the conjugate gradient process should converge in at most NEQ iterations (124 in this case) the amount of computational effort in large problems may seem to be daunting. Fortunately, as the set of equations to be solved grows larger, the proportion of iterations to converge to number of equations (ITERS/NEQ) usually decreases dramatically. It does, however, depend crucially on the "condition number" (nature of the eigenvalue spectrum) of the assembled stiffness matrix.

```
Global coordinates
Node    1          0.0000E+00  0.0000E+00  0.0000E+00
Node    2          0.2500E00  0.0000E+00  0.0000E+00
Node    3          0.5000E00  0.0000E+00  0.0000E+00
Node    4          0.0000E+00  0.0000E+00 -0.5000E00
Node    5          0.5000E00  0.0000E+00 -0.5000E00
.............................................................

Node   67          0.5000E00  0.3000E+01 -0.1500E+01
Node   68          0.0000E+00  0.3000E+01 -0.2000E+01
Node   69          0.2500E00  0.3000E+01 -0.2000E+01
Node   70          0.5000E00  0.3000E+01 -0.2000E+01
Global node numbers
Element    1      6  4  1  2  3  5  8  7 16 14 15 17 25 23 20 21 22 24 27 26
Element    2     11  9  6  7  8 10 13 12 18 16 17 19 30 28 25 26 27 29 32 31
Element    3     25 23 20 21 22 24 27 26 35 33 34 36 44 42 39 40 41 43 46 45
Element    4     30 28 25 26 27 29 32 31 37 35 36 38 49 47 44 45 46 48 51 50
Element    5     44 42 39 40 41 43 46 45 54 52 53 55 63 61 58 59 60 62 65 64
Element    6     49 47 44 45 46 48 51 50 56 54 55 57 68 66 63 64 65 67 70 69
The number of unknowns is  124
The total load is -0.5000E00
The number of iterations to convergence was     56
The nodal displacements are    :
    1      0.0000E+00  0.0000E+00 -0.1606E-01
    2      0.1344E-02  0.0000E+00 -0.1616E-01
    3      0.2687E-02  0.0000E+00 -0.1658E-01
    4      0.0000E+00  0.0000E+00 -0.1168E-01
    5      0.1703E-02  0.0000E+00 -0.1197E-01
    6      0.0000E+00  0.0000E+00 -0.7423E-02
    7      0.6783E-03  0.0000E+00 -0.7406E-02
    8      0.1345E-02  0.0000E+00 -0.7490E-02
    9      0.0000E+00  0.0000E+00 -0.3078E-02
   10      0.1220E-02  0.0000E+00 -0.3328E-02
   11      0.0000E+00  0.0000E+00  0.0000E+00
   12      0.0000E+00  0.0000E+00  0.0000E+00
   13      0.0000E+00  0.0000E+00  0.0000E+00
   14      0.0000E+00 -0.1678E-02 -0.1475E-01
   15      0.2249E-02 -0.1379E-02 -0.1510E-01
   16      0.0000E+00  0.1049E-02 -0.6449E-02
   17      0.1249E-02  0.1161E-02 -0.6516E-02
   18      0.0000E+00  0.0000E+00  0.0000E+00
   19      0.0000E+00  0.0000E+00  0.0000E+00
   20      0.0000E+00 -0.2881E-02 -0.9123E-02
   21      0.5035E-03 -0.2800E-02 -0.9143E-02
   22      0.1012E-02 -0.2388E-02 -0.9149E-02
The Centroid point stresses for element    1  are :
Point      1
  0.4610E-02 -0.1648E00 -0.8883E00  0.2476E-02  0.8305E-01  0.2309E-02
The Centroid point stresses for element    2  are :
Point      1
  0.2581E-02 -0.6851E-01 -0.6645E00 -0.6375E-03  0.8299E-01  0.6888E-02
The Centroid point stresses for element    3  are :
Point      1
 -0.6057E-02 -0.9576E-01 -0.1171E00  0.1026E-02  0.8193E-01 -0.2191E-02
The Centroid point stresses for element    4  are :
Point      1
  0.2350E-02 -0.9292E-01 -0.2653E00 -0.1321E-02  0.1252E00  0.3126E-02
The Centroid point stresses for element    5  are :
Point      1
 -0.5703E-03 -0.1013E-01 -0.2528E-02 -0.1237E-02 -0.5974E-02  0.6605E-03
The Centroid point stresses for element    6  are :
Point      1
  0.2178E-02 -0.4532E-01 -0.2885E-01 -0.1753E-03  0.5599E-01  0.2917E-04
```

Figure 5.38 Results from Program 5.10

PROGRAM 5.11: THREE-DIMENSIONAL ANALYSIS USING 20-NODE BRICK ELEMENTS : MESH-FREE STRATEGY VECTORISED VERSION

```
program p511
!-----------------------------------------------------------------------
!     program 5.11 three dimensional analysis of an elastic
!     solid using 20-node brick elements
!     preconditioned conjugate gradient solver  ;  only integrate one element
!     diagonal preconditioner diag_precon    ;  vectorised version
!-----------------------------------------------------------------------
 use new_library  ;  use  geometry_lib    ;     implicit none
 integer::nxe,nze,neq,nn,nr,nip,nodof=3,nod=20,nst=6,ndof,loaded_nodes,     &
         i,k,ndim=3,iters,limit,iel,nels
 real::aa,bb,cc,e,v,det,tol,up,alpha,beta,big
 logical :: converged  ; character(len=15) :: element = 'hexahedron'
!------------------------- dynamic arrays-------------------------------
 real     ,allocatable :: points(:,:),dee(:,:),coord(:,:), weights(:),        &
                          g_coord(:,:), jac(:,:), der(:,:), deriv(:,:),       &
                          bee(:,:), km(:,:),eld(:),eps(:),sigma(:),           &
                          diag_precon(:),p(:),r(:),x(:),xnew(:),              &
                          u(:),g_pmul(:,:),g_utemp(:,:),d(:)
 integer, allocatable :: nf(:,:), g(:), num(:), g_num(:,:) ,g_g( : , :)
!------------------------input and initialisation--------------------------
  open (10,file='p511.dat',status=    'old',action='read')
  open (11,file='p511.res',status='replace',action='write')
  read (10,*) nels,nxe,nze,nn,nip,aa,bb,cc,e,v,   tol,limit ;  ndof=nod*nodof
  allocate ( nf(nodof,nn), points(nip,ndim),dee(nst,nst),coord(nod,ndim),    &
           jac(ndim,ndim),der(ndim,nod),deriv(ndim,nod),                     &
           bee(nst,ndof),km(ndof,ndof),eld(ndof),eps(nst),sigma(nst),        &
           g(ndof),g_pmul(ndof,nels),g_utemp(ndof,nels), g_coord(ndim,nn),   &
           g_num(nod,nels),weights(nip),num(nod),g_g(ndof,nels))
   nf=1; read(10,*) nr ; if(nr>0) read(10,*)(k,nf(:,k),i=1,nr)
         call formnf(nf);neq=maxval(nf)
  allocate(p(0:neq),r(0:neq),x(0:neq),xnew(0:neq),u(0:neq),&
          diag_precon(0:neq),d(0:neq))
       r=0.; p=0.; x=0.; xnew=0.  ; diag_precon=0.
   call deemat(dee,e,v);    call sample(element,points,weights)
! ----------------- single element stiffness integration ---------------------
            iel=1
            call geometry_20bxz(iel,nxe,nze,aa,bb,cc,coord,num)
            km=0.0
     gauss_pts_1:  do i=1,nip
                call shape_der (der,points,i) ; jac = matmul(der,coord)
                det = determinant(jac)        ; call invert(jac)
                deriv = matmul(jac,der) ;call beemat (bee,deriv)
                km=km+matmul(matmul(transpose(bee),dee),bee)*det*weights(i)
     end do gauss_pts_1
! -------------- store global arrays and build the preconditioner -------------
    elements_1: do iel = 1,nels
                  call geometry_20bxz(iel,nxe,nze,aa,bb,cc,coord,num)
                  g_num(:, iel) = num; g_coord(: ,num) = transpose(coord)
                  call num_to_g(num,nf,g);  g_g( : , iel) = g
               do k=1,ndof;diag_precon(g(k))=diag_precon(g(k))+km(k,k);end do
    end do elements_1
   write(11,'(a)') "Global coordinates "
   do k=1,nn;write(11,'(a,i5,a,3e12.4)')"Node",k,"       ",g_coord(:,k);end do
   write(11,'(a)') "Global node numbers "
   do k = 1 , nels; write(11,'(a,i5,a,27i3)')                                   &
                              "Element ",k,"    ",g_num(:,k); end do
   write(11,'(a,i5)') "The number of equations is  ",neq
!--------------------invert the preconditioner and get starting r--------------
        read(10,*) loaded_nodes,(k,r(nf(:,k)),i=1,loaded_nodes)
        write(11,'(a,e12.4)') "The total load is", sum(r)
       diag_precon(1:neq)=1./ diag_precon(1:neq)  ; diag_precon(0) = .0
                d=diag_precon*r  ; p = d
!----------------------preconditioned c. g. iterations-------------------------
      iters = 0
```

```
    iterations  :       do
            iters = iters + 1      ;      u = 0.
      elements_2 : do iel = 1 , nels                 ! gather
                    g_pmul(: , iel) = p( g_g( : , iel))
      end do elements_2
!------------------------- global matrix multiply ----------------------------
         g_utemp = matmul( km , g_pmul )
!dir$ ivdep
      elements_2a : do iel = 1 , nels                 ! scatter
                    u(g_g(:,iel))=u(g_g(:,iel))+g_utemp(:,iel)
      end do elements_2a      ! let this vectorise by compiler directive
!---------------------------pcg equation solution-----------------------------
   up=dot_product(r,d); alpha= up/ dot_product(p,u)
   xnew = x + p* alpha ; r=r - u*alpha;  d = diag_precon*r
   beta=dot_product(r,d)/up; p=d+p*beta  ; p(0) = .0
   big=0.; converged = .true.
   do i=1,neq; if(abs(xnew(i))>big) big= abs(xnew(i)) ; end do
   do i=1,neq; if(abs(xnew(i)-x(i))/big>tol)converged=.false.; end do; x=xnew
   if(converged .or. iters==limit) exit
                  end do iterations
      write(11,'(a,i5)') "The number of iterations to convergence was  ",iters
      write(11,'(a)')    "The nodal displacements are   :"
  do k=1,22; write(11,'(i5,a,3e12.4)') k,"     ",xnew(nf(:,k)); end do
!-------------------recover stresses at centroidal gauss-point-----------------
  nip=1; deallocate(points,weights); allocate(points(nip,ndim),weights(nip))
  elements_3:do iel = 1, nels
                 num = g_num(: ,iel)   ; coord =transpose( g_coord(:,num))
                 g = g_g( : , iel) ; eld=xnew(g)
                 write(11,'(a,i5,a)')                                      &
                        "The Gauss point stresses for element",iel,"  are :"
    gauss_pts_2: do i= 1 , nip
      call shape_der(der,points,i); jac= matmul(der,coord)
      call invert (jac);   deriv= matmul(jac,der)
      bee= 0.;call beemat(bee,deriv); sigma = matmul (dee,matmul(bee,eld))
      write(11,'(a,i5)') "Point  ",i   ; write(11,'(6e12.4)') sigma
    end do gauss_pts_2
  end do elements_3
 end program p511
```

Table 5.1 Timings of vectorised programs

Original code (Program 5.10)	44.7 seconds
No dependency	25.3 seconds
Replace MAXVAL	21.6 seconds
Matrix–matrix (Program 5.11)	9.3 seconds

This program is used to solve exactly the same problem as was detailed for Programs 5.8 and 5.10, using the mesh-free strategy of Program 5.10. However, it is used as an example of some of the issues which arise when programming for vector computers (see section 1.3).

All practical vector computers enable code to be analysed to see where most time is being used and where there are features of the program inhibiting most effective use of vectorising compilers.

When Program 5.10 was run through such an analysis program, three main points arose:

1. There is potential "dependency" in the scatter operation `U(G) = U(G) + UTEMP`. The compiler does not know whether there may be repeated entries in `G` and so does not vectorise this statement.
2. A surprising amount of time was spent in the Fortran 90 intrinsic `MAXVAL` (for testing convergence).
3. The most time-consuming operation is `MATMUL`, and on the particular vector computer, it was running considerably slower than the peak machine speed.

Program 5.11 addresses all of these issues. First, unless freedoms are "tied" together (a device not used in this book) we can be sure that entries in `G` are not duplicated and so the scatter operation can be vectorised. A "compiler directive" is inserted before the loop "`ELEMENTS_2A`" enabling the loop to be vectorised. Second, `MAXVAL` is replaced by its longhand equivalent. This is obviously a problem with the particular vendor whose implementation of `MAXVAL` could be much improved. Third, and this is probably also a vendor problem, the `MATMUL` operation is changed from matrix–vector to matrix–matrix by collecting all the `PMUL` vectors into a global matrix `G_PMUL`. Otherwise the program is the same as Program 5.10 and of course produces the same results. However, Table 5.1 shows the progressive effects of making changes to the coding in Program 5.10.

The speed-up of Program 5.11 over Program 5.10 on this particular vector computer was by a factor of about 5 and illustrates the importance of code analysis when using such machines.

5.1 PARALLELISATION STRATEGIES

Effective parallelisation of programs is a complex matter and all that is given here is a description of some of the issues involved. The target machine is of the MIMD (multiple instruction stream, multiple data stream) architecture with distributed memory (DM-MIMD). It is assumed that an identical program runs on several processors of a MIMD machine (SPMD) with conditional control. As was described in Chapter 1, although programs like Program 5.10 are tailor-made for HPF (High Performance Fortran) compilers, the practical implementation of these is just beginning at the time of writing. Therefore the environment envisaged is one of "message passing" where the problem is decomposed and data is distributed over the multiple processors by the programmer. The specific library used is MPI (message passing interface).

Unlike in HPF, each processor knows only about its own local data. If it requires data from the local memory of a different processor, it must send a message to that processor asking for it. Much more time is taken in accessing "remote" data than local data so a major objective in a message-passing environment is to distribute data in a way that minimises remote communications.

5.1.1 CODE STRUCTURE

Program 5.10 can be considered to consist of the following steps:

1. Input and initialisation;
2. Stiffness integration, building and inverting the preconditioner;
3. Preconditioned conjugate gradient iterations;
4. Post-processing with the solution vector.

On a serial computer, 90% of the time is spent in the iterations section and it is this time which can be significantly reduced by vectorisation (Program 5.11).

The iterations section can itself be split into the following steps:

1. Matrix multiplication, including gather and scatter;
2. Vector operations, including dot products;
3. Convergence testing.

The serial version is dominated by the matrix multiplication. Even the most modest parallelisation of this section leads to the vector operations section being dominant. Initially, convergence testing can be ignored in a parallel code.

5.1.2 MATRIX MULTIPLICATION SECTION

The code is

```
PMUL = P(G)!gather
UTEMP = MATMUL(KM,PMUL)! matrix multiply
U(G) = U(G)+UTEMP! scatter
```

In a test problem similar to that shown in Figure 5.30, but with 8000 elements, `NEQ` is about 100 000. The `KM` matrix is (60×60) and `G_PMUL` is (60×8000). The objective is to distribute this data so that calculation requires a minimum number of remote accesses.

In the very simple case when there is a single `KM`, it can be replicated on every processor. The large arrays `PMUL` and `UTEMP` can be chopped into block columns and each processor performs the matrix multiplication on its own data. This assumes communication associated with gather/scatter is acceptable (see below).

5.1.3 VECTOR OPERATIONS SECTION

This involves the dot products and whole array operations described by equations (3.25). It is easy to chop the vectors over the processors. Simple vector operations involve no communication, but dot products can be split into local dot products, for which tuned basic linear algebra subroutines (BLAS) routines can be used (no communication again) and global sum routines which are tuned to minimise communication.

5.1.4 GATHER AND SCATTER

The distributions described in 5.1.2 and 5.1.3 are incompatible, and a decision has to be taken as to which will lead to the least communication if implemented

Number of Processors

		1	2	4	8	16	32
Total time	cpu	28.1	15.6	8.8	4.86	3.1	2.37
Host time	cpu		0.81	0.83	0.79	0.80	1.03
Serial time	cpu	0.67	0.66	0.70	0.65	0.63	0.86
Slave time	cpu		0.97	0.97	0.82	0.74	0.97
Common set-up time	cpu		0.50	0.33	0.25	0.11	0.22
Iterations	cpu	27.4	14.1	7.4	3.73	2.11	1.11
	spdup	1.00	1.94	3.71	7.35	13.0	24.6
	Mflps	64	123	235	468	825	1562
	% Pk	43	41	39	39	34	33
Matrix multiply	cpu	22.9	11.8	6.3	3.06	1.68	0.87
	spdup	1.00	1.94	3.67	7.45	13.6	26.3
	Mflps	75	145	272	557	1014	1969
Vector operations	cpu	4.5	2.34	1.2	0.66	0.43	0.25
	spdup	1.00	1.94	3.75	6.9	10.70	18.4
	Mflps	9	17	36	63	97	170
Peak	Mflps	150	300	600	1200	2400	4800

Spdup—speed-up from single node version.
Mflps—Mflop rate.

Figure 5.39 Parallel processing statistics

throughout the code. The approach adopted here is the following:

1. Chop the vectors for the vector operations;
2. Broadcast `KM` and distribute the `G` index matrix in blocked columns;
3. Generate local vectors, say `P_LOCAL` which contain all elements of `P` required for local assignments of `PMUL` and corresponding local index matrix, say `G_LOCAL` so that `P_LOCAL(G_LOCAL) = P(G)`.

To achieve this, each processor must identify the elements it requires (a once only start-up cost) and arrange for "remote" elements to be transferred (to be done every iteration.)

4. A similar process to scatter back to `U`.

There are many other alternatives but this one leads to the performance on the test problem (`NELS` = 8000, `NEQ` = 100 000) shown in Figure 5.39.

It can be seen that this program is reasonably scalable up to 32 processors on which it is running at about 1.5 Gflps and 30% of peak. On this particular system the single processor speed is disappointing as is the vector operation Mflps rate. Nevertheless the scope for very significant gains through parallelism of finite element code has been demonstrated.

5.2 EXAMPLES

1. Derive in terms of local coordinates, any shape function of the following elements:
   (a) 6-node triangle
   (b) 8-node quadrilateral
   (c) 9-node quadrilaterlal
2. Given that the "first" shape function of a four-node rectangular element of width a and height b is

$$N_1 = \left(1 - \frac{x}{a}\right)\left(1 - \frac{y}{b}\right)$$

   use analytical integration to show that the element stiffness (plane stress) and mass matrices include the following terms:

$$k_{11} = \frac{E}{1 - \nu^2}\left(\frac{b}{3a} + \frac{1 - \nu}{2}\frac{a}{3b}\right)$$

$$m_{11} = \frac{\rho ab}{9}$$

3. Assuming "small" strains, make sure you can derive the differential equations of 2-d elastic equilibrium for conditions of plane stress.

Given the square element with the local freedom numbering in Figure 5.40, show that the element stiffness matrix contains the term

$$K_{12} = \frac{E\nu}{6(1-\nu^2)}$$

4. Selective reduced integration (SRI) is a way of improving the performance of four-node plane elements as Poisson's ratio approaches 0.5 (e.g. undrained clay-incompressibility). The **D** matrix is split into volumetric and deviatoric components and the element stiffness matrix is constructed in two stages, thus

$$\mathbf{k} = \int \mathbf{B}^{\mathrm{T}}\mathbf{D}\nu\mathbf{B}\,\mathrm{dVol} + \int \mathbf{B}^{\mathrm{T}}\mathbf{D}^{\mathrm{d}}\mathbf{B}\,\mathrm{dVol}$$

where

$$\mathbf{D}^{\mathrm{d}} = \mu \begin{bmatrix} 2 & 0 & 0 \\ 0 & 2 & 0 \\ 0 & 0 & 1 \end{bmatrix} \tag{1}$$

$$\mathbf{D}^{\mathrm{v}} = \lambda \begin{bmatrix} 1 & 1 & 0 \\ 1 & 1 & 0 \\ 0 & 0 & 0 \end{bmatrix} \tag{2}$$

and Lamé's parameters are defined:

$$\mu = \frac{E}{2(1+\nu)} \quad \text{and} \quad \lambda = \frac{E\nu}{(1+\nu)(1-2\nu)} \tag{3}$$

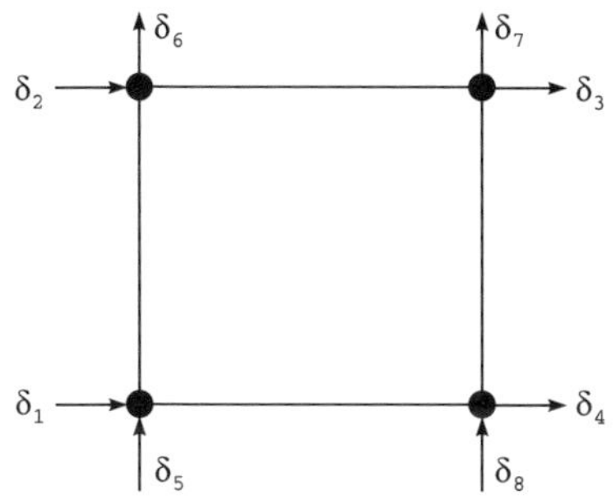

Figure 5.40

In computer programs, the SRI approach involves exact integration (4 Gauss point per element) of the $\mathbf{D}^d$ term and "reduced" integration (1 Gauss point per element) of the $\mathbf{D}^v$ term. For your hand calculation of this rectangular element however, you can use analytical integration for the $\mathbf{D}^d$ term, but you will need to use numerical integration for the $\mathbf{D}^v$ term. Use this technique to estimate the displacement of the loaded node in Figure 5.41.

5. Given the following shape functions for the eight-node plane quadrilateral element shown in Figure 5.42.

$$N_1 = {}^1\!/_4(1-\xi)(1-\eta)(-\xi-\eta-1)$$
$$N_2 = {}^1\!/_2(1-\xi)(1-\eta^2)$$
$$N_3 = {}^1\!/_4(1-\xi)(1+\eta)(-\xi+\eta-1)$$
$$N_4 = {}^1\!/_2(1-\xi^2)(1+\eta)$$
$$N_5 = {}^1\!/_4(1+\xi)(1+\eta)(\xi+\eta-1)$$
$$N_6 = {}^1\!/_2(1+\xi)(1-\eta^2)$$
$$N_7 = {}^1\!/_4(1+\xi)(1-\eta)(\xi-\eta-1)$$
$$N_8 = {}^1\!/_2(1-\xi^2)(1-\eta)$$

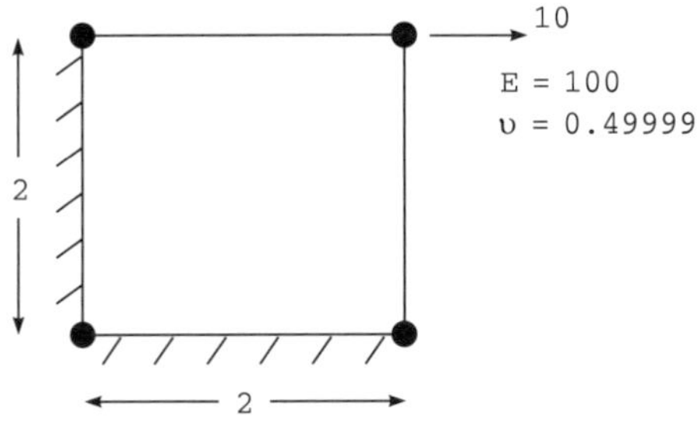

Figure 5.41

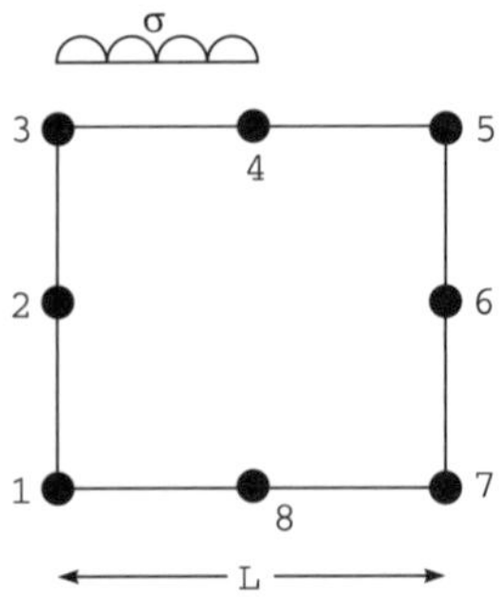

Figure 5.42

compute the equivalent nodal forces for the loading case in which a uniformly distributed load is applied over half of the top surface of the element.

6. Using Pascal's pyramid, suggest the terms that might appear in a typical shape function of a 14-noded cubic element which has nodes at each corner and in the middle of each face.
7. For the problem shown in Figure 5.43, estimate the force necessary to displace the loaded node horizontally by 0.015 units.
8. Derive the nodal forces that are equivalent to the triangular stress distribution acting on the four-node element shown in Figure 5.44. Given that the stiffness matrix of this element (assuming local freedom numbering

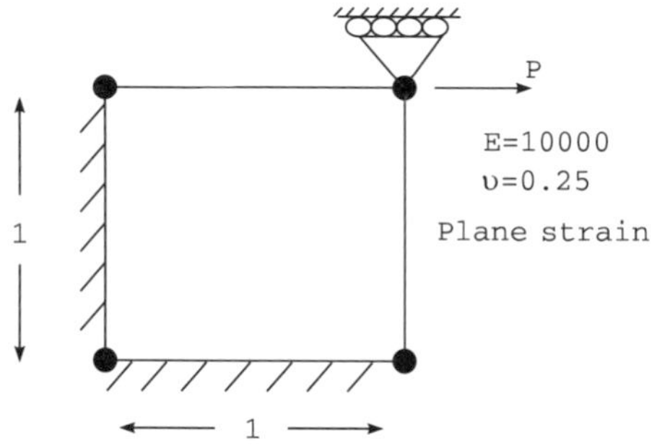

Figure 5.43

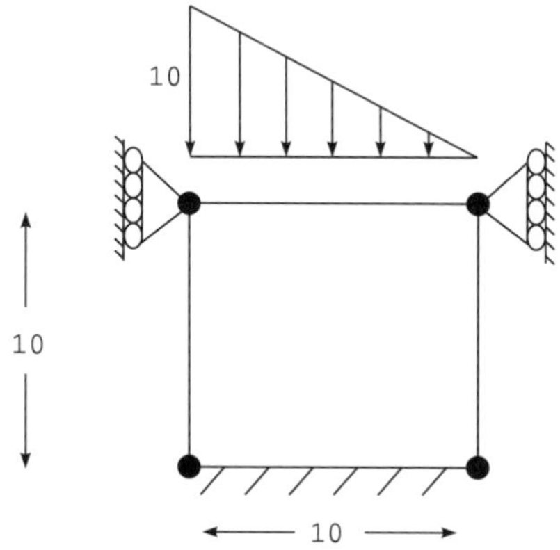

Figure 5.44

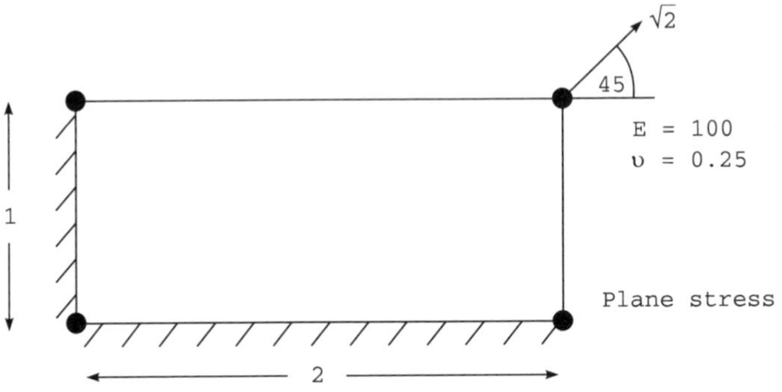

Figure 5.45

in the order $u_1v_1u_2v_2u_3v_3u_4v_4$) is

57.69	24.04	9.62	−4.81	−28.85	−24.04	−38.46	4.81
	57.69	4.81	−38.46	−24.04	−28.85	−4.81	9.62
		57.69	−24.04	−38.46	−4.81	−28.85	24.04
			57.69	4.81	9.62	24.04	−28.85
				57.69	24.04	9.62	−4.81
					57.69	4.81	−38.46
						57.69	−24.04
							57.69

compute the vertical displacement of the top two nodes.

9. A rectangular, blinear, four-noded element under plane stress conditions is subjected to the corner loading shown in Figure 5.45. By deriving the appropriate terms in the stiffness matrix, estimate the corner displacements.
10. A planar square eight-node quadrilateral of unit side length and unit mass density has the following shape functions at nodes 1 and 2.

$$N_1 = 0.25(1-\xi)(1-\eta)(-\eta-\xi-1)$$
$$N_2 = 0.5(1-\xi)(1-\eta^2)$$

    Use two-point Gaussian integration to compute the terms m_{11} and m_{12} of the element consistent mass matrix.
11. The four-node axisymmetric element shown in Figure 5.46 is subjected to the distributed vertical load on its top surface which varies from one unit at the centreline to zero at the outer boundary of the element. Compute:
    (a) The equivalent nodal forces F_1 and F_3.
    (b) Tht **B** matrix at the centre of the element.
    (c) The strains at the centre of the element given that the nodal displacements under this loading system are: $u_1 = -0.0550, u_2 = 0.0059, u_3 = -0.0225, u_4 = 0.0141$. Check your answer using Program 5.4.

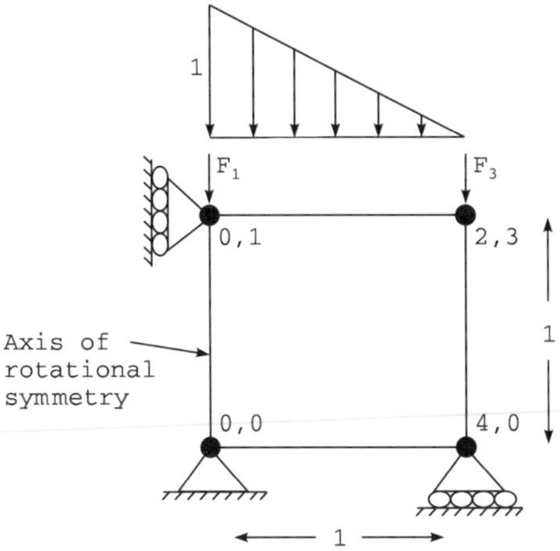

Figure 5.46

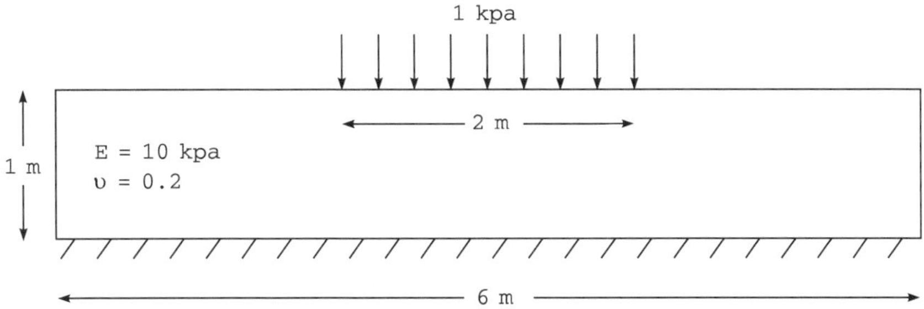

Figure 5.47

12. Use Program 5.3 to compute the vertical displacement at the edge of the flexible strip footing shown in Figure 5.47. You should use symmetry to reduce the number of elements in the discretisation.

5.3 REFERENCES

Cuthill, E. and McKee, J. (1969) Reducing the bandwidth of sparse symmetric matrices. *ACM Proceedings of 24th National Conference*, New York.

Griffiths, D.V. (1986) HARMONY—a program for analysing elasto-plastic axisymmetric bodies subjected to non-axisymmetric loads. Internal report, Engineering Dept, University of Manchester.

Griffiths, D.V. (1994) Stiffness matrix of the four-node quadrilateral element in closed form. *Int. J. Num. Meth. Eng.*, **37**, 1027–1038.

Hicks, M. and Mar, A. (1996) A benchmark computational study of finite element error estimation. *IJNME*, **39**(23), 3969–3983.

Irons, B.M. (1971) Quadrature rules for brick-based finite elements. *Int. J. Num. Meth. Eng.*, **3**, 293–294.

Peano, A. (1987) Inadmissible distortion of solid elements and patch test results. *Commun. Appl. Numer. Methods*, **5**, 97–101.

Poulos, H.G. and Davies, E.H. (1974) *Elastic Solutions for Soil and Rock Mechanics.* John Wiley, New York.

Smith, I.M. and Kidger, D.J. (1992) Elastoplastic analysis using the 14-node brick element family. *Int. J. Num. Meth. Eng.*, **35**, 1263–1275.

Wilson, E.L. (1965) Structural analysis of axisymmetric solids. JAIAA, **3**, 2269–2274.

Zienkiewicz, O.C. (1988) *The Finite Element Method*, 4th ed., p. 85. McGraw-Hill, New York.

6 Material Nonlinearity

6.0 INTRODUCTION

Nonlinear processes pose very much greater analytical problems than do the linear processes so far considered in this book. The nonlinearity may be found in the dependence of the equation coefficients on the solution itself or in the appearance of powers and products of the unknowns or their derivatives.

Two main types of nonlinearity can manifest themselves in finite element analysis of solids: material nonlinearity, in which the relationship between stresses and strains (or other material properties) are complicated functions which result in the equation coefficients depending on the solution, and geometric nonlinearity (otherwise known as "large strain" or "large displacement" analysis), which leads to products of the unknowns in the equations.

In order to keep the present book to a manageable size, the 11 programs described in this chapter deal only with material nonlinearity. As far as the organisation of computer programs is concerned, material nonlinearity is simpler to implement than geometric nonlinearity. However, programs can be adapted to cope with geometric nonlinearity as well (see, for example, Smith, 1997), and Programs 6.4, 6.5.

In practical finite element analysis two main types of solution procedure can be adopted to model material nonlinearity. The first approach involves "constant stiffness" iterations in which nonlinearity is introduced by iteratively modifying the right-hand side "loads" vector. The (usually elastic) global stiffness matrix in such an analysis is formed once only. Each iteration thus represents an elastic analysis of the type described in Chapter 5. Convergence is said to occur when stresses generated by the loads satisfy some stress–strain law or yield or failure criterion within prescribed tolerances. The loads vector at each iteration consists of externally applied loads and self-equilibrating "bodyloads". The body-loads have the effect of redistributing stresses within the system, but as they are self-equilibrating, they do not alter the net loading on the system. The "constant stiffness" method is shown diagrammatically in Figure 6.1. For load-controlled problems, many iterations may be required as failure is approached, because the elastic (constant) global stiffness matrix starts to seriously overestimate the actual material stiffness. The numbers in parentheses on the figure indicate the number of iterations that might typically be required to reach convergence.

Less iterations per load step are required if the second approach, the "variable" or "tangent" stiffness method, is adopted. This method, shown in Figure 6.2, takes account of the reduction in stiffness of the material as failure is approached. If small enough load steps are taken, the method can become

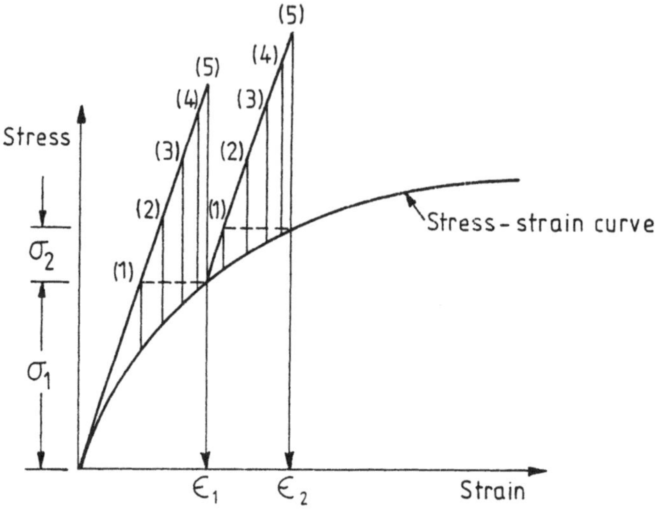

Figure 6.1 Constant stiffness method

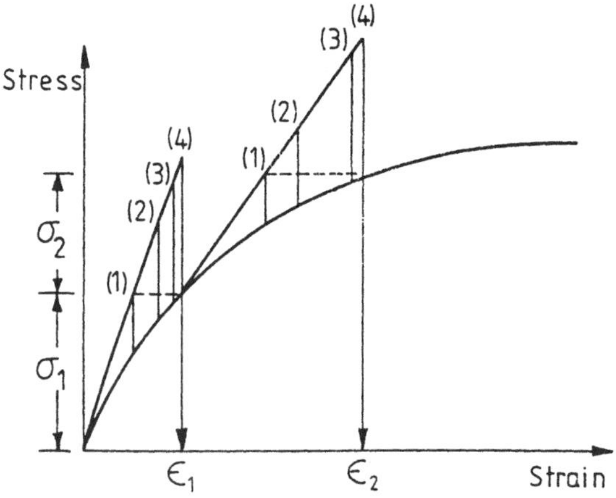

Figure 6.2 Variable stiffness method

equivalent to a simple Euler "explicit" method. In practice, the global stiffness matrix may be updated periodically and "body-loads" iterations employed to achieve convergence. In contrasting the two methods, the extra cost of re-forming and factorising the global stiffness matrix in the "variable stiffness" method is offset by reduced numbers of iterations, especially as failure is approached.

A further possibility, introduced in later programs, is "implicit" integration of the rate equations rather than the "explicit" methods just described. This helps to further reduce the number of iterations to convergence.

The first four programs in this chapter employ the "constant stiffness" approach and explicit integration for the sake of simplicity, and are similar in structure to Program 4.5 described previously for plastic analysis of frames.

Before describing the programs, some discussion is necessary regarding the form of the stress–strain laws that are to be adopted. In addition, two popular methods of generating body-loads for "constant stiffness" methods, namely "visco plasticity" and "initial stress", are described.

6.1 STRESS–STRAIN BEHAVIOUR

Although nonlinear elastic constitutive relations have been applied in finite element analyses and especially soil mechanics applications (e.g. Duncan and Chang, 1970), the main physical feature of nonlinear material behaviour is usually the irrecoverability of strain. A convenient mathematical framework for describing this phenomenon is to be found in the theory of plasticity (e.g. Hill, 1950). The simplest stress–strain law of this type that could be implemented in a finite element analysis involves elastic–perfectly plastic material behaviour (Figure 6.3). Although a simple law of this type was described in Chapter 4 (Figure 4.27), it is convenient in solid mechanics to introduce a "yield" surface in principal stress space which separates stress states that give rise to elastic and to plastic (irrecoverable) strains. To take account of complicated processes like cyclic loading, the yield surface may move in stress space "kinematically" (e.g. Molenkamp, 1983), but in this book only immovable surfaces are considered.

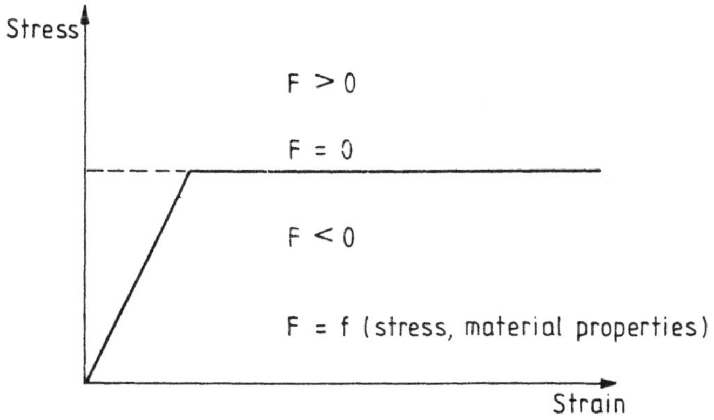

Figure 6.3 Elastic–perfectly plastic stress–strain behaviour

An additional simplification introduced here is that the yield and ultimate "failure" surfaces are identical.

Algebraically, the surfaces are expressed in terms of a yield or failure function F. This function, which has units of stress, depends on the material strength and invariant combinations of the stress components. The function is designed such that it is negative within the yield or failure surface and zero on the yield or failure surface. Positive values of F imply stresses lying outside the yield or failure surface which are illegal and which must be redistributed via the iterative process described previously.

During plastic straining, the material may flow in an "associated" manner, that is the vector of plastic strain increment may be normal to the yield or failure surface. Alternatively, normality may not exist and the flow may be "non-associated". Associated flow leads to various mathematically attractive simplifications and, when allied to the von Mises or Tresca failure criterion, correctly predicts zero plastic volume change during yield for undrained clays. For frictional materials, whose ultimate state is described by the Mohr–Coulomb criterion, associated flow leads to physically unrealistic volumetric expansion or dilation during yield. In such cases, non-associated flow rules must be preferred in which plastic straining is described by a plastic potential function Q. This function is often geometrically similar to the failure function F but with the friction angle ϕ replaced by a dilation angle ψ. The implementation of the plastic potential function will be described further in a later section.

Before outlining some commonly used failure criteria and their representations in principal stress space, some useful stress-invariant expressions are briefly reviewed.

6.2 STRESS INVARIANTS

The Cartesian stress tensor defining the stress conditions at a point within a loaded body is given by

$$\{\sigma_x \sigma_y \sigma_z \tau_{xy} \tau_{yz} \tau_{zx}\} \tag{6.1}$$

which can be shown to be equivalent to three principal stresses acting on orthogonal planes

$$\{\sigma_1 \sigma_2 \sigma_3\} \tag{6.2}$$

Principal stress space is obtained by treating the principal stresses as 3-d coordinates and such a plot represents a useful means of defining the stresses acting at a point. It may be noted that although principal stress space defines the magnitudes of the principal stresses, no indication is given of their orientation in physical space.

Instead of defining a point in principal stress space with coordinates $(\sigma_1, \sigma_2, \sigma_3)$ it is often more convenient to use invariants (s, t, θ), where

$$s = \frac{1}{\sqrt{3}}(\sigma_x + \sigma_y + \sigma_z)$$

$$t = \frac{1}{\sqrt{3}}[(\sigma_x - \sigma_y)^2 + (\sigma_y - \sigma_z)^2 + (\sigma_z - \sigma_x)^2 + 6\tau_{xy}^2 + 6\tau_{yz}^2 + 6\tau_{zx}^2]^{1/2} \quad (6.3)$$

$$\theta = 1/3 \ \arcsin\left(\frac{-3\sqrt{6}\,J_3}{t^3}\right)$$

where

$$J_3 = s_x s_y s_z - s_x \tau_{yz}^2 - s_y \tau_{zx}^2 - s_z \tau_{xy}^2 + 2\tau_{xy}\tau_{yz}\tau_{zx}$$

and

$$s_x = (2\sigma_x - \sigma_y - \sigma_z)/3, \text{etc.}$$

As shown in Figure 6.4, s gives the distance from the origin to the π plane in which the stress point lies and t represents the perpendicular distance of the stress point from the space diagonal. The Lode angle θ, called `LODE_THETA` in the programs, is a measure of the angular position of the stress point within the π plane.

It may be noted that in some soil mechanics applications, plane strain conditions apply and equations (6.3) are simplified because $\tau_{yz} = \tau_{zx} = 0$.

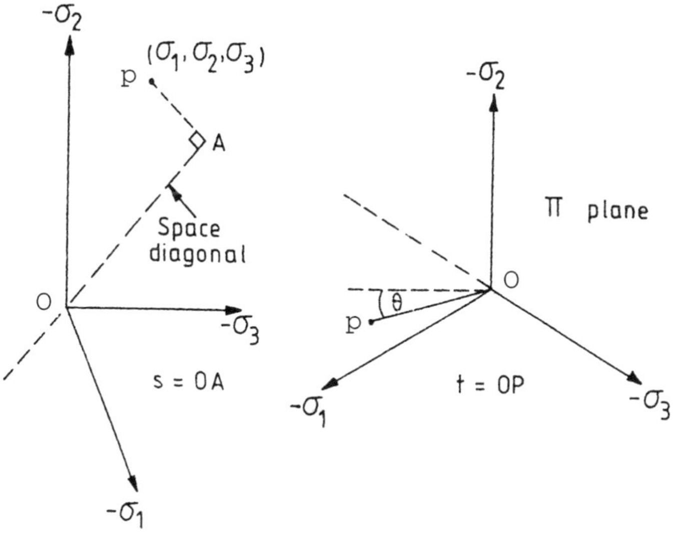

Figure 6.4 Principal stress space

In the programs described later in this chapter, the invariants that are used are slightly different from these defined in (6.3) whence

$$\sigma_m = s/\sqrt{3}$$
$$\bar{\sigma} = t\sqrt{(3/2)} \tag{6.4}$$

These expressions, called `SIGM` and `DSBAR` in program terminology, have more physical meaning than s and t in that they represent respectively the "mean stress" and "deviator stress" in a triaxial test. The relationship between principal stresses and invariants is given as follows:

$$\sigma_1 = \sigma_m + 2/3\bar{\sigma}\sin\left(\theta - \frac{2\pi}{3}\right)$$
$$\sigma_2 = \sigma_m + 2/3\bar{\sigma}\sin\theta \tag{6.5}$$
$$\sigma_3 = \sigma_m + 2/3\bar{\sigma}\sin\left(\theta + \frac{2\pi}{3}\right)$$

6.3 FAILURE CRITERIA

Several failure criteria have been proposed as suitable for representing the strength of soils as engineering materials. For soils possessing both frictional and cohesive components of shear strength, the best known criterion is undoubtedly that called "Mohr–Coulomb" and takes the form of an irregular hexagonal cone in principal stress space.

For metals or undrained clays which behave in a "frictionless" ($\phi_u = 0$) manner, cylindrical failure criteria are appropriate. These are the simplest criteria, which do not depend on the first invariant s (or σ_m). The Tresca criterion in fact does not require separate treatment mathematically because it is a special case of the Mohr–Coulomb criterion, as will be shown. Alternatively, the von Mises criterion can be used. The difference in strengths predicted by the two criteria does not exceed about 15%.

6.3.1 VON MISES

As shown in Figure 6.5, this criterion takes the form of a right circular cylinder lying along the space diagonal. Only one of the three invariants, namely t (or $\bar{\sigma}$), is of any significance when determining whether a stress state has reached the limit of elastic behaviour. The onset of yield in a von Mises material is not dependent upon invariants s or θ.

The symmetry of the von Mises criterion when viewed in the π plane indicates why it is not ideally suited to correlations with traditional soil

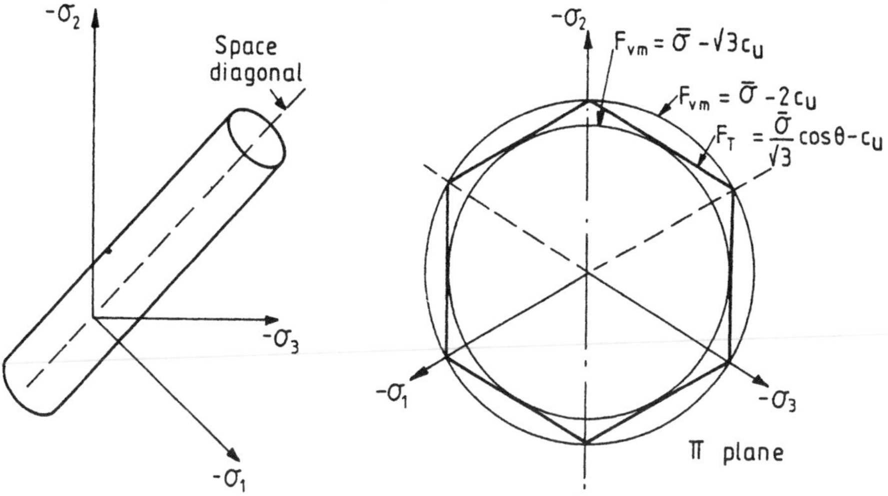

Figure 6.5 Von Mises and Tresca failure criteria

mechanics concepts of strength. The criterion gives equal weighting to all three principal stresses, so if it is to be used to model undrained clay behaviour, consideration must be given to the value of the intermediate principal stress, σ_2, at failure.

For plane strain applications it can be shown that at failure

$$\sigma_2 = \frac{\sigma_1 + \sigma_3}{2} \tag{6.6}$$

Hence, the failure criterion is given by

$$F = \bar{\sigma} - \sqrt{3}c_u \tag{6.7}$$

where c_u = undrained "cohesion" of the soil.

Under triaxial conditions, where at all times

$$\sigma_2 = \sigma_3 \tag{6.8}$$

the criterion is given by

$$F = \bar{\sigma} - 2c_u \tag{6.9}$$

Both of these expressions ensure that at failure

$$c_u = \frac{\sigma_1 - \sigma_3}{2} \tag{6.10}$$

6.3.2 MOHR–COULOMB

In principal stress space, this criterion takes the form of an irregular hexagonal cone, as shown in Figure 6.6. The irregularity is due to the fact that σ_2 is not

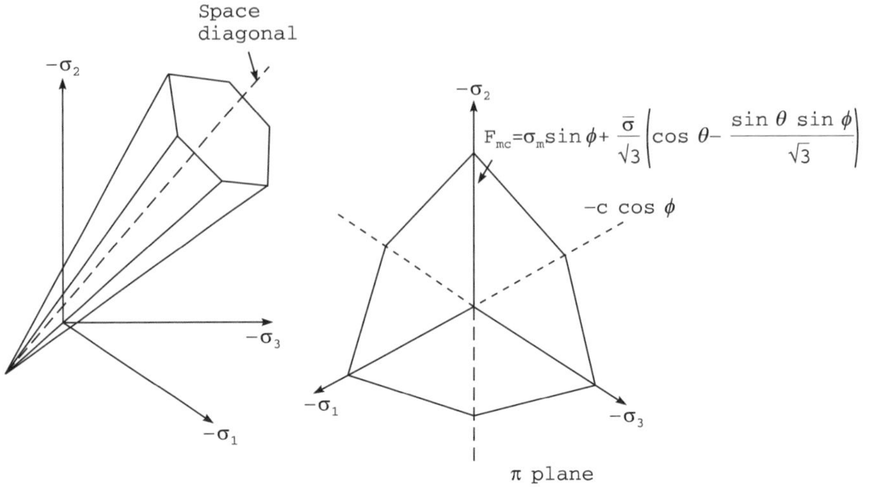

Figure 6.6 Mohr–Coulomb failure criterion

taken into account. In order to derive the invariant form of this criterion, it should first be written in terms of principal stresses from the geometry of Mohr's circle, thus

$$\frac{\sigma_1 + \sigma_3}{2}\sin\phi - \frac{\sigma_1 - \sigma_3}{2} - c\cos\phi = 0 \tag{6.11}$$

Substituting for σ_1 and σ_3 from equations (6.5) gives the function

$$F = \sigma_m \sin\phi + \bar{\sigma}\left(\frac{\cos\theta}{\sqrt{3}} - \frac{\sin\theta\,\sin\phi}{3}\right) - c\cos\phi \tag{6.12}$$

which is clearly dependent upon all three invariants $(\sigma_m, \bar{\sigma}, \theta)$.

The Tresca criterion is obtained from (6.12) by putting $\phi = 0$ to give

$$F = \frac{\bar{\sigma}\cos\theta}{\sqrt{3}} - c_u \tag{6.13}$$

This criterion is preferred to von Mises for applications involving undrained clays because (6.10) is always satisfied at failure, regardless of the value of σ_2. In principal stress space the Tresca criterion is a regular hexagonal cylinder tangential to the von Mises cylinder defined by equation (6.7), as shown in Figure 6.5.

6.4 SIMPLE GENERATION OF BODY-LOADS

Constant stiffness methods of the type described in this chapter use repeated elastic solutions to achieve convergence by iteratively varying the loads on the system. Within each load increment, the system of equations

$$\mathbf{K}\boldsymbol{\delta}^i = \mathbf{p}^i \tag{6.14}$$

must be solved for the displacement increments $\boldsymbol{\delta}^i$, where i represents the iteration number, $\mathbf{K}$ the global stiffness and $\mathbf{p}^i$ the external and internal loads.

The element displacement increments $\mathbf{u}^i$ are extracted from $\boldsymbol{\delta}^i$, and these lead to total strain increments via the element strain-displacement relationships:

$$\Delta\boldsymbol{\varepsilon}^i = \mathbf{B}\mathbf{u}^i \tag{6.15}$$

Assuming the material is yielding, the strains will contain both elastic and (visco) plastic components; thus

$$\Delta\boldsymbol{\varepsilon}^i = (\Delta\boldsymbol{\varepsilon}^{\mathrm{e}} + \Delta\boldsymbol{\varepsilon}^{\mathrm{p}})^i \tag{6.16}$$

It is only the elastic strain increments $\Delta\varepsilon^{\mathrm{e}}$ that generate stresses through the elastic stress–strain matrix; hence

$$\Delta\boldsymbol{\sigma}^i = \mathbf{D}^{\mathrm{e}}(\Delta\boldsymbol{\varepsilon}^{\mathrm{e}})^i \tag{6.17}$$

These stress increments are added to stresses already existing from the previous load step and the updated stresses substituted into the failure criterion.

If stress redistribution is necessary, this is done by altering the load increment vector $\mathbf{p}^i$ in equation (6.14). In general, this vector holds two types of load, as given by

$$\mathbf{p}^i = \mathbf{p}_{\mathrm{a}} + \mathbf{p}_{\mathrm{b}}^i \tag{6.18}$$

where $\mathbf{p}_{\mathrm{a}}$ is the actual applied load increment that is required and $\mathbf{p}_b^i$ is the body-loads vector that varies from one iteration to the next. The $\mathbf{p}_{\mathrm{b}}^i$ vector must be self-equilibrating so that the net loading on the system is not affected by it. Two simple methods for generating body-loads are now described briefly.

6.5 VISCO-PLASTICITY

In this method (Zienkiewicz and Cormeau, 1974), the material is allowed to sustain stresses outside the failure criterion for finite "periods". Overshoot of the failure criterion, as signified by a positive value of F, is an integral part of the method and is actually used to drive the algorithm.

Instead of plastic strains, we now refer to visco-plastic strains and these are generated at a rate that is related to the amount by which yield has been violated through the expression

$$\dot{\boldsymbol{\varepsilon}}^{\mathrm{VP}} = F\frac{\partial Q}{\partial \boldsymbol{\sigma}} \tag{6.19}$$

It should be noted that a pseudo-viscosity property equal to unity is implied on the right-hand side of equation (6.19) from dimensional considerations.

Multiplication of the visco-plastic strain rate by a pseudo-time step gives an increment of visco-plastic strain which is accumulated from one "time step" or iteration to the next; thus

$$(\boldsymbol{\delta\varepsilon}^{\mathrm{VP}})^i = \Delta t(\dot{\boldsymbol{\varepsilon}}^{\mathrm{VP}})^i \tag{6.20}$$

and

$$(\boldsymbol{\Delta\varepsilon}^{\mathrm{VP}})^i = (\boldsymbol{\Delta\varepsilon}^{\mathrm{VP}})^{i-1} + (\boldsymbol{\delta\varepsilon}^{\mathrm{VP}})^i \tag{6.21}$$

The "time step" for unconditional numerical stability has been derived by Cormeau (1975) and depends on the assumed failure criterion. Thus, for von Mises materials:

$$\Delta t = \frac{4(1+\nu)}{3E} \tag{6.22}$$

and for Mohr–Coulomb materials:

$$\Delta t = \frac{4(1+\nu)(1-2\nu)}{E(1-2\nu+\sin^2\phi)} \tag{6.23}$$

The derivatives of the plastic potential function Q with respect to stresses are conveniently expressed through the chain rule; thus

$$\frac{\partial Q}{\partial \boldsymbol{\sigma}} = \frac{\partial Q}{\partial \sigma_{\mathrm{m}}}\frac{\partial \sigma_{\mathrm{m}}}{\partial \boldsymbol{\sigma}} + \frac{\partial Q}{\partial J_2}\frac{\partial J_2}{\partial \boldsymbol{\sigma}} + \frac{\partial Q}{\partial J_3}\frac{\partial J_3}{\partial \boldsymbol{\sigma}} \tag{6.24}$$

where $J_2 = 1/2t^2$ and the visco-plastic strain rate given by equation (6.19) is evaluated numerically by an expression of the form

$$\dot{\boldsymbol{\varepsilon}}^{\mathrm{VP}} = F(\mathtt{DQ1}\,\mathbf{M}^1 + \mathtt{DQ2}\,\mathbf{M}^2 + \mathtt{DQ3}\,\mathbf{M}^3)\boldsymbol{\sigma} \tag{6.25}$$

where `DQ1`, `DQ2` and `DQ3` are scalars equal to $\partial Q/\partial \sigma_{\mathrm{m}}$, $\partial Q/\partial J_2$ and $\partial Q/\partial J_3$ respectively, and $\mathbf{M}^1\boldsymbol{\sigma}$, $\mathbf{M}^2\boldsymbol{\sigma}$ and $\mathbf{M}^3\boldsymbol{\sigma}$ are vectors representing $\partial \sigma_{\mathrm{m}}/\partial \boldsymbol{\sigma}$, $\partial J_2/\partial \boldsymbol{\sigma}$ and $\partial J_3/\partial \boldsymbol{\sigma}$ respectively. This is essentially the notation used by Zienkiewicz (1991) and these quantities are given in more detail in Appendix 2.

The body-loads $\mathbf{p}_{\mathrm{b}}^i$ are accumulated at each "time step" within each load step by summing the following integrals for all elements containing a yielding Gauss point:

$$\mathbf{p}_{\mathrm{b}}^i = \mathbf{p}_{\mathrm{b}}^{i-1} + \sum_{\text{element}}^{\text{all}} \int \mathbf{B}^{\mathrm{T}}\mathbf{D}^{\mathrm{e}}(\boldsymbol{\delta\varepsilon}^{\mathrm{VP}})^i \mathrm{d(element)} \tag{6.26}$$

This process is repeated at each "time step" iteration until no Gauss point stresses violate the failure criterion to within a certain tolerance. The convergence criterion is based on a dimensionless measure of the amount by which the displacement increment vector $\boldsymbol{\delta}^i$ changes from one iteration to the next. The convergence checking process is identical to that used in Program 4.5.

6.6 INITIAL STRESS

The visco-plastic algorithm is often referred to as an "initial strain" method to distinguish it from the more widely used "initial stress" approaches (e.g. Zienkiewicz et al, 1969).

Initial stress methods involve an explicit relationship between increments of stress and increments of strain. Thus, whereas linear elasticity was described by

$$\boldsymbol{\Delta\sigma} = \mathbf{D}^{\mathrm{e}}\boldsymbol{\Delta\varepsilon} \tag{6.27}$$

elasto-plasticity is described by

$$\boldsymbol{\Delta\sigma} = \mathbf{D}^{\mathrm{PL}}\boldsymbol{\Delta\varepsilon}$$

where

$$\mathbf{D}^{\mathrm{PL}} = \mathbf{D}^{\mathrm{e}} - \mathbf{D}^{\mathrm{P}} \tag{6.29}$$

For perfect plasticity in the absence of hardening or softening it is assumed that once a stress state reaches a failure surface, subsequent changes in stress may shift the stress state to a different position on the failure surface, but not outside it; thus

$$\frac{\partial F}{\partial \boldsymbol{\sigma}}\boldsymbol{\Delta\sigma} = 0 \tag{6.30}$$

Allowing for the possibility of non-associated flow, plastic strain increments occur normal to a plastic potential surface; thus

$$\boldsymbol{\Delta\varepsilon}^{\mathrm{P}} = \lambda\frac{\partial Q}{\partial \boldsymbol{\sigma}} \tag{6.31}$$

Assuming stress changes are generated by elastic strain components only gives

$$\boldsymbol{\Delta\sigma} = \mathbf{D}^{\mathrm{e}}\left(\boldsymbol{\Delta\sigma} - \lambda\frac{\partial Q}{\partial \boldsymbol{\sigma}}\right) \tag{6.32}$$

Substitution of equation (6.32) into (6.30) leads to

$$\mathbf{D}^{\mathrm{P}} = \frac{\mathbf{D}^{\mathrm{e}}\dfrac{\partial Q}{\partial \boldsymbol{\sigma}}\left(\dfrac{\partial F}{\partial \boldsymbol{\sigma}}\right)^{\mathrm{T}}\mathbf{D}^{\mathrm{e}}}{\left(\dfrac{\partial F}{\partial \boldsymbol{\sigma}}\right)^{\mathrm{T}}\mathbf{D}^{\mathrm{e}}\dfrac{\partial Q}{\partial \boldsymbol{\sigma}}} \tag{6.33}$$

Explicit versions of $\mathbf{D}^{\mathrm{P}}$ may be obtained for simple failure and potential functions and these are given for von Mises (Yamada et al, 1968) and Mohr–Coulomb (Griffiths and Willson, 1986) in Appendix 2.

The body-loads $\mathbf{p}_{\mathrm{b}}^{i}$ in the stress redistribution process are re-formed at each iteration by summing the following integral for all elements that possess

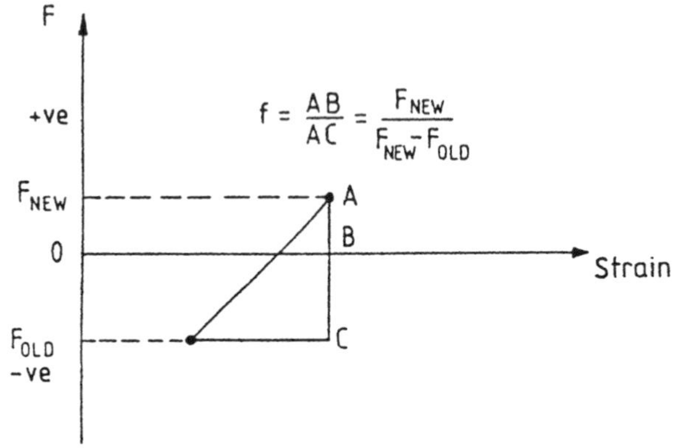

Figure 6.7 Factoring process for just yielding elements

yielding Gauss points; thus

$$\mathbf{p}_b^i = \sum_{\text{elements}}^{\text{all}} \int \mathbf{B}^{\mathrm{T}}(\mathbf{D}^{\mathrm{P}}\boldsymbol{\Delta\varepsilon})^i \mathrm{d}(\text{element}) \tag{6.34}$$

In the event of a loading increment causing a Gauss point to go plastic for the first time, it may be necessary to factor the matrix $\mathbf{D}^{\mathrm{P}}$ in (6.34). A linear interpolation can be used as indicated in Figure 6.7. Thus, instead of using $\mathbf{D}^{\mathrm{P}}$ we use $f\mathbf{D}^{\mathrm{P}}$ where

$$f = \frac{F_{\text{NEW}}}{F_{\text{NEW}} - F_{\text{OLD}}} = \texttt{FAC} \tag{6.35}$$

This simple method represents a forward Euler approach to integrating the elasto-plastic rate equations, extrapolating from the point at which the yield surface is crossed. More complicated integrations, which are mainly relevant to tangent stiffness methods, are given later in section 6.8.

It may be noted that in the visco-plastic algorithm elastic overshoot of the yield or failure criterion is an integral part of the solution process.

6.7 CORNERS ON THE FAILURE AND POTENTIAL SURFACES

For non-continuous F and Q surfaces, the derivatives in equations (6.19) and (6.33) become indeterminate. In the case of the Mohr–Coulomb (or Tresca) surface, this occurs when the angular invariant $\theta = \pm 30°$. The method used in the programs to overcome this difficulty is to replace the hexagonal surface by a smooth conical surface if

$$|\sin\theta| > 0.49 \tag{6.36}$$

The conical surfaces are those obtained by substituting either $\theta = 30°$ or $\theta = -30°$ into (6.12), depending upon the sign of θ as it approaches $\pm 30°$. It should be noted that in the initial stress approach, both F and Q functions must be approximated in this way due to the inclusion of both $\partial F/\partial \boldsymbol{\sigma}$ and $\partial Q/\partial \boldsymbol{\sigma}$ terms in (6.33). The visco-plastic algorithm, however, only requires the potential function Q to be smoothed from (6.19), as derivatives of F with respect to stresses are not required.

All the programs in this chapter for solving 2-d problems have been based on the eight-node quadrilateral element, together with reduced integration (four Gauss points per element). This particular combination has been chosen for its simplicity, and also its well-known ability to compute collapse loads accurately (e.g. Zienkiewicz, 1991; Griffiths, 1982). Of course, other element types could be used if required, and this would involve making similar subroutine changes to those illustrated in Chapter 5. To alleviate "locking" problems the order of integration of the "volumetric" components can also be reduced even further.

The first four programs involve computation of collapse loads in problems which have known closed form solutions. In addition to the different boundary value problems tackled, both the visco-plastic method and the simple initial stress method are implemented. Both assembly and mesh-free strategies are described.

PROGRAM 6.0: PLANE STRAIN BEARING CAPACITY ANALYSIS — VISCO-PLASTIC METHOD USING THE VON MISES CRITERION

The first four programs in this chapter are derived from Program 5.3 for linear elastic analysis, which used eight-node quadrilateral elements. Program 6.0 employs the visco-plastic method to compute the response to loading of an elastic–perfectly plastic (von Mises) material. Plane strain conditions are enforced and, in order to monitor the load–displacement response, the loads are applied incrementally.

In the first part of the program, however, the global elastic stiffness matrix is formed in the usual way and the resulting matrix, `KB`, is then factorised using (in this case) Choleski's method. An obvious advantage of constant stiffness methods is that the time-consuming reduction process is only performed once. An outline of the visco-plastic algorithm which comes after the stiffness matrix formation is given in the structure chart in Figure 6.8.

New variable and array names that have not already been identified in Chapter 5 are listed as follows:

`CU`	undrained "cohesion"
`INCS`	number of load increments
`LIMIT`	maximum number of iterations

```
   program p60
!-------------------------------------------------------------------------
!      program 6.0 plane strain of an elastic-plastic(Von Mises) solid
!      using 8-node quadrilateral elements; viscoplastic strain method
!-------------------------------------------------------------------------
 use new_library      ; use geometry_lib  ;     implicit none
 integer::nels,nxe,nye,neq,nband,nn,nr,nip,nodof=2,nod=8,nst=4,ndof,         &
          loaded_nodes,i,k,iel,iters,limit,incs,iy,ndim=2
 character (len = 15) :: element = 'quadrilateral';   logical::converged
 real::e,v,det,cu,dt,ptot,f,dsbar,dq1,dq2,dq3,lode_theta,sigm,tol
!------------------------- dynamic arrays-----------------------------------
 real    ,allocatable :: kb(:,:),loads(:),points(:,:),totd(:),bdylds(:),      &
                         evpt(:,:,:),oldis(:),width(:),depth(:),              &
                         tensor(:,:,:),val(:,:),stress(:),qinc(:),            &
                         dee(:,:),coord(:,:),jac(:,:),weights(:),             &
                         der(:,:),deriv(:,:),bee(:,:),km(:,:),eld(:),eps(:),  &
                         sigma(:),bload(:),eload(:),erate(:),g_coord(:,:),    &
                         evp(:),devp(:),m1(:,:),m2(:,:),m3(:,:),flow(:,:)
 integer, allocatable :: nf(:,:) , g(:), no(:) ,num(:), g_num(:,:) ,g_g(:,:)
!----------------input and initialisation---------------------
  open (10,file='p60.dat',status=    'old',action='read')
  open (11,file='p60.res',status='replace',action='write')
  read (10,*) cu,e,v,nels,nxe,nye,nn,nip,tol,limit
  ndof=nod*nodof
  allocate (nf(nodof,nn), points(nip,ndim),weights(nip),g_coord(ndim,nn),    &
           width(nxe+1),depth(nye+1),num(nod),dee(nst,nst),g_g(ndof,nels),   &
           evpt(nst,nip,nels), tensor(nst,nip,nels),coord(nod,ndim),         &
           jac(ndim,ndim),der(ndim,nod),deriv(ndim,nod),g_num(nod,nels),     &
           bee(nst,ndof),km(ndof,ndof),eld(ndof),eps(nst),sigma(nst),        &
           bload(ndof),eload(ndof),erate(nst),evp(nst),devp(nst),g(ndof),    &
           m1(nst,nst),m2(nst,nst),m3(nst,nst),flow(nst,nst),stress(nst))
  nf=1; read(10,*) nr ; if(nr>0) read(10,*)(k,nf(:,k),i=1,nr)
   call formnf(nf); neq=maxval(nf)     ;    read(10,*) width, depth
!---------- loop the elements to find nband and set up global arrays ----------
     nband = 0
       elements_1:   do iel = 1 , nels
                        call geometry_8qyv(iel,nye,width,depth,coord,num)
                        call num_to_g(num,nf,g); g_num(:,iel)=num
                        g_coord(:,num)=transpose(coord);  g_g(:,iel) = g
                        if (nband<bandwidth(g)) nband = bandwidth(g)
     end do elements_1
   write(11,'(a)') "Global coordinates "
   do k=1,nn;write(11,'(a,i5,a,2e12.4)')"node",k,"       ",g_coord(:,k);end do
   write(11,'(a)') "Global node numbers "
   do k=1,nels; write(11,'(a,i5,a,8i5)')                                      &
                         "Element ",k,"        ",g_num(:,k); end do
   write(11,'(a,i5,a,i5)')                                                    &
           "There are ",neq ," equations and the half-bandwidth is ",nband
  allocate(kb(neq,nband+1),loads(0:neq),bdylds(0:neq),oldis(0:neq),totd(0:neq))
  kb=0.0; oldis=0.0; totd=0.0 ; tensor = 0.0
  call deemat(dee,e,v); call sample(element,points,weights)
  dt=4.*(1.+v)/(3.*e)
!------------------ element stiffness integration and assembly-----------------
 elements_2: do iel = 1 , nels
               num = g_num(:,iel) ; coord = transpose(g_coord(: ,num))
               g = g_g(: ,iel ) ;          km=0.0
             gauss_pts_1:  do i =1 , nip
               call shape_der (der,points,i);  jac = matmul(der,coord)
               det = determinant(jac)  ;   call invert(jac)
               deriv = matmul(jac,der) ;  call beemat (bee,deriv)
               km = km + matmul(matmul(transpose(bee),dee),bee) *det* weights(i)
            end do gauss_pts_1
   call formkb (kb,km,g)
 end do elements_2
```

(continued)

```
!------------------read load weightings and factorise equations----------------
    read(10,*)loaded_nodes; allocate(no(loaded_nodes),val(loaded_nodes,ndim))
    read(10,*)(no(i),val(i,:),i=1,loaded_nodes)
    call cholin(kb)
!-------------------load increment loop--------------------------------------
    read(10,*)incs;  allocate(qinc(incs));  read(10,*) qinc
    ptot=.0
   load_increments: do iy=1,incs
   write(11,'(a,i5)') 'increment',iy
    ptot=ptot+qinc(iy) ; iters=0;  bdylds=.0;  evpt=.0
!--------------------   iteration loop  --------------------------------------
   iterations: do
    iters=iters+1;  loads=.0
      do i=1,loaded_nodes ; loads(nf(:,no(i)))=val(i,:)*qinc(iy) ;  end do
      loads=loads+bdylds     ;    call chobac(kb,loads)
!----------------------   check convergence  ---------------------------------
      call checon(loads,oldis,tol,converged)
      if(iters==1)converged=.false. ;  if(converged.or.iters==limit)bdylds=.0
!------------------- go round the Gauss Points --------------------------------
      elements_3: do iel = 1 , nels
       bload=.0
       num = g_num( : ,iel ) ; coord = transpose (g_coord( : , num))
       g = g_g( : , iel ) ;        eld = loads ( g )
       gauss_points_2 : do i = 1 , nip
          call shape_der ( der,points,i); jac=matmul(der,coord)
          det = determinant(jac)  ;   call invert(jac)
          deriv = matmul(jac,der) ; call beemat (bee,deriv);eps=matmul(bee,eld)
          eps = eps - evpt( : ,i , iel) ;   sigma=matmul(dee,eps)
          stress = sigma + tensor( : ,i , iel)
          call invar(stress,sigm,dsbar,lode_theta)
!-----------------  check whether yield is violated  -------------------------
         f=dsbar-sqrt(3.)*cu
         if(converged.or.iters==limit) then
         devp=stress
           else
           if(f>=.0) then
           dq1=.0; dq2=1.5/dsbar; dq3=.0     ;   call formm(stress,m1,m2,m3)
           flow=f*(m1*dq1+m2*dq2+m3*dq3)     ;   erate=matmul(flow,stress)
           evp = erate*dt; evpt(:,i,iel)=evpt(:,i,iel) + evp
           devp=matmul(dee,evp)
         end if; end if
      if(f>=.0) then
        eload=matmul(transpose(bee),devp) ; bload=bload+eload*det*weights(i)
      end if
      if(converged.or.iters==limit)then
!------------------------- update the Gauss Point stresses -------------------
          tensor( : , i , iel) = stress
      end if
    end do gauss_points_2
!         compute the total bodyloads vector
    bdylds( g ) = bdylds( g ) + bload      ; bdylds(0) = .0
  end do elements_3
  if(converged.or.iters==limit)exit
 end do iterations
 totd=totd+loads
 write(11,'(a,e12.4)') "The total load is ", ptot
 write(11,'(a,10e12.4)')"Displacements are",                               &
                        (totd(nf(2,no(i))),i=1,loaded_nodes)
 write(11,'(a,i5,a)')"It took ",iters ,"   iterations to converge"
 if(iters==limit)stop
end do load_increments
end program p60
```

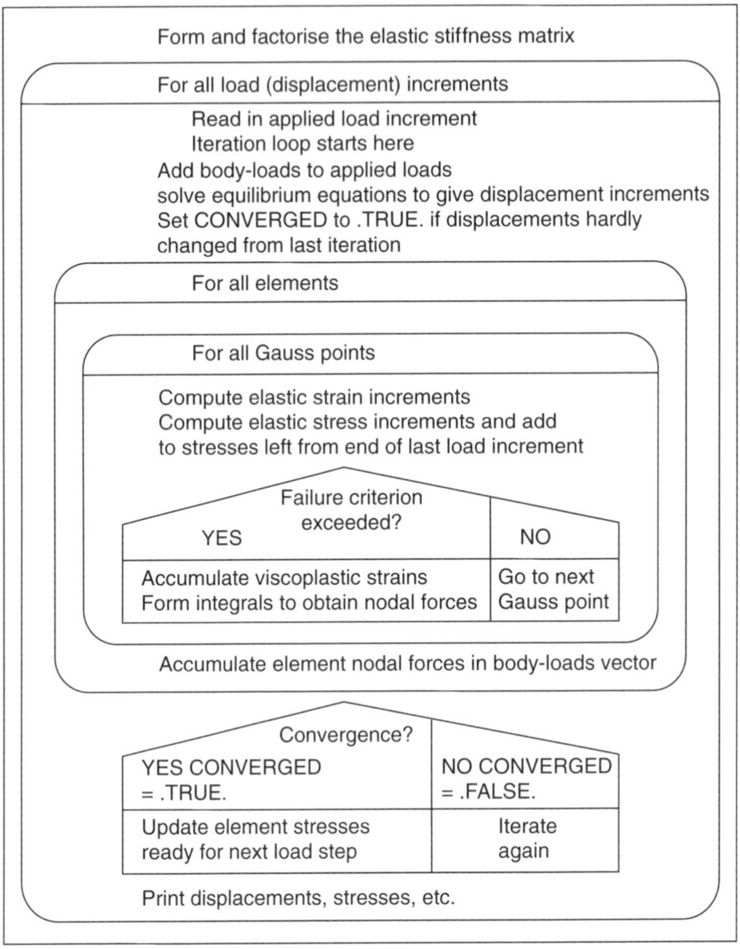

Figure 6.8 Structure chart for visco-plastic algorithm

`DT`	critical time step
`ITERS`	iteration counter
`PTOT`	keeps running total of loading
`CONVERGED`	checks convergence (`.TRUE.`=convergence)
`SIGM`	mean stress invariant σ_m
`DSBAR`	deviatoric stress invariant $\bar{\sigma}$
`LODE_THETA`	Lode angle θ
`F`	value of failure function
`DQ1 DQ2 DQ3`	$\partial Q/\partial\sigma_m$, $\partial Q/\partial J_2$, $\partial Q/\partial J_3$
`ELOAD`	product $\mathbf{B}^T\mathbf{D}\delta\boldsymbol{\varepsilon}^{VP}$ at each Gauss point
`BLOAD`	accumulates `ELOAD` from each Gauss point

BDYLDS	accumulates BLOAD from each element
OLDIS	displacements from previous iteration
QINC	load increments
WIDTH	x coordinates of elements
DEPTH	y coordinates of elements
TOTD	accumulated displacements
TENSOR	accumulates $\sigma_x, \sigma_y, \tau_{xy}$ and σ_z at each Gauss point
STRESS	vector holding $\sigma_x, \sigma_y, \tau_{xy}$ and σ_z
NO	loaded freedom numbers
VAL	load weighting at each freedom
ERATE	visco-plastic strain rate $\boldsymbol{\varepsilon}^{\mathrm{VP}}$
EVP	visco-plastic strain increment $\boldsymbol{\delta\varepsilon}^{\mathrm{VP}}$
DEVP	product $\mathbf{D}^{\mathrm{e}}\boldsymbol{\delta\varepsilon}^{\mathrm{VP}}$
EVPT	accumulated visco-plastic strain at each Gauss point
M1,M2,M3	arrays holding $\partial\sigma_{\mathrm{m}}/\partial\boldsymbol{\sigma}$, $\partial J_2/\partial\boldsymbol{\sigma}$, $\partial J_3/\partial\boldsymbol{\sigma}$
FLOW	used to calculate $\partial Q/\partial\boldsymbol{\sigma}$

Several subroutines appear for the first time in Program 6.0. All elements are rectangular with nodes counted in the y direction; thus the nodal coordinates and steering vector are provided by the eight-node variable geometry subroutine GEOMETRY_8QYV (Appendix 3). Suboutine INVAR forms the three invariants $\sigma_{\mathrm{m}}, \bar{\sigma}$ and θ (Lode angle) from the four Cartesian stress components held in the 1-d array STRESS. It should be noted that in plane strain plasticity applications, it is necessary to retain four components of stress and strain. Although, by definition, one of the strains (ε_z) must equal zero, the elastic strain in that direction may be non-zero, provided

$$\varepsilon_z^{\mathrm{e}} = -\varepsilon_z^{\mathrm{VP}} \tag{6.37}$$

is always true. For this reason, the 4×4 elastic stress–strain matrix is provided by the routine DEEMAT with NST $= 4$.

The only other subroutine not encountered before is FORMM, which creates arrays M1,M2 and M3 used in the calculation of the visco-plastic strain rate from equation (6.25).

The example shown in Figure 6.9 is of a flexible strip footing at the surface of a layer of uniform undrained clay. The footing supports a uniform stress, q, which is increased incrementally to failure. The elasto-plastic soil is described by three parameters, namely the elastic properties, E, ν and the undrained "cohesion" c_{u}. Bearing failure in this problem occurs when q reaches the "Prandtl" load given by

$$q_{\mathrm{ULT}} = (2 + \pi)c_{\mathrm{u}} \tag{6.38}$$

The data follows the familiar pattern established in Chapter 5. The "load weightings" provide a uniform stress of 1 kN/m^2 across the footing semi-width of 2 m (Appendix 1). These "weightings" are then increased proportionally by

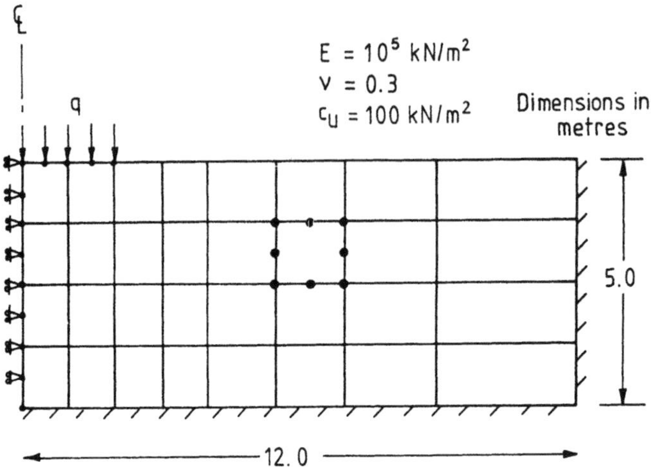


```
Soil                CU      E       V
properties          100.    1.E5    0.3

Structure, mesh     NELS    NXE     NYE     NN      NIP     TOL     LIMIT
and iteration data  32      8       4       121     4       0.001   250

Node freedom        NR
data                33
                    K,NF(:,K),I=1,NR
                    1 0 1 2 0 1 3 0 1 4 0 1 5 0 1 6 0 1
                    7 0 1 8 0 1 9 0 0 14 0 0 23 0 0 28 0 0 37 0 0
                    42 0 0 51 0 0 56 0 0 65 0 0 70 0 0 79 0 0 84 0 0
                    93 0 0 98 0 0 107 0 0 112 0 0 113 0 0 114 0 0 115 0 0
                    116 0 0 117 0 0 118 0 0 119 0 0 120 0 0 121 00
                    WIDTH
Geometry data       0.  1.  2.  3.  4.  5.5  7.  9.  12
                    DEPTH
                    0.  -1.25  -2.5  -3.75  -5.

Load                LOADED_NODES
weightings          5
                    (NO(I),VAL(I,NDIM),I=1,LOADED_NODES)
                    1  .0  -.166667  10  .0  -.666667  15  .0  -.333333
                    24  .0  -.666667  29  .0  -.166667

Load                INCS
increments          10
                    QINC(I),I=1,INCS
                    200.  100.  50.  50.  50.  30.  20.  10.  5.  5.
```

Figure 6.9 Mesh and data for Program 6.0

```
Global coordinates
node    1          0.0000E+00  0.0000E+00
node    2          0.0000E+00 -0.6250E+00
node    3          0.0000E+00 -0.1250E+01
node    4          0.0000E+00 -0.1875E+01
node    5          0.0000E+00 -0.2500E+01
.............................................................................

node  119          0.1200E+02 -0.3750E+01
node  120          0.1200E+02 -0.4375E+01
node  121          0.1200E+02 -0.5000E+01
Global node numbers
Element     1           3    2    1   10   15   16   17   11
Element     2           5    4    3   11   17   18   19   12
Element     3           7    6    5   12   19   20   21   13
Element     4           9    8    7   13   21   22   23   14
Element     5          17   16   15   24   29   30   31   25
.............................................................................
Element    30         103  102  101  109  115  116  117  110
Element    31         105  104  103  110  117  118  119  111
Element    32         107  106  105  111  119  120  121  112
There are   184   equations and the half-bandwidth is    29
increment    1
The total load is    0.2000E+03
Displacements are -0.6592E-02 -0.6486E-02 -0.6116E-02 -0.5418E-02 -0.3849E-02
It took     2   iterations to converge
increment    2
The total load is    0.3000E+03
Displacements are -0.1155E-01 -0.1128E-01 -0.1051E-01 -0.9099E-02 -0.6005E-02
It took    11   iterations to converge
increment    3
The total load is    0.3500E+03
Displacements are -0.1630E-01 -0.1596E-01 -0.1512E-01 -0.1327E-01 -0.7557E-02
It took    20   iterations to converge
increment    4
The total load is    0.4000E+03
Displacements are -0.2316E-01 -0.2283E-01 -0.2217E-01 -0.1986E-01 -0.9360E-02
It took    33   iterations to converge
increment    5
The total load is    0.4500E+03
Displacements are -0.3317E-01 -0.3285E-01 -0.3242E-01 -0.2958E-01 -0.1150E-01
It took    45   iterations to converge
increment    6
The total load is    0.4800E+03
Displacements are -0.4227E-01 -0.4193E-01 -0.4173E-01 -0.3852E-01 -0.1314E-01
It took    65   iterations to converge
increment    7
The total load is    0.5000E+03
Displacements are -0.5084E-01 -0.5041E-01 -0.5034E-01 -0.4677E-01 -0.1455E-01
It took    81   iterations to converge
increment    8
The total load is    0.5100E+03
Displacements are -0.5665E-01 -0.5611E-01 -0.5600E-01 -0.5209E-01 -0.1543E-01
It took    99   iterations to converge
increment    9
The total load is    0.5150E+03
Displacements are -0.6093E-01 -0.6026E-01 -0.5998E-01 -0.5569E-01 -0.1594E-01
It took   159   iterations to converge
increment   10
The total load is    0.5200E+03
Displacements are -0.7459E-01 -0.7317E-01 -0.7122E-01 -0.6463E-01 -0.1654E-01
It took   250   iterations to converge
```

Figure 6.10 Results from Program 6.0

the load increment values held in the vector QINC. It is usual in problems of this type to make the load increments smaller as the failure load is approached. At load levels well below failure, convergence should occur in relatively few iterations. In the data provided in this example LIMIT is set to 250, and this represents the maximum number of iterations that will be allowed within any load increment. If ITERS ever becomes equal to LIMIT, the algorithm stops and no more load increments are applied.

The computed results for this example are given in Figure 6.10 and show the applied stress, the vertical displacement under the loaded nodes and the number of iterations at each stage of the calculation. These results have been plotted in Figure 6.11 in the form of a dimensionless bearing capacity factor q/c_u versus centreline displacement. The number of iterations to achieve convergence for each load increment is also shown. It is seen that convergence was achieved in 159 iterations when $q/c_u = 5.1$, but convergence could not be achieved within the upper limit of 250 when $q/c_u = 5.2$. In addition, the displacements are also increasing rapidly at this level of loading, indicating that bearing failure is taking place at a value very close to the "Prandtl" load of 5.14.

Although Program 6.0 does not incorporate any graphical output capability, it is a simple matter for users to write their own from commercially available libraries. Figure 6.12 shows a plot of the displacement pattern for this example at failure. Although the finite element mesh is constrained to remain a continuum, the picture is still able to give an indication of the form of the failure mechanism.

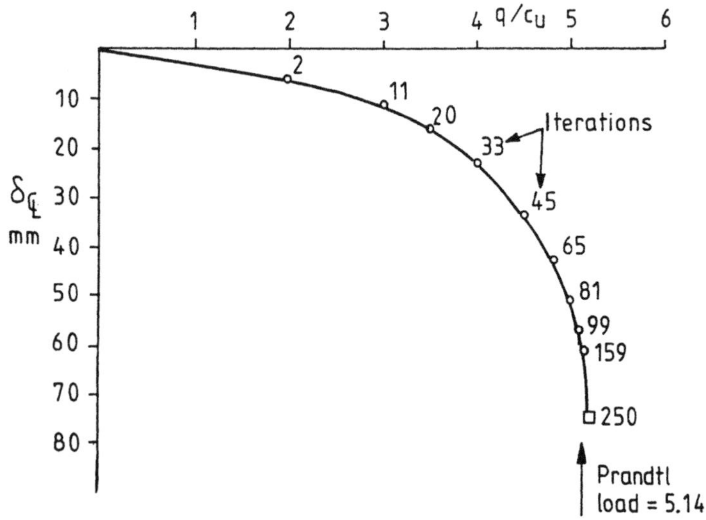

Figure 6.11 Plot of bearing stress versus centreline displacement

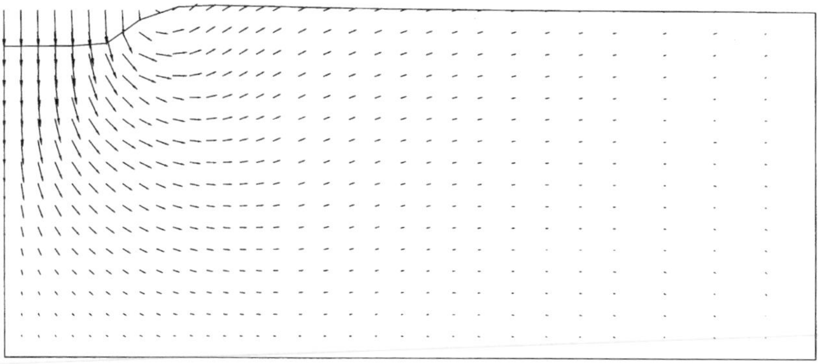

Figure 6.12 Displacement pattern at failure from Program 6.0

PROGRAM 6.1: PLANE STRAIN OF AN ELASTIC–PLASTIC (VON MISES) SOLID USING EIGHT-NODE QUADRILATERAL ELEMENTS; VISCO-PLASTIC STRAIN METHOD, MESH-FREE

For nonlinear problems, computation times are now more demanding. For example Program 6.0 carries out some 765 linear elastic solutions to reach the nonlinear solution and clearly vectorised and parallelised algorithms will become essential for much larger problems. Therefore the mesh-free approach begun in Program 5.10 will be attractive for nonlinear problems and is implemented in Program 6.1. The data in Figure 6.13 is identical to that used in Program 6.0 with the addition of conjugate gradient solution tolerance `CJTOL` and iteration limit `CJITS` added to the plasticity tolerance and plasticity iteration limit, now called `PLASTOL` and `PLASITS` respectively. The results are shown in Figure 6.14, and are nearly identical to those in Figure 6.10. The number of plastic iterations per load step is identical in Programs 6.0 and 6.1 and it took approximately 50 conjugate gradient equation solutions per plastic iteration. In scalar computations, therefore, this algorithm will not be attractive, but it does have some strong attractions for parallel computation. Because a "constant stiffness" approach is being used, groups of elements in Figure 6.9 have constant properties throughout the calculation (there are only 4 distinct element types in this case). This feature can be exploited in parallel versions of Program 6.1.

```
   program p61
!------------------------------------------------------------------------
!      program 6.1 plane strain of an elastic-plastic(Von Mises) solid
!      using 8-node quadrilateral elements; viscoplastic strain method
!------------------------------------------------------------------------
 use new_library    ;    use  geometry_lib  ;        ; implicit none
 integer::nels,nxe,nye,neq,nn,nr,nip,nodof=2,nod=8,nst=4,ndof,loaded_nodes,   &
          i,k,iel,plasiters,plasits,cjiters,cjits,incs,iy,ndim=2,cjtot
 logical:: plastic_converged, cj_converged
 real::e,v,det,cu,dt,ptot,f,dsbar,dq1,dq2,dq3,lode_theta,sigm,                &
     up,alpha,beta,big,plastol,cjtol;character(len=15)::element='quadrilateral'
!--------------------------- dynamic arrays-----------------------------------
 real    ,allocatable :: loads(:),points(:,:),totd(:),bdylds(:),pmul(:),       &
                         evpt(:,:,:),oldis(:),width(:),depth(:),qinc(:),        &
                         tensor(:,:,:),val(:,:),stress(:),storkm(:,:,:),        &
                         dee(:,:),coord(:,:),jac(:,:),weights(:),               &
                         der(:,:),deriv(:,:),bee(:,:),km(:,:),eld(:),eps(:),    &
                         sigma(:),bload(:),eload(:),erate(:),g_coord(:,:),      &
                         evp(:),devp(:),m1(:,:),m2(:,:),m3(:,:),flow(:,:),      &
                         p(:),x(:),xnew(:),u(:),diag_precon(:),d(:),utemp(:)
 integer, allocatable :: nf(:,:) , g(:), no(:) ,num(:), g_num(:,:) ,g_g(:,:)
!-----------------------input and initialisation------------------------------
  open (10,file='p61.dat',status=    'old',action='read')
  open (11,file='p61.res',status='replace',action='write')
  read (10,*) cu,e,v,  nels,nxe,nye,nn,nip,    plasits,cjits,plastol,cjtol
  ndof=nod*nodof
  allocate (nf(nodof,nn), points(nip,ndim),weights(nip),g_coord(ndim,nn),      &
           width(nxe+1),depth(nye+1),num(nod),dee(nst,nst),coord(nod,ndim),    &
           evpt(nst,nip,nels), tensor(nst,nip,nels),pmul(ndof),utemp(ndof),    &
           jac(ndim,ndim),der(ndim,nod),deriv(ndim,nod),g_num(nod,nels),       &
           bee(nst,ndof),km(ndof,ndof),eld(ndof),eps(nst),sigma(nst),          &
           bload(ndof),eload(ndof),erate(nst),evp(nst),devp(nst),g(ndof),      &
           m1(nst,nst),m2(nst,nst),m3(nst,nst),flow(nst,nst),stress(nst),      &
           g_g(ndof,nels),storkm(ndof,ndof,nels))
      nf=1; read(10,*) nr ; if(nr>0) read(10,*)(k,nf(:,k),i=1,nr)
      call formnf(nf); neq=maxval(nf)   ;      read(10,*) width, depth
! ---------------loop the elements to set up global arrays --------------------
       elements_1:   do iel = 1 , nels
                         call geometry_8qyv(iel,nye,width,depth,coord,num)
                         call num_to_g( num , nf , g );  g_num(:,iel)=num
                         g_coord(:,num)=transpose(coord);   g_g(:,iel) = g
       end do elements_1
    write(11,'(a)') "Global coordinates "
    do k=1,nn;write(11,'(a,i5,a,2e12.4)')"node",k,"       ",g_coord(:,k);end do
    write(11,'(a)') "Global node numbers "
    do k = 1 , nels; write(11,'(a,i5,a,8i5)')                                  &
                                "Element ",k,"        ",g_num(:,k); end do
    write(11,'(a,i5)') "The number of equations to be solved is",neq
 allocate(loads(0:neq),bdylds(0:neq),oldis(0:neq),totd(0:neq),p(0:neq),        &
         x(0:neq),xnew(0:neq),u(0:neq),diag_precon(0:neq),d(0:neq))
   oldis=0.0; totd=0.0 ; tensor = 0.0
   p = .0;  xnew = .0; diag_precon = .0
  call deemat(dee,e,v); call sample(element,points,weights)
  dt=4.*(1.+v)/(3.*e)
!---------- element stiffness integration,storage and preconditioner --------
 elements_2: do iel = 1 , nels
                num = g_num(:,iel) ; coord = transpose(g_coord(:  , num ))
                g = g_g(:,iel)     ;            km=0.0
              gauss_pts_1:  do i =1 , nip
                call shape_der (der,points,i);  jac = matmul(der,coord)
                det = determinant(jac)  ;   call invert(jac)
                deriv = matmul(jac,der) ;  call beemat (bee,deriv)
                km = km + matmul(matmul(transpose(bee),dee),bee) *det* weights(i)
             end do gauss_pts_1
       storkm(:,:,iel) = km
       do k=1,ndof; diag_precon(g(k))=diag_precon(g(k))+km(k,k); end do
  end do elements_2
       diag_precon(1:neq)=1./diag_precon(1:neq) ; diag_precon(0)=.0
```

```
!-----------------------read load weightings ---------------------------------
    read(10,*)loaded_nodes  ; allocate(no(loaded_nodes),val(loaded_nodes,ndim))
    read(10,*)(no(i),val(i,:),i=1,loaded_nodes)
!--------------------------load increment loop-------------------------------
    read(10,*)incs  ; allocate(qinc(incs));  read(10,*)qinc
    ptot=.0
   load_increments: do iy=1,incs
    write (11,'(a,i5)') "Load increment" ,iy
    ptot=ptot+qinc(iy) ; plasiters=0;bdylds=.0;evpt=.0  ;  cjtot = 0
!-------------------------   plastic iteration loop   -----------------------
   plastic_iterations: do
    plasiters=plasiters+1;  loads=.0
      do i=1,loaded_nodes ; loads(nf(:,no(i)))=val(i,:)*qinc(iy) ;  end do
      loads=loads+bdylds      ; d=diag_precon*loads ; p = d   ; x = .0
!-------------------   solve the simultaneous equations by pcg ----------------
       cjiters = 0
      conjugate_gradients:  do
       cjiters = cjiters + 1 ; u = .0
      elements_3 : do iel = 1 , nels
                   g = g_g( : , iel ); km = storkm( : , : ,iel)
                   pmul = p(g); utemp=matmul(km,pmul)
!dir$ ivdep
          do i = 1 , ndof
             u(g(i)) = u(g(i)) +  utemp(i)
          end do
      end do elements_3
!------------------------------pcg process ----------------------------------
    up =dot_product(loads,d); alpha=up/dot_product(p,u)
    xnew = x + p* alpha; loads = loads - u*alpha; d = diag_precon*loads
    beta = dot_product(loads,d)/up; p = d + p * beta
    big = .0; cj_converged = .true.
    do i = 1,neq; if(abs(xnew(i))>big)big=abs(xnew(i)); end do
    do i = 1,neq; if(abs(xnew(i)-x(i))/big>cjtol)cj_converged=.false.;end do
    x = xnew
    if(cj_converged.or.cjiters==cjits) exit
      end do conjugate_gradients
      cjtot = cjtot + cjiters
!--------------------------- end of pcg process -----------------------------
    loads = xnew          ; loads(0) = .0
!-----------------------   check plastic convergence   -----------------------
      call checon(loads,oldis,plastol,plastic_converged)
      if(plasiters==1)plastic_converged=.false.
      if(plastic_converged.or.plasiters==plasits)bdylds=.0
!----------------------- go round the Gauss Points --------------------------
      elements_4: do iel = 1 , nels
       bload=.0
       num = g_num( : ,iel) ; coord =transpose( g_coord( : , num ))
       g = g_g( : , iel )    ; eld = loads ( g )
       gauss_points_2 : do i = 1 , nip
          call shape_der ( der,points,i); jac=matmul(der,coord)
          det = determinant(jac);   call invert(jac) ; deriv = matmul(jac,der)
          call beemat (bee,deriv);eps=matmul(bee,eld)
          eps = eps - evpt(: ,i ,iel) ;sigma=matmul(dee,eps)
          stress = sigma+tensor(: , i, iel)
          call invar(stress,sigm,dsbar,lode_theta)
!---------------------  check whether yield is violated  --------------------
         f=dsbar-sqrt(3.)*cu
         if(plastic_converged.or.plasiters==plasits) then
         devp=stress
           else
           if(f>=.0) then
           dq1=.0; dq2=1.5/dsbar; dq3=.0     ;   call formm(stress,m1,m2,m3)
           flow=f*(m1*dq1+m2*dq2+m3*dq3)     ;   erate=matmul(flow,stress)
           evp=erate*dt; evpt(:,i,iel)=evpt(:,i,iel)+evp; devp=matmul(dee,evp)
         end if; end if
      if(f>=.0) then
        eload=matmul(transpose(bee),devp) ; bload=bload+eload*det*weights(i)
      end if
      if(plastic_converged.or.plasiters==plasits)then
```

```
!---------------------- update the Gauss Point stresses ----------------------
          tensor(: , i , iel) = stress
      end if
    end do gauss_points_2
!-------------compute the total bodyloads vector ; dependency if vectorised----
!dir$ ivdep
    do i=1,ndof
         bdylds( g(i) ) = bdylds( g(i) ) + bload(i)
    end do  ;    bdylds(0) = .0
  end do elements_4
  if(plastic_converged.or.plasiters==plasits)exit
 end do plastic_iterations
 totd=totd+loads
 write(11,'(a,e12.4)')"The total load is  ",ptot
 write(11,'(a,10e12.4)')"Displacements are",(totd(nf(2,no(i))),i=1,loaded_nodes)
 write(11,'(a,i12)')"The total number of cj iterations was    ",cjtot
 write(11,'(a,i12)')"The number of plastic iterations was     ",plasiters
 write(11,'(a,f11.2)')"cj iterations per plastic iteration were   ", &
  & real(cjtot)/real(plasiters)
 if(plasiters==plasits)stop
end do load_increments
end program p61
```

Soil properties	CU E V 100. 1.E5 0.3
Structure and mesh data	NELS NXE NYE NN NIP 32 8 4 121 4
Iteration data	PLASITS CJITS PLASTOL CJTOL 250 100 0.001 0.0001
Node freedom data	NR 33 K,NF(:,K),I=1,NR 1 0 1 2 0 1 3 0 1 4 0 1 5 0 1 6 0 1 7 0 1 8 0 1 9 0 0 14 0 0 23 0 0 28 0 0 37 0 0 42 0 0 51 0 0 56 0 0 65 0 0 70 0 0 79 0 0 84 0 0 93 0 0 98 0 0 107 0 0 112 0 0 113 0 0 114 0 0 115 0 0 116 0 0 117 0 0 118 0 0 119 0 0 120 0 0 121 00
Geometry data	WIDTH 0. 1. 2. 3. 4. 5.5 7. 9. 12 DEPTH 0. -1.25 -2.5 -3.75 -5.
Load weightings	LOADED_NODES 5 (NO(I),VAL(I,NDIM),I=1,LOADED_NODES) 1 .0 -.166667 10 .0 -.666667 15 .0 -.333333 24 .0 -.666667 29 .0 -.166667
Load increments	INCS 10 QINC(I),I=1,INCS 200. 100. 50. 50. 50. 30. 20. 10. 5. 5.

Figure 6.13 Data for Program 6.1

```
Global coordinates
node    1          0.0000E+00  0.0000E+00
node    2          0.0000E+00 -0.6250E+00
node    3          0.0000E+00 -0.1250E+01
node    4          0.0000E+00 -0.1875E+01
node    5          0.0000E+00 -0.2500E+01
.........................................................................

node  118          0.1200E+02 -0.3125E+01
node  119          0.1200E+02 -0.3750E+01
node  120          0.1200E+02 -0.4375E+01
node  121          0.1200E+02 -0.5000E+01
Global node numbers
Element     1            3    2    1   10   15   16   17   11
Element     2            5    4    3   11   17   18   19   12
..........................................................................
Element    30          103  102  101  109  115  116  117  110
Element    31          105  104  103  110  117  118  119  111
Element    32          107  106  105  111  119  120  121  112
The number of equations to be solved is  184
Load increment     1
The total load is    0.2000E+03
Displacements are -0.6593E-02 -0.6487E-02 -0.6116E-02 -0.5418E-02 -0.3849E-02
The total number of cj iterations was              92
The number of plastic iterations was                2
cj iterations per plastic iteration were        46.00
Load increment     2
The total load is    0.3000E+03
Displacements are -0.1155E-01 -0.1128E-01 -0.1051E-01 -0.9096E-02 -0.6004E-02
The total number of cj iterations was             515
The number of plastic iterations was               11
cj iterations per plastic iteration were        46.82
..............................................................................

Load increment     7
The total load is    0.5000E+03
Displacements are -0.5090E-01 -0.5047E-01 -0.5038E-01 -0.4681E-01 -0.1455E-01
The total number of cj iterations was            4346
The number of plastic iterations was               82
cj iterations per plastic iteration were        53.00
Load increment     8
The total load is    0.5100E+03
Displacements are -0.5674E-01 -0.5619E-01 -0.5605E-01 -0.5214E-01 -0.1542E-01
The total number of cj iterations was            5207
The number of plastic iterations was               98
cj iterations per plastic iteration were        53.13
Load increment     9
The total load is    0.5150E+03
Displacements are -0.6110E-01 -0.6042E-01 -0.6010E-01 -0.5580E-01 -0.1594E-01
The total number of cj iterations was            8485
The number of plastic iterations was              159
cj iterations per plastic iteration were        53.36
Load increment    10
The total load is    0.5200E+03
Displacements are -0.7516E-01 -0.7370E-01 -0.7165E-01 -0.6496E-01 -0.1652E-01
The total number of cj iterations was           13782
The number of plastic iterations was              250
cj iterations per plastic iteration were        55.13
```

Figure 6.14 Results from Program 6.1

PROGRAM 6.2: PLANE STRAIN SLOPE STABILITY ANALYSIS—VISCO-PLASTIC METHOD USING THE MOHR–COULOMB CRITERION

```
    program p62
!----------------------------------------------------------------------
!      program 6.2 plane strain of an elastic-plastic(Mohr-Coulomb) solid
!      using 8-node quadrilateral elements; viscoplastic strain method
!----------------------------------------------------------------------
 use new_library   ;   use geometry_lib     ; implicit none
 integer::nels,nxe,nye,neq,nband,nn,nr,nip,nodof=2,nod=8,nst=4,ndof,          &
         i,k,iel,iters,limit,incs,iy,ndim=2
 logical::converged      ;    character(len=15) :: element='quadrilateral'
 real::e,v,det,phi,c,psi,gama,dt,f,dsbar,dq1,dq2,dq3,lode_theta,              &
          sigm,pi,tnph,phif,snph,cf,tol
!--------------------------- dynamic arrays------------------------------------
 real    ,allocatable :: kb(:,:),loads(:),points(:,:),bdylds(:),bot(:),fos(:),&
                         evpt(:,:,:),oldis(:),top(:),depth(:),gravlo(:),      &
                         dee(:,:),coord(:,:),fun(:),jac(:,:),weights(:),      &
                         der(:,:),deriv(:,:),bee(:,:),km(:,:),eld(:),eps(:),  &
                         sigma(:),bload(:),eload(:),erate(:),g_coord(:,:),    &
                         evp(:),devp(:),m1(:,:),m2(:,:),m3(:,:),flow(:,:)
 integer, allocatable :: nf(:,:) , g(:), no(:) ,num(:), g_num(:,:) ,g_g(:,:)
!------------------------input and initialisation-----------------------------
  open (10,file='p62.dat',status=    'old',action='read')
  open (11,file='p62.res',status='replace',action='write')
  read (10,*) phi,c,psi,gama,e,v,    nels,nxe,nye,nn,nip,tol,limit
  ndof=nod*nodof
  allocate (nf(nodof,nn), points(nip,ndim),weights(nip),g_coord(ndim,nn),     &
           top(nxe+1),depth(nye+1),num(nod),dee(nst,nst),evpt(nst,nip,nels), &
           bot(nxe+1),coord(nod,ndim),fun(nod),g_g(ndof,nels),              &
           jac(ndim,ndim),der(ndim,nod),deriv(ndim,nod),g_num(nod,nels),     &
           bee(nst,ndof),km(ndof,ndof),eld(ndof),eps(nst),sigma(nst),        &
           bload(ndof),eload(ndof),erate(nst),evp(nst),devp(nst),g(ndof),    &
           m1(nst,nst),m2(nst,nst),m3(nst,nst),flow(nst,nst))
  nf=1; read(10,*) nr ; if(nr>0) read(10,*)(k,nf(:,k),i=1,nr)
  call formnf(nf); neq=maxval(nf)     ;  read(10,*) top , bot , depth
!---------- loop the elements to find nband and set up global arrays ----------
     nband = 0
       elements_1:   do iel = 1 , nels
                       call slope_geometry(iel,nye,top,bot,depth,coord,num)
                       call num_to_g(num,nf,g);      g_num(:,iel)=num
                       g_coord(:,num)=transpose(coord);   g_g(:,iel) = g
                       if (nband<bandwidth(g)) nband = bandwidth(g)
      end do elements_1
    write(11,'(a)') "Global coordinates "
    do k=1,nn;write(11,'(a,i5,a,2e12.4)')"Node",k,"        ",g_coord(:,k);end do
    write(11,'(a)') "Global node numbers "
    do k = 1 , nels; write(11,'(a,i5,a,8i5)')                                  &
                            "Element ",k,"        ",g_num(:,k); end do
   write(11,'(a,i5,a,i5)')                                                     &
           "The system has",neq," equations and the half-bandwidth is ",nband
allocate(kb(neq,nband+1),loads(0:neq),bdylds(0:neq),oldis(0:neq),gravlo(0:neq))
           kb=0.0; oldis=0.0; gravlo=0.0
  call deemat(dee,e,v);       call sample(element,points,weights)
  pi = acos( -1. ); tnph = tan(phi*pi/180.)
!----------------- element stiffness integration and assembly------------------
 elements_2: do iel = 1 , nels
                num = g_num(:,iel) ; coord = transpose(g_coord( : , num ))
                g = g_g( : , iel )  ;         km=0.0    ;  eld = .0
             gauss_pts_1:  do i =1 , nip     ; call shape_fun(fun,points,i)
               call shape_der (der,points,i);  jac = matmul(der,coord)
               det = determinant(jac)  ;    call invert(jac)
               deriv = matmul(jac,der) ;  call beemat (bee,deriv)
               km = km + matmul(matmul(transpose(bee),dee),bee) *det* weights(i)
               do k=2,ndof,2;eld(k)=eld(k)+fun(k/2)*det*weights(i);end do
             end do gauss_pts_1
   call formkb (kb,km,g)
```

```
   gravlo ( g ) = gravlo ( g ) - eld * gama ; gravlo(0) = .0
 end do elements_2
!------------------------ factorise left hand side-----------------------------
          call cholin(kb)
!------------------------trial factor of safety loop---------------------------
    read(10,*) incs;   allocate ( fos (incs )) ;    read(10,*) fos
    load_increments: do iy=1,incs
    phif = atan(tnph/fos(iy))*180./pi; snph = sin(phif*pi/180.)
    dt = 4.*(1.+v)*(1.-2.*v)/(e*(1.-2.*v+snph**2)) ; cf = c/fos(iy)
    write(11,'(a,i5)') "Load increment",iy ;  iters=0;  bdylds=.0;  evpt=.0
!----------------------------  iteration loop  ------------------------------
   iterations: do
    iters=iters+1;  loads = gravlo + bdylds   ;  call chobac(kb,loads)
!-------------------------  check convergence --------------------------------
      call checon(loads,oldis,tol,converged)
      if(iters==1)converged=.false. ;  if(converged.or.iters==limit)bdylds=.0
!----------------------- go round the Gauss Points  --------------------------
      elements_3: do iel = 1 , nels
       bload=.0
       num = g_num( : , iel ) ; coord =transpose( g_coord( : ,num ))
       g = g_g( : , iel )  ;       eld = loads ( g )
       gauss_points_2 : do i = 1 , nip
          call shape_der ( der,points,i); jac=matmul(der,coord)
          det = determinant(jac); call invert(jac) ; deriv = matmul(jac,der)
          call beemat (bee,deriv);   eps=matmul(bee,eld)
          eps = eps -evpt( : , i , iel)   ;          sigma=matmul(dee,eps)
          call invar(sigma,sigm,dsbar,lode_theta)
!------------------ check whether yield is violated  -------------------------
         call mocouf (phif, cf , sigm, dsbar , lode_theta , f )
         if(converged.or.iters==limit) then
         devp=sigma
           else
           if(f>=.0) then
           call mocouq(psi,dsbar,lode_theta,dq1,dq2,dq3)
           call formm(sigma,m1,m2,m3)   ;    flow=f*(m1*dq1+m2*dq2+m3*dq3)
           erate=matmul(flow,sigma)     ;    evp=erate*dt
           evpt(:,i,iel)=evpt(:,i,iel)+evp;  devp=matmul(dee,evp)
         end if; end if
      if(f>=.0) then
        eload=matmul(devp,bee) ; bload=bload+eload*det*weights(i)
      end if
    end do gauss_points_2
!------------------- compute the total bodyloads vector ---------------------
    bdylds( g ) = bdylds( g ) + bload     ; bdylds(0) = .0
  end do elements_3
  if(converged.or.iters==limit)exit
 end do iterations
 write(11,'(a)') "    fos    max displacement"
 write(11,'(2e12.4)')fos(iy),maxval(abs(loads))
 write(11,'(a,i5,a)') "It took",iters,"   iterations to converge"
 if(iters==limit)stop
end do load_increments
end program p62
```

This program is, in many ways, similar to Program 6.0. The problem to be analysed is a slope of Mohr–Coulomb material subjected to gravity loading. The factor of safety (`FOS`) of the slope is to be assessed, and this quantity is defined as the proportion by which tan ϕ and c must be reduced in order to cause failure. This is in contrast to the previous program in which failure was induced by increasing the loads with the material properties remaining constant.

The gravity loading vector $\mathbf{p}_a$ for a material with unit weight γ is accumulated for each element from integrals of the type

$$\mathbf{p}_a = \gamma \sum_{\text{elements}}^{\text{all}} \int \mathbf{N}^T \text{d(element)} \tag{6.39}$$

and these calculations are performed in the same part of the program that forms the global stiffness matrix. It may be noted that only those freedoms corresponding to vertical movement are incorporated in the integrals.

At the element level, the 1-d array ELD is used to gather the contributions from each Gauss point. The global gravity loads vector GRAVLO accumulates ELD from each element after multiplication by the unit weight GAMA (γ).

The gravity loads vector in this program is applied to the slope in a single increment, and what was previously called the "load increment loop" is now called the "trial factor of safety loop". Each entry into this loop corresponds to a different factor of safety (FOS) on the soil strength parameters. The factored soil strength parameters that go into the elasto-plastic analysis are obtained from

$$\begin{aligned} \phi_f &= \arctan(\tan\phi/\texttt{FOS}) \\ c_f &= c/\texttt{FOS} \end{aligned} \tag{6.40}$$

Keeping the loads constant, several (usually increasing) values of the factor of safety are attempted until the algorithm fails to converge. The actual factor of safety of the slope is the value to cause failure.

The following variables that are new to this program are now defined:

PHI	friction angle (ϕ)
C	cohesion (c)
PSI	dilation angle (ψ)
PHIF	factored friction angle (ϕ_f)
CF	factored cohesion (c_f)
TNPH	tan ϕ_f
SNPH	sin ϕ_f
FOS	1-d array holding trial factors of safety

The first subroutine that is new to this program is the geometry subroutine SLOPE_GEOMETRY. This subroutine generates a mesh shaped like a trapezium with the restriction that the top and bottom boundaries are parallel to the x axis. Hence, instead of a single 1-d array to hold the x coordinates of the element boundaries, the following two 1-d arrays must be read in as data:

TOP	element x coordinates at top of mesh
BOT	element x coordinates at bottom of mesh

The subroutine assumes that these coordinates are connected by straight lines. The element y coordinates are read into the array `DEPTH`, as was done in Program 6.0.

The subroutine `MOCOUF` forms the Mohr–Coulomb failure function F from the current stress state and the operating shear strength parameters. The subroutine `MOCOUQ` forms the derivatives of the Mohr–Coulomb potential function Q with respect to the three stress invariants and these values are held in `DQ1`, `DQ2` and `DQ3`.

In Program 6.0, similar subroutines corresponding to the von Mises criterion could have been used (`VMF`, `VMQ`, etc.), but the required expressions were so trivial that they were written directly into the main program.

Figure 6.15 shows the mesh and data for a typical slope stability analysis. The parameters are given as $\phi = 40°$, $c = 1\,\text{kN/m}^3$ and the dilation angle ψ is put equal to zero. The unit weight of the material is given as $\gamma = 20\,\text{kN/m}^3$. The "structure data" and "node freedom data" follow a familiar pattern in which

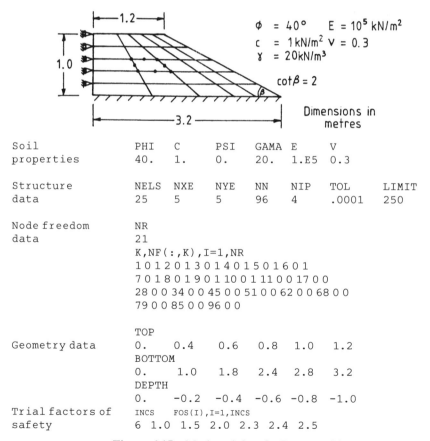

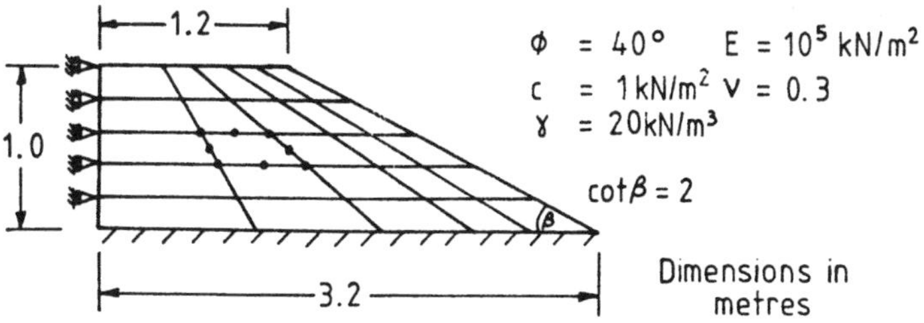


```
Soil             PHI   C    PSI   GAMA  E     V
properties       40.   1.   0.    20.   1.E5  0.3

Structure        NELS  NXE  NYE   NN    NIP   TOL     LIMIT
data             25    5    5     96    4     .0001   250

Node freedom     NR
data             21
                 K,NF(:,K),I=1,NR
                 1 0 1 2 0 1 3 0 1 4 0 1 5 0 1 6 0 1
                 7 0 1 8 0 1 9 0 1 10 0 1 11 0 0 17 0 0
                 28 0 0 34 0 0 45 0 0 51 0 0 62 0 0 68 0 0
                 79 0 0 85 0 0 96 0 0

                 TOP
Geometry data    0.    0.4   0.6   0.8   1.0   1.2
                 BOTTOM
                 0.    1.0   1.8   2.4   2.8   3.2
                 DEPTH
                 0.    -0.2  -0.4  -0.6  -0.8  -1.0
Trial factors of INCS  FOS(I),I=1,INCS
safety           6 1.0 1.5 2.0 2.3 2.4 2.5
```

Figure 6.15 Mesh and data for Program 6.2

```
Global coordinates
Node    1          0.0000E+00  0.0000E+00
Node    2          0.0000E+00 -0.1000E+00
Node    3          0.0000E+00 -0.2000E+00
Node    4          0.0000E+00 -0.3000E+00
Node    5          0.0000E+00 -0.4000E+00
.................................................................

Node   93          0.2600E+01 -0.7000E+00
Node   94          0.2800E+01 -0.8000E+00
Node   95          0.3000E+01 -0.9000E+00
Node   96          0.3200E+01 -0.1000E+01
Global node numbers
Element     1            3    2    1   12   18   19   20   13
Element     2            5    4    3   13   20   21   22   14
Element     3            7    6    5   14   22   23   24   15
Element     4            9    8    7   15   24   25   26   16
Element     5           11   10    9   16   26   27   28   17
Element     6           20   19   18   29   35   36   37   30
Element     7           22   21   20   30   37   38   39   31
Element     8           24   23   22   31   39   40   41   32
Element     9           26   25   24   32   41   42   43   33
Element    10           28   27   26   33   43   44   45   34
Element    11           37   36   35   46   52   53   54   47
Element    12           39   38   37   47   54   55   56   48
Element    13           41   40   39   48   56   57   58   49
Element    14           43   42   41   49   58   59   60   50
Element    15           45   44   43   50   60   61   62   51
Element    16           54   53   52   63   69   70   71   64
Element    17           56   55   54   64   71   72   73   65
Element    18           58   57   56   65   73   74   75   66
Element    19           60   59   58   66   75   76   77   67
Element    20           62   61   60   67   77   78   79   68
Element    21           71   70   69   80   86   87   88   81
Element    22           73   72   71   81   88   89   90   82
Element    23           75   74   73   82   90   91   92   83
Element    24           77   76   75   83   92   93   94   84
Element    25           79   78   77   84   94   95   96   85
The system has  160  equations and the half-bandwidth is    35
Load increment     1
    fos    max displacement
  0.1000E+01  0.7580E-04
It took    2   iterations to converge
Load increment     2
    fos    max displacement
  0.1500E+01  0.7580E-04
It took    2   iterations to converge
Load increment     3
    fos    max displacement
  0.2000E+01  0.7748E-04
It took   16   iterations to converge
Load increment     4
    fos    max displacement
  0.2300E+01  0.8495E-04
It took   29   iterations to converge
Load increment     5
    fos    max displacement
  0.2400E+01  0.9024E-04
It took  140   iterations to converge
Load increment     6
    fos    max displacement
  0.2500E+01  0.1200E-03
It took  250   iterations to converge
```

Figure 6.16 Results from Program 6.2

nodes are counted in the y direction. Three 1-d arrays read in the coordinate information, and six trial factors of safety ranging from 1.0 to 2.5 are to be attempted.

The output in Figure 6.16 gives the factor of safety, the maximum displacement at convergence and the number of iterations to achieve convergence. The results have also been plotted in Figure 6.17, and these indicate that the factor of safety of the slope is around 2.5. Bishop and Morgenstern (1960) produced charts for slope stability analysis using slip circle techniques, and these give a factor of safety of 2.505 for the slope considered in this example. The displaced mesh after 250 iterations with a factor of safety of 2.5 are given in Figure 6.18 to indicate the nature of the failure mechanism. This was produced from the coarse 5 × 5 mesh results by the technique of "regridding" using the shape functions (Kidger, 1994).

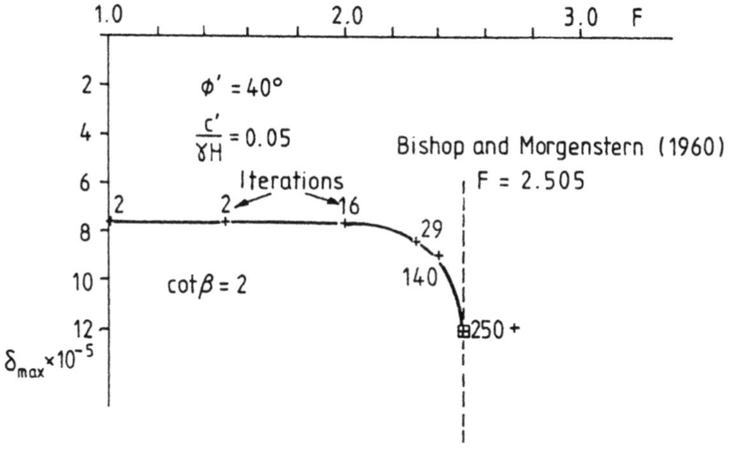

Figure 6.17 Plot of maximum displacement versus factor of safety

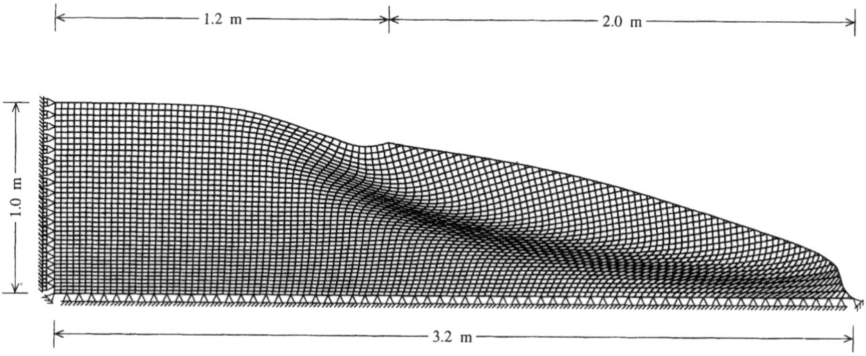

Figure 6.18 Displaced mesh at failure from Program 6.1

PROGRAM 6.3: PLANE STRAIN PASSIVE EARTH PRESSURE ANALYSIS—INITIAL STRESS METHOD USING THE MOHR–COULOMB CRITERION

```
    program p63
!-----------------------------------------------------------------------
!      program 6.3 plane strain of an elastic-plastic(Mohr-Coulomb) solid
!      using 8-node quadrilateral elements; initial stress method
!-----------------------------------------------------------------------
 use new_library ;   use geometry_lib  ; implicit none
 integer::nels,nxe,nye,neq,nband,nn,nr,nip,nodof=2,nod=8,nst=4,ndof,         &
          loaded_nodes, i,k,iel,iters,limit,incs,iy,ndim=2
 logical::converged        ;  character (len=15) :: element='quadrilateral'
 real::e,v,det,c,phi,psi,epk0,ptot,fac,f,fnew,dsbar,lode_theta,sigm,presc,    &
       pav,gama,tol
!---------------------------- dynamic arrays---------------------------------
 real    ,allocatable :: kb(:,:),loads(:),points(:,:),totd(:),bdylds(:),       &
                         width(:),depth(:),tensor(:,:,:),                      &
                         dee(:,:),coord(:,:),fun(:),jac(:,:),weights(:),       &
                         der(:,:),deriv(:,:),bee(:,:),km(:,:),eld(:),eps(:),  &
                         sigma(:),bload(:),eload(:),elso(:),g_coord(:,:),      &
                         oldis(:),storkb(:),stress(:),pl(:,:),gc(:)
 integer, allocatable :: nf(:,:) , g(:), no(:) ,num(:), g_num(:,:) ,g_g(:,:)
!-----------------------input and initialisation-----------------------------
  open (10,file='p63a.dat',status=    'old',action='read')
  open (11,file='p63a.res',status='replace',action='write')
  read (10,*) phi,c,psi,gama,epk0,e,v,    nels,nxe,nye,nn,nip
  ndof=nod*nodof
  allocate (nf(nodof,nn), points(nip,ndim),weights(nip),g_coord(ndim,nn),     &
            width(nxe+1),depth(nye+1),num(nod),dee(nst,nst),fun(nod),gc(ndim),&
            tensor(nst,nip,nels),g_g(ndof,nels),coord(nod,ndim),stress(nst),  &
            jac(ndim,ndim),der(ndim,nod),deriv(ndim,nod),g_num(nod,nels),     &
            bee(nst,ndof),km(ndof,ndof),eld(ndof),eps(nst),sigma(nst),        &
            bload(ndof),eload(ndof),pl(nst,nst),elso(nst),g(ndof))
  nf=1; read(10,*) nr ; if(nr>0) read(10,*)(k,nf(:,k),i=1,nr)
  call formnf(nf); neq=maxval(nf)     ;  read(10,*) width, depth
!----------- loop the elements to find nband and set up global arrays ---------
     nband = 0
       elements_1:   do iel = 1 , nels
                        call geometry_8qyv(iel,nye,width,depth,coord,num)
                        call num_to_g(num,nf,g);          g_num(:,iel)=num
                        g_coord(:,num)=transpose(coord);  g_g(:,iel) = g
                        if (nband<bandwidth(g)) nband = bandwidth(g)
      end do elements_1
    write(11,'(a)') "Global coordinates "
    do k=1,nn;write(11,'(a,i5,a,2e12.4)')"Node",k,"       ",g_coord(:,k);end do
    write(11,'(a)') "Global node numbers "
    do k = 1 , nels; write(11,'(a,i5,a,8i5)')                                  &
                              "Element ",k,"        ",g_num(:,k); end do
   write(11,'(a,i5,a,i5)')                                                     &
           "The system has ",neq," equations and the half-bandwidth is ",nband
  allocate(kb(neq,nband+1),loads(0:neq),bdylds(0:neq),oldis(0:neq),totd(0:neq))
  kb=0.0; oldis=0.0; totd=0.0 ;bdylds=.0;  tensor = 0.0
  call deemat(dee,e,v);     call sample(element,points,weights)
!---------------- element stiffness integration and assembly-------------------
 elements_2: do iel = 1 , nels
                num = g_num(:, iel) ; coord =transpose( g_coord(: ,num))
                g = g_g( : , iel )  ;         km=0.0
             gauss_pts_1:  do i =1 , nip
                call shape_fun(fun,points,i);  gc = matmul(fun,coord)
!--------------------starting stress state  -----------------------------------
         tensor(2,i,iel)=gc(2)*gama;tensor(1,i,iel)=gc(2)*gama*epk0
         tensor(4,i,iel)=tensor(1,i,iel)
                call shape_der (der,points,i);  jac = matmul(der,coord)
                det = determinant(jac)  ;    call invert(jac)
                deriv = matmul(jac,der) ;  call beemat (bee,deriv)
                km = km + matmul(matmul(transpose(bee),dee),bee) *det* weights(i)
              end do gauss_pts_1
```

```
   call formkb (kb,km,g)
 end do elements_2
!---------------read prescribed displacements and factorise l.h.s.-------------
    read(10,*) loaded_nodes ; allocate (no(loaded_nodes),storkb(loaded_nodes))
    read(10,*)(no(i),i=1,loaded_nodes),presc,incs,tol,limit
    do i=1,loaded_nodes
            kb(nf(1,no(i)),nband+1)=kb(nf(1,no(i)),nband+1)+1.e20
            storkb(i) = kb(nf(1,no(i)),nband+1)
    end do;          call cholin(kb)
!------------------displacement increment loop--------------------------------
      write(11,'(a)') "Displacement    Force         Iterations"
     displacement_increments: do iy=1,incs
            ptot=presc*iy; iters=0
!------------------------   iteration loop   ---------------------------------
   iterations: do
    iters=iters+1;  loads=.0   ;loads = loads + bdylds
      do i=1,loaded_nodes ; loads(nf(1,no(i)))=storkb(i)*presc ;  end do
        call chobac(kb,loads)  ; bdylds = .0
!------------------------   check convergence   ------------------------------
      call checon(loads,oldis,tol,converged)
      if(iters==1)converged=.false.
!------------------------ go round the Gauss Points --------------------------
      elements_3: do iel = 1 , nels
       bload=.0
       num = g_num( : ,iel ) ; coord = transpose(g_coord( : , num ))
       g = g_g( : , iel )    ;    eld = loads ( g )
       gauss_points_2 : do i = 1 , nip
          elso = .0 ; call shape_der ( der,points,i); jac=matmul(der,coord)
          det = determinant(jac)  ;   call invert(jac)
          deriv = matmul(jac,der) ; call beemat (bee,deriv)
          eps = matmul(bee,eld);    sigma = matmul(dee,eps)
          stress = sigma + tensor(:,i,iel)
          call invar(stress,sigm,dsbar,lode_theta)
!-------------------  check whether yield is violated ------------------------
         call mocouf(phi,c,sigm,dsbar,lode_theta,fnew)
         if (fnew>.0) then
           stress = tensor(:,i,iel) ;  call invar(stress,sigm,dsbar,lode_theta)
           call mocouf(phi,c,sigm,dsbar,lode_theta,f)
           fac = fnew / (fnew - f);    stress = (1.-fac)*sigma+tensor(:,i,iel)
           call mocopl(phi,psi,e,v,stress,pl); pl = fac * pl
           elso =  matmul(pl,eps) ;  eload = matmul(elso,bee)
           bload = bload + eload * det * weights(i)
         end if
       if(converged.or.iters==limit)then
!-------------------    update the Gauss Point stresses  ----------------------
          tensor(:,i,iel) = tensor(:,i,iel) + sigma - elso
       end if
    end do gauss_points_2
!--------------------- compute the total bodyloads vector --------------------
    bdylds( g ) = bdylds( g ) + bload  ; bdylds(0) = .0
  end do elements_3
  if(converged.or.iters==limit)exit
 end do iterations
 totd=totd+loads
 pav = .5*((depth(1)-depth(2))*(tensor(1,1,1)+tensor(1,3,1)) &
          +(depth(2)-depth(3))*(tensor(1,1,2)+tensor(1,3,2)) &
          +(depth(3)-depth(4))*(tensor(1,1,3)+tensor(1,3,3)) &
          +(depth(4)-depth(5))*(tensor(1,1,4)+tensor(1,3,4)))
 write(11,'(2e12.4,i12)') ptot,pav,iters
 if(iters==limit)stop
end do displacement_increments
end program p63
```

The initial stress method of stress redistribution is demonstrated in a problem of passive earth pressure, in which a smooth wall is translated into a bed of "sand". As in Program 6.0, a rectangular mesh of eight-noded elements is generated with nodes and freedoms counted in the y direction. An additional feature of this program which appears in the element integration and assembly section is the generation of starting self-weight stresses. The coordinates of each Gauss point are calculated using the isoparametric property

$$x = \sum_{i=1}^{8} N_i x_i$$
$$y = \sum_{i=1}^{8} N_i y_i \tag{6.41}$$

The x and y coordinates that result are held in the 1-d array GC, in positions 1 and 2 respectively. Only the y coordinate is required in this case and the vertical stress σ_y is obtained after multiplication by the unit weight γ held in GAMA. The normal effective stresses σ_x and σ_z are obtained by multiplying σ_y by the "at rest" earth pressure coefficient K_0 held in EPK0.

After the stiffness matrix formulation, the freedoms which are to receive prescribed displacements are read, followed by the magnitude of the displacement increment held in PRESC. In this program, INCS represents the number of constant displacement increments that are to be applied. The upper limit on iterations held in LIMIT does not need to be as large as it was in load-controlled problems. Convergence is quicker when using displacement control, especially as failure conditions are approached, since unconfined flow cannot occur.

The "penalty" technique is used to implement the prescribed displacements, as was first demonstrated in Program 4.0 and described more fully in section 3.8.

The program follows a familiar course until the calculation of the failure function. Initially, the failure function FNEW is obtained after adding the full elastic stress increment to those stresses existing previously. If FNEW is positive, indicating a yielding Gauss point, then the failure function F is obtained using just those stresses existing previously. The scaling parameter FAC is then calculated as described in equation (6.35). The plastic stress–strain matrix $\mathbf{D}^p$ for a Mohr–Coulomb material is formed by the routine MOCOPL (if implementing the von Mises criterion, the subroutine VMPL should be substituted) using stresses that have been factored to ensure they lie on the failure surface. The resulting matrix PL is multiplied by the scaling parameter FAC and then by the total strain increment array EPS to yield the "plastic" stress increment array ELSO. This is simple "forward Euler" integration of the rate equations. "Implicit" versions are described in the next section. Integrals of the type described by equation (6.34) then follow and the array BDYLDS is accumulated from each element. It may be noted that in the algorithm presented here, the body-loads vector is completely reformed at each iteration. This is in contrast to the visco-plasticity algorithm presented in Programs 6.0 and 6.1 in which the body-loads vector was accumulated at each iteration.

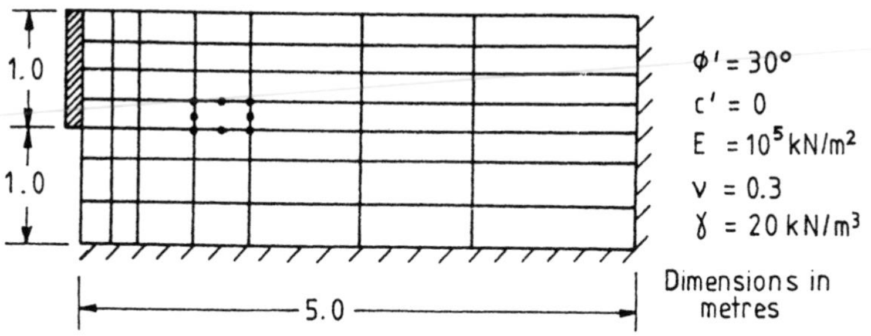

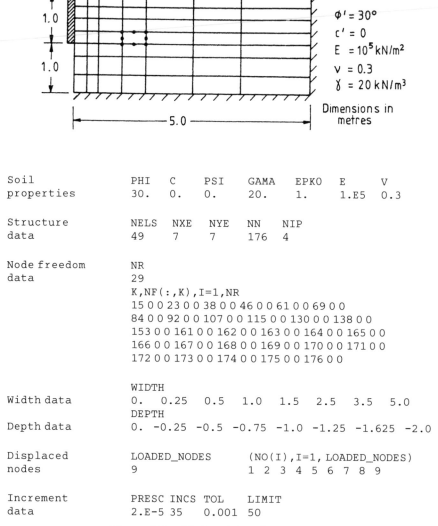

Soil properties	PHI C PSI GAMA EPKO E V 30. 0. 0. 20. 1. 1.E5 0.3
Structure data	NELS NXE NYE NN NIP 49 7 7 176 4
Node freedom data	NR 29 K,NF(:,K),I=1,NR 15 0 0 23 0 0 38 0 0 46 0 0 61 0 0 69 0 0 84 0 0 92 0 0 107 0 0 115 0 0 130 0 0 138 0 0 153 0 0 161 0 0 162 0 0 163 0 0 164 0 0 165 0 0 166 0 0 167 0 0 168 0 0 169 0 0 170 0 0 171 0 0 172 0 0 173 0 0 174 0 0 175 0 0 176 0 0
Width data	WIDTH 0. 0.25 0.5 1.0 1.5 2.5 3.5 5.0
Depth data	DEPTH 0. -0.25 -0.5 -0.75 -1.0 -1.25 -1.625 -2.0
Displaced nodes	LOADED_NODES (NO(I),I=1, LOADED_NODES) 9 1 2 3 4 5 6 7 8 9
Increment data	PRESC INCS TOL LIMIT 2.E-5 35 0.001 50

Figure 6.19 Mesh and data for Program 6.3

At convergence, the stresses must be updated ready for the next displacement (load) increment. This involves adding, to the stresses remaining from the previous increment, the 1-d array of total stress increments (SIGMA) minus the 1-d array of corrective "plastic" stresses (ELSO).

```
Global coordinates
Node    1          0.0000E+00  0.0000E+00
Node    2          0.0000E+00 -0.1250E+00
Node    3          0.0000E+00 -0.2500E+00
Node    4          0.0000E+00 -0.3750E+00
Node    5          0.0000E+00 -0.5000E+00
..................................................................

Node  174          0.5000E+01 -0.1625E+01
Node  175          0.5000E+01 -0.1813E+01
Node  176          0.5000E+01 -0.2000E+01
Global node numbers
Element     1            3    2    1   16   24   25   26   17
Element     2            5    4    3   17   26   27   28   18
Element     3            7    6    5   18   28   29   30   19
Element     4            9    8    7   19   30   31   32   20
..................................................................

Element    46          147  146  145  157  168  169  170  158
Element    47          149  148  147  158  170  171  172  159
Element    48          151  150  149  159  172  173  174  160
Element    49          153  152  151  160  174  175  176  161
The system has   294  equations and the half-bandwidth is    47
Displacement     Force          Iterations
  0.2000E-04 -0.1097E+02           2
  0.4000E-04 -0.1194E+02           2
  0.6000E-04 -0.1292E+02           4
  0.8000E-04 -0.1385E+02          12
  0.1000E-03 -0.1477E+02          10
  0.1200E-03 -0.1568E+02           6
  0.1400E-03 -0.1659E+02           8
  0.1600E-03 -0.1750E+02           3
  0.1800E-03 -0.1841E+02           7
  0.2000E-03 -0.1932E+02           3
  0.2200E-03 -0.2019E+02          15
  0.2400E-03 -0.2106E+02          10
  0.2600E-03 -0.2192E+02           6
  0.2800E-03 -0.2277E+02           9
  0.3000E-03 -0.2361E+02           9
  0.3200E-03 -0.2441E+02          16
  0.3400E-03 -0.2529E+02          15
  0.3600E-03 -0.2616E+02           7
  0.3800E-03 -0.2698E+02          15
  0.4000E-03 -0.2770E+02          21
  0.4200E-03 -0.2834E+02          19
  0.4400E-03 -0.2891E+02          18
  0.4600E-03 -0.2941E+02          16
  0.4800E-03 -0.2989E+02           9
  0.5000E-03 -0.3034E+02          14
  0.5200E-03 -0.3058E+02          30
  0.5400E-03 -0.3061E+02          38
  0.5600E-03 -0.3063E+02          10
  0.5800E-03 -0.3065E+02           9
  0.6000E-03 -0.3065E+02          11
  0.6200E-03 -0.3065E+02          12
  0.6400E-03 -0.3066E+02           3
  0.6600E-03 -0.3066E+02           3
  0.6800E-03 -0.3067E+02           7
  0.7000E-03 -0.3068E+02           3
```

Figure 6.20 Results from Program 6.3

The example problem shown in Figure 6.19 represents a sand with strength parameters $\phi = 30°$, $c = 0$ and $\psi = 0°$, subjected to prescribed displacements along the left face. The boundary condition is applied to the x components of displacement at the nine nodes adjacent to the hypothetical smooth, rigid wall shown hatched. The initial stresses in the ground are calculated assuming the unit weight $\gamma = 20\,\text{kN/m}^3$ and "at rest" earth pressure coefficient $K_0 = 1$.

Following each displacement increment, and after numerical convergence, the resultant force, PAV, acting on the wall is computed by averaging the σ_x stresses at the eight Gauss points closest to the "wall".

The output shown in Figure 6.20 gives the wall displacement, PTOT, the resultant force and the number of iterations to convergence at each step, and these are plotted in Figure 6.21. The force is seen to build up to a maximum value of around 31 kN/m. This is in close agreement with the closed form Rankine solution of 30 kN/m, despite the relatively crude mesh. Contours of plastic shear strain increment at failure are shown in Figure 6.22. The classical "Coulomb" wedge is apparent.

The initial stress algorithm presented in this program will tend to overestimate collapse loads, especially if the displacement (load) steps are made too big. Users are recommended to try one or two different increment sizes to test the sensitivity of the solutions. The problem is caused by incremental "drift" of the stress state at individual Gauss points into illegal stress space in spite of apparent numerical convergence. Although not included in the present work, various strategies are available (e.g. Nayak and Zienkiewicz, 1972) for drift correction. In the next section, more complicated

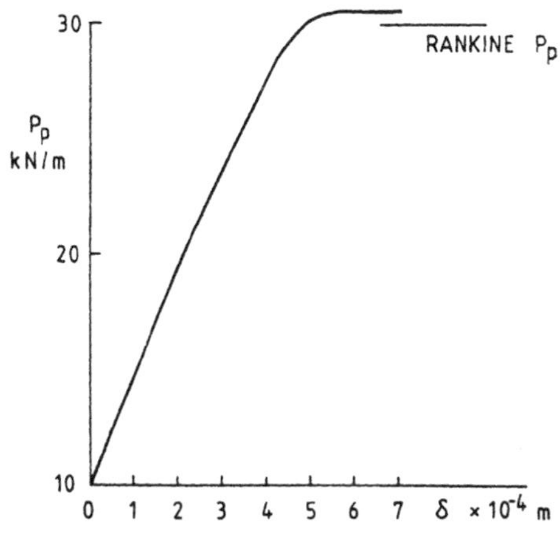

Figure 6.21 Passive force versus horizontal movement

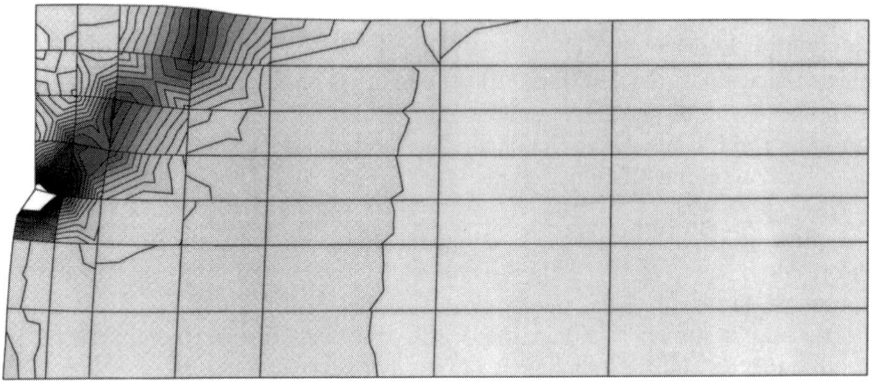

Figure 6.22 Plastic shear strain increment contours

"stress-return" procedures are illustrated which ensure stresses at each Gauss point return accurately to the yield surface.

6.8 ELASTO-PLASTIC RATE INTEGRATION

For the purposes of this description, we return to elastic–perfectly plastic materials obeying the von Mises yield and failure criterion. Similar, if more complicated, arguments apply to Mohr–Coulomb materials.

Using the notation previously developed in sections 6.1 and 6.2, F is the yield function and J_2 the second invariant of the stress tensor (like t):

$$J_2 = \frac{1}{6}[(\sigma_1 - \sigma_2)^2 + (\sigma_2 - \sigma_3)^2 + (\sigma_3 - \sigma_1)^2] \tag{6.42}$$

The first derivative of F with respect to the stresses is

$$\mathbf{a} = \frac{\partial F}{\partial \boldsymbol{\sigma}} = \frac{1.5}{\sqrt{3J_2}} \begin{Bmatrix} s_x \\ s_y \\ s_z \\ 2\tau_{xy} \\ 2\tau_{yz} \\ 2\tau_{zx} \end{Bmatrix} \tag{6.43}$$

where S are deviatoric components as before. The second derivative of F with respect to stress is

$$\frac{\partial \mathbf{a}}{\partial \boldsymbol{\sigma}} = \frac{1}{2\sqrt{3J_2}}\mathbf{A} - \frac{1}{\sqrt{3J_2}}\mathbf{a}\mathbf{a}^{\mathrm{T}} \tag{6.44}$$

where

$$\mathbf{A} = \begin{bmatrix} 2 & -1 & -1 & & & \\ -1 & 2 & -1 & \mathbf{0} & & \\ -1 & -1 & 2 & & & \\ & & & 6 & & \\ & \mathbf{0} & & & 6 & \\ & & & & & 6 \end{bmatrix} \tag{6.45}$$

Ortiz and Popov (1985) described various methods of elasto-plastic rate integration, which essentially consist of an (elastic) predictor, followed by a plastic corrector to ensure the final stress is (nearly) on the yield surface.

Referring to Figure 6.23 let $\boldsymbol{\sigma}_x$ refer to the unyielded stress at the start of a step and $\boldsymbol{\Delta\sigma}_e$ the (elastic) increment. The stress crosses the yield surface at $\boldsymbol{\sigma}_\mathbf{A}$ while the elastic increment ends up at $\boldsymbol{\sigma}_\mathbf{B}$. We wish to "return" to the "correct" stresses on the yield surface at $\boldsymbol{\sigma}_\mathbf{C}$.

If $\boldsymbol{\Delta\varepsilon}$ is the total incremental strain, $\boldsymbol{\Delta\varepsilon}_\mathbf{P}$ the incremental plastic strain, and λ the scalar multiplier (see equation (6.31)) an elastic stress–strain matrix $\mathbf{D}$ will lead to

$$\boldsymbol{\sigma}_\mathbf{C} = \boldsymbol{\sigma}_\mathbf{A} + \mathbf{D}(\boldsymbol{\Delta\varepsilon} - \boldsymbol{\Delta\varepsilon}_\mathbf{p}) \tag{6.46}$$

$$\boldsymbol{\Delta\varepsilon}_\mathbf{p} = \lambda[(1-\beta)\boldsymbol{\sigma}_\mathbf{A} + \beta\boldsymbol{\sigma}_\mathbf{C}] \tag{6.47}$$

$$\boldsymbol{\Delta\varepsilon}_\mathbf{p} = \lambda[\mathbf{a}\{(1-\beta)\boldsymbol{\sigma}_\mathbf{A} + \beta\boldsymbol{\sigma}_\mathbf{C}\}] \tag{6.48}$$

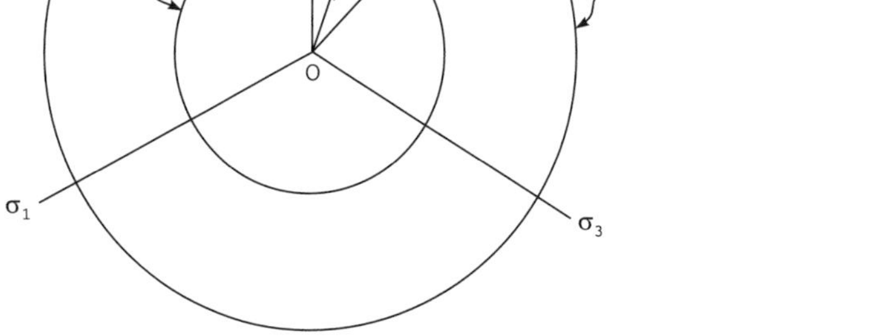

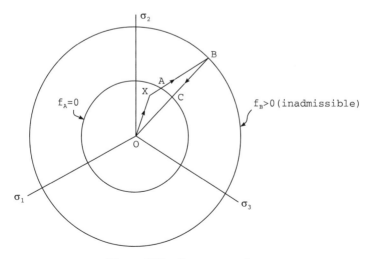

Figure 6.23 Stress corrections

where β is a scalar interpolating parameter (see also Chapter 8) such that $0 \leq \beta \leq 1$.

6.8.1 FORWARD EULER METHOD

Here $\beta = 0$ and the rate equation is integrated at the point at which the yield surface is crossed ($\boldsymbol{\sigma}_\mathbf{A}$ in Figure 6.21). Thus

$$F(\boldsymbol{\sigma}_\mathbf{X} + \alpha \boldsymbol{\Delta\sigma}_\mathbf{e}) = 0 \tag{6.49}$$

For a non-hardening von-Mises material the point can be found explicitly:

$$\alpha = \frac{-3\mathrm{B} \pm \sqrt{9\mathrm{B}^2 - 12\mathrm{A}(3\mathrm{C} - 3\mathrm{c}_\mathrm{u}^2)}}{6\mathrm{A}} \tag{6.50}$$

$$\mathrm{A} = \frac{1}{2}(\Delta s_{xe}^2 + \Delta s_{ye}^2 + \Delta s_{ze}^2) + \Delta \tau_{xye}^2 \tag{6.51}$$

$$\mathrm{B} = s_x \Delta s_{xe} + s_y \Delta s_{ye} + s_z \Delta s_{ze} + 2\tau_{xy} \Delta \tau_{xye} \tag{6.52}$$

$$\mathrm{C} = \frac{1}{2}(s_x^2 + s_y^2 + s_z^2) + \tau_{xy}^2 \tag{6.53}$$

but an approximate α can be found by linearly interpolating between points X and B:

$$\alpha \simeq \frac{-F(\boldsymbol{\sigma}_\mathbf{X})}{F(\boldsymbol{\sigma}_\mathbf{B}) - F(\boldsymbol{\sigma}_\mathbf{X})} \tag{6.54}$$

The remaining stress $(1 - \alpha)\mathbf{D\Delta\varepsilon}$ causes the "illegal" stress state outside the yield surface and for non-hardening plasticity it is assumed that once a stress state reaches a yield surface, subsequent changes in stress may shift the stress state to a different position on the yield surface but not outside it (see equation (6.30)). Thus

$$\Delta F = \mathbf{a}^\mathrm{T} \boldsymbol{\Delta\sigma} = 0 \tag{6.55}$$

$$\mathbf{a}_\mathrm{A}^\mathrm{T}(\mathbf{D\Delta\varepsilon} - (1 - \alpha)\lambda \mathbf{Da}_\mathrm{A}) = 0 \tag{6.56}$$

$$\lambda = \frac{\mathbf{a}_\mathrm{A}^\mathrm{T}\mathbf{D\Delta\varepsilon}}{(1 - \alpha)\mathbf{a}_\mathrm{A}^\mathrm{T}\mathbf{Da}_\mathrm{A}} \tag{6.57}$$

The final stress is then

$$\boldsymbol{\sigma}_\mathrm{C} = \boldsymbol{\sigma}_\mathrm{X} + \boldsymbol{\Delta\sigma}_\mathrm{e} - \lambda_\mathrm{A}\mathbf{Da}_\mathrm{A} \tag{6.58}$$

This is the method used in equation (6.35) in which $(1 - \alpha)$ is called `FAC`.

6.8.2 BACKWARD EULER METHOD

Here the rate equation is integrated at the "illegal" state B ($\beta = 1$). This results in a simple evaluation of the plastic multiplier λ for non-hardening von Mises materials. A first-order Taylor expansion of the yield function at B gives

$$F(\boldsymbol{\sigma}_{\mathrm{C}}) = F(\boldsymbol{\sigma}_{\mathbf{B}}) + \left(\frac{\partial F}{\partial \boldsymbol{\sigma}}\right)^{\mathrm{T}} \boldsymbol{\Delta\sigma} \tag{6.59}$$

By enforcing consistency of the yield function at point C,

$$0 = F(\boldsymbol{\sigma}_{\mathbf{B}}) - \lambda a_{\mathrm{B}}^{\mathrm{T}} \mathbf{D} \mathbf{a}_{\mathrm{B}} \tag{6.60}$$

so that

$$\lambda_{\mathrm{B}} = \frac{F(\boldsymbol{\sigma}_{\mathbf{B}})}{\mathbf{a}_{\mathrm{B}}^{\mathrm{T}} \mathbf{D} \mathbf{a}_{\mathrm{B}}} \tag{6.61}$$

The increase in stress is

$$\boldsymbol{\Delta\sigma} = \boldsymbol{\Delta\sigma}_{\mathrm{e}} - \frac{F(\boldsymbol{\sigma}_{\mathbf{B}}) \mathbf{D} \mathbf{a}_{\mathrm{B}}}{\mathbf{a}_{\mathbf{B}}^{\mathrm{T}} \mathbf{D} \mathbf{a}_{\mathbf{B}}} \tag{6.62}$$

The final stress is

$$\boldsymbol{\sigma} = \boldsymbol{\sigma}_{\mathbf{X}} + \boldsymbol{\Delta\sigma}_{\mathrm{e}} - \lambda_{\mathrm{B}} \mathbf{D} \mathbf{a}_{\mathbf{B}} \tag{6.63}$$

Rice and Tracey (1973) advocated a mean normal method for a von Mises yield criterion so that

$$\frac{(\mathbf{a}_{\mathrm{A}} + \mathbf{a}_{\mathrm{B}})^{\mathrm{T}} \boldsymbol{\Delta\sigma}_{\mathrm{e}}}{2} = 0 \tag{6.64}$$

In a von Mises yield criterion under plane strain and 3-d stress states, the yield surface appears as a circle on the deviatoric plane. Any "illegal" stress can be corrected along a radial path directed from the hydrostatic stress axis. The final deviatoric stress at point C is

$$\mathbf{s}_{\mathrm{C}} = \frac{\sqrt{3}c_{\mathrm{u}}}{\sqrt{3J_2}} \mathbf{s}_{\mathbf{B}} \tag{6.65}$$

and the components of $\boldsymbol{\sigma}_{\mathrm{C}}$ can then be determined by superimposing the hydrostatic stress from point B.

In practice, it has been found that this method offers no advantages over forward Euler in constant stiffness algorithms. The same is not true for tangent stiffness methods as is shown in the next paragraph.

6.9 TANGENT STIFFNESS APPROACHES

The difference between constant stiffness and tangent stiffness methods was discussed in section 6.0. In general, constant stiffness methods can be attractive in displacement-controlled situations (see Figure 6.20 where the number of iterations per displacement increment is modest) but in load-controlled situations, particularly close to collapse loads, large numbers of iterations tend to arise (see for example Figure 6.16). These counterbalance the speed of constant stiffness methods in which the global stiffness matrix is only factorised once. If convergence in cases like Figure 6.16 is monitored, it will be found that most Gauss points have converged to the yield surface, leaving only a few Gauss points responsible for the lack of convergence. Tangent stiffness methods, with backward Euler integration, can significantly improve the convergence properties of algorithms and the cost of re-forming and refactorising the global stiffness can be justified.

6.9.1 "INCONSISTENT" TANGENT MATRIX

The change in stress is composed of two parts, the elastic predictor $\mathbf{D}\Delta\boldsymbol{\varepsilon}$ and a plastic corrector $\mathbf{D}\lambda\mathbf{a}$, i.e.

$$\Delta\boldsymbol{\sigma} = \mathbf{D}(\Delta\boldsymbol{\varepsilon} - \lambda\mathbf{a}) \tag{6.66}$$

Substituting λ into the above equation,

$$\Delta\boldsymbol{\sigma} = \mathbf{D}\left(\Delta\boldsymbol{\varepsilon} - \frac{\mathbf{a}^{\mathrm{T}}\mathbf{D}\Delta\boldsymbol{\varepsilon}}{\mathbf{a}^{\mathrm{T}}\mathbf{D}\mathbf{a}}\mathbf{a}\right) \tag{6.67}$$

and hence

$$\Delta\boldsymbol{\sigma} = \left(\mathbf{D} - \frac{\mathbf{D}\mathbf{a}\mathbf{a}^{\mathrm{T}}\mathbf{D}}{\mathbf{a}^{\mathrm{T}}\mathbf{D}\mathbf{a}}\right)\Delta\boldsymbol{\varepsilon} \tag{6.68}$$

$$\Delta\boldsymbol{\sigma} = \mathbf{D}_{\mathrm{ep}}\Delta\boldsymbol{\varepsilon} \tag{6.69}$$

The term $\mathbf{D}_{\mathrm{ep}}$ is known as the standard or "inconsistent" tangent matrix.

6.9.2 CONSISTENT TANGENT MATRIX

With the backward Euler integration scheme a consistent tangent modular matrix can be formed:

$$\boldsymbol{\sigma} = \boldsymbol{\sigma}_{\mathbf{B}} - \lambda_{\mathrm{B}}\mathbf{D}\mathbf{a}_{\mathbf{B}} = (\boldsymbol{\sigma}_{\mathbf{x}} + \mathbf{D}\Delta\boldsymbol{\varepsilon}) - \lambda_{\mathrm{B}}\mathbf{D}\mathbf{a}_{\mathbf{B}} \tag{6.70}$$

On differentiation, we get

$$\delta\boldsymbol{\sigma} = \mathbf{D}\delta\boldsymbol{\varepsilon} - \Delta\lambda\mathbf{D}\mathbf{a} - \lambda_{\mathrm{B}}\mathbf{D}\frac{\partial\mathbf{a}}{\partial\boldsymbol{\sigma}}\Big|_{\mathrm{B}}\delta\boldsymbol{\sigma} \tag{6.71}$$

or

$$\delta\boldsymbol{\sigma} = \left(\mathbf{I} + \lambda\mathbf{D}\frac{\partial\mathbf{a}}{\partial\boldsymbol{\sigma}}\Big|_{\mathrm{B}}\right)^{-1}\mathbf{D}(\delta\boldsymbol{\varepsilon} - \Delta\lambda\mathbf{a}) \tag{6.72}$$

$$= \mathbf{R}(\delta\boldsymbol{\varepsilon} - \Delta\lambda\mathbf{a}) \tag{6.73}$$

and hence

$$\delta\boldsymbol{\sigma} = \left(\mathbf{R} - \frac{\mathbf{R}\mathbf{a}\mathbf{a}^{\mathrm{T}}\mathbf{R}}{\mathbf{a}^{\mathrm{T}}\mathbf{R}\mathbf{a}}\right)\delta\boldsymbol{\varepsilon} \tag{6.74}$$

or

$$\boldsymbol{\Delta\sigma} = \mathbf{D}_{\mathrm{epc}}\boldsymbol{\Delta\varepsilon} \tag{6.75}$$

The term $\mathbf{D}_{\mathrm{epc}}$ is known as the "consistent" **D** matrix and its use in element matrix construction leads to the "consistent" tangent stiffness matrix.

6.9.3 CONVERGENCE CRITERION

Programs 6.0–6.3 use a very simple convergence criterion, based on the fact that the out-of-balance stresses lead to a convergent vector of nodal loads called `BDYLDS`. Amongst other advantages, this leads to ideas of acceleration of convergence. The criterion used is that the maximum change in any component of `BDYLDS`, as a fraction of the maximum absolute number in `BDYLDS`, is less than, say, 0.1%.

When the convergence of `BDYLDS` is examined, it is found that in typical problems nearly all Gauss points converge early, and the time is taken in the analysis as a few points struggle towards the yield surface.

In the consistent tangent method, the efficiency of the return algorithm is such that all Gauss points converge much faster to the yield surface. On the other hand, there can be no concept of a converging `BDYLDS`. Rather, the residual remaining in `BDYLDS` tends to zero as convergence is approached (this is actually how many codes operate for the constant stiffness, forward Euler, case as well). A criterion based on the size of the maximum component of the reducing `BDYLDS` as a percentage of, say, the root mean square (RMS) of `BDYLDS` can be used.

In practice, when an element assembly technique is chosen, the strategy for constant stiffness is simple to implement and just as efficient computationally as the consistent tangent approach. This is because the plastic iterations involve only backsubstitutions in a direct equation-solving process.

When a tangent stiffness method is used, the extra time involved in re-forming the stiffness matrices and completely resolving the equilibrium equations can more than compensate for the reduced iteration counts.

However, when iterative strategies are adopted for equilibrium equation solution (see Program 6.1) the equilibrium equations have to be reassembled and solved on every iteration anyway. In these circumstances, the consistent tangent stiffness with backward Euler return, leading to low iteration counts, is essential.

PROGRAM 6.4: PLANE STRAIN OF AN ELASTIC–PLASTIC VON MISES SOLID USING EIGHT-NODE QUADRILATERAL ELEMENTS; INITIAL STRESS METHOD WITH TANGENT STIFFNESS AND CONSISTENT RETURN

```
 program p64
!----------------------------------------------------------------------
!      program 6.4 plane strain of an elastic-plastic(Von Mises) solid
!      using 8-node quadrilateral elements; initial stress method  with
!      tangent stiffness ; consistent return algorithm for problem of p60
!----------------------------------------------------------------------
 use new_library  ;     use geometry_lib ;       implicit none
 integer::nels,nxe,nye,neq,nband,nn,nr,nip,nodof=2,nod=8,nst=4,ndof,          &
          loaded_nodes,i,k,iel,iters,limit,incs,iy,ndim=2
 logical::converged       ; character(len=15) :: element='quadrilateral'
 real::e,v,det,cu,ptot,fnew,ff,fstiff,dlam,dslam,dsbar,lode_theta,sigm,        &
       top,bot,tload,tloads,residual,tol,fftol,ltol
!-------------------------- dynamic arrays----------------------------------
 real    ,allocatable :: kb(:,:),loads(:),points(:,:),totd(:),bdylds(:),      &
                         width(:),depth(:),tensor(:,:,:),                     &
                         dee(:,:),coord(:,:),fun(:),jac(:,:),weights(:),      &
                         der(:,:),deriv(:,:),bee(:,:),km(:,:),eld(:),eps(:),  &
                         sigma(:),bload(:),eload(:),elso(:),g_coord(:,:),     &
                         oldis(:),val(:,:),stress(:),qinc(:),ddylds(:),       &
                         dl(:,:),dload(:),vmfl(:),caflow(:),dsigma(:),        &
                         ress(:),rmat(:,:),acat(:,:),acatc(:,:),qmat(:,:),    &
                         qinva(:),daatd(:,:),temp(:,:),vmflq(:),              &
                         vmfla(:),qinvr(:),vmtemp(:,:)
 integer, allocatable :: nf(:,:) , g(:), no(:) ,num(:), g_num(:,:) ,g_g(:,:)
!------------------------input and initialisation----------------------------
  open (10,file='p64.dat',status=    'old',action='read')
  open (11,file='p64.res',status='replace',action='write')
  read (10,*) cu,e,v,    nels,nxe,nye,nn,nip,tol,fftol,ltol,limit
  ndof=nod*nodof
  allocate (nf(nodof,nn), points(nip,ndim),weights(nip),g_coord(ndim,nn),      &
            width(nxe+1),depth(nye+1),num(nod),dee(nst,nst),                   &
            tensor(nst,nip,nels),g_g(ndof,nels),coord(nod,ndim),stress(nst),   &
            jac(ndim,ndim),der(ndim,nod),deriv(ndim,nod),g_num(nod,nels),      &
            bee(nst,ndof),km(ndof,ndof),eld(ndof),eps(nst),sigma(nst),         &
            bload(ndof),eload(ndof),elso(nst),g(ndof),vmfl(nst),qinvr(nst),    &
            temp(nst,nst),dl(nip,nels),                                        &
            dload(ndof),caflow(nst),dsigma(nst),ress(nst),rmat(nst,nst),       &
            acat(nst,nst),acatc(nst,nst),qmat(nst,nst),qinva(nst),             &
            daatd(nst,nst),vmflq(nst),vmfla(nst),vmtemp(1,nst))
  nf=1; read(10,*) nr ; if(nr>0) read(10,*)(k,nf(:,k),i=1,nr)
  call formnf(nf); neq=maxval(nf)      ;  read(10,*) width, depth
  temp = .0; temp(1,1)=1.;temp(2,2)=1.;temp(3,3)=3.;temp(4,4)=1.
  temp(1,2)=-.5;temp(2,1)=-.5;temp(1,4)=-.5
  temp(2,4)=-.5;temp(4,1)=-.5;temp(4,2)=-.5
!--------------- loop the elements to find nband and set up global arrays -----
     nband = 0
       elements_1:   do iel = 1 , nels
                        call geometry_8qyv(iel,nye,width,depth,coord,num)
                        call num_to_g(num,nf,g);             g_num(:,iel)=num
                        g_coord(:,num)=transpose(coord)  ;  g_g(: , iel ) = g
                        if (nband<bandwidth(g)) nband = bandwidth(g)
     end do elements_1
    write(11,'(a)') "Global coordinates "
    do k=1,nn;write(11,'(a,i5,a,2e12.4)')"Node",k,"       ",g_coord(:,k);end do
    write(11,'(a)') "Global node numbers "
    do k = 1 , nels; write(11,'(a,i5,a,8i5)')                                 &
                              "Element ",k,"        ",g_num(:,k); end do
    write(11,'(a,i5,a,i5)')                                                    &
            "The number of equations is ",neq," with half-bandwidth ",nband
  allocate(kb(neq,nband+1),loads(0:neq),bdylds(0:neq),oldis(0:neq),           &
           totd(0:neq),ddylds(0:neq))
  kb=0.0; oldis=0.0; totd=0.0
  call deemat(dee,e,v); call sample(element,points,weights);tensor=.0;  dl=.0
!------------- starting element stiffness integration and assembly-------------
```

```
 elements_2: do iel = 1 , nels
               num = g_num(: , iel) ; coord =transpose( g_coord( : , num ))
                g = g_g( : , iel )   ;       km=0.0
            gauss_pts_1:  do i =1 , nip
              call shape_der (der,points,i);  jac = matmul(der,coord)
              det = determinant(jac)  ;   call invert(jac)
              deriv = matmul(jac,der) ;  call beemat (bee,deriv)
              km = km + matmul(matmul(transpose(bee),dee),bee) *det* weights(i)
           end do gauss_pts_1
   call formkb (kb,km,g)
 end do elements_2
!--------------------read load weightings and factorise l.h.s. --------------
 read(10,*) loaded_nodes ; allocate(no(loaded_nodes),val(loaded_nodes,ndim))
 read(10,*)(no(i),val(i,:),i=1,loaded_nodes)
         call cholin(kb)
!------------------------- load increment loop-------------------------------
 read(10,*) incs ; allocate(qinc(incs)) ;   read(10,*)qinc ;   ptot = .0
     load_increments: do iy=1,incs
         write(11,'(/,a,i5)') ' Load increment  ',iy
         ptot=ptot + qinc(iy) ;   iters = 0 ;  bdylds=.0 ;  loads=.0
      do i=1,loaded_nodes ; loads(nf(:,no(i)))=val(i,:)*qinc(iy) ;  end do
!----------------------   iteration loop  ----------------------------------
   iterations: do
    iters=iters+1;  if(iters/=1)loads=.0   ;loads = loads + bdylds
      write(11,'(a,i5)') "Iteration number",iters
       call chobac(kb,loads)  ; bdylds = .0    ; ddylds = .0
!---------------------- go round the elements -------------------------------
      kb = .0
      elements_3: do iel = 1 , nels
       bload=.0   ;   dload = .0
       num = g_num( : , iel ) ; coord = transpose(g_coord( : , num ))
       g = g_g( : , iel )  ;   km = .0 ;    eld = loads(g)
!---------------------- go round the Gauss points --------------------------
       gauss_points_2 : do i = 1 , nip
          elso = .0 ; call shape_der ( der,points,i); jac=matmul(der,coord)
          det = determinant(jac)  ;   call invert(jac)
          deriv = matmul(jac,der) ; call beemat (bee,deriv)
          eps = matmul(bee,eld);    call deemat(dee,e,v)
          stress = tensor(: , i , iel)
          call invar(stress,sigm,dsbar,lode_theta) ; ff = dsbar-sqrt(3.)*cu
           if(ff>fftol) then
             dlam = dl(i,iel) ; call vmflow(stress,dsbar,vmfl)
             call fmrmat(vmfl,dsbar,dlam,dee,temp,rmat)
             caflow = matmul(rmat,vmfl); bot=dot_product(vmfl,caflow)
             call formaa(vmfl,rmat,daatd); dee = rmat - daatd/bot
           end if
          sigma = matmul(dee,eps)   ;  stress = sigma + tensor( : , i , iel)
          call invar(stress,sigm,dsbar,lode_theta)
!--------------------  check whether yield is violated ----------------------
          fnew = dsbar - sqrt(3.)*cu ; fstiff = fnew
          if (fnew>=.0) then
              call deemat(dee,e,v) ; call vmflow(stress,dsbar,vmfl)
              caflow = matmul(dee,vmfl); bot=dot_product(vmfl,caflow)
              dlam = fnew/bot; elso = caflow*dlam
              stress = tensor( : , i , iel) + sigma - elso
              call invar(stress,sigm,dsbar,lode_theta);fnew= dsbar-sqrt(3.)*cu
             iterate_on_fnew : do
               call vmflow(stress,dsbar,vmfl); caflow = matmul(dee,vmfl)*dlam
               ress = stress - (tensor(: , i , iel) +sigma - caflow)
               call fmacat(vmfl,temp,acat); acat = acat / dsbar
               acatc = matmul(dee,acat); qmat = acatc*dlam
               do k=1,4; qmat(k,k)=qmat(k,k)+1.; end do; call invert(qmat)
               vmtemp(1,:)=vmfl; vmtemp = matmul(vmtemp,qmat);vmflq=vmtemp(1,:)
               top = dot_product(vmflq,ress)
```

```
                vmtemp = matmul(vmtemp,dee);vmfla=vmtemp(1,:)
                bot = dot_product(vmfla,vmfl) ; dslam = (fnew - top)/bot
                qinvr = matmul(qmat,ress); qinva=matmul(matmul(qmat,dee),vmfl)
                dsigma = -qinvr - qinva*dslam; stress = stress + dsigma
                call invar(stress,sigm,dsbar,lode_theta)
                fnew = dsbar - sqrt(3.)*cu;  dlam = dlam + dslam
                if (fnew<tol) exit
              end do iterate_on_fnew
              dl(i,iel) = dlam
              elso = tensor( : , i , iel) + sigma - stress
              eload=matmul(elso,bee);bload=bload+eload*det*weights(i)
              call vmflow(stress,dsbar,vmfl)
              call fmrmat(vmfl,dsbar,dlam,dee,temp,rmat)
              caflow=matmul(rmat,vmfl);bot = dot_product(vmfl,caflow)
              call formaa(vmfl,rmat,daatd)
              dee = rmat - daatd/bot
           end if
           if(fstiff<.0) call deemat(dee,e,v)
               km = km + matmul(matmul(transpose(bee),dee),bee) *det* weights(i)
!----------------------- update the Gauss Point stresses ----------------------
           tensor( : , i , iel) = tensor( : , i , iel) + sigma - elso
           stress = tensor ( : , i , iel)
           eload=matmul(stress,bee); dload=dload+eload*det*weights(i)
     end do gauss_points_2
!-------------------     compute the total bodyloads vector  -------------------
     bdylds( g ) = bdylds( g ) + bload  ; bdylds(0) = .0
     ddylds( g ) = ddylds( g ) + dload  ; ddylds(0) = .0
     call formkb (kb,km,g)
   end do elements_3
    call cholin(kb)  ; tload = sum(ddylds)  ; tloads = sum(bdylds)
    if(iters==1)converged=.false.;if(iters/=1.and.tloads<ltol)converged=.true.
    residual = (2.*ptot+tload-tloads)/(2.*ptot)
    write(11,'(a,10e12.4)')"tloads,tload,residual are",tloads,tload,residual
    totd = totd + loads
   if(converged.or.iters==limit)exit
  end do iterations
  totd=totd+loads
  write(11,'(a,e12.4)') "The total load is  ",ptot
  write(11,'(a,10e12.4)') "Displacements are",totd(nf(2,no))
  write(11,'(a,i5,a)') "It took ",iters,"   iterations to converge"
  if(iters==limit)stop
 end do load_increments
 end program p64
```

The structure chart for a typical tangent stiffness approach is shown in Figure 6.24. Preliminary element loops, elements_1 and elements_2 set up the geometry and the starting tangent matrix respectively.

The load increment loop is entered and new subroutines encountered are FMRMAT and FORMAA which are used to form the matrices needed in equations (6.44) and (6.45) at every Gauss point in the elements_3 loop. Plastic multiplier λ is called DLAM. The 4×4 matrix QMAT has to be inverted to complete the return strategy at each Gauss point and ultimately the consistent tangent DEE matrix is obtained, leading to the consistent KM.

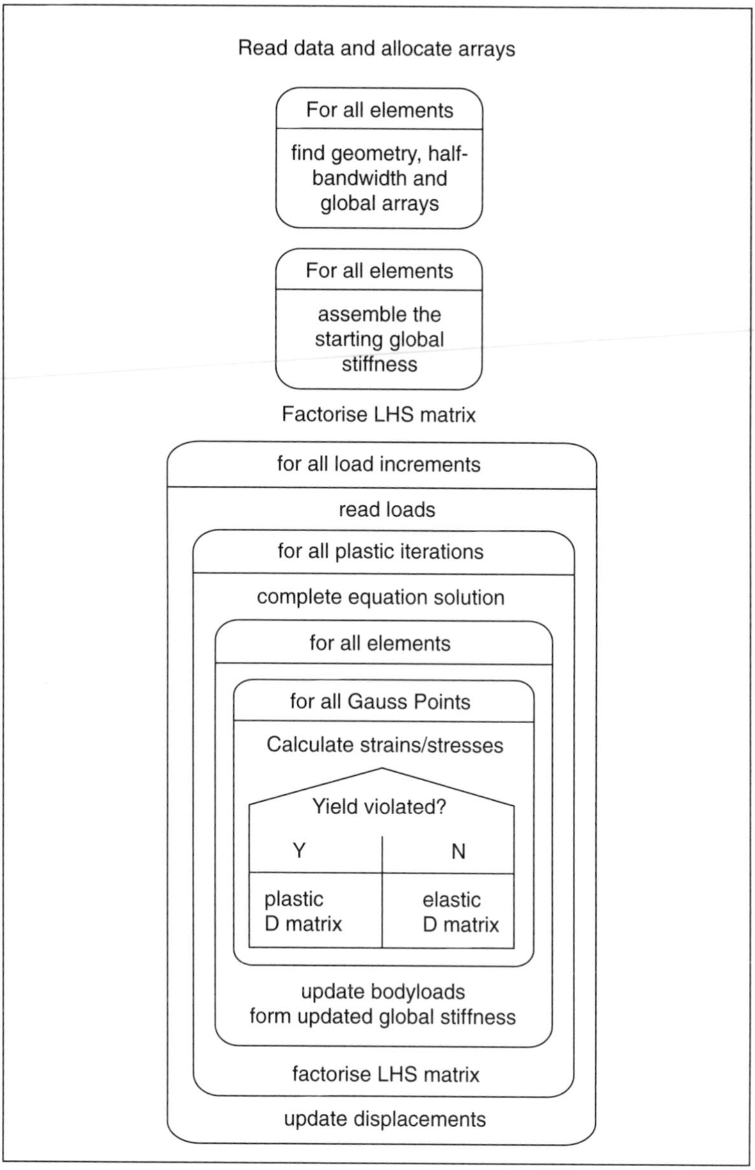

Figure 6.24 Structure chart for Program 6.4

```
Soil                    CU       E        V
properties              100.     1.E5     0.3

Structure, mesh         NELS     NXE      NYE     NN      NIP
and iteration data      32       8        4       121     4

                        TOL      FFTOL    LTOL      LIMIT   Program 6.4
                        1.E-6    -1.E-6   0.00005   50

                        PLASITS  CJITS  PLASTOL  CJTO     FFTOL  LTOL
                        50       100    1.E-6    0.0001   -1.E6  0.00005
                                                              Program 6.5
Node freedom            NR
data                    33
                        K,NF(:,K),I=1,NR
                        1 0 1 2 0 1 3 0 1 4 0 1 5 0 1 6 0 1
                        7 0 1 8 0 1 9 0 0 14 0 0 23 0 0 28 0 0 37 0 0
                        42 0 0 51 0 0 56 0 0 65 0 0 70 0 0 79 0 0 84 0 0
                        93 0 0 98 0 0 107 0 0 112 0 0 113 0 0 114 0 0 115 0 0
                        116 0 0 117 0 0 118 0 0 119 0 0 120 0 0 121 00

                        WIDTH
Geometry data           0. 1. 2. 3. 4. 5.5 7. 9. 12.
                        DEPTH
                        0. -1.25 -2.5 -3.75 -5.

Load                    LOADED_NODES
weightings              5
                        (NO(I),VAL(I,NDIM),I=1,LOADED_NODES)
                        1  .0 -.166667 10 .0 -.666667 15 .0 -.333333
                        24 .0 -.666667 29 .0 -.166667

Load                    INCS
increments              10
                        QINC(I),I=1,INCS
                        200. 100. 50. 50. 50. 30. 20. 10. 5. 5.
```

Figure 6.25 Data for Programs 6.4 and 6.5

Figure 6.25 shows the data for the problem analysed, for which we return to the original problem of Figure 6.9. The extra information required is confined to the tolerances.

$$\text{TOL} = 1*10^{-6}$$

$$\text{FFTOL} = 1*10^{-6}$$

$$\text{LTOL} = 0.00005$$

while `LIMIT` can be assigned a much more economic value of, say, 50.

The results are listed as Figure 6.26 and are found to be very similar to those in Figure 6.10 up to the ninth load increment. Up until then, the consistent return algorithm leads to convergence within four iterations on every increment. On the final increment (at "failure") Program 6.4 took 28 iterations compared to Program 6.0 which took (at least) 250.

```
Global coordinates
Node    1          0.0000E+00  0.0000E+00
Node    2          0.0000E+00 -0.6250E+00
Node    3          0.0000E+00 -0.1250E+01
Node    4          0.0000E+00 -0.1875E+01
Node    5          0.0000E+00 -0.2500E+01
.........................................................................

Node  118          0.1200E+02 -0.3125E+01
Node  119          0.1200E+02 -0.3750E+01
Node  120          0.1200E+02 -0.4375E+01
Node  121          0.1200E+02 -0.5000E+01
Global node numbers
Element     1            3    2    1   10   15   16   17   11
Element     2            5    4    3   11   17   18   19   12
Element     3            7    6    5   12   19   20   21   13
Element     4            9    8    7   13   21   22   23   14
.........................................................................
Element    30          103  102  101  109  115  116  117  110
Element    31          105  104  103  110  117  118  119  111
Element    32          107  106  105  111  119  120  121  112
The number of equations is   184 with half-bandwidth    29

 Load increment      1
Iteration number    1
tloads,tload,residual are  0.0000E+00 -0.4000E+03 -0.5000E-06
Iteration number    2
tloads,tload,residual are  0.0000E+00 -0.4000E+03 -0.5000E-06
The total load is    0.2000E+03
Displacements are -0.6592E-02 -0.6486E-02 -0.6116E-02 -0.5418E-02 -0.3849E-02
It took     2   iterations to converge

 Load increment      2
Iteration number    1
tloads,tload,residual are  0.3494E+02 -0.6349E+03 -0.1165E+00
Iteration number    2
tloads,tload,residual are  0.7230E+00 -0.6007E+03 -0.2410E-02
Iteration number    3
tloads,tload,residual are  0.2092E-02 -0.6000E+03 -0.7474E-05
Iteration number    4
tloads,tload,residual are  0.3277E-06 -0.6000E+03 -0.5011E-06
The total load is    0.3000E+03
Displacements are -0.1154E-01 -0.1127E-01 -0.1052E-01 -0.9110E-02 -0.6008E-02
It took     4   iterations to converge
.............................................................................

 Load increment      7
Iteration number    1
tloads,tload,residual are  0.2992E+01 -0.1003E+04 -0.5985E-02
Iteration number    2
tloads,tload,residual are  0.2670E+00 -0.1000E+04 -0.5345E-03
Iteration number    3
tloads,tload,residual are  0.6363E-03 -0.1000E+04 -0.1775E-05
Iteration number    4
tloads,tload,residual are  0.1418E-08 -0.1000E+04 -0.5027E-06
The total load is    0.5000E+03
Displacements are -0.5031E-01 -0.5002E-01 -0.5063E-01 -0.4768E-01 -0.1452E-01
It took     4   iterations to converge
.............................................................................

 Load increment     10
Iteration number    1
tloads,tload,residual are  0.7399E+00 -0.1041E+04 -0.1423E-02
Iteration number    2
```

Figure 6.26 *(continued)*

```
tloads,tload,residual are  0.7215E+00 -0.1041E+04 -0.1388E-02
.....................................................................

Iteration number   25
tloads,tload,residual are  0.6542E-01 -0.1040E+04 -0.1263E-03
Iteration number   26
tloads,tload,residual are  0.1177E-01 -0.1040E+04 -0.2313E-04
Iteration number   27
tloads,tload,residual are  0.5084E-03 -0.1040E+04 -0.1480E-05
Iteration number   28
tloads,tload,residual are  0.5855E-06 -0.1040E+04 -0.5037E-06
The total load is     0.5200E+03
Displacements are -0.3307E+00 -0.3117E+00 -0.2643E+00 -0.2047E+00 -0.1556E-01
It took    28    iterations to converge
```

Figure 6.26 Results from Program 6.4

PROGRAM 6.5: PLANE STRAIN OF AN ELASTIC–PLASTIC VONMISES SOLID USING EIGHT-NODE QUADRILATERAL ELEMENTS; INITIAL STRESS METHOD, TANGENT STIFFNESS WITH CONSISTENT STRESS RETURN; MESH-FREE APPROACH USING PCG

The mesh-free version of the constant stiffness method, Program 6.1, was inefficient (at least on a scalar computer) due to the large number of repeated equation solutions. Program 6.4 has shown that a tangent stiffness approach, with consistent return, has reduced the number of equation solutions to four per load increment. This implies that a mesh-free approach could be appropriate, and this is the subject of Program 6.5. It should be seen as a merging of Programs 6.1 and 6.4. The data for the problem solved, again the original one of Figure 6.9, is shown in Figure 6.25.

Extra information is limited the conjugate gradient tolerance and iteration limit, `CJTOL` and `CJITS`, of 0.0001 and 100 respectively. The plasticity tolerance `TOL` is renamed `PLASTOL` and the plasticity iteration limit `LIMIT` is `PLASITS`.

The output is listed as Figure 6.27 which can be compared with Figure 6.26. Between 62 and 100 conjugate gradient iterations per plastic iteration were required but the program ran faster than Program 6.0 for the solution of this problem even in scalar mode.

In parallel implementations there is a trade-off between constant stiffness and tangent stiffness methods because for the latter, all yielded elements are different although their geometries and elastic properties may be identical.

```
    program p65
!-------------------------------------------------------------------------
!      program 6.5 plane strain of an elastic-plastic(Von Mises)
!      solid using 8-node quadrilateral elements; initial stress method
!      consistent return algorithm for problem of p60 ; pcg version
!-------------------------------------------------------------------------
 use new_library  ; use geometry_lib    ;   implicit none
 integer::nels,nxe,nye,neq,nn,nr,nip,nodof=2,nod=8,nst=4,ndof,loaded_nodes,  &
          i,k,iel,plasiters,plasits,cjiters,cjits,incs,iy,ndim=2,cjtot
 logical::plastic_converged,cj_converged
 character(len=15)::element='quadrilateral'
 real::e,v,det,cu,ptot,fnew,ff,fstiff,dlam,dslam,dsbar,lode_theta,sigm,top,  &
       bot,tload,tloads,residual,up,alpha,beta,big,plastol,cjtol,fftol,ltol
!------------------------------- dynamic arrays--------------------------------
 real    ,allocatable :: loads(:),points(:,:),totd(:),bdylds(:),             &
                         width(:),depth(:),tensor(:,:,:),pmul(:),utemp(:),   &
                         dee(:,:),coord(:,:),fun(:),jac(:,:),weights(:),     &
                         der(:,:),deriv(:,:),bee(:,:),km(:,:),eld(:),eps(:), &
                         sigma(:),bload(:),eload(:),elso(:),g_coord(:,:),    &
                         oldis(:),val(:,:),stress(:),qinc(:),ddylds(:),      &
                         dl(:,:),dload(:),vmfl(:),caflow(:),dsigma(:),       &
                         ress(:),rmat(:,:),acat(:,:),acatc(:,:),qmat(:,:),   &
                         qinva(:),daatd(:,:),temp(:,:),vmflq(:),storkm(:,:,:),&
                         vmfla(:),qinvr(:),vmtemp(:,:),                      &
                         p(:),x(:),xnew(:),u(:),diag_precon(:),d(:)
 integer, allocatable :: nf(:,:) , g(:), no(:) ,num(:), g_num(:,:) ,g_g(:,:)
!------------------------input and initialisation-----------------------------
  open (10,file='p65.dat',status=    'old',action='read')
  open (11,file='p65.res',status='replace',action='write')
  read (10,*) cu,e,v, nels,nxe,nye,nn,nip,                                  &
              plasits,cjits,plastol,cjtol,fftol,ltol
  ndof=nod*nodof
  allocate (nf(nodof,nn), points(nip,ndim),weights(nip),g_coord(ndim,nn),    &
            width(nxe+1),depth(nye+1),num(nod),dee(nst,nst),pmul(ndof),      &
            tensor(nst,nip,nels),g_g(ndof,nels),storkm(ndof,ndof,nels),      &
            coord(nod,ndim),stress(nst),temp(nst,nst),dl(nip,nels),          &
            jac(ndim,ndim),der(ndim,nod),deriv(ndim,nod),g_num(nod,nels),    &
            bee(nst,ndof),km(ndof,ndof),eld(ndof),eps(nst),sigma(nst),       &
            bload(ndof),eload(ndof),elso(nst),g(ndof),vmfl(nst),qinvr(nst),  &
            dload(ndof),caflow(nst),dsigma(nst),ress(nst),rmat(nst,nst),     &
            acat(nst,nst),acatc(nst,nst),qmat(nst,nst),qinva(nst),           &
            daatd(nst,nst),vmflq(nst),vmfla(nst),vmtemp(1,nst),utemp(ndof))
  nf=1; read(10,*) nr ; if(nr>0) read(10,*)(k,nf(:,k),i=1,nr)
  call formnf(nf); neq=maxval(nf)    ;   read(10,*) width, depth
  temp = .0; temp(1,1)=1.;temp(2,2)=1.;temp(3,3)=3.;temp(4,4)=1.
  temp(1,2)=-.5;temp(2,1)=-.5;temp(1,4)=-.5
  temp(2,4)=-.5;temp(4,1)=-.5;temp(4,2)=-.5
!---------------- loop the elements to set up global arrays -------------------
       elements_1:   do iel = 1 , nels
                        call geometry_8qyv(iel,nye,width,depth,coord,num)
                        call num_to_g(num,nf,g);          g_num(:,iel)=num
                        g_coord(:,num)=transpose(coord);  g_g( : , iel ) = g
       end do elements_1
    write(11,'(a)') "Global coordinates "
    do k=1,nn;write(11,'(a,i5,a,2e12.4)')"Node",k,"        ",g_coord(:,k);end do
    write(11,'(a)') "Global node numbers "
    do k = 1 , nels; write(11,'(a,i5,a,8i5)')                                 &
                              "Element ",k,"        ",g_num(:,k); end do
  allocate(loads(0:neq),bdylds(0:neq),oldis(0:neq),totd(0:neq),ddylds(0:neq), &
           p(0:neq),x(0:neq),xnew(0:neq),u(0:neq),diag_precon(0:neq),d(0:neq))
  oldis=0.0; totd=0.0 ; p = .0; xnew = .0; diag_precon = .0
  call deemat(dee,e,v); call sample(element,points,weights); tensor= .0; dl=.0
!---------- starting element stiffness integration,storage,preconditioner------
 elements_2: do iel = 1 , nels
```

```
                num = g_num(: , iel ) ; coord = transpose(g_coord(: , num ))
                g = g_g( : , iel )    ;      km=0.0
             gauss_pts_1:  do i =1 , nip
               call shape_der (der,points,i);  jac = matmul(der,coord)
               det = determinant(jac)  ;   call invert(jac)
               deriv = matmul(jac,der) ;  call beemat (bee,deriv)
               km = km + matmul(matmul(transpose(bee),dee),bee) *det* weights(i)
            end do gauss_pts_1
   storkm(:,:,iel) = km
   do k=1,ndof; diag_precon(g(k))=diag_precon(g(k)) + km(k,k); end do
 end do elements_2
   diag_precon(1:neq) = 1./diag_precon(1:neq) ; diag_precon(0) = .0
!---------------read load weightings --------------------------------------
  read(10,*) loaded_nodes ; allocate (no(loaded_nodes),val(loaded_nodes,ndim))
  read(10,*)(no(i),val(i,:),i=1,loaded_nodes)
!----------------- load increment loop--------------------------------------
      read(10,*) incs ; allocate(qinc(incs));  read(10,*)qinc ;   ptot = .0
     load_increments: do iy=1,incs
        write(11,'(/,a,i5)') ' Load increment  ',iy
        ptot=ptot + qinc(iy) ; plasiters = 0;bdylds=.0 ;loads=.0; cjtot=0
      do i=1,loaded_nodes ; loads(nf(:,no(i)))=val(i,:)*qinc(iy) ;  end do
!---------------------   plastic iteration loop  ----------------------------
   plastic_iterations: do
    plasiters=plasiters+1;  if(plasiters/=1)loads=.0   ;loads = loads + bdylds
    if(abs(sum(loads))<1.e-5) then; plasiters=plasiters-1;exit ; end if
       write(11,'(a,i5)') "Plastic iteration number",plasiters
        bdylds = .0    ; ddylds = .0 ; d=diag_precon*loads ; p = d   ; x = .0
!---------------  solve the simultaneous equations by pcg  ------------------
       cjiters = 0
      conjugate_gradients:  do
       cjiters = cjiters + 1 ; u = .0
      elements_3 : do iel = 1 , nels
                       g = g_g( : , iel ); km = storkm( : , : ,iel)
                       pmul = p(g); utemp=matmul(km,pmul)
!dir$ ivdep
          do i = 1 , ndof
             u(g(i)) = u(g(i)) +  utemp(i)
          end do
      end do elements_3
!---------------------------pcg process ------------------------------------
    up =dot_product(loads,d); alpha=up/dot_product(p,u)
    xnew = x + p* alpha; loads = loads - u*alpha; d = diag_precon*loads
    beta = dot_product(loads,d)/up; p = d + p * beta
    big = .0; cj_converged = .true.
    do i = 1,neq; if(abs(xnew(i))>big)big=abs(xnew(i)); end do
    do i = 1,neq; if(abs(xnew(i)-x(i))/big>cjtol)cj_converged=.false.;end do
    x = xnew
    if(cj_converged.or.cjiters==cjits) exit
      end do conjugate_gradients
    cjtot = cjtot + cjiters
!------------------------ end of pcg process --------------------------------
    loads = xnew           ; loads(0) = .0    ; diag_precon = .0
!---------------------- go round the elements--------------------------------
      elements_4: do iel = 1 , nels
       bload=.0   ;   dload = .0
       num = g_num( : , iel ) ; coord = transpose(g_coord( : , num ))
       g = g_g( : , iel )     ;    km = .0; eld = loads(g)
!----------------------- go round the Gauss points --------------------------
       gauss_points_2 : do i = 1 , nip
          elso = .0 ; call shape_der ( der,points,i); jac=matmul(der,coord)
          det = determinant(jac)  ;   call invert(jac)
          deriv = matmul(jac,der) ;   call beemat (bee,deriv)
          eps = matmul(bee,eld)   ;   call deemat(dee,e,v)
          stress = tensor(: , i , iel)
```

```
         call invar(stress,sigm,dsbar,lode_theta) ; ff = dsbar-sqrt(3.)*cu
          if(ff>fftol) then
            dlam = dl(i,iel) ; call vmflow(stress,dsbar,vmfl)
            call fmrmat(vmfl,dsbar,dlam,dee,temp,rmat)
            caflow = matmul(rmat,vmfl); bot=dot_product(vmfl,caflow)
            call formaa(vmfl,rmat,daatd); dee = rmat - daatd/bot
          end if
         sigma = matmul(dee,eps)
         stress = sigma + tensor( : , i , iel)
         call invar(stress,sigm,dsbar,lode_theta)
!--------------------- check whether yield is violated ---------------------
         fnew = dsbar - sqrt(3.)*cu ; fstiff = fnew
         if (fnew>=.0) then
             call deemat(dee,e,v) ; call vmflow(stress,dsbar,vmfl)
             caflow = matmul(dee,vmfl); bot=dot_product(vmfl,caflow)
             dlam = fnew/bot; elso = caflow*dlam
             stress = tensor( : , i , iel) + sigma - elso
             call invar(stress,sigm,dsbar,lode_theta);fnew=dsbar-sqrt(3.)*cu
            iterate_on_fnew : do
              call vmflow(stress,dsbar,vmfl); caflow = matmul(dee,vmfl)*dlam
              ress = stress - (tensor(: , i , iel) +sigma - caflow)
              call fmacat(vmfl,temp,acat); acat = acat / dsbar
              acatc = matmul(dee,acat); qmat = acatc*dlam
              do k=1,4; qmat(k,k)=qmat(k,k)+1.; end do; call invert(qmat)
              vmtemp(1,:)=vmfl; vmtemp = matmul(vmtemp,qmat);vmflq=vmtemp(1,:)
              top = dot_product(vmflq,ress)
              vmtemp = matmul(vmtemp,dee);vmfla=vmtemp(1,:)
              bot = dot_product(vmfla,vmfl) ; dslam = (fnew - top)/bot
              qinvr = matmul(qmat,ress); qinva=matmul(matmul(qmat,dee),vmfl)
              dsigma = -qinvr - qinva*dslam; stress = stress + dsigma
              call invar(stress,sigm,dsbar,lode_theta)
              fnew = dsbar - sqrt(3.)*cu;  dlam = dlam + dslam
              if (fnew<plastol) exit
            end do iterate_on_fnew
            dl(i,iel) = dlam
            elso = tensor( : , i , iel) + sigma - stress
            eload=matmul(elso,bee);bload=bload+eload*det*weights(i)
            call vmflow(stress,dsbar,vmfl)
            call fmrmat(vmfl,dsbar,dlam,dee,temp,rmat)
            caflow=matmul(rmat,vmfl);bot = dot_product(vmfl,caflow)
            call formaa(vmfl,rmat,daatd)
            dee = rmat - daatd/bot
         end if
         if(fstiff<.0) call deemat(dee,e,v)
             km = km + matmul(matmul(transpose(bee),dee),bee) *det* weights(i)
!--------------------- update the Gauss Point stresses -----------------------
         tensor( : , i , iel) = tensor( : , i , iel) + sigma - elso
         stress = tensor ( : , i , iel)
         eload=matmul(stress,bee); dload=dload+eload*det*weights(i)
    end do gauss_points_2
!        compute the total bodyloads vector
    bdylds( g ) = bdylds( g ) + bload  ; bdylds(0) = .0
    ddylds( g ) = ddylds( g ) + dload  ; ddylds(0) = .0
    storkm(:,:,iel) = km
   do k =1,ndof; diag_precon(g(k))=diag_precon(g(k))+km(k,k); end do
  end do elements_4
    diag_precon(1:neq)=1./diag_precon(1:neq) ; diag_precon(0) = .0
         tload = sum(ddylds)  ; tloads = sum(bdylds)
   if(plasiters==1)plastic_converged=.false.
   if(plasiters/=1.and.tloads<ltol)plastic_converged=.true.
   residual = (2.*ptot+tload-tloads)/(2.*ptot)
   write(11,'(a,10e12.4)')"tloads,tload,residual are",tloads,tload,residual
   totd = totd + loads
  if(plastic_converged.or.plasiters==plasits)exit
 end do plastic_iterations
 totd=totd+loads
 write(11,'(a,e12.4)')"The total load is   ",ptot
 write(11,'(a,10e12.4)')"Displacements are",totd(nf(2,no))
 write(11,'(a,i12)')"The number of cj iterations was    ",cjtot
 write(11,'(a,i12)')"The number of plastic iterations was  ",plasiters
 write(11,'(a,f11.2)')"cj iterations per plastic iteration were   ",           &
       &  real(cjtot)/real(plasiters)
 if(plasiters==plasits)stop
end do load_increments
end program p65
```

```
Global coordinates
Node    1         0.0000E+00  0.0000E+00
Node    2         0.0000E+00 -0.6250E+00
Node    3         0.0000E+00 -0.1250E+01
Node    4         0.0000E+00 -0.1875E+01
Node    5         0.0000E+00 -0.2500E+01
...........................................................................

Node  118         0.1200E+02 -0.3125E+01
Node  119         0.1200E+02 -0.3750E+01
Node  120         0.1200E+02 -0.4375E+01
Node  121         0.1200E+02 -0.5000E+01
Global node numbers
Element     1          3    2    1   10   15   16   17   11
Element     2          5    4    3   11   17   18   19   12
Element     3          7    6    5   12   19   20   21   13
Element     4          9    8    7   13   21   22   23   14
...........................................................................

Element    30        103  102  101  109  115  116  117  110
Element    31        105  104  103  110  117  118  119  111
Element    32        107  106  105  111  119  120  121  112

 Load increment      1
Plastic iteration number    1
tloads,tload,residual are  0.0000E+00 -0.4001E+03 -0.1496E-03
The total load is     0.2000E+03
Displacements are -0.6593E-02 -0.6487E-02 -0.6116E-02 -0.5418E-02 -0.3849E-02
The number of cj iterations was              46
The number of plastic iterations was            1
cj iterations per plastic iteration were        46.00

 Load increment      2
Plastic iteration number    1
tloads,tload,residual are  0.3491E+02 -0.6350E+03 -0.1165E+00
Plastic iteration number    2
tloads,tload,residual are  0.7229E+00 -0.6006E+03 -0.2230E-02
Plastic iteration number    3
tloads,tload,residual are  0.2032E-02 -0.5999E+03  0.1980E-03
Plastic iteration number    4
tloads,tload,residual are  0.3066E-06 -0.5999E+03  0.2052E-03
The total load is     0.3000E+03
Displacements are -0.1154E-01 -0.1127E-01 -0.1051E-01 -0.9106E-02 -0.6007E-02
The number of cj iterations was             206
The number of plastic iterations was            4
cj iterations per plastic iteration were        51.50
...........................................................................

 Load increment      7
Plastic iteration number    1
tloads,tload,residual are  0.3031E+01 -0.1003E+04 -0.6097E-02
Plastic iteration number    2
tloads,tload,residual are  0.2658E+00 -0.1000E+04 -0.5747E-03
Plastic iteration number    3
tloads,tload,residual are  0.6272E-03 -0.1000E+04 -0.3411E-04
Plastic iteration number    4
tloads,tload,residual are  0.1375E-08 -0.1000E+04 -0.3285E-04
The total load is     0.5000E+03
Displacements are -0.5031E-01 -0.5002E-01 -0.5061E-01 -0.4763E-01 -0.1452E-01
The number of cj iterations was             313
The number of plastic iterations was            4
cj iterations per plastic iteration were        78.25
...........................................................................
```

Figure 6.27 *(continued)*

```
  Load increment      10
 Plastic iteration number    1
 tloads,tload,residual are  0.7241E+00 -0.1041E+04 -0.1475E-02
 Plastic iteration number    2
 tloads,tload,residual are  0.7550E+00 -0.1038E+04  0.1521E-02
 ...............................................................

 Plastic iteration number   15
 tloads,tload,residual are  0.7798E-02 -0.1046E+04 -0.6033E-02
 Plastic iteration number   16
 tloads,tload,residual are  0.6485E-04 -0.1046E+04 -0.6033E-02
 Plastic iteration number   17
 tloads,tload,residual are  0.6052E-08 -0.1046E+04 -0.6033E-02
 The total load is      0.5200E+03
 Displacements are -0.2046E+00 -0.1934E+00 -0.1680E+00 -0.1331E+00 -0.1520E-01
 The number of cj iterations was             1695
 The number of plastic iterations was              17
 cj iterations per plastic iteration were          99.71
```

Figure 6.27 Results from Program 6.5

6.10 UNDRAINED ANALYSES

Little mention has been made so far of the role of the dilation angle ψ on the calculation of collapse loads in Mohr–Coulomb materials. The reason is that the dilation angle governs volumetric strains during plastic yield and will have little influence on collapse loads in "unconfined" problems. The examples considered so far in this chapter have been relatively unconfined (e.g. slope stability, earth pressures).

"Undrained" soils are two-phase particulate materials in which the voids between the particles are full of water. In addition, the permeability of the material may be sufficiently low or the loads applied so quickly that porewater pressures that are generated have no time to dissipate during the time-scale of the analysis.

In the case of undrained clays that have soft soil skeletons, the shear strength appears to be constant and given by an undrained "cohesion" c_u and $\phi_u = 0$. In such materials, the von Mises or Tresca failure criterion can be successfully applied, as was demonstrated in Program 6.0.

In the case of saturated soils with hard skeletons, such as dense quartz sand, shear stresses will tend to cause dilation which will be resisted by tensile water pressures in the voids of the soil. In turn, the effective stresses between particles will rise and, in a frictional material, the shear stresses necessary to cause failure will also rise. Thus, a dense sand, far from exhibiting a constant shear strength when sheared undrained, would have infinite strength provided the pore fluid could sustain infinite suction and the grains did not crush. In reality, a finite shear strength is recorded due to either grain crushing or pore fluid cavitation.

PROGRAM 6.6: AXISYMMETRIC "UNDRAINED" ANALYSIS—VISCO-PLASTIC METHOD USING THE MOHR-COULOMB CRITERION

```
   program p66
!-----------------------------------------------------------------------
!      program 6.6 axisymmetric 'undrained' strain of an elastic-plastic
!      (Mohr-Coulomb) solid
!      using 8-node quadrilateral elements; viscoplastic strain method
!-----------------------------------------------------------------------
 use new_library    ;   use  geometry_lib  ;      implicit none
 integer::nels,nxe,nye,neq,nband,nn,nr,nip,nodof=2,nod=8,nst=4,ndof,        &
         i,j,k,iel,iters,limit,incs,iy,ndim=2,loaded_nodes
 logical::converged      ; character (len=15) :: element='quadrilateral'
 real::e,v,det,phi,c,psi,dt,f,dsbar,dq1,dq2,dq3,lode_theta,                 &
          sigm,pi,snph,bulk,cons,presc,ptot,radius,tol
!---------------------------- dynamic arrays--------------------------------
 real    ,allocatable :: kv(:),loads(:),points(:,:),bdylds(:),totd(:),      &
                         evpt(:,:,:),oldis(:),width(:),depth(:),stress(:),   &
                         dee(:,:),coord(:,:),jac(:,:),weights(:),storkv(:),  &
                         der(:,:),deriv(:,:),bee(:,:),km(:,:),eld(:),eps(:), &
                         sigma(:),bload(:),eload(:),erate(:),g_coord(:,:),   &
                         evp(:),devp(:),m1(:,:),m2(:,:),m3(:,:),flow(:,:),   &
                         tensor(:,:,:),etensor(:,:,:),pore(:,:),fun(:)
 integer, allocatable :: nf(:,:) , g(:), no(:) ,num(:), g_num(:,:) ,g_g(:,:)
!-------------------------input and initialisation---------------------------
  open (10,file='p66.dat',status=    'old',action='read')
  open (11,file='p66.res',status='replace',action='write')
  read (10,*) phi,c,psi,e,v,bulk,cons,     nels,nxe,nye,nn,nip
  ndof=nod*nodof
  allocate (nf(nodof,nn), points(nip,ndim),weights(nip),g_coord(ndim,nn),    &
           width(nxe+1),depth(nye+1),num(nod),evpt(nst,nip,nels),            &
           coord(nod,ndim),g_g(ndof,nels),tensor(nst,nip,nels),fun(nod),     &
           etensor(nst,nip,nels),dee(nst,nst),pore(nip,nels),stress(nst),    &
           jac(ndim,ndim),der(ndim,nod),deriv(ndim,nod),g_num(nod,nels),     &
           bee(nst,ndof),km(ndof,ndof),eld(ndof),eps(nst),sigma(nst),        &
           bload(ndof),eload(ndof),erate(nst),evp(nst),devp(nst),g(ndof),    &
           m1(nst,nst),m2(nst,nst),m3(nst,nst),flow(nst,nst))
  nf=1; read (10,*) nr ; if(nr>0) read(10,*)(k,nf(:,k),i=1,nr)
  call formnf(nf); neq=maxval(nf);read(10,*) width , depth
!------------ loop the elements to find nband and set up global arrays --------
     nband = 0
       elements_1:   do iel = 1 , nels
                       call geometry_8qyv(iel,nye,width,depth,coord,num)
                       call num_to_g(num,nf,g) ;    g_num(:,iel)=num
                       g_coord(: , num )=transpose(coord); g_g( : , iel ) = g
                       if (nband<bandwidth(g)) nband = bandwidth(g)
       end do elements_1
    write(11,'(a)') "Global coordinates "
    do k=1,nn;write(11,'(a,i5,a,2e12.4)')"Node",k,"       ",g_coord(:,k);end do
    write(11,'(a)') "Global node numbers "
    do k = 1 , nels; write(11,'(a,i5,a,8i5)')                                 &
                           "Element ",k,"        ",g_num(:,k); end do
   write(11,'(a,i5,a,i5)')                                                    &
          "The system has ",neq," equations and the half-bandwidth is",nband
allocate(kv(neq*(nband+1)),loads(0:neq),bdylds(0:neq),oldis(0:neq),totd(0:neq))
          kv=0.0; oldis=0.0; totd=0.0 ; tensor = 0.0; etensor = 0.0
  call deemat(dee,e,v); call sample(element,points,weights)
!------------------    fluid bulk modulus is "bulk" -------------------------
  do i=1,nst; do j=1,nst;if(i/=3.and.j/=3)dee(i,j)=dee(i,j)+bulk; end do; end do
  pi = acos( -1. ); snph = sin(phi*pi/180.)
  dt = 4.*(1.+ v)*(1.-2.*v)/(e*(1.-2.*v+snph*snph))
!---------- element stiffness integration and assembly & initial conditions----
 elements_2: do iel = 1 , nels
               num = g_num(: ,iel ) ; coord = transpose (g_coord(: ,num ))
               g = g_g( : ,iel )     ;       km=0.0
           gauss_pts_1:  do i =1 , nip    ; call shape_fun(fun,points,i)
             call shape_der (der,points,i);  jac = matmul(der,coord)
```

```
            det = determinant(jac)  ;   call invert(jac)
            deriv=matmul(jac,der);call bmataxi(bee,radius,coord,deriv,fun)
          km=km+matmul(matmul(transpose(bee),dee),bee)*det*weights(i)*radius
            tensor(1:2,i,iel)=cons; tensor(4,i,iel)=cons
          end do gauss_pts_1
   call formkv (kv,km,g,neq)
 end do elements_2
!--------------- prescribe displacements and factorise l.h.s. ---------------
      read(10,*) loaded_nodes ; allocate(no(loaded_nodes),storkv(loaded_nodes))
          read(10,*)no , presc  , incs , tol , limit
          do i=1,loaded_nodes
             kv(nf(2,no(i)))=kv(nf(2,no(i)))+1.e20 ; storkv(i)=kv(nf(2,no(i)))
          end do               ;      call banred(kv,neq)
!------------------displacement increment loop--------------------------------
    call deemat(dee,e,v)
   load_increments: do iy=1,incs
    ptot = presc * iy
    write(11,'(/,a,i5)') 'Load increment',iy ;  iters=0;  bdylds=.0;  evpt=.0
!--------------------------    iteration loop  -------------------------------
   iterations: do
    iters=iters+1;  loads = .0
     do i=1,loaded_nodes;loads(nf(2,no(i)))=storkv(i)*presc; end do
     loads = loads + bdylds  ;  call bacsub(kv,loads)
!------------------------   check convergence --------------------------------
      call checon(loads,oldis,tol,converged)
      if(iters==1)converged=.false.
!---------------------- go round the Gauss Points ----------------------------
      elements_3: do iel = 1 , nels
       bload=.0
       num = g_num( : , iel ) ; coord = transpose( g_coord( : , num ))
       g = g_g( : , iel )     ; eld = loads ( g )
       gauss_points_2 : do i = 1 , nip
          call shape_fun(fun,points,i)
          call shape_der ( der,points,i); jac=matmul(der,coord)
          det = determinant(jac)  ;  call invert(jac)
          deriv = matmul(jac,der) ;  call bmataxi (bee,radius,coord,deriv,fun)
          eps=matmul(bee,eld); det = det * radius;  eps=eps-evpt(:,i,iel)
          sigma=matmul(dee,eps)  ;    stress=sigma+tensor(: , i , iel)
          call invar(stress,sigm,dsbar,lode_theta)
!--------------------  check whether yield is violated -----------------------
          call mocouf (phi, c , sigm, dsbar , lode_theta , f )
          if(f>=.0) then
           call mocouq(psi,dsbar,lode_theta,dq1,dq2,dq3)
           call formm(stress,m1,m2,m3)
           flow=f*(m1*dq1+m2*dq2+m3*dq3)     ;   erate=matmul(flow,stress)
           evp=erate*dt; evpt(:,i,iel)=evpt(:,i,iel)+evp;devp=matmul(dee,evp)
           eload=matmul(devp,bee)     ; bload=bload+eload*det*weights(i)
          end if
        if(converged.or.iters==limit)  then
!----------------   update stresses and calculate porepressures  -------------
          tensor(:,i,iel)=stress
          etensor(:,i,iel)=etensor(:,i,iel)+eps+evpt(:,i,iel)
          pore(i,iel)=(etensor(1,i,iel)+etensor(2,i,iel)+etensor(4,i,iel))*bulk
        end if
    end do gauss_points_2
!         compute the total bodyloads vector
    bdylds( g ) = bdylds( g ) + bload      ; bdylds(0) = .0
  end do elements_3
  if(converged.or.iters==limit)exit
 end do iterations
  totd = totd + loads
 write(11,'(a,e12.4)') "     Displacement" , ptot
 write(11,'(a,3e12.4)')                                                      &
        "     Effective stresses",tensor(1,1,1),tensor(2,1,1),tensor(4,1,1)
 write(11,'(a,2e12.4)') "     Deviator stress and porepressure",dsbar,pore(1,1)
 write(11,'(a,i5,a)') "It took",iters," iterations to converge"
 if(iters==limit)stop
end do load_increments
end program p66
```

To perform analyses of this type it is necessary to separate stresses into porewater pressures (isotropic) and effective interparticle stresses (isotropic + shear). Such a treatment has already been described in Chapter 2 in terms of time-dependent "consolidation" properties of two-phase materials (Biot's poro-elastic theory) and programs to deal with this will be found in Chapter 9. However, the undrained problem pertaining at the beginning of the Biot process is so important in soil mechanics that it merits special treatment.

Naylor (1974) has described a method of separating the stresses into pore pressures and effective stresses. The method uses as its basis the concept of effective stress in matrix notation; thus

$$\boldsymbol{\sigma} = \boldsymbol{\sigma}' + \mathbf{u} \tag{6.76}$$

in which the 1-d array $\mathbf{u}$ contains no shear terms. The stress–strain relationships can be written as

$$\boldsymbol{\sigma}' = \mathbf{D}'\boldsymbol{\varepsilon} \tag{6.77}$$

and

$$\mathbf{u} = \mathbf{D}_{\mathrm{u}}\boldsymbol{\varepsilon} \tag{6.78}$$

which combine to give

$$\boldsymbol{\sigma} = \mathbf{D}\boldsymbol{\varepsilon} \tag{6.79}$$

where

$$\mathbf{D} = \mathbf{D}' + \mathbf{D}_{\mathrm{u}} \tag{6.80}$$

The matrix $\mathbf{D}'$ is the familiar elastic stress–strain matrix in terms of effective Young's modulus (E') and Poisson's ratio (ν'). The matrix $\mathbf{D}_{\mathrm{u}}$ contains the apparent bulk modulus of the fluid K_{e} in the following locations:

$$\mathbf{D}_{\mathrm{u}} = \begin{bmatrix} K_{\mathrm{e}} & K_{\mathrm{e}} & 0 & K_{\mathrm{e}} \\ & K_{\mathrm{e}} & 0 & K_{\mathrm{e}} \\ & & 0 & 0 \\ \text{symmetrical} & & & K_{\mathrm{e}} \end{bmatrix} \tag{6.81}$$

assuming that the third column corresponds to the shear terms in a 2-d plane strain analysis.

To implement this method in the programs described in this chapter, it is necessary to form the global stiffness matrix using the total stress–strain matrix $\mathbf{D}$, while effective stresses for use in the failure function are computed from total strains using the effective stress–strain matrix $\mathbf{D}'$. Pore pressures are simply computed from

$$u = K_{\mathrm{e}}(\varepsilon_r + \varepsilon_z + \varepsilon_\theta) \tag{6.82}$$

For relatively large values of K_e, the analysis is insensitive to the exact magnitude of K_e. For axisymmetric analyses, Griffiths (1985) defined the dimensionless grouping

$$\beta = \frac{(1 - 2\nu')K_e}{E'} \tag{6.83}$$

and showed that results were essentially constant for $\beta \geq 20$. The example shown in Figure 6.28 represents a single axisymmetric eight-node element subjected to vertical compressive displacement increments along its top face. In

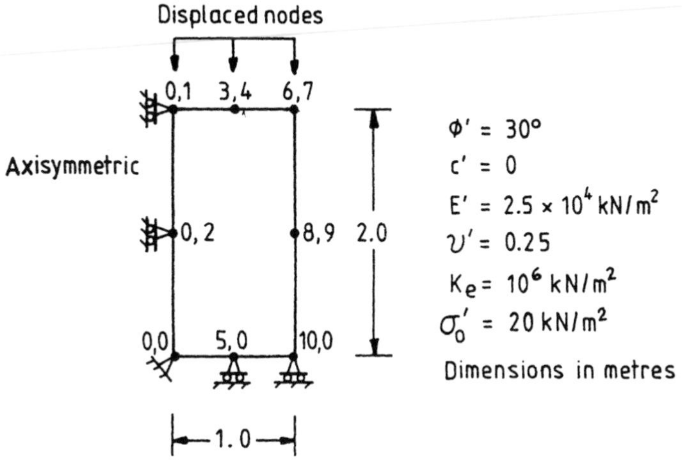

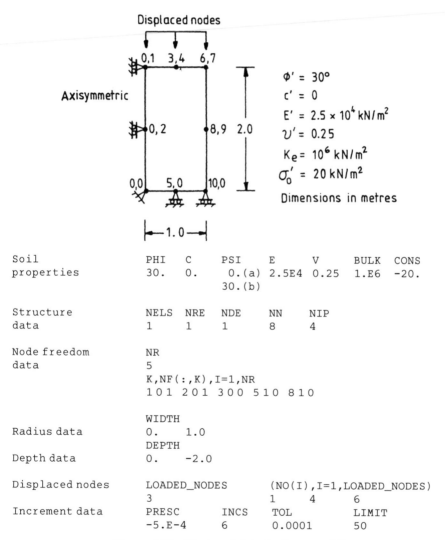

Soil properties	PHI	C	PSI	E	V	BULK	CONS
	30.	0.	0.(a) 30.(b)	2.5E4	0.25	1.E6	-20.
Structure data	NELS	NRE	NDE	NN	NIP		
	1	1	1	8	4		

Node freedom data	NR
	5
	K,NF(:,K),I=1,NR
	1 0 1 2 0 1 3 0 0 5 1 0 8 1 0
	WIDTH
Radius data	0. 1.0
	DEPTH
Depth data	0. -2.0

Displaced nodes	LOADED_NODES		(NO(I),I=1,LOADED_NODES)		
	3		1	4	6
Increment data	PRESC	INCS	TOL		LIMIT
	-5.E-4	6	0.0001		50

Figure 6.28 Mesh and data for Program 6.6

order to compute pore pressures during undrained loading, it is necessary to update strains as well as stresses after each increment; hence the following additional 3-d arrays are declared:

`ETENSOR`	accumulates ε_r, ε_z, γ_{rz} and ε_θ at each Gauss point
`PORE`	pore pressure at each Gauss point

Additional input parameters are as follows:

`BULK`	apparent pore fluid bulk modulus K_e
`CONS`	initial effective stress σ'_r, σ'_z, $\sigma'_\theta(\tau_{rz} = 0)$

The coordinates of the elements are read into 1-d arrays `WIDTH` and `DEPTH` following the notation established for axisymmetric problems by Program 5.4.

After the effective stress–strain matrix has been augmented by the fluid bulk modulus, the global stiffness matrix is formed in the usual way. It may be noted that in this listing the stiffness matrix is stored as a 1-d array (`KV`) in preparation for Gaussian direct methods of solution. Information about the magnitude of the prescribed displacement increments and the freedoms to which they are to be applied is read and the "penalty" technique duly implemented.

Just before the displacement increment loop begins, the routine `DEEMAT` is called to form the effective stress–strain matrix. The only other difference from earlier programs is that both stresses and strains are updated after each increment. The pore pressure is also computed from equation (6.82).

The data shown in Figure 6.28 is for an undrained sand with the following properties:
Effective soil properties:

$$\phi' = 30^\circ$$
$$c' = 0$$
$$E' = 2.5 \times 10^4\,\text{kN/m}^2$$
$$\nu' = 0.25$$

Bulk modulus:

$$K_e = 10^6\,\text{kN/m}^2$$

The triaxial specimen has been consolidated under a cell pressure of $20\,\text{kN/m}^2$ before undrained loading commences.

The output of an analysis is presented in Figure 6.29. In this analysis (a), $\psi = 0$ while in analysis (b), $\psi = 30^\circ$. As expected, the inclusion of dilation has a considerable impact on the response in this "confined" problem. The deviator stress and pore pressure versus axial strain have been plotted for both cases in Figure 6.30. Case (a), in which there is no plastic volume change ($\psi = 0$),

```
Global coordinates
Node     1          0.0000E+00  0.0000E+00
Node     2          0.0000E+00 -0.1000E+01
Node     3          0.0000E+00 -0.2000E+01
Node     4          0.5000E+00  0.0000E+00
Node     5          0.5000E+00 -0.2000E+01
Node     6          0.1000E+01  0.0000E+00
Node     7          0.1000E+01 -0.1000E+01
Node     8          0.1000E+01 -0.2000E+01
Global node numbers
Element      1           3    2    1    4    6    7    8    5
The system has    10  equations and the half-bandwidth is    9

Load increment    1
     Displacement -0.5000E-03
     Effective stresses -0.1755E+02 -0.2502E+02 -0.1755E+02
     Deviator stress and porepressure  0.7475E+01 -0.2451E+01
It took    2  iterations to converge

Load increment    2
     Displacement -0.1000E-02
     Effective stresses -0.1510E+02 -0.3005E+02 -0.1510E+02
     Deviator stress and porepressure  0.1495E+02 -0.4902E+01
It took    2  iterations to converge

Load increment    3
     Displacement -0.1500E-02
     Effective stresses -0.1265E+02 -0.3507E+02 -0.1265E+02
     Deviator stress and porepressure  0.2243E+02 -0.7353E+01
It took    2  iterations to converge

Load increment    4
     Displacement -0.2000E-02
     Effective stresses -0.1207E+02 -0.3626E+02 -0.1207E+02
     Deviator stress and porepressure  0.2419E+02 -0.7931E+01
It took    4  iterations to converge

Load increment    5
     Displacement -0.2500E-02
     Effective stresses -0.1207E+02 -0.3626E+02 -0.1207E+02
     Deviator stress and porepressure  0.2420E+02 -0.7934E+01
It took    4  iterations to converge

Load increment    6
     Displacement -0.3000E-02
     Effective stresses -0.1207E+02 -0.3626E+02 -0.1207E+02
     Deviator stress and porepressure  0.2420E+02 -0.7934E+01
It took    4  iterations to converge
```

Figure 6.29 Results from Program 6.6 ($\psi = 0$)

reaches a peak deviator stress that is less than the drained failure load. This is due to the increase in pore pressure occurring in the elastic phase of compression. The closed form solution (Griffiths, 1985) for this case is seen to be in close agreement for $\beta = 20$. Case (b), on the other hand, shows no sign of failure due to the tendency for dilation. In this case, the pore pressures would continue to fall and the deviator stress continue to rise indefinitely unless some additional criterion was introduced.

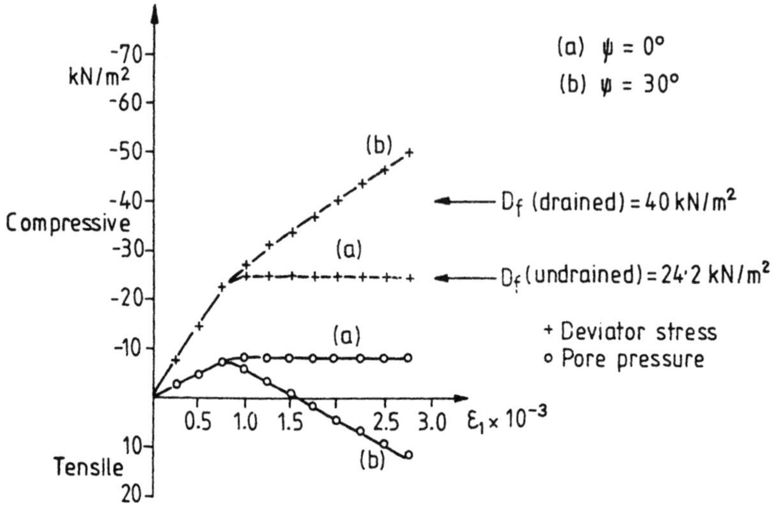

Figure 6.30 Deviator stress and pore pressure versus axial strain

6.11 THREE-DIMENSIONAL ANALYSES AND GENERAL PROGRAMS

Although storage requirements increase particularly if an assembly strategy is used it is simple to extend any of the programs in this chapter to deal with 3-d strain conditions. The program described here as Program 6.7 is based on Program 5.8, and modified to acount for elasto-plastic material behaviour using the visco-plastic algorithm of Program 6.0.

Indeed, the program can scarcely be distinguished from Program 6.0, although it economises on storage while still using an assembly strategy.

As in Program 5.8, a skyline strategy is adopted for storage of the global stiffness matrix. As the length of the resulting 1-d array `KV` is not known in advance, for large problems, a preliminary run of the program can be made to obtain this length held in `KDIAG(NEQ)`. After formation of the global stiffness matrix, the "penalty" method is implemented and it is worth noting how the locations of the diagonal terms in `KV` are found using the 1-d array `KDIAG`.

The example shown in Figure 6.31 represents a cube of material subjected to prescribed displacements along its top face. The symmetry of the boundary conditions implies that the element represents one-eighth of a cube of side length 2 m. The material properties are those of a cohesionless material ($\phi' = 30°, c' = 0$) and an initial isotropic stress state of 20 kN/m^2 is imposed. The output given in Figure 6.32 shows the axial displacement, the three

PROGRAM 6.7: THREE-DIMENSIONAL ELASTO-PLASTIC ANALYSIS—VISCO-PLASTIC METHOD USING THE MOHR–COULOMB CRITERION

```
    program p67
!-----------------------------------------------------------------------
!      program 6.7 three-d strain of an elastic-plastic(Mohr-Coulomb) solid
!      using 20-node brick elements; viscoplastic strain method
!-----------------------------------------------------------------------
 use new_library  ;   use geometry_lib  ;      implicit none
 integer::nels,nxe,nze,neq,nn,nr,nip,nodof=3,nod=20,nst=6,ndof,            &
         i,k,iel,iters,limit,incs,iy,ndim=3,loaded_nodes
 logical::converged   ; character (len=15) :: element='hexahedron'
 real::e,v,det,phi,c,psi,dt,f,dsbar,dq1,dq2,dq3,lode_theta,presc,          &
          sigm,pi,snph,cons,ptot,aa,bb,cc,tol
!--------------------------- dynamic arrays-----------------------------------
 real   ,allocatable :: kv(:),loads(:),points(:,:),bdylds(:),evpt(:,:,:),   &
                        dee(:,:),coord(:,:),jac(:,:),weights(:),oldis(:),   &
                        der(:,:),deriv(:,:),bee(:,:),km(:,:),eld(:),eps(:), &
                        sigma(:),bload(:),eload(:),erate(:),g_coord(:,:),   &
                        evp(:),devp(:),m1(:,:),m2(:,:),m3(:,:),flow(:,:),   &
                        val(:),storkv(:),tensor(:,:,:),stress(:),totd(:)
 integer, allocatable :: nf(:,:) , g(:), no(:) ,num(:), g_num(:,:) ,g_g(:,:), &
                        kdiag(:)
!-------------------------input and initialisation----------------------------
  open (10,file='p67.dat',status=    'old',action='read')
  open (11,file='p67.res',status='replace',action='write')
  read (10,*) phi,c,psi,e,v,cons,    nels,nxe,nze,nn,nip,aa,bb,cc
  ndof=nod*nodof
  allocate (nf(nodof,nn), points(nip,ndim),weights(nip),g_coord(ndim,nn),   &
           num(nod),dee(nst,nst),evpt(nst,nip,nels),tensor(nst,nip,nels),   &
           coord(nod,ndim),g_g(ndof,nels), stress(nst),                     &
           jac(ndim,ndim),der(ndim,nod),deriv(ndim,nod),g_num(nod,nels),    &
           bee(nst,ndof),km(ndof,ndof),eld(ndof),eps(nst),sigma(nst),       &
           bload(ndof),eload(ndof),erate(nst),evp(nst),devp(nst),g(ndof),   &
           m1(nst,nst),m2(nst,nst),m3(nst,nst),flow(nst,nst))
  nf=1; read(10,*) nr ; if(nr>0) read(10,*)(k,nf(:,k),i=1,nr)
  call formnf(nf); neq=maxval(nf)  ;   allocate(kdiag(neq))     ; kdiag = 0
!---------- loop the elements to set up global arrays and kdiag --------------
      elements_1:   do iel = 1 , nels
                       call geometry_20bxz(iel,nxe,nze,aa,bb,cc,coord,num)
                       call num_to_g ( num , nf , g )
                       g_num(:,iel)=num; g_coord(:,num)=transpose(coord)
                       g_g( : , iel ) = g   ;  call fkdiag(kdiag,g)
      end do elements_1
    write(11,'(a)') "Global coordinates "
    do k=1,nn;write(11,'(a,i5,a,3e12.4)')"Node",k,"       ",g_coord(:,k);end do
    write(11,'(a)') "Global node numbers "
    do k = 1 , nels; write(11,'(a,i5,a,20i3)')                                &
                          "Element ",k,"        ",g_num(:,k); end do
      kdiag(1)=1; do i=2,neq; kdiag(i)=kdiag(i)+kdiag(i-1); end do
   write(11,'(a,i5,a,i5)')                                                    &
          "There are ",neq," equations and the skyline storage is",kdiag(neq)
 allocate(kv(kdiag(neq)),loads(0:neq),bdylds(0:neq),oldis(0:neq),totd(0:neq))
          kv=0.0; oldis=0.0; totd=0.0
  call deemat(dee,e,v); call sample(element,points,weights)
  pi = acos( -1. ); snph = sin(phi*pi/180.)
  dt=4.*(1.+v)*(1.-2.*v)/(e*(1.-2.*v+snph*snph))    ; tensor = .0
  write(11,'(a,e12.4)') "The critical timestep is   ",dt
!---------- element stiffness integration and assembly & set initial stress---
 elements_2: do iel = 1 , nels
               num = g_num(: , iel ) ; coord = transpose (g_coord(:,num ))
               g = g_g( : , iel )    ;       km=0.0
            gauss_pts_1:  do i =1 , nip
              tensor(1:3,i,iel)=cons
              call shape_der (der,points,i);  jac = matmul(der,coord)
              det = determinant(jac)  ;   call invert(jac)
              deriv = matmul(jac,der) ;   call beemat (bee,deriv)
```

```
            km = km + matmul(matmul(transpose(bee),dee),bee) *det* weights(i)
          end do gauss_pts_1
          call fsparv (kv,km,g,kdiag)
 end do elements_2
!--------------- read prescribed displacements and factorise l.h.s. -----------
    read(10,*) loaded_nodes ; allocate(no(loaded_nodes),storkv(loaded_nodes))
    read(10,*)no ,presc, incs, tol, limit
    do i=1,loaded_nodes
         kv(kdiag(nf(3,no(i))))=kv(kdiag(nf(3,no(i)))) + 1.e20
         storkv(i)=kv(kdiag(nf(3,no(i))))
    end do
         call sparin (kv,kdiag)
!-------------------displacement increment loop-------------------------------
    load_increments: do iy=1,incs
    ptot = presc * iy
    write(11,'(/,a,i5)')"Load increment ",iy ;  iters=0;  bdylds=.0;  evpt=.0
!-----------------------   iteration loop   ---------------------------------
   iterations: do
    iters=iters+1; loads =.0
      do i=1,loaded_nodes;loads(nf(3,no(i)))=storkv(i)*presc; end do
      loads=loads+bdylds; call spabac(kv,loads,kdiag)
!---------------------   check convergence ----------------------------------
      call checon(loads,oldis,tol,converged)
      if(iters==1)converged=.false. ;  if(converged.or.iters==limit)bdylds=.0
!-------------------- go round the Gauss Points  ----------------------------
      elements_3: do iel = 1 , nels
       bload=.0
       num = g_num( : ,iel ) ; coord = transpose(g_coord( : , num ))
       g = g_g( : , iel )    ;    eld = loads ( g )
       gauss_points_2 : do i = 1 , nip
          call shape_der ( der,points,i); jac=matmul(der,coord)
          det = determinant(jac);call invert(jac) ;  deriv = matmul(jac,der)
          call beemat (bee,deriv);eps=matmul(bee,eld); eps=eps-evpt(:,i,iel)
          sigma=matmul(dee,eps) ;  stress = sigma + tensor(:,i,iel)
          call invar(stress,sigm,dsbar,lode_theta)
!-------------------  check whether yield is violated ------------------------
         call mocouf (phi, c , sigm, dsbar , lode_theta , f )
         if(converged.or.iters==limit) then
         devp=stress
           else
           if(f>=.0) then
           call mocouq(psi,dsbar,lode_theta,dq1,dq2,dq3)
           call formm(stress,m1,m2,m3);  flow=f*(m1*dq1+m2*dq2+m3*dq3)
           erate=matmul(flow,stress)
           evp=erate*dt; evpt(:,i,iel)=evpt(:,i,iel)+evp; devp=matmul(dee,evp)
         end if; end if
      if(f>=.0) then
        eload=matmul(devp,bee) ; bload=bload+eload*det*weights(i)
      end if
    if(converged.or.iters==limit) then
!-----------------------update the Gauss point stresses -----------------------
                tensor(:,i,iel) = stress
    end if
    end do gauss_points_2
!-------------------  compute the total bodyloads vector-----------------------
    bdylds( g ) = bdylds( g ) + bload      ; bdylds(0) = .0
  end do elements_3
  if(converged.or.iters==limit)exit
 end do iterations
 totd = totd + loads
 write(11,'(a,e12.4)') "The displacement is ",totd(1)
 write(11,'(a)') "   sigma z      sigma x      sigma y"
 write(11,'(3e12.4)') tensor(3,1,1),tensor(1,1,1),tensor(2,1,1)
 write(11,'(a,i5,a)') "It took",iters,"  iterations to converge"
  if(iters==limit)stop
 end do load_increments
 end program p67
```

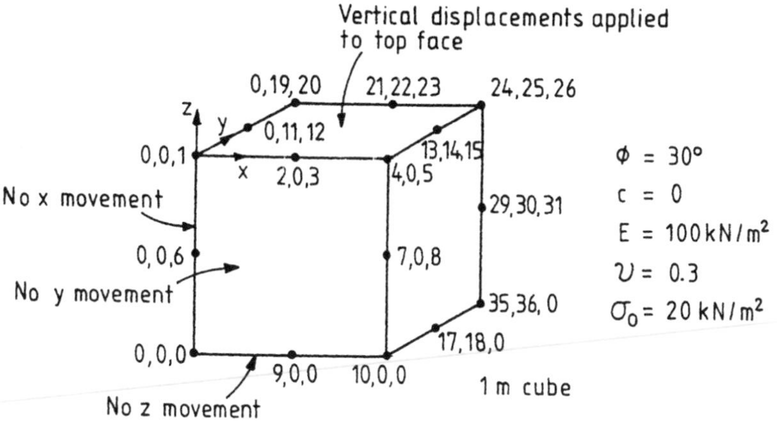


```
Soil properties      PHI    C      PSI    E      V      CONS
                     30.    0.     0.     100.   0.3    -20.

Structure            NELS   NXE    NZE    NN     NIP
data                 1      1      1      20     27

Element data         AA     BB     CC
                     1.     1.     1.

Node freedom         NR
data                 16
                     K,NF(:,K),I=1,NR
                     1 0 0 1 2 1 0 1 3 1 0 1 4 0 0 1 5 1 0 1
                     6 0 0 0 7 1 0 0 8 1 0 0 9 0 1 1 11 1 0 1 0
                     12 1 1 0 13 0 1 1 16 0 1 1 18 0 1 0 19 1 1 0
                     20 1 1 0

Displaced            LOADED_NODES       (NO(I),I=1,LOADED_NODES)
freedoms             8                  1 2 3 9 10 13 14 15
Increment data       PRESC      INCS    TOL           LIMIT
                     -0.05      10      0.0001        50
```

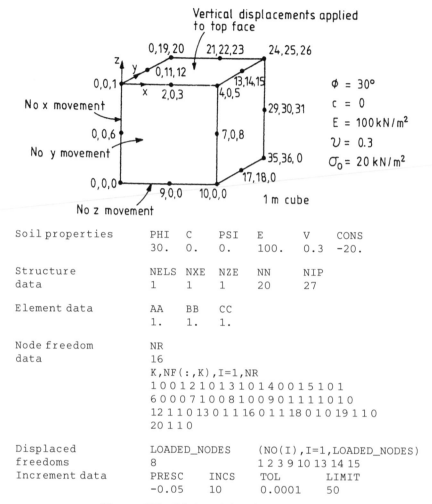

Figure 6.31 Mesh and data for Program 6.7

normal stresses, σ_x, σ_y and σ_z (in this case equal to the principal stresses), and the number of iterations to reach convergence after each displacement increment.

As loads are applied axially and the cube is free to expand in the other orthogonal directions, two of the principal stresses remain constant at 20 kN/m^2 throughout. As expected for a Mohr–Coulomb material with $\phi' = 30°$ and $c' = 0$, the axial stress (σ_z or σ_1) builds up to a maximum of 60 kN/m^2 and remains constant at that value.

```
Global coordinates
Node    1          0.0000E+00  0.0000E+00  0.0000E+00
Node    2          0.5000E+00  0.0000E+00  0.0000E+00
Node    3          0.1000E+01  0.0000E+00  0.0000E+00
Node    4          0.0000E+00  0.0000E+00 -0.5000E+00
Node    5          0.1000E+01  0.0000E+00 -0.5000E+00
Node    6          0.0000E+00  0.0000E+00 -0.1000E+01
Node    7          0.5000E+00  0.0000E+00 -0.1000E+01
Node    8          0.1000E+01  0.0000E+00 -0.1000E+01
Node    9          0.0000E+00  0.5000E+00  0.0000E+00
Node   10          0.1000E+01  0.5000E+00  0.0000E+00
Node   11          0.0000E+00  0.5000E+00 -0.1000E+01
Node   12          0.1000E+01  0.5000E+00 -0.1000E+01
Node   13          0.0000E+00  0.1000E+01  0.0000E+00
Node   14          0.5000E+00  0.1000E+01  0.0000E+00
Node   15          0.1000E+01  0.1000E+01  0.0000E+00
Node   16          0.0000E+00  0.1000E+01 -0.5000E+00
Node   17          0.1000E+01  0.1000E+01 -0.5000E+00
Node   18          0.0000E+00  0.1000E+01 -0.1000E+01
Node   19          0.5000E+00  0.1000E+01 -0.1000E+01
Node   20          0.1000E+01  0.1000E+01 -0.1000E+01
Global node numbers
Element     1            6  4  1  2  3  5  8  7 11  9 10 12 18 16 13 14 15 17 20 19
There are    36  equations and the skyline storage is   666
The critical timestep is      0.3200E-01

Load increment      1
The displacement is  -0.5000E-01
   sigma z      sigma x      sigma y
 -0.2500E+02 -0.2000E+02 -0.2000E+02
It took    2  iterations to converge

Load increment      2
The displacement is  -0.1000E+00
   sigma z      sigma x      sigma y
 -0.3000E+02 -0.2000E+02 -0.2000E+02
It took    2  iterations to converge
...........................................................................

Load increment      8
The displacement is  -0.4000E+00
   sigma z      sigma x      sigma y
 -0.6000E+02 -0.2000E+02 -0.2000E+02
It took    2  iterations to converge

Load increment      9
The displacement is  -0.4500E+00
   sigma z      sigma x      sigma y
 -0.6000E+02 -0.2000E+02 -0.2000E+02
It took   16  iterations to converge

Load increment     10
The displacement is  -0.5000E+00
   sigma z      sigma x      sigma y
 -0.6000E+02 -0.2000E+02 -0.2000E+02
It took   16  iterations to converge
```

Figure 6.32 Results from Program 6.7

PROGRAM 6.8: GENERAL PURPOSE ELASTO-PLASTIC ANALYSIS OF MOHR–COULOMB SOLIDS USING A VISCO-PLASTIC ALGORITHM

Since Programs 6.0 and 6.7 are almost identical, except for the specification of the respective geometries, general programs can be prepared, in the way that

```
    program p68
!-----------------------------------------------------------------------
!      program 6.8 general strain of an elastic-plastic(Mohr-Coulomb)
!      solid --- viscoplastic strain method
!-----------------------------------------------------------------------
 use new_library  ;  use  geometry_lib       ; implicit none
 integer::nels,neq,nn,nr,nip,nodof,nod,nst,ndof,nprops,np_types,           &
          i,k,iel,iters,limit,incs,iy,ndim,loaded_nodes,fixed_nodes
 logical::converged   ;  character (len=15) :: element
 real::e,v,phi,c,psi,                                                     &
       det,dt,ddt,f,dsbar,dq1,dq2,dq3,lode_theta,sigm,pi,snph,ptot,tol
!--------------------------- dynamic arrays----------------------------------
 real   ,allocatable :: kv(:),loads(:),points(:,:),bdylds(:),oldis(:),      &
                        dee(:,:),coord(:,:),fun(:),jac(:,:),weights(:),     &
                        der(:,:),deriv(:,:),bee(:,:),km(:,:),eld(:),eps(:), &
                        sigma(:),bload(:),eload(:),erate(:),g_coord(:,:),   &
                        evp(:),devp(:),m1(:,:),m2(:,:),m3(:,:),flow(:,:),   &
                        val(:),storkv(:),tensor(:,:,:),stress(:),totd(:),   &
                        evpt(:,:,:),value(:),load_store(:),prop(:,:)
 integer, allocatable :: nf(:,:) , g(:), no(:) ,num(:), g_num(:,:) ,g_g(:,:), &
                        kdiag(:), sense(:), node(:) , etype(:)
!--------------------------input and initialisation--------------------------
  open (10,file='p68.dat',status=    'old',action='read')
  open (11,file='p68.res',status='replace',action='write')
  read (10,*) element,nels,nn,nip,nodof,nod,nst,ndim,incs,tol,limit
  ndof=nod*nodof
  allocate (nf(nodof,nn), points(nip,ndim),weights(nip),g_coord(ndim,nn),   &
            num(nod),dee(nst,nst),evpt(nst,nip,nels),tensor(nst,nip,nels),  &
            coord(nod,ndim),g_g(ndof,nels),stress(nst),etype(nels),         &
            jac(ndim,ndim),der(ndim,nod),deriv(ndim,nod),g_num(nod,nels),   &
            bee(nst,ndof),km(ndof,ndof),eld(ndof),eps(nst),sigma(nst),      &
            bload(ndof),eload(ndof),erate(nst),evp(nst),devp(nst),g(ndof),  &
            m1(nst,nst),m2(nst,nst),m3(nst,nst),flow(nst,nst))
   read(10,*) nprops , np_types
    allocate(prop(nprops,np_types)) ; read(10,*) prop
    etype = 1 ; if(np_types>1) read(10,*) etype
!--------------------- read geometry and connectivity ------------------------
  read(10,*) g_coord; read(10,*) g_num
  nf=1; read(10,*) nr ; if(nr>0) read(10,*)(k,nf(:,k),i=1,nr)
  call formnf(nf); neq=maxval(nf)    ;  allocate(kdiag(neq))     ; kdiag = 0
!--------- loop the elements to set up global arrays and kdiag ----------------
      elements_1:   do iel = 1 , nels
                       num = g_num(:,iel); coord = transpose(g_coord(:,num))
                       call num_to_g( num , nf , g )
                       g_g( : , iel ) = g ;   call fkdiag(kdiag,g)
      end do elements_1
    write(11,'(a)') "Global coordinates "
    do k=1,nn;write(11,'(a,i5,a,3e12.4)')"Node",k,"       ",g_coord(:,k);end do
    write(11,'(a)') "Global node numbers "
    do k = 1 , nels; write(11,'(a,i5,a,27i3)')                              &
                            "Element ",k,"   ",g_num(:,k); end do
      kdiag(1)=1; do i=2,neq; kdiag(i)=kdiag(i)+kdiag(i-1); end do
 write(11,'(a,i5,a,i5,a)')                                                  &
        "The skyline storage is", kdiag(neq),"and there are",neq," equations"
 allocate(kv(kdiag(neq)),loads(0:neq),bdylds(0:neq),oldis(0:neq),totd(0:neq), &
          load_store(0:neq))
           kv=0.0; oldis=0.0; totd=0.0
  call sample(element,points,weights) ;  pi = acos( -1. )
  dt = 100.
  do iel = 1 , nels
   e = prop(1,etype(iel)); v = prop(2,etype(iel))
   snph = sin(prop(3,etype(iel))*pi/180.)
   ddt=4.*(1.+v)*(1.-2.*v)/(e*(1.-2.*v+snph*snph))
   if(ddt<dt)dt=ddt
```

```
  end do
 write(11,'(a,e12.4)') "The critical timestep is ",dt
!---------- element stiffness integration and assembly & set stresses--------
 elements_2: do iel = 1 , nels
               num = g_num(: , iel) ; coord = transpose(g_coord(: , num ))
               g = g_g( : , iel )    ;  km=0.0   ; tensor = .0
            gauss_pts_1:  do i =1 , nip
              tensor(1:3 , i , iel) = prop( 6 , etype(iel) )
              e=prop(1,etype(iel)); v=prop(2,etype(iel)); call deemat(dee,e,v)
              call shape_der (der,points,i);  jac = matmul(der,coord)
              det = determinant(jac)  ;   call invert(jac)
              deriv = matmul(jac,der) ;   call beemat (bee,deriv)
              km = km + matmul(matmul(transpose(bee),dee),bee) *det* weights(i)
            end do gauss_pts_1
            call fsparv (kv,km,g,kdiag)
 end do elements_2
!------------ read prescribed loads/displacements and factorise l.h.s. -------
    read(10,*) loaded_nodes
    if(loaded_nodes/=0) then
     read(10,*)(k,loads(nf(:,k)),i=1,loaded_nodes); load_store = loads
    end if
    read(10,*) fixed_nodes
    if(fixed_nodes /=0) then
     allocate(node(fixed_nodes),sense(fixed_nodes),value(fixed_nodes),        &
              no(fixed_nodes),storkv(fixed_nodes))
     read(10,*) (node(i), sense(i), value(i),i=1,fixed_nodes)
     do i=1,fixed_nodes; no(i)=nf(sense(i),node(i)); end do
     kv(kdiag(no)) = kv(kdiag(no)) + 1.e20  ; storkv = kv(kdiag(no))
    end if
         call sparin (kv,kdiag)
!-----------------displacement increment loop-------------------------------
    load_increments: do iy=1,incs
    ptot = value(1) * iy
    write(11,'(a,i5)')"Load increment",iy ;  iters=0;  bdylds=.0;  evpt=.0
!-------------------------   iteration loop  -------------------------------
   iterations: do
    iters=iters+1; loads =.0
      if(loaded_nodes/=0) loads = load_store
      if(fixed_nodes/=0) loads(no) = storkv * value
      loads=loads+bdylds; call spabac(kv,loads,kdiag)
!-------------------------   check convergence -------------------------------
      call checon(loads,oldis,tol,converged)
      if(iters==1)converged=.false. ;  if(converged.or.iters==limit)bdylds=.0
!----------------------- go round the Gauss Points ---------------------------
      elements_3: do iel = 1 , nels
       bload=.0
       num = g_num( : , iel ) ; coord = transpose(g_coord( : , num ))
       g = g_g( : , iel )     ; eld = loads ( g )
       gauss_points_2 : do i = 1 , nip
          call shape_der ( der,points,i); jac=matmul(der,coord)
          det = determinant(jac)  ; call invert(jac)
          deriv = matmul(jac,der) ; call beemat (bee,deriv);eps=matmul(bee,eld)
          eps=eps-evpt(:,i,iel);  sigma=matmul(dee,eps)
          stress = sigma + tensor(:,i,iel)
          call invar(stress,sigm,dsbar,lode_theta)
!------------------- check whether yield is violated -------------------------
        phi = prop(3,etype(iel));c=prop(4,etype(iel)); psi=prop(5,etype(iel))
        call mocouf (phi, c , sigm, dsbar , lode_theta , f )
        if(converged.or.iters==limit) then
        devp=stress
          else
          if(f>=.0) then
          call mocouq(psi,dsbar,lode_theta,dq1,dq2,dq3)
          call formm(stress,m1,m2,m3)
```

```
           flow=f*(m1*dq1+m2*dq2+m3*dq3)      ;   erate=matmul(flow,stress)
           evp=erate*dt; evpt(:,i,iel)=evpt(:,i,iel)+evp; devp=matmul(dee,evp)
         end if; end if
      if(f>=.0) then
        eload=matmul(devp,bee) ; bload=bload+eload*det*weights(i)
      end if
    if(converged.or.iters==limit) then
!----------------------- update the Gauss point stresses ----------------------
                tensor ( : , i , iel) = stress
    end if
    end do gauss_points_2
!----------------------- compute the total bodyloads vector -------------------
    bdylds( g ) = bdylds( g ) + bload      ; bdylds(0) = .0
  end do elements_3
  if(converged.or.iters==limit)exit
 end do iterations
 totd = totd + loads
 write(11,'(a,e12.4)') "The displacement is ",totd(1)
 write(11,'(a)')"   The  stresses  are  "
 write(11,'(6e12.4)') tensor(: , 1 , 1)
 write(11,'(a,i5,a)') "It took",iters,"  iterations to converge"
 if(iters==limit)stop
end do load_increments
end program p68
```

Mesh and	ELEMENT	NELS	NN	NIP	NODOF	NOD	NST	NDIM
element data	'HEXAHEDRON'	1	20	27	3	20	6	3

Property	INCS	TOL	LIMIT	NPROPS	NP_TYPES
data	10	0.0001	50	6	1
	PROP				
	100. .3 30. 0. 0. -20.				

Geometry	G_COORD		
	0.	0.	0.
	.5	0.	0.
	1.	0.	0.
	0.	0.	-.5
	1.	0.	-.5
	0.	0.	-1.
	.5	0.	-1.
	1.	0.	-1.
	0.	.5	0.
	1.	.5	0.
	0.	.5	-1.
	1.	.5	-1.
	0.	1.	0.
	.5	1.	0.
	1.	1.	0.
	0.	1.	-.5
	1.	1.	-.5
	0.	1.	-1.
	.5	1.	-1.
	1.	1.	-1.

Figure 6.33 *(continued)*

```
Connectivity  G_NUM
              6 4 1 2 3 5 8 7 11 9 10 12 18 16 13 14 15 17 20 19
Node          NR
freedom data  16
              K,NF(:K,)I=1,NR
              1 0 0 1 2 1 0 1 3 1 0 1 4 0 0 1 5 1 0 1
              6 0 0 0 7 1 0 0 8 1 0 0 9 0 1 1 11 0 1 0
              12 1 1 0 13 0 1 1 16 0 1 1 18 0 1 0 19 1 1 0 20 1 1 0
Loading       LOADED NODES
data          0
              FIXED_NODES
              8
              NODE   SENSE  VALUE
              1      3      -.05
              2      3      -.05
              3      3      -.05
              9      3      -.05
              10     3      -.05
              13     3      -.05
              14     3      -.05
              15     3      -.05
```

Figure 6.33 Data for Program 6.8

was done for elasticity analyses in Program 5.9. The expectation would be that global geometry G_COORD and global connectivity G_NUM would be provided by a pre-processing mesh generator.

Program 6.8 (visco-plastic method) illustrates this feature, and uses a skyline storage assembly strategy.

Figure 6.33 shows input to Program 6.8 for the problem previously analysed by Program 6.7 and Figure 6.34 lists the output which is effectively identical to Figure 6.32.

6.12 THE GEOTECHNICAL PROCESSES OF EMBANKING AND EXCAVATION

One of the main features of analyses of geotechnical problems is the need to model construction processes. Gravity is one of the main agencies causing deformations and it used to be common to employ “gravity turn-on” as the loading mechanism in embankments for example. The final geometry of the embankment was modelled by a finite element mesh as usual and gravity was then gradually applied up to or beyond its terrestrial value (see for example Program 6.2). Alternatively, or simultaneously, the soil’s strength parameters could be reduced.

```
Global coordinates
Node    1        0.0000E+00  0.0000E+00  0.0000E+00
Node    2        0.5000E+00  0.0000E+00  0.0000E+00
Node    3        0.1000E+01  0.0000E+00  0.0000E+00
Node    4        0.0000E+00  0.0000E+00 -0.5000E+00
Node    5        0.1000E+01  0.0000E+00 -0.5000E+00
..................................................................

Node   15        0.1000E+01  0.1000E+01  0.0000E+00
Node   16        0.0000E+00  0.1000E+01 -0.5000E+00
Node   17        0.1000E+01  0.1000E+01 -0.5000E+00
Node   18        0.0000E+00  0.1000E+01 -0.1000E+01
Node   19        0.5000E+00  0.1000E+01 -0.1000E+01
Node   20        0.1000E+01  0.1000E+01 -0.1000E+01
Global node numbers
Element     1     6  4  1  2  3  5  8  7 11  9 10 12 18 16 13 14 15 17 20 19
The skyline storage is 666 and there are   36 equations
The critical timestep is   0.3200E-01
Load increment    1
The displacement is  -0.5000E-01
   The  stresses  are
 -0.2000E+02 -0.2000E+02 -0.2500E+02  0.1068E-13 -0.2135E-13 -0.3416E-13
It took    2  iterations to converge
Load increment    2
The displacement is  -0.1000E+00
   The  stresses  are
 -0.2000E+02 -0.2000E+02 -0.3000E+02  0.2135E-13 -0.4270E-13 -0.6832E-13
It took    2  iterations to converge
Load increment    3
The displacement is  -0.1500E+00
   The  stresses  are
 -0.2000E+02 -0.2000E+02 -0.3500E+02  0.3203E-13 -0.6405E-13 -0.1025E-12
It took    2  iterations to converge
Load increment    4
The displacement is  -0.2000E+00
   The  stresses  are
 -0.2000E+02 -0.2000E+02 -0.4000E+02  0.4270E-13 -0.8540E-13 -0.1366E-12
It took    2  iterations to converge
..................................................................

Load increment    8
The displacement is  -0.4000E+00
   The  stresses  are
 -0.2000E+02 -0.2000E+02 -0.6000E+02 -0.7757E-08 -0.1307E-07 -0.6972E-08
It took    2  iterations to converge
Load increment    9
The displacement is  -0.4500E+00
   The  stresses  are
 -0.2000E+02 -0.2000E+02 -0.6000E+02 -0.4022E-07 -0.1409E-07 -0.2354E-07
It took   16  iterations to converge
Load increment   10
The displacement is  -0.5000E+00
   The  stresses  are
 -0.2000E+02 -0.2000E+02 -0.6000E+02 -0.8898E-08 -0.5979E-08  0.6053E-09
It took   16  iterations to converge
```

Figure 6.34 Results from Program 6.8

Although it has been known for a long time (Smith and Hobbs, 1974) that this method can capture some of the realities of a construction process which takes place piece by piece (in layers for example) it is far more appropriate to be able to build up a mesh in stages using the true value of gravity at each stage.

PROGRAM 6.9: PLANE STRAIN OF AN ELASTIC–PLASTIC (MOHR–COULOMB) SOLID USING EIGHT-NODE QUADRILATERAL ELEMENTS : VISCO-PLASTIC STRAIN METHOD: CONSTRUCTION OF AN EMBANKMENT IN LAYERS ON A FOUNDATION

Program 6.9 illustrates such an analysis. Figure 6.35 shows a typical staged construction of an embankment (plane strain) constructed on a rectangular foundation block of soil. The embankment is assumed to be raised in a series of "lifts" the first of which merely stresses the foundation block gravitationally. The "soil" is assumed to be an elastic–plastic (Mohr–Coulomb) material and a visco-plastic strain algorithm is used, so the program can be considered to be a development of Program 6.2. Figure 6.36 shows the final mesh in more detail with the input data for the problem.

Alternatives, controlled by `ITYPE` as shown in the figure, allow quadrilaterals to be degenerated into triangles (on the face of the embankment slope) by two different methods.

Quantities not encountered in previous programs are:

`EF VF CF PHIF PSIF GAMAF`	foundation soil properties: Young's modulus, Poisson's ratio, cohesion, angle of friction, dilation angle, weight
`ES VS CS PHIS PSIS GAMAS`	embankment soil properties
`FNXE, FNYE`	number of foundation elements in x, y directions
`LNXE, LNYE`	number of embankment elements in x, y directions
`TOTELE`	total number of elements
`LIFTS`	Number of "lifts" with foundation counted as 1
`FWIDTH` (:)	positions of vertical foundation mesh lines
`FDEPTH` (:)	positions of horizontal foundation mesh lines
`WIDTH` (:)	positions of vertical embankment mesh lines
`DEPTH` (:)	positions of horizontal embankment mesh lines
`LNF` (:,:)	local node freedom array

Because the mesh is updated at every "lift" there is a need for a "local" node freedom array `LNF` which is found from the final `NF` at every stage. `LNF` is allocated and deallocated at each lift. Subroutines `FMGLEM` and `FMCOEM` set up the global node numbers and element nodal coordinates `G_NUM` and `G_COORD` respectively. Then, for each lift the geometry and connectivity can be calculated and hence the number of equations, `NEQ` and half-bandwidth, `NBAND`, operating at that stage of the construction process. Stiffness matrix and load vectors can then be set by an `ALLOCATE` statement and the visco-plastic algorithm initiated. The gravity loads are applied in `INCS` increments (five in this case).

The results are shown as Figure 6.37.

```
    program p69
!-----------------------------------------------------------------------
!       program 6.9 plane strain of an elastic-plastic(Mohr-Coulomb) solid
!       using 8-node quadrilateral elements; viscoplastic strain method
!       construction of an embankment in layers on a foundation
!-----------------------------------------------------------------------
 use new_library      ; use geometry_lib   ; implicit none
 integer::nels,lnxe,lnye,neq,nband,nn,nr,nip,nodof=2,nod=8,nst=4,ndof,       &
          i,k,iel,iters,limit,incs,iy,ndim=2,fnxe,fnye,lifts,oldele,newele,  &
          oldnn,lnn,ii,itype
 logical::converged   ; character (len=15) :: element = 'quadrilateral'
 real::ef,es,vf,vs,det,phif,phis,cf,cs,psif,psis,gamaf,gamas,dt,f,dsbar,     &
       dq1,dq2,dq3,lode_theta,sigm,pi,snph,e,v,c,phi,psi,gama,epk0,tol
!--------------------------- dynamic arrays-------------------------------------
 real    ,allocatable :: kb(:,:),loads(:),points(:,:),bdylds(:),              &
                         evpt(:,:,:),oldis(:),width(:),depth(:),gravlo(:),    &
                         dee(:,:),coord(:,:),fun(:),jac(:,:),weights(:),      &
                         der(:,:),deriv(:,:),bee(:,:),km(:,:),eld(:),eps(:),  &
                         sigma(:),bload(:),eload(:),erate(:),g_coord(:,:),    &
                         evp(:),devp(:),m1(:,:),m2(:,:),m3(:,:),flow(:,:),    &
                         totd(:),fwidth(:),fdepth(:),tensor(:,:,:),gc(:),s(:)
 integer, allocatable :: nf(:,:) , g(:),  num(:), g_num(:,:) ,g_g(:,:),       &
                         prop(:) , lnf(:,:)
!------------------------input and initialisation------------------------------
  open (10,file='p69.dat',status=    'old',action='read')
  open (11,file='p69.res',status='replace',action='write')
  read (10,*) fnxe,fnye,nn,nip,incs,limit,tol,                                &
              lifts,lnxe,lnye,itype,epk0,                                     &
              ef,vf,cf,phif,psif,gamaf,                                       &
              es,vs,cs,phis,psis,gamas
  ndof=nod*nodof   ; pi = acos( -1. )
!------------------ calculate the total number of elements --------------------
   k=0;do i=1,lnye-1;k=i+k;end do ; nels=fnxe*fnye+(lnxe*lnye-k)
        write(11,'(a,i5)') "The total number of elements is ",nels
  allocate (nf(nodof,nn), points(nip,ndim),weights(nip),g_coord(ndim,nn),     &
            depth(lnye+1),num(nod),dee(nst,nst),evpt(nst,nip,nels),           &
            width(lnxe+1),coord(nod,ndim),fun(nod),prop(nels),g_g(ndof,nels), &
            jac(ndim,ndim),der(ndim,nod),deriv(ndim,nod),g_num(nod,nels),     &
            bee(nst,ndof),km(ndof,ndof),eld(ndof),eps(nst),sigma(nst),        &
            bload(ndof),eload(ndof),erate(nst),evp(nst),devp(nst),g(ndof),    &
            m1(nst,nst),m2(nst,nst),m3(nst,nst),flow(nst,nst),s(nst),         &
            fwidth(fnxe+1),fdepth(fnye+1),gc(ndim),tensor(nst,nip,nels))
  nf=1; read(10,*) nr ; if(nr>0) read(10,*)(k,nf(:,k),i=1,nr)
  call formnf(nf); neq=maxval(nf)
        write(11,'(a,i5)') "The final number of equations is:",neq
  allocate(totd(0:neq))
  read(10,*) fwidth , fdepth , width , depth
!------------------------- set the element type ------------------------------
   prop(1:fnxe*fnye)=1 ; prop(fnxe*fnye+1:nels)=2
!----------- set up the global node numbers and element nodal coordinates -----
   call fmglem(fnxe,fnye,lnxe,1,g_num,lifts)
   call fmcoem(g_num,g_coord,fwidth,fdepth,width,depth,         &
               lnxe,lifts,fnxe,fnye,itype)
    write(11,'(a)') "Global coordinates "
    do k=1,nn;write(11,'(a,i5,a,2e12.4)')"Node",k,"        ",g_coord(:,k);end do
    write(11,'(a)') "Global node numbers "
    do k = 1 , nels; write(11,'(a,i5,a,8i5)')                                  &
                             "Element ",k,"        ",g_num(:,k); end do
    tensor = .0; totd = .0; call sample(element,points,weights)
!------------- loop the elements to find the global g -------------------------
       elements_1:  do iel = 1 , nels    ; num = g_num(:,iel)
                       call num_to_g (num,nf,g) ;   g_g(:,iel) = g
       end do elements_1
! ------------------ construct another lift --------------------------------
```

```
  lift_number : do ii = 1 , lifts
! ------------      calculate how many elements there are -------------------
    if (ii<=lifts) then
        if(ii==1) then
           newele=fnxe*fnye; oldele = newele
        else
           newele = lnxe - (ii -2); oldele = oldele + newele
        end if
!--------- go round the elements and get nband from the g vectors ------------
       nband = 0
       elements_2 : do iel = 1 , oldele
                      g=g_g( : , iel )
                      if(nband<bandwidth(g)) nband = bandwidth( g )
       end do elements_2
! -------------      calculate how many nodes there are    --------------------
        if(ii==1) then
           lnn=(fnxe*2+1)*(fnye+1)+(fnxe+1)*fnye ; oldnn = lnn
        end if
        if(ii>1) then
           lnn=oldnn+(lnxe-(ii-2))*2+1+(lnxe-(ii-2)+1) ; oldnn = lnn
        end if
!------------------- now get the new node freedom array -----------------------
       allocate(lnf(nodof,lnn))     ; lnf = nf(:,1:lnn)
!----------------- recalculate the number of freedoms neq --------------------
          neq = maxval(lnf)
     write(11,'(/,3(a,i5))')                                                 &
             "There are",neq," freedoms and",lnn," nodes in lift",ii
     write(11,'(a,i5,a,i5,a)')                                               &
             "There are ",oldele," elements and",newele," were added"
    end if
allocate(kb(neq,nband+1),loads(0:neq),bdylds(0:neq),oldis(0:neq),gravlo(0:neq))
           kb=0.0;  gravlo=0.0  ; loads = .0
!----------------- element stiffness integration and assembly-----------------
 elements_3: do iel = 1 , oldele
             if(prop(iel)==1)then
                gama = gamaf; e = ef ; v = vf
             else
                gama = gamas; e = es ; v = vs
             end if
             if(iel<=(oldele-newele)) gama = .0
               num = g_num(: , iel) ; coord = transpose(g_coord(:,num ))
               g = g_g(:,iel);  km=0.0   ;  call deemat(dee,e,v); eld = .0
            gauss_pts_1:  do i =1 , nip
              call shape_fun(fun,points,i)  ;  gc = matmul ( fun , coord )
!-------------------- initial stress in foundation ---------------------------
             if(ii==1) then
              tensor(2,i,iel)=-1.*(fdepth(fnye+1)-gc(2))*gama
              tensor(1,i,iel)=epk0*tensor(2,i,iel)
              tensor(4,i,iel)=tensor(1,i,iel);tensor(3,i,iel)=.0
             end if
              call shape_der (der,points,i);  jac = matmul(der,coord)
              det = determinant(jac)  ;   call invert(jac)
              deriv = matmul(jac,der) ;   call beemat (bee,deriv)
              km = km + matmul(matmul(transpose(bee),dee),bee) *det* weights(i)
              do k=2,ndof,2;eld(k)=eld(k)+fun(k/2)*det*weights(i);end do
            end do gauss_pts_1
   call formkb (kb,km,g)
   if(ii<=lifts) gravlo ( g ) = gravlo ( g ) - eld * gama ; gravlo(0) = .0
 end do elements_3
!------------------------ factorise equations---------------------------------
         call cholin(kb)
!------------------ factor gravlo by incs------------------------------------
    write(11,'(a,i5,a,e12.4)')                                               &
            "The total gravity load in lift", ii, " is" , sum(gravlo)
```

```
       gravlo = gravlo / incs
!-------------------- apply gravity loads incrementally -----------------------
    load_increments: do iy=1,incs
  write(11,'(a,i5)')                                                          &
         "Increment",iy ; iters=0; oldis =.0; bdylds=.0; evpt(:,:,1:oldele)=.0
!-------------------------   iteration loop  ---------------------------------
   iterations: do
    iters=iters+1;  loads = .0; loads= gravlo+bdylds ;  call chobac(kb,loads)
!------------------------   check convergence --------------------------------
      call checon(loads,oldis,tol,converged)
      if(iters==1)converged=.false. ;  if(converged.or.iters==limit)bdylds=.0
!---------------------- go round the Gauss Points ----------------------------
      elements_4: do iel = 1 , oldele
                  if(prop(iel)==1)then
                     phi = phif; c = cf;  e = ef ; v = vf; psi = psif
                  else
                     phi = phis; c = cs;  e = es;  v = vs; psi = psis
                  end if
             snph=sin(phi*pi/180.);dt=4.*(1.+v)*(1.-2.*v)/(e*(1.-2.*v+snph**2))
             call deemat(dee,e,v);     bload=.0
             num = g_num( : , iel ) ; coord = transpose(g_coord( : , num ))
             g = g_g( : , iel )      ; eld = loads ( g )
       gauss_points_2 : do i = 1 , nip
          call shape_der ( der,points,i); jac=matmul(der,coord)
          det = determinant(jac)  ;   call invert(jac)
          deriv = matmul(jac,der) ; call beemat (bee,deriv);eps=matmul(bee,eld)
          eps = eps -evpt( : , i , iel)    ;        sigma=matmul(dee,eps)
          if(ii==1)then;s=tensor(:,i,iel);else;s=tensor(:,i,iel)+sigma;end if
          call invar(s,sigm,dsbar,lode_theta)
!------------------  check whether yield is violated -------------------------
         call mocouf (phi, c , sigm, dsbar , lode_theta , f )
         if(converged.or.iters==limit) then
         devp=s
           else
           if(f>=.0) then
           call mocouq(psi,dsbar,lode_theta,dq1,dq2,dq3);call formm(s,m1,m2,m3)
           flow=f*(m1*dq1+m2*dq2+m3*dq3)     ;   erate=matmul(flow,s)
           evp=erate*dt; evpt(:,i,iel)=evpt(:,i,iel)+evp; devp=matmul(dee,evp)
         end if; end if
      if(f>=.0) then
        eload=matmul(devp,bee) ; bload=bload+eload*det*weights(i)
      end if
!---------------- if appropriate update the Gauss point stresses -------------
      if(converged.or.iters==limit) then
        if(ii/=1) tensor(:,i,iel) = s
      end if
    end do gauss_points_2
!---------------- compute the total bodyloads vector -------------------------
    bdylds( g ) = bdylds( g ) + bload      ; bdylds(0) = .0
  end do elements_4
  if(converged.or.iters==limit)exit
 end do iterations
 if(ii/=1) totd(:neq) = totd(:neq) + loads(:neq)
 write(11,'(2(a,i5))') "Lift number",ii," gravity load increment",iy
 write(11,'(a,i5,a)') "It took ",iters, " iterations to converge"
 if(iy==incs.or.iters==limit)write(11,'(a,e12.4)')                            &
                                    "Max displacement is",maxval(abs(loads))
 if(iters==limit)stop
end do load_increments
deallocate(lnf,kb,loads,bdylds,oldis,gravlo)
end do lift_number
end program p69
```

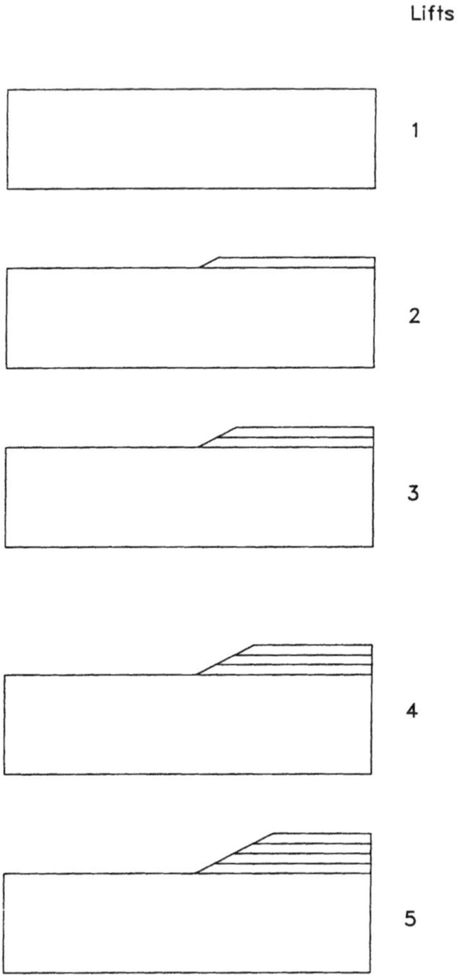

Figure 6.35 Staged construction of an embankment

Results for this kind of analysis are summarised in Figures 6.38(a) and (b). In the first of these, embankments with a slope angle of 26.5°, a soil weight of 20 kN/m^3 and a foundation depth of 10m are constructed assuming a frictionless "soil" with various c_u values. Limit equilibrium (slip circle) analyses indicate that this slope is just stable at $c_u \simeq 14\,\text{kN/m}^2$. In the finite element analyses the displacement of the toe of the slope accelerates as c_u is reduced from 16 to 14 kN/m^2 indicating failure. In Figure 6.38(b) the progress of toe displacement as the embankment is raised with $c_u = 14\,\text{kN/m}^2$ is shown. Failure is clearly occurring at the final height of 4 m.

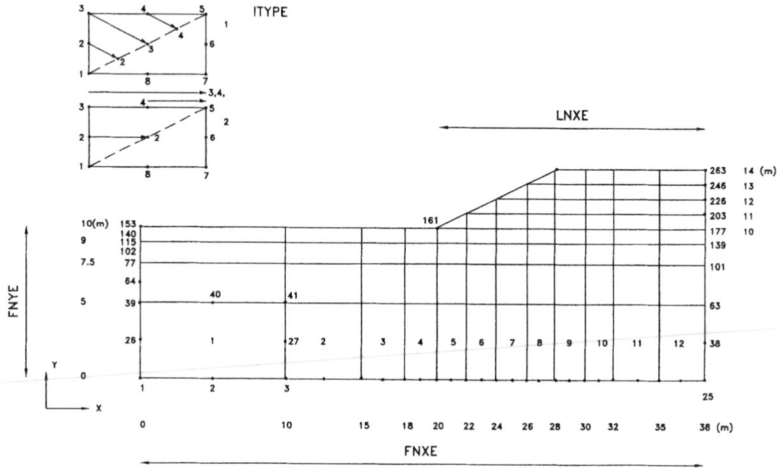

Geometry

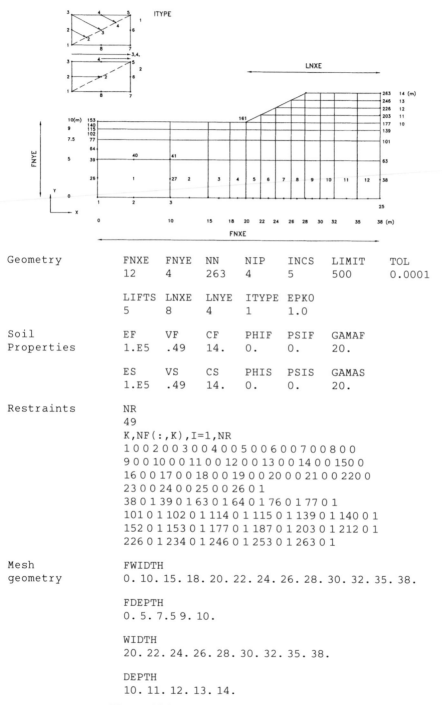

FNXE	FNYE	NN	NIP	INCS	LIMIT	TOL
12	4	263	4	5	500	0.0001

LIFTS	LNXE	LNYE	ITYPE	EPKO
5	8	4	1	1.0

Soil Properties

EF	VF	CF	PHIF	PSIF	GAMAF
1.E5	.49	14.	0.	0.	20.

ES	VS	CS	PHIS	PSIS	GAMAS
1.E5	.49	14.	0.	0.	20.

Restraints

NR
49
K,NF(:,K),I=1,NR
1 0 0 2 0 0 3 0 0 4 0 0 5 0 0 6 0 0 7 0 0 8 0 0
9 0 0 10 0 0 11 0 0 12 0 0 13 0 0 14 0 0 15 0 0
16 0 0 17 0 0 18 0 0 19 0 0 20 0 0 21 0 0 22 0 0
23 0 0 24 0 0 25 0 0 26 0 1
38 0 1 39 0 1 63 0 1 64 0 1 76 0 1 77 0 1
101 0 1 102 0 1 114 0 1 115 0 1 139 0 1 140 0 1
152 0 1 153 0 1 177 0 1 187 0 1 203 0 1 212 0 1
226 0 1 234 0 1 246 0 1 253 0 1 263 0 1

Mesh geometry

FWIDTH
0. 10. 15. 18. 20. 22. 24. 26. 28. 30. 32. 35. 38.

FDEPTH
0. 5. 7.5 9. 10.

WIDTH
20. 22. 24. 26. 28. 30. 32. 35. 38.

DEPTH
10. 11. 12. 13. 14.

Figure 6.36 Mesh and data for Program 6.9

```
The total number of elements is    74
The final number of equations is:  452
Global coordinates
Node    1          0.0000E+00  0.0000E+00
Node    2          0.5000E+01  0.0000E+00
Node    3          0.1000E+02  0.0000E+00
Node    4          0.1250E+02  0.0000E+00
Node    5          0.1500E+02  0.0000E+00
.................................................................

Node  259          0.3200E+02  0.1400E+02
Node  260          0.3350E+02  0.1400E+02
Node  261          0.3500E+02  0.1400E+02
Node  262          0.3650E+02  0.1400E+02
Node  263          0.3800E+02  0.1400E+02
Global node numbers
Element     1            1   26   39   40   41   27    3    2
Element     2            3   27   41   42   43   28    5    4
Element     3            5   28   43   44   45   29    7    6
Element     4            7   29   45   46   47   30    9    8
Element     5            9   30   47   48   49   31   11   10
.................................................................

Element    70          236  247  253  254  255  248  238  237
Element    71          238  248  255  256  257  249  240  239
Element    72          240  249  257  258  259  250  242  241
Element    73          242  250  259  260  261  251  244  243
Element    74          244  251  261  262  263  252  246  245

There are  288 freedoms and  177 nodes in lift    1
There are    48 elements and   48 were added
The total gravity load in lift    1 is -0.6967E+04
Increment    1
Lift number    1 gravity load increment    1
It took    2 iterations to converge
Increment    2
Lift number    1 gravity load increment    2
It took    2 iterations to converge
Increment    3
Lift number    1 gravity load increment    3
It took    2 iterations to converge
Increment    4
Lift number    1 gravity load increment    4
It took    2 iterations to converge
Increment    5
Lift number    1 gravity load increment    5
It took    2 iterations to converge
Max displacement is  0.1169E-03
.................................................................

There are  452 freedoms and  263 nodes in lift    5
There are    74 elements and    5 were added
The total gravity load in lift    5 is -0.2200E+03
Increment    1
Lift number    5 gravity load increment    1
It took   18 iterations to converge
Increment    2
Lift number    5 gravity load increment    2
It took   25 iterations to converge
Increment    3
Lift number    5 gravity load increment    3
It took   42 iterations to converge
Increment    4
Lift number    5 gravity load increment    4
It took   77 iterations to converge
Increment    5
Lift number    5 gravity load increment    5
It took  500 iterations to converge
Max displacement is  0.3451E-01
```

Figure 6.37 Output from Program 6.9

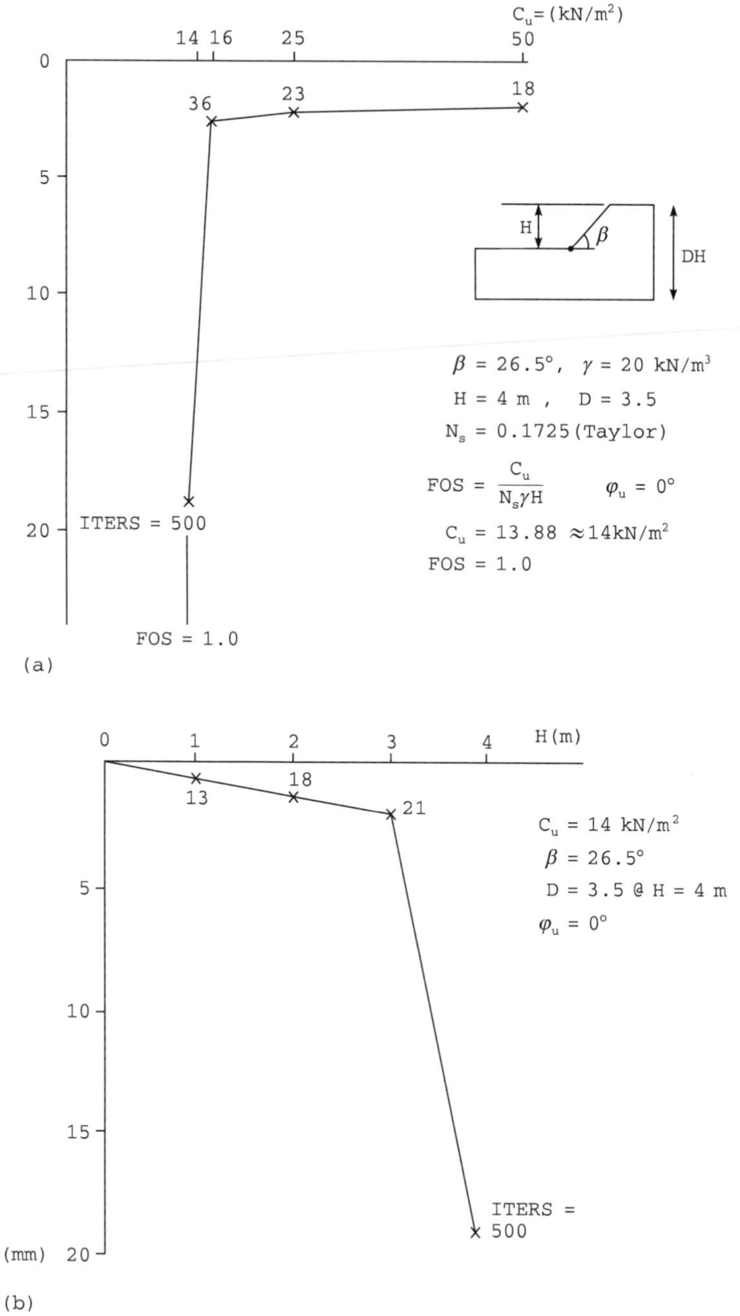

Figure 6.38 (a) Constant embankment height constructed, varying soil properties. (b) Constant soil properties, embankment constructed to 4 m

6.12.1 EXCAVATION

The second important geotechnical construction process involving change of geometry occurs when material is removed from the ground, either in open excavations ("cuts") or in enclosed tunnels. The ground is stressed prior to removal of part of it and this starting stress state may be difficult to infer from the known history.

The aim in an analysis is that, when a portion of material is excavated, and forces are applied along the excavated surface, the remaining material should experience the correct stress relief so that the new "free surface" is indeed stress-free. Suppose body A is to be removed from body B as shown in Figure 6.39. The stresses to begin with are σ_{AO} and σ_{BO}, respectively. Any external loads are taken into consideration in forming these stresses prior to the removal of A. Since both bodies are in equilibrium, forces F_{AB} must be applied to body B due to body A to maintain σ_{BO} and, similarly, F_{BA} must act on body A. Forces F_{AB} and F_{BA} are equal in magnitude and opposite in sign. In general, therefore, the excavation forces acting on a boundary depend on the stress state in the excavated material and on the self-weight of that material. It can be shown that

$$F_{\mathrm{BA}} = -\int_{V_{\mathrm{A}}} \mathbf{B}^{\mathrm{T}} \sigma_{\mathrm{AO}} \, \mathrm{d}V_{\mathrm{A}} + \int_{V_{\mathrm{A}}} \mathbf{N}^{\mathrm{T}} \gamma \, \mathrm{d}V_{\mathrm{A}} \tag{6.84}$$

where $\mathbf{B}$ is the strain–displacement matrix, V_{A} the excavated volume, $\mathbf{N}$ the element shape functions and γ the soil self-weight.

Single-stage and multi-stage excavations give the same results, and in a 1-d situation a stress-free excavated surface results, and the expected stresses are

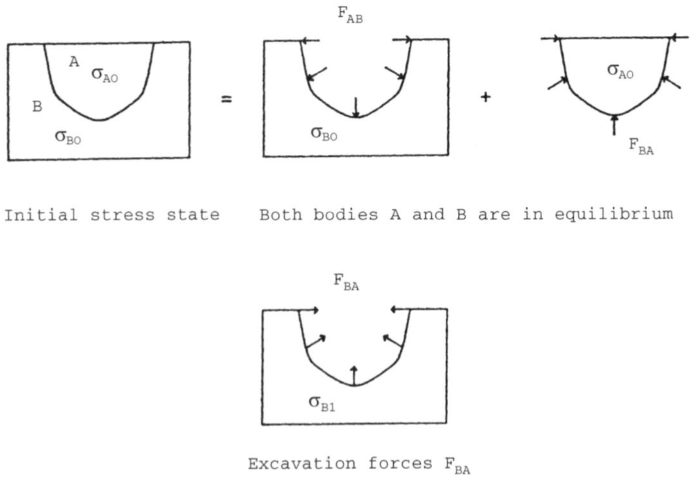

Figure 6.39 Excavation: force formulation

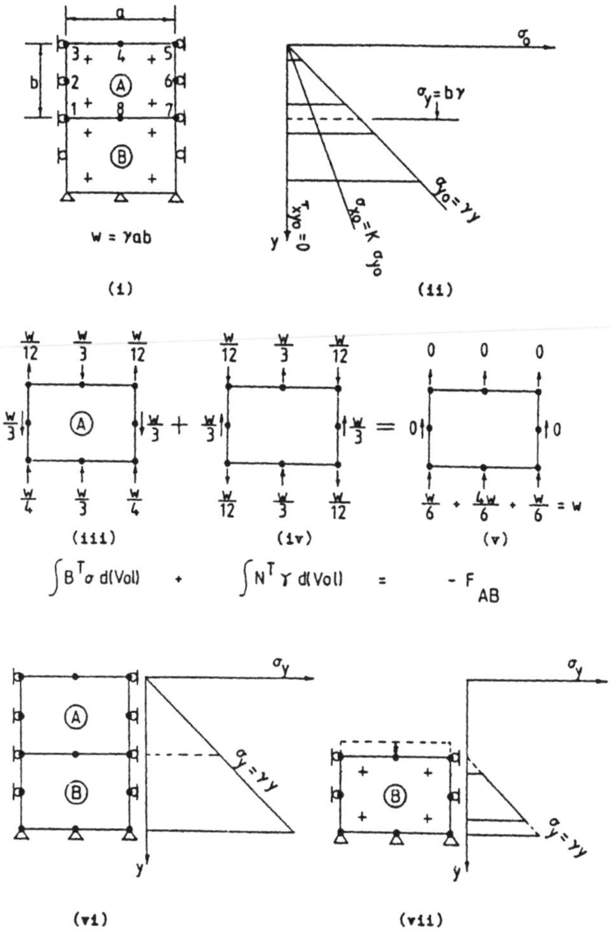

Figure 6.40 One-dimensional excavation

found in the remaining material, as shown in Figure 6.40 for eight-node elastic elements with stress evaluation at 2×2 Gauss points, w being the weight of element A.

When the situation is 2-d or 3-d, equation (6.84) applies again, but in the corners of excavations a rather complex stress concentration exists. This means that the finite element results will be mesh-dependent (Smith and Ho, 1992).

A program illustrating the analysis of excavation processes is listed as Program 6.10. As with the previous program, it can be derived from Program 6.2.

PROGRAM 6.10: PLANE STRAIN OF AN ELASTIC–PLASTIC (MOHR–COULOMB) SOLID USING EIGHT-NODE QUADRILATERAL ELEMENTS : VISCO-PLASTIC STRAIN METHOD: CONSTRUCTION OF AN EXCAVATION IN STAGES INTRODUCING AIR ELEMENTS

```
  program p610
!-------------------------------------------------------------------------
!     program 6.10 plane strain of an elastic-plastic(Mohr-Coulomb) solid
!     using 8-node quadrilateral elements; viscoplastic strain method
!     construction of an excavation in stages introducing air elements
!-------------------------------------------------------------------------
 use new_library      ; use geometry_lib   ;     implicit none
 integer::nels,nxe,nye,neq,nband,nn,nr,nip,nodof=2,nod=8,nst=4,ndof,       &
         i,k,iel,iters,limit,incs,iy,ndim=2,layers,ii,iq,nofix,noexe
 logical::converged          ; character (len=15) :: element='quadrilateral'
 real::es,ea,det,phi,psi,c,gama,                                            &
      dt,f,dsbar,dq1,dq2,dq3,lode_theta,sigm,pi,snph,e,v,epk0,tol
!------------------------------ dynamic arrays---------------------------------
 real    ,allocatable :: kb(:,:),loads(:),points(:,:),bdylds(:),disps(:,:),   &
                         evpt(:,:,:),oldis(:),width(:),depth(:),dloads(:),    &
                         dee(:,:),coord(:,:),fun(:),jac(:,:),weights(:),      &
                         der(:,:),deriv(:,:),bee(:,:),km(:,:),eld(:),eps(:),  &
                         sigma(:),bload(:),eload(:),erate(:),g_coord(:,:),    &
                         evp(:),devp(:),m1(:,:),m2(:,:),m3(:,:),flow(:,:),    &
                         totd(:),tensor(:,:,:),gc(:),s(:),lod(:,:)
 integer, allocatable :: nf(:,:) , g(:),  num(:), g_num(:,:) ,               &
                         prop(:) , lnf(:,:) , fixnod(:) , exele(:)
!------------------------input and initialisation------------------------------
  open (10,file='p610.dat',status=    'old',action='read')
  open (11,file='p610.res',status='replace',action='write')
  read (10,*) nxe,nye,nn,nip,incs,limit,tol,layers,epk0,                     &
             ea,es,v,c,phi,psi,gama
  ndof=nod*nodof    ; pi = acos( -1. )     ; snph = sin(phi*pi/180.)
 ! calculate the total number of elements
   nels = nxe*nye;   write(11,'(a,i5)')"The total number of elements is ",nels
  allocate (nf(nodof,nn), points(nip,ndim),weights(nip),g_coord(ndim,nn),     &
            depth(nye+1),num(nod),dee(nst,nst),evpt(nst,nip,nels),            &
            width(nxe+1),coord(nod,ndim),fun(nod),prop(nels),                  &
            jac(ndim,ndim),der(ndim,nod),deriv(ndim,nod),g_num(nod,nels),     &
            bee(nst,ndof),km(ndof,ndof),eld(ndof),eps(nst),sigma(nst),        &
            bload(ndof),eload(ndof),erate(nst),evp(nst),devp(nst),g(ndof),    &
            m1(nst,nst),m2(nst,nst),m3(nst,nst),flow(nst,nst),s(nst),         &
            disps(nodof,nn),gc(ndim),tensor(nst,nip,nels),lnf(nodof,nn))
!------ nf is an index array of 1s and 0s : lnf is the local nf  --------------
            nf=1; read(10,*) nr; if(nr>0) read(10,*)(k,nf(:,k),i=1,nr)
            lnf = nf;   call formnf(lnf); neq=maxval(lnf)
            write(11,'(a,i5)')"The total possible number of equations is:",neq
  allocate(totd(0:neq))          ;  read(10,*)  width , depth
!--------------------- set the element type -----------------------------------
   prop = 1
!-------- set up the global node numbers and global nodal coordinates ---------
   call fmglob(nxe,nye,g_num)
   call fmcoco(g_num,g_coord,width,depth,nxe,nye)
    write(11,'(a)') "Global coordinates "
    do k=1,nn;write(11,'(a,i5,a,2e12.4)')"Node",k,"       ",g_coord(:,k);end do
    write(11,'(a)') "Global node numbers "
    do k = 1 , nels; write(11,'(a,i5,a,8i5)')                                 &
                           "Element ",k,"        ",g_num(:,k); end do
    disps = .0;     call sample(element,points,weights)
!--------------- loop the elements to set starting stresses -------------------
       elements_1:   do iel = 1 , nels
                       num = g_num(:,iel); coord=transpose(g_coord(:,num))
              gauss_points_1: do i=1,nip
                       call shape_fun(fun,points,i); gc = matmul(fun,coord)
                       tensor(2,i,iel)=gc(2)*gama
                       tensor(1,i,iel)=epk0*tensor(2,i,iel)
                       tensor(4,i,iel)=tensor(1,i,iel) ;  tensor(3,i,iel)=.0
             end do gauss_points_1
       end do elements_1
```

```
! ----------------------- excavate another layer -----------------------------
  layer_number : do ii = 1 , layers ;write(11,'(a,i5)') "Layer no",ii
!--------- recalculate the number of freedoms neq and half-bandwidth nband ----
          read(10,*)nofix ;   allocate(fixnod(nofix))
             read(10,*)fixnod ; nf(:,fixnod) = 0 ; lnf = nf
             call formnf(lnf) ;    neq = maxval(lnf)
             nband = 0
             elements_1a : do iel = 1 , nels
                          num = g_num(:,iel);call num_to_g(num,lnf,g)
                          if(nband<bandwidth(g))nband = bandwidth(g)
             end do elements_1a
      write(11,'(/,3(a,i5))')                                                &
              "There are ",neq, " freedoms and nband is",nband," in step",ii
allocate(kb(neq,nband+1),loads(0:neq),bdylds(0:neq),oldis(0:neq),dloads(0:neq))
          kb=0.0;   loads = .0
!------------------- specify the elements to be removed ----------------------
    read(10,*)noexe  ; allocate(exele(noexe),lod(ndof,noexe))
    read(10,*)exele; prop(exele)=0
!---------------------- calculate excavation load ----------------------------
    s = .0;  lod = .0
    elements_2 : do iel = 1 , noexe
       iq = exele(iel) ; bload = .0; eld = .0 ;  num = g_num(:,iq)
       call num_to_g(num,lnf,g);coord = transpose(g_coord(:,num))
       gauss_points_2 : do i = 1 , nip
          call shape_fun(fun,points,i)
          call shape_der (der,points,i);  jac = matmul(der,coord)
          det = determinant(jac)  ;   call invert(jac)
          deriv = matmul(jac,der) ;   call beemat (bee,deriv)
          s = tensor(:,i,iq) ; eload = matmul(s,bee)
          bload = bload + eload * det * weights(i)
          do k=2,ndof,2;eld(k)=eld(k)+fun(k/2)*det*weights(i)*gama;end do
       end do gauss_points_2
          lod(:,iel) = eld + bload; loads(g)=loads(g)+lod(:,iel);loads(0)=.0
    end do elements_2
!------------------ element stiffness integration and assembly-----------------
 elements_3: do iel = 1 , nels
              if(prop(iel)==0) e = ea
              if(prop(iel)==1) e = es
              km = .0; eld = .0; call deemat(dee,e,v)
              num=g_num(:,iel);call num_to_g(num,lnf,g)
              coord = transpose(g_coord(: , num))
              gauss_points_3: do i = 1 , nip
               call shape_der (der,points,i);  jac = matmul(der,coord)
               det = determinant(jac)  ;   call invert(jac)
               deriv = matmul(jac,der) ;   call beemat (bee,deriv)
               km = km + matmul(matmul(transpose(bee),dee),bee) *det* weights(i)
              end do gauss_points_3
   call formkb (kb,km,g)
 end do elements_3
!--------------- factorise l.h.s. ----------------------------------------
          call cholin(kb)
!------------------ factor excavation load by incs----------------------------
     write(11,'(a,i5,a,e12.4)')                                              &
             "The total gravity load in lift", ii, " is" , sum(loads)
        loads = loads / incs
!------------ apply excavation loads incrementally ----------------------------
    load_increments: do iy=1,incs
              iters=0; oldis =.0; bdylds=.0; evpt=.0
!------------------------   iteration loop -----------------------------------
   iterations: do
    iters=iters+1;  dloads = .0; dloads= loads+bdylds ;  call chobac(kb,dloads)
!------------------------   check convergence ------------------------------
      call checon(dloads,oldis,tol,converged)
      if(iters==1)converged=.false.
```

```
      if(converged.or.iters==limit) then
        bdylds=.0
          do iq=1,nn;do i=1,nodof
           if(lnf(i,iq)/=0)disps(i,iq) = disps(i,iq) + dloads(lnf(i,iq))
          end do;end do
      end if
!------------------------ go round the Gauss Points --------------------------
 elements_4: do iel = 1 , nels
              if(prop(iel)==0) then;e = ea;dt=1.e10; end if
              if(prop(iel)==1) then
                e = es;dt=(4.*(1.+v)*(1.-2.*v))/(e*(1.-2.*v+snph*snph))
              end if
              bload = .0; call deemat(dee,e,v)
              num=g_num(:,iel);call num_to_g(num,lnf,g)
              coord = transpose(g_coord(: , num)) ;   eld = dloads(g)
              gauss_points_4: do i = 1 , nip
               call shape_der (der,points,i);  jac = matmul(der,coord)
               det = determinant(jac)  ;   call invert(jac)
               deriv = matmul(jac,der) ;  call beemat (bee,deriv)
               eps = matmul(bee,eld); eps=eps-evpt(:,i,iel)
               s = tensor(:,i,iel) + matmul(dee,eps)
!---------------------- air element stresses are zero ------------------------
               if(prop(iel)==0) s = .0
               call invar(s,sigm,dsbar,lode_theta)
!-------------------- check whether yield is violated ------------------------
         call mocouf (phi, c , sigm, dsbar , lode_theta , f )
         if(converged.or.iters==limit) then
         devp=s
           else
           if(f>=.0) then
           call mocouq(psi,dsbar,lode_theta,dq1,dq2,dq3);call formm(s,m1,m2,m3)
           flow=f*(m1*dq1+m2*dq2+m3*dq3)     ;   erate=matmul(flow,s)
           evp=erate*dt; evpt(:,i,iel)=evpt(:,i,iel)+evp; devp=matmul(dee,evp)
         end if; end if
      if(f>=.0) then
        eload=matmul(devp,bee)   ; bload=bload+eload*det*weights(i)
      end if
!--------------- if appropriate update the Gauss point stresses ---------------
      if(converged.or.iters==limit) tensor(:,i,iel) = s
    end do gauss_points_4
!-------------------   compute the total bodyloads vector ---------------------
    bdylds( g ) = bdylds( g ) + bload      ; bdylds(0) = .0
  end do elements_4
  if(converged.or.iters==limit)exit
 end do iterations
 write(11,'(2(a,i5))') "Lift number",ii," gravity load increment",iy
 write(11,'(a,i5,a)') "It took ",iters, "  iterations to converge"
 if(iy==incs.or.iters==limit)then
    write(11,'(a)') "The displacements are :"
    write(11,'(5e12.4)')disps(1,61),disps(2,61),disps(1,37),disps(2,37)
 end if
 if(iters==limit)stop
end do load_increments
deallocate(kb,loads,bdylds,oldis,dloads,fixnod,exele,lod)
end do layer_number
end program p610
```

Figure 6.41 shows a square block of "soil" in a state of plane strain. The upper right-hand corner of the block is to be removed ("excavated") and five possible sequences for reaching the final excavated geometry are shown as cases A to E. The specific data is for case B.

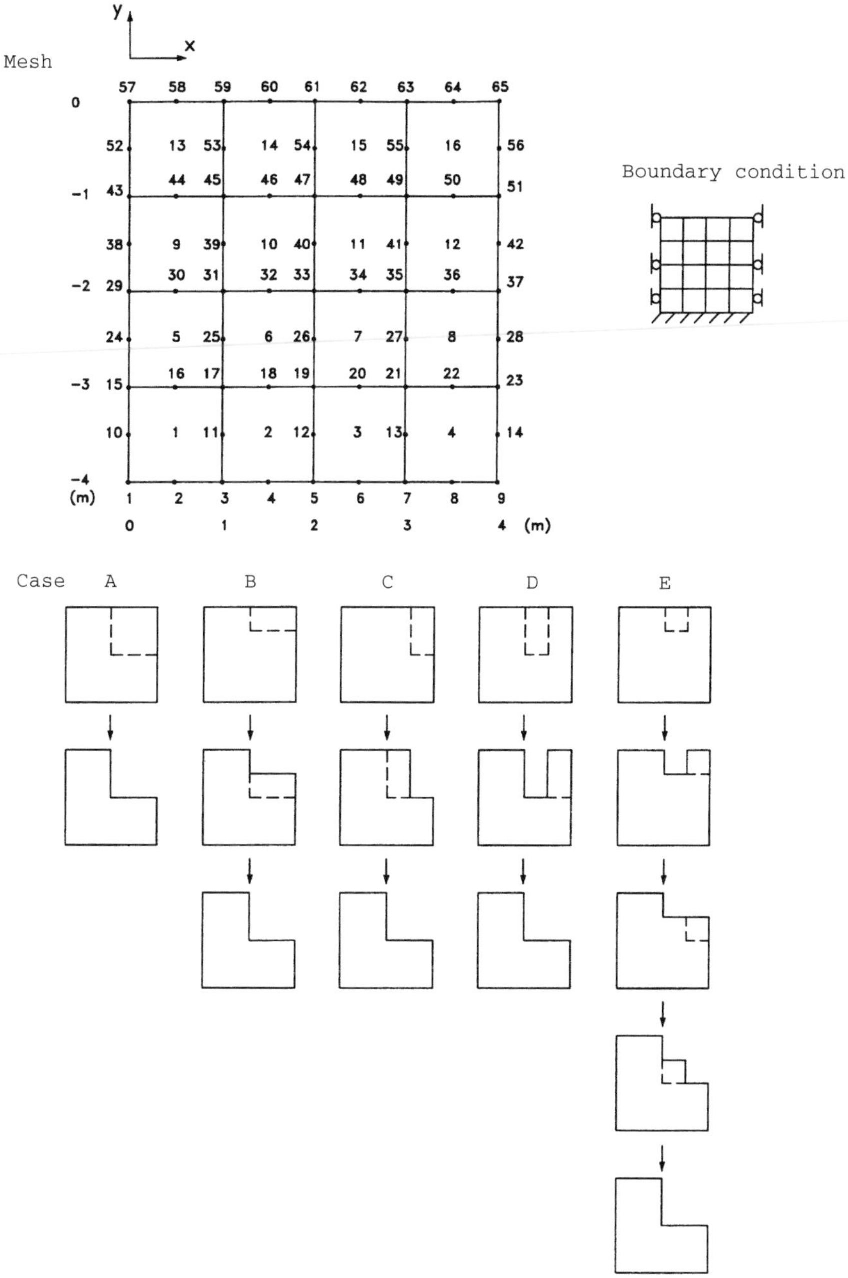

Figure 6.41 (*continued*)

```
Mesh          NXE    NYE    NN     NIP    INCS   LIMIT   TOL
              4      4      65     4      5      250     .0001

              LAYERSEPKO
              2      1.

Soil          EA     ES     V      C      PHI    PSI     GAMA
Properties    1.     1.E5   .49    10.    .0     .0      20.

Restraints    NR
              25
              K,NF(:,K),I=1,NR
              1 0 0 2 0 0 3 0 0 4 0 0 5 0 0 6 0 0 7 0 0
              8 0 0 9 0 0 10 0 1 14 0 1 15 0 1 23 0 1 24 0 1
              28 0 1 29 0 1 37 0 1 38 0 1 42 0 1 43 0 1 51 0 1
              52 0 1 56 0 1 57 0 1 65 0 1

Geometry      WIDTH
              0. 1. 2. 3. 4.

              DEPTH
              -4. -3. -2. -1. 0.

              For each excavation step "LAYERS"
              Number of nodes to be "fixed", and their values

              NOFIX
              6 (6)

              FIXNOD(:)
              55 56 62 63 64 65 (41 42 48 49 50 51)

              Number of elements to be excavated and their values

              NOEXE
              2 (2)
              EXELE(:)
              15 16 (11 12)

              (the data for the second of the two lifts are shown
                in parentheses)
```

Figure 6.41 Mesh and input data for Program 6.10

Variables not encountered in previous programs are:

`LOD(:)`	excavation loads
`FIXNOD(:)`	values of nodes to be "fixed" or removed in an excavation stage
`EXELE(:)`	values of elements to be removed in an excavation stage
`LAYERS`	number of stages of excavation

The same concept of a "local" node freedom array LNF as was used in the previous program is used again.

The geometry node numbers G_NUM and G_COORD of the starting geometry, before excavation commences, are formed by subroutines FMGLOB and FMCOCO

```
The total number of elements is     16
The total possible number of equations is:   96
Global coordinates
Node     1          0.0000E+00 -0.4000E+01
Node     2          0.5000E+00 -0.4000E+01
Node     3          0.1000E+01 -0.4000E+01
Node     4          0.1500E+01 -0.4000E+01
Node     5          0.2000E+01 -0.4000E+01
.............................................................................

Node    60          0.1500E+01  0.0000E+00
Node    61          0.2000E+01  0.0000E+00
Node    62          0.2500E+01  0.0000E+00
Node    63          0.3000E+01  0.0000E+00
Node    64          0.3500E+01  0.0000E+00
Node    65          0.4000E+01  0.0000E+00
Global node numbers
Element      1             1    10    15    16    17    11     3     2
Element      2             3    11    17    18    19    12     5     4
Element      3             5    12    19    20    21    13     7     6
Element      4             7    13    21    22    23    14     9     8
Element      5            15    24    29    30    31    25    17    16
Element      6            17    25    31    32    33    26    19    18
.............................................................................

Element     13            43    52    57    58    59    53    45    44
Element     14            45    53    59    60    61    54    47    46
Element     15            47    54    61    62    63    55    49    48
Element     16            49    55    63    64    65    56    51    50
Layer no     1

There are     86 freedoms and nband is     29 in step     1
The total gravity load in lift     1 is  0.5000E+02
Lift number     1 gravity load increment     1
It took      2  iterations to converge
Lift number     1 gravity load increment     2
It took      2  iterations to converge
Lift number     1 gravity load increment     3
It took      2  iterations to converge
Lift number     1 gravity load increment     4
It took      2  iterations to converge
Lift number     1 gravity load increment     5
It took      2  iterations to converge
The displacements are :
  0.7626E-05 -0.5872E-04  0.0000E+00  0.1223E-03
Layer no     2

There are     76 freedoms and nband is     29 in step     2
The total gravity load in lift     2 is  0.5848E+02
Lift number     2 gravity load increment     1
It took      2  iterations to converge
Lift number     2 gravity load increment     2
It took      2  iterations to converge
Lift number     2 gravity load increment     3
It took      4  iterations to converge
Lift number     2 gravity load increment     4
It took      6  iterations to converge
Lift number     2 gravity load increment     5
It took     30  iterations to converge
The displacements are :
  0.8686E-04 -0.3944E-03  0.0000E+00  0.2323E-03
```

Figure 6.42 Output from Program 6.10

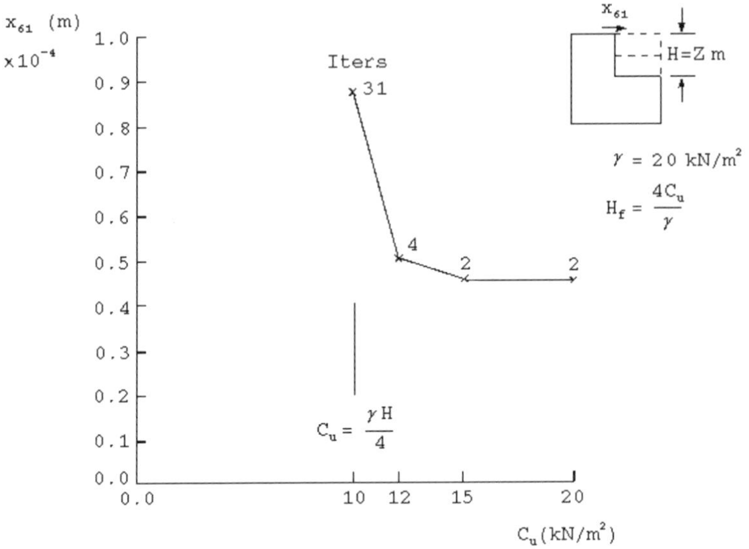

Figure 6.43 Crest displacement as a function of c_u

respectively. In a preliminary loop, labelled `ELEMENTS_1`, the ground is stressed by its own weight. Note that due to the simplicity of the constitutive model, a very limited range of `EPK0` could be achieved automatically and so `EPK0` is input as data.

For each "layer" the geometry is modified and a new stiffness matrix and load vectors formed by `ALLOCATE` statements. Then the standard visco-plastic analysis is carried out. The loads are removed in `INCS` increments (five in the case listed). The output (for case B) is shown as Figure 6.42.

Some results are plotted graphically in Figure 6.43.

The "soil" is a frictionless one and the displacement at the crest of the excavation when it reaches the final depth of 2 m is shown as a function of undrained strength c_u. The classical limit plasticity solution gives collapse of this excavation for an undrained strength of 10 kN/m^2. The finite element results show a sharp acceleration in the crest displacement as c_u is reduced from 12 to 10 kN/m^2 indicating "collapse".

6.13 REFERENCES

Bishop, A.W. and Morgenstern, N. (1960) Stability coefficients for earth slopes. *Géotechnique*, **10**(4), 129–150.

Cormeau, I.C. (1975) Numerical stability in quasi-static elasto-viscoplasticity. *Int. J. Num. Meth. Eng.*, **9**(1), 109–127.

Duncan, J.M. and Chang, C.Y. (1970) Nonlinear analysis of stress and strain in soils. *Soil Mech. Found. Eng. Div., ASCE*, **96**, No. SM5, 1629–1653.

Griffiths, D.V. (1982) Computation of bearing capacity factors using finite elements. *Géotechnique*, **32**(3), 195–202.

Griffiths, D.V. (1985) The effect of pore fluid compressibility on failure loads in elastic plastic soil. *Int. J. Num. Anal. Meth. Geomech.*, **9**, 253–259.

Griffiths, D.V. and Willson, S.M. (1986) An explicit form of the plastic matrix for Mohr–Coulomb materials. *Comm. App. Num. Meths.*, **2** 523–529.

Hill, R. (1950) *The Mathematical Theory of Plasticity.* Oxford University Press.

Kidger, D.J. (1994) Visualisation of three-dimensional processes in geomechanics computations. *Proc 8th IACMAG Conference*, West Virginia, pp. 453–457 Balkema.

Molenkamp, F. (1983) Kinematic Elastoplastic Model ALTERNAT. Delft Soil Mechanics Laboratory Report.

Nayak, G.C. and Zienkiewicz, O.C. (1972) Elasto/plastic stress analysis. A generalisation for various constitutive relationships including strain softening. *Int. J. Num. Meth. Eng.*, **5**, 113–135.

Naylor, D.J. (1974) Stresses in nearly incompressible materials by finite elements with application to the calculation of excess pore pressures. *Int. J. Num. Meth. Eng.*, **8**, 443–460.

Ortiz, M. and Popov, E.P. (1985) Accuracy and stability integration algorithms for elastoplastic constitutive relations, *IJNME*, 21, 1561–1576.

Rice, J.R. and Tracey, D.M. (1973) Computational fracture mechanics, *Proc. Symp. Num. Meth. Struct. Mech.*, ed S.J. Fenves, Academic Press.

Smith, I.M. and Hobbs, R. (1974) Finite element analysis of centrifuged and built-up slopes. *Géotechnique*, **24**(4), 531–559.

Smith, I.M. and Ho, D.K.H. (1992) Influence of construction technique on performance of a braced excavation in marine clay. *IJNAMG*, **16**(12), 845–867.

Smith, I.M. (1997) Computation of Large Scale Viscoplastic Flows of Frictional Geotechnical Materials in 'Dynamics of Complex Fluids', eds M.J. Adams, R.A. Mashelkar and A.R. Rennie. Imperial College Press – The Royal Society.

Yamada, Y.,Yoshimura, N. and Sakurai, T. (1968) Plastic stress–strain matrix and its application for the solution of elastic plastic problems by the finite element method. *Int. J. Mech. Sci.*, **10**, 343–354.

Zienkiewicz, O.C. and Cormeau, I.C. (1974) Viscoplasticity, plasticity and creep in elastic solids. A unified numerical solution approach. *Int. J. Num. Meth. Eng.*, **8**, 821–845.

Zienkiewicz, O.C., Valliappan, S. and King, I.P. (1969) Elasto-plastic solutions of engineering problems, "initial-stress" finite element approach. *Int. J. Num. Meth. Eng.*, **1**, 75–100.

Zienkiewicz, O.C. (1991) *The Finite Element Method*, 4th edn. McGraw-Hill, New York.

7 Steady-state Flow

7.0 INTRODUCTION

The four programs presented in this chapter solve steady-state problems governed by Laplace's equation (2.99). Typical examples include steady seepage through soils or steady heat flow through a conductor. Solutions are presented for plane, axisymmetric and 3-d flow. Unlike the problems solved in Chapters 5 and 6 which gave vector fields of displacements, the dependent variable in these problems is a scalar, generically called the "potential" which may represent the total head in a seepage problem or the temperature in a heat flow analysis. Each node therefore has only one degree of freedom associated with it.

Systems that are governed by Laplace's equation require boundary conditions to be prescribed at all points around a closed domain. These boundary conditions commonly take the form of fixed values of the potential or its first derivative normal to the boundary. The problem amounts to finding the values of the potential at points within the closed domain.

Being "elliptic" in character, solution of Laplace's equation quite closely resembles the solution of equilibrium equations (2.54) in solid elasticity. Both methods ultimately require the solution of a set of linear simultaneous equations. The element conductivity or "stiffness" matrices can be formed numerically, as described by equations (3.57)–(3.61) or "analytically" as discussed in section 3.3. Either way, if an assembly strategy is chosen, the element matrices can be assembled into a global "stiffness" matrix which is symmetrical and banded. Taking the analogy with Chapter 5 one stage further, "displacements" now become potentials and "loads" become net inflow or outflow.

Program 7.0 describes the solution of Laplace's equation over a 2-d plane domain. Program 7.1 describes the rather more specialised nonlinear problem of free surface flow in which the mesh is allowed to deform iteratively until it assumes the shape of the free surface at convergence. Program 7.2 is the axisymmetric counterpart of Program 7.0, and Program 7.3 is a general program for the solution of Laplace's equation in 2-d or 3-d domains with variable material properties.

PROGRAM 7.0: PLANE STEADY FLOW

Figure 7.1 shows a typical problem of steady seepage beneath an impermeable sheet pile wall. The total head loss across the wall has been normalised to 100 units, but due to symmetry only half the problem needs to be analysed. In this

```
program p70
!---------------------------------------------------------------------------
!      program 7.0 solution of Laplace's equation
!      over a plane area using 4-node quadrilaterals
!---------------------------------------------------------------------------
use new_library   ;      use geometry_lib;         implicit none
integer::nels,nxe,neq,nn,nr,nip=4,nodof=1,nod=4,ndof=4,loaded_nodes,i,k,  &
         iel,ndim=2,fixed_nodes
real::det,aa,bb    ; character (len=15) :: element = 'quadrilateral'
!----------------------------- dynamic arrays---------------------------------
real,allocatable::kv(:),kvh(:),loads(:),points(:,:),coord(:,:),jac(:,:),  &
                  der(:,:),deriv(:,:),weights(:),kp(:,:),g_coord(:,:),    &
                  value(:),kay(:,:),disps(:),perms(:)
integer,allocatable::nf(:,:),g(:),num(:),g_num(:,:),g_g(:,:),kdiag(:),    &
                     node(:),no(:)
!------------------------input and initialisation---------------------------
open(10,file='p70.dat',status =   'old',    action = 'read')
open(11,file='p70.res',status='replace',    action = 'write')
read (10,*)nels,nxe,aa,bb; nn=(nxe+1)*(nels/nxe+1)
allocate(nf(nodof,nn),points(nip,ndim),g(ndof),g_coord(ndim,nn),           &
         coord(nod,ndim),jac(ndim,ndim),weights(nip),der(ndim,nod),       &
         deriv(ndim,nod),kp(ndof,ndof),num(nod),g_num(nod,nels),          &
         g_g(ndof,nels),kay(ndim,ndim),perms(ndim))
read(10,*)perms
kay=0.0; do i=1,ndim; kay(i,i)=perms(i); end do
read(10,*)nr
nf=1; if(nr>0)read(10,*)(k,nf(:,k),i=1,nr); call formnf(nf); neq=maxval(nf)
allocate (kdiag(neq))
! ----------loop the elements to set up global geometry and kdiag -------------
kdiag=0
elements_1: do iel=1,nels
              call geometry_4qx(iel,nxe,aa,bb,coord,num)
              g_num(:,iel)=num  ; g_coord(:,num)=transpose(coord)
              call num_to_g (num, nf , g ); g_g(:,iel)=g; call fkdiag(kdiag,g)
end do elements_1
write(11,'(a)')"Global coordinates"
do k = 1 , nn
  write(11,'(a,i5,a,3e12.4)')"Node     ",k,"     ",g_coord(:,k); end do
write(11,'(a)')"Global node numbers"
do k=1,nels
  write(11,'(a,i5,a,27i3)')"Element ",k,"      ",g_num(:,k); end do
kdiag(1)=1; do i=2,neq; kdiag(i)=kdiag(i)+kdiag(i-1); end do
write(11,'(2(a,i5),/)')                                                    &
  "There are ",neq," equations and the skyline storage is ",kdiag(neq)
allocate(kv(kdiag(neq)),kvh(kdiag(neq)),loads(0:neq),disps(0:neq))
kv=0.0; loads=0.0
call sample(element,points,weights)
!------- element conductivity integration and assembly--------------------
elements_2: do iel=1,nels
        kp= 0.0
        num=g_num(:,iel); coord=transpose(g_coord(:,num)); g=g_g(:,iel)
        integrating_pts_1: do i=1,nip
            call shape_der(der,points,i); jac=matmul(der,coord)
            det=determinant(jac); call invert(jac); deriv=matmul(jac,der)
            kp=kp+matmul(matmul(transpose(deriv),kay),deriv)*det*weights(i)
        end do integrating_pts_1
        call fsparv (kv,kp,g,kdiag)
end do elements_2
kvh=kv
read (10,*)loaded_nodes
if(loaded_nodes/=0)read(10,*)(k,loads(nf(:,k)),i=1,loaded_nodes)
read (10,*)fixed_nodes
if(fixed_nodes/=0)then
     allocate(node(fixed_nodes),no(fixed_nodes),value(fixed_nodes))
```

```
      read(10,*)(node(i),value(i),i=1,fixed_nodes)
      do i=1,fixed_nodes; no(i)=nf(1,node(i)); end do
      kv(kdiag(no))=kv(kdiag(no))+1.e20; loads(no)=kv(kdiag(no))*value
  end if
!--------------------------equation solution--------------------------------
  call sparin(kv,kdiag); call spabac(kv,loads,kdiag)
!------------------------- retrieve flow rate ------------------------------
  call linmul_sky(kvh,loads,disps,kdiag)
  write(11,'(a)')"The nodal values are:"
  write(11,'(a)')"                 Potentials   Flow rates"
  do k=1,nn
    write(11,'(i5,a,2f12.2)')k,"    ",loads(nf(1,k)),disps(nf(1,k)); end do
  write(11,'(a)')"       Inflow      Outflow"
  write(11,'(2f12.2)')sum(disps,mask=disps>0.),sum(disps,mask=disps<0.)
end program p70
```

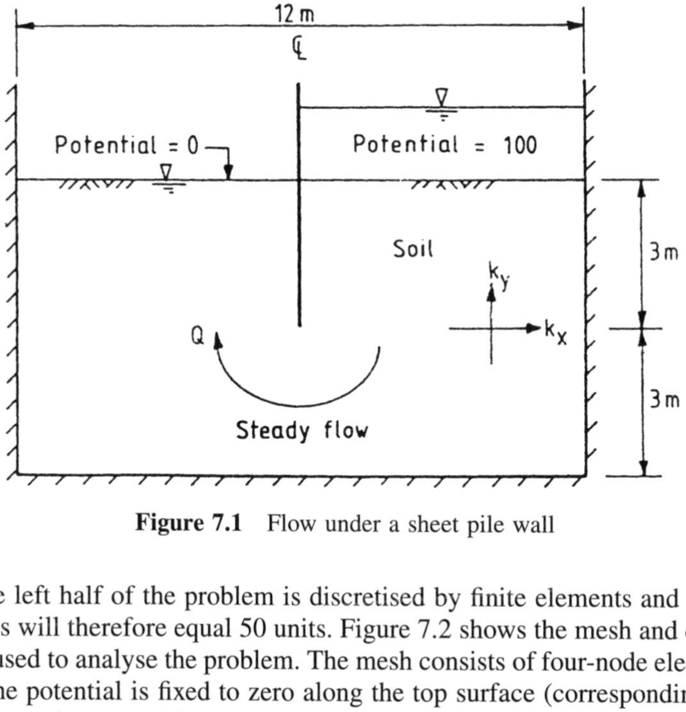

Figure 7.1 Flow under a sheet pile wall

case, the left half of the problem is discretised by finite elements and the total head loss will therefore equal 50 units. Figure 7.2 shows the mesh and data that will be used to analyse the problem. The mesh consists of four-node elements in which the potential is fixed to zero along the top surface (corresponding to the downstream face), and fixed to 50 along the lower half of the right-hand face (beneath the wall). In earlier chapters of this book, "zero freedoms" were achieved by reading in nodal freedom data which resulted in those equations being eliminated from the assembly process. Fixing freedoms at any value, including zero, can also be achieved using the "penalty" technique first described in section 3.8. This is the method used in the current chapter. The price paid for including the zero freedoms is that a few more equations have to be solved. In seepage problems it is useful to retain all the fixed freedoms in the analysis in case the flow rate is required at those locations. All other boundaries are impermeable, so a "natural" boundary condition of $\partial\phi/\partial n = 0$ is required

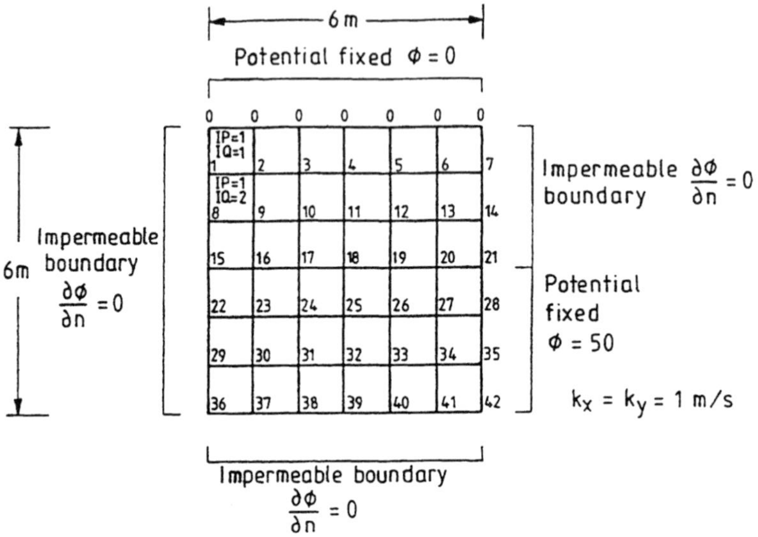

Mesh and element data	NELS	NXE	AA	BB					
	36	6	1.0	1.0					
	PERMS (kx ky)								
	1.0		1.0						
Node program data	NR								
	0								
''Loading'' data	LOADED_NODES								
	0								
	FIXED_NODES								
	11								
	NODE(I), VALUE(I), I=1,FIXED_NODES								
	1	0.	2	0.	3	0.	4	0.	5 0.
	6	0.							
	7	0.	28	50.	35	50.	42	50.	49 50.

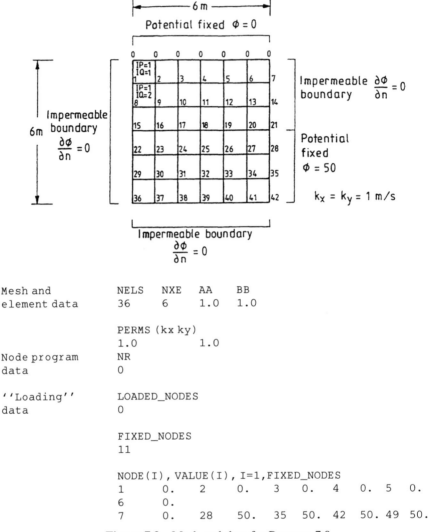

Figure 7.2 Mesh and data for Program 7.0

which is obtained at the boundaries of the mesh by taking no further action (see section 3.8).

The program assumes a rectangular mesh made up of rectangular constant size elements, with nodal coordinates and geometry to be generated by a library subroutine GEOMETRY_4QX. This geometry subroutine assumes that nodes are numbered in the x direction. The data is similar to that read in the programs of Chapter 5. The total number of elements and the number of elements in the x direction (NELS, NXE) are read followed by the size of each rectangular element in the x and y directions as AA and BB respectively. The number of nodes (NN) can be internally calculated for the rectangular mesh and the number of integrating points per elements (NIP) is set to four in the declaration statement.

After allocation of array sizes, the permeabilities in the x and y directions are read into the vector PERMS. The program assumes homogeneous properties throughout although anisotropy can be included.

Since all the fixed freedoms are to be retained in the assembly process, there is no nodal freedom data and NR is read as zero.

After scanning the elements to determine the storage requirements, the program uses numerical integration to form the element conductivity matrices KP which are then assembled into a global conductivity matrix KV. The sequence of operations described by the ELEMENTS_2 loop bears a striking similarity to the integration of an element stiffness matrix used, for example, in Program 5.2. Program 7.0 is even simpler because the derivative array DERIV is used directly in the products described by equation (3.58). The first program in this chapter uses the skyline storage strategy described in Figure 3.19.

Returning to the main program, the meaning of the variable names used in Program 7.0 are given as follows:

Scalar integers:

NELS	number of elements
NXE	number of elements in the x direction
NEQ	number of degrees of freedom in the mesh
NN	number of nodes in the mesh
NR	number of restrained nodes
NIP	number of integrating points per element
NOD	number of nodes per element
NODOF	number of degrees of freedom per node
NDOF	number of degrees of freedom per element
NDIM	number of dimensions
LOADED_NODES	number of loaded nodes
FIXED_NODES	number of fixed displacements
IEL, I, K	simple counters

Scalar reals:

DET	determinant of the Jacobian matrix
AA	width of each element
BB	height of each element

Allocatable real arrays:

KV	global conductivity matrix (stored as a skyline vector)
KVH	copy of KV

LOADS	input flow rates and potentials
COORD	element nodal coordinates
JAC	Jacobian matrix
ELD	element displacement vector
DER	derivatives of shape functions w.r.t. local coordinates
DERIV	derivatives of shape functions w.r.t. global coordinates
WEIGHTS	quadrature weights
KP	element conductivity matrix
G_COORD	global nodal coordinates matrix
VALUE	fixed displacements vector
KAY	element property (permeability) matrix
DISPS	output flow rates
PERMS	permeabilities in the x and y directions

Allocatable integer arrays:

NF	nodal freedom matrix
G	element steering vector
NUM	element node numbers vector
G_NUM	global element node numbers matrix
NO	fixed displacement freedom numbers vector
G_G	global element steering matrix
NODE	fixed displacement nodes vector
KDIAG	diagonal addresses of vector KV

Following formation of the global conductivity matrix KV, a copy KVH is made and the remaining data read. First the LOADED_NODES data indicates whether any sources or sinks are required at nodes internal to the mesh. In this case the only net inflow or outflow occurs at the up- and downstream boundaries and LOADED_NODES is set to zero. Finally the FIXED_NODES data is read which corresponds to the 11 nodes at which the potential is fixed to either zero or 50.

Following solution of the "equilibrium" equations which is performed by library subroutines SPARIN and SPABAC the nodal potentials are held in the vector LOADS and printed. In order to retrieve the nodal flow rates, the matrix KVH is multiplied by the nodal potentials by library subroutine LINMUL_SKY to give a vector of net nodal inflow/outflows held in vector DISPS. Examination of vector DISPS reveals that the majority of flow rates corresponding to internal nodes are zero, the only non-zero values occurring at the boundary nodes that had their potentials fixed. If we had chosen to include an internal source or sink as data using LOADED_NODES, this would have appeared at the appropriate node in the DISPS vector.

Finally the net inflow and outflow through the system is computed by summing, respectively, the positive and negative terms in DISPS. The output from Program 7.0 is shown in Figure 7.3. As expected, the inflow and outflow values are virtually identical and give a steady-state flow rate of 48.6. The method of fragments for this constrained seepage problem (Griffiths, 1984) would predict a flow rate of around 47. The theoretical solution for a sheet pile wall embedded to half the depth of a stratum of similar soil in a domain which extends to infinity laterally would be exactly 50.0.

A way to visualise the results of a seepage analysis such as this is to draw a contour map of the nodal potentials. Figure 7.4 shows a contour map of the potentials and the stream functions that would be computed by the mesh of

```
Global coordinates
Node        1        0.0000E+00  0.0000E+00
Node        2        0.1000E+01  0.0000E+00
Node        3        0.2000E+01  0.0000E+00
Node        4        0.3000E+01  0.0000E+00
Node        5        0.4000E+01  0.0000E+00
Node        6        0.5000E+01  0.0000E+00
Node        7        0.6000E+01  0.0000E+00
Node        8        0.0000E+00 -0.1000E+01
Node        9        0.1000E+01 -0.1000E+01
Node       10        0.2000E+01 -0.1000E+01
.............................................................

Node       44        0.1000E+01 -0.6000E+01
Node       45        0.2000E+01 -0.6000E+01
Node       46        0.3000E+01 -0.6000E+01
Node       47        0.4000E+01 -0.6000E+01
Node       48        0.5000E+01 -0.6000E+01
Node       49        0.6000E+01 -0.6000E+01
Global node numbers
Element     1          8  1  2  9
Element     2          9  2  3 10
Element     3         10  3  4 11
Element     4         11  4  5 12
.............................................................

Element    33         45 38 39 46
Element    34         46 39 40 47
Element    35         47 40 41 48
Element    36         48 41 42 49
There are     49  equations and the skyline storage is    385

The nodal values are:
                 Potentials   Flow rates
    1               0.00         -2.93
    2               0.00         -6.06
    3               0.00         -6.71
    4               0.00         -7.81
    5               0.00         -9.18
.............................................................

   42              50.00          9.20
   43              21.96          0.00
   44              22.72          0.00
   45              25.02          0.00
   46              28.97          0.00
   47              34.64          0.00
   48              41.85          0.00
   49              50.00          4.26
       Inflow       Outflow
       -48.57        48.57
```

Figure 7.3 Output from Program 7.0

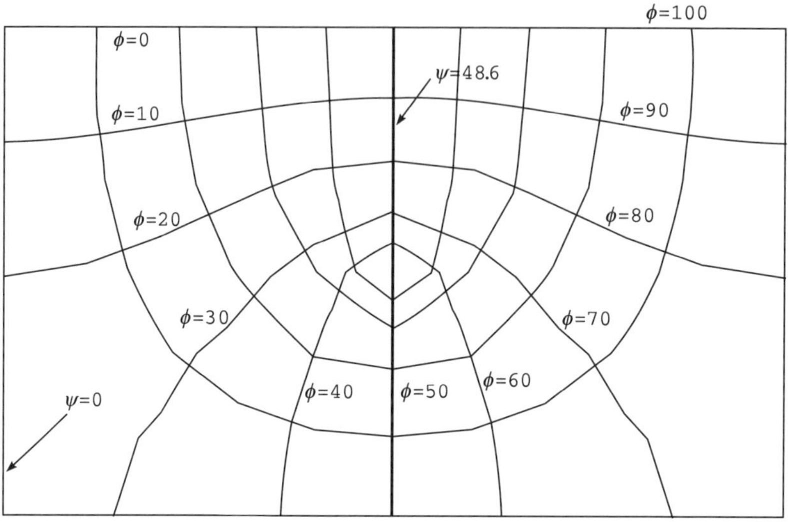

Figure 7.4 A flow net consisting of contours of the potential ϕ and stream function ψ

Figure 7.2. Both sides of the wall are shown for clarity, although only half of the problem was actually analysed. The contours are a little ragged due to the crude discretisation. The stream function problem has not been solved in this example, but it could easily be included by solving the "inverse" problem:

$$\frac{1}{k_y}\frac{\partial^2\psi}{\partial x^2}+\frac{1}{k_x}\frac{\partial^2\psi}{\partial y^2}=0 \tag{7.1}$$

where ψ is the stream function. The boundary conditions are "inverted", hence those boundaries that had fixed values in the potential problem such as the up- and downstream boundaries, have $\partial\psi/\partial n=0$ boundary conditions in the stream problem, and boundaries that had $\partial\phi/\partial n=0$ conditions in the potential problem such as the impermeable boundaries and the sheet pile wall, are given fixed values of the stream function. In order to choose a contour interval which satisfies the usual flow-net "rules", it is suggested that when solving the stream problem, the uppermost streamline (in this case the wall) is fixed equal to the flow rate ($\psi=48.6$), and the lowest streamline (the impermeable boundary) is set equal to zero ($\psi=0$).

PROGRAM 7.1: PLANE FREE-SURFACE FLOW

In this program we consider a boundary condition frequently met in geomechanics in relation to flow of water through dams. Free-surface problems involve an upper boundary, the location of which is not known a priori. This

```
program p71
!-----------------------------------------------------------------------
!      program 7.1 solution of Laplace's equation
!      for plane free-surface flow using 4-node quadrilaterals
!-----------------------------------------------------------------------
 use new_library   ;   use geometry_lib ; use vlib   ;  implicit none
 integer::nels,nxe,neq,nn,nr,nodof=1,nod=4,ndof=4,np_types,i,k,iel,ndim=2,&
         fixed_up,fixed_down,fixed_seep,iters,limit,nband
 real::tol,upstream,downstream
 logical::converged
!-------------------------- dynamic arrays-------------------------------------
 real,allocatable::kv(:),kvh(:),loads(:),coord(:,:),kp(:,:),g_coord(:,:), &
                   kay(:,:),disps(:),oldpot(:),width(:),surf(:),prop(:,:),&
                   angs(:)
 integer,allocatable::nf(:,:),g(:),num(:),g_num(:,:),g_g(:,:),node_up(:), &
                      no_up(:),etype(:),node_down(:),no_down(:),          &
                      node_seep(:),no_seep(:)
!-----------------------input and initialisation------------------------------
 open (10,file='p71a.dat',status='old'    , action ='read' )
 open (11,file='p71a.res',status='replace', action = 'write')
 read(10,*) nels,nxe,tol,limit,np_types; nn=(nxe+1)*(nels/nxe+1)
 allocate(nf(nodof,nn),g(ndof),g_coord(ndim,nn),coord(nod,ndim),           &
         width(nxe+1),surf(nxe+1),angs(nxe+1),kp(ndof,ndof),num(nod),      &
         g_num(nod,nels),prop(ndim,np_types),g_g(ndof,nels),               &
         kay(ndim,ndim),etype(nels))
 read(10,*)prop; etype=1; if(np_types>1)read(10,*)etype
 read(10,*)width; read(10,*)angs; read(10,*)surf
 read(10,*)nr
 nf=1; if(nr>0)read(10,*)(k,nf(:,k),i=1,nr); call formnf(nf); neq=maxval(nf)
 read(10,*)upstream,fixed_up ; allocate (node_up(fixed_up),no_up(fixed_up))
 read(10,*) node_up          ; read (10,*)downstream,fixed_down
 allocate (node_down(fixed_down),no_down(fixed_down))
 read(10,*)node_down      ;  fixed_seep=nels/nxe-fixed_down
 allocate (node_seep(fixed_seep),no_seep(fixed_seep))
 do i=1,fixed_seep; node_seep(i)=i*(nxe+1)+1; end do
!--------- loop the elements to find nband and set up global arrays ------
 nband=0
 elements_1: do iel=1,nels
      call geometry_freesurf(iel,nxe,fixed_seep,fixed_down,downstream,&
                             width,angs,surf,coord,num)
      call num_to_g ( num , nf , g )
      g_coord(:,num)=transpose(coord); g_num(:,iel)=num
      if(nband<bandwidth(g))nband=bandwidth(g)
 end do elements_1
 write(11,'(a)')"Initial global coordinates"
 do k=1,nn
   write(11,'(a,i5,a,3e12.4)')"Node    ",k,"    ",g_coord(:,k); end do
 write(11,'(2(a,i5),/)')                                                     &
     "There are ",neq," equations and the half bandwidth is ",nband
 allocate(kv(neq*(nband+1)),kvh(neq*(nband+1)),loads(0:neq),disps(0:neq), &
         oldpot(0:neq)); oldpot=0.0
!--------------- element conductivity integration and assembly-----------------
 iters=0
 iterations: do
    iters=iters+1; kv=0.0
    elements_2: do iel=1,nels
         kay=0.0; do i=1,ndim; kay(i,i)=prop(i,etype(iel)); end do
         call geometry_freesurf(iel,nxe,fixed_seep,fixed_down,downstream, &
                                width,angs,surf,coord,num)
         call num_to_g ( num , nf , g )
         g_coord(:,num)=transpose(coord)
         call seep4(coord,kay,kp)  ;   call formkv(kv,kp,g,neq)
    end do elements_2
    kvh=kv
```

```
!----------   specify fixed potentials and factorise equations
    loads=0.0
    do i=1,fixed_up; no_up(i)=nf(1,node_up(i)); end do
    kv(no_up)=kv(no_up)+1.e20; loads(no_up)=kv(no_up)*upstream
    do i=1,fixed_down; no_down(i)=nf(1,node_down(i)); end do
    kv(no_down)=kv(no_down)+1.e20; loads(no_down)=kv(no_down)*downstream
    do i=1,fixed_seep
      no_seep(i)=nf(1,node_seep(i))
      kv(no_seep(i))=kv(no_seep(i))+1.e20
      loads(no_seep(i))=kv(no_seep(i))*                                   &
      (downstream+(surf(1)-downstream)*(fixed_seep+1-i)/(fixed_seep+1))
    end do
!-----------------------------equation solution-------------------------------
    call banred(kv,neq);call bacsub(kv,loads)
    surf(1:nxe)=loads(1:nxe)
!------------------------------check convergence-----------------------------
    call checon(loads,oldpot,tol,converged)
    if(converged.or.iters==limit)exit
 end do iterations
! ------------------- write out the results -------------------------------
 write(11,'(a)')"Final global coordinates for plotting"
 do k=1,nn
    write(11,'(a,i5,a,3e12.4)')"Node     ",k,"     ",g_coord(:,k); end do
 write(11,'(a)')"Global node numbers"
 do k=1,nels
   write(11,'(a,i5,a,27i3)')"Element ",k,"      ",g_num(:,k); end do
 call linmul(kvh,loads,disps)
 write(11,'(a)')"The nodal values are:"
 write(11,'(a)')"            Potentials  Flow rate"
 do k=1,nn
   write(11,'(i5,a,2e12.4)')k,"    ",loads(nf(1,k)),disps(nf(1,k)); end do
 write(11,'(a)')"  Inflow      Outflow"
 write(11,'(2e12.4)')sum(disps,mask=disps>0.),sum(disps,mask=disps<0.)
 write(11,'(a,i5)')"Number of iterations =",iters
end program p71
```

boundary can be found by iteration in several ways; for example, a fixed mesh can be used and nodes separated into "active" and "inactive" ones depending upon whether fluid exists at that point. An alternative strategy is to use the present program, whereby the mesh is deformed so that its upper surface ultimately coincides with the free surface. A summary of the boundary conditions is given in Figure 7.5.

The analysis starts by assuming an initial position for the free surface. Solution of Laplace's equation gives values of the fluid potential which will not in general equal the elevation of the upper surface of the mesh. The elevations of the nodes along the upper surface are therefore adjusted to equal the potential values just calculated at those locations. In order to avoid distorted elements, the subroutine GEOMETRY_FREESURF ensures that the nodes beneath the top surface are evenly distributed. The analysis is then repeated with the new mesh. Since many of the coordinates have changed, the conductivity matrices of all the elements must be recomputed and assembled into the global system. In order to avoid the need for numerical integration of the element conductivity matrices at each iteration, library subroutine SEEP4 computes the element conductivity matrices "analytically". This subroutine could of course have been used in Program 7.0 also. The assembly is made into a global conductivity matrix KV which in this case is stored using a constant bandwidth strategy.

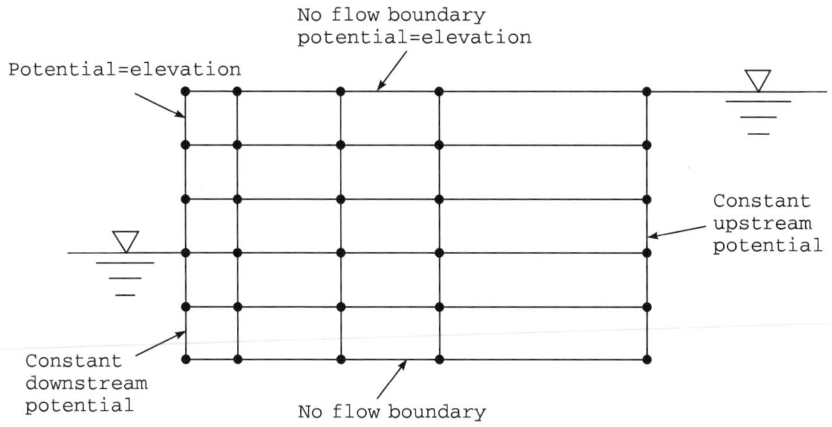

Figure 7.5 Boundary conditions for free surface flow

This process is repeated until the change in computed potential values from one iteration to the next is less than a tolerance value `TOL`. The convergence check is performed by library subroutine `CHECON` and depends on the value of the logical variable `CONVERGED` which is `.TRUE.` if the solution has converged.

Referring to the program the following new variables are required:

Scalar integers:

`NP_TYPES`	number of different property types
`FIXED_UP`	number of upstream nodes to be fixed
`FIXED_DOWN`	number of downstream nodes to be fixed
`FIXED_SEEP`	number of nodes to be fixed on the seepage surface
`ITERS`	iteration counter
`LIMIT`	iteration ceiling
`NBAND`	half-bandwidth of mesh

Scalar reals:

`TOL`	convergence tolerance
`UPSTREAM`	upstream potential
`DOWNSTREAM`	downstream potential

Scalar logicals:

`CONVERGED`	convergence check (`.TRUE.` = converged)

Allocatable real arrays:

`OLDPOT`	potential values from the previous iteration
`WIDTH`	x coordinates of nodes at the base of the mesh

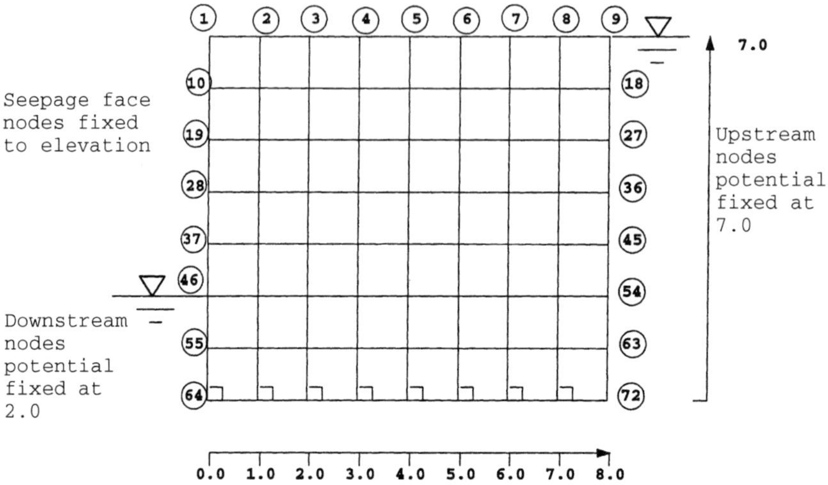

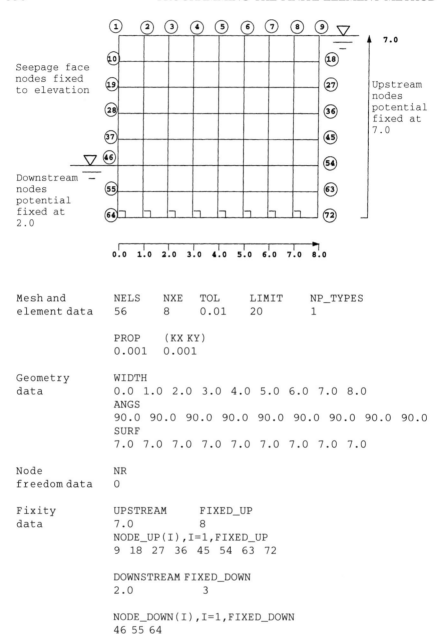

```
Mesh and        NELS   NXE  TOL    LIMIT  NP_TYPES
element data    56     8    0.01   20     1

                PROP   (KX KY)
                0.001  0.001

Geometry        WIDTH
data            0.0 1.0 2.0 3.0 4.0 5.0 6.0 7.0 8.0
                ANGS
                90.0 90.0 90.0 90.0 90.0 90.0 90.0 90.0 90.0
                SURF
                7.0 7.0 7.0 7.0 7.0 7.0 7.0 7.0 7.0

Node            NR
freedom data    0

Fixity          UPSTREAM     FIXED_UP
data            7.0          8
                NODE_UP(I),I=1,FIXED_UP
                9 18 27 36 45 54 63 72

                DOWNSTREAM FIXED_DOWN
                2.0          3

                NODE_DOWN(I),I=1,FIXED_DOWN
                46 55 64
```

Figure 7.6 Mesh and data for vertical faced dam with Program 7.1

```
Initial global coordinates
Node        1      0.0000E+00  0.7000E+01
Node        2      0.1000E+01  0.7000E+01
Node        3      0.2000E+01  0.7000E+01
Node        4      0.3000E+01  0.7000E+01
Node        5      0.4000E+01  0.7000E+01
.................................................................

Node       70      0.6000E+01  0.0000E+00
Node       71      0.7000E+01  0.0000E+00
Node       72      0.8000E+01  0.0000E+00
There are   72  equations and the half bandwidth is    10

Final global coordinates for plotting
Node        1      0.0000E+00  0.2806E+01
Node        2      0.1000E+01  0.3757E+01
Node        3      0.2000E+01  0.4508E+01
Node        4      0.3000E+01  0.5060E+01
Node        5      0.4000E+01  0.5571E+01
.................................................................

Node       68      0.4000E+01  0.0000E+00
Node       69      0.5000E+01  0.0000E+00
Node       70      0.6000E+01  0.0000E+00
Node       71      0.7000E+01  0.0000E+00
Node       72      0.8000E+01  0.0000E+00
Global node numbers
Element     1       10  1  2 11
Element     2       11  2  3 12
Element     3       12  3  4 13
Element     4       13  4  5 14
Element     5       14  5  6 15
.................................................................

Element    52       67 58 59 68
Element    53       68 59 60 69
Element    54       69 60 61 70
Element    55       70 61 62 71
Element    56       71 62 63 72
The nodal values are:
          Potentials  Flow rate
    1     0.2759E+01 -0.5551E-16
    2     0.3784E+01 -0.6939E-16
    3     0.4498E+01  0.4857E-16
    4     0.5062E+01  0.4510E-16
    5     0.5571E+01 -0.1561E-16
    6     0.6016E+01 -0.3469E-16
    7     0.6409E+01 -0.9194E-16
.................................................................

   66     0.3619E+01  0.2776E-16
   67     0.4310E+01  0.0000E+00
   68     0.4925E+01  0.0000E+00
   69     0.5485E+01 -0.2776E-16
   70     0.6009E+01 -0.2776E-16
   71     0.6510E+01 -0.8327E-16
   72     0.7000E+01  0.2432E-03
  Inflow       Outflow
 -0.2813E-02  0.2813E-02
Number of iterations =    10
```

Figure 7.7 Output from vertical face dam analysis with Program 7.1

SURF	y coordinates of initial "guess" of free surface
PROP	holds the permeabilities for each property type
ANGS	angle (in degrees) of line rising from the base of the mesh

Allocatable integer arrays:

NODE_UP	nodes to be fixed on upstream side
NODE_DOWN	nodes to be fixed on downstream side
NODE_SEEP	nodes to be fixed on seepage face
NO_UP	freedoms to be fixed on upstream side
NO_DOWN	freedoms to be fixed on downstream side
NO_SEEP	freedoms to be fixed on seepage face
ETYPE	element property type vector

The subroutine GEOMETRY_FREESURF is designed for solving free-surface problems with initially trapezoidal meshes and counts nodes in the x direction. The rectangular undeformed mesh shown in Figure 7.6 represents the starting point for analysis of free-surface seepage through a vertical faced dam. The free surface described by nodes 1–9 is initially assumed to be horizontal. Following the initial data relating to the number of elements in the x and y directions, the tolerance TOL is set equal to 0.01 and the iteration ceiling LIMIT is set equal to

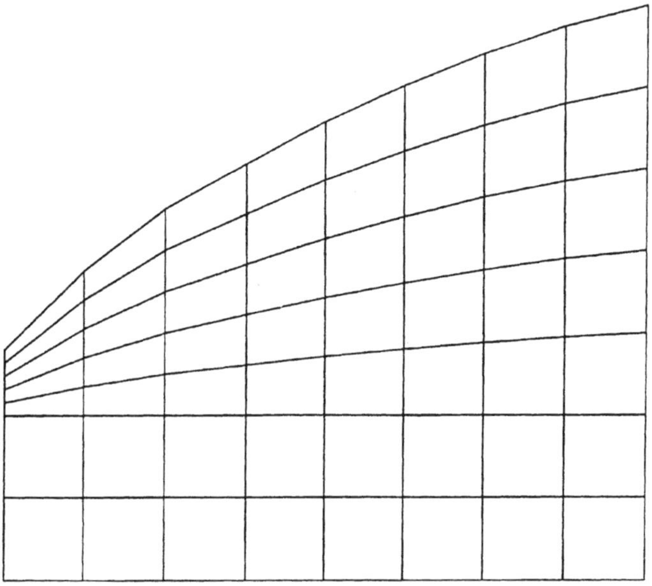

Figure 7.8 Computed free surface at convergence in vertical face dam analysis

20. The program can analyse free surface problems in layered soils, but in this example the dam is homogeneous, so NP_TYPES is set equal to unity. If there were several material property groups (NP_TYPES > 1), additional data would be read into the integer vector ETYPE identifying which element was to be assigned to which property group.

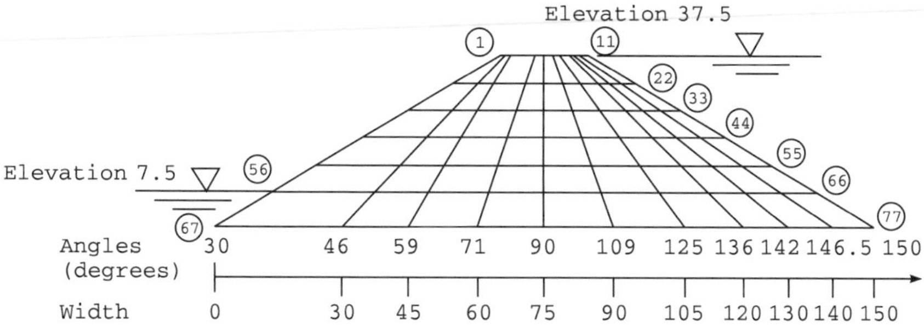

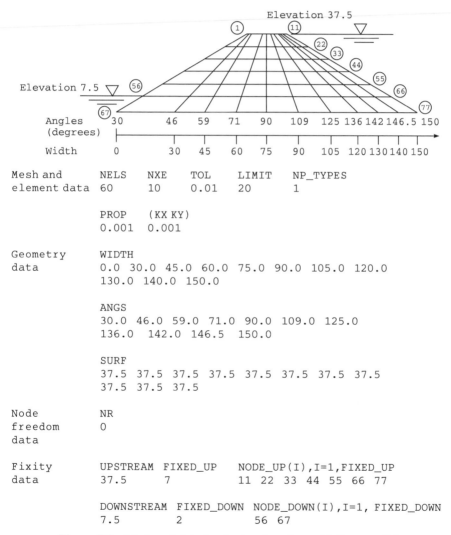

Mesh and element data
NELS NXE TOL LIMIT NP_TYPES
60 10 0.01 20 1

PROP (KX KY)
0.001 0.001

Geometry data
WIDTH
0.0 30.0 45.0 60.0 75.0 90.0 105.0 120.0
130.0 140.0 150.0

ANGS
30.0 46.0 59.0 71.0 90.0 109.0 125.0
136.0 142.0 146.5 150.0

SURF
37.5 37.5 37.5 37.5 37.5 37.5 37.5 37.5
37.5 37.5 37.5

Node freedom data
NR
0

Fixity data
UPSTREAM FIXED_UP NODE_UP(I),I=1,FIXED_UP
37.5 7 11 22 33 44 55 66 77

DOWNSTREAM FIXED_DOWN NODE_DOWN(I),I=1, FIXED_DOWN
7.5 2 56 67

Figure 7.9 Mesh and data for sloping face dam with Program 7.1

The permeabilities in the x and y directions are read followed by the x coordinates of the bottom nodes `WIDTH`, the angles `ANGS` (all 90° in this case) at which the mesh lines rise from the base to the free surface and the initial height of the free surface `SURF`. The value of the upstream potential `UPSTREAM` is then read, followed by the number of nodes to be fixed on the upstream side

```
Initial global coordinates
Node        1       0.6495E+02  0.3750E+02
Node        2       0.6621E+02  0.3750E+02
Node        3       0.6753E+02  0.3750E+02
Node        4       0.7291E+02  0.3750E+02
Node        5       0.7500E+02  0.3750E+02
.....................................................................

Node       74       0.1200E+03  0.0000E+00
Node       75       0.1300E+03  0.0000E+00
Node       76       0.1400E+03  0.0000E+00
Node       77       0.1500E+03  0.0000E+00
There are     77  equations and the half bandwidth is     12

Final global coordinates for plotting
Node        1       0.2958E+02  0.1708E+02
Node        2       0.5706E+02  0.2802E+02
Node        3       0.6270E+02  0.2946E+02
Node        4       0.7097E+02  0.3186E+02
Node        5       0.7500E+02  0.3313E+02
Node        6       0.7821E+02  0.3424E+02
.....................................................................

Node       75       0.1300E+03  0.0000E+00
Node       76       0.1400E+03  0.0000E+00
Node       77       0.1500E+03  0.0000E+00
Global node numbers
Element     1        12  1  2 13
Element     2        13  2  3 14
Element     3        14  3  4 15
Element     4        15  4  5 16
Element     5        16  5  6 17
.....................................................................

Element    56        72 61 62 73
Element    57        73 62 63 74
Element    58        74 63 64 75
Element    59        75 64 65 76
Element    60        76 65 66 77
The nodal values are:
          Potentials   Flow rate
    1     0.1672E+02  0.0000E+00
    2     0.2805E+02  0.1332E-14
    3     0.2947E+02 -0.7216E-15
    4     0.3186E+02 -0.3469E-15
    5     0.3313E+02  0.0000E+00
.....................................................................

   74     0.3709E+02 -0.6661E-15
   75     0.3738E+02 -0.8882E-15
   76     0.3749E+02  0.8882E-15
   77     0.3750E+02 -0.2318E-05
  Inflow      Outflow
 -0.1155E-01  0.1155E-01
Number of iterations =    16
```

Figure 7.10 Output from sloping face dam analysis with Program 7.1

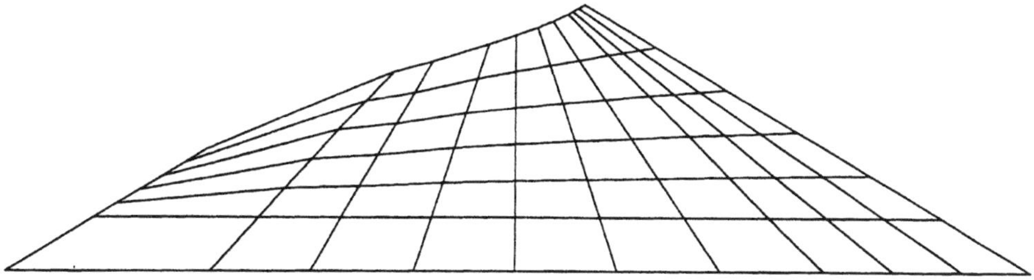

Figure 7.11 Computed free surface at convergence in sloping face dam analysis

FIXED_UP and their node numbers NODE_UP. The same is then read for the downstream side, namely DOWNSTREAM, FIXED_DOWN and NODE_DOWN.

The output for this example is shown in Figure 7.7. Both the initial nodal coordinates and the final values are printed to enable plotting of the deformed mesh at convergence as shown in Figure 7.8. In addition, the potentials and flow rates at each node are also output.

The case of free-surface flow through a dam with vertical faces is a classical problem for which the Dupuit formula (see e.g. Verruijt, 1970) predicts a flow rate given by

$$Q = \frac{k(H_1^2 - H_2^2)}{2D} \tag{7.2}$$

where $H_1 = 7.0$ and $H_2 = 2.0$ refer to the up- and downstream water elevations, $k = 0.001$ refers to the permeability (assumed isotropic and homogeneous) and $D = 8.0$ refers to the width of the dam. In this case the formula gives a flow rate of 0.00281 which is in very good agreement with the computed value.

A second example of a dam with sloping sides is presented in Figure 7.9. The initial mesh is trapezoidal with an initially horizontal free surface. The upstream nodes are all fixed at an elevation of 37.5 while the bottom two nodes on the downstream side are fixed at an elevation of 7.5. The initial mesh is defined by the x coordinates of the nodes at the base, and the angles of inclination of the lines moving towards the top of the embankment. During the mesh iteration stage of the algorithm, the nodes are constrained to remain on the sloping lines and maintain even spacing in the y direction. The output from this example is shown in Figure 7.10 indicating a steady flow rate of 0.0116. The deformed mesh, which was reached after 16 iterations, is shown in Figure 7.11.

PROGRAM 7.2: AXISYMMETRIC STEADY FLOW

```
 program p72
!-------------------------------------------------------------------------
!      program 7.2 solution of Laplace's equation
!      over an axisymmetric region using 4-node quadrilaterals
!-------------------------------------------------------------------------
 use new_library   ;    use   geometry_lib ;   implicit none
 integer::nels,nxe,neq,nn,nr,nip,nodof=1,nod=4,ndof=4,lOaded_nodes,i,k,    &
              iel,ndim=2,fixed_nodes
 real::det,aa,bb,radius  ;  character (len=15) :: element='quadrilateral'
!--------------------------- dynamic arrays---------------------------------
 real,allocatable::kv(:),kvh(:),loads(:),points(:,:),coord(:,:),jac(:,:), &
                   der(:,:),deriv(:,:),weights(:),kp(:,:),g_coord(:,:),   &
                   value(:),kay(:,:),disps(:),perms(:),fun(:)
 integer,allocatable::nf(:,:),g(:),num(:),g_num(:,:),g_g(:,:),kdiag(:),    &
                      node(:),no(:)
!-----------------------input and initialisation--------------------------
 open(10,file='p72.dat',status='old',     action='read')
 open(11,file='p72.res',status='replace',action='write')
 read (10,*)nels,nxe,nip,aa,bb; nn=(nxe+1)*(nels/nxe+1)
 allocate(nf(nodof,nn),points(nip,ndim),g(ndof),g_coord(ndim,nn),          &
          coord(nod,ndim),jac(ndim,ndim),fun(ndof),weights(nip),           &
          der(ndim,nod),deriv(ndim,nod),kp(ndof,ndof),num(nod),            &
          g_num(nod,nels),g_g(ndof,nels),kay(ndim,ndim),perms(ndim))
 read(10,*)perms
 kay=0.0; do i=1,ndim; kay(i,i)=perms(i); end do
 read(10,*)nr
 nf=1; if(nr>0)read(10,*)(k,nf(:,k),i=1,nr); call formnf(nf); neq=maxval(nf)
 allocate (kdiag(neq))
!------- loop the elements to set up global geometry and kdiag ---------------
 kdiag=0
 elements_1: do iel=1,nels
              call geometry_4qx(iel,nxe,aa,bb,coord,num)
              g_num(:,iel)=num; g_coord(:,num)=transpose(coord)
              call num_to_g(num,nf,g);   g_g(:,iel)=g  ; call fkdiag(kdiag,g)
 end do elements_1
 write(11,'(a)')"Global coordinates"
 do k=1,nn
   write(11,'(a,i5,a,3e12.4)')"Node    ",k,"    ",g_coord(:,k); end do
 write(11,'(a)')"Global node numbers"
 do k=1,nels
   write(11,'(a,i5,a,27i3)')"Element ",k,"      ",g_num(:,k); end do
 kdiag(1)=1; do i=2,neq; kdiag(i)=kdiag(i)+kdiag(i-1); end do
 write(11,'(2(a,i5),/)')                                                  &
   "There are ",neq," equations and the skyline storage is ",kdiag(neq)
 allocate(kv(kdiag(neq)),kvh(kdiag(neq)),loads(0:neq),disps(0:neq))
 kv=0.0; loads=0.0   ; call sample(element,points,weights)
!--------------- element conductivity integration and assembly----------------
 elements_2: do iel=1,nels
        kp=0.0
        num=g_num(:,iel); coord=transpose(g_coord(:,num)); g=g_g(:,iel)
        integrating_pts_1: do i=1,nip
           call shape_der(der,points,i); jac=matmul(der,coord)
           det=determinant(jac); call invert(jac)
           deriv=matmul(jac,der); call shape_fun(fun,points,i)
           radius=sum(fun(:)*coord(:,1))
           kp=kp+matmul(matmul(transpose(deriv),kay),deriv) &
              *radius*det*weights(i)
        end do integrating_pts_1
        call fsparv(kv,kp,g,kdiag)
 end do elements_2
 kvh=kv
 read (10,*)loaded_nodes
 if(loaded_nodes/=0)read(10,*)(k,loads(nf(:,k)),i=1,loaded_nodes)
 read (10,*) fixed_nodes
```

```
if(fixed_nodes/=0)then
    allocate( node(fixed_nodes),no(fixed_nodes),value(fixed_nodes))
    read(10,*) (node(i),value(i),i=1,fixed_nodes)
    do i=1,fixed_nodes; no(i)=nf(1,node(i)); end do
    kv(kdiag(no))=kv(kdiag(no))+1.e20; loads(no)=kv(kdiag(no))*value
end if
!-----------------------equation solution-------------------------------
call sparin(kv,kdiag); call spabac(kv,loads,kdiag)
!----------------------- retrieve flow rate ----------------------------
call linmul_sky(kvh,loads,disps,kdiag)
write(11,'(a)')"The nodal values are:"
write(11,'(a)')"                Potentials   Flow rate"
do k=1,nn
  write(11,'(i5,a,2f12.2)')k,"    ",loads(nf(1,k)),disps(nf(1,k)); end do
write(11,'(a)')"      Inflow      Outflow"
write(11,'(2f12.2)')sum(disps,mask=disps>.0),sum(disps,mask=disps<.0)
end program p72
```

Program 7.2 is the axisymmetric counterpart of Program 7.0. The x axis has become a radial r axis, and the y axis has become a "depth" z axis. The only new variable introduced into the program is `RADIUS`, which represents the radial coordinate of a Gauss point used during numerical integration in cylindrical coordinates (see e.g. equation (3.56)). The number of integrating points `NIP` is included as data in this program and is set to 9 (3 points in each direction). In contrast to the plane case in Program 7.0, axisymmetric integration is never "exact" using conventional Gaussian quadrature and slightly different results can be expected as `NIP` is increased.

The simple mesh shown in Figure 7.12 represents a radial plane of a cylinder of porous material. The model subtends one radian of the angular coordinate θ at the axis of symmetry. The boundary conditions consist of a fixed potential of 100 units on the top of the cylinder and on the central axis ($r = 0$). The outer surface of the cylinder and the bottom surface are fixed to zero. In order to compute flow rates, the "penalty" method has again been used for fixing freedoms, even when they are fixed to zero.

In this example, a point sink is applied to node 10 where -25 m^3/(s radian) (assuming metric units) of fluid is being removed from the system. The computed results are shown in Figure 7.13. In addition to the usual flow rates recorded at the boundary nodes, the fluid removed from the system at node 10 also appears in the "flow rates" column. The net inflow (outflow) from the system is around 363 m^3/(s radian).

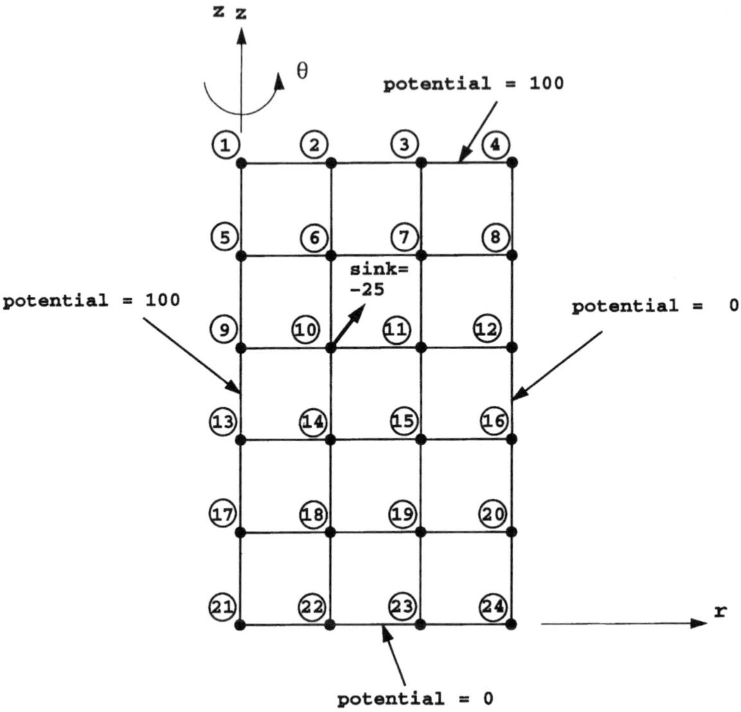

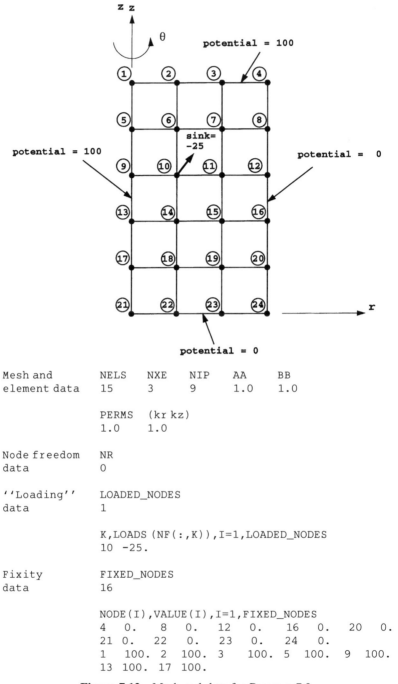

```
Mesh and        NELS   NXE   NIP   AA    BB
element data    15     3     9     1.0   1.0

                PERMS  (kr kz)
                1.0    1.0

Node freedom    NR
data            0

''Loading''     LOADED_NODES
data            1

                K,LOADS(NF(:,K)),I=1,LOADED_NODES
                10 -25.

Fixity          FIXED_NODES
data            16

                NODE(I),VALUE(I),I=1,FIXED_NODES
                4  0.    8  0.    12  0.    16  0.   20  0.
                21 0.    22  0.   23  0.    24  0.
                1  100.  2  100.  3  100.   5  100.  9  100.
                13 100.  17 100.
```

Figure 7.12 Mesh and data for Program 7.2

```
Global coordinates
Node        1      0.0000E+00  0.0000E+00
Node        2      0.1000E+01  0.0000E+00
Node        3      0.2000E+01  0.0000E+00
Node        4      0.3000E+01  0.0000E+00
Node        5      0.0000E+00 -0.1000E+01
.....................................................

Node       20      0.3000E+01 -0.4000E+01
Node       21      0.0000E+00 -0.5000E+01
Node       22      0.1000E+01 -0.5000E+01
Node       23      0.2000E+01 -0.5000E+01
Node       24      0.3000E+01 -0.5000E+01
Global node numbers
Element     1        5  1  2  6
Element     2        6  2  3  7
Element     3        7  3  4  8
Element     4        9  5  6 10
Element     5       10  6  7 11
Element     6       11  7  8 12
Element     7       13  9 10 14
Element     8       14 10 11 15
Element     9       15 11 12 16
Element    10       17 13 14 18
Element    11       18 14 15 19
Element    12       19 15 16 20
Element    13       21 17 18 22
Element    14       22 18 19 23
Element    15       23 19 20 24
There are    24  equations and the skyline storage is    122

The nodal values are:
             Potentials   Flow rate
    1          100.00        6.07
    2          100.00       45.58
    3          100.00      187.80
    4            0.00      -69.25
    5          100.00       17.26
    6           63.59        0.00
    7           33.10        0.00
    8            0.00     -125.52
    9          100.00       28.63
   10           32.83      -25.00
   11           17.52        0.00
   12            0.00      -51.24
   13          100.00       35.95
   14           31.79        0.00
   15           10.86        0.00
   16            0.00      -28.80
   17          100.00       41.42
   18           19.67        0.00
   19            6.18        0.00
   20            0.00      -14.21
   21            0.00       -3.28
   22            0.00      -26.31
   23            0.00      -13.96
   24            0.00       -5.15
      Inflow      Outflow
     -362.72       362.72
```

Figure 7.13 Output from Program 7.2

PROGRAM 7.3: A GENERAL PROGRAM FOR 2-D OR 3-D STEADY FLOW

```
program p73
!-----------------------------------------------------------------------
!      program 7.3 general program for two- or three-dimensional
!                  analysis of Laplace's equation
!-----------------------------------------------------------------------
use new_library   ;    use  geometry_lib    ;  implicit none
integer::nels,neq,nband,nn,nr,nip,nodof,nod,ndof,i,k,iel,ndim,            &
          loaded_nodes,fixed_nodes,np_types
real::det    ;     character (len=15) :: element
!---------------------------- dynamic arrays---------------------------------
real,allocatable::kv(:),kvh(:),loads(:),disps(:),points(:,:),             &
                  coord(:,:),jac(:,:),der(:,:),deriv(:,:),weights(:),     &
                  prop(:,:),kp(:,:),g_coord(:,:),value(:),kay(:,:)
integer,allocatable::nf(:,:),g(:),num(:),g_num(:,:),g_g( :, :),no(:),     &
                     node(:),etype(:)
!----------------------input and initialisation------------------------------
open (10, file = 'p73.dat' , status = 'old' , action = 'read')
open (11, file = 'p73.res' , status='replace',action = 'write')

read(10,*)element,nels,nn,nip,nodof,nod,ndim,np_types; ndof=nod*nodof
allocate(nf(nodof,nn),points(nip,ndim),g_coord(ndim,nn),coord(nod,ndim), &
         etype(nels),jac(ndim,ndim),weights(nip),num(nod),                &
         g_num(nod,nels),der(ndim,nod),deriv(ndim,nod),kp(ndof,ndof),    &
         g(ndof),g_g(ndof,nels),kay(ndim,ndim),prop(ndim,np_types))
read(10,*)prop
etype=1; if(np_types>1)read(10,*)etype
read(10,*)g_coord; read(10,*)g_num
nf=1; read(10,*)nr; if(nr>0)read(10,*)(k,nf(:,k),i=1,nr)
call formnf(nf); neq=maxval(nf); call sample(element,points,weights)
!------------- loop the elements to find nband and store steering vectors ---
nband=0
elements_1: do iel =1,nels
       num=g_num(:,iel) ; call num_to_g(num,nf,g); g_g(:,iel)=g
       if(nband<bandwidth(g))nband=bandwidth(g)
end do elements_1
write(11,'(a)')"Global coordinates"
do k=1,nn
  write(11,'(a,i5,a,3e12.4)')"Node    ",k,"    ",g_coord(:,k); end do
write(11,'(a)')"Global node numbers"
do k=1,nels
  write(11,'(a,i5,a,20i4)')"Element ",k,"      ",g_num(:,k); end do
write(11,'(2(a,i5),/)')                                                   &
  "There are ",neq,"  equations and the half bandwidth is ",nband
allocate( kv(neq*(nband+1)),kvh(neq*(nband+1)),loads(0:neq),disps(0:neq))
kv=0.0; loads =0.0
!------------- element stiffness integration and assembly-------------------
elements_2: do iel=1,nels
           kay=0.0; do i=1,ndim; kay(i,i)=prop(i,etype(iel)); end do
           num=g_num(:,iel); coord=transpose(g_coord(:,num))
           g=g_g(:,iel); kp=0.0
           integrating_pts_1: do i=1,nip
                 call shape_der(der,points,i); jac=matmul(der,coord)
                 det=determinant(jac); call invert(jac)
                 deriv=matmul(jac,der)
                 kp= kp+matmul(matmul(transpose(deriv),kay),deriv)       &
                     *det*weights(i)
           end do integrating_pts_1
           call formkv(kv,kp,g,neq)
end do elements_2
kvh=kv
read(10,*)loaded_nodes
if(loaded_nodes/=0)read(10,*)(k,loads(nf(:,k)),i=1,loaded_nodes)
read(10,*)fixed_nodes
if(fixed_nodes/=0)then
```

```
        allocate(node(fixed_nodes),no(fixed_nodes),value(fixed_nodes))
        read(10,*)(node(i),value(i),i=1,fixed_nodes)
        do i=1,fixed_nodes; no(i)=nf(1,node(i)); end do
        kv(no)=kv(no)+1.e20; loads(no)=kv(no)*value
  end if
!-------------------------equation solution------------------------------------
  call banred(kv,neq); call bacsub(kv,loads)

!-------------------------retrieve flow rates----------------------------------
  call linmul(kvh,loads,disps)
  write(11,'(a)')"The nodal values are:"
  write(11,'(a)')"                Potentials  Flow rates"
  do k=1,nn
     write(11,'(i5,a,2f12.2)')k,"   ",loads(nf(1,k)),disps(nf(1,k)); end do
  write(11,'(a)')"        Inflow      Outflow"
  write(11,'(2f12.2)')sum(disps,mask=disps<0.),sum(disps,mask=disps>0.)
end program p73
```

The final program in this chapter, Program 7.3, can analyse steady seepage over any 2-d or 3-d domain with non-homogeneous and anisotropic material properties. The program can use any of the 2-d or 3-d elements referred to in this book. For most 3-d applications, the simple eight-node hexahedral element will usually be adequate.

The program is very similar to the earlier programs in the chapter. Some of the variables that were previously fixed in the declaration statements must now be read as data in order to identify the dimensionality of the problem and the type of element required. The first line of data requires the following information:

NELS	number of elements
NN	number of nodes in the mesh
NIP	number of integrating points per element
NODOF	number of degrees of freedom per node
NOD	number of nodes per element
NDIM	number of dimensions
NP_TYPES	number of property types
ELEMENT	element type (character variable)

The next data to be read is the property values for each of the NP_TYPES groups. If the problem is 3-d (NDIM=3), then three permeabilities (k_x, k_y and k_z) must be read for each group. In this chapter it is always assumed that the principal axes of the permeability tensor coincide with the global Cartesian coordinate axes leading to a diagonal property array KAY. If this is not the case, the KAY matrix will be fully populated with the appropriate $x, y[, z]$ components.

If NP_TYPES is greater than 1, the ETYPE vector is read next assigning different property groups to the appropriate elements.

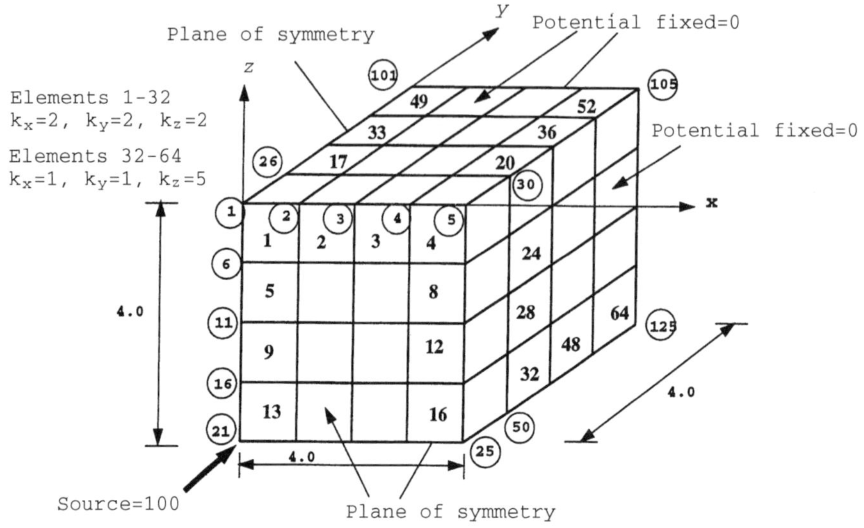

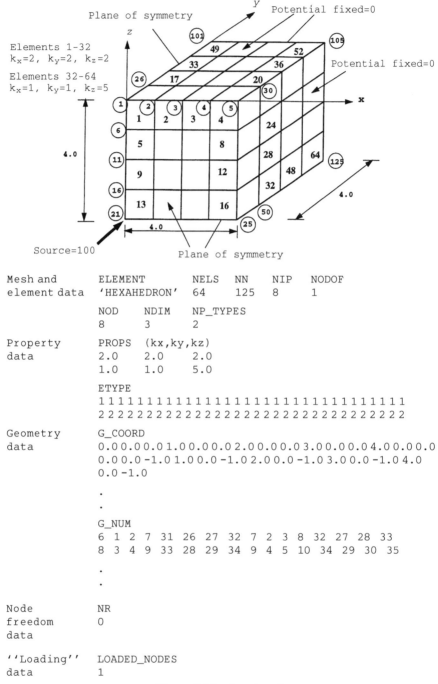

Mesh and element data
ELEMENT 'HEXAHEDRON' — NELS 64 — NN 125 — NIP 8 — NODOF 1

NOD	NDIM	NP_TYPES
8	3	2

Property data
PROPS (kx,ky,kz)

2.0	2.0	2.0
1.0	1.0	5.0

ETYPE
1 1 1 1 1 1 1 1 1 1 1 1 1 1 1 1 1 1 1 1 1 1 1 1 1 1 1 1 1 1 1 1
2 2 2 2 2 2 2 2 2 2 2 2 2 2 2 2 2 2 2 2 2 2 2 2 2 2 2 2 2 2 2 2

Geometry data
G_COORD
0.0 0.0 0.0 1.0 0.0 0.0 2.0 0.0 0.0 3.0 0.0 0.0 4.0 0.0 0.0
0.0 0.0 -1.0 1.0 0.0 -1.0 2.0 0.0 -1.0 3.0 0.0 -1.0 4.0
0.0 -1.0

.
.

G_NUM
6 1 2 7 31 26 27 32 7 2 3 8 32 27 28 33
8 3 4 9 33 28 29 34 9 4 5 10 34 29 30 35

.
.

Node freedom data
NR
0

''Loading'' data
LOADED_NODES
1

Figure 7.14 (*continued*)

```
                K,LOADS(NF(:,K)),I=1,LOADED_NODES
                21   100.0

Fixity          FIXED_NODES
data            61

                NODE(I),VALUE(I),I=1,FIXED_NODES
                 1 0. 2 0. 3 0. 4 0. 5 0. 10 0. 15 0. 20 0. 25 0. 26 0. 27 0.
                28 0. 29 0. 30 0. 35 0. 40 0. 45 0. 50 0. 51 0. 52 0. 53 0. 54 0.
                55 0. 60 0. 65 0. 70 0. 75 0 76 0. 77 0. 78 0. 79 0. 80 0. 85 0.
                76 0. 77 0. 78 0. 79 0. 80 0. 85 0. 90 0. 95 0. 100 0. 101 0.
                102 0.
                103 0. 104 0. 105 0. 106 0. 107 0. 108 0. 109 0. 110 0. 111 0.
                112 0. 113 0.
                110 0. 111 0. 112 0. 113 0. 114 0. 115 0. 116 0. 117 0. 118 0.
                119 0. 120 0.
                119 0. 120 0. 121 0. 122 0. 123 0. 124 0. 125 0.
```

Figure 7.14 Mesh and data for 3-d seepage example with Program 7.3

The next data to be read is the $x, y[, z]$ coordinates of the NN nodes in the mesh followed by the global node numbers of each the NELS elements. If dealing with a 3-d eight-node element for example, the order in which the node numbers are read must follow the sequence described in Figure 3.12.

Next comes the nodal freedom data starting with NR, which indicates the number of "restricted" freedoms in the mesh. In line with the other programs in this chapter, NR is set to zero as all freedoms are kept active in seepage problems.

Next comes the LOADED_NODES data which indicates which nodes if any are to be sources or sinks. Finally FIXED_NODES data is read which gives the boundary conditions where the potential is to be fixed.

A 3-d seepage example is shown in Figure 7.14. The model represents one-eighth of a symmetrical cube with a point source of 100 units at its centroid with all outside faces maintained at a potential of zero. Referring to the figure, node numbers are indicated in circles and some of the element numbers have also been included. The example has 125 nodes and 64 elements. To numerically integrate an 8-node hexahedral element, 8 Gauss points are required (2 in each of the three coordinate directions), so NIP is read as 8.

In this example, there are two material property groups (NP_TYPES=2) in which elements 1 to 32 are isotropic with $k_x = k_y = 2$, and elements 33 to 64 anisotropic with $k_x = k_y = 1$ and $k_z = 5$. The ETYPE vector reads the information required to match elements with property groups. There is one source at node 21 equal to 100.0 indicated in the LOADED_NODES data. All the outside faces of the cube are fixed to zero which requires 61 FIXED_NODES data.

A truncated version of the output from the program is shown in Figure 7.15. The potential is greatest at the central node and computed to be about 169.

```
Global coordinates
Node       1       0.0000E+00  0.0000E+00  0.0000E+00
Node       2       0.1000E+01  0.0000E+00  0.0000E+00
Node       3       0.2000E+01  0.0000E+00  0.0000E+00
Node       4       0.3000E+01  0.0000E+00  0.0000E+00
Node       5       0.4000E+01  0.0000E+00  0.0000E+00
Node       6       0.0000E+00  0.0000E+00 -0.1000E+01
Node       7       0.1000E+01  0.0000E+00 -0.1000E+01
..........................................................
Node     120       0.4000E+01  0.4000E+01 -0.3000E+01
Node     121       0.0000E+00  0.4000E+01 -0.4000E+01
Node     122       0.1000E+01  0.4000E+01 -0.4000E+01
Node     123       0.2000E+01  0.4000E+01 -0.4000E+01
Node     124       0.3000E+01  0.4000E+01 -0.4000E+01
Node     125       0.4000E+01  0.4000E+01 -0.4000E+01
Global node numbers
Element    1          6   1   2   7  31  26  27  32
Element    2          7   2   3   8  32  27  28  33
Element    3          8   3   4   9  33  28  29  34
Element    4          9   4   5  10  34  29  30  35
..........................................................
Element   63         98  93  94  99 123 118 119 124
Element   64         99  94  95 100 124 119 120 125
There are   125  equations and the half bandwidth is     31

The nodal values are:
          Potentials   Flow rates
    1          0.00          -1.54
    2          0.00          -2.73
    3          0.00          -2.03
    4          0.00          -1.03
    5          0.00          -0.35
..............................................................
  124          0.00          -0.35
  125          0.00          -0.13
      Inflow       Outflow
      100.00      -100.00
```

Figure 7.15 Output from Program 7.3

Outflow occurs at all the outside nodes of the mesh (those nodes set initially to a potential of zero). For example, the outflow at node number 20 is −2.68.

7.1 EXAMPLES

1. Derive the element conductivity matrix for a square four-node element suitable for solving Laplace's equation for an isotropic material of permeability k.
2. Using the matrix from the previous question, assemble the global conductivity matrix for the heat conduction problem shown in Figure 7.16 and hence solve for the steady-state internal temperatures.
3. Derive the conductivity matrix of a three-noded, right-angled isosceles triangular element suitable for discretisation of Laplace's equation. Use your element to estimate the steady-state value of the potential at the central node of the mesh with the boundary conditions given in Figure 7.17.
4. A square four-node plane element of side length unity as shown in Figure 7.18 is to be used in the solution of Laplace's equation over a 2-d isotropic

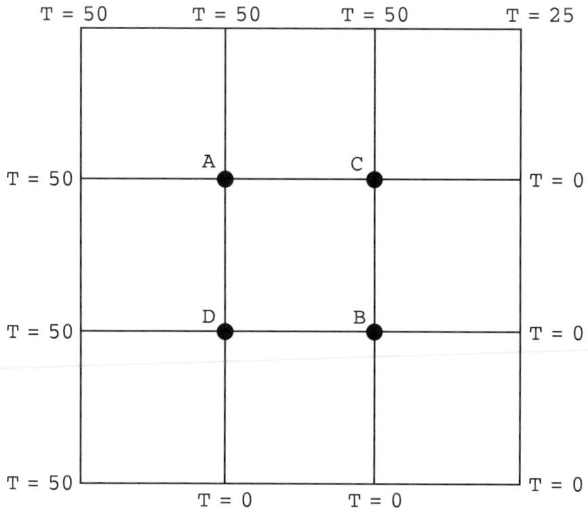

Figure 7.16

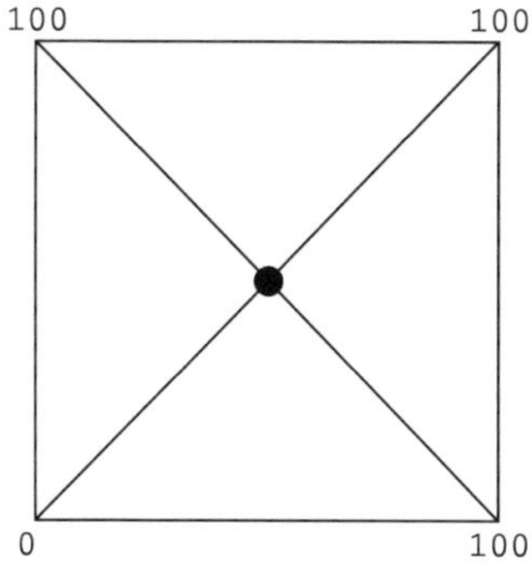

Figure 7.17

medium. If the terms of the element conductivity matrix can be expressed in the form

$$k_{ij} = \int_0^1 \int_0^1 f_{ij}(x, y) \mathrm{d}x \, \mathrm{d}y \quad i, j = 1, 2, 3, 4$$

then find the functions f_{ij} and evaluate k_{11} explicitly.

5. The square region in Figure 7.19 has anisotropic conductivity properties and boundary temperatures fixed at the values indicated. Estimate the steady-state temperature at point A.
6. Steady seepage is taking place along a 1-d pipe containing three porous materials with different permeabilities as indicated in Figure 7.20. The total

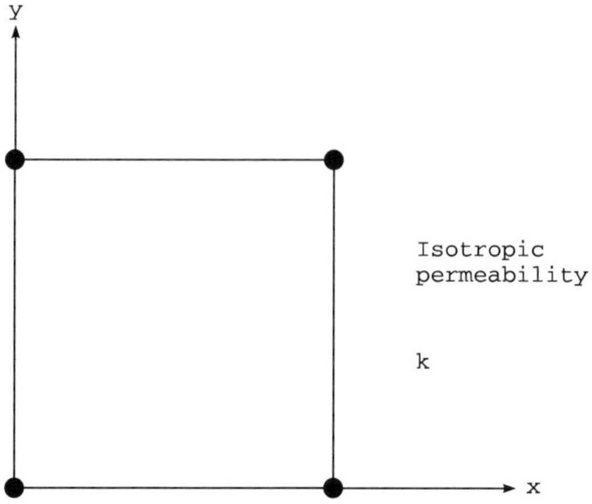

Figure 7.18

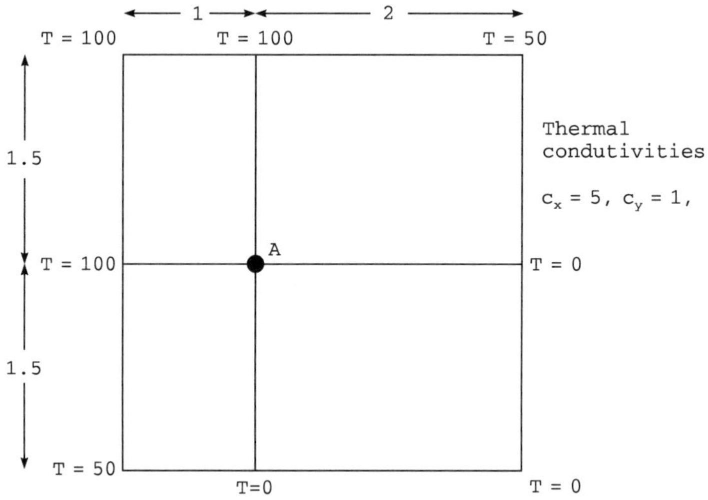

Figure 7.19

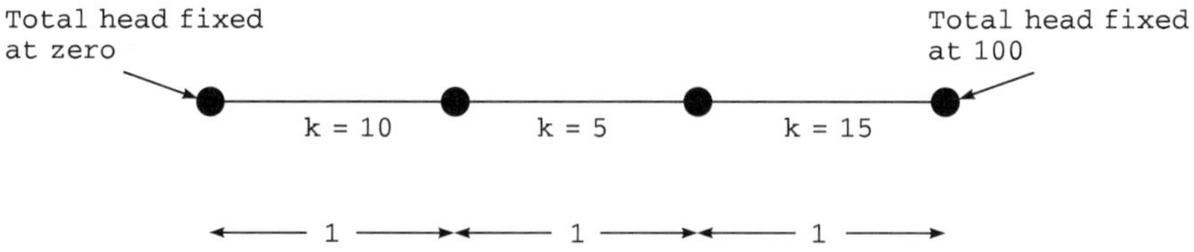

Figure 7.20

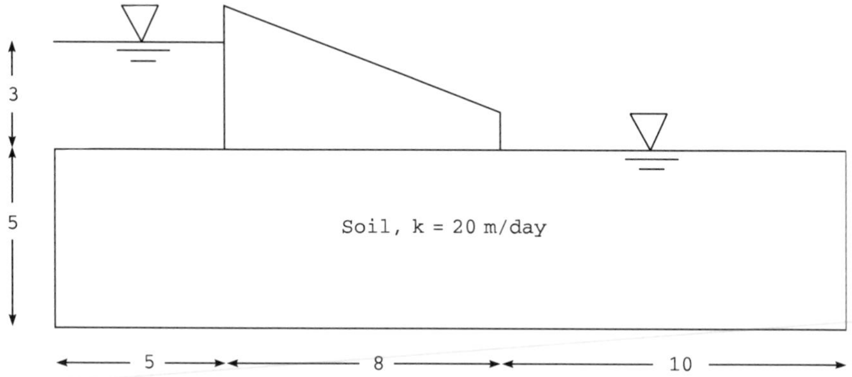

Figure 7.21

head difference between the ends of the pipe is 100 units. Use three 1-d "rod" elements to discretise the steady flow problem and hence compute the total head values at the two intermediate locations along the pipe and the steady flow rate through the pipe.

7. Use Program 7.0 to estimate the steady flow rate under the impermeable dam shown in Figure 7.21.

7.2 REFERENCES

Griffiths, D.V. (1984) Rationalised charts for the method of fragments applied to confined seepage. *Géotechnique*, **34**(2), 229–238.

Verruijt, A. (1970) *Theory of Groundwater Flow.* Macmillan, London.

8 Transient Problems: First Order (Uncoupled)

8.0 INTRODUCTION

In the previous chapter, programs for the solution of steady-state potential flow problems were described. Typically, Laplace's equation (2.99) was discretised into an equilibrium equation (2.100) involving the solution of a set of simultaneous equations. For well-posed problems there are usually no associated numerical difficulties.

When a flow process is transient, or time dependent, the simplest extension of equation (2.99), or reduction of the Navier–Stokes equations, is provided by equations like (2.107). There is still a single dependent variable (for example potential), and so the analysis is "uncoupled". After discretisation in space, a typical element equation is given by equation (2.108). This is a set of first-order ordinary differential equations, the solution of which is no longer a simple numerical task for large numbers of elements.

Some of the many solution techniques available were described in Chapter 3. Possibly the simplest, and most robust, are the "implicit" methods described by equation (3.95) and by the structure chart in Figure 3.20. These θ methods form the basis of Program 8.0. Different strategies, involving "mesh-free" and "explicit" techniques, will be described in later programs.

PROGRAM 8.0: SOLUTION OF THE CONDUCTION EQUATION OVER A RECTANGULAR AREA

In the absence of sources or sinks, equation (3.95) reduces to

$$(\mathbf{PM} + \theta\Delta t\mathbf{KP})\boldsymbol{\phi}_1 = [\mathbf{PM} - (1 - \theta)\Delta t\mathbf{KP}]\boldsymbol{\phi}_0 \tag{8.1}$$

or

$$(\mathbf{PM} + \theta\Delta t\mathbf{KP})\boldsymbol{\phi}_1 = [\mathbf{PM} - \Delta t\mathbf{KP} + \theta\Delta t\mathbf{KP}]\boldsymbol{\phi}_0 \tag{8.2}$$

which is the form used in the present program. The element **KP** matrices are assembled, including the multiple $\theta\Delta t$ (`THETA*DTIM` in program terminology), into a global matrix **BK** and the **PM** matrices are assembled into the global matrix **BP**. Thus, the global system to be solved is

$$(\mathbf{BP} + \mathbf{BK})\boldsymbol{\phi}_1 = \left(\mathbf{BP} + \mathbf{BK} - \frac{1}{\theta}\mathbf{BK}\right)\boldsymbol{\phi}_0 \tag{8.3}$$

with **BP** and **BK** being stored as vectors using the assembly routine `FORMKV`. For constant θ and Δt the left-hand side of (8.3) is constant and so the strategy

```
   program p80
!-----------------------------------------------------------------------
!      program 8.0 conduction equation on rectangular
!      area using 4-node quadrilateral elements
!      implicit integration in time using 'theta' method
!-----------------------------------------------------------------------
 use new_library    ;  use geometry_lib  ;      implicit none
 integer::nels,nxe,neq,nband,nn,nr,nip,nodof=1,nod=4,ndof,ndim=2,        &
          i,j,k,iel,nstep,npri,nres
 real::aa,bb,permx,permy,det,theta,dtim,val0,time
 character(len=15) :: element = 'quadrilateral'
!------------------------------ dynamic arrays-------------------------------
 real ,allocatable :: bp(:),bk(:),loads(:),points(:,:),kay(:,:),coord(:,:),  &
                      fun(:),jac(:,:),der(:,:),deriv(:,:),weights(:),        &
                      kp(:,:), pm(:,:), newlo(:) ,funny(:,:),g_coord(:,:)
 integer, allocatable :: nf(:,:), g(:) , num(:) , g_num(:,:) ,g_g(:,:)
!-----------------------input and initialisation---------------------------
  open (10,file='p80.dat',status=    'old',action='read')
  open (11,file='p80.res',status='replace',action='write')
  read (10,*) nels,nxe,nn,nip,aa,bb,permx,permy ,                          &
              dtim,nstep,theta,npri,nres
  ndof=nod*nodof
  allocate ( nf(nodof,nn), points(nip,ndim),weights(nip),kay(ndim,ndim),   &
            coord(nod,ndim), fun(nod), jac(ndim,ndim),g_coord(ndim,nn),     &
            der(ndim,nod), deriv(ndim,nod), pm(ndof,ndof),g_num(nod,nels),  &
            kp(ndof,ndof), g(ndof),funny(1,nod),num(nod),g_g(ndof,nels))
  kay=0.0 ; kay(1,1)=permx; kay(2,2)=permy
  call sample (element,points,weights)
  nf=1; read(10,*) nr ; if(nr>0)read(10,*)(k,nf(:,k),i=1,nr)
  call formnf(nf)  ;    neq=maxval(nf)
!-------------loop the elements to find nband and set up global arrays---------
   nband = 0
   elements_1: do iel = 1 , nels
               call geometry_4qx(iel,nxe,aa,bb,coord,num)
               g_num(:,iel) = num; g_coord(: , num ) = transpose(coord)
               call num_to_g (num,nf,g);   g_g( : , iel ) = g
               if(nband<bandwidth(g)) nband = bandwidth(g)
   end do elements_1
    write(11,'(a)') "Global coordinates "
    do k=1,nn;write(11,'(a,i5,a,2e12.4)')"Node",k,"        ",g_coord(:,k);end do
    write(11,'(a)') "Global node numbers "
    do k = 1 , nels; write(11,'(a,i5,a,4i5)')                                &
                              "Element ",k,"        ",g_num(:,k); end do
    allocate(bp(neq*(nband+1)),bk(neq*(nband+1)),loads(0:neq),newlo(0:neq))
      bp = 0.;  bk = 0.
      write(11,'(2(a,i5))')                                                  &
              "There are ",neq,"  equations and the half-bandwidth is ",nband
!-------------------- element integration and assembly-------------------------
 elements_2: do iel = 1 , nels
             num = g_num(:,iel) ; coord = transpose( g_coord( : , num ))
             g = g_g( : , iel )      ;      kp=0.0 ; pm=0.0
      gauss_pts:  do i =1 , nip
               call shape_der (der,points,i) ; call shape_fun(fun,points,i)
               funny(1,:)=fun(:) ; jac = matmul(der,coord)
               det=determinant(jac); call invert(jac);deriv = matmul(jac,der)
               kp = kp + matmul(matmul(transpose(deriv),kay),deriv) &
                    *det*weights(i)*theta*dtim
               pm  =  pm + matmul( transpose(funny),funny)*det*weights(i)
      end do gauss_pts
   call formkv (bk,kp,g,neq) ; call formkv(bp,pm,g,neq)
 end do elements_2
!-----------------------factorise left hand side-----------------------------
          bp=bp+bk; bk=bp-bk/theta ; call banred(bp,neq)
    read(10,*) val0; loads=val0
```

```
!--------------------time stepping recursion-----------------------------------
   write(11,'(a)') "    Time      Pressure"
 timesteps: do j=1,nstep
               time=j*dtim ; call linmul(bk,loads,newlo)
               call bacsub(bp,newlo) ;  loads=newlo
               if(j/npri*npri==j)write(11,'(2e12.4)')time,loads(nres)
            end do timesteps
end program p80
```

will be to form **BP** + **BK**, factorise the resulting matrix once only and then for each Δt to carry out the matrix-by-vector multiplication on the right-hand side of (8.3) followed by a forward and backward substitution. The process is described in detail by the structure chart in Figure 8.1. Note, however, that the

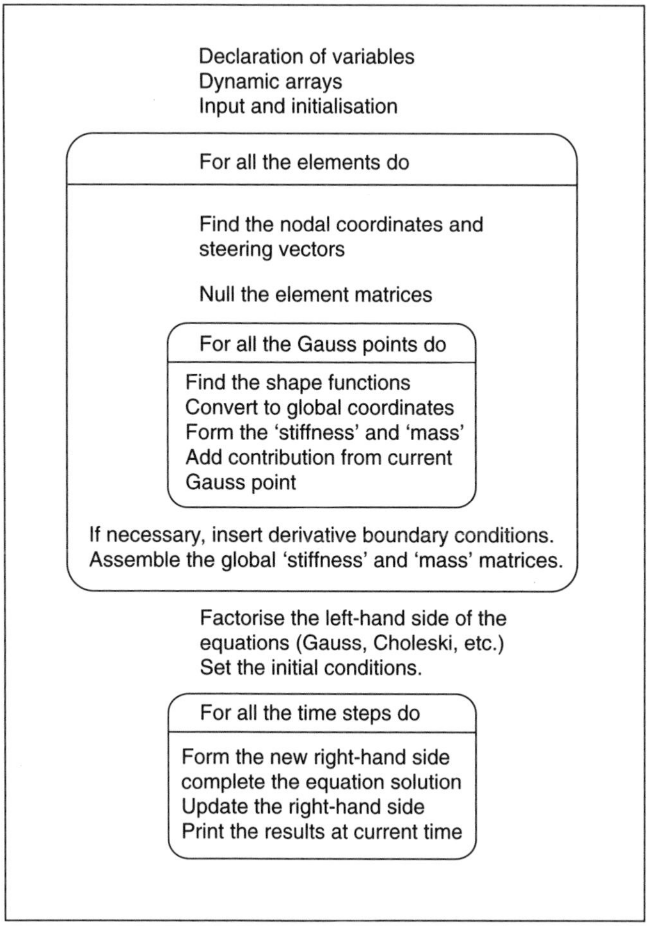

Figure 8.1 Structure chart for implicit analyses of transient problems

matrix-by-vector multiplication on the right-hand side could be done using element-by-element summation, avoiding storage of one large matrix.

The example chosen is shown in Figure 8.2 and could represent dissipation of excess porewater pressure from a rectangle of soil. Due to fourfold symmetry

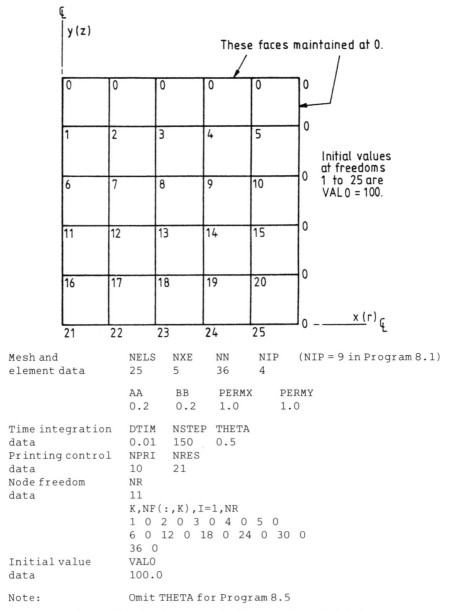

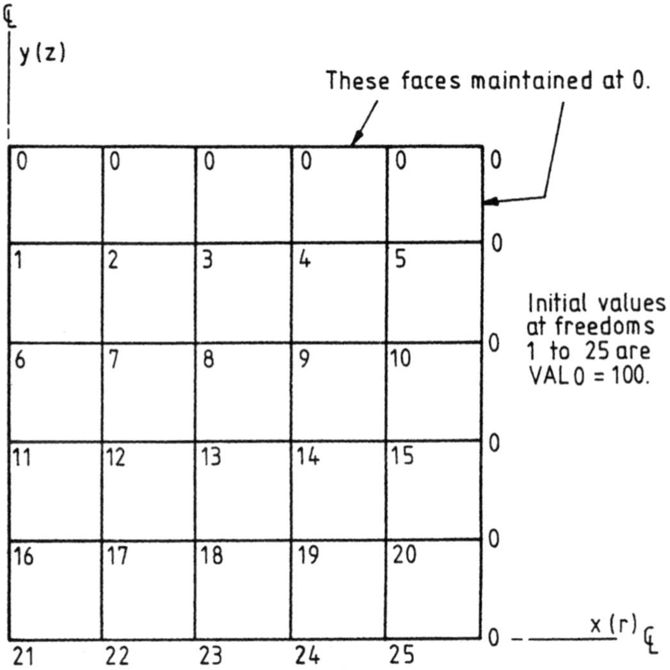

Mesh and element data	NELS 25	NXE 5	NN 36	NIP 4	(NIP = 9 in Program 8.1)
	AA 0.2	BB 0.2	PERMX 1.0	PERMY 1.0	
Time integration data	DTIM 0.01	NSTEP 150	THETA 0.5		
Printing control data	NPRI 10	NRES 21			
Node freedom data	NR 11				
	K,NF(:,K),I=1,NR 1 0 2 0 3 0 4 0 5 0 6 0 12 0 18 0 24 0 30 0 36 0				
Initial value data	VAL0 100.0				
Note:	Omit THETA for Program 8.5				

Figure 8.2 Mesh and data for Programs 8.0, 8.1, 8.5, 8.6

two boundaries have zero pressure while the remainder of the soil has constant initial pressure VAL0. The problem is to compute the pressures as time passes. The early part of the program, involving element integration and assembly, closely resembles Program 7.0. Since four-noded elements are used, with

```
Global coordinates
Node    1          0.0000E+00  0.0000E+00
Node    2          0.2000E00   0.0000E+00
Node    3          0.4000E00   0.0000E+00
Node    4          0.6000E00   0.0000E+00
Node    5          0.8000E00   0.0000E+00
Node    6          0.1000E+01  0.0000E+00
...................................................................

Node   30          0.1000E+01 -0.8000E00
Node   31          0.0000E+00 -0.1000E+01
Node   32          0.2000E00  -0.1000E+01
Node   33          0.4000E00  -0.1000E+01
Node   34          0.6000E00  -0.1000E+01
Node   35          0.8000E00  -0.1000E+01
Node   36          0.1000E+01 -0.1000E+01
Global node numbers
Element     1            7    1    2    8
Element     2            8    2    3    9
Element     3            9    3    4   10
Element     4           10    4    5   11
Element     5           11    5    6   12
Element     6           13    7    8   14
Element     7           14    8    9   15
Element     8           15    9   10   16
Element     9           16   10   11   17
Element    10           17   11   12   18
Element    11           19   13   14   20
Element    12           20   14   15   21
Element    13           21   15   16   22
Element    14           22   16   17   23
Element    15           23   17   18   24
Element    16           25   19   20   26
Element    17           26   20   21   27
Element    18           27   21   22   28
Element    19           28   22   23   29
Element    20           29   23   24   30
Element    21           31   25   26   32
Element    22           32   26   27   33
Element    23           33   27   28   34
Element    24           34   28   29   35
Element    25           35   29   30   36
There are    25  equations and the half-bandwidth is      6
    Time      Pressure
  0.1000E00   0.9009E+02
  0.2000E00   0.5845E+02
  0.3000E00   0.3580E+02
  0.4000E00   0.2178E+02
  0.5000E00   0.1324E+02
  0.6000E00   0.8051E+01
  0.7000E00   0.4895E+01
  0.8000E00   0.2976E+01
  0.9000E00   0.1809E+01
  0.1000E+01   0.1100E+01
  0.1100E+01   0.6687E00
  0.1200E+01   0.4065E00
  0.1300E+01   0.2471E00
  0.1400E+01   0.1503E00
  0.1500E+01   0.9135E-01
```

Figure 8.3 Results from Program 8.0

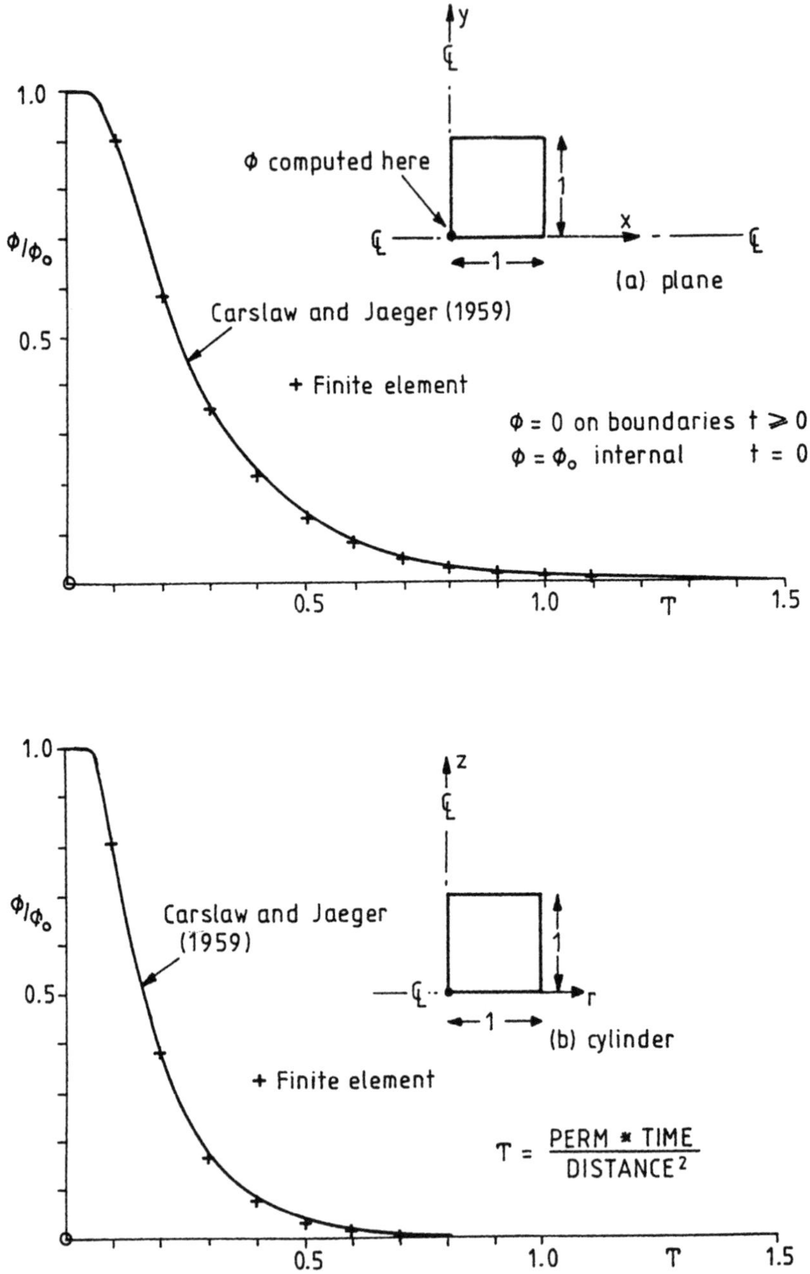

Figure 8.4 Comparisons of finite element results with series solutions

numbering in the x direction, the appropriate geometry subroutine is GEOMETRY_4QX. The elements have coefficients of consolidation in the x and y directions, denoted by PERMX and PERMY (assumed constant for all elements). At the end of the element assembly **BP** and **BK**, as required by (8.3), have been stored.

In the section headed "factorisation of the left-hand side", **BP** is reset to **BP** + **BK**, and **BK** is reset to (new) **BP**-$(1/\theta)$**BK**. The matrix **BP** is then factorised using BANRED and the non-zero initial pressures set to VAL0 (in this case 100.0 units).

The final section of the program consists of the time-stepping loop completed NSTEP times. The matrix-by-vector multiplication is carried out by LINMUL and forward and backward substitution by BACSUB. The pressure at node 25 (NRES), the centre of the mesh, is output every NPRI (10 in this case) time steps, that is at time intervals of 0.1, and is listed in Figure 8.3. The results are plotted in Figure 8.4(a), where they are compared with series solution values obtained by Carslaw and Jaeger (1959). The crude finite element idealisation gives excellent results.

PROGRAM 8.1: SOLUTION OF THE CONDUCTION EQUATION OVER A CYLINDER

In the same way as a few modifications to Program 5.3 for plane elasticity led to Program 5.4 for axisymmetric conditions, Program 8.0 is readily modified to cylindrical coordinates. The coordinate x is associated with the radial coordinate r while y is associated with the axial coordinate z.

The mesh in Figure 8.2 now represents a symmetrical quarter of a cross-section through a right cylinder. In soil mechanics, the physical analogue would be a "triaxial" specimen of soil draining from all its boundaries. However, the data is identical to that for the plane case. Integration in the circumferential direction is carried out over 1 radian and the only additional code needed is to form RADIUS which is the radius of the current Gauss point (contained in the axisymmetric integrals, e.g. equation (3.55)).

Program 8.1 differs from Program 8.0 only in the four lines from RADIUS= .0 onwards.

The results are listed in Figure 8.5 and compared with series solutions in Figure 8.4(b). The agreement is rather poorer than for the plane case due to the approximate integration as r tends to zero. However, the results are perfectly acceptable. Note that nine-point Gaussian integration was used while in Program 8.0 four-point is sufficient.

To give an idea of the accuracy obtained for points other than the centre of a cylinder, Figure 8.6 shows results after 100 steps with DTIM= 0.01 for a 5×10 quadrant measuring 2.5 units radially and 5.0 units vertically. Otherwise, parameters are those of Figure 8.2.

```
   program p81
!-------------------------------------------------------------------------
!      program 8.1 conduction equation on rectangular section of a
!      cylindrical area using 4-node quadrilateral elements
!      implicit integration in time using 'theta' method
!-------------------------------------------------------------------------
 use new_library    ;  use geometry_lib  ;      implicit none
 integer::nels,nxe,neq,nband,nn,nr,nip,nodof=1,nod=4,ndof,ndim=2,           &
          i,j,k,iel,nstep,npri,nres
 real::aa,bb,permx,permy,det,theta,dtim,val0,time,radius
 character (len=15) :: element = 'quadrilateral'
!------------------------- dynamic arrays-------------------------------------
 real ,allocatable :: bp(:),bk(:),loads(:),points(:,:),kay(:,:),coord(:,:),  &
                      fun(:),jac(:,:),der(:,:),deriv(:,:),weights(:),        &
                      kp(:,:), pm(:,:), newlo(:) ,funny(:,:),g_coord(:,:)
 integer, allocatable :: nf(:,:), g(:) , num(:) , g_num(:,:) ,g_g(:,:)
!------------------------input and initialisation---------------------------
  open (10,file='p81.dat',status=    'old',action='read')
  open (11,file='p81.res',status='replace',action='write')
  read (10,*) nels,nxe,nn,nip,aa,bb,permx,permy ,                            &
              dtim,nstep,theta,npri,nres
  ndof=nod*nodof
  allocate ( nf(nodof,nn), points(nip,ndim),weights(nip),kay(ndim,ndim),     &
           coord(nod,ndim), fun(nod), jac(ndim,ndim),g_coord(ndim,nn),       &
           der(ndim,nod), deriv(ndim,nod), pm(ndof,ndof),g_num(nod,nels),    &
           kp(ndof,ndof), g(ndof),funny(1,nod),num(nod),g_g(ndof,nels))
  kay=0.0 ; kay(1,1)=permx; kay(2,2)=permy
  call sample(element,points,weights)
  nf=1; read(10,*) nr ; if(nr>0)read(10,*)(k,nf(:,k),i=1,nr)
  call formnf(nf);neq=maxval(nf)
!------------loop the elements to find nband and set up global arrays----------
   nband = 0
   elements_1: do iel = 1 , nels
                call geometry_4qx(iel,nxe,aa,bb,coord,num)
                g_num(:,iel) = num; g_coord(: , num ) = transpose(coord)
                call num_to_g(num,nf,g);  g_g( : , iel ) = g
                if(nband<bandwidth(g)) nband = bandwidth(g)
   end do elements_1
    write(11,'(a)') "Global coordinates "
    do k=1,nn;write(11,'(a,i5,a,2e12.4)')"Node",k,"       ",g_coord(:,k);end do
    write(11,'(a)') "Global node numbers "
    do k = 1 , nels; write(11,'(a,i5,a,4i5)')                                   &
                            "Element ",k,"        ",g_num(:,k); end do
    allocate(bp(neq*(nband+1)),bk(neq*(nband+1)),loads(0:neq),newlo(0:neq))
      bp = 0.;  bk = 0.
      write(11,'(2(a,i5))')                                                       &
               "There are ",neq," equations and the half-bandwidth is ",nband
!-------------------- element integration and assembly-----------------------
 elements_2: do iel = 1 , nels
              num = g_num(:,iel) ; coord = transpose( g_coord( : , num ))
              g = g_g( : , iel )      ;      kp=0.0 ; pm=0.0
        gauss_pts:  do i =1 , nip
                call shape_der (der,points,i) ; call shape_fun(fun,points,i)
                funny(1,:)=fun(:) ; jac = matmul(der,coord)
                det=determinant(jac); call invert(jac);deriv = matmul(jac,der)
                radius = .0
                do k=1,nod; radius=radius+fun(k)*coord(k,1); end do
                kp = kp + matmul(matmul(transpose(deriv),kay),deriv) &
                     *det*weights(i)*theta*dtim*radius
                pm  = pm + matmul(transpose(funny),funny)*det*weights(i)*radius
        end do gauss_pts
   call formkv (bk,kp,g,neq) ; call formkv(bp,pm,g,neq)
 end do elements_2
!-----------------------factorise left hand side-----------------------------
```

```
          bp=bp+bk; bk=bp-bk/theta ; call banred(bp,neq)
    read(10,*) val0; loads=val0
!------------------time stepping recursion---------------------------------
   write(11,'(a)') "    Time      Pressure"
  timesteps: do j=1,nstep
               time=j*dtim ; call linmul(bk,loads,newlo)
               call bacsub(bp,newlo) ;  loads=newlo
               if(j/npri*npri==j)write(11,'(2e12.4)')time,loads(nres)
            end do timesteps
 end program p81
```

```
Global coordinates
Node    1          0.0000E+00  0.0000E+00
Node    2          0.2000E00  0.0000E+00
Node    3          0.4000E00  0.0000E+00
Node    4          0.6000E00  0.0000E+00
Node    5          0.8000E00  0.0000E+00
Node    6          0.1000E+01  0.0000E+00
.....................................................................

Node   32          0.2000E00 -0.1000E+01
Node   33          0.4000E00 -0.1000E+01
Node   34          0.6000E00 -0.1000E+01
Node   35          0.8000E00 -0.1000E+01
Node   36          0.1000E+01 -0.1000E+01
Global node numbers
Element    1            7    1    2    8
Element    2            8    2    3    9
Element    3            9    3    4   10
Element    4           10    4    5   11
Element    5           11    5    6   12
Element    6           13    7    8   14
Element    7           14    8    9   15
Element    8           15    9   10   16
Element    9           16   10   11   17
Element   10           17   11   12   18
Element   11           19   13   14   20
Element   12           20   14   15   21
Element   13           21   15   16   22
Element   14           22   16   17   23
Element   15           23   17   18   24
Element   16           25   19   20   26
Element   17           26   20   21   27
Element   18           27   21   22   28
Element   19           28   22   23   29
Element   20           29   23   24   30
Element   21           31   25   26   32
Element   22           32   26   27   33
Element   23           33   27   28   34
Element   24           34   28   29   35
Element   25           35   29   30   36
There are    25 equations and the half-bandwidth is     6
    Time     Pressure
  0.1000E00  0.8096E+02
  0.2000E00  0.3798E+02
  0.3000E00  0.1667E+02
  0.4000E00  0.7280E+01
  0.5000E00  0.3176E+01
  0.6000E00  0.1386E+01
  0.7000E00  0.6046E00
  0.8000E00  0.2638E00
  0.9000E00  0.1151E00
  0.1000E+01  0.5022E-01
  0.1100E+01  0.2191E-01
  0.1200E+01  0.9560E-02
  0.1300E+01  0.4171E-02
  0.1400E+01  0.1820E-02
  0.1500E+01  0.7940E-03
```

Figure 8.5 Results from Program 8.1

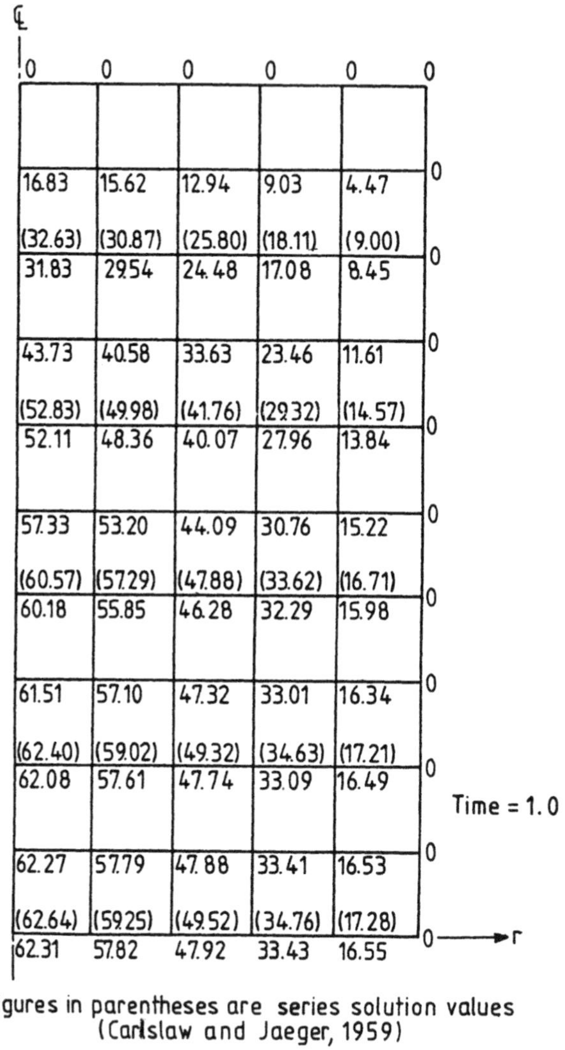
Time = 1.0
Figures in parentheses are series solution values
(Carlslaw and Jaeger, 1959)

Figure 8.6 Results from triaxial consolidation analysis

PROGRAM 8.2: DIFFUSION–CONVECTION EQUATION OVER A RECTANGLE—TRANSFORMED ANALYSIS

When convection terms are retained in the simplified flow equations, (2.94) or (2.109) have to be solved. Again many techniques could be employed but, in the present program, an implicit algorithm based on equation (3.94) is used. Thus this program is an extension of Program 8.0.

```
    program p82
!------------------------------------------------------------------------
!      program 8.2 diffusion - convection equation on rectangular
!      area using 4-node quadrilateral elements
!      self-adjoint transformation
!      implicit integration in time using 'theta' method
!------------------------------------------------------------------------
 use new_library   ;  use geometry_lib    ;    implicit none
 integer::nels,nxe,neq,nband,nn,nr,nip,nodof=1,nod=4,ndof,ndim=2,          &
         i,j,k,iel,nstep,npri
 real::aa,bb,permx,permy,det,theta,dtim,ux,uy,time,f1,f2
 character (len=15) :: element = 'quadrilateral'
!--------------------------- dynamic arrays-------------------------------
real ,allocatable ::kb(:,:),pb(:,:),loads(:),points(:,:),kay(:,:),coord(:,:),&
                    fun(:),jac(:,:),der(:,:),deriv(:,:),weights(:),         &
                    kp(:,:), pm(:,:), ans(:) ,funny(:,:),g_coord(:,:)
 integer, allocatable :: nf(:,:), g(:) , num(:) , g_num(:,:) ,g_g(:,:)
!------------------------input and initialisation-------------------------
  open (10,file='p82.dat',status=    'old',action='read')
  open (11,file='p82.res',status='replace',action='write')
  read (10,*) nels,nxe,nn,nip,aa,bb,permx,permy,ux,uy,                     &
              dtim,nstep,theta,npri
  ndof=nod*nodof
  allocate ( nf(nodof,nn), points(nip,ndim),weights(nip),kay(ndim,ndim),    &
           coord(nod,ndim), fun(nod), jac(ndim,ndim),g_coord(ndim,nn),       &
           der(ndim,nod), deriv(ndim,nod), pm(ndof,ndof),g_num(nod,nels),    &
           kp(ndof,ndof), g(ndof),funny(1,nod),num(nod),g_g(ndof,nels))
  kay=0.0 ; kay(1,1)=permx; kay(2,2)=permy
  call sample(element,points,weights)
  nf=1;  read(10,*) nr ; if(nr>0)read(10,*)(k,nf(:,k),i=1,nr)
  call formnf(nf);neq=maxval(nf)
!----------loop the elements to find nband and set up global arrays ----------
  nband = 0
   elements_1: do iel = 1 , nels
             call geometry_4qx(iel,nxe,aa,bb,coord,num)
             g_num(:,iel) = num; g_coord(:,num) = transpose(coord)
             call num_to_g(num,nf,g);  g_g( : , iel ) = g
             if(nband<bandwidth(g)) nband = bandwidth(g)
   end do elements_1
    write(11,'(a)') "Global coordinates "
    do k=1,nn;write(11,'(a,i5,a,2e12.4)')"Node",k,"       ",g_coord(:,k);end do
    write(11,'(a)') "Global node numbers "
    do k = 1 , nels; write(11,'(a,i5,a,4i5)')                                 &
                        "Element ",k,"        ",g_num(:,k); end do
    allocate(kb(neq,nband+1),pb(neq,nband+1),loads(0:neq),ans(0:neq))
      kb = 0.;  pb = 0. ; loads = .0
      write(11,'(2(a,i5))')                                                   &
             "There are ",neq," equations and the half-bandwidth is ",nband
!------- element integration and assembly-----------------------------------
elements_2: do iel = 1 , nels
           num = g_num(:,iel) ; coord =  transpose(g_coord(: , num ))
           g = g_g( : , iel )    ;     kp=0.0 ; pm=0.0
      gauss_pts:  do i =1 , nip
             call shape_der (der,points,i) ; call shape_fun(fun,points,i)
             funny(1,:)=fun(:) ; jac = matmul(der,coord)
             det=determinant(jac); call invert(jac); deriv = matmul(jac,der)
             kp = kp + matmul(matmul(transpose(deriv),kay),deriv) &
                  *det*weights(i)
             pm  =  pm + matmul( transpose(funny),funny)*det*weights(i)
      end do gauss_pts
             kp = kp + pm*(ux*ux/permx+uy*uy/permy)*.25
             pm = pm/(theta*dtim)
!------------------- derivative boundary conditions --------------------------
      if(iel==1) then
```

```
          kp(2,2)=kp(2,2)+uy*aa/6.; kp(2,3)=kp(2,3)+uy*aa/12.
          kp(3,2)=kp(3,2)+uy*aa/12.; kp(3,3)=kp(3,3)+uy*aa/6.
       else if(iel==nels) then
         kp(1,1)=kp(1,1)+uy*aa/6.; kp(1,4)=kp(1,4)+uy*aa/12.
         kp(4,1)=kp(4,1)+uy*aa/12.; kp(4,4)=kp(4,4)+uy*aa/6.
       end if
    call formkb (kb,kp,g) ; call formkb(pb,pm,g)
 end do elements_2
!-----------------------factorise left hand side----------------------------
          f1=uy*aa/(2.*theta); f2 = f1
          pb = pb + kb; kb = pb - kb/theta ; call cholin(pb)
!------------------time stepping recursion-----------------------------------
   write(11,'(a)') "    Time      Concentration"
 timesteps: do j=1,nstep
              time=j*dtim ; call banmul(kb,loads,ans)
              ans(neq)=ans(neq)+f1; ans(neq-1) = ans(neq-1)+f2
              call chobac(pb,ans) ;  loads=ans
              if(j/npri*npri==j)write(11,'(2e12.4)')  &
                           time,loads(neq)*exp(ux/2./permx)*exp(uy/2./permy)
         end do timesteps
end program p82
```

When the transformation of equation (2.111) is employed, the equation to be solved becomes

$$c_x \frac{\partial^2 h}{\partial x^2} + c_y \frac{\partial^2 h}{\partial y^2} - \left(\frac{u^2}{4c_x} + \frac{v^2}{4c_y}\right) h = \frac{\partial h}{\partial t} \tag{8.4}$$

Thus the extra term involving h distinguishes the process from a simple diffusion one. However, reference to Table 2.1 shows that the semi-discretised "stiffness" matrix for this problem will still be symmetrical, the h term involving an element matrix of the "mass matrix" type $\int\int N_i N_j \, dx \, dy$.

Comparison with Program 8.0 will show essentially the same array declarations and input parameters, although Choleski rather than Gauss factorisation is used. Extra variables are the velocities in the x and y directions, `UX` and `UY` respectively.

The problem chosen is the 1-d example shown in Figure 8.7. The dependent variable ϕ refers to concentration of sediment picked up by the flow from the base of the mesh, and distributed with time in the y direction. Thus, `UX` is zero and for numerical reasons `PERMX` is set to a small number, 1×10^{-6}, which is effectively zero. The equation to be solved is in effect

$$c_y \frac{\partial^2 h}{\partial y^2} - \frac{v^2}{4c_y} h = \frac{\partial h}{\partial t} \tag{8.5}$$

subject to the boundary conditions

$$\frac{\partial \phi}{\partial y} = \frac{v}{c_y} = C_2 \quad \text{(constant)} \tag{8.6}$$

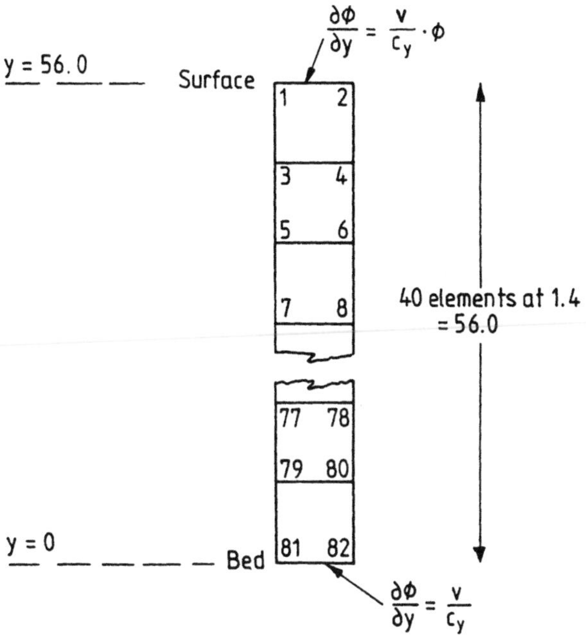

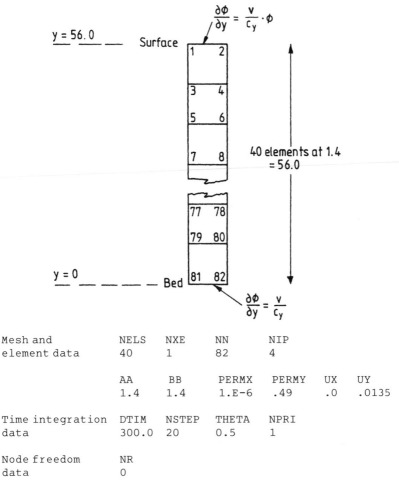

Mesh and element data	NELS	NXE	NN	NIP		
	40	1	82	4		
	AA	BB	PERMX	PERMY	UX	UY
	1.4	1.4	1.E-6	.49	.0	.0135
Time integration data	DTIM	NSTEP	THETA	NPRI		
	300.0	20	0.5	1		
Node freedom data	NR					
	0					

Figure 8.7 Mesh and data for Program 8.2

at $y = 0$ and

$$\frac{\partial \phi}{\partial y} = \frac{v}{c_y}\phi = C_1\phi \quad \text{where } C_1 \text{ is constant} \tag{8.7}$$

at $y = 56.0$.

After transformation, these conditions become

$$\frac{\partial h}{\partial y} = \frac{-v}{2c_y}h + \frac{v}{c_y} \tag{8.8}$$

and

$$\frac{\partial h}{\partial y} = +\frac{v}{2c_y}h \tag{8.9}$$

Boundary condition (8.9) is clearly of the type described in section 3.8, equation (3.29). Therefore, at that boundary, the element matrix will have to be

```
Global coordinates
Node    1          0.0000E+00  0.0000E+00
Node    2          0.1400E+01  0.0000E+00
Node    3          0.0000E+00 -0.1400E+01
Node    4          0.1400E+01 -0.1400E+01
Node    5          0.0000E+00 -0.2800E+01
..............................................................

Node   76          0.1400E+01 -0.5180E+02
Node   77          0.0000E+00 -0.5320E+02
Node   78          0.1400E+01 -0.5320E+02
Node   79          0.0000E+00 -0.5460E+02
Node   80          0.1400E+01 -0.5460E+02
Node   81          0.0000E+00 -0.5600E+02
Node   82          0.1400E+01 -0.5600E+02
Global node numbers
Element     1            3    1    2    4
Element     2            5    3    4    6
Element     3            7    5    6    8
Element     4            9    7    8   10
Element     5           11    9   10   12
Element     6           13   11   12   14
Element     7           15   13   14   16
Element     8           17   15   16   18
Element     9           19   17   18   20
Element    10           21   19   20   22
..............................................................

Element    34           69   67   68   70
Element    35           71   69   70   72
Element    36           73   71   72   74
Element    37           75   73   74   76
Element    38           77   75   76   78
Element    39           79   77   78   80
Element    40           81   79   80   82
There are     82 equations and the half-bandwidth is     3
     Time      Concentration
  0.3000E+03  0.2867E00
  0.6000E+03  0.4067E00
  0.9000E+03  0.4862E00
  0.1200E+04  0.5464E00
  0.1500E+04  0.5949E00
  0.1800E+04  0.6356E00
  0.2100E+04  0.6708E00
  0.2400E+04  0.7016E00
  0.2700E+04  0.7292E00
  0.3000E+04  0.7539E00
  0.3300E+04  0.7762E00
  0.3600E+04  0.7965E00
  0.3900E+04  0.8149E00
  0.4200E+04  0.8318E00
  0.4500E+04  0.8472E00
  0.4800E+04  0.8612E00
  0.5100E+04  0.8741E00
  0.5400E+04  0.8859E00
  0.5700E+04  0.8966E00
  0.6000E+04  0.9065E00
```

Figure 8.8 Results from Program 8.2

augmented by the matrix shown in equation (3.33). The multiple $C_1 c_y(x_k - x_j)/6$ in (3.33) is just $v(x_k - x_j)/12$ or `UY*AA/12` in the program. This is carried out in the section of program headed "insert derivative boundary conditions".

The condition (8.8) contains a similar contribution, but in addition the term v/c_y is of the type described by the equation (3.30). Thus, an addition must be made to the right-hand side of the equations at such a boundary in accordance with equation (3.35). In this case the terms in equation (3.35) are $v(x_k - x_j)/2$ and are incorporated in the program immediately after the comment "factorisation of left-hand side".

This example shows that quite complicated coding would be necessary to permit very general boundary conditions to be specified in all problems. The authors prefer to write specialised code when necessary.

Comparing Program 8.2 with Program 8.0, it can be seen that the array storages `KB` and `PB` are used for global matrices in place of vectors `BK` and `BP`. In the present example there are no zero freedoms, `NR` = 0. In the element integration and assembly loop `PART1` accumulates the diffusive part of the element "stiffness" and `PART2` the convective part.

After insertion of boundary conditions the (constant) global left-hand side is factorised using `CHOLIN`. The time-stepping loop is as before. Output for 20 steps is listed as Figure 8.8, while Figure 8.9 shows how the finite element solution compares with an "analytical" one due to Dobbins (1944).

It should be remembered that solutions are in terms of the transformed variable h, and the true solution ϕ has been recovered using (2.111).

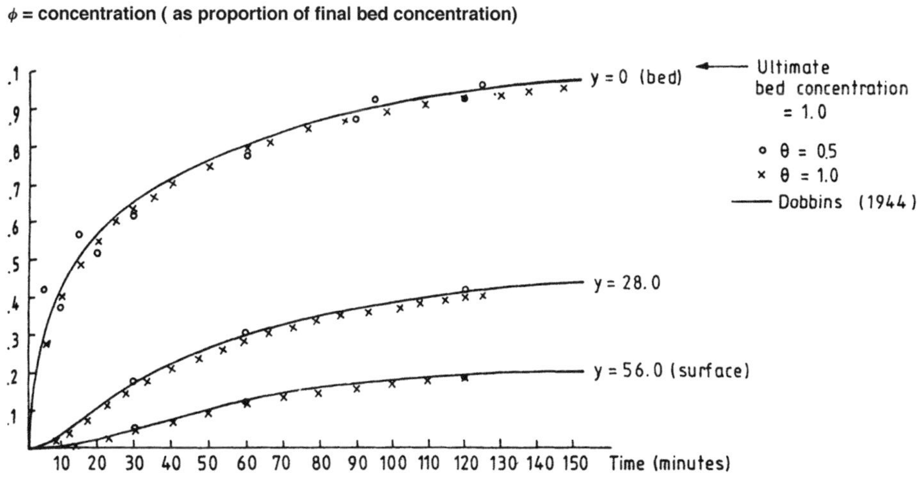

Figure 8.9 Graph of concentration versus time for Program 8.2

PROGRAM 8.3: DIFFUSION–CONVECTION EQUATION OVER A RECTANGLE—UNTRANSFORMED ANALYSIS

```
     program p83
!-----------------------------------------------------------------------
!      program 8.3 diffusion - convection equation on rectangular
!      area using 4-node quadrilateral elements
!      untransformed solution by Galerkin's method
!      implicit integration in time using 'theta' method
!-----------------------------------------------------------------------
 use new_library        ;  use  geometry_lib        ;    implicit none
 integer::nels,nxe,neq,nband,nn,nr,nip,nodof=1,nod=4,ndof,ndim=2,          &
         i,j,k,l,iel,nstep,npri,nfix
 real::aa,bb,permx,permy,det,theta,dtim,ux,uy,time,part1,part2
 character (len=15) :: element = 'quadrilateral'
!------------------------- dynamic arrays------------------------------------
real ,allocatable ::kb(:,:),pb(:,:),loads(:),points(:,:),kay(:,:),coord(:,:),&
                    fun(:),jac(:,:),der(:,:),deriv(:,:),weights(:),            &
                    kp(:,:), pm(:,:), ans(:) ,funny(:,:),g_coord(:,:),         &
                    storpb(:),work(:,:),copy(:,:) , dtkd(:,:)
integer, allocatable :: nf(:,:), g(:) , num(:) , g_num(:,:) ,g_g(:,:) , no(:)
!------------------------input and initialisation-----------------------------
  open (10,file='p83.dat',status=    'old',action='read')
  open (11,file='p83.res',status='replace',action='write')
  read (10,*) nels,nxe,nn,nip,aa,bb,permx,permy,ux,uy,                      &
             dtim,nstep,theta,npri,nfix
  ndof=nod*nodof
  allocate ( nf(nodof,nn), points(nip,ndim),weights(nip),kay(ndim,ndim),    &
            coord(nod,ndim), fun(nod),jac(ndim,ndim),g_coord(ndim,nn)      ,&
            der(ndim,nod), deriv(ndim,nod), pm(ndof,ndof),g_num(nod,nels), &
            kp(ndof,ndof), g(ndof),funny(1,nod),num(nod),g_g(ndof,nels),   &
            no(nfix),storpb(nfix),dtkd(ndof,ndof))
  kay=0.0 ; kay(1,1)=permx; kay(2,2)=permy
  call sample(element,points,weights)
  nf=1;  read(10,*) nr ; if(nr>0)read(10,*)(k,nf(:,k),i=1,nr)
  call formnf(nf);neq=maxval(nf)
!-------------loop the elements to find nband and set up global arrays---------
   nband = 0
   elements_1: do iel = 1 , nels
               call geometry_4qx(iel,nxe,aa,bb,coord,num)
               g_num( : , iel ) = num; g_coord(:,num) = transpose(coord)
               call num_to_g(num,nf,g)  ;   g_g( : , iel ) = g
               if(nband<bandwidth(g)) nband = bandwidth(g)
   end do elements_1
    write(11,'(a)') "Global coordinates "
    do k=1,nn;write(11,'(a,i5,a,2e12.4)')"Node",k,"       ",g_coord(:,k);end do
    write(11,'(a)') "Global node numbers "
    do k = 1 , nels; write(11,'(a,i5,a,4i5)')                                 &
                          "Element ",k,"        ",g_num(:,k); end do
    allocate(kb(neq,2*nband+1),pb(neq,2*nband+1),loads(0:neq),ans(0:neq),&
            work(nband+1,neq),copy(nband+1,neq))
      kb = 0.;  pb = 0. ; work = .0 ; loads = .0
      write(11,'(2(a,i5))')                                                    &
              "There are ",neq," equations and the half-bandwidth is",nband
!------------------ element integration and assembly------------------------
 elements_2: do iel = 1 , nels
             num = g_num(: , iel ) ; coord =  transpose(g_coord( : , num ))
             g = g_g( : , iel )     ;     kp=0.0 ; pm=0.0
       gauss_pts:  do i =1 , nip
               call shape_der (der,points,i) ; call shape_fun(fun,points,i)
               funny(1,:)=fun(:) ; jac = matmul(der,coord)
               det=determinant(jac); call invert(jac); deriv = matmul(jac,der)
               do k=1,nod;do l=1,nod
                 part1=permx*deriv(1,k)*deriv(1,l)+permy*deriv(2,k)*deriv(2,l)
                 part2=ux*fun(k)*deriv(1,l)+uy*fun(k)*deriv(2,l)
                 dtkd(k,l)=(part1-part2)*det*weights(i)
               end do; end do
```

```
            kp = kp + dtkd
            pm  =  pm + matmul( transpose(funny),funny)*det*weights(i)
     end do gauss_pts
            pm = pm/(theta*dtim)
   call formtb (kb,kp,g) ; call formtb(pb,pm,g)
 end do elements_2
!------------------------specify fixed nodal values --------------------------
          pb = pb + kb; kb = pb - kb / theta    ; read(10,*) no
          pb(no,nband+1) = pb(no,nband+1) + 1.e20 ; storpb = pb(no,nband+1)
!------------------------factorise left hand side-----------------------------
          call gauss_band(pb,work)
!------------------time stepping recursion-----------------------------------
   write(11,'(a)') "    Time      Concentration"
 timesteps: do j=1,nstep
            time=j*dtim ; copy = work ; call bantmul(kb,loads,ans); ans(0)=.0
            if(time<=.2) then ;ans(no)=storpb; else; ans(no) = .0; end if
            call solve_band(pb,copy,ans) ;  ans(0)=.0; loads=ans
            if(j/npri*npri==j)write(11,'(2e12.4)') time,loads(3)
          end do timesteps
end program p83
```

In the previous example, difficulties would have arisen had the ratio of u to c_x been large, because the transformation involving exp (u/c_x) would not have been numerically feasible (Smith et al, 1973). Under these circumstances (and even when $c_x = c_y = 0$) equation (2.109) can still be solved, but with the drawback that the element and system matrices become unsymmetrical, although the latter are still banded.

Program 8.3 will be used to solve a purely convective problem, that is with $c_x = c_y = 0$. The problem chosen is 1-d (see Figure 8.10) so the equation is effectively

$$-v\frac{\partial \phi}{\partial y} = \frac{\partial \phi}{\partial t} \tag{8.10}$$

in the region $0 \le y \le 2$, subject to the boundary conditions $\phi = 1$ at $y = 0$ for $0 \le t \le 0.2, \phi = 0$ at $y = 0$ for $t > 0.2$ and $\partial\phi/\partial y = 0$ at $y = 2$ for all t.

Comparing with Program 8.2, the usual geometry routine for four-noded elements numbered in the x direction, GEOMETRY_4QX, is used. Arrays NO and STORPB are used to read in the numbers of fixed freedoms and to store information about them during the time-stepping process. The system KB matrix is now unsymmetrical and is formed as a total band using FORMTB. Although PB is symmetrical it too is stored as a full band to be compatible with KB.

The PART2 contribution to the KP matrix is the only one in the present example and it is unsymmetrical (skew-symmetrical in fact).

The structure of the program is modelled on the previous one. There are no zero freedoms and the solution routines are GAUSS_BAND and SOLVE_BAND which use an extra array WORK as working space. Matrix-by-vector multiplication needs the subroutine BANTMUL (see Chapter 3).

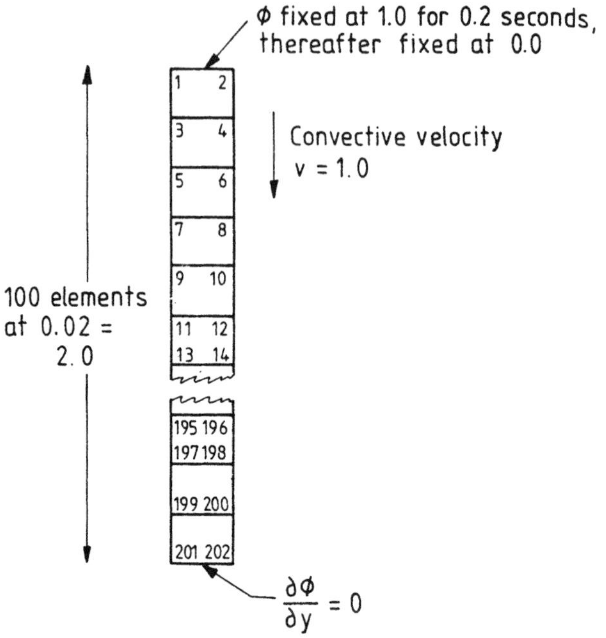


```
Mesh and            NELS    NXE     NN      NIP
element data        100     1       202     4

                    AA      BB      PERMX   PERMY   UX      UY
                    .02     .02     .0      .0      .0      1.0

Time integration    DTIM    NSTEP   THETA
data                .04     1       .5

Fixed freedom       NR
data                0
                    NFIX    (NO(I),I=1,NFIX)
                    2       1 2
```

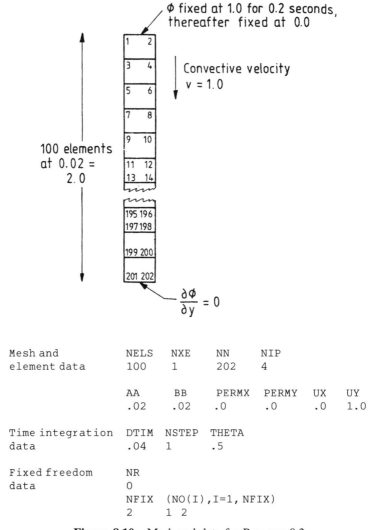

Figure 8.10 Mesh and data for Program 8.3

In the section 'time-stepping recursion' it can be seen that the solution at nodes 1 and 2 is held at the value 1.0 for the first 0.2 seconds of convection.

The variation of concentration with time at node 3 is shown in Figure 8.11. The correct solution to the problem is described by a rectangular pulse moving with unit velocity in the *y* direction. Figure 8.12 shows the computed

```
Global coordinates
Node    1          0.0000E+00  0.0000E+00
Node    2          0.2000E-01  0.0000E+00
Node    3          0.0000E+00 -0.2000E-01
Node    4          0.2000E-01 -0.2000E-01
Node    5          0.0000E+00 -0.4000E-01
..................................................................

Node  198          0.2000E-01 -0.1960E+01
Node  199          0.0000E+00 -0.1980E+01
Node  200          0.2000E-01 -0.1980E+01
Node  201          0.0000E+00 -0.2000E+01
Node  202          0.2000E-01 -0.2000E+01
Global node numbers
Element     1            3    1    2    4
Element     2            5    3    4    6
Element     3            7    5    6    8
Element     4            9    7    8   10
Element     5           11    9   10   12
Element     6           13   11   12   14
..................................................................

Element    95          191  189  190  192
Element    96          193  191  192  194
Element    97          195  193  194  196
Element    98          197  195  196  198
Element    99          199  197  198  200
Element   100          201  199  200  202
There are   202  equations and the half-bandwidth is     3
    Time      Concentration
  0.4000E-01  0.3660E00
  0.8000E-01  0.1116E+01
  0.1200E00  0.1066E+01
  0.1600E00  0.9285E00
  0.2000E00  0.1017E+01
  0.2400E00  0.6577E00
  0.2800E00 -0.1436E00
  0.3200E00 -0.5885E-01
  0.3600E00  0.8405E-01
  0.4000E00 -0.3271E-01
  0.4400E00 -0.1985E-01
  0.4800E00  0.3561E-01
  0.5200E00 -0.1689E-01
  0.5600E00 -0.1002E-01
  0.6000E00  0.2089E-01
  0.6400E00 -0.1101E-01
  0.6800E00 -0.6154E-02
  0.7200E00  0.1419E-01
  0.7600E00 -0.7981E-02
  0.8000E00 -0.4226E-02
  0.8400E00  0.1046E-01
  0.8800E00 -0.6148E-02
  0.9200E00 -0.3116E-02
  0.9600E00  0.8139E-02
  0.1000E+01 -0.4933E-02
```

Figure 8.11 Results from Program 8.3

solution after one second. Spurious spatial oscillations are seen to have been introduced by the numerical solution. Measures to improve the solutions are beyond the scope of the present treatment, but see Smith (1976, 1979). Of course, in the present case improvements can be achieved by simply reducing Δy and Δt.

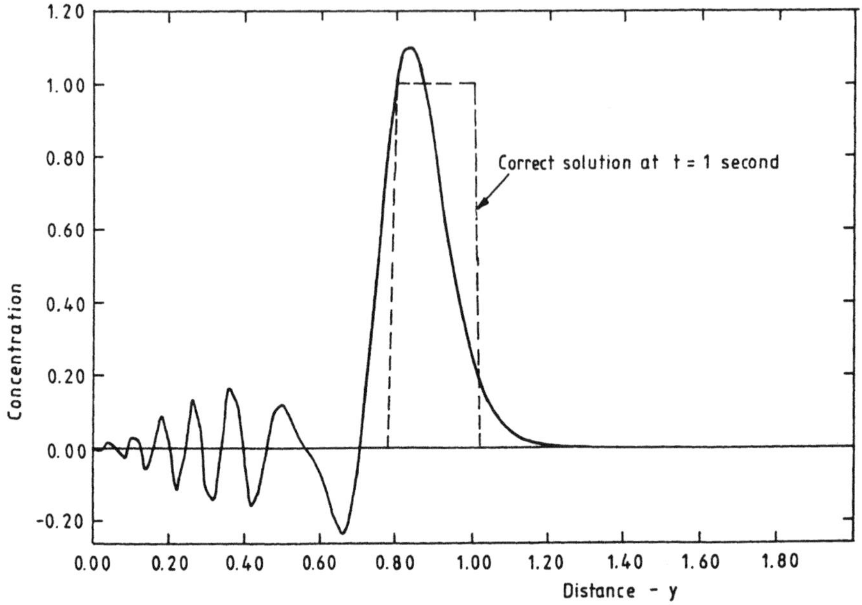

Figure 8.12 Solution from Program 8.3 after 1 second

8.1 MESH-FREE STRATEGIES

The first four programs in this chapter have used implicit integration in time and assembly strategies for mesh "stiffness" and "mass" matrices. For very large problems, the demands on computer storage become significant and, as was the case in previous chapters, it is natural to seek "mesh-free" strategies which avoid the need to store system matrices at all.

Possibly the simplest (and oldest) mesh-free strategy is based on equation (3.98) which shows that, for lumped "mass" matrix **PM**, the solution at a new time can be found from the solution at the previous time by a simple matrix–vector multiplication which can be done element-by-element. Such "explicit" techniques suffer from numerical stability difficulties and are described later. First we consider the natural extension to Program 8.0 which was seen to consist of a linear equation solution on every time step. Clearly this solution can be accomplished iteratively using p.c.g. or some similar technique, and this is done in Program 8.4.

PROGRAM 8.4: SOLUTION OF THE CONDUCTION EQUATION OVER A RECTANGLE USING AN IMPLICIT METHOD: MESH-FREE PCG VERSION

In this program the element **PM** and **KP** matrices are stored in the forms $\mathbf{PM} + \theta\Delta t\mathbf{KP}$ and $\mathbf{PM} - (1 - \theta)\Delta t\mathbf{KP}$ as required by equation (3.95). The

```
   program p84
!-------------------------------------------------------------------------
!      program 8.4 conduction equation on rectangular
!      area using 4-node quadrilateral elements : pcg version
!      implicit integration in time using 'theta' method
!-------------------------------------------------------------------------
 use new_library    ;  use geometry_lib  ;      implicit none
 integer::nels,nxe,neq,nn,nr,nip,nodof=1,nod=4,ndof,ndim=2,              &
         i,j,k,iel,nstep,npri,nres,iters,limit
 real::aa,bb,permx,permy,det,theta,dtim,val0,time,tol,alpha,beta,up,big
 logical::converged      ; character(len=15) :: element='quadrilateral'
!------------------------ dynamic arrays-----------------------------------
 real ,allocatable :: loads(:),u(:),p(:),points(:,:),kay(:,:),coord(:,:),   &
                      fun(:),jac(:,:),der(:,:),deriv(:,:),weights(:),d(:),   &
                      kp(:,:), pm(:,:), funny(:,:),g_coord(:,:),             &
                      storka(:,:,:),storkb(:,:,:),x(:),xnew(:),pmul(:),      &
                      utemp(:),diag_precon(:)
 integer, allocatable :: nf(:,:), g(:) , num(:) , g_num(:,:) ,g_g(:,:)
!----------------------input and initialisation--------------------------------
  open (10,file='p84.dat',status=    'old',action='read')
  open (11,file='p84.res',status='replace',action='write')
  read (10,*) nels,nxe,nn,nip,aa,bb,permx,permy ,                          &
             dtim,nstep,theta,npri,nres,tol,limit
  ndof=nod*nodof
  allocate ( nf(nodof,nn), points(nip,ndim),weights(nip),kay(ndim,ndim),   &
           coord(nod,ndim), fun(nod), jac(ndim,ndim),g_coord(ndim,nn),     &
           der(ndim,nod), deriv(ndim,nod), pm(ndof,ndof),g_num(nod,nels),  &
           kp(ndof,ndof), g(ndof),funny(1,nod),num(nod),g_g(ndof,nels),    &
           storka(ndof,ndof,nels),storkb(ndof,ndof,nels),utemp(ndof),      &
           pmul(ndof))
  kay=0.0 ; kay(1,1)=permx; kay(2,2)=permy
  call sample (element,points,weights)
  nf=1; read(10,*) nr ; if(nr>0)read(10,*)(k,nf(:,k),i=1,nr)
  call formnf(nf);neq=maxval(nf)
!-------------loop the elements to  set up global arrays ---------------------
   elements_1: do iel = 1 , nels
             call geometry_4qx(iel,nxe,aa,bb,coord,num)
             g_num(:,iel) = num; g_coord(: , num ) = transpose(coord)
             call num_to_g (num,nf,g);    g_g( : , iel ) = g
   end do elements_1
   write(11,'(a)') "Global coordinates "
   do k=1,nn;write(11,'(a,i5,a,2e12.4)')"Node",k,"       ",g_coord(:,k);end do
   write(11,'(a)')  "Global node numbers "
   do k = 1 , nels; write(11,'(a,i5,a,4i5)')                                 &
                             "Element ",k,"        ",g_num(:,k); end do
   allocate(loads(0:neq),diag_precon(0:neq),u(0:neq),d(0:neq),p(0:neq),      &
           x(0:neq),xnew(0:neq)) ; storka = .0; storkb = .0
     write(11,'(a,i5,a)') "There are ",neq," equations to be solved"
     p = .0; diag_precon = .0; xnew = .0
!----------- element integration ,storage and build preconditioner -----------
   elements_2: do iel = 1 , nels
           num = g_num(:,iel) ; coord = transpose( g_coord( : , num ))
           g = g_g( : , iel )     ;     kp=0.0 ; pm=0.0
      gauss_pts:  do i =1 , nip
             call shape_der (der,points,i) ; call shape_fun(fun,points,i)
             funny(1,:)=fun(:) ; jac = matmul(der,coord)
             det=determinant(jac); call invert(jac); deriv = matmul(jac,der)
             kp = kp + matmul(matmul(transpose(deriv),kay),deriv) &
                  *det*weights(i)
             pm  =  pm + matmul( transpose(funny),funny)*det*weights(i)
      end do gauss_pts
      storka(:,:,iel)=pm+kp*theta*dtim;storkb(:,:,iel)=pm-kp*(1.-theta)*dtim
    do k=1,ndof; diag_precon(g(k))=diag_precon(g(k))+ storka(k,k,iel); end do
   end do elements_2
```

```
    diag_precon(1:neq) = 1./diag_precon(1:neq) ; diag_precon(0) = .0
!---------------------------initial conditions --------------------------------
      read(10,*) val0; loads=val0   ; loads(0) = .0
!---------------------time stepping recursion ---------------------------------
   write(11,'(a)') "    Time     Pressure  Iterations"
  timesteps: do j=1,nstep
              time=j*dtim
    u = .0
    elements_3 : do iel = 1 , nels    ! gather for rhs multiply
                    g = g_g( : , iel ) ; kp = storkb(:, :, iel)
                    pmul = loads(g) ; utemp = matmul(kp , pmul)
                    u ( g ) = u ( g ) + utemp   ! scatter
    end do elements_3
    u(0) = .0 ; loads = u
!------------------- solve simultaneous equations by pcg ----------------------
      d = diag_precon*loads; p = d; x = .0
      iters = 0
     iterations  :       do
           iters = iters + 1     ;    u = 0.
      elements_4 : do iel = 1, nels
                      g = g_g( : , iel ) ; kp = storka(:,:,iel);  pmul = p(g)
                      utemp = matmul(kp,pmul); u(g) = u(g)+ utemp
      end do elements_4    ; u(0) = .0
!-------------------------pcg equation solution-------------------------------
         up=dot_product(loads,d); alpha= up/ dot_product(p,u)
         xnew = x + p* alpha ; loads=loads - u*alpha;  d = diag_precon*loads
         beta=dot_product(loads,d)/up; p=d+p*beta
         big = .0; converged = .true.
         u = xnew ; where(u<.0) u = -u; big = maxval(u)
         u = (xnew - x)/big ;where(u<.0) u=-u ; big = maxval(u)
         if(big>tol) converged=.false.   ;     x=xnew
         if(converged .or. iters==limit) exit
     end do iterations
              loads=xnew
              if(j/npri*npri==j) then
               write(11,'(2e12.4,i5)')time,loads(nres),iters
              end if
  end do timesteps
end program p84
```

Mesh and element data	NELS 25	NXE 5	NN 36	NIP 4
	AA 0.2	BB 0.2	PERMX 1.0	PERMY 1.0
Time integration data	DTIM 0.01	NSTEP 150	THETA 0.5	
Printing control data	NPRI 10	NRES 21		
Iteration data	TOL .0001	LIMIT 100		
Node freedom data	NR 11 K,NF,(:,K),I=1,NR 1 0 2 0 3 0 4 0 5 0 6 0 12 0 18 0 24 0 30 0 36 0			
Initial value data	VAL0 100.0			

Figure 8.13 Input data for Program 8.4

```
Global coordinates
Node    1          0.0000E+00  0.0000E+00
Node    2          0.2000E00  0.0000E+00
Node    3          0.4000E00  0.0000E+00
Node    4          0.6000E00  0.0000E+00
Node    5          0.8000E00  0.0000E+00
Node    6          0.1000E+01  0.0000E+00
.....................................................................

Node   32          0.2000E00 -0.1000E+01
Node   33          0.4000E00 -0.1000E+01
Node   34          0.6000E00 -0.1000E+01
Node   35          0.8000E00 -0.1000E+01
Node   36          0.1000E+01 -0.1000E+01
Global node numbers
Element      1           7    1    2    8
Element      2           8    2    3    9
Element      3           9    3    4   10
Element      4          10    4    5   11
Element      5          11    5    6   12
Element      6          13    7    8   14
Element      7          14    8    9   15
Element      8          15    9   10   16
Element      9          16   10   11   17
Element     10          17   11   12   18
Element     11          19   13   14   20
Element     12          20   14   15   21
Element     13          21   15   16   22
Element     14          22   16   17   23
Element     15          23   17   18   24
Element     16          25   19   20   26
Element     17          26   20   21   27
Element     18          27   21   22   28
Element     19          28   22   23   29
Element     20          29   23   24   30
Element     21          31   25   26   32
Element     22          32   26   27   33
Element     23          33   27   28   34
Element     24          34   28   29   35
Element     25          35   29   30   36
There are     25  equations to be solved
    Time        Pressure   Iterations
  0.1000E00   0.9009E+02     3
  0.2000E00   0.5845E+02     3
  0.3000E00   0.3580E+02     2
  0.4000E00   0.2178E+02     2
  0.5000E00   0.1324E+02     2
  0.6000E00   0.8051E+01     2
  0.7000E00   0.4895E+01     2
  0.8000E00   0.2976E+01     2
  0.9000E00   0.1809E+01     2
  0.1000E+01   0.1100E+01     2
  0.1100E+01   0.6687E00     2
  0.1200E+01   0.4065E00     2
  0.1300E+01   0.2472E00     2
  0.1400E+01   0.1503E00     2
  0.1500E+01   0.9135E-01     2
```

Figure 8.14 Output from Program 8.4

diagonal preconditioner is formed from the diagonal terms of the first of these as it would have been assembled. The only additional inputs compared to Program 8.0 are the iteration tolerance, `TOL`, and the limiting number of p.c.g. iterations, `LIMIT`. The data is shown as Figure 8.13 with output as Figure 8.14. For an iteration tolerance of 0.0001 and the same time step as was used in Program 8.0, the p.c.g. process converges in at most three iterations and leads to the same solution as given in Figure 8.3.

PROGRAM 8.5: SOLUTION OF THE CONDUCTION EQUATION OVER A RECTANGLE USING AN EXPLICIT METHOD

```
    program p85
!-------------------------------------------------------------------------
!      program 8.5 conduction equation on rectangular area using 4-node
!      quadrilateral elements and a simple explicit algorithm
!-------------------------------------------------------------------------
 use new_library     ; use  geometry_lib      ;     implicit none
 integer::nels,nxe,neq,nn,nr,nip,nodof=1,nod=4,ndof,ndim=2,               &
          i,j,k,iel,nstep,npri,nres
 real::aa,bb,permx,permy,det,dtim,val0,time
 character (len=15) :: element = 'quadrilateral'
!------------------------- dynamic arrays---------------------------------
 real ,allocatable :: loads(:),points(:,:),kay(:,:),coord(:,:),mass(:),   &
                      jac(:,:),der(:,:),deriv(:,:),weights(:),kp(:,:),    &
                      pm(:,:), funny(:,:),g_coord(:,:),globma(:),fun(:),  &
                      store_pm(:,:,:), newlo(:)
 integer, allocatable :: nf(:,:), g(:) , num(:) , g_num(:,:) ,g_g(:,:)
!-----------------------input and initialisation--------------------------
  open (10,file='p85.dat',status=    'old',action='read')
  open (11,file='p85.res',status='replace',action='write')
  read (10,*) nels,nxe,nn,nip,aa,bb,permx,permy                           &
             dtim,nstep,npri,nres
  ndof=nod*nodof
  allocate ( nf(nodof,nn), points(nip,ndim),weights(nip),kay(ndim,ndim),  &
            coord(nod,ndim), fun(nod), jac(ndim,ndim),g_coord(ndim,nn),   &
            der(ndim,nod), deriv(ndim,nod), pm(ndof,ndof),g_num(nod,nels),&
            kp(ndof,ndof), g(ndof),funny(1,nod),num(nod), g_g(ndof,nels), &
            globma(0:nn),store_pm(ndof,ndof,nels),mass(ndof))
  kay=0.0 ; kay(1,1)=permx; kay(2,2)=permy
  globma = .0   ;  do i = 1,nn; nf(nodof,i)=i; end do
  call sample(element,points,weights)
!------------ loop the elements for integration and to store globals  --------
 elements_1: do iel = 1 , nels
             call geometry_4qx(iel,nxe,aa,bb,coord,num)
             g_num(:,iel) = num;  g_coord(:,num) =transpose( coord )
             call num_to_g(num,nf,g);     kp=0.0 ; pm=0.0
      gauss_pts:  do i =1 , nip
               call shape_der (der,points,i) ; call shape_fun(fun,points,i)
               funny(1,:)=fun(:) ; jac = matmul(der,coord)
               det=determinant(jac); call invert(jac); deriv = matmul(jac,der)
               kp=kp+matmul(matmul(transpose(deriv),kay),deriv)*det*weights(i)
               pm  =  pm + matmul( transpose(funny),funny)*det*weights(i)
      end do gauss_pts
     do i=1,ndof; mass(i) = sum(pm(i,:)); end do
     pm = .0 ; do i = 1 , ndof; pm(i,i) = mass(i); end do
     store_pm(:,:,iel) = pm - kp*dtim  ; globma(g) = globma(g) + mass
 end do elements_1
   write(11,'(a)') "Global coordinates "
   do k=1,nn;write(11,'(a,i5,a,2e12.4)')"Node",k,"       ",g_coord(:,k);end do
   write(11,'(a)') "Global node numbers "
   do k = 1 , nels; write(11,'(a,i5,a,4i5)')                              &
                         "Element ",k,"       ",g_num(:,k); end do
!--------------take account of initial and boundary conditions----------------
  nf=1; read(10,*) nr ;if(nr>0)read(10,*)(k,nf(:,k),i=1,nr)
  call formnf(nf);neq=maxval(nf)
  write(11,'(a,i5)') "The number of equations is : ",   neq
  allocate(loads(0:neq), newlo(0:neq))
  j = 0        ;      globma(0) = .0
  do i=1,nn; if(nf(1,i)/=0)then;j=j+1;globma(j)=1./globma(i);end if ; end do
  read(10,*) val0; loads=val0   ; loads(0) = .0
!--------------- go round the elements  for revised g -----------------------
   elements_2 : do iel = 1 , nels   ! g is different
                   call geometry_4qx(iel,nxe,aa,bb,coord,num)
                   call num_to_g(num,nf,g);   g_g ( : , iel) = g
   end do elements_2
```

```
!-------------------time stepping recursion-----------------------------------
   write(11,'(a)') "    Time      Pressure"
  timesteps: do j=1,nstep
                time=j*dtim         ; newlo = .0
!---------------- go round the elements  ------------------------------------
    elements_3 : do iel = 1 , nels   ! g is new one
                    g = g_g ( : , iel) ;   pm = store_pm(: , : , iel)
                    loads(0) = .0 ; mass = loads(g) ; fun = matmul(pm , mass)
                    newlo ( g ) = newlo ( g ) + fun
    end do elements_3
    newlo(0) = .0;     loads = newlo * globma
  if(j/npri*npri==j)write(11,'(2e12.4)')time,loads(nres)
 end do timesteps
end program p85
```

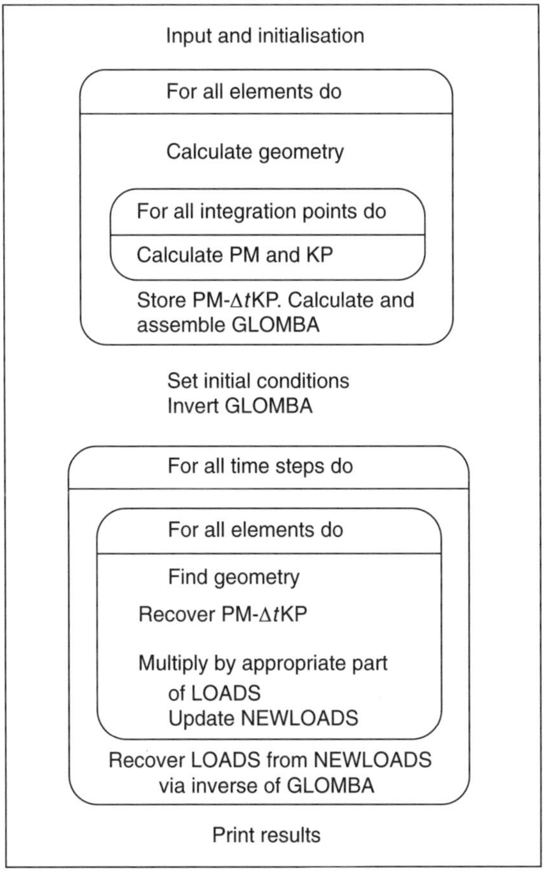

Figure 8.15 Structure chart for explicit time integration

```
Global coordinates
Node    1          0.0000E+00  0.0000E+00
Node    2          0.2000E00  0.0000E+00
Node    3          0.4000E00  0.0000E+00
Node    4          0.6000E00  0.0000E+00
Node    5          0.8000E00  0.0000E+00
Node    6          0.1000E+01  0.0000E+00
..............................................................

Node   32          0.2000E00 -0.1000E+01
Node   33          0.4000E00 -0.1000E+01
Node   34          0.6000E00 -0.1000E+01
Node   35          0.8000E00 -0.1000E+01
Node   36          0.1000E+01 -0.1000E+01
Global node numbers
Element     1             7    1    2    8
Element     2             8    2    3    9
Element     3             9    3    4   10
Element     4            10    4    5   11
Element     5            11    5    6   12
Element     6            13    7    8   14
Element     7            14    8    9   15
Element     8            15    9   10   16
Element     9            16   10   11   17
Element    10            17   11   12   18
Element    11            19   13   14   20
Element    12            20   14   15   21
Element    13            21   15   16   22
Element    14            22   16   17   23
Element    15            23   17   18   24
Element    16            25   19   20   26
Element    17            26   20   21   27
Element    18            27   21   22   28
Element    19            28   22   23   29
Element    20            29   23   24   30
Element    21            31   25   26   32
Element    22            32   26   27   33
Element    23            33   27   28   34
Element    24            34   28   29   35
Element    25            35   29   30   36
The number of equations is :     25
    Time       Pressure
  0.1000E00   0.8972E+02
  0.2000E00   0.5881E+02
  0.3000E00   0.3624E+02
  0.4000E00   0.2215E+02
  0.5000E00   0.1353E+02
  0.6000E00   0.8258E+01
  0.7000E00   0.5042E+01
  0.8000E00   0.3078E+01
  0.9000E00   0.1879E+01
  0.1000E+01  0.1147E+01
  0.1100E+01  0.7005E00
  0.1200E+01  0.4277E00
  0.1300E+01  0.2611E00
  0.1400E+01  0.1594E00
  0.1500E+01  0.9733E-01
```

Figure 8.16 Results from Program 8.5

In nonlinear problems which are "ill-conditioned", p.c.g. iterations could become significant despite the unconditional stability exhibited by Program 8.4. The oldest and simplest mesh-free process is the "explicit" method, based on equation (3.98). This was given as

$$\mathbf{PM}\boldsymbol{\phi}_1 = (\mathbf{PM} - \Delta t\mathbf{KP})\boldsymbol{\phi}_0 \tag{8.11}$$

If **PM** is lumped (diagonalised) this can be written as

$$\boldsymbol{\phi}_1 = \mathbf{PM}^{-1}(\mathbf{PM} - \Delta t\mathbf{KP})\boldsymbol{\phi}_0 \tag{8.12}$$

Since $\mathbf{PM} - \Delta t\mathbf{KP}$ are element matrices which when summed give global matrices, the summation can as well be done of the vectors $(\mathbf{PM} - \Delta t\mathbf{KP})\boldsymbol{\phi}_0$ without there being any need to store a global matrix at all. This having been done, the global $\boldsymbol{\phi}_1$ is computed by multiplying by the inverse of the global mass matrix $\sum \mathbf{PM}^{-1}$ (called GLOBMA in the program). The process is illustrated by the structure chart in Figure 8.15.

The program is readily derived from Program 8.0 and contains only three new arrays: the global mass vector GLOBMA, the element lumped mass vector MASS and a storage array for element matrices STORE_PM.

The element integration loop follows the now standard course but the element matrix $\mathbf{PM} - \Delta t\mathbf{KP}$ is stored for each element rather than assembled as would be the case for a traditional implicit technique.

After the initial conditions have been established, GLOBMA is redefined as the inverse of the global mass vector required by equation (8.5). Then in the time-stepping loop, element matrices are recovered from STORE_PM and the multiplication and summation required by (8.5) completed for all elements.

After the elements have been inspected, the "loads" are updated by multiplying by the inverse of the global mass vector to give the solution at the new time.

The problem shown on Figure 8.2 has been analysed again and the results are listed as Figure 8.16.

PROGRAM 8.6: SOLUTION OF THE CONDUCTION EQUATION OVER A RECTANGLE USING AN EBE PRODUCT ALGORITHM

The motivation in using this algorithm is to preserve the storage economy achieved by the previous explicit technique while attaining the stability properties enjoyed by implicit methods typified by Programs 8.0 and 8.4 without involving solution of sets of global equations. The process is described in Chapter 3 by equation (3.102) and in structure by Figure 3.21. The program bears a strong resemblance to Program 8.5. The element integration loop is employed to store the element **KP** matrices in a storage array STORE_KP. In

```
   program p86
!------------------------------------------------------------------------
!      program 8.6 conduction equation on rectangular area using 4-node
!      quadrilateral elements and an ebe product algorithm
!------------------------------------------------------------------------
 use new_library     ;  use  geometry_lib      ;     implicit none
 integer::nels,nxe,neq,nn,nr,nip,nodof=1,nod=4,ndof,ndim=2,              &
          i,j,k,iel,nstep,npri,nres
 real::aa,bb,permx,permy,det,dtim,theta,val0,time
 character (len=15) :: element = 'quadrilateral'
!--------------------------- dynamic arrays-----------------------------
 real ,allocatable :: loads(:),points(:,:),kay(:,:),coord(:,:),mass(:),fun(:),&
                      jac(:,:),der(:,:),deriv(:,:),weights(:),kp(:,:),      &
                      pm(:,:), funny(:,:),g_coord(:,:),globma(:),store_kp(:,:,:)
 integer, allocatable :: nf(:,:), g(:) , num(:) , g_num(:,:) ,g_g(:,:)
!-------------------------input and initialisation----------------------
  open (10,file='p86.dat',status=    'old',action='read')
  open (11,file='p86.res',status='replace',action='write')
  read (10,*) nels,nxe,nn,nip,aa,bb,permx,permy ,                       &
              dtim,nstep,theta,npri,nres
  ndof=nod*nodof
  allocate ( nf(nodof,nn), points(nip,ndim),weights(nip),kay(ndim,ndim),     &
             coord(nod,ndim), fun(nod), jac(ndim,ndim),g_coord(ndim,nn),     &
             der(ndim,nod), deriv(ndim,nod), pm(ndof,ndof),g_num(nod,nels),  &
             kp(ndof,ndof), g(ndof),funny(1,nod),num(nod),g_g(ndof,nels),    &
             globma(0:nn),store_kp(ndof,ndof,nels),mass(ndof))
  kay=0.0 ; kay(1,1)=permx; kay(2,2)=permy
  globma = .0    ;  do i = 1,nn; nf(nodof,i)=i; end do
  call sample(element,points,weights)
!----------- loop the elements for integration and to store globals  ----------
 elements_1: do iel = 1 , nels
             call geometry_4qx(iel,nxe,aa,bb,coord,num)
             g_num(:,iel) = num;  g_coord(:,num) =transpose( coord)
             call num_to_g(num,nf,g) ;g_g( : , iel ) = g ;  kp=0.0 ; pm=0.0
       gauss_pts:  do i =1 , nip
               call shape_der (der,points,i) ; call shape_fun(fun,points,i)
               funny(1,:)=fun(:) ; jac = matmul(der,coord)
               det=determinant(jac); call invert(jac); deriv = matmul(jac,der)
               kp=kp+matmul(matmul(transpose(deriv),kay),deriv)*det*weights(i)
               pm   =  pm + matmul( transpose(funny),funny)*det*weights(i)
       end do gauss_pts
      store_kp(:,:,iel) = kp
      do i=1,ndof ; globma(g(i))=globma(g(i))+sum(pm(i,:)); end do
      globma ( 0 ) = .0
 end do elements_1
   write(11,'(a)') "Global coordinates "
   do k=1,nn;write(11,'(a,i5,a,2e12.4)')"Node",k,"       ",g_coord(:,k);end do
   write(11,'(a)') "Global node numbers "
   do k = 1 , nels; write(11,'(a,i5,a,4i5)')                                &
                        "Element ",k,"        ",g_num(:,k); end do
!-------------- recover  element A and B matrices  ---------------------------
 elements_2 : do iel = 1 , nels        ;            g = g_g( : , iel )
              kp = - store_kp(:,:,iel) * (1. - theta) * dtim * .5
              pm =  store_kp(:,:,iel) * theta * dtim * .5
              do i = 1,ndof
               pm (i,i) = pm (i,i) + globma ( g(i))
               kp (i,i) = kp (i,i) + globma ( g(i))
              end do
              call invert ( pm )  ; pm = matmul( pm , kp)
              store_kp ( : , : , iel) =  pm
 end do elements_2
!--------------take account of initial and boundary conditions----------------
  nf=1; read(10,*) nr ;if(nr>0)read(10,*)(k,nf(:,k),i=1,nr)
  call formnf(nf);neq=maxval(nf)
```

```
   write(11,'(a,i5)') "The number of equations is : ", neq
   allocate(loads(0:neq))
   read(10,*) val0; loads=val0    ; loads(0) = .0
!-------------------time stepping recursion-----------------------------------
   write(11,'(a)') "    Time      Pressure"
 timesteps: do j=1,nstep
               time=j*dtim
!---------------  first pass 1 to nels -----------------------------------------
    elements_3 : do iel = 1 , nels   ! g is different
                   call geometry_4qx(iel,nxe,aa,bb,coord,num)
                   call num_to_g(num,nf,g) ; g_g( : , iel ) = g
                   pm = store_kp(: , : , iel) ; mass = loads(g)
                   fun = matmul(pm , mass); loads(g) = fun; loads(0) = .0
    end do elements_3
!---------------  second pass nels to 1 ---------------------------------------
    elements_4 : do iel = nels , 1 , -1 ;    g = g_g( : , iel )
                   pm = store_kp(: , : , iel) ; mass = loads(g)
                   fun = matmul(pm , mass); loads(g) = fun; loads(0) = .0
    end do elements_4
  if(j/npri*npri==j)write(11,'(2e12.4)')time,loads(nres)
 end do timesteps
end program p86
```

addition, the element consistent mass matrices **PM** are diagonalised and the global mass vector, GLOBMA, assembled. A second loop over the elements is then made, headed "recover element **A** and **B** matrices". These are the matrices given in Figure 3.21 by $[\mathbf{M} - (1-\theta)\Delta t\mathbf{KP}/2]$ and $[\mathbf{M} + \theta\Delta t\mathbf{KP}/2]$ respectively, and they are called KP and PM respectively in the program. The algorithm calls for **B**(PM) to be inverted, which is done using the library subroutine INVERT. Then **A** is formed as $\mathbf{B}^{-1}[\mathbf{M} - (1-\theta)\Delta t\mathbf{KP}/2]$, that is by multiplying PM and KP. The result, called PM in the program, is re-stored as STORE_KP.

Initial conditions can then be prescribed and the time-stepping loop entered. Within that loop, two passes are made over the elements from first to last and back again. Half of the total $\Delta t\mathbf{KP}$ increment operates on each pass, and this has been accounted for already in forming **A** and **B**. The essential coding recovers each element $\mathbf{B}^{-1}\mathbf{A}$ matrix from STORE_KP and multiplies it by the appropriate part of the solution LOADS. Note that in this product algorithm the solution is continually being updated so there is no need for any "new loads" vector such as had to be employed in the explicit summation algorithm.

The process is fully detailed in Figure 3.21. Results for the problem described by Figure 8.2 are shown in Figure 8.17.

8.2 COMPARISON OF PROGRAMS 8.0, 8.4, 8.5 AND 8.6

These four programs already described in this chapter can all be used to solve plane conduction or uncoupled consolidation problems. Comparison of Figures

```
Global coordinates
Node    1          0.0000E+00  0.0000E+00
Node    2          0.2000E00  0.0000E+00
Node    3          0.4000E00  0.0000E+00
Node    4          0.6000E00  0.0000E+00
Node    5          0.8000E00  0.0000E+00
Node    6          0.1000E+01  0.0000E+00
...............................................................

Node   32          0.2000E00 -0.1000E+01
Node   33          0.4000E00 -0.1000E+01
Node   34          0.6000E00 -0.1000E+01
Node   35          0.8000E00 -0.1000E+01
Node   36          0.1000E+01 -0.1000E+01
Global node numbers
Element     1            7    1    2    8
Element     2            8    2    3    9
Element     3            9    3    4   10
Element     4           10    4    5   11
Element     5           11    5    6   12
Element     6           13    7    8   14
Element     7           14    8    9   15
Element     8           15    9   10   16
Element     9           16   10   11   17
Element    10           17   11   12   18
Element    11           19   13   14   20
Element    12           20   14   15   21
Element    13           21   15   16   22
Element    14           22   16   17   23
Element    15           23   17   18   24
Element    16           25   19   20   26
Element    17           26   20   21   27
Element    18           27   21   22   28
Element    19           28   22   23   29
Element    20           29   23   24   30
Element    21           31   25   26   32
Element    22           32   26   27   33
Element    23           33   27   28   34
Element    24           34   28   29   35
Element    25           35   29   30   36
The number of equations is :    25
    Time       Pressure
  0.1000E00   0.8912E+02
  0.2000E00   0.6107E+02
  0.3000E00   0.3894E+02
  0.4000E00   0.2450E+02
  0.5000E00   0.1537E+02
  0.6000E00   0.9644E+01
  0.7000E00   0.6050E+01
  0.8000E00   0.3795E+01
  0.9000E00   0.2380E+01
  0.1000E+01   0.1493E+01
  0.1100E+01   0.9366E00
  0.1200E+01   0.5875E00
  0.1300E+01   0.3685E00
  0.1400E+01   0.2312E00
  0.1500E+01   0.1450E00
```

Figure 8.17 Results from Program 8.6

8.3, 8.14, 8.16 and 8.17 shows that for the chosen problem—at the time step used (that is 0.01)—all solutions are accurate, and indeed the explicit solution (Figure 8.16) is probably as accurate as any despite being the simplest and cheapest to obtain.

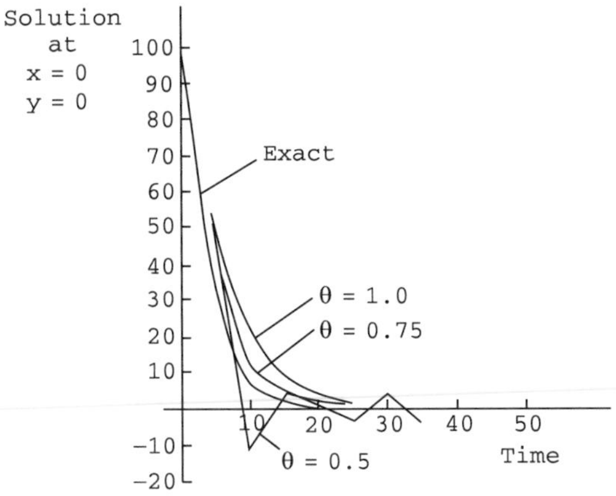

Figure 8.18 Typical solutions from Program 8.0 with varying θ

It must, however, be remembered that as the time step Δt is increased, the explicit algorithm will lead to unstable results (the stability limit for the chosen problem is about $\Delta t = 0.02$). At that time step, Program 8.0 with $\theta = 0.5$ will tend to produce oscillatory results, which can be damped, at the expense of average accuracy, by increasing θ towards 1.0. Typical behaviour of the implicit algorithm is illustrated in Figure 8.18.

Program 8.6, while retaining the storage economies of Program 8.5, allows the time step to be increased well beyond the explicit limit. For example, in the selected problem, reasonable results are still produced at $\Delta t = 10\Delta t_{\text{crit}}$. However, as Δt is increased still further, accuracy becomes poorer and Program 8.0 yields the best solutions for very large Δt.

It will be clear that algorithm choice in this area is not a simple one and depends on the nature of the problem (degree of nonlinearity, etc.) and on the hardware employed. All the mesh-free methods afford much scope for parallelisation.

8.3 EXAMPLES

1. Show that the 1-d "consolidation" equation

$$c_v \frac{\partial^2 u}{\partial x^2} = \frac{\partial u}{\partial t}$$

becomes after discretisation by a finite element in space and linear finite differences in time:

$$(\mathbf{m} + \theta \Delta t \mathbf{k})\mathbf{u}_1 = (\mathbf{m} - (1 - \theta)\Delta t \mathbf{k})\mathbf{u}_0$$

A layer of clay of thickness $2D$, free draining at its top and bottom surfaces, is subjected to a suddenly applied distributed load of one unit. Working in terms of a dimensionless time factor given by

$$T = \frac{c_v t}{D^2}$$

and using a single finite element, use the Crank–Nicolson approach ($\theta = 0.5$) with a time step of $\Delta T = 0.1$ to estimate the mid-depth pore pressure when $T = 0.3$. Compare this result with the analytical solution to your equation.

2. A rectangle of soil with x dimension 4 units and y dimension 2 units consolidates by drainage in the x and y directions according to Terzaghi's 2-d equation:

$$c_x \frac{\partial^2 u}{\partial x^2} + c_y \frac{\partial^2 u}{\partial y^2} = \frac{\partial u}{\partial t}$$

where c_x and c_y are the coefficients of consolidation and u is the excess pore pressure. If drainage is allowed from all four faces of the rectangle and the initial excess pore pressure is set at all points to $u = 1.0$, use a single finite element to estimate the variation of u with time at the centre of the rectangle if $c_x = 10$ and $c_y = 1$.

3. A rod of length 1 unit and thermal conductivity 1 unit is initially at a temperature of zero degrees along its entire length. One end of the rod is then suddenly subjected to a temperature of 100° and is maintained at that value. You may assume that the other end of the rod is perfectly insulated (i.e. there is no temperature gradient at that point). Using two 1-d "rod" elements, and assuming time-stepping parameters Δt and θ, set up (but do not solve) the recursive matrix equations that will model the change in temperature along the rod as a function of time.

4. A layer of saturated soil is subjected at time $t = 0$ to a triangular excess pore pressure distribution varying from 60 units at the top to zero at the bottom. During the subsequent dissipation phase, the top and bottom of the layer can be considered to be fully drained. Using the three-element discretisation shown in Figure 8.19 and a finite difference scheme with $\theta = 0.5$, estimate the pore pressures at the nodes after 0.1s using a single time step of $\Delta t = 0.1$

5. A layer of saturated clay of depth D and coefficient of consolidation c_v is drained at its top surface only. The layer is subjected to a sudden excess

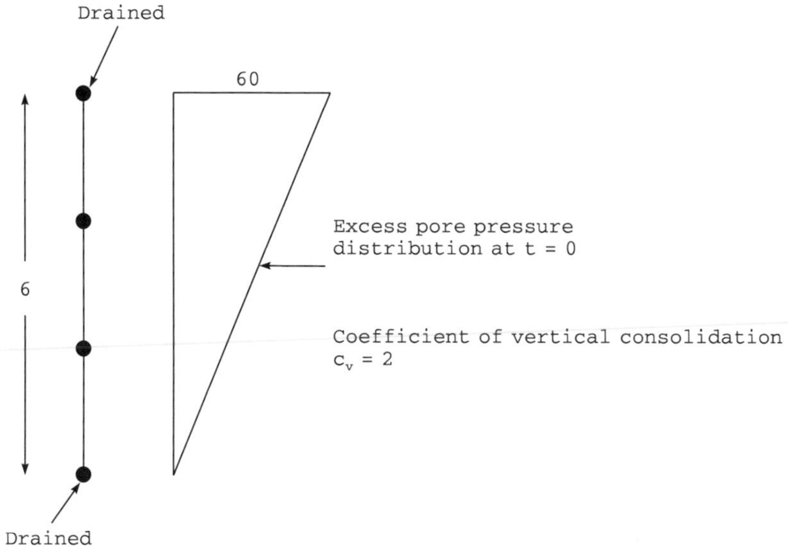

Figure 8.19

pore pressure which varies linearly from zero at the surface to U_0 at the base. Using two 1-d elements across the soil depth and a dimensionless time step of $\Delta T = 0.1$, estimate the average degree of consolidation when $T = 0.2$.

8.4 REFERENCES

Carslaw, H.S., and Jaeger, J.C. (1959) *Conduction of Heat in Solids*, 2nd edn. Clarendon Press, Oxford.

Dobbins, W.E. (1944) Effect of turbulence on sedimentation. *Trans. Am. Soc. Civ. Eng.*, **109**, 629–656.

Smith, I.M. (1976) Integration in time of diffusion and diffusion–convection equations. In *Finite Elements in Water Resources*, eds W.G. Gay, G. Pinder and C. Brebbia, pp. 1.3–1.20. Pentech Press, Plymouth.

Smith, I.M. (1979) The diffusion–convection equation. In *Summary of Numerical Methods for Partial Differential Equations*, eds I. Gladwell and R. Wait, Chapter 11. Oxford University Press.

Smith, I.M., Farraday, R.V. and O'Connor, B.A. (1973) Rayleigh–Ritz and Galerkin finite elements for diffusion–convection problems. *Water Resources Research*, **9** (3), 593-606.

9 "Coupled" Problems

9.0 INTRODUCTION

In the previous chapter, flow problems were treated in terms of a single dependent variable, for example the "potential" ϕ, and solutions involved only one degree of freedom per node in the finite element mesh. While this simplification may be adequate in some cases, it may be necessary to solve problems in which several degrees of freedom exist at the nodes of the mesh and the several dependent variables—for example velocities and pressures, or displacements and pressures—are "coupled" in the differential equations. Strictly speaking, the equations of 2-d and 3-d elasticity involve coupling between the various components of displacement, but the term "coupled problems" is really reserved for those in which variables of entirely different types are interdependent.

Both steady-state and transient problems are considered in this chapter. As usual, the former involve the solution of sets of simultaneous equations, as in Chapters 4–7. For the first problem chosen—solution of the Navier–Stokes equations—the simultaneous equations are, however, nonlinear and involve unsymmetrical coefficient matrices, and an iterative strategy.

The subsequent transient problems, although described by coupled systems of algebraic and differential equations, (the "Biot" equations) are cast as (linear) first-order differential equations in the time variable, and solved by the techniques introduced in Chapter 8—specifically by implicit integration, as illustrated by Program 8.0.

As in Chapter 8, a "mesh-free" alternative, using p.c.g., is given. When the Biot equations are nonlinear (see section 3.16.1) an incremental approach is necessary.

The (steady state) Navier–Stokes equations were developed in section 2.16. The equilibrium equations to be solved are (2.91), whose coefficients are themselves functions of the velocities $\mathbf{u}$ and $\mathbf{v}$ so that the equations are nonlinear. Furthermore, the coefficient submatrices $\mathbf{C}$ are not, in general, symmetrical and reference to section 3.10 shows that the subroutines `GAUSS_BAND` and `SOLVE_BAND` will be required to operate on the banded equation coefficients.

Section 3.15 illustrates how the element submatrices $\mathbf{C}$ are assembled and uses much of the program terminology already developed for uncoupled flow problems in Chapters 7 and 8.

PROGRAM 9.0: SOLUTION OF THE STEADY-STATE NAVIER–STOKES EQUATIONS OVER A RECTANGLE

```
program p90
!-------------------------------------------------------------------------
!      program 9.0 steady state Navier-Stokes equation
!      using 8-node velocity quadrilateral elements
!      coupled to 4-node pressure quadrilateral elements ; u-p-v order
!-------------------------------------------------------------------------
 use new_library     ; use geometry_lib     ;   implicit none
 integer::nels,nxe,nye,neq,nband,nn,nr,nip,nodof=3,nod=8,nodf=4,ndim=2,       &
         i,k,iel,ntot,limit ,fixed_nodes ,iters , inc
 real::visc, rho, det ,ubar, vbar , tol  ; logical :: converged
 character (len=15) :: element = 'quadrilateral'
!--------------------------- dynamic arrays---------------------------------
 real    ,allocatable :: points(:,:), coord(:,:),derivf(:,:),fun(:),work(:,:),&
                         jac(:,:),kay(:,:),der(:,:),deriv(:,:),weights(:)  ,  &
                         derf(:,:),funf(:), coordf(:,:),ke(:,:), g_coord(:,:),&
                         width(:), depth(:),c11(:,:),c21(:,:),c12(:,:),val(:),&
                         c23(:,:),c32(:,:), pb(:,:), loads(:), oldlds(:) ,    &
                         funny(:,:),row1(:,:),row2(:,:),uvel(:),vvel(:)  ,    &
                         funnyf(:,:),rowf(:,:)
 integer, allocatable :: nf(:,:),g(:),num(:),g_num(:,:) , g_g(:,:) ,no(:),    &
                         sense(:), node(:)
!---------------------------input and initialisation------------------------
  open (10,file='p90.dat',status=    'old',action='read')
  open (11,file='p90.res',status='replace',action='write')
  read (10,*) nels,nxe,nye,nn,nip,visc,rho,tol,limit
        ntot=nod+nodf+nod
  allocate (points(nip,ndim),coord(nod,ndim),derivf(ndim,nodf), &
           jac(ndim,ndim),kay(ndim,ndim),der(ndim,nod),deriv(ndim,nod),      &
           derf(ndim,nodf),funf(nodf),coordf(nodf,ndim),funny(nod,1),        &
           g_g(ntot,nels),c11(nod,nod),c12(nod,nodf),c21(nodf,nod),g(ntot),  &
           ke(ntot,ntot),fun(nod),width(nxe+1),depth(nye+1),nf(nodof,nn),    &
           g_coord(ndim,nn),g_num(nod,nels),num(nod),weights(nip),           &
           c32(nod,nodf),c23(nodf,nod),uvel(nod),vvel(nod),                  &
           row1(1,nod),row2(1,nod),funnyf(nodf,1),rowf(1,nodf))
      read(10,*) width , depth
      uvel =.0; vvel =.0 ; kay=0.0; kay(1,1)=visc/rho; kay(2,2)=visc/rho
  nf=1; read(10,*) nr ; if(nr>0) read(10,*)(k,nf(:,k),i=1,nr)
  call formnf(nf);neq=maxval(nf)  ;  call sample(element,points,weights)
!------- loop the elements to find nband and set up global arrays------------
  nband = 0
 elements_1: do iel = 1 , nels
             call geometry_8qxv(iel,nxe,width,depth,coord,num)
             inc=0
             do i=1,8;inc=inc+1;g(inc)=nf(1,num(i));end do
             do i=1,7,2;inc=inc+1;g(inc)=nf(2,num(i));end do
             do i=1,8;inc=inc+1;g(inc)=nf(3,num(i));end do
             g_num(:,iel )=num; g_coord(:,num)=transpose(coord); g_g(:,iel)=g
             if(nband<bandwidth(g))nband=bandwidth(g)
 end do elements_1
    write(11,'(a)') "Global coordinates "
    do k=1,nn;write(11,'(a,i5,a,2e12.4)')"Node",k,"       ",g_coord(:,k);end do
    write(11,'(a)') "Global node numbers "
    do k = 1 , nels; write(11,'(a,i5,a,8i5)')                                &
                                "Element ",k,"       ",g_num(:,k); end do
 write(11,'(2(a,i5))')                                                        &
         "There are ",neq," equations and the half-bandwidth is   ",nband
  allocate(pb(neq,2*(nband+1)-1),loads(0:neq),oldlds(0:neq),work(nband+1,neq))
    loads = .0  ; oldlds =.0   ; iters = 0
    read(10,*) fixed_nodes
       allocate(node(fixed_nodes),sense(fixed_nodes),val(fixed_nodes),        &
               no(fixed_nodes))
       read(10,*) (node(i),sense(i),val(i),i=1,fixed_nodes )
!-------------------iteration loop  ----------------------------------------
  iterations: do
```

```
             iters = iters + 1    ; converged = .false.
     pb = .0; work = .0; ke = .0
!------------ element stiffness integration and assembly---------------------

      elements_2:  do iel = 1 , nels
             num = g_num(: , iel ); coord=transpose(g_coord(:,num))
             g = g_g( : , iel )    ; coordf = coord(1 : 7 : 2, : )
             uvel = (loads(g(1:nod))+oldlds(g(1:nod)))*.5
             do i = nod + nodf + 1 , ntot
                vvel(i-nod-nodf) = (loads(g(i))+oldlds(g(i)))*.5
             end do
             c11 = .0; c12 = .0; c21 = .0; c23 = .0; c32 = .0
          gauss_points_1: do i = 1 , nip
!--------------------- velocity contribution ---------------------------------
            call shape_fun(fun,points,i) ;funny(:,1) = fun
            ubar = dot_product(fun,uvel);vbar = dot_product(fun,vvel)
            if(iters==1) then; ubar = 1.; vbar = 0.; end if
            call shape_der(der,points,i);  jac = matmul(der,coord)
            det = determinant(jac )      ; call invert(jac)
            deriv = matmul(jac,der);row1(1,:)=deriv(1,:);row2(1,:)=deriv(2,:)
            c11 = c11 + matmul(matmul(transpose(deriv),kay),deriv) &
                   *det* weights(i) + &
                        matmul(funny,row1)*det*weights(i)*ubar + &
                        matmul(funny,row2)*det*weights(i)*vbar
!----------------------now the pressure contribution--------------------------
             call shape_fun(funf,points,i); funnyf(:,1)=funf
             call shape_der(derf,points,i)  ;jac=matmul(derf,coordf)
             det=determinant(jac)       ;     call invert(jac)
             derivf=matmul(jac,derf)
             rowf(1,:) = derivf(1,:)
             c12 = c12 + matmul(funny,rowf)*det*weights(i)/rho
             rowf(1,:) = derivf(2,:)
             c32 = c32 + matmul(funny,rowf)*det*weights(i)/rho
             c21 = c21 + matmul(funnyf,row1)*det*weights(i)
             c23 = c23 + matmul(funnyf,row2)*det*weights(i)
          end do gauss_points_1
        call formupv(ke,c11,c12,c21,c23,c32) ; call formtb(pb,ke,g)
      end do elements_2
!----------- prescribed values of velocity and pressure ----------------------
       loads = .0
       do i=1, fixed_nodes; no(i) = nf(sense(i),node(i))  ; end do
         pb( no ,nband+1)=pb( no ,nband+1) + 1.e20
         loads(no) = pb(no,nband+1) * val
!----------------------- solve the simultaneous equations -------------------
    call gauss_band(pb,work); call solve_band(pb,work,loads); loads(0) = .0
    call checon(loads,oldlds,tol,converged);if(converged.or.iters==limit) exit
  end do iterations
    write(11,'(a,i5,a)')"The solution took",iters,"  iterations to converge"
       write(11,'(a)') " The nodal velocities and porepressures are     :"
       write(11,'(a)')"   Node    u - velocity    pressure     v - velocity"
          do k=1,nn; write(11,'(i5,a,3e12.4)')k,"     ",loads(nf(:,k));end do
 end program p90
```

The simple problem chosen to illustrate this program is shown in Figure 9.1. Flow is confined to a rectangular cavity and driven by a uniform horizontal velocity at the top. The boundary velocities at the other three boundaries are zero. Note that a dummy freedom has been inserted at mid-side nodes where there is no **p** variable. Thus, the second freedom at all mid-side nodes is zero.

The meanings of the variables in the program (excluding simple counters) are as follows:

NELS	number of elements in the problem (25)
NXE	number of elements in x direction (5)
NYE	number of elements in y direction (5)
NN	total number of nodes in mesh (96)
NIP	number of Gaussian integrating points in each element (4)
VISC	molecular viscosity, $\mu(0.01)$
RHO	density of the fluid, $\rho(1.0)$
TOL	iteration tolerance (0.001)
LIMIT	maximum number of iterations allowed (30)
FIXED_NODES	number of nodes with prescribed non-zero values (11)
NDIM	dimension of the problem (2)
NOD	number of nodes for velocity shape functions (8)
NR	number of restrained nodes in mesh (96)

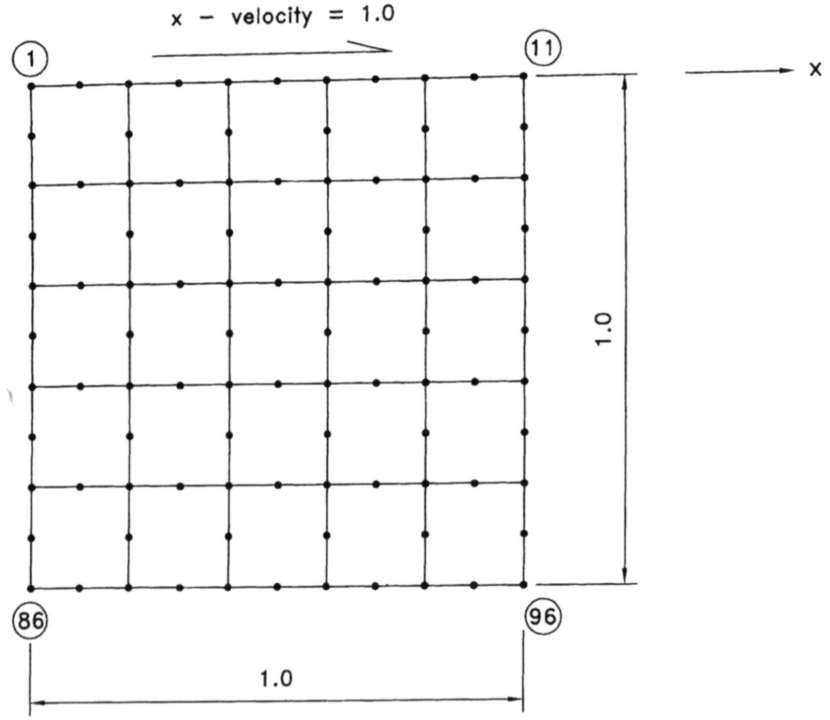

Figure 9.1 *(continued)*

```
Mesh and           NELS    NXE   NYE   NN    NIP
element data       25      5     5     96    4

                   VISC    RHO   TOL   LIMIT
                   .01     1.0   .001  30

Variable geometry  WIDTH(I),I=1,NXE+1
data               .0 .2 .4 .6 .8 1.0

                   DEPTH(I),I=1,NYE+1
                   .0 -.2 -.4 -.6 -.8 -1.0
Node freedom       NR
data               96
                   K,NF(:,K),K=1,NR
                    1 1 0 0  2 1 0 0  3 1 1 0  4 1 0 0  5 1 1 0  6 1 0 0
                    7 1 1 0  8 1 0 0  9 1 1 0 10 1 0 0 11 1 1 0 12 0 0 0
                   13 1 0 1 14 1 0 1 15 1 0 1 16 1 0 1 17 0 0 0 18 0 1 0
                   19 1 0 1 20 1 1 1 21 1 0 1 22 1 1 1 23 1 0 1 24 1 1 1
                   25 1 0 1 26 1 1 1 27 1 0 1 28 0 1 0 29 0 0 0 30 1 0 1
                   31 1 0 1 32 1 0 1 33 1 0 1 34 0 0 0 35 0 1 0 36 1 0 1
                   37 1 1 1 38 1 0 1 39 1 1 1 40 1 0 1 41 1 1 1 42 1 0 1
                   43 1 1 1 44 1 0 1 45 0 1 0 46 0 0 0 47 1 0 1 48 1 0 1
                   49 1 0 1 50 1 0 1 51 0 0 0 52 0 1 0 53 1 0 1 54 1 1 1
                   55 1 0 1 56 1 1 1 57 1 0 1 58 1 0 1 59 1 0 1 60 1 1 1
                   61 1 0 1 62 0 1 0 63 0 0 0 64 1 0 1 65 1 0 1 66 1 0 1
                   67 1 0 1 68 0 0 0 69 0 1 0 70 1 0 1 71 1 1 1 72 1 0 1
                   73 1 1 1 74 1 0 1 75 1 1 1 76 1 0 1 77 1 1 1 78 1 0 1
                   79 0 1 0 80 0 0 0 81 1 0 1 82 1 0 1 83 1 0 1 84 1 0 1
                   85 0 0 0 86 0 1 0 87 0 0 0 88 0 1 0 89 0 0 0 90 0 1 0
                   91 0 0 0 92 0 1 0 93 0 0 0 94 0 1 0 95 0 0 0 96 0 1 0

Fixed freedom      FIXED_NODES
data               11
                   NODE    SENSE  VALUE

                    1      1      1.0
                    2      1      1.0
                    3      1      1.0
                    4      1      1.0
                    5      1      1.0
                    6      1      1.0
                    7      1      1.0
                    8      1      1.0
                    9      1      1.0
                   10      1      1.0
                   11      1      1.0
```

Figure 9.1 Mesh and data for Program 9.0

Space is reserved for the following arrays:

POINTS	quadrature abscissae and weights
COORD	coordinates of "velocity" nodes

UVEL	element nodal u velocity values
VVEL	element nodal v velocity values
JAC	Jacobian matrix
KAY	viscosity matrix
DER	derivatives of velocity shape functions in local coordinates
DERIV	derivatives of velocity shape functions in global coordinates
KE	element "stiffness" matrix
FUN	velocity shape functions in local coordinates
FUNF	pressure shape functions in local coordinates
COORDF	coordinates of "pressure" nodes
DERF	derivatives of pressure shape functions in local coordinates
DERIVF	derivatives of pressure shape functions in global coordinates
ROW1,ROW2	temporary storage of rows of DERIV, etc.
C11, C12, C21, C23, C32	element submatrices (see equation 2.91)
G	element steering vector

The following arrays have their dimensions adjusted after the problem data has been specified to reflect the size of problem being analysed:

OLDLDS	solution at the previous iteration
WIDTH	x coordinates of the mesh lines
DEPTH	y coordinates of the mesh lines
VAL	values of fixed non-zero freedoms
PB	unsymmetrical band global "stiffness" matrix
WORK	working space
LOADS	solution at the current iteration
NO	numbers of freedoms to be fixed at non-zero values
NF	node freedom array

The structure of the program is described by the structure chart in Figure 9.2. After the data has been read in, various arrays must be initialised to zero. Note that the KAY matrix which, in the previous chapter, held coefficients of consolidation c_x etc., now holds viscosities in the form μ/ρ.

The iteration loop is then entered, controlled by the counter ITERS. System arrays PB and WORK must be nulled together with the element "stiffness" matrix KE. Element matrix integration and assembly then proceeds as usual. The nodal coordinates and steering vector are formed by the geometry library

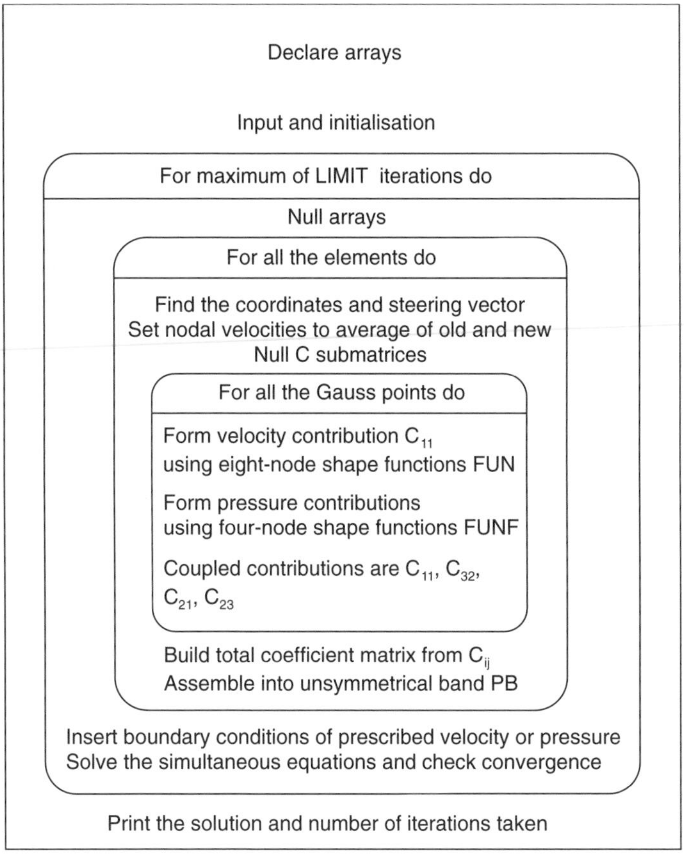
Declare arrays
Input and initialisation
For maximum of LIMIT iterations do
Null arrays
For all the elements do
Find the coordinates and steering vector
Set nodal velocities to average of old and new
Null C submatrices
For all the Gauss points do
Form velocity contribution C11
using eight-node shape functions FUN
Form pressure contributions
using four-node shape functions FUNF
Coupled contributions are C11, C32,
C21, C23
Build total coefficient matrix from Cij
Assemble into unsymmetrical band PB
Insert boundary conditions of prescribed velocity or pressure
Solve the simultaneous equations and check convergence
Print the solution and number of iterations taken

Figure 9.2 Structure chart for Program 9.0

routine GEOMETRY_8QXV. Nodal velocities used to form $\bar{u}$ and $\bar{v}$ in equation (2.91) are taken to be the average of those computed in the last two iterations. Element submatrices C11, etc., are set to zero and the numerical integration loop entered. Average velocities $\bar{u}$ and $\bar{v}$ are recovered form UVEL and VVEL, except in the first iteration where the guess $\bar{u} = 1.0, \bar{v} = 0.0$ is used. The submatrix C11 is formed as required by equation (3.104). One component is DERIV^{T}*KAY*DERIV which is formed as usual. However, there are two extra components calculated by the two calls to the intrinsic subroutine MATMUL.

Finally, submatrices C12, C32, C21, and C23 are computed as demanded by equation (3.105). The element "stiffness" KE is built up by the subroutine FORMUPV and assembled into the global unsymmetrical band matrix by the assembly routine FORMTB.

It remains only to specify the fixed freedoms by the "penalty" technique (see section 3.8) and to complete the equation solution using `GAUSS_BAND` and `SOLVE_BAND`. The maximum number of iterations allowed is 30 but a convergence check of 0.1% is invoked by `CHECON`.

The results are listed as Figure 9.3 and illustrated in Figure 9.4 for a "regridded" mesh (Kidger, 1994). As the Reynolds number increases, convergence of this algorithm will be slow and more specialised techniques are necessary. Due to the unsymmetrical properties of `KE` no p.c.g. mesh-free technique is presented here, although iterative strategies for unsymmetrical systems are available.

```
Global coordinates
Node    1          0.0000E+00  0.0000E+00
Node    2          0.1000E00  0.0000E+00
Node    3          0.2000E00  0.0000E+00
Node    4          0.3000E00  0.0000E+00
Node    5          0.4000E00  0.0000E+00
Node    6          0.5000E00  0.0000E+00
Node    7          0.6000E00  0.0000E+00
Node    8          0.7000E00  0.0000E+00
..........................................................................

Node   93          0.7000E00 -0.1000E+01
Node   94          0.8000E00 -0.1000E+01
Node   95          0.9000E00 -0.1000E+01
Node   96          0.1000E+01 -0.1000E+01
Global node numbers
Element    1          18   12    1    2    3   13   20   19
Element    2          20   13    3    4    5   14   22   21
Element    3          22   14    5    6    7   15   24   23
Element    4          24   15    7    8    9   16   26   25
Element    5          26   16    9   10   11   17   28   27
..........................................................................

Element   23          90   82   73   74   75   83   92   91
Element   24          92   83   75   76   77   84   94   93
Element   25          94   84   77   78   79   85   96   95
There are   158  equations and the half-bandwidth is       39
The solution took     9  iterations to converge
 The nodal velocities and porepressures are      :
   Node     u - velocity    pressure      v - velocity
     1      0.1000E+01  0.0000E+00  0.0000E+00
     2      0.1000E+01  0.0000E+00  0.0000E+00
     3      0.1000E+01  0.2429E00  0.0000E+00
     4      0.1000E+01  0.0000E+00  0.0000E+00
     5      0.1000E+01  0.1704E00  0.0000E+00
     6      0.1000E+01  0.0000E+00  0.0000E+00
     7      0.1000E+01  0.2096E00  0.0000E+00
..........................................................................

    91      0.0000E+00  0.0000E+00  0.0000E+00
    92      0.0000E+00  0.2236E00  0.0000E+00
    93      0.0000E+00  0.0000E+00  0.0000E+00
    94      0.0000E+00  0.2275E00  0.0000E+00
    95      0.0000E+00  0.0000E+00  0.0000E+00
    96      0.0000E+00  0.2202E00  0.0000E+00
```

Figure 9.3 Results from Program 9.0

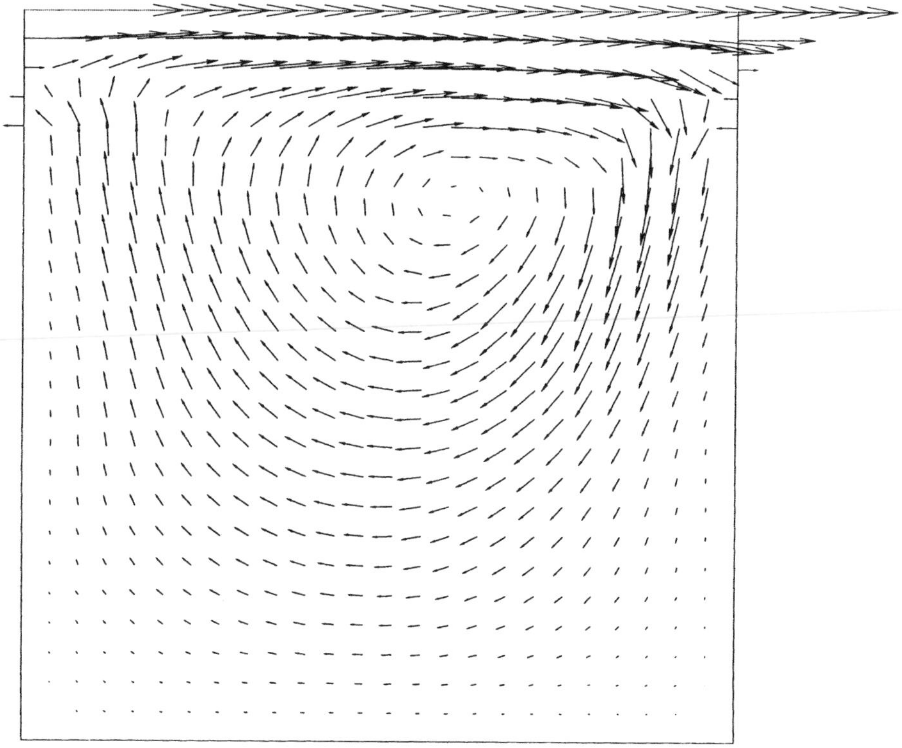

Figure 9.4 Distribution of velocity from Program 9.0

9.1 ANALYSIS OF SOIL CONSOLIDATION USING BIOT'S THEORY

The analysis of the behaviour of porous elastic solids under load is in many ways analogous to the coupled flow analysis described in the previous section. The displacements of the soil skeleton take over the role of the velocities **u** and **v**, and the excess porewater pressure the role of the fluid pressure **p**.

The differential equations to be solved are (2.113), (2.114) and (2.117). Due to the coupling of fluid and solid phases there arises the complication that the applied "total" stresses, $\boldsymbol{\sigma}$, are divided between a portion carried by the soil skeleton, called "effective" stress $\boldsymbol{\sigma}'$, and a portion carried by the porewater, called in soil mechanics the "pore pressure" and denoted in Chapter 2 by $\mathbf{u_w}$ to distinguish it from the mean total stress **p**.

After discretisation in space by finite elements, the coupled equations in **u**,**v** and $\mathbf{u_w}$ are given by (2.119). These can be seen to be partly algebraic equations and partly first-order differential equations in time. Common techniques for discretisation in the time domain are shown by (3.113), in which θ has the same role as it had in Program 8.0, controlling the implicit integration.

Thus, equations (3.113) or (3.114) are in principle no different from (3.95) or (3.96) for uncoupled problems. Solutions will involve setting up the coupled "stiffness" matrix on the left-hand side of these equations, followed by an equation solution for every time step Δt to advance the solution from time 0 to time 1. For constant element properties and time step Δt, the left hand side needs to be factorised only once, the remainder of the solution involving matrix-by-vector multiplication on the right hand side, followed by back-substitution. The process may be written:

$$\mathbf{KE}\boldsymbol{\delta}_1 = \mathbf{KD}\boldsymbol{\delta}_0 + \mathbf{f} \tag{9.1}$$

PROGRAM 9.1: BIOT CONSOLIDATION OF A RECTANGULAR SOLID IN PLANE STRAIN (EIGHT-NODE/FOUR-NODE ELEMENTS)

The problem chosen is of a plane strain "oedometer" specimen subjected to a ramp loading at its surface as shown in Figure 9.5. The structure chart describing the process of the program is shown in Figure 9.6 and the mesh and input data in Figure 9.7. There are roller boundaries at both sides of the mesh.

Since the shape functions governing the variation of the unknowns (u and v displacement and pore pressure u_w) are the same, the program is very similar to Program 9.0. The variables have the same significance, with the following additions:

`PERMX`	permeability (D'Arcy) in the x direction/γ_w } (see equation (2.117))
`PERMY`	permeability in the y direction/γ_w } (see equation (2.117))
`E`	effective Young's modulus
`V`	effective Poisson's ratio
`DTIM`	time step
`NSTEP`	number of time steps
`THETA`	parameter in implicit time-stepping process $(1/2 \leq \theta \leq 1)$

Arrays not encountered in Program 9.0 are:

`DEE`	effective stress–strain matrix (see equation (2.114))
`BEE`	strain–displacement matrix
`KM`	element (solid) stiffness matrix
`ELD`	element nodal displacements
`EPS`	element strain vector
`SIGMA`	element effective stress vector
`KP`	element (fluid) stiffness matrix
`KE,KD`	element matrices (see equation (9.1))

```
program p91
!-------------------------------------------------------------------------
!      program 9.1 plane strain consolidation of a Biot elastic
!      solid using 8-node solid quadrilateral elements
!      coupled to 4-node fluid elements
!-------------------------------------------------------------------------
 use new_library       ; use geometry_lib   ;   implicit none
 integer::nels,nxe,nye,neq,nband,nn,nr,nip,nodof=3,nod=8,nodf=4,nst=3,     &
         ndim=2,ndof, i,k,l,iel,ns,nstep,ntot, nodofs=2 ,inc
 real::permx,permy,e,v,det,dtim,theta,x1,x2,time
 character (len=15) :: element = 'quadrilateral'
!--------------------------- dynamic arrays-------------------------------
 real    ,allocatable :: dee(:,:), points(:,:), coord(:,:), derivf(:,:),    &
                         jac(:,:),kay(:,:),der(:,:),deriv(:,:),weights(:),  &
                         derf(:,:),funf(:), coordf(:,:), bee(:,:), km(:,:), &
                         eld(:), sigma(:), kp(:,:), ke(:,:), g_coord(:,:),  &
                         kd(:,:),fun(:), c(:,:), width(:), depth(:), bk(:), &
                         vol(:), pb(:,:), loads(:), ans(:) ,volf(:,:)
 integer, allocatable :: nf(:,:),g(:),num(:),g_num(:,:) , g_g(:,:)
!-------------------------input and initialisation-------------------------
  open (10,file='p91.dat',status=    'old',action='read')
  open (11,file='p91.res',status='replace',action='write')
  read (10,*) nels,nxe,nye,nn,nip,                                          &
            permx, permy, e,v, dtim, nstep, theta
  ndof=nod*2; ntot=ndof+nodf
  allocate (dee(nst,nst),points(nip,ndim),coord(nod,ndim),derivf(ndim,nodf), &
           jac(ndim,ndim),kay(ndim,ndim),der(ndim,nod),deriv(ndim,nod),      &
           derf(ndim,nodf),funf(nodf),coordf(nodf,ndim),bee(nst,ndof),       &
           km(ndof,ndof),eld(ndof),sigma(nst),kp(nodf,nodf),g_g(ntot,nels), &
           ke(ntot,ntot),kd(ntot,ntot),fun(nod),c(ndof,nodf),width(nxe+1),   &
           depth(nye+1),vol(ndof),nf(nodof,nn), g(ntot), volf(ndof,nodf),    &
           g_coord(ndim,nn),g_num(nod,nels),num(nod),weights(nip))
           kay=0.0; kay(1,1)=permx; kay(2,2)=permy
           read (10,*)width , depth
  nf=1; read(10,*) nr; if(nr>0) read(10,*)(k,nf(:,k),i=1,nr)
  call formnf(nf);neq=maxval(nf)
  call deemat (dee,e,v); call sample(element,points,weights)
!--------- loop the elements to find nband and set up global arrays------------
  nband = 0
 elements_1: do iel = 1 , nels
             call geometry_8qxv(iel,nxe,width,depth,coord,num)
             inc=0
             do i=1,8; do k=1,2; inc=inc+1;g(inc)=nf(k,num(i));end do;end do
             do i=1,7,2; inc=inc+1;g(inc)=nf(3,num(i)); end do
             g_num(:,iel)=num; g_coord(:,num)=transpose(coord); g_g(:,iel)= g
             if(nband<bandwidth(g))nband=bandwidth(g)
 end do elements_1
    write(11,'(a)') "Global coordinates "
    do k=1,nn;write(11,'(a,i5,a,2e12.4)')"Node",k,"        ",g_coord(:,k);end do
    write(11,'(a)') "Global node numbers "
    do k = 1 , nels; write(11,'(a,i5,a,8i5)')                                  &
                              "Element ",k,"         ",g_num(:,k); end do
  write(11,'(2(a,i5))')                                                         &
         "There are ",neq,"  equations and the half-bandwidth is   ",nband
  allocate(bk(neq*(nband+1)),pb(neq,2*(nband+1)-1),loads(0:neq),ans(0:neq))
    pb = .0 ; bk = .0 ; loads = .0
!--------------- element stiffness integration and assembly---------------------

      elements_2:  do iel = 1 , nels
              num = g_num(: , iel ); coord=transpose(g_coord(:,num))
              g = g_g( : , iel )   ; coordf = coord(1 : 7 : 2, : )
              km = .0; c = .0; kp = .0
          gauss_points_1: do i = 1 , nip
             call shape_der(der,points,i);  jac = matmul(der,coord)
```

```
            det = determinant(jac ); call invert(jac);deriv = matmul(jac,der)
            call beemat(bee,deriv); vol(:)=bee(1,:)+bee(2,:)
            km = km + matmul(matmul(transpose(bee),dee),bee) *det* weights(i)
!--------------------------now the fluid contribution--------------------------
             call shape_fun(funf,points,i)
             call shape_der(derf,points,i)  ; derivf=matmul(jac,derf)
        kp=kp+matmul(matmul(transpose(derivf),kay),derivf)*det*weights(i)*dtim
             do l=1,nodf; volf(:,l)=vol(:)*funf(l); end do
             c= c+volf*det*weights(i)
          end do gauss_points_1
 call fmkdke(km,kp,c,ke,kd,theta);call formkv(bk,ke,g,neq);call formtb(pb,kd,g)
     end do elements_2
!------------------------factorise left hand side------------------------------
   call banred(bk,neq)
! ----------------------- enter the time-stepping loop------------------------
     time = .0
  time_steps:  do ns = 1 , nstep
     time=time+dtim;write(11,'(a,e12.4)')"The time is",time
    call bantmul(pb,loads,ans)
!   ramp loading
       x1=(.1*ns+.1*(theta-1.))/6.; x2=x1*4.
       if(ns>10) then
          ans(1)=ans(1)-1./6.; ans(3)=ans(3)-2./3.
          ans(4)=ans(4)-1./6.
       else if(ns<10) then
          ans(1)=ans(1)-x1;ans(3)=ans(3)-x2; ans(4)=ans(4)-x1
       end if
       call bacsub(bk,ans) ; loads=ans
       write(11,'(a)') " The nodal displacements and porepressures are    :"
       do k=1,23,22; write(11,'(i5,a,3e12.4)')k,"     ",ans(nf(:,k)) ; end do
!-------------------recover stresses at  Gauss-points--------------------------
     elements_3 :  do iel = 1 , nels
             num = g_num(:,iel); coord=transpose(g_coord(:,num))
             g = g_g( : , iel );  eld = ans( g ( 1 : ndof ) )
        !    print*,"The Gauss Point effective stresses for element",iel,"are"
           gauss_points_2: do i = 1,nip
             call shape_der (der,points,i);  jac= matmul(der,coord)
             call invert ( jac );    deriv= matmul(jac,der)
             bee= 0.;call beemat(bee,deriv);sigma= matmul(dee,matmul(bee,eld))
        !     print*,"Point    ",i       ;!  print*,sigma
          end do gauss_points_2
     end do elements_3
  end do time_steps
end program p91
```

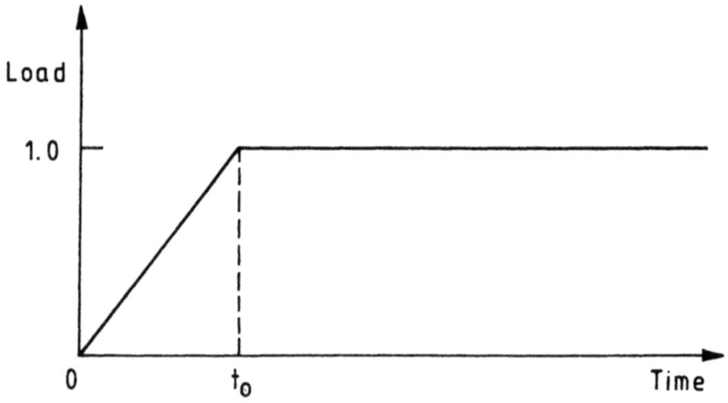

Figure 9.5 Ramp loading

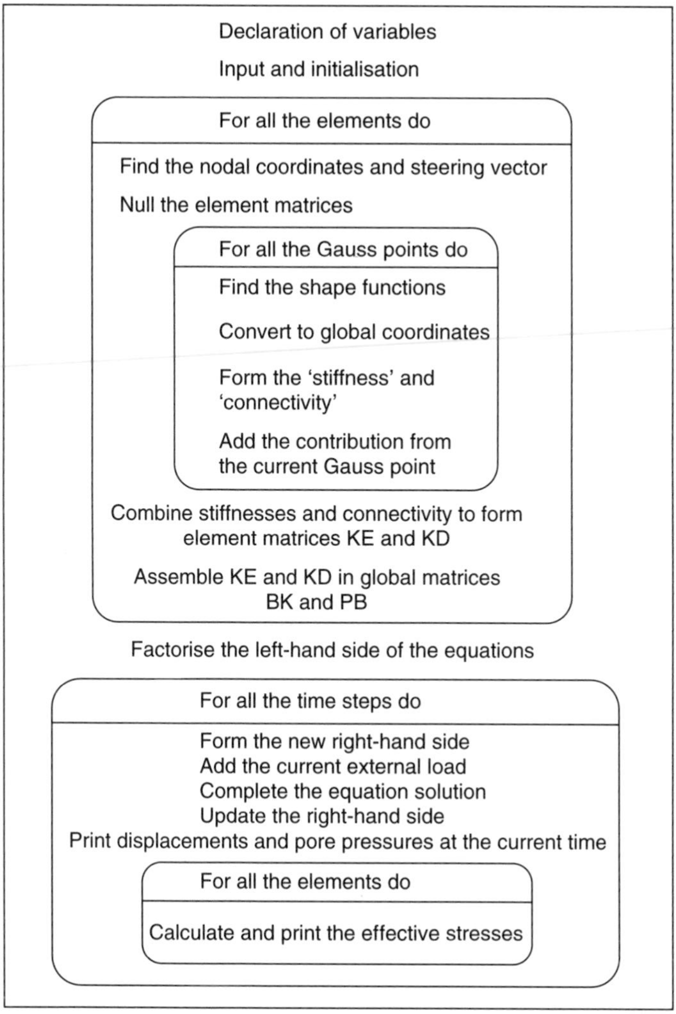

Figure 9.6 Structure chart for Program 9.1

`C`	coupling matrix
`BK`	global left-hand side matrix
`VOL`	volumetric strain vector
`PB`	global right-hand side matrix
`ANS`	solution (answer) at current time step

Turning to the program, the nodal coordinates and steering vector are again provided by subroutine `GEOMETRY_8QXV` and the subroutine `FMKDKE` builds up the `KD` and `KE` matrices required by equation (9.1). Then `KE` is assembled into

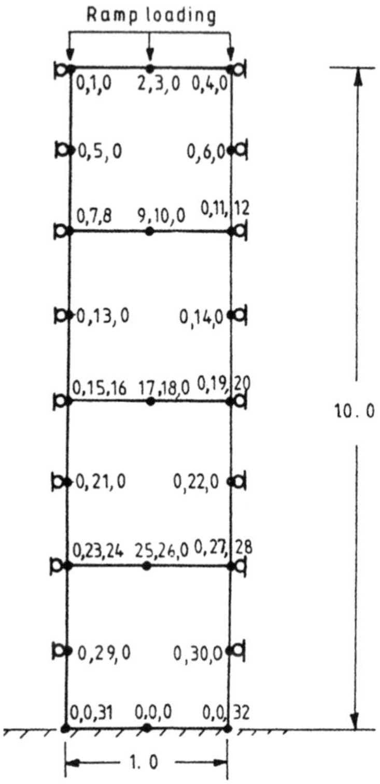

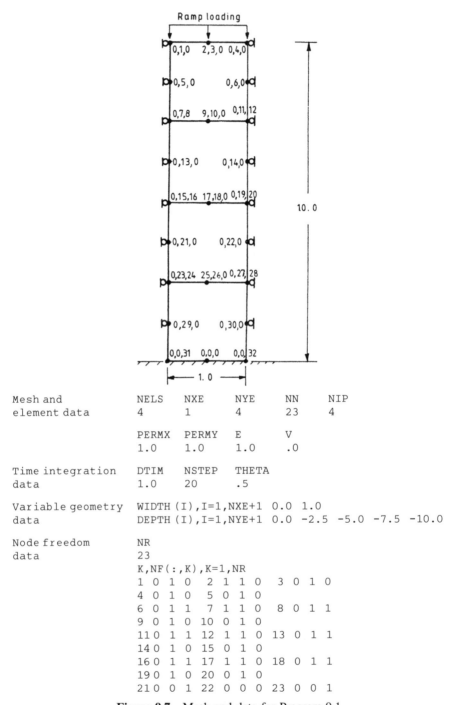

Mesh and element data	NELS	NXE	NYE	NN	NIP
	4	1	4	23	4
	PERMX	PERMY	E	V	
	1.0	1.0	1.0	.0	
Time integration data	DTIM	NSTEP	THETA		
	1.0	20	.5		

Variable geometry data

```
WIDTH(I),I=1,NXE+1  0.0  1.0
DEPTH(I),I=1,NYE+1  0.0  -2.5  -5.0  -7.5  -10.0
```

Node freedom data

```
NR
23
K,NF(:,K),K=1,NR
1  0 1 0    2  1 1 0    3  0 1 0
4  0 1 0    5  0 1 0
6  0 1 1    7  1 1 0    8  0 1 1
9  0 1 0   10  0 1 0
11 0 1 1   12  1 1 0   13  0 1 1
14 0 1 0   15  0 1 0
16 0 1 1   17  1 1 0   18  0 1 1
19 0 1 0   20  0 1 0
21 0 0 1   22  0 0 0   23  0 0 1
```

Figure 9.7 Mesh and data for Program 9.1

the (symmetric) global matrix (actually vector) BK while KD is assembled into the unsymmetrical band matrix PB. When using (3.115) the presence of many zeros in KD means that element-by-element summation to avoid assembling KD is a better strategy, and is used in the remaining programs in this chapter.

The subroutine BANRED is used to factorise the symmetric left-hand side global matrix BK summed from KE and the time-stepping loop is entered. The right-hand side global matrix, PB, summed from KD, is multiplied by the old solution LOADS using library subroutine BANTMUL to give ANS, which is then modified for ramp loading. The subroutine BACSUB completes the solution and the element effective stresses can be recovered at the element Gauss points if required.

```
Global coordinates
Node    1          0.0000E+00  0.0000E+00
Node    2          0.5000E00   0.0000E+00
Node    3          0.1000E+01  0.0000E+00
Node    4          0.0000E+00 -0.1250E+01
Node    5          0.1000E+01 -0.1250E+01
Node    6          0.0000E+00 -0.2500E+01
Node    7          0.5000E00  -0.2500E+01
.....................................................................
Node   18          0.1000E+01 -0.7500E+01
Node   19          0.0000E+00 -0.8750E+01
Node   20          0.1000E+01 -0.8750E+01
Node   21          0.0000E+00 -0.1000E+02
Node   22          0.5000E00  -0.1000E+02
Node   23          0.1000E+01 -0.1000E+02
Global node numbers
Element     1           6    4    1    2    3    5    8    7
Element     2          11    9    6    7    8   10   13   12
Element     3          16   14   11   12   13   15   18   17
Element     4          21   19   16   17   18   20   23   22
There are   32  equations and the half-bandwidth is       13
The time is  0.1000E+01
 The nodal displacements and porepressures are    :
    1       0.0000E+00 -0.1233E00  0.0000E+00
   23       0.0000E+00  0.0000E+00 -0.1000E00
The time is  0.2000E+01
 The nodal displacements and porepressures are    :
    1       0.0000E+00 -0.2872E00  0.0000E+00
   23       0.0000E+00  0.0000E+00 -0.2000E00
The time is  0.3000E+01
 The nodal displacements and porepressures are    :
    1       0.0000E+00 -0.4849E00  0.0000E+00
   23       0.0000E+00  0.0000E+00 -0.3000E00
.....................................................................
The time is  0.1800E+02
 The nodal displacements and porepressures are    :
    1       0.0000E+00 -0.3932E+01  0.0000E+00
   23       0.0000E+00  0.0000E+00 -0.9092E00
The time is  0.1900E+02
 The nodal displacements and porepressures are    :
    1       0.0000E+00 -0.4095E+01  0.0000E+00
   23       0.0000E+00  0.0000E+00 -0.8942E00
The time is  0.2000E+02
 The nodal displacements and porepressures are    :
    1       0.0000E+00 -0.4251E+01  0.0000E+00
   23       0.0000E+00  0.0000E+00 -0.8783E00
```

Figure 9.8 Results from Program 9.1

The results are shown as Figure 9.8 and the mid-plane pore pressure is plotted against time in Figure 9.9 for two different rise-times, $t_0 = 10$ and $t_0 = 50$. The "time-factor" in Figure 9.9 is the dimensionless number

$$T = \frac{c_{\mathrm{v}} t}{H^2}$$

where H is the "drainage path distance" of 10.0 in the present instance. The coefficient of consolidation c_v is found from

$$c_{\mathrm{v}} = \frac{k}{m_{\mathrm{v}} \gamma_{\mathrm{w}}} \tag{9.3}$$

where

$$m_{\mathrm{v}} = \frac{(1+\nu')(1-2\nu')}{E'(1-\nu')} \tag{9.4}$$

and k is the "permeability" (Chapter 7) with γ_{w} being the unit weight of water.

In the present example ν' is 0 and $E' = 1.0$, so m_{v} is just 1.0. Similarly, k/γ_{w} is 1.0 so that $T = t/100$. Thus, for DTIM equal to 1.0, the solution at the first step is for $T = 0.01$ and the rise-time is $T_0 = 0.1$. The results will be found to agree exactly with those of Schiffman (1960), and problems of practical importance have been solved since Smith and Hobbs (1976).

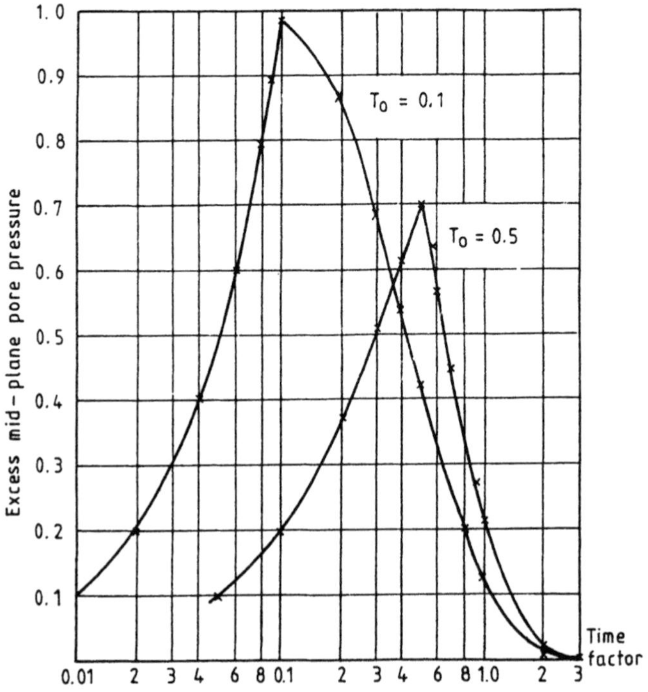

Figure 9.9 Mid-plane pore pressure response to ramp loading (Program 9.1)

PROGRAM 9.2: BIOT CONSOLIDATION OF A RECTANGULAR SOLID IN PLANE STRAIN (EIGHT-NODE/FOUR-NODE ELEMENTS) INCREMENTAL VERSION

```
program p92
!-------------------------------------------------------------------------
!      program 9.2 plane strain consolidation of a Biot elastic
!      solid using 8-node solid quadrilateral elements
!      coupled to 4-node fluid elements : incremental version
!-------------------------------------------------------------------------
 use new_library       ; use geometry_lib    ;   implicit none
 integer::nels,nxe,nye,neq,nband,nn,nr,nip,nodof=3,nod=8,nodf=4,nst=3,     &
          ndim=2,ndof, i,k,l,iel,ns,nstep,ntot, nodofs=2,inc
 real::permx,permy,e,v,det,dtim,theta,time
 character(len=15)::element='quadrilateral'
!---------------------------- dynamic arrays------------------------------------
 real    ,allocatable :: dee(:,:), points(:,:), coord(:,:), derivf(:,:),    &
                         jac(:,:),kay(:,:),der(:,:),deriv(:,:),weights(:),  &
                         derf(:,:),funf(:), coordf(:,:), bee(:,:), km(:,:), &
                         eld(:), sigma(:), kp(:,:), ke(:,:), g_coord(:,:),  &
                         fun(:), c(:,:), width(:), depth(:), bk(:),         &
                         vol(:), loads(:), ans(:) ,volf(:,:),               &
                         store_kp(:,:,:),phi0(:),phi1(:)
 integer, allocatable :: nf(:,:),g(:),num(:),g_num(:,:) , g_g(:,:)
!---------------------------input and initialisation--------------------------
  open (10,file='p92.dat',status='old',action='read')
  open (11,file='p92.res',status='replace',action='write')
  read (10,*) nels,nxe,nye,nn,nip,                                          &
              permx, permy, e,v, dtim, nstep, theta
  ndof=nod*2; ntot=ndof+nodf
  allocate (dee(nst,nst),points(nip,ndim),coord(nod,ndim),derivf(ndim,nodf), &
           jac(ndim,ndim),kay(ndim,ndim),der(ndim,nod),deriv(ndim,nod),     &
           derf(ndim,nodf),funf(nodf),coordf(nodf,ndim),bee(nst,ndof),      &
           km(ndof,ndof),eld(ndof),sigma(nst),kp(nodf,nodf),g_g(ntot,nels), &
           ke(ntot,ntot),fun(nod),c(ndof,nodf),width(nxe+1),                &
           depth(nye+1),vol(ndof),nf(nodofs,nn), g(ntot), volf(ndof,nodf),  &
           g_coord(ndim,nn),g_num(nod,nels),num(nod),weights(nip),          &
           store_kp(nodf,nodf,nels),phi0(nodf),phi1(nodf))
           kay=0.0; kay(1,1)=permx; kay(2,2)=permy
           read (10,*)width , depth
  nf=1; read(10,*) nr ; if(nr>0) read(10,*)(k,nf(:,k),i=1,nr)
  call formnf(nf);neq=maxval(nf)
  call deemat (dee,e,v); call sample(element,points,weights)
!--------- loop the elements to find nband and set up global arrays------------
  nband = 0
 elements_1: do iel = 1 , nels
             call geometry_8qxv(iel,nxe,width,depth,coord,num)
             inc=0
             do i=1,8; do k=1,2; inc=inc+1; g(inc)=nf(k,num(i));end do;end do
             do i=1,7,2;inc=inc+1;g(inc)=nf(3,num(i)); end do
             g_num(:,iel)=num; g_coord(:,num)=transpose(coord); g_g(:,iel)= g
             if(nband<bandwidth(g))nband=bandwidth(g)
 end do elements_1
    write(11,'(a)') "Global coordinates "
    do k=1,nn;write(11,'(a,i5,a,2e12.4)')"Node",k,"        ",g_coord(:,k);end do
    write(11,'(a)') "Global node numbers "
    do k = 1 , nels; write(11,'(a,i5,a,8i5)')                               &
                            "Element ",k,"        ",g_num(:,k); end do
  write(11,'(2(a,i5))')                                                     &
          "There are  ",neq, "  equations and the half-bandwidth is   ",nband
  allocate(bk(neq*(nband+1)),loads(0:neq),ans(0:neq))
            bk = .0 ; loads = .0
!---------------- element stiffness integration and assembly------------------
      elements_2:  do iel = 1 , nels
               num = g_num( : ,iel ); coord= transpose(g_coord(:,num))
               g = g_g ( : , iel )  ; coordf = coord(1 : 7 : 2, : )
               km = .0; c = .0; kp = .0
           gauss_points_1: do i = 1 , nip
```

```
              call shape_der(der,points,i);  jac = matmul(der,coord)
              det = determinant(jac); call invert(jac);deriv = matmul(jac,der)
              call beemat(bee,deriv); vol(:)=bee(1,:)+bee(2,:)
              km = km + matmul(matmul(transpose(bee),dee),bee) *det* weights(i)
!-----------------------now the fluid contribution----------------------------
               call shape_fun(funf,points,i)
               call shape_der(derf,points,i)  ; derivf=matmul(jac,derf)
         kp=kp+matmul(matmul(transpose(derivf),kay),derivf)*det*weights(i)*dtim
               do l=1,nodf; volf(:,l)=vol(:)*funf(l); end do
               c= c+volf*det*weights(i)
           end do gauss_points_1
         store_kp( : , : , iel) = kp
         call formke(km,kp,c,ke,theta);call formkv(bk,ke,g,neq)
      end do elements_2
!------------------------factorise left hand side-----------------------------
    call banred(bk,neq)
! --------- enter the time-stepping loop--------------------------------------
    time = .0
  time_steps:  do ns = 1 , nstep
                   time = time +dtim     ;    ans = .0
    write(11,'(a,e12.4)') "The time is  ",time
      elements_3 : do iel = 1 , nels
                      g = g_g(: , iel ) ; kp = store_kp( : , : , iel)
                      phi0 = loads ( g (ndof + 1  : ))  ! gather
                      phi1 = matmul(kp,phi0)
                      ans(g(ndof+1:))=ans(g(ndof+1:))+ phi1;ans(0)=.0;! scatter
      end do elements_3
!   ramp loading
        if(ns<=10) then
           ans(1)=ans(1)-.1/6.; ans(3)=ans(3)-.2/3.
           ans(4)=ans(4)-.1/6.
        end if
        call bacsub(bk,ans) ; loads = loads + ans
        write(11,'(a)') " The nodal displacements and porepressures are    :"
        do k=1,23,22; write(11,'(i5,a,3e12.4)')k,"    ",loads(nf(:,k)) ; end do
!-------------------recover stresses at  Gauss-points-------------------------
      elements_4 :  do iel = 1 , nels
               num = g_num(: , iel ); coord=transpose(g_coord(:,num))
               g = g_g( : , iel )   ;  eld = loads( g ( 1 : ndof ) )
!             print*,"The Gauss Point effective stresses for element",iel,"are"
            gauss_pts_2: do i = 1,nip
              call shape_der (der,points,i);  jac= matmul(der,coord)
              call invert ( jac );    deriv= matmul(jac,der)
              bee= 0.;call beemat(bee,deriv);sigma= matmul(dee,matmul(bee,eld))
!              print*,"Point    ",i       ;!  print*,sigma
           end do gauss_pts_2
      end do elements_4
   end do time_steps
end program p92
```

In the previous program, solutions were obtained in terms of the total load on the system. As a prelude to the solution of nonlinear problems, where it will be essential to be able to apply loads incrementally, the next program, Program 9.2, repeats the analysis done by Program 9.1 but in an incremental form.

The formulation is described in section 3.16.1 with the resulting set of equations to be solved listed as equations (3.121). Comparison with, for example, (3.115) will show that the algorithms involved will be almost identical. The presence of the $\mathbf{KPu}_w$ term on the right-hand side of (3.121) encourages an element-wise approach in which the **KP** matrices are stored and a gather–multiply–scatter operation employed to compute the right-hand side vector in (3.121). Thus in this program the large global matrix **PB** which was present in Program 9.1 does not appear.

Otherwise, Programs 9.2 and 9.1 are very similar and operate on identical data. The results (Figure 9.10) are also very similar to those in Figure 9.8 with the small differences being due to differences in the time integration algorithm.

```
Global coordinates
Node     1            0.0000E+00  0.0000E+00
Node     2            0.5000E00   0.0000E+00
Node     3            0.1000E+01  0.0000E+00
Node     4            0.0000E+00 -0.1250E+01
Node     5            0.1000E+01 -0.1250E+01
Node     6            0.0000E+00 -0.2500E+01
Node     7            0.5000E00  -0.2500E+01
....................................................................
Node    20            0.1000E+01 -0.8750E+01
Node    21            0.0000E+00 -0.1000E+02
Node    22            0.5000E00  -0.1000E+02
Node    23            0.1000E+01 -0.1000E+02
Global node numbers
Element       1                6    4    1    2    3    5    8    7
Element       2               11    9    6    7    8   10   13   12
Element       3               16   14   11   12   13   15   18   17
Element       4               21   19   16   17   18   20   23   22
There are       32  equations and the half-bandwidth is           13
The time is      0.1000E+01
 The nodal displacements and porepressures are         :
    1        0.0000E+00 -0.1233E00   0.0000E+00
   23        0.0000E+00  0.0000E+00 -0.1000E00
The time is      0.2000E+01
 The nodal displacements and porepressures are         :
    1        0.0000E+00 -0.2872E00   0.0000E+00
   23        0.0000E+00  0.0000E+00 -0.2000E00
....................................................................
The time is      0.1700E+02
 The nodal displacements and porepressures are         :
    1        0.0000E+00 -0.3966E+01  0.0000E+00
   23        0.0000E+00  0.0000E+00 -0.9144E00
The time is      0.1800E+02
 The nodal displacements and porepressures are         :
    1        0.0000E+00 -0.4125E+01  0.0000E+00
   23        0.0000E+00  0.0000E+00 -0.8981E00
The time is      0.1900E+02
 The nodal displacements and porepressures are         :
    1        0.0000E+00 -0.4277E+01  0.0000E+00
   23        0.0000E+00  0.0000E+00 -0.8811E00
The time is      0.2000E+02
 The nodal displacements and porepressures are         :
    1        0.0000E+00 -0.4424E+01  0.0000E+00
   23        0.0000E+00  0.0000E+00 -0.8636E00
```

Figure 9.10 Results from Program 9.2

PROGRAM 9.3: PLANE STRAIN CONSOLIDATION OF A BIOT ELASTO-PLASTIC SOLID USING EIGHT-NODE SOLID QUADRILATERAL ELEMENTS COUPLED TO FOUR-NODE FLUID ELEMENTS: INCREMENTAL VERSION: MOHR–COULOMB FAILURE CRITERION: VISCO–PLASTIC STRAIN METHOD

```
program p93
!-----------------------------------------------------------------------
!      program 9.4 plane strain consolidation of a Biot elasto-plastic
!      solid using 8-node solid quadrilateral elements
!      coupled to 4-node fluid elements : incremental version
!      Mohr-Coulomb failure criterion : viscoplastic strain method
!-----------------------------------------------------------------------
 use new_library       ; use geometry_lib    ;    implicit none
 integer::nels,nxe,nye,neq,nband,nn,nr,nip,nodof=3,nod=8,nodf=4,nst=4,       &
          ndim=2,ndof, i,k,l,iel,ns,nstep,ntot, nodofs=2 ,iters,limit,inc
 real    ::permx,permy,e,v,det,dtim,theta,phi,snph,coh,psi,cons,p0,tol,      &
          dt,f,dsbar,dq1,dq2,dq3,lode_theta,sigm,pi,dpore,time
 logical::converged    ;  character(len=15) :: element = 'quadrilateral'
!--------------- dynamic arrays---------------------------
 real    ,allocatable :: dee(:,:), points(:,:), coord(:,:), derivf(:,:),    &
                         jac(:,:),kay(:,:),der(:,:),deriv(:,:),weights(:),  &
                         derf(:,:),funf(:), coordf(:,:), bee(:,:), km(:,:), &
                         eld(:), sigma(:), kp(:,:), ke(:,:), g_coord(:,:),  &
                         fun(:),c(:,:), width(:), depth(:), bk(:),stress(:),&
                         vol(:), loads(:), ans(:) ,volf(:,:),tensor(:,:,:), &
                         store_kp(:,:,:),phi0(:),phi1(:),bdylds(:),eps(:),  &
                         evpt(:,:,:),bload(:),eload(:),erate(:),            &
                         evp(:),devp(:),m1(:,:),m2(:,:),m3(:,:),flow(:,:),  &
                         tempdis(:),newdis(:),displ(:)
 integer, allocatable :: nf(:,:),g(:),num(:),g_num(:,:) , g_g(:,:)
!-----------------------input and initialisation------------------------------
  open (10,file='p93.dat',status=    'old',action='read')
  open (11,file='p93.res',status='replace',action='write')
  read (10,*) nels,nxe,nye,nn,nip,                                          &
              permx, permy, phi, coh, psi, e, v, dtim, nstep, theta ,       &
              cons , p0 , tol , limit
  ndof=nod*2; ntot=ndof+nodf
  allocate (dee(nst,nst),points(nip,ndim),coord(nod,ndim),derivf(ndim,nodf), &
            jac(ndim,ndim),kay(ndim,ndim),der(ndim,nod),deriv(ndim,nod),     &
            derf(ndim,nodf),funf(nodf),coordf(nodf,ndim),bee(nst,ndof),      &
            km(ndof,ndof),eld(ndof),sigma(nst),kp(nodf,nodf),g_g(ntot,nels), &
            ke(ntot,ntot),fun(nod),c(ndof,nodf),width(nxe+1),phi0(nodf),     &
            depth(nye+1),vol(ndof),nf(nodof,nn), g(ntot), volf(ndof,nodf),   &
            g_coord(ndim,nn),g_num(nod,nels),num(nod),weights(nip),phi1(nodf),&
            store_kp(nodf,nodf,nels),tensor(nst+1,nip,nels),eps(nst),evp(nst),&
            evpt(nst,nip,nels),bload(ndof),eload(ndof),erate(nst),devp(nst), &
            m1(nst,nst),m2(nst,nst),m3(nst,nst),flow(nst,nst),stress(nst))
        kay=0.0; kay(1,1)=permx; kay(2,2)=permy
        read (10,*)width , depth  ; pi = acos( -1.); snph=sin(phi*pi/180.)
        dt = 4.*(1.+v)*(1.-2.*v)/(e*(1.-2.*v*snph*snph))
  write(11,'(a,e12.4)') "The viscoplastic timestep is ", dt
  nf=1; read(10,*) nr ; if(nr>0) read(10,*)(k,nf(:,k),i=1,nr)
  call formnf(nf);neq=maxval(nf)
  call deemat (dee,e,v); call sample(element,points,weights)
!--------- loop the elements to find nband and set up global arrays------------
  nband = 0
 elements_1: do iel = 1 , nels
              call geometry_8qxv(iel,nxe,width,depth,coord,num)
              inc = 0
              do i=1,8; do k=1,2; inc=inc+1;g(inc)=nf(k,num(i));end do;end do
              do i=1,7,2 ; inc=inc+1 ; g(inc) = nf(3,num(i)); end do
              g_num(:,iel)=num; g_coord(:,num)=transpose(coord); g_g(:,iel) = g
              if(nband<bandwidth(g))nband=bandwidth(g)
 end do elements_1
    write(11,'(a)') "Global coordinates "
    do k=1,nn;write(11,'(a,i5,a,2e12.4)')"Node",k,"       ",g_coord(:,k);end do
    write(11,'(a)') "Global node numbers "
    do k = 1 , nels; write(11,'(a,i5,a,8i5)')                                &
                          "Element ",k,"        ",g_num(:,k); end do
```

```
  write(11,'(2(a,i5))')                                                       &
         "There are ",neq ," equations and the half-bandwidth is    ",nband
  allocate(bk(neq*(nband+1)),loads(0:neq),ans(0:neq),bdylds(0:neq),          &
           displ(0:neq),newdis(0:neq),tempdis(0:neq))
            bk = .0 ; loads = .0 ; displ = .0 ;  tensor = .0
!-------------- element stiffness integration and assembly---------------------
      elements_2:  do iel = 1 , nels
              num = g_num(: , iel ); coord=transpose(g_coord(:,num))
              g = g_g( : , iel )   ; coordf = coord(1 : 7 : 2, : )
              km = .0; c = .0; kp = .0
          gauss_points_1: do i = 1 , nip
             call shape_der(der,points,i);  jac = matmul(der,coord)
             det = determinant(jac)  ; call invert(jac)
             deriv = matmul(jac,der) ; tensor(1:2,i,iel) = cons
             tensor(4,i,iel) = cons; tensor(5,i,iel) = p0
             call beemat(bee,deriv); vol(:)=bee(1,:)+bee(2,:)
             km = km + matmul(matmul(transpose(bee),dee),bee) *det* weights(i)
!----------------------now the fluid contribution------------------------------
              call shape_fun(funf,points,i)
              call shape_der(derf,points,i)  ; derivf=matmul(jac,derf)
        kp=kp+matmul(matmul(transpose(derivf),kay),derivf)*det*weights(i)*dtim
              do l=1,nodf; volf(:,l)=vol(:)*funf(l); end do
              c= c+volf*det*weights(i)
          end do gauss_points_1
        store_kp( : , : , iel) = kp
        call formke(km,kp,c,ke,theta);call formkv(bk,ke,g,neq)
      end do elements_2
!-----------------------reduce left hand side--------------------------------
    call banred(bk,neq)  ; bdylds = .0 ; evpt = .0      ; tempdis = .0
! --------- enter the time-stepping (load increment) loop---------------------
  time = .0
  time_steps:  do ns = 1 , nstep
       time = time + dtim ; write(11,'(a,e12.4)') "The time is    ", time
                   ans = .0 ; bdylds = .0 ; evpt = .0 ;newdis = .0
      elements_3 : do iel = 1 , nels
                   g = g_g(: , iel ) ; kp = store_kp( : , : , iel)
                   phi0 = loads ( g (ndof + 1  : ))  ! gather
                   phi1 = matmul(kp,phi0),
                   ans(g(ndof+1:))=ans(g(ndof+1:))+ phi1;ans(0)=.0;! scatter
      end do elements_3
!-------------------------- constant loading --------------------------------
          ans(1)=ans(1)-1./24.; ans(3)=ans(3)-1./6.
          ans(5)=ans(5)-1./12. ; ans(7)=ans(7)-1./6. ; ans(9) = ans(9)-1./24.
!-----------------------     iteration loop ---------------------------------
        iters = 0
    iterations: do
     iters=iters+1;  loads = ans + bdylds ;   call bacsub(bk,loads)
!-----------------------     check convergence -------------------------------
      newdis = loads ; newdis(nf(3,:))=.0
      call checon(newdis,tempdis,tol,converged)
      if(iters==1)converged=.false. ;  if(converged.or.iters==limit)bdylds=.0
!-------------------- go round the Gauss Points------------------------------
      elements_4: do iel = 1 , nels
       num = g_num( : , iel ) ; coord = transpose(g_coord( : , num ))
       g = g_g( : , iel )  ;  eld = loads ( g( 1 : ndof) )    ;  bload = .0
       gauss_points_2 : do i = 1 , nip
          call shape_der ( der,points,i); jac=matmul(der,coord)
          det = determinant(jac)  ;   call invert(jac)
          deriv = matmul(jac,der) ; call beemat (bee,deriv);eps=matmul(bee,eld)
          eps = eps -evpt( : , i , iel)    ;        sigma=matmul(dee,eps)
          stress = sigma + tensor( 1:4 , i , iel )
          call invar(stress,sigm,dsbar,lode_theta)
!------------------ check whether yield is violated -------------------------
call mocouf (phi, coh , sigm, dsbar , lode_theta , f )
```

```
        if(converged.or.iters==limit) then
        devp=stress
          else
          if(f>=.0) then
          call mocouq(psi,dsbar,lode_theta,dq1,dq2,dq3)
          call formm(stress,m1,m2,m3)
          flow=f*(m1*dq1+m2*dq2+m3*dq3)      ;    erate=matmul(flow,stress)
          evp=erate*dt; evpt(:,i,iel)=evpt(:,i,iel)+evp; devp=matmul(dee,evp)
        end if; end if
     if(f>=.0) then
       eload=matmul(transpose(bee),devp) ; bload=bload+eload*det*weights(i)
     end if
    if(converged.or.iters==limit) then
! ------------------- update stresses and porepressures -----------------------
      tensor (1:4 , i , iel ) = stress   ;    dpore = .0
      call shape_fun( funf , points , i )
      do k = 1 , nodf; dpore = dpore + funf(k)*loads(g(k+ndof)) ; end do
      tensor( 5 , i , iel ) = tensor ( 5 , i , iel ) + dpore
    end if
   end do gauss_points_2
!-------------------- compute the total bodyloads vector ---------------------
   bdylds(g( 1 : ndof)) = bdylds( g( 1 : ndof) ) + bload  ; bdylds(0) = .0
  end do elements_4
  if(converged.or.iters==limit)exit
 end do iterations
 displ = displ + loads
 write(11,'(a,i5,a)') "It took",iters,"  iterations to converge"
 write(11,'(a,e12.4)') "The displacements are : ", displ(1)
 if(iters==limit)stop
end do time_steps
end program p93
```

For nonlinear problems involving elasto-plastic solid skeletons in a Biot analysis, the incremental form of the equations becomes essential (the algorithm of Program 9.2 rather than that of Program 9.1). The resulting program is an amalgamation of Program 9.2 and Program 6.2 which used the visco-plastic strain method for redistributing excess internal stresses (Griffiths, 1994). Terminology to note includes:

`NODOF`	total number of degrees of freedom per node (3)
`NODOFS`	number of solid degrees of freedom per node (2)
`NOD`	total number of nodes per element (8)
`NODF`	number of fluid nodes per element (4)
`THETA`	time-stepping parameter θ
`LODE_THETA`	Lode angle (see Figure 6.4)
`DPORE`	increment of pore pressure

The 3-d array `TENSOR` is used to store the element integrating point stresses with the four effective stress components coming first, followed by the porewater stress as the fifth component.

The illustrative problem shown in Figure 9.11 is a very simple one involving compression of a block of saturated elasto-plastic material by a “deviator” stress D_f which is the difference between the vertical and (constant) horizontal

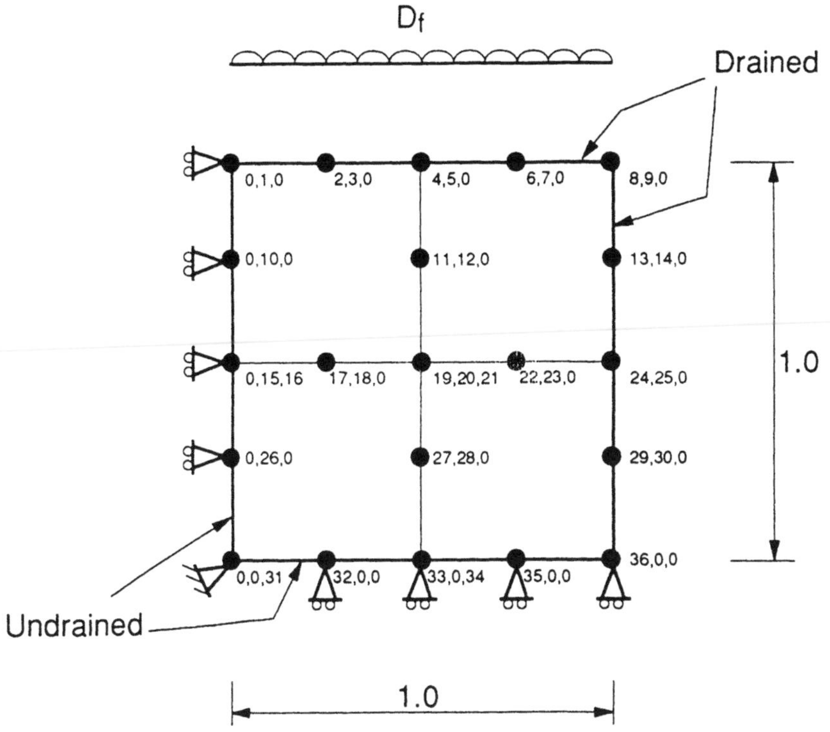

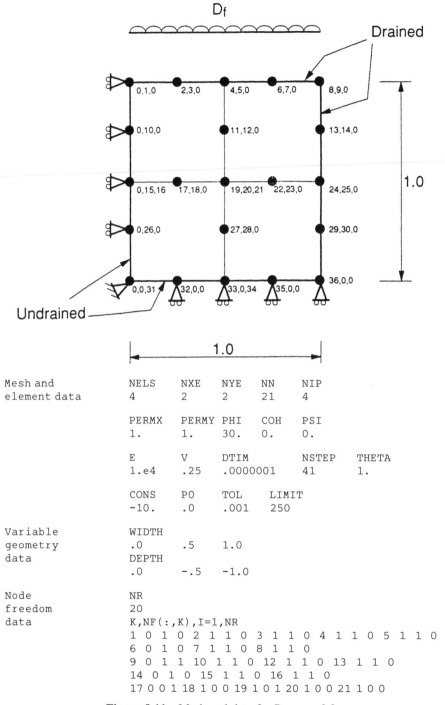

Mesh and element data	NELS	NXE	NYE	NN	NIP	
	4	2	2	21	4	
	PERMX	PERMY	PHI	COH	PSI	
	1.	1.	30.	0.	0.	
	E	V	DTIM		NSTEP	THETA
	1.e4	.25	.0000001		41	1.
	CONS	PO	TOL	LIMIT		
	-10.	.0	.001	250		
Variable geometry data	WIDTH					
	.0	.5	1.0			
	DEPTH					
	.0	-.5	-1.0			

Node freedom data

```
NR
20
K,NF(:,K),I=1,NR
1 0 1 0 2 1 1 0 3 1 1 0 4 1 1 0 5 1 1 0
6 0 1 0 7 1 1 0 8 1 1 0
9 0 1 1 10 1 1 0 12 1 1 0 13 1 1 0
14 0 1 0 15 1 1 0 16 1 1 0
17 0 0 1 18 1 0 0 19 1 0 1 20 1 0 0 21 1 0 0
```

Figure 9.11 Mesh and data for Program 9.3

stresses on the "soil". These have been set initially as `CONS` to be -10 kN/m^2 (compression negative).

The results as D_f is increased are listed as Figure 9.12 for the case of dilation angle $\psi = 0$ and a "fast" rate of application of D_f such that

$$\frac{dD_f}{dt} = 5 \times 10^6$$

The effects of changing the loading rate are shown in Figure 9.13.

```
The viscoplastic timestep is     0.2857E-03
Global coordinates
Node     1            0.0000E+00   0.0000E+00
Node     2            0.2500E00   0.0000E+00
Node     3            0.5000E00   0.0000E+00
Node     4            0.7500E00   0.0000E+00
Node     5            0.1000E+01   0.0000E+00
Node     6            0.0000E+00  -0.2500E00
Node     7            0.5000E00  -0.2500E00
.....................................................................

Node    17            0.0000E+00  -0.1000E+01
Node    18            0.2500E00  -0.1000E+01
Node    19            0.5000E00  -0.1000E+01
Node    20            0.7500E00  -0.1000E+01
Node    21            0.1000E+01  -0.1000E+01
Global node numbers
Element      1               9     6     1     2     3     7    11    10
Element      2              11     7     3     4     5     8    13    12
Element      3              17    14     9    10    11    15    19    18
Element      4              19    15    11    12    13    16    21    20
There are     36  equations and the half-bandwidth is          21
The time is          0.1000E-06
It took     2  iterations to converge
The displacements are :    -0.3342E-04
The time is          0.2000E-06
It took     2  iterations to converge
The displacements are :    -0.6689E-04
The time is          0.3000E-06
It took     2  iterations to converge
The displacements are :    -0.1004E-03
.....................................................................

The time is          0.1600E-05
It took     2  iterations to converge
The displacements are :    -0.5355E-03
The time is          0.1700E-05
It took     2  iterations to converge
The displacements are :    -0.5690E-03
The time is          0.1800E-05
It took    16  iterations to converge
The displacements are :    -0.6176E-03
The time is          0.1900E-05
It took    37  iterations to converge
The displacements are :    -0.6856E-03
The time is          0.2000E-05
It took   250  iterations to converge
The displacements are :    -0.6581E-02
```

Figure 9.12 Results from Program 9.3

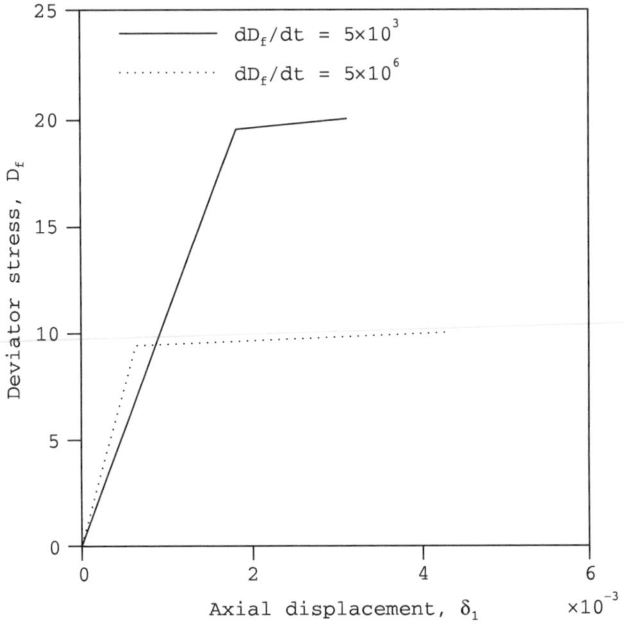

Figure 9.13 Response for different loading rates

For the faster loading rate, collapse occurs at about half the D_f for the slow loading rate due to the build-up of excess porewater pressures in the "soil". Griffiths (1985) has produced an analytical solution which gives

$$\frac{D_f}{\sigma_3} = \frac{(K_p - 1)(2\beta_{ps} + 1)}{(K_p + 1)\beta_{ps} + 1} \tag{9.5}$$

In this expression, $K_p = \tan^2(45^\circ + \phi'/2) = 3$ in the present example and $\beta_{ps} \to 0$ for very slow loading "drained" conditions while $\beta_{ps} \to \infty$ for very fast loading "undrained" conditions. Equation (9.5) leads to expected values of D_f/σ_3 of 2 and 1 respectively and these are closely approximated by the numerical solutions shown in Figure 9.13.

PROGRAM 9.4: PLANE STRAIN CONSOLIDATION OF A BIOT ELASTIC SOLID USING EIGHT-NODE SOLID ELEMENTS COUPLED TO FOUR-NODE FLUID ELEMENTS: MESH-FREE PCG VERSION

For the final program in this chapter, Program 9.1 is recast in mesh-free form using the p.c.g. algorithm. Inspection of equations (3.114)–(3.116) shows that

```
program p94
!-------------------------------------------------------------------------
!      program 9.4 plane strain consolidation of a Biot elastic
!      solid using 8-node solid quadrilateral elements
!      coupled to 4-node fluid elements  -  pcg version
!-------------------------------------------------------------------------
 use new_library       ; use geometry_lib   ;   implicit none
 integer::nels,nxe,nye,neq,nn,nr,nip,nodof=3,nod=8,nodf=4,nst=3,ndim=2,     &
         ndof,i,k,l,iel,ns,nstep,ntot,nodofs=2,cjiters,cjits,inc
 real::permx,permy,e,v,det,dtim,theta,x1,x2,time,up,alpha,beta,big,cjtol
 logical :: cj_converged   ;     character(len=15) :: element='quadrilateral'
!--------------------------- dynamic arrays-------------------------------------
 real    ,allocatable :: dee(:,:),points(:,:),coord(:,:),derivf(:,:),pmul(:),&
                         jac(:,:),kay(:,:),der(:,:),deriv(:,:),weights(:),   &
                         derf(:,:),funf(:), coordf(:,:), bee(:,:), km(:,:),  &
                         eld(:), sigma(:), kp(:,:), ke(:,:), g_coord(:,:),   &
                         kd(:,:),fun(:), c(:,:), width(:), depth(:),loads(:),&
                         vol(:), storke(:,:,:), ans(:) ,volf(:,:) ,           &
                         p(:),x(:),xnew(:),u(:),diag_precon(:),d(:),          &
                         utemp(:), storkd(:,:,:)
 integer, allocatable :: nf(:,:),g(:),num(:),g_num(:,:) , g_g(:,:)
!------------------------input and initialisation--------------------------------
  open (10,file='p94.dat',status=    'old',action='read')
  open (11,file='p94.res',status='replace',action='write')
  read (10,*) nels,nxe,nye,nn,nip,                                          &
              permx, permy, e,v, dtim, nstep, theta , cjits , cjtol
  ndof=nod*2; ntot=ndof+nodf
  allocate (dee(nst,nst),points(nip,ndim),coord(nod,ndim),derivf(ndim,nodf), &
           jac(ndim,ndim),kay(ndim,ndim),der(ndim,nod),deriv(ndim,nod),      &
           derf(ndim,nodf),funf(nodf),coordf(nodf,ndim),bee(nst,ndof),       &
           km(ndof,ndof),eld(ndof),sigma(nst),kp(nodf,nodf),g_g(ntot,nels),  &
           ke(ntot,ntot),kd(ntot,ntot),fun(nod),c(ndof,nodf),width(nxe+1),   &
           depth(nye+1),vol(ndof),nf(nodof,nn), g(ntot), volf(ndof,nodf),    &
           g_coord(ndim,nn),g_num(nod,nels),num(nod),weights(nip),           &
           storke(ntot,ntot,nels),storkd(ntot,ntot,nels),                    &
           pmul(ntot),utemp(ntot))
           kay=0.0; kay(1,1)=permx; kay(2,2)=permy
           read (10,*)width , depth
  nf=1; read(10,*) nr ; if(nr>0) read(10,*)(k,nf(:,k),i=1,nr)
  call formnf(nf);neq=maxval(nf)
  call deemat (dee,e,v); call sample(element,points,weights)
!----------------- loop the elements to  set up global arrays------------------
 elements_1: do iel = 1 , nels
             call geometry_8qxv(iel,nxe,width,depth,coord,num)
             inc = 0
             do i=1,8; do k=1,2;inc=inc+1;g(inc)=nf(k,num(i)); end do; end do
             do i=1,7,2;inc=inc+1;g(inc)=nf(3,num(i)); end do
             g_num(:,iel)=num; g_coord(:,num)=transpose(coord); g_g(:,iel) = g
 end do elements_1
    write(11,'(a)') "Global coordinates "
    do k=1,nn;write(11,'(a,i5,a,2e12.4)')"Node",k,"        ",g_coord(:,k);end do
    write(11,'(a)') "Global node numbers "
    do k = 1 , nels; write(11,'(a,i5,a,8i5)')                                  &
                               "Element ",k,"        ",g_num(:,k); end do
  write(11,'(a,i5,a)') "There are  ",neq, "  equations to be solved"
  allocate(loads(0:neq),ans(0:neq),p(0:neq),x(0:neq),xnew(0:neq),u(0:neq),     &
           diag_precon(0:neq),d(0:neq))
    loads = .0 ; p = .0; xnew = .0; diag_precon = .0
!------------ element stiffness integration , storage and preconditioner -----
      elements_2:  do iel = 1 , nels
              num = g_num(:,iel); coord=transpose(g_coord(:,num))
              g = g_g( :, iel ) ; coordf = coord(1 : 7 : 2, : )
              km = .0; c = .0; kp = .0
         gauss_points_1: do i = 1 , nip
```

```
           call shape_der(der,points,i);  jac = matmul(der,coord)
           det = determinant(jac);call invert(jac); deriv = matmul(jac,der)
           call beemat(bee,deriv); vol(:)=bee(1,:)+bee(2,:)
           km = km + matmul(matmul(transpose(bee),dee),bee) *det* weights(i)
!-------------------------now the fluid contribution-------------------------
            call shape_fun(funf,points,i)
            call shape_der(derf,points,i)  ; derivf=matmul(jac,derf)
      kp=kp+matmul(matmul(transpose(derivf),kay),derivf)*det*weights(i)*dtim
            do l=1,nodf; volf(:,l)=vol(:)*funf(l); end do
            c= c+volf*det*weights(i)
         end do gauss_points_1
        call fmkdke(km,kp,c,ke,kd,theta)
        storke(: , : , iel) = ke   ;    storkd ( : , : , iel ) = kd
        do k=1,ndof;diag_precon(g(k))=diag_precon(g(k))+theta*km(k,k);end do
        do k=1 , nodf
         diag_precon(g(ndof+k))=diag_precon(g(ndof+k))-theta*theta*kp(k,k)
        end do
    end do elements_2
        diag_precon(1:neq) = 1./ diag_precon(1:neq) ; diag_precon(0) = .0
! ----------------------- enter the time-stepping loop-----------------------
    time = .0
  time_steps:  do ns = 1 , nstep
    ans = .0    ;   time=time+dtim;write(11,'(a,e12.4)')"The time is",time
    elements_3 : do iel = 1 , nels
                 g = g_g( : , iel );  kd = storkd ( : , : , iel )
                 pmul = loads ( g ) ; utemp = matmul ( kd , pmul )
!dir$ ivdep
        do i = 1 , ntot
         ans( g ( i )) = ans( g ( i )) + utemp ( i )
        end do
    end do elements_3          ;  ans(0) = .0
!-------------------------   ramp loading  ----------------------------------
      x1=(.1*ns+.1*(theta-1.))/6.; x2=x1*4.
      if(ns>10) then
        ans(1)=ans(1)-1./6.; ans(3)=ans(3)-2./3.
        ans(4)=ans(4)-1./6.
      else if(ns<10) then
        ans(1)=ans(1)-x1;ans(3)=ans(3)-x2; ans(4)=ans(4)-x1
      end if              ;
    d = diag_precon*ans    ;   p = d ; x = .0  ! depends on starting x = .0
!--------------------   solve the simultaneous equations by pcg --------------
     cjiters = 0
    conjugate_gradients:  do
     cjiters = cjiters + 1 ; u = .0
    elements_4 : do iel = 1 , nels
                  g = g_g( : , iel ); ke = storke( : , : ,iel)
                  pmul = p(g); utemp=matmul(ke,pmul)
!dir$ ivdep
        do i = 1 , ntot
           u(g(i)) = u(g(i)) +  utemp(i)
        end do
    end do elements_4
!---------------------------pcg process -------------------------------------
  up =dot_product(ans,d); alpha=up/dot_product(p,u)
  xnew = x + p* alpha; ans = ans - u*alpha; d = diag_precon*ans
  beta = dot_product(ans,d)/up; p = d + p * beta
  big = .0; cj_converged = .true.
  do i = 1,neq; if(abs(xnew(i))>big)big=abs(xnew(i)); end do
  do i = 1,neq; if(abs(xnew(i)-x(i))/big>cjtol)cj_converged=.false.;end do
  x = xnew
  if(cj_converged.or.cjiters==cjits) exit
    end do conjugate_gradients
!----------- end of pcg process--------------------------------------------
  write(11,'(a,i5,a)')                                                   &
```

```
            "Conjugate gradients took ",cjiters, "  iterations to converge"
   ans = xnew           ; ans(0) = .0          ; loads = ans

       write(11,'(a)') " The nodal displacements and porepressures are    :"
       do k=1,23,22; write(11,'(i5,a,3e12.4)')k,"     ",ans(nf(:,k)) ; end do
!-------------------recover stresses at centroidal gauss-point------------
     elements_5 :  do iel = 1 , nels
              num = g_num(:,iel); coord=transpose(g_coord(:,num))
              g = g_g( : , iel ); eld = ans( g ( 1 : ndof ) )
        !    print*,"The Gauss Point effective stresses for element",iel,"are"
            gauss_pts_2: do i = 1,nip
              call shape_der (der,points,i);  jac= matmul(der,coord)
              call invert ( jac );     deriv= matmul(jac,der)
              bee= 0.;call beemat(bee,deriv);sigma= matmul(dee,matmul(bee,eld))
        !     print*,"Point    ",i        ;!  print*,sigma
           end do gauss_pts_2
     end do elements_5
  end do time_steps
 end program p94
```

the left-hand side matrix in Biot analyses is not positive definite due to the negative **KP** terms. Simple diagonal preconditioning does yield a symmetric positive definite preconditioned matrix and this is the method used in Program 9.4. It is recognised that alternative iterative strategies may well be more efficient. The **KE** and **KD** element matrices are stored as STORKE and STORKD

```
Mesh, element          NELS     NXE      NYE      NN      NIP
and iteration data     4        1        4        23      4

                       PERMX     PERMY     E       V
                       1.0       1.0       1.0     .0

Time integration       DTIM      NSTEP     THETA     CJITS     CJTOL
data                   1.0       20        1.0       100       .00001

Variable geometry      WIDTH(I),I=1,NXE+1 0.0  1.0
data                   DEPTH(I),I=1,NYE+1 0.0 -2.5 -5.0 -7.5 -10.0

Node freedom           NR
data                   23
                       K,NF(:,K),K=1,NR
                        1  0  1  0   2  1  1  0   3  0  1  0
                        4  0  1  0   5  0  1  0
                        6  0  1  1   7  1  1  0   8  0  1  1
                        9  0  1  0  10  0  1  0
                       11  0  1  1  12  1  1  0  13  0  1  1
                       14  0  1  0  15  0  1  0
                       16  0  1  1  17  1  1  0  18  0  1  1
                       19  0  1  0  20  0  1  0
                       21  0  0  1  22  0  0  0  23  0  0  1
```

Figure 9.14 Data for Program 9.4

```
Global coordinates
Node     1           0.0000E+00  0.0000E+00
Node     2           0.5000E+00  0.0000E+00
Node     3           0.1000E+01  0.0000E+00
Node     4           0.0000E+00 -0.1250E+01
Node     5           0.1000E+01 -0.1250E+01
Node     6           0.0000E+00 -0.2500E+01
Node     7           0.5000E+00 -0.2500E+01
.............................................................
Node    20           0.1000E+01 -0.8750E+01
Node    21           0.0000E+00 -0.1000E+02
Node    22           0.5000E+00 -0.1000E+02
Node    23           0.1000E+01 -0.1000E+02
Global node numbers
Element      1            6    4    1    2    3    5    8    7
Element      2           11    9    6    7    8   10   13   12
Element      3           16   14   11   12   13   15   18   17
Element      4           21   19   16   17   18   20   23   22
There are     32  equations to be solved
The time is  0.1000E+01
Conjugate gradients took     18     iterations to converge
 The nodal displacements and porepressures are     :
    1       0.0000E+00 -0.1233E+00  0.0000E+00
   23       0.0000E+00  0.0000E+00 -0.1000E+00
The time is  0.2000E+01
Conjugate gradients took     19     iterations to converge
 The nodal displacements and porepressures are     :
    1       0.0000E+00 -0.2872E+00  0.0000E+00
   23       0.0000E+00  0.0000E+00 -0.2000E+00
.............................................................

The time is  0.1800E+02
Conjugate gradients took     23     iterations to converge
 The nodal displacements and porepressures are     :
    1       0.0000E+00 -0.3932E+01  0.0000E+00
   23       0.0000E+00  0.0000E+00 -0.9092E+00
The time is  0.1900E+02
Conjugate gradients took     21     iterations to converge
 The nodal displacements and porepressures are     :
    1       0.0000E+00 -0.4095E+01  0.0000E+00
   23       0.0000E+00  0.0000E+00 -0.8942E+00
The time is  0.2000E+02
Conjugate gradients took     22     iterations to converge
 The nodal displacements and porepressures are     :
    1       0.0000E+00 -0.4251E+01  0.0000E+00
   23       0.0000E+00  0.0000E+00 -0.8783E+00
```

Figure 9.15 Results from Program 9.4

respectively but otherwise the terminology is that of Program 9.1 and of previous p.c.g. programs such as Program 8.4. The problem solved is the same as was solved by Program 9.1, and data is listed as Figure 9.14.

The extra data items are just the conjugate gradient iteration tolerance and iteration limit, `CJTOL` and `CJITS` respectively. The results are listed as Figure 9.15 which should be compared with Figure 9.8.

The two sets of results are in agreement but the conjugate gradient process takes an average 0.8 ∗ `NEQ` iterations to converge and sometimes > `NEQ`.

9.2 REFERENCES

Griffiths, D.V. (1985) The effects of pore fluid compressibility on failure loads in elastoplastic soils. *Int.J.Num. Analytical Meth. Geomech.* **9**, 253–259.

Griffiths, D.V. (1994) Coupled analyses in geomechanics. In *Viscoplastic Behaviour of Geomaterials*, ed. Cristescu, N.D. and Gioda, G. Chapter 5, Springer-Verlag, Vienna and New York.

Kidger, D. J. (1994) Visualisation of three-dimensional processes in geomechanics computations. *Proc. 8th IACMAG Conference*, West Virginia, pp. 453–457, Balkema.

Schiffman, R. L. (1960) Field applications of soil consolidation time-dependent loading and variable permeability. *Highway Research Board, Bulletin 248*, Washington, USA.

Smith, I. M. and Hobbs, R. (1976) Biot analysis of consolidation beneath embankments. *Géotechnique*, **26**, (1), 149–171.

10 Eigenvalue Problems

10.0 INTRODUCTION

The ability to solve eigenvalue problems is important in many aspects of finite element work. For example, the number of zero eigenvalues of an element "stiffness" matrix (its rank deficiency) is an important guide to the suitability of that element. In that context, the problem to be solved is just

$$\mathbf{Ax} = \lambda \mathbf{x} \tag{10.1}$$

which is the eigenvalue problem in "standard form". More often, the eigenvalue equation will describe a physical situation such as free vibration of a solid or fluid. For example, equation (2.19) for a freely vibrating elastic solid was

$$\mathbf{KMa} = \omega^2 \mathbf{MMa} \tag{10.2}$$

which, although not in standard form, can readily be converted to it. Section 3.13.1 describes the conversion process for the cases in which **MM** is diagonalised ("lumped") or not ("consistent").

The present chapter describes five programs for the determination of eigenvalues and eigenvectors of such elastic solids. Different algorithms and storage strategies are employed in the various cases. Since elastic solids are treated, the programs can be viewed as extensions of the programs described in Chapters 4 and 5. The same terminology is used.

PROGRAM 10.0: EIGENVALUES AND EIGENVECTORS OF A STRING OF BEAM ELEMENTS, LUMPED MASS

This program illustrates a free vibration analysis of a typical line structure and can be thought of as an extension to Program 4.2. Only "string" topologies are allowed and Figure 10.1 shows a simple cantilever beam made up from five slender beam elements, with two degrees of freedom per node. The beam has a length of 4.0, a flexural stiffness *EI* of 0.08333 and a mass per unit length ρA of 1.0. The natural frequencies and mode shapes of the cantilever are to be found.

The meanings of the variable names used in Program 10.0 are given as follows:

Scalar integers:

`NELS`	number of elements
`NEQ`	number of degrees of freedom in the mesh
`NN`	number of nodes

```
program p100
!-------------------------------------------------------------------------
! program 10.0 eigenvalues and eigenvectors of a string of beam elements
!-------------------------------------------------------------------------
use new_library      ;    use geometry_lib ;        implicit none
integer::nels,neq,nn,nband,nr,nod=2,nodof=2,ndof=4,iel,i,j,k,ndim=1,      &
         np_types,ifail,icount,nmodes
real::tol=1.e-30, el_ei , el_ell
!-------------------------dynamic arrays-----------------------------------
real,allocatable::km(:,:),ku(:,:),loads(:),coord(:,:),g_coord(:,:),ei(:), &
                  rhoa(:),ell(:),diag(:),udiag(:),emm(:,:),kv(:),kh(:),   &
                  rrmass(:)
integer,allocatable::nf(:,:),g(:),num(:),g_num(:,:),g_g(:,:),etype(:)
!--------------------input and initialisation------------------------------
open(10,file='p100.dat'); open(11,file='p100.res')
read(10,*)nels,np_types,nmodes; nn=nels+1
allocate(nf(nodof,nn),km(ndof,ndof),coord(nod,ndim);g_coord(ndim,nn),     &
        g_num(nod,nels),num(nod),g(ndof),emm(ndof,ndof),ei(np_types),     &
        rhoa(np_types),ell(nels),g_g(ndof,nels),etype(nels))
read(10,*)(ei(i),rhoa(i),i=1,np_types)
etype=1; if(np_types>1)read(10,*)etype
read(10,*)ell,nr
nf=1; if(nr>0)read(10,*)(k,nf(:,k),i=1,nr); call formnf(nf); neq=maxval(nf)
!-------------loop the elements to find global array sizes-------------------
nband=0
elements_1: do iel=1,nels
              el_ell = ell(iel) ; call geometry_2l(iel,el_ell,coord,num)
              call num_to_g ( num , nf , g )
              g_num(:,iel)=num; g_coord(:,num)=transpose(coord)
              g_g(:,iel)=g; if(nband<bandwidth(g))nband=bandwidth(g)
end do elements_1
allocate(ku(neq,nband+1),kv(neq*(nband+1)),kh(neq*(nband+1)),              &
        loads(0:neq),diag(0:neq),udiag(0:neq),rrmass(0:neq))
write(11,'(a)')"Global coordinates"
do k=1,nn; write(11,'(a,i5,a,3e12.4)')                                     &
    "Node     ",k,"      ",g_coord(:,k); end do
write(11,'(a)')"Global node numbers"
do k=1,nels; write(11,'(a,i5,a,27i3)')                                     &
    "Element ",k,"       ",g_num(:,k); end do
write(11,'(2(a,i5),/)')                                                    &
    "There are ",neq,"  equations and the half-bandwidth is ",nband
!-------------global stiffness and (lumped) mass matrix assembly-------------
diag=0.0; ku=0.0
elements_2: do iel=1,nels
              emm=0.0;   el_ei = ei(etype(iel)); el_ell = ell(iel)
              emm(1,1)=0.5*rhoa(etype(iel))*el_ell; emm(3,3)=emm(1,1)
              emm(2,2)=emm(1,1)*el_ell**2/12.; emm(4,4)=emm(2,2)
              call beam_km(km,el_ei,el_ell); g=g_g(:,iel)
              call formku(ku,km,g); call formlump(diag,emm,g)
end do elements_2
write(11,*)"The global mass diagonal is:"; write(11,'(6e12.4)')diag(1:)
!----------------------reduce to standard eigenvalue problem-----------------
rrmass(1:)=1./sqrt(diag(1:))
do i=1,neq
  if(i<=neq-nband)then; k=nband+1; else; k=neq-i+1; end if
  do j=1,k; ku(i,j)=ku(i,j)*rrmass(i)*rrmass(i+j-1); end do
end do
icount=0
do j=1,nband+1; do i=1,neq
icount=icount+1; kh(icount)=ku(i,j)
end do; end do
!------------------------extract the eigenvalues-----------------------------
call bandred(ku,diag,udiag,loads);ifail=1; call bisect(diag,udiag,tol,ifail)
write(11,*)"The eigenvalues are:"; write(11,'(6e12.4)')diag(1:)
```

```
!----------------------extract the eigenvectors-----------------------------
do i=1,nmodes
  kv=kh; kv(:neq)=kv(:neq)-diag(i); kv(1)=kv(1)+1.e20
  udiag=0.0; udiag(1)=kv(1)
  call banred(kv,neq); call bacsub(kv,udiag); udiag=rrmass*udiag
  write(11,'("Eigenvector number",i3," is:")')i
  write(11,'(6e12.4)')udiag(1:)/maxval(abs(udiag))
end do
end program p100
```

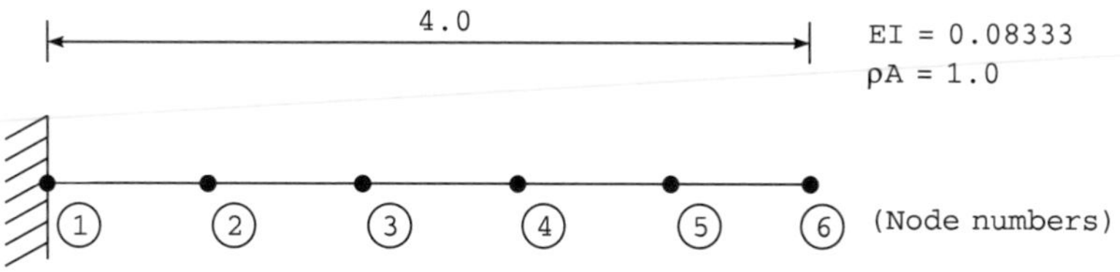

Structure and element data	NELS	NP_TYPES	NMODES
	5	1	3
	(EI(I),RHOA(I),I=1,NP_TYPES)		
	0.08333 1.0		
	ELL		
	0.8 0.8 0.8 0.8 0.8		
Node freedom data	NR	K,NF(:,K),I=1,NR	
	1	1 0 0	

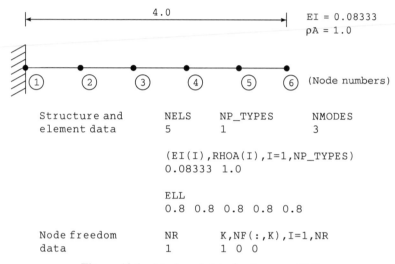

Figure 10.1 Mesh and data for Program 10.0

NBAND	half-bandwidth of mesh
NR	number of restrained nodes
NOD	number of nodes per element
NODOF	number of degrees of freedom per node
NDOF	number of degrees of freedom per element
NDIM	number of dimensions
NP_TYPES	number of different property types
IEL,I,J,K,ICOUNT	simple counters
IFAIL	eigensolver flag
NMODES	number of eigenvectors required

Scalar reals:

TOL	eigensolver tolerance
EL_ELL	element length
EL_EI	element stiffness

Allocatable real arrays:

KM	element stiffness matrix
KU	global system matrix (upper triangle stored)
LOADS	working space vector
COORD	element nodal coordinates matrix
G_COORD	global nodal coordinates matrix
EI	element flexural rigidities vector
RHOA	mass per unit length vector
ELL	element lengths vector
DIAG	lumped mass matrix vector (overwritten by eigenvalues)
UDIAG	working space vector
EMM	element (lumped) mass matrix
KV	KU in vector form
KH	copy of KV
RRMASS	copy of square root of reciprocal of DIAG

Allocatable integer arrays:

NF	nodal freedom matrix
G	element steering vector
NUM	element node numbers vector
G_NUM	global element node numbers matrix
G_G	global element steering matrix
ETYPE	element property type vector

After the usual preliminary steps to establish the number of freedoms and bandwidth of the problem, the elements are assembled into global stiffness and mass matrices. In this case, the global stiffness is stored in KU as an upper band rectangle by library subroutine FORMKU, and the global lumped mass matrix is stored as a vector in DIAG by library subroutine FORMLUMP and printed. The structure of the program is shown in Figure 10.2.

The diagonal element mass matrices are held in EMM and use the following lumping (see e.g. Cook et al, 1989):

$$\mathbf{M} = \frac{\rho AL}{2}\begin{bmatrix} 1 & 0 & 0 & 0 \\ 0 & L^2/12 & 0 & 0 \\ 0 & 0 & 1 & 0 \\ 0 & 0 & 0 & L^2/12 \end{bmatrix} \tag{10.3}$$

By factorising DIAG and altering the appropriate terms in KU (see section 3.13.1), the symmetrical matrix for the standard eigenvalue problem is retrieved (still called KU). The eigenvalues of this band matrix (ω^2) are then calculated using Jacobi's method which employs library subroutines BANDRED and BISECT. The eigenvalues are listed in Figure 10.3.

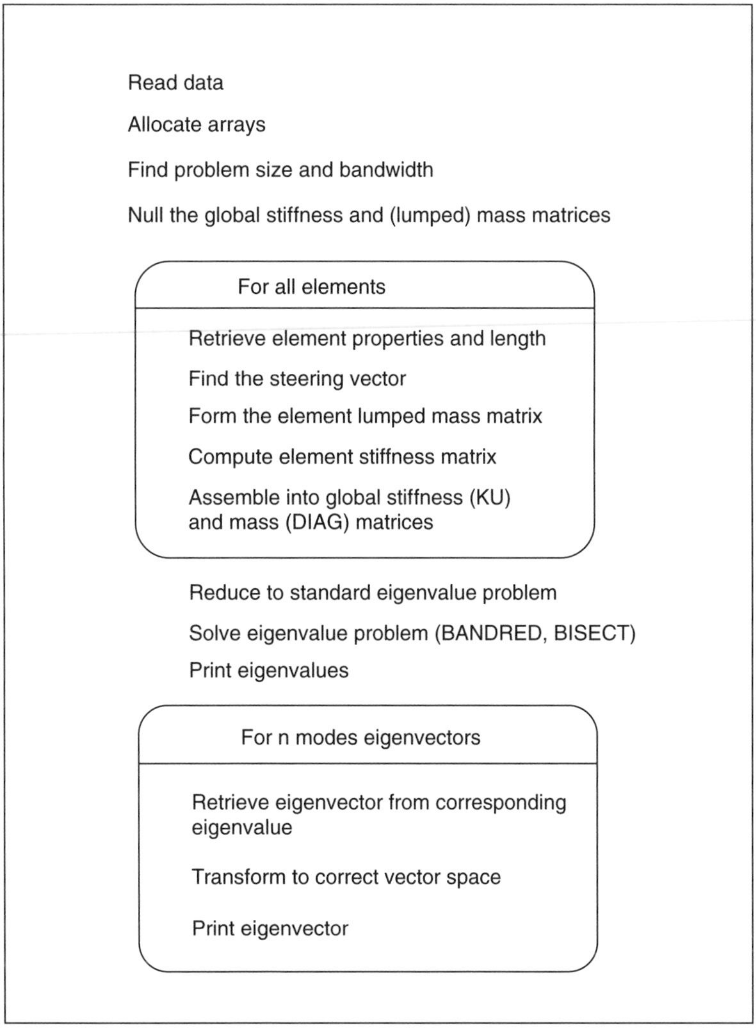
Read data
Allocate arrays
Find problem size and bandwidth
Null the global stiffness and (lumped) mass matrices
For all elements
Retrieve element properties and length
Find the steering vector
Form the element lumped mass matrix
Compute element stiffness matrix
Assemble into global stiffness (KU)
and mass (DIAG) matrices
Reduce to standard eigenvalue problem
Solve eigenvalue problem (BANDRED, BISECT)
Print eigenvalues
For n modes eigenvectors
Retrieve eigenvector from corresponding
eigenvalue
Transform to correct vector space
Print eigenvector

Figure 10.2 Structure chart for Program 10.0

The computed results indicate a fundamental frequency of $\omega = \sqrt{0.0038} = 0.062$ which should be compared with the exact result for a slender beam of

$$\frac{3.516}{L^2}\sqrt{\frac{EI}{\rho A}} = 0.063$$

If required, the first `NMODES` eigenvectors can be extracted by finding the relevant non-trivial solutions to the homogeneous equations and transforming them back to the correct vector space. In this example the first three

```
Global coordinates
Node        1       0.0000E+00
Node        2       0.8000E+00
Node        3       0.1600E+01
Node        4       0.2400E+01
Node        5       0.3200E+01
Node        6       0.4000E+01
Global node numbers
Element     1         1  2
Element     2         2  3
Element     3         3  4
Element     4         4  5
Element     5         5  6
There are     10  equations and the half-bandwidth is      3

 The global mass diagonal is:
  0.8000E+00  0.4267E-01  0.8000E+00  0.4267E-01  0.8000E+00  0.4267E-01
  0.8000E+00  0.4267E-01  0.4000E+00  0.2133E-01
 The eigenvalues are:
  0.3823E-02  0.1278E+00  0.8511E+00  0.2765E+01  0.6323E+01  0.1147E+02
  0.1748E+02  0.2324E+02  0.2756E+02  0.3225E+02
Eigenvector number  1 is:
  0.6303E-01  0.1499E+00  0.2275E+00  0.2539E+00  0.4579E+00  0.3154E+00
  0.7229E+00  0.3423E+00  0.1000E+01  0.3485E+00
Eigenvector number  2 is:
  0.2307E+00  0.4535E+00  0.5431E+00  0.2347E+00  0.5094E+00 -0.3411E+00
  0.2270E-01 -0.8308E+00 -0.7287E+00 -0.1000E+01
Eigenvector number  3 is:
  0.2888E+00  0.4155E+00  0.3179E+00 -0.3976E+00 -0.1579E+00 -0.5696E+00
 -0.2684E+00  0.3779E+00  0.3425E+00  0.1000E+01
```

Figure 10.3 Results from beam example with Program 10.0

eigenvectors are printed as shown in Figure 10.3. The eigenvectors are normalised so that the largest component is equal to ± 1.

PROGRAM 10.1: EIGENVALUES AND EIGENVECTORS OF A RECTANGULAR SOLID IN PLANE STRAIN—FOUR-NODE QUADRILATERALS, LUMPED MASS

This program is an extension of Program 5.2 for the analysis of elastic solids in plane strain using four-node quadrilaterals. The element coordinates and steering information are produced by the geometry subroutine `GEOMETRY_4QY`, otherwise the structure of the program has much in common with Program 10.0 (see Figure 10.2). Only those variables that have not already been defined under the previous program are listed as follows:

Scalar integers:

`NYE`	number of elements in the y direction

Scalar reals:

`AA`	x dimension of elements
`BB`	y dimension of elements

```
program p101
!-------------------------------------------------------------------------
!      program 10.1 eigenvalues and eigenvectors of a rectangular
!      elastic solid in plane strain using uniform 4-node
!      quadrilateral elements  :     lumped mass
!-------------------------------------------------------------------------
 use new_library  ;   use geometry_lib  ;  use vlib   ;  implicit none
 integer::nels,nye,neq,nband,nn,nr,nodof=2,nod=4,ndof,                   &
          i,j,k,iel,ndim=2,ifail,icount,nmodes
 real::aa,bb,e,v,rho,tol=1.e-30
!----------------------------- dynamic arrays--------------------------------
 real    ,allocatable :: ku(:,:),loads(:),coord(:,:),km(:,:),g_coord(:,:),  &
                         diag(:),udiag(:),emm(:,:),kv(:),kh(:),rrmass(:)
 integer, allocatable :: nf(:,:), g(:)  , num(:)  , g_num(:,:) , g_g (:,:)
!-----------------------input and initialisation-----------------------------
  open (10,file='p101.dat',status=    'old',action='read')
  open ( 6,file='p101.res',status='replace',action='write')
  read (10,*) nels,nye,nn,aa,bb,rho,e,v,nmodes
  ndof=nod*nodof
  allocate ( nf(nodof,nn), g_coord(ndim,nn),coord(nod,ndim),emm(ndof,ndof),  &
           g_num(nod,nels),num(nod),km(ndof,ndof),g(ndof),g_g(ndof,nels))
  nf=1; read(10,*) nr ; if(nr>0) read(10,*) (k,nf(:,k),i=1,nr)
  call formnf(nf); neq=maxval(nf)
!------- loop the elements to find nband and set up global arrays -----------
      nband=0
 elements_1   : do iel =1,nels
                 call geometry_4qy(iel,nye,aa,bb,coord,num)
                 call num_to_g ( num , nf , g )
                 g_num(:,iel)=num;g_coord(:,num)=transpose(coord);g_g(:,iel)=g
                 if(nband<bandwidth(g))nband=bandwidth(g)
 end do elements_1
        write(6,'(a)') "Global coordinates"
        do k=1,nn;write(6,'(a,i5,a,2e12.4)')"Node",k,"    ",g_coord(:,k);end do
        write(6,'(a)') "Global node numbers"
        do k=1,nels;write(6,'(a,i5,a,4i5)')                                  &
                         "Element",k,"      ",g_num(:,k) ; end do
    write(6,'(2(a,i5))')                                                     &
            "There are ",neq,"  equations and the half-bandwidth is", nband
   allocate( ku(neq,nband+1),loads(0:neq),diag(0:neq),udiag(0:neq),          &
            kv(neq*(nband+1)),kh(neq*(nband+1)),rrmass(0:neq))
        emm = .0; diag = .0; ku = .0
!--------element mass matrix is lumped----------------------------------------
        do i=1,ndof; emm(i,i)=.25*aa*bb*rho; end do
!------- element stiffness and mass integration and assembly------------------
 elements_2: do iel=1,nels
                num = g_num(:,iel); coord =transpose(g_coord(:,num))
                g = g_g( : , iel );   call analy4(km,coord,e,v)
                call formku (ku,km,g)   ;  call formlump(diag,emm,g)
  end do elements_2
  write(6,'(a)') "The global mass diagonal is:"
  write(6,'(6e12.4)') diag(1:neq)
!----------------------reduce to standard eigenvalue problem-----------------
  rrmass(1:neq) = 1./sqrt(diag(1:neq))
  do i=1,neq
     if(i<=neq-nband)then;k=nband+1;else;k=neq-i+1;end if
     do j=1,k; ku(i,j)=ku(i,j)*rrmass(i)*rrmass(i+j-1); end do
  end do
  icount=0
  do j=1,nband+1;do i=1,neq; icount=icount+1;kh(icount)=ku(i,j);end do; end do
!----------------------extract the eigenvalues--------------------------------
  call bandred(ku,diag,udiag,loads);ifail=1; call bisect(diag,udiag,tol,ifail)
  write(6,'(a)') "The eigenvalues are:"    ;  write(6,'(6e12.4)') diag(1:neq)
  do i = 1 , nmodes
    kv = kh; kv(:neq)=kv(:neq)-diag(i);kv(1)=kv(1)+1.e20
             udiag=0.0; udiag(1)=kv(1)
             call banred(kv,neq);call bacsub(kv,udiag);udiag=rrmass*udiag
             write(6,'("Eigenvector number ",i3," is: ")')i
             write(6,'(6e12.4)')udiag(1:)/maxval(abs(udiag))
  end do
  end program p101
```

E	Young's modulus
V	Poisson's ratio
RHO	mass density

The example problem shown in Figure 10.4 is nominally the same as the beam analysed in Figure 10.1, namely an elastic solid cantilever 4.0 units long in the x direction with a flexural rigidity of 0.08333. The solid modelled by the five four-node elements in Figure 10.4 is truly 2-d, however, and has the additional property of Poisson's ratio, set in this example to 0.3.

The lumped mass matrix `EMM` is readily formed with eight diagonal terms, in which one-quarter of the total mass of each element (`AA*BB*RHO*`0.25) is lumped at each node in each direction. The stiffness and mass matrix assembly loop is then entered for each element as was done for stiffness in Chapter 5. The analytical formulation for the stiffness matrix of a four-node quadrilateral is implemented using library subroutine `ANALY4` and the global stiffness matrix is assembled as an upper triangle band rectangle by library subroutine `FORMKU`. As in Program 10.0, the assembled global lumped mass matrix is held in the

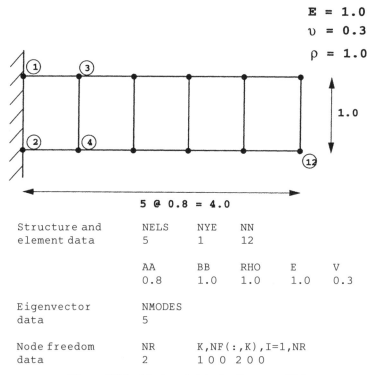

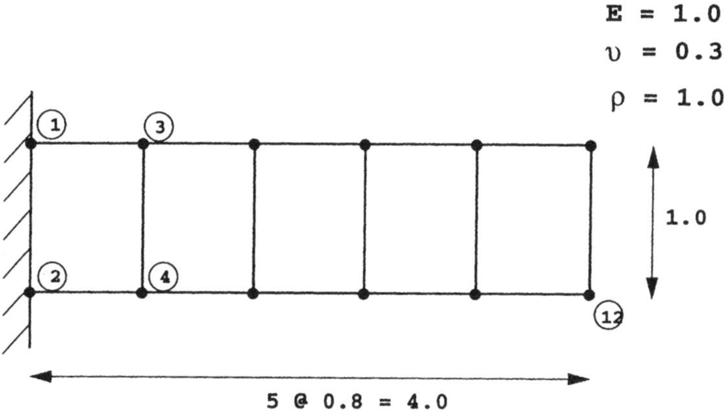

Structure and element data	NELS 5	NYE 1	NN 12		
	AA 0.8	BB 1.0	RHO 1.0	E 1.0	V 0.3
Eigenvector data	NMODES 5				
Node freedom data	NR 2	K,NF(:,K),I=1,NR 1 0 0 2 0 0			

Figure 10.4 Mesh and data for Program 10.1

vector DIAG and printed in Figure 10.5. The remainder of the program is identical to Program 10.0.

It can be seen that the fundamental frequency, printed as ω^2, is 0.00538, significantly higher than that calculated by Program 10.0 for a slender beam. Thus, the solid in plane strain, represented by the four-node elements, is a poor

```
Global coordinates
Node    1        0.0000E+00  0.0000E+00
Node    2        0.0000E+00 -0.1000E+01
Node    3        0.8000E00   0.0000E+00
Node    4        0.8000E00  -0.1000E+01
Node    5        0.1600E+01  0.0000E+00
Node    6        0.1600E+01 -0.1000E+01
Node    7        0.2400E+01  0.0000E+00
Node    8        0.2400E+01 -0.1000E+01
Node    9        0.3200E+01  0.0000E+00
Node   10        0.3200E+01 -0.1000E+01
Node   11        0.4000E+01  0.0000E+00
Node   12        0.4000E+01 -0.1000E+01
Global node numbers
Element    1           2    1    3    4
Element    2           4    3    5    6
Element    3           6    5    7    8
Element    4           8    7    9   10
Element    5          10    9   11   12
There are    20  equations and the half-bandwidth is      7
The global mass diagonal is:
  0.4000E00  0.4000E00  0.4000E00  0.4000E00  0.4000E00  0.4000E00
  0.4000E00  0.4000E00  0.4000E00  0.4000E00  0.4000E00  0.4000E00
  0.4000E00  0.4000E00  0.4000E00  0.4000E00  0.2000E00  0.2000E00
  0.2000E00  0.2000E00
The eigenvalues are:
  0.5382E-02  0.1117E00  0.1721E00  0.5182E00  0.1153E+01  0.1378E+01
  0.1882E+01  0.1934E+01  0.2579E+01  0.2617E+01  0.2665E+01  0.3028E+01
  0.3088E+01  0.3332E+01  0.3547E+01  0.4047E+01  0.5336E+01  0.5765E+01
  0.7284E+01  0.8693E+01
Eigenvector number   1 is:
  0.7034E-01 -0.7440E-01 -0.7031E-01 -0.7439E-01  0.1195E00 -0.2437E00
 -0.1195E00 -0.2437E00  0.1489E00 -0.4738E00 -0.1489E00 -0.4738E00
  0.1621E00 -0.7336E00 -0.1621E00 -0.7336E00  0.1653E00 -0.1000E+01
 -0.1652E00 -0.1000E+01
Eigenvector number   2 is:
  0.1973E00 -0.4765E00 -0.1973E00 -0.4765E00  0.4990E-01 -0.9043E00
 -0.4987E-01 -0.9043E00 -0.2736E00 -0.7874E00  0.2736E00 -0.7874E00
 -0.5473E00 -0.4639E-01  0.5473E00 -0.4639E-01 -0.6445E00  0.1000E+01
  0.6446E00  0.1000E+01
Eigenvector number   3 is:
  0.2874E00 -0.9586E-01  0.2874E00  0.9589E-01  0.5752E00 -0.6990E-01
  0.5752E00  0.6995E-01  0.8012E00 -0.5258E-01  0.8012E00  0.5263E-01
  0.9465E00 -0.2598E-01  0.9465E00  0.2600E-01  0.1000E+01 -0.1074E-01
  0.1000E+01  0.1070E-01
Eigenvector number   4 is:
  0.1000E00 -0.1000E+01 -0.1000E00 -0.1000E+01 -0.4292E00 -0.7942E00
  0.4292E00 -0.7943E00 -0.3600E00  0.4644E00  0.3600E00  0.4644E00
  0.4293E00  0.6358E00 -0.4293E00  0.6358E00  0.8955E00 -0.7455E00
 -0.8955E00 -0.7455E00
Eigenvector number   5 is:
  0.3961E00  0.1000E+01 -0.3961E00  0.1000E+01  0.6559E00 -0.4435E00
 -0.6559E00 -0.4435E00 -0.2008E00 -0.5586E00  0.2008E00 -0.5586E00
  0.1317E00  0.8174E00 -0.1317E00  0.8174E00  0.9032E00 -0.2581E00
 -0.9032E00 -0.2581E00
```

Figure 10.5 Results from Program 10.1

representation of a slender beam, at least in the flexural modes. The elements are too "stiff". The longitudinal modes as indicated by the third eigenvalue and eigenvector are in better agreement, however; the computed eigenvalue equals 0.1721 as compared with the analytical solution of $\omega^2 = \pi^2 E/(4L^2\rho) = 0.1542$. Better agreement still would be obtained if Poisson's ratio was set to zero, completely removing the lateral stiffening effect also absent from the slender beam solutions.

PROGRAM 10.2: EIGENVALUES AND EIGENVECTORS OF A RECTANGULAR SOLID IN PLANE STRAIN—EIGHT-NODE QUADRILATERALS, LUMPED MASS

A much better representation of flexural modes of "beams" made up of solid elements is achieved by the use of eight-node quadrilaterals and this program uses these elements with a "lumped" mass approximation. The process of mass lumping is not obvious in these elements. For example, the summation of rows of the consistent matrix leads to negative values at the corners. It can be shown, however (e.g. Smith, 1977), that a reasonable approximation is to lump the mass to the mid-point and corner nodes in the ratio 4 : 1, thus `AA*BB*RHO/20` is assigned to each corner node and `AA*BB*RHO/5` is assigned to each mid-side node.

Apart from the different lumping of the element mass matrix, the program is virtually identical to Program 10.1. The only differences occur in the geometry routine which has been changed to `GEOMETRY_8QY` to reflect the different element type, and in the element stiffness matrix calculation which is now computed using (reduced) numerical integration (`NIP=4`) using an identical sequence of subroutines as Program 5.3.

Variables not used in the previous programs in this chapter are as follows:

Scalar integers:

`NST`	number of stress components
`NIP`	number of Gauss points in each element

Scalar real:

`DET`	determinant of the Jacobian matrix

Allocatable real arrays:

`DER`	derivatives of shape functions w.r.t. local coordinates
`DERIV`	derivatives of shape functions w.r.t. global coordinates
`WEIGHTS`	quadrature weights
`POINTS`	local coordinates of integrating points
`DEE`	elastic plane strain stress–strain matrix

```
program p102
!-----------------------------------------------------------------------
!      program 10.2 eigenvalues and eigenvectors of a rectangular
!      elastic solid in plane strain using uniform 8-node
!      quadrilateral elements   :      lumped mass
!-----------------------------------------------------------------------
 use new_library     ;   use geometry_lib      ;  implicit none
 integer::nels,nye,neq,nband,nn,nr,nip,nodof=2,nst=3,nod=8,ndof,          &
          i,j,k,iel,ndim=2,ifail,icount,nmodes
 real::aa,bb,e,v,det,rho,tol=1.e-30;   character(len=15)::element='quadrilateral'
!----------------------------- dynamic arrays-----------------------------------
 real    ,allocatable :: ku(:,:),loads(:),coord(:,:),km(:,:),g_coord(:,:),   &
                         points(:,:),dee(:,:),jac(:,:),der(:,:),deriv(:,:),  &
                         diag(:),udiag(:),emm(:,:),kv(:),kh(:),rrmass(:) ,   &
                         weights(:),bee(:,:)
 integer, allocatable :: nf(:,:), g(:)  , num(:)  , g_num(:,:) , g_g (:,:)
!------------------------input and initialisation-------------------------------
  open (10,file='p102.dat',status=    'old',action='read')
  open ( 6,file='p102.res',status='replace',action='write')
  read (10,*) nels,nye,nn,nip,aa,bb,rho,e,v,nmodes
  ndof=nod*nodof
  allocate ( nf(nodof,nn), g_coord(ndim,nn),coord(nod,ndim),emm(ndof,ndof),  &
           g_num(nod,nels),der(ndim,nod),deriv(ndim,nod),bee(nst,ndof),      &
           num(nod),km(ndof,ndof),g(ndof),g_g(ndof,nels),points(nip,ndim),   &
           dee(nst,nst),jac(ndim,ndim),weights(nip))
  nf=1; read(10,*) nr ; if(nr>0) read(10,*) (k,nf(:,k),i=1,nr)
  call formnf(nf); neq=maxval(nf)
  call deemat(dee,e,v)  ;  call sample (element , points, weights)
!------- loop the elements to find nband and set up global arrays ------------
      nband=0
 elements_1   : do iel =1,nels
                 call geometry_8qy(iel,nye,aa,bb,coord,num)
                 call num_to_g ( num , nf , g )
                 g_num(:,iel)=num;g_coord(:,num)=transpose(coord);g_g(:,iel)=g
                 if(nband<bandwidth(g))nband=bandwidth(g)
 end do elements_1
        write(6,'(a)') "Global coordinates"
        do k=1,nn;write(6,'(a,i5,a,2e12.4)')"Node",k,"     ",g_coord(:,k);end do
        write(6,'(a)') "Global node numbers"
        do k=1,nels;write(6,'(a,i5,a,8i5)')                                   &
                       "Element",k,"      ",g_num(:,k) ; end do
    write(6,'(2(a,i5))')                                                      &
             "There are ",neq," equations and the half-bandwidth is", nband
   allocate( ku(neq,nband+1),loads(0:neq),diag(0:neq),udiag(0:neq),          &
             kv(neq*(nband+1)),kh(neq*(nband+1)),rrmass(0:neq))
       emm = .0; diag = .0; ku = .0
       call sample(element,points,weights); call deemat(dee,e,v)
!--------element mass matrix is lumped------------------------------------------
    emm = .0; do i=1,ndof; emm(i,i)=.2*aa*bb*rho; end do
             do i=1,13,4; emm(i,i)=.25*emm(3,3); end do
             do i=2,14,4; emm(i,i)=.25*emm(3,3); end do
!------- element stiffness and mass integration and assembly-----------------
 elements_2: do iel=1,nels
               num = g_num(:,iel); coord =transpose(g_coord(:,num))
               g = g_g( : , iel );    km=0.0
               integrating_pts_1:  do i=1,nip
                 call shape_der(der,points,i); jac=matmul(der,coord)
                 det= determinant(jac)  ; call invert(jac)
                 deriv = matmul(jac,der);call beemat(bee,deriv)
                 km= km+matmul(matmul(transpose(bee),dee),bee)*det*weights(i)
               end do integrating_pts_1
             call formku (ku,km,g)   ;  call formlump(diag,emm,g)
  end do elements_2
  write(6,'(a)') "The global mass diagonal is:"
```

```
  write(6,'(6e12.4)') diag(1:neq)
!----------------------reduce to standard eigenvalue problem-----------------
  rrmass(1:neq) = 1./sqrt(diag(1:neq))
  do i=1,neq
     if(i<=neq-nband)then;k=nband+1;else;k=neq-i+1;end if
     do j=1,k; ku(i,j)=ku(i,j)*rrmass(i)*rrmass(i+j-1); end do
  end do
  icount=0
  do j=1,nband+1;do i=1,neq; icount=icount+1;kh(icount)=ku(i,j);end do; end do
!----------------------extract the eigenvalues--------------------------------
  call bandred(ku,diag,udiag,loads);ifail=1; call bisect(diag,udiag,tol,ifail)
  write(6,'(a)') "The eigenvalues are:"    ;  write(6,'(6e12.4)') diag(1:neq)
 do i = 1 , nmodes
    kv = kh; kv(:neq)=kv(:neq)-diag(i);kv(1)=kv(1)+1.e20
    udiag=0.0; udiag(1)=kv(1)
    call banred(kv,neq);call bacsub(kv,udiag);udiag=rrmass*udiag
    write(6,'("Eigenvector number ",i3," is: ")')i
    write(6,'(6e12.4)')udiag(1:)/maxval(abs(udiag))
 end do
end program p102
```

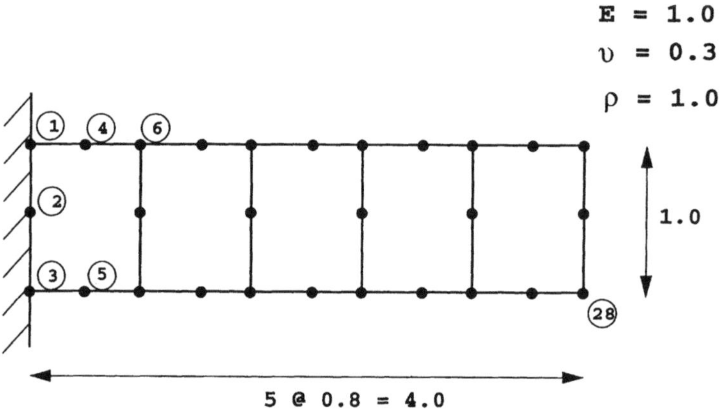

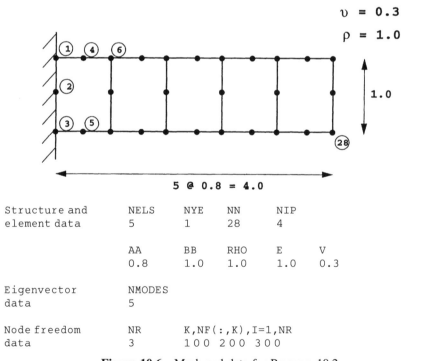

Structure and element data	NELS	NYE	NN	NIP	
	5	1	28	4	
	AA	BB	RHO	E	V
	0.8	1.0	1.0	1.0	0.3
Eigenvector data	NMODES				
	5				
Node freedom data	NR	K,NF(:,K),I=1,NR			
	3	1 0 0 2 0 0 3 0 0			

Figure 10.6 Mesh and data for Program 10.2

```
Global coordinates
Node    1       0.0000E+00  0.0000E+00
Node    2       0.0000E+00 -0.5000E00
Node    3       0.0000E+00 -0.1000E+01
Node    4       0.4000E00   0.0000E+00
.....................................................................

Node   26       0.4000E+01  0.0000E+00
Node   27       0.4000E+01 -0.5000E00
Node   28       0.4000E+01 -0.1000E+01
Global node numbers
Element    1           3    2    1    4    6    7    8    5
Element    2           8    7    6    9   11   12   13   10
Element    3          13   12   11   14   16   17   18   15
Element    4          18   17   16   19   21   22   23   20
Element    5          23   22   21   24   26   27   28   25
There are     50  equations and the half-bandwidth is    15
The global mass diagonal is:
  0.1600E00  0.1600E00  0.1600E00  0.1600E00  0.8000E-01  0.8000E-01
  0.3200E00  0.3200E00  0.8000E-01  0.8000E-01  0.1600E00  0.1600E00
.....................................................................

  0.1600E00  0.1600E00  0.4000E-01  0.4000E-01  0.1600E00  0.1600E00
  0.4000E-01  0.4000E-01
The eigenvalues are:
  0.4086E-02  0.9840E-01  0.1706E00  0.5002E00  0.1240E+01  0.1337E+01
  0.1606E+01  0.2232E+01  0.3035E+01  0.3071E+01  0.3617E+01  0.4170E+01
  0.4351E+01  0.5301E+01  0.5505E+01  0.6241E+01  0.6451E+01  0.6959E+01
  0.7002E+01  0.8034E+01  0.8105E+01  0.8305E+01  0.8730E+01  0.8908E+01
  0.9296E+01  0.9747E+01  0.9865E+01  0.1021E+02  0.1043E+02  0.1063E+02
  0.1114E+02  0.1193E+02  0.1223E+02  0.1313E+02  0.1452E+02  0.1611E+02
  0.2046E+02  0.2224E+02  0.2460E+02  0.2610E+02  0.2764E+02  0.3194E+02
  0.3224E+02  0.4417E+02  0.4521E+02  0.5879E+02  0.6235E+02  0.6759E+02
  0.8007E+02  0.9102E+02
Eigenvector number   1 is:
  0.3441E-01 -0.2442E-01 -0.3448E-01 -0.2443E-01  0.6816E-01 -0.7611E-01
 -0.2051E-04 -0.6465E-01 -0.6827E-01 -0.7611E-01  0.9810E-01 -0.1474E00
 -0.9818E-01 -0.1474E00  0.1211E00 -0.2397E00 -0.3943E-04 -0.2351E00
 -0.1212E00 -0.2398E00  0.1384E00 -0.3492E00 -0.1384E00 -0.3492E00
  0.1510E00 -0.4703E00 -0.3260E-04 -0.4673E00 -0.1511E00 -0.4703E00
  0.1595E00 -0.5985E00 -0.1596E00 -0.5985E00  0.1643E00 -0.7313E00
 -0.3513E-04 -0.7305E00 -0.1644E00 -0.7313E00  0.1664E00 -0.8656E00
 -0.1665E00 -0.8655E00  0.1671E00 -0.1000E+01 -0.3370E-04 -0.9999E00
 -0.1671E00 -0.1000E+01
Eigenvector number   2 is:
  0.1321E00 -0.1834E00 -0.1321E00 -0.1834E00  0.1872E00 -0.4372E00
  0.1601E-05 -0.4192E00 -0.1872E00 -0.4372E00  0.1606E00 -0.6530E00
 -0.1606E00 -0.6530E00  0.5414E-01 -0.7995E00  0.3084E-05 -0.8395E00
.....................................................................

  0.2766E-05  0.8700E-02  0.5182E00  0.3221E-01 -0.5800E00  0.5061E00
  0.5800E00  0.5061E00 -0.5995E00  0.1000E+01  0.3059E-05  0.9983E00
  0.5995E00  0.1000E+01
.....................................................................

Eigenvector number   5 is:
  0.4003E-01 -0.6070E00 -0.4031E-01 -0.6071E00 -0.3091E00 -0.9026E00
 -0.1491E-03 -0.1000E+01  0.3086E00 -0.9026E00 -0.6174E00 -0.3276E00
  0.6169E00 -0.3276E00 -0.5015E00  0.4257E00 -0.1856E-03  0.5116E00
.....................................................................

  0.7460E00 -0.2448E00 -0.9509E00  0.5518E00  0.7999E-04  0.5574E00
  0.9518E00  0.5523E00
```

Figure 10.7 Results from Program 10.2

JAC	Jacobian matrix
BEE	strain/displacement matrix

The example chosen is the same as before, namely a cantilever of length four units, modelled using five eight-node elements as shown in Figure 10.6. The output is shown in Figure 10.7 and consists of the diagonal global lumped mass matrix followed by the eigenvalues (as usual in the form ω^2) and the eigenvectors. Five eigenvectors were requested in the data, partially shown in Figure 10.7.

The square of the fundamental frequency is now $\omega^2 = 0.004086$, in much closer agreement with the value of 0.003823 produced by Program 10.0.

PROGRAM 10.3: EIGENVALUES AND EIGENVECTORS OF A RECTANGULAR SOLID IN PLANE STRAIN: FOUR-NODE QUADRILATERALS, CONSISTENT MASS

In Programs 10.0–10.2 Jacobi transformation was used to solve the eigenvalue problem. Although reliable and robust this method is time-consuming for large problems. For such problems, iterative methods are more attractive, for example the Lanczos method (e.g. Parlett and Reid, 1981). The process is described in section 3.13.2, and is used in the present program, whose structure chart is given as Figure 10.8. It should be noted that for the first time in this chapter, consistent mass has been used. The global mass matrix is therefore symmetric and banded and stored in array MB in the same way that the global stiffness matrix is stored in array KB.

Variables new to this program are listed as follows:

Scalar integers:

JFLAG	equals zero if eigenvectors computed properly
IFLAG	failure option
ITAPE	scratch tape number
LP	output device
LALFA	limit of Lanczos steps
LEIG	required number of eigenvalues in range
LX	set to at least 3*LEIG
LZ	holds the first dimension of array Z
ITERS	iteration counter
NEIG	set to the number of eigenvalues in EIG

Scalar reals:

EL	lower limit of eigenvalue spectrum
ER	upper limit of eigenvalue spectrum
ACC	tolerance for iterations

```
program p103                          ! kind=1 precision
!-------------------------------------------------------------------------
!      program 10.3 eigenvalues and eigenvectors of a
!      rectangular elastic solid in plane strain using
!      uniform 4-node quadrilateral elements : consistent mass
!-------------------------------------------------------------------------
 use libks    ; use new_library  ; use geometry_lib ; implicit none
 integer::nels,nye,neq,nband,nn,nr,nip,nodof=2,nod=4,nst=3,ndof,           &
         i,k,iel,ndim=2,nmodes,jflag,iflag=-1,itape=1,lp=6 ,               &
         lalfa=500,leig=20, lx=80, lz=500   ,iters   ,neig = 0
 real :: aa,bb,rho,e,v,det   , el,er,  acc = 1.e-6
 character (len=15) :: element = 'quadrilateral'
!--------------------------- dynamic arrays--------------------------------
 real,    allocatable :: kb(:,:),mb(:,:),points(:,:),dee(:,:),coord(:,:),   &
                         fun(:),jac(:,:),der(:,:),deriv(:,:),weights(:),    &
                         bee(:,:),km(:,:),emm(:,:),ecm(:,:),g_coord(:,:),   &
                         ua(:),va(:),eig(:),x(:),del(:), udiag(:),diag(:),  &
                         alfa(:),beta(:),w1(:),y(:,:),z(:,:)
 integer, allocatable :: nf(:,:), g(:)   , num(:)   , g_num(:,:) , g_g (:,:), &
                         nu(:),jeig(:,:)
!--------------------------input and initialisation-------------------------
  open (10,file='p103.dat',status=    'old',action='read')
  open (11,file='p103.res',status='replace',action='write')
  open ( 1,file='p103.tem',form='unformatted')
  read (10,*) nels,nye,nn,nip,aa,bb,rho,e,v,nmodes,el,er
  ndof=nod*nodof
  allocate ( nf(nodof,nn), points(nip,ndim),dee(nst,nst), g_coord(ndim,nn),  &
           coord(nod,ndim),fun(nod),jac(ndim,ndim), weights(nip),            &
           g_num(nod,nels),der(ndim,nod),deriv(ndim,nod),bee(nst,ndof),      &
           num(nod),km(ndof,ndof),g(ndof),g_g(ndof,nels),emm(ndof,ndof),     &
           ecm(ndof,ndof),eig(leig),x(lx),del(lx),nu(lx),jeig(2,leig),       &
           alfa(lalfa),beta(lalfa),z(lz,leig))
  nf=1; read(10,*) nr ; if(nr>0) read(10,*) (k,nf(:,k),i=1,nr)
  call formnf(nf); neq=maxval(nf)
!-------- loop the elements to find nband and set up global arrays -----------
      nband=0
 elements_1   : do iel =1,nels
                 call geometry_4qy(iel,nye,aa,bb,coord,num)
                 call num_to_g ( num , nf , g )
                 g_num(:,iel)=num;g_coord(:,num)=transpose(coord);g_g(:,iel)=g
                 if(nband<bandwidth(g))nband=bandwidth(g)
 end do elements_1
    write(11,'(a)') "Global coordinates "
    do k=1,nn;write(11,'(a,i5,a,2e12.4)')"Node",k,"       ",g_coord(:,k);end do
    write(11,'(a)') "Global node numbers "
    do k = 1 , nels; write(11,'(a,i5,a,4i5)')                                 &
                               "Element ",k,"        ",g_num(:,k); end do
   write(11,'(2(a,i5))')                                                      &
           "There are ",neq," equations and the half-bandwidth is", nband
   allocate  ( kb(neq,nband+1),mb(neq,nband+1),ua(0:neq),va(0:neq),           &
               diag(0:neq),udiag(0:neq),w1(0:neq), y(0:neq,leig))
    kb=0.0 ; mb=0.0  ; ua = .0 ; va = .0  ; eig = .0
    jeig = 0;  x=.0; del=.0; nu=0; alfa=.0; beta=.0
    diag = .0 ; udiag = .0 ; w1 = .0 ; y=.0; z=.0
    call sample(element,points,weights); call deemat(dee,e,v)
!----------------- element stiffness integration and assembly------------------
 elements_2: do iel=1,nels
                num = g_num(:,iel); coord =transpose( g_coord(:,num))
                g = g_g( : , iel ); km=0.0  ; emm=0.0
                integrating_pts_1:  do i=1,nip
                  call shape_fun(fun,points,i); call shape_der(der,points,i)
                  jac=matmul(der,coord);det= determinant(jac);call invert(jac)
                  deriv = matmul(jac,der);call beemat(bee,deriv)
                  km= km+matmul(matmul(transpose(bee),dee),bee)*det*weights(i)
```

```
                call ecmat(ecm,fun,ndof,nodof);emm=emm+ecm*det*weights(i)*rho
              end do integrating_pts_1
             call formkb (kb,km,g)  ; call formkb (mb,emm,g)
  end do elements_2
!------------------------------find eigenvalues---------------------------------
    call cholin(mb)
    do iters = 1 , lalfa
       call lancz1(neq,el,er,acc,leig,lx,lalfa,lp,itape,iflag,ua,va,&
                    eig,jeig,neig,x,del,nu,alfa,beta)
       if(iflag==0) exit
       if(iflag>1) then
          write(11,'(a,i5)')                                              &
                   " Lancz1 is signalling failure, with iflag = ",   iflag
          stop
       end if
!--- iflag = 1 therefore form u + a * v  ( candidate for ebe ) ---------------
        udiag = va ; call chobk2(mb,udiag); call banmul(kb,udiag,diag)
        call chobk1(mb,diag); ua = ua + diag
    end do
!--- iflag = 0 therefore write out the spectrum  ------------------------------
      write(11,'(2(a,e12.4))') "The range was",el,"  to",er
      write(11,'(a,i5,a)') "It took ",iters,"  iterations"
      write(11,'(a)')"The eigenvalues are  :"
      write(11,'(6e12.4)') eig(1:neig)
!------------------- calculate the eigenvectors ------------------------------
  if(neig>10)neig = 10
  call lancz2(neq,lalfa,lp,itape,eig,jeig,neig,alfa,beta,lz,jflag,y,w1,z)
!-------------------if jflag is zero  calculate the eigenvectors -------------
  if (jflag==0) then
  write(11,'(a)') "The eigenvectors are"
    do i = 1 , nmodes
       udiag(:) = y(:,i)  ; call chobk2(mb,udiag)
       write(11,'("Eigenvector  number  ",i4," is: ")') i
       write(11,'(11e12.4)') udiag(1:)
    end do
  else
! lancz2 fails
    write(11,'(a,i5)')" Lancz2 is signalling failure with jflag = ",  jflag
  end if
end program p103
```

Allocatable real arrays:

KB	global stiffness matrix stored as lower rectangle
MB	global mass matrix stored as lower rectangle
FUN	shape functions
ECM	used to compute element consistent mass matrix
EMM	element consistent mass matrix
UA	Lanczos iteration vector
VA	Lanczos iteration vector
UDIAG	Lanczos iteration vector
DIAG	Lanczos iteration vector
EIG	holds eigenvalues in increasing order
X	working array
DEL	working array
ALFA	working array

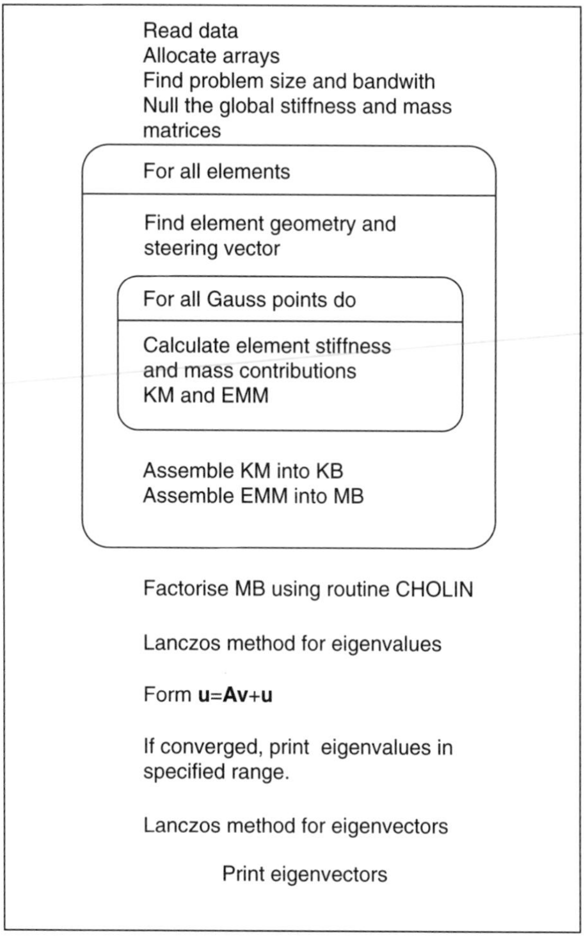

Figure 10.8 Structure chart for Program 10.3

`BETA`	working array
`W1`	working vector
`Y`	set to the required eigenvectors
`Z`	working vector

Allocatable integer arrays:

`NU`	working array
`JEIG`	Lanczos iteration array

The principal new vectors are **u** and **v** (called `UA` and `VA` in the program) as described in the structure chart of Figure 10.8. As seen in the above list, many of the arrays are used for working space. Library subroutines `LANCZ1` and `LANCZ2` are used to retrieve the eigenvalues and eigenvectors respectively (see also Smith, 1984).

The same problem considered previously is used to demonstate this program as shown in Figure 10.9. The only difference is the additional data read as `EL` and `ER` to set the limits on the eigenvalue search.

The output from the program is listed as Figure 10.10. It consistes of the number of Lanczos iterations (20 in this case) and the range of the eigenvalue spectrum (0.0–20.0). Then follows the eigenvalues and the first five eigenvectors.

Since the consistent mass assumption was made, the eigenvalues differ somewhat from those computed by Program 10.1. For example the first eigenvalue is computed as $\omega^2 = 0.00578$ as compared with 0.00538 previously computed for a similar problem with lumped mass.

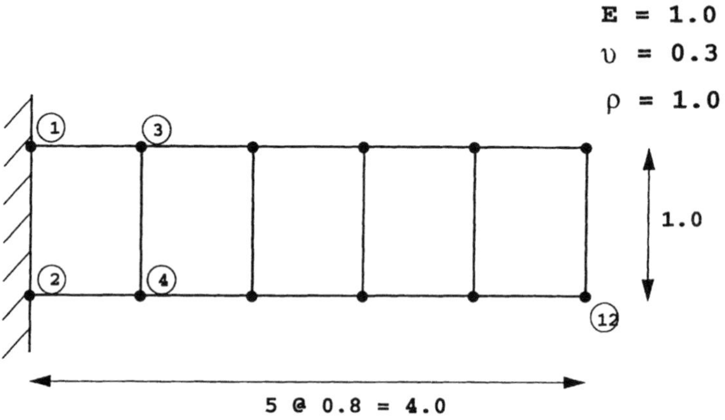

Structure and element data	NELS 5	NYE 1	NN 12	NIP 4	
	AA 0.8	BB 1.0	RHO 1.0	E 1.0	V 0.3
Eigenvector data	NMODES 5				
	EL 0.0	ER 20.0			
Node freedom data	NR 2	K,NF(:,K),I=1,NR 1 0 0 2 0 0			

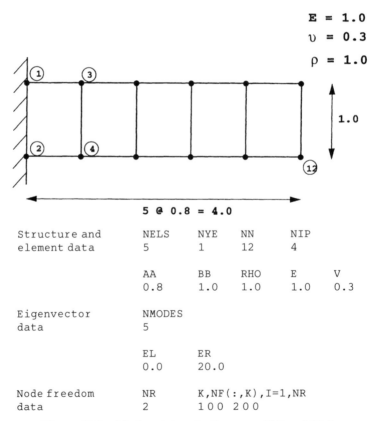

Figure 10.9 Mesh and data for Programs 10.3 and 10.4

```
Global coordinates
Node   1          0.0000E+00  0.0000E+00
Node   2          0.0000E+00 -0.1000E+01
Node   3          0.8000E+00  0.0000E+00
Node   4          0.8000E+00 -0.1000E+01
Node   5          0.1600E+01  0.0000E+00
Node   6          0.1600E+01 -0.1000E+01
Node   7          0.2400E+01  0.0000E+00
Node   8          0.2400E+01 -0.1000E+01
Node   9          0.3200E+01  0.0000E+00
Node  10          0.3200E+01 -0.1000E+01
Node  11          0.4000E+01  0.0000E+00
Node  12          0.4000E+01 -0.1000E+01
Global node numbers
Element     1             2    1    3    4
Element     2             4    3    5    6
Element     3             6    5    7    8
Element     4             8    7    9   10
Element     5            10    9   11   12
There are    20  equations and the half-bandwidth is     7
The range was  0.0000E+00  to  0.5000E+01
It took    22  iterations
The eigenvalues are  :
  0.5757E-02  0.1544E+00  0.1758E+00  0.9066E+00  0.1690E+01  0.2665E+01
The eigenvectors are
Eigenvector  number     1 is:
 -0.6889E-01  0.7395E-01  0.6889E-01  0.7395E-01 -0.1163E+00  0.2401E+00  0.1163E+00
0.2401E+00 -0.1440E+00  0.4639E+00  0.1440E+00
  0.4639E+00 -0.1559E+00  0.7149E+00  0.1559E+00  0.7149E+00 -0.1585E+00  0.9706E+00
0.1585E+00  0.9706E+00
Eigenvector  number     2 is:
 -0.1456E+00  0.3984E+00  0.1456E+00  0.3984E+00 -0.1520E-01  0.7104E+00  0.1520E-01
0.7104E+00  0.2332E+00  0.5567E+00 -0.2332E+00
  0.5567E+00  0.4207E+00 -0.6842E-01 -0.4207E+00 -0.6842E-01  0.4797E+00 -0.8658E+00 -
0.4797E+00 -0.8658E+00
Eigenvector  number     3 is:
  0.2071E+00 -0.6698E-01  0.2071E+00  0.6692E-01  0.4138E+00 -0.4936E-01  0.4138E+00
0.4924E-01  0.5762E+00 -0.3700E-01  0.5763E+00
  0.3690E-01  0.6807E+00 -0.1834E-01  0.6807E+00  0.1830E-01  0.7191E+00 -0.7630E-02
0.7192E+00  0.7678E-02
Eigenvector  number     4 is:
  0.1911E-01 -0.7993E+00 -0.1911E-01 -0.7993E+00 -0.3275E+00 -0.4257E+00  0.3275E+00 -
0.4257E+00 -0.1579E+00  0.5743E+00  0.1579E+00
  0.5743E+00  0.4220E+00  0.4028E+00 -0.4220E+00  0.4028E+00  0.7114E+00 -0.8100E+00 -
0.7114E+00 -0.8100E+00
Eigenvector  number     5 is:
  0.5787E+00 -0.1552E+00  0.5787E+00  0.1552E+00  0.7299E+00  0.5737E-01  0.7299E+00 -
0.5737E-01  0.2581E+00  0.1947E+00  0.2581E+00
 -0.1947E+00 -0.4276E+00  0.1597E+00 -0.4276E+00 -0.1597E+00 -0.7774E+00  0.8630E-01 -
0.7774E+00 -0.8630E-01
```

Figure 10.10 Results from Program 10.3

PROGRAM 10.4: EIGENVALUES AND EIGENVECTORS OF A RECTANGULAR SOLID IN PLANE STRAIN: FOUR-NODE QUADRILATERALS, LUMPED MASS ELEMENT-BY-ELEMENT FORMULATION

The Lanczos algorithm relies on iterations which involve matrix–vector multiplication followed by vector addition. In the previous program, the matrix–

```
program p104
!-------------------------------------------------------------------------
!      program 10.4 eigenvalues and eigenvectors of a
!      rectangular elastic solid in plane strain using
!      uniform 4-node quadrilateral elements : kind = 1  precision
!      for lumped mass this is done element by element
!-------------------------------------------------------------------------
 use libks  ;  use new_library ; use geometry_lib ;  implicit none
 integer::nels,nye,neq,nn,nr,nip,nodof=2,nod=4,nst=3,ndof,               &
          i,k,iel,ndim=2,nmodes,jflag,iflag=-1,itape=1,lp=6 ,            &
          lalfa=500,leig=20, lx=80, lz=500  ,iters  ,neig = 0
 real::aa,bb,rho,e,v,det  , el,er,  acc = 1.e-6
 character (len=15) :: element = 'quadrilateral'
!-------------------------- dynamic arrays-------------------------------
 real,allocatable :: points(:,:),dee(:,:),coord(:,:),vdiag(:),     &
                     fun(:),jac(:,:),der(:,:),deriv(:,:),weights(:),  &
                     bee(:,:),km(:,:),emm(:,:),ecm(:,:),g_coord(:,:), &
                     ua(:),va(:),eig(:),x(:),del(:), udiag(:),diag(:),&
                     alfa(:),beta(:),w1(:),y(:,:),z(:,:),pmul(:),utemp(:)
 integer, allocatable :: nf(:,:), g(:)  , num(:)  , g_num(:,:) , g_g (:,:),&
                     nu(:),jeig(:,:)
!----------------------input and initialisation--------------------------
  open (10,file='p104.dat',status=    'old',action='read')
  open (11,file='p104.res',status='replace',action='write')
  open ( 1,file='p104.tem',form='unformatted')
  read (10,*) nels,nye,nn,nip,aa,bb,rho,e,v,nmodes,el,er
  ndof=nod*nodof
  allocate ( nf(nodof,nn), points(nip,ndim),dee(nst,nst), g_coord(ndim,nn),  &
           coord(nod,ndim),fun(nod),jac(ndim,ndim), weights(nip),          &
           g_num(nod,nels),der(ndim,nod),deriv(ndim,nod),bee(nst,ndof),    &
           num(nod),km(ndof,ndof),g(ndof),g_g(ndof,nels),emm(ndof,ndof),   &
           ecm(ndof,ndof),eig(leig),x(lx),del(lx),nu(lx),jeig(2,leig),     &
           alfa(lalfa),beta(lalfa),z(lz,leig),pmul(ndof),utemp(ndof))
  nf=1; read(10,*) nr ; if(nr>0) read(10,*) (k,nf(:,k),i=1,nr)
        call formnf(nf); neq=maxval(nf)
!------------------ loop the elements to set up global arrays  ----------------
 elements_1   : do iel =1,nels
                 call geometry_4qy(iel,nye,aa,bb,coord,num)
                 call num_to_g ( num , nf , g )
                 g_num(:,iel)=num;g_coord(:,num)=transpose(coord);g_g(:,iel)=g
                end do elements_1
    write(11,'(a)') "Global coordinates "
    do k=1,nn;write(11,'(a,i5,a,2e12.4)')"Node",k,"       ",g_coord(:,k);end do
    write(11,'(a)') "Global node numbers "
    do k = 1 , nels; write(11,'(a,i5,a,4i5)')                                 &
                              "Element ",k,"        ",g_num(:,k); end do
       write(11,'(a,i5,a)') "There are ",neq,"  equations to be solved"
   allocate  ( ua(0:neq),va(0:neq),vdiag(0:neq),                             &
               diag(0:neq),udiag(0:neq),w1(0:neq), y(0:neq,leig))
    ua = .0 ; va = .0  ; eig = .0
    jeig = 0;  x=.0; del=.0; nu=0; alfa=.0; beta=.0
    diag = .0 ; udiag = .0 ; w1 = .0 ; y=.0; z=.0
    call sample( element, points, weights); call deemat(dee,e,v)
!--------------- element stiffness integration and assembly--------------------
 elements_2: do iel=1,nels
                num = g_num(:,iel); coord =transpose( g_coord(:, num ))
                g = g_g( : , iel );     km=0.0    ; emm=0.0
                integrating_pts_1:  do i=1,nip
                  call shape_fun(fun,points,i)
                  call shape_der(der,points,i); jac=matmul(der,coord)
                  det= determinant(jac) ; call invert(jac)
                  deriv = matmul(jac,der);call beemat(bee,deriv)
                  km= km+matmul(matmul(transpose(bee),dee),bee)*det*weights(i)
                  call ecmat(ecm,fun,ndof,nodof);emm=emm+ecm*det*weights(i)*rho
```

```
               end do integrating_pts_1
               do i=1,ndof; diag(g(i))=diag(g(i))+sum(emm(i,:));end do
  end do elements_2
!------------------------------find eigenvalues--------------------------------
  diag = 1. / sqrt(diag) ; diag(0) = .0 ! diag holds l**(-1/2)
    do iters = 1 , lalfa
       call lancz1(neq,el,er,acc,leig,lx,lalfa,lp,itape,iflag,ua,va,          &
                    eig,jeig,neig,x,del,nu,alfa,beta)
       if(iflag==0) exit
       if(iflag>1) then
          write(11,'(a,i5)')                                                  &
                  " Lancz1 is signalling failure, with iflag = ",   iflag
          stop
       end if
!----- iflag = 1 therefore form u + a * v  ( done element by element )---------
      vdiag = va ;    vdiag = vdiag * diag ! vdiag is l**(-1/2).va
      udiag = .0  ; vdiag(0)=.0
      elements_3 : do iel = 1 , nels
                      g = g_g( : , iel )
                      pmul = vdiag (g); utemp = matmul(km,pmul)
                      udiag(g) = udiag(g) + utemp     ! udiag is A.l**(-1/2).va
      end do elements_3
      udiag = udiag *  diag  ; ua = ua + udiag
    end do
!-------------- iflag = 0 therefore write out the spectrum --------------------
      write(11,'(2(a,e12.4))') "The range is",el,"  to ",er
      write(11,'(a,i5,a)') "It took ",iters,"  iterations"
      write(11,'(a)') "The eigenvalues are    :"
      write(11,'(6e12.4)') eig(1:neig)
!  calculate the eigenvectors
   if(neig>10)neig = 10
   call lancz2(neq,lalfa,lp,itape,eig,jeig,neig,alfa,beta,lz,jflag,y,w1,z)
!------------------if jflag is zero  calculate the eigenvectors ---------------
   if (jflag==0) then
     write(11,'(a)') "The eigenvectors are  :"
     do i = 1 , nmodes
        udiag(:) = y(:,i)  ; udiag = udiag * diag
        write(11,'("Eigenvector number  ",i4," is: ")')  i
        write(11,'(6e12.4)') udiag(1:)
     end do
   else
! lancz2 fails
     write(11,'(a,i5)')" Lancz2 is signalling failure with jflag = ",  jflag
   end if
end program p104
```

vector product was performed using library subroutine BANMUL which took account of the storage strategy used to hold the global stiffness matrix KB.

Program 10.4 performs the same operations, but without the need to assemble and store a global stiffness matrix. The matrix–vector product described above can be achieved using element-by-element products as has been demonstrated in earlier programs in this book (e.g. Program 5.10).

Variables new to this program are listed as follows:

Dynamic real arrays:

VDIAG	temporary store of VA
PMUL	relevant terms gathered from VDIAG
UTEMP	element product of KM and PMUL

The example and data are exactly the same as in Program 10.3 (Figure 10.9). The results are listed as Figure 10.11 which essentially reproduce the results from Program 10.1 which also assumed lumped mass. The Lanczos process took 23 iterations to converge in this case. Note the economy in storage requirements compared with Program 10.1.

```
Global coordinates
Node    1          0.0000E+00  0.0000E+00
Node    2          0.0000E+00 -0.1000E+01
Node    3          0.8000E+00  0.0000E+00
Node    4          0.8000E+00 -0.1000E+01
Node    5          0.1600E+01  0.0000E+00
Node    6          0.1600E+01 -0.1000E+01
Node    7          0.2400E+01  0.0000E+00
Node    8          0.2400E+01 -0.1000E+01
Node    9          0.3200E+01  0.0000E+00
Node   10          0.3200E+01 -0.1000E+01
Node   11          0.4000E+01  0.0000E+00
Node   12          0.4000E+01 -0.1000E+01
Global node numbers
Element     1            2     1     3     4
Element     2            4     3     5     6
Element     3            6     5     7     8
Element     4            8     7     9    10
Element     5           10     9    11    12
There are    20  equations to be solved
The range is  0.0000E+00  to   0.2000E+02
It took    23  iterations
The eigenvalues are   :
  0.5383E-02  0.1117E+00  0.1721E+00  0.5182E+00  0.1153E+01  0.1378E+01
  0.1882E+01  0.1934E+01  0.2579E+01  0.2617E+01  0.2665E+01  0.3028E+01
  0.3088E+01  0.3332E+01  0.3547E+01  0.4047E+01  0.5336E+01  0.5765E+01
  0.7284E+01  0.8693E+01
The eigenvectors are  :
Eigenvector number     1 is:
 -0.6622E-01  0.7007E-01  0.6622E-01  0.7007E-01 -0.1126E+00  0.2296E+00
  0.1126E+00  0.2296E+00 -0.1403E+00  0.4462E+00  0.1403E+00  0.4462E+00
 -0.1527E+00  0.6910E+00  0.1527E+00  0.6910E+00 -0.1556E+00  0.9419E+00
  0.1556E+00  0.9419E+00
Eigenvector number     2 is:
 -0.1320E+00  0.3189E+00  0.1320E+00  0.3189E+00 -0.3339E-01  0.6052E+00
  0.3339E-01  0.6052E+00  0.1831E+00  0.5270E+00 -0.1831E+00  0.5270E+00
  0.3663E+00  0.3105E-01 -0.3663E+00  0.3105E-01  0.4314E+00 -0.6693E+00
 -0.4314E+00 -0.6693E+00
Eigenvector number     3 is:
  0.2045E+00 -0.6822E-01  0.2045E+00  0.6822E-01  0.4093E+00 -0.4976E-01
  0.4093E+00  0.4976E-01  0.5701E+00 -0.3743E-01  0.5701E+00  0.3743E-01
  0.6735E+00 -0.1849E-01  0.6735E+00  0.1849E-01  0.7116E+00 -0.7628E-02
  0.7116E+00  0.7627E-02
Eigenvector number     4 is:
  0.6030E-01 -0.6030E+00 -0.6030E-01 -0.6030E+00 -0.2588E+00 -0.4789E+00
  0.2588E+00 -0.4789E+00 -0.2171E+00  0.2801E+00  0.2171E+00  0.2801E+00
  0.2589E+00  0.3834E+00 -0.2589E+00  0.3834E+00  0.5400E+00 -0.4496E+00
 -0.5400E+00 -0.4496E+00
Eigenvector number     5 is:
  0.2451E+00  0.6190E+00 -0.2451E+00  0.6190E+00  0.4060E+00 -0.2745E+00
 -0.4060E+00 -0.2745E+00 -0.1243E+00 -0.3457E+00  0.1243E+00 -0.3457E+00
  0.8153E-01  0.5060E+00 -0.8153E-01  0.5060E+00  0.5590E+00 -0.1598E+00
 -0.5590E+00 -0.1598E+00
```

Figure 10.11 Results from Program 10.4

10.1 EXAMPLES

1. Use Program 10.0 to evaluate the lowest two natural frequencies of the beam shown in Figure 10.12 with the boundary conditions:
   (a) pinned at both ends;
   (b) fixed at both ends;
   (c) fixed at one end and pinned at the other;
   (d) fixed at one end and free at the other.
2. Repeat question 1 using Program 10.2.
3. Use Program 10.1 to estimate the first two axial natural frequencies and mode shapes of the rod shown in Figure 10.13.
4. Repeat question 3 using rod elements. Program 10.0 can be easily modified to analyse axial vibrations of rods. Make the following changes:

   On line 6 of the program change

   `nodof=2` to `nodof=1` and `ndof=4` to `ndof=2`

   On lines 3–5 in the `element_2` do-loop change:

```
emm(1,1)=0.5*rhoa(etype(iel))*el_ell;
   emm(3,3)=emm(1,1)
emm(2,2)=emm(1,1)*ell(iel)**2/12.;
   emm(4,4)=emm(2,2)
call beam_km (km,ei(etype(iel)),el_ell));
   g=g_g(:,iel)
```

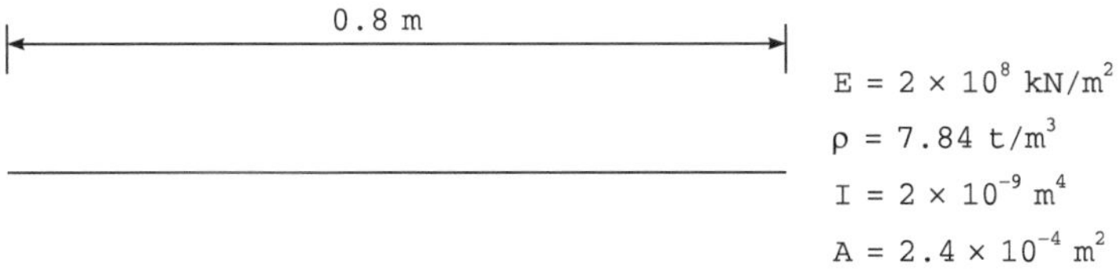

Figure 10.12

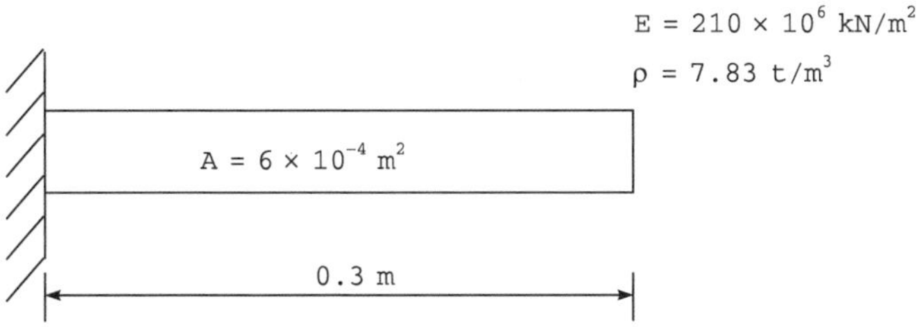

Figure 10.13

to

```
emm(1,1)=0.5*rhoa(etype(iel))*el_ell;
emm(2,2)=emm(1,1)
call rod_km (km,ei(etype(iel)),el_ell));
  g=g_g(:,iel)
```

The array `ei` will hold the values of the axial stiffness EA of each element.

5. Using the modified program from the previous question determine the first two natural frequencies and eigenvectors for the stepped bar shown in Figure 10.14.
6. A vital attribute of an element "stiffness matrix" is that it should possess the right number of "rigid body modes", i.e. zero eigenvalues of the stiffness matrix. Test the two-node elastic rod element stiffness matrix and prove that it has only one zero eigenvalue, corresponding to one rigid body mode (translation). If the rod has length L, cross-sectional area A, modulus E and mass per unit length ρ, assume "lumped mass" and calculate its non-zero eigenvalue and hence natural frequency of free vibration. Compare with the "exact" value of

$$\frac{\pi}{L}\sqrt{\frac{E}{\rho}}$$

and also with the finite element solution adopting "consistent mass".

7. Show how dynamic equilibrium of a multi-degree of freedom system vibrating at a resonant frequency leads to an equation of the form

$$\mathbf{Ka} = \omega^2 \mathbf{Ma}$$

where **K**, **M** = system stiffness and lumped mass matrices
a = displacement amplitudes
ω = angular frequency

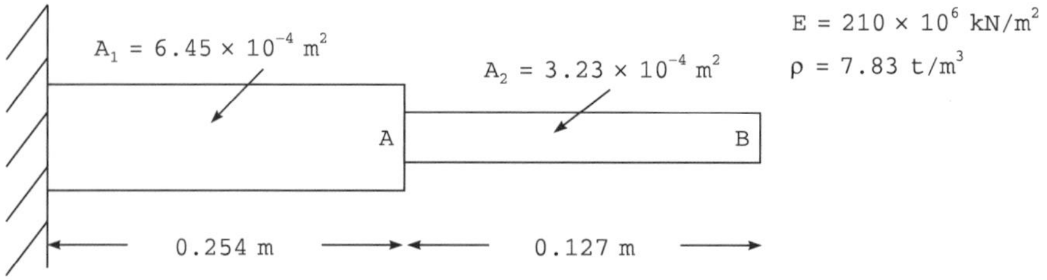

Figure 10.14

Describe a method for reducing this equation to a standard eigenvalue equation of the form

$$\mathbf{Ax} = \lambda\mathbf{x}$$

where $\mathbf{A}$ is symmetrical.

10.2 REFERENCES

Cook, R. D., Malkus, D. S. and Plesha, M. E. (1989) *Concepts and Applications of Finite Element Analysis*, 3rd edn. John Wiley & Sons, Chichester, New York.

Parlett, B. N. and Reid, J. K. (1981) Tracking the progress of the Lanczos algorithm for large eigenproblems. *IMA Journal of Numerical Analysis*, **1**, 135–155.

Smith, I. M. (1977) Transient phenomena of offshore foundations. In *Numerical Methods in Offshore Engineering*, Chapter 14. John Wiley & Sons, Chichester, New York.

Smith, I. M. (1984) Adaptability of truly modular software. *Eng. Comput. (Swansea, Wales)*, **1**(1), 25–35.

11 Forced Vibrations

11.0 INTRODUCTION

In the previous chapter, programs were described which enable the calculation of the intrinsic dynamic properties of systems—their undamped natural frequencies and mode shapes. The next stage in a dynamic analysis is usually the calculation of the response of the system to an imposed time dependent disturbance. This chapter describes six programs which enable such calculations to be made.

The type of equations to be solved was derived early in the book (see, for example, (2.13)). After semi-discretisation in space using finite elements, the resulting matrix equations are typified by (2.17), a set of second-order ordinary differential equations in the time variable. On inclusion of damping, the relevant equations become (3.123) and section 3.17 describes the principles behind the various solution procedures used below. For example, the first program described involves a forced vibration analysis of a beam structure similar to the one whose natural frequencies were found by Program 10.0.

PROGRAM 11.0: FORCED VIBRATION OF A BEAM: CONSISTENT MASS

This program computes the response of a string of beam elements to a combination of time-dependent applied nodal loads. The program allows for (Rayleigh) damping and uses a consistent mass matrix (see equation (2.29)). Time stepping is achieved using a direct Newmark method involving time-stepping parameters β and γ, the values of which determine the accuracy and stability characteristics of the algorithm (see e.g. Bathe, 1996). For the special case of $\beta = 0.25$ and $\gamma = 0.5$, the method is identical to the Crank–Nicolson $\theta = 0.5$ method described in section 3.17.1.

The algorithm used by Program 11.0 is as follows. Given the equation of motion in matrix form and initial conditions on displacements and velocities:

$$\mathbf{M}\ddot{\mathbf{u}} + \mathbf{C}\dot{\mathbf{u}} + \mathbf{K}\mathbf{u} = \mathbf{f} \tag{11.1}$$

with $\mathbf{u}(0) = \mathbf{u}_0$ and $\dot{\mathbf{u}}(0) = \dot{\mathbf{u}}_0$.

Let $\mathbf{f}_i$, $\mathbf{u}_i$, $\dot{\mathbf{u}}_i$ and $\ddot{\mathbf{u}}_i$ represent conditions at time $t = i\Delta t$ where $i = 0, 1, 2, \ldots$. Assuming that the following quantities are known:

$\mathbf{M}$, $\mathbf{C}$, $\mathbf{K}$, $\mathbf{u}_0$, $\dot{\mathbf{u}}_0$, β, γ Δt and $\mathbf{f}_i$ for all i,

1. Compute: $\mathbf{K}' = \mathbf{K} + \frac{\gamma}{\beta\Delta t}\mathbf{C} + \frac{1}{\beta\Delta t^2}\mathbf{M}$
2. Factorise $\mathbf{K}'$ to facilitate step 8.

```
program p110
!-------------------------------------------------------------
! program 11.0 forced vibration analysis of beams
!-------------------------------------------------------------
 use new_library      ;  use  geometry_lib    ;  implicit none
 integer::nels,neq,nn,nband,nr,nod=2,nodof=2,ndof=4,iel,i,j,k,ndim=1,      &
  np_types,nln,lnode,lsense,np,ntp,non
 real::beta,gamma,fm,fk,dt,f1,f2,el_ei,el_ell
!--------------------------dynamic arrays---------------------------------------

 real,allocatable::km(:,:),mm(:,:),kb(:,:),mb(:,:),ek(:,:),em(:,:),cb(:,:),&
  kp(:,:),a(:),d(:),v(:),a1(:),b1(:),vc(:),kd(:),coord(:,:),g_coord(:,:),  &
  ei(:),rhoa(:),ell(:),rt(:),rl(:),al(:,:),dis(:,:),vel(:,:),acc(:,:)
 integer,allocatable::nf(:,:),g(:),num(:),g_num(:,:),g_g(:,:),etype(:),    &
  lp(:),lf(:)

!------------------------input and initialisation------------------------------

 open(10,file='p110.dat'); open(11,file='p110.res')
 read(10,*)nels,np_types; nn=nels+1
 allocate(nf(nodof,nn),km(ndof,ndof),mm(ndof,ndof),coord(nod,ndim),          &
  g_coord(ndim,nn),g_num(nod,nels),num(nod),g(ndof),&
  ei(np_types),rhoa(np_types),ell(nels),g_g(ndof,nels),etype(nels))
 read(10,*)(ei(i),rhoa(i),i=1,np_types)
 etype=1; if(np_types>1)read(10,*)etype
 read(10,*)ell,nr
 nf=1; if(nr>0)read(10,*)(k,nf(:,k),i=1,nr); call formnf(nf); neq=maxval(nf)

!--------------loop the elements to find global array sizes--------------------

 nband=0
 elements_1: do iel=1,nels
              el_ell = ell(iel)
              call geometry_2l(iel,el_ell,coord,num);  call num_to_g(num,nf,g)
              g_num(:,iel)=num; g_coord(:,num)=transpose(coord)
              g_g(:,iel)=g; if(nband<bandwidth(g))nband=bandwidth(g)
 end do elements_1
 allocate(kb(neq,nband+1),mb(neq,nband+1),ek(neq,nband+1),em(neq,nband+1), &
  cb(neq,nband+1),kp(neq,nband+1),a1(0:neq),b1(0:neq),                       &
  vc(0:neq),kd(0:neq),a(0:neq),d(0:neq),v(0:neq))
 kb=0.0; mb=0.0

!--------global stiffness and mass matrix assembly------------------------

 elements_2: do iel=1, nels
              el_ei = ei(etype(iel)) ; el_ell = ell(iel)
              call beam_km(km,el_ei,el_ell); g=g_g(:,iel)
              call beam_mm(mm,rhoa(etype(iel)),ell(iel))
              call formkb(kb,km,g); call formkb(mb,mm,g)
 end do elements_2
 ek=kb; em=mb

!--------read newmark time-stepping parameters-----------------------------
!--------rayleigh damping parameters and initial conditions----------------

 read(10,*)beta,gamma,dt,fm,fk; d=0.0; v=0.0

! read initial conditions if /= 0.0 d(1:),v(1:)
!--------read number of loaded freedoms
!--------for each of these read number of points in load/time history
!--------and coordinates of load/time history

 read(10,*)nln; allocate(lf(nln))
 do k=1,nln
```

```
      read(10,*)lnode,lsense,np;  allocate(rt(np),rl(np))
      lf(k)=nf(lsense,lnode)
      do j=1,np; read(10,*)rt(j),rl(j); end do
      if(k==1)then
        ntp=nint((rt(np)-rt(1))/dt)+1; allocate(al(ntp,nln))
      end if
      call interp(k,dt,rt,rl,al,ntp)
    end do

    f1=beta*dt**2; f2=beta*dt
    cb=fm*mb+fk*kb; kp=mb/f1+gamma*cb/f2+kb
    call cholin(em); call cholin(kp); a=0.0
    a(lf(:))=al(1,:)
    call banmul(cb,v,vc); call banmul(kb,d,kd)
    a=a-vc-kd; call chobac(em,a)

    read(10,*)non
    allocate(lp(non),dis(ntp,non),vel(ntp,non),acc(ntp,non))

    do k=1,non; read(10,*)lnode,lsense; lp(k)=nf(lsense,lnode); end do
    dis(1,:)=d(lp); vel(1,:)=v(lp); acc(1,:)=a(lp)

   !---------Time stepping------------------------------------------------

    do j=2,ntp
      a1=d/f1+v/f2+a*(0.5/beta-1.)
      b1=gamma*d/f2-v*(1.-gamma/beta)-dt*a*(1.-0.5*gamma/beta)
      call banmul(mb,a1,vc); call banmul(cb,b1,kd)
      d=vc+kd; d(lf(:))=d(lf(:))+al(j,: )
      call chobac(kp,d); v=gamma*d/f2-b1; a=d/f1-a1
      dis(j,:)=d(lp); vel(j,:)=v(lp); acc(j,:)=a(lp)
    end do
    write(11,'(//,a)')"Computed time histories   "
    do i=1,non
   !   write(11,'(/,a,i4)')"          Freedom",lp(i)
   !   write(11,'(a)')"        Time     Disp          Velo          Accel"
      do j=1,ntp
        write(11,'(f10.2,4e12.4)')(j-1)*dt,dis(j,i),vel(j,i),acc(j,i)
      end do
    end do
   end program p110
```

3. Solve the linear equations: $\mathbf{M}\ddot{\mathbf{u}}_0 = \mathbf{f}_0 - \mathbf{C}\dot{\mathbf{u}}_0 - \mathbf{K}\mathbf{u}_0$
4. Set $i = 0$
5. Compute: $\mathbf{a}_i = \frac{1}{\beta\Delta t^2}\mathbf{u}_i + \frac{1}{\beta\Delta t}\dot{\mathbf{u}}_i + (\frac{1}{2\beta} - 1)\ddot{\mathbf{u}}_i$
6. Compute: $\mathbf{b}_i = \frac{\gamma}{\beta\Delta t}\mathbf{u}_i - (1 - \frac{\gamma}{\beta})\dot{\mathbf{u}}_i - (1 - \frac{\gamma}{2\beta})\Delta t\ddot{\mathbf{u}}_i$
7. Compute: $\mathbf{f}'_{i+1} = \mathbf{f}_{i+1} + \mathbf{M}\mathbf{a}_i + \mathbf{C}\mathbf{b}_i$
8. Solve the linear equations: $\mathbf{K}'\mathbf{u}_{i+1} = \mathbf{f}'_{i+1}$
9. Compute: $\dot{\mathbf{u}}_{i+1} = \frac{\gamma}{\beta\Delta t}\mathbf{u}_{i+1} - \mathbf{b}_i$
10. Compute: $\ddot{\mathbf{u}}_{i+1} = \frac{1}{\beta\Delta t^2}\mathbf{u}_{i+1} - \mathbf{a}_i$
11. Increment i and repeat from 5.

The meanings of the variable names used in Program 11.0 are given as follows:

Scalar integers:

`NELS`	number of elements
`NEQ`	number of degrees of freedom in the mesh

NN	number of nodes
NBAND	half-bandwidth of mesh
NR	number of restrained nodes
NOD	number of nodes per element
NODOF	number of degrees of freedom per node
NDOF	number of degrees of freedom per element
NDIM	number of dimensions
NP_TYPES	number of different property types
IEL,I,J,K	simple counters
NLN	number of loaded nodes
LNODE	loaded node number (and for output)
LSENSE	sense of loading on LNODE (and for output)
NP	number of points in load/time function
NTP	total number of computation time steps
NON	number of output nodes

Scalar reals:

BETA	Newmark time stepping parameter
GAMMA	Newmark time stepping parameter
FM	Rayleigh damping factor on mass matrix (conventionally α)
FK	Rayleigh damping factor on stiffness matrix (conventionally β)
DT	computation time step
F1	working value
F2	working value
EL_ELL	element length
EL_EI	element stiffness

Allocatable real arrays:

KM	element stiffness matrix
MM	element mass matrix
KB	global stiffness matrix (lower triangle stored)
MB	global mass matrix (lower triangle stored)
EK	copy of KB
EM	copy of MB
CB	global (Rayleigh) damping matrix
KP	modified stiffness matrix
A	accelerations at each time step
V	velocities at each time step
D	displacements at each time step
A1	working vector
B1	working vector
VC	working vector
KD	working vector
COORD	element nodal coordinates matrix

`G_COORD`	global nodal coordinates matrix
`EI`	element flexural rigidities vector
`RHOA`	mass per unit length vector
`ELL`	element lengths vector
`RT`	time coordinates of load/time functions
`RL`	load coordinates of load/time functions
`AL`	load values at each calculation time step
`ACC`	full computed accelerations history
`VEL`	full computed velocities history
`DIS`	full computed displacements history

Allocatable integer arrays:

`NF`	nodal freedom matrix
`G`	element steering vector
`NUM`	element node numbers vector
`G_NUM`	global element node numbers matrix
`G_G`	global element steering matrix
`ETYPE`	element property type vector
`LP`	freedoms for output history print out
`LF`	freedoms receiving applied load/time histories

Library subroutines that have not been seen before in this text include `BEAM_MM`, which forms the beam element consistent mass matrix and `INTERP` which takes the load/time function data points and interpolates linearly to give load/time function values at the resolution of the calculation time step.

The example and data shown in Figure 11.1 are of a cantilever of unit length, subjected to a tip loading given by a half-sine pulse with an amplitude of *EI* and a duration equal to *T*, the fundamental period of the cantilever. The properties of the beam are given by an *EI* of 3.1941 and a mass per unit length ρA of 1.0. The fundamental frequency of the beam is therefore given by $\omega = 3.165\sqrt{3.1941} = 6.284$ (see discussion of Program 10.0), and the fundamental period by $T = 2\pi/\omega = 1.00$.

As shown in Figure 11.1, the beam has been discretised using a single beam element fully fixed at its left node. Conventional Newmark time-stepping parameters are used ($\beta = 0.25$ and $\gamma = 0.5$) with a calculation time step of 0.05. The beam is undamped, thus the Rayleigh damping parameters are both set to zero. There is one load time function (`NLN` $= 1$) applied to node 2 in sense 1. The sinusoidal loading function has been input using a total of 12 coordinates comprising 11 coordinates at 0.1 s intervals up to 1 s followed by a "quiet zone" involving a single interval of 0.8 s up to 1.8 s. The program linearly interpolates this load/time function at the calculation step length of 0.05 s for the Newmark algorithm.

The output from Program 11.0 is shown in Figure 11.2 and a plot of the computed displacement/time history of the cantilever tip is shown in Figure

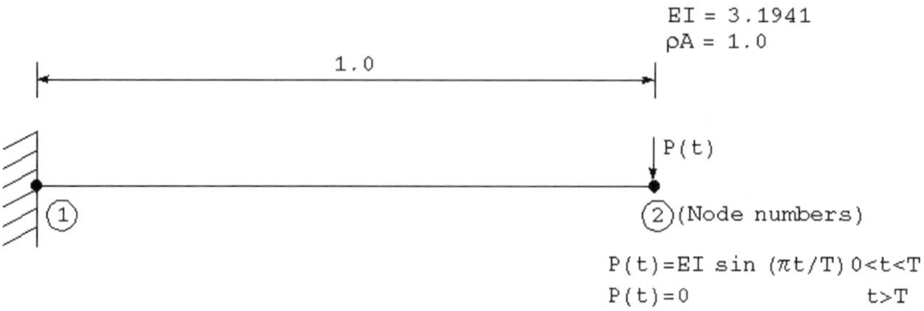

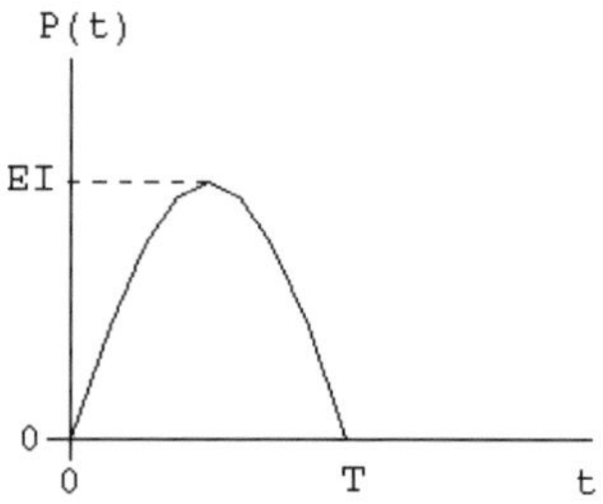

Structure and mesh data	NELS	NP_TYPES			
	1	1			
	(EI(I),RHOA(I),I=1,NP_TYPES)				
	3.1941	1.0			
	ELL				
	1.0				
Node freedom data	NR				
	1				
	K,NF(:,K),I=1,NR				
	1 0 0				
Time integration data	BETA	GAMMA	DT	FM	FK
	0.25	0.5	0.05	0.0	0.0
Load data	NLN				
	1				
	LNODE	LSENSE	NP		
	2	1	12		
	(RT(J),RL(J),J=1,NP)				
	.000	.000			
	.100	.987			
	.200	1.877			
	.300	2.584			
	.400	3.038			
	.500	3.194			

```
.600     3.038
.700     2.584
.800     1.877
.900      .987
1.000     .000
1.800     .000
NON
1

LNODE          LSENSE
2              1
```

Figure 11.1 Mesh and data for Program 11.0

```
Computed time histories
     0.00  0.0000E+00  0.0000E+00  0.0000E+00
     0.05  0.1942E-02  0.7768E-01  0.3107E+01
     0.10  0.9511E-02  0.2251E00  0.2788E+01
     0.15  0.2531E-01  0.4069E00  0.4485E+01
     0.20  0.5285E-01  0.6946E00  0.7026E+01
     0.25  0.9488E-01  0.9866E00  0.4652E+01
     0.30  0.1497E00  0.1207E+01  0.4159E+01
     0.35  0.2153E00  0.1415E+01  0.4184E+01
     0.40  0.2892E00  0.1541E+01  0.8436E00
     0.45  0.3650E00  0.1493E+01 -0.2780E+01
     0.50  0.4361E00  0.1352E+01 -0.2843E+01
     0.55  0.4976E00  0.1104E+01 -0.7080E+01
     0.60  0.5420E00  0.6757E00 -0.1005E+02
     0.65  0.5633E00  0.1745E00 -0.9996E+01
     0.70  0.5589E00 -0.3524E00 -0.1108E+02
     0.75  0.5261E00 -0.9574E00 -0.1312E+02
     0.80  0.4638E00 -0.1534E+01 -0.9945E+01
     0.85  0.3757E00 -0.1993E+01 -0.8412E+01
     0.90  0.2659E00 -0.2397E+01 -0.7734E+01
     0.95  0.1393E00 -0.2670E+01 -0.3187E+01
     1.00  0.4293E-02 -0.2729E+01  0.8098E00
     1.05 -0.1284E00 -0.2581E+01  0.5127E+01
     1.10 -0.2487E00 -0.2229E+01  0.8928E+01
     1.15 -0.3455E00 -0.1641E+01  0.1458E+02
     1.20 -0.4081E00 -0.8624E00  0.1658E+02
     1.25 -0.4308E00 -0.4544E-01  0.1610E+02
     1.30 -0.4123E00  0.7824E00  0.1701E+02
     1.35 -0.3534E00  0.1575E+01  0.1468E+02
     1.40 -0.2597E00  0.2174E+01  0.9284E+01
     1.45 -0.1415E00  0.2554E+01  0.5934E+01
     1.50 -0.9312E-02  0.2733E+01  0.1195E+01
     1.55  0.1244E00  0.2614E+01 -0.5936E+01
     1.60  0.2452E00  0.2221E+01 -0.9788E+01
     1.65  0.3423E00  0.1660E+01 -0.1265E+02
     1.70  0.4067E00  0.9180E00 -0.1702E+02
     1.75  0.4310E00  0.5502E-01 -0.1750E+02
     1.80  0.4132E00 -0.7675E00 -0.1540E+02
```

Figure 11.2 Results from Program 11.0

11.3. For comparison, the analytical solution (e.g. Warburton, 1964) for this problem during the loading phase when $0 \leq t \leq T$ is given by

$$v(t) = 0.441 \sin \frac{\pi t}{T} - 0.216 \sin \frac{2\pi t}{T} \tag{11.2}$$

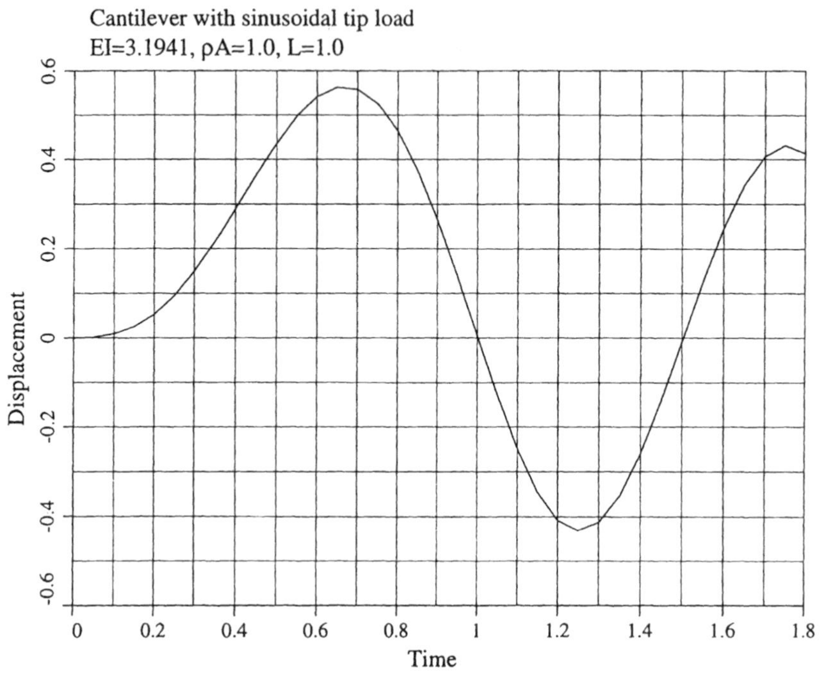

Figure 11.3 Computed tip displacement–time history

and for the free vibration phase when $t > T$ by

$$v(t) = -0.432 \sin 6.284\,(t - T) \tag{11.3}$$

This analytical solution is virtually indistinguishable from the computed result shown in Figure 11.3.

The structure chart is essentially the same as that shown in Figure 11.5.

PROGRAM 11.1: FORCED VIBRATION OF A RECTANGULAR SOLID IN PLANE STRAIN USING EIGHT-NODE QUADRILATERALS—LUMPED OR CONSISTENT MASS, IMPLICIT INTEGRATION BY THETA METHOD

In this program whose structure chart is given as Figure 11.5, a similar problem to that previously analysed using line beam elements is solved again using plane, solid elements and a direct integration procedure. In this example the tip loading takes the form of a cosine function. The specific method is the same implicit technique as was used for first-order problems in Program 8.0,

```
program p111
!------------------------------------------------------------------------------
!      program 11.1 forced vibration of a rectangular elastic
!      solid in plane strain using uniform 8-node quadrilateral elements
!      numbered in the y direction - lumped or consistent mass
!      implicit integration by theta method
!------------------------------------------------------------------------------
use new_library;     use  geometry_lib   ;      implicit none
integer::nels,nye,neq,nband,nn,nr,nip,nodof=2,nod=8,nst=3,ndof,           &
         i,k,iel,ndim=2,nstep,npri
real    ::aa,bb,e,v,det,rho,alpha,beta,omega,theta,period,pi,dtim,area,     &
         c1,c2,c3,c4,tim
logical :: consistent = .false. ; character(len=15):: element='quadrilateral'
!---------------------------- dynamic arrays-----------------------------------
real    ,allocatable :: kv(:),mv(:),loads(:),points(:,:),dee(:,:),coord(:,:),&
                        fun(:),jac(:,:), der(:,:),deriv(:,:), weights(:),   &
                        bee(:,:),km(:,:),g_coord(:,:),                      &
                        emm(:,:),ecm(:,:),f1(:),x0(:),d1x0(:),d2x0(:),       &
                        x1(:),d1x1(:),d2x1(:)
integer, allocatable :: nf(:,:), g(:) , num(:)  , g_num(:,:) , g_g(:,:)
!-----------------------input and initialisation-------------------------------
 open (10,file='p111.dat',status=    'old',action='read')
 open (11,file='p111.res',status='replace',action='write')
 read (10,*) nels,nye,nn,nip,aa,bb,rho,e,v,                                  &
             alpha,beta,nstep,npri,theta,omega
 ndof=nod*nodof
 allocate ( nf(nodof,nn), points(nip,ndim),g(ndof), g_coord(ndim,nn),        &
           dee(nst,nst),coord(nod,ndim),jac(ndim,ndim),weights(nip),        &
           der(ndim,nod), deriv(ndim,nod), bee(nst,ndof), km(ndof,ndof),     &
           num(nod),g_num(nod,nels),g_g(ndof,nels),                          &
           emm(ndof,ndof),ecm(ndof,ndof),fun(nod))
 nf=1; read(10,*) nr ; if(nr>0)read(10,*)(k,nf(:,k),i=1,nr)
 call formnf (nf);neq=maxval(nf)
 nband = 0       ; pi=acos(-1.)    ; period = 2.*pi/omega ; dtim =period/20.
 call deemat (dee,e,v); call sample( element ,points , weights)
!---------------loop the elements to find bandwidth and neq--------------------
 elements_1: do iel = 1 , nels
             call geometry_8qy(iel,nye,aa,bb,coord,num)
             call num_to_g ( num , nf , g
             g_num(:,iel)=num; g_coord(:,num)=transpose(coord);g_g(:,iel) = g
             if(nband<bandwidth(g))nband=bandwidth(g)
            end do elements_1
   write(11,'(a)') "Global coordinates "
   do k=1,nn;write(11,'(a,i5,a,2e12.4)')"Node",k,"        ",g_coord(:,k);end do
   write(11,'(a)') "Global node numbers "
   do k = 1 , nels; write(11,'(a,i5,a,8i5)')                                 &
                              "Element ",k,"        ",g_num(:,k); end do
   write(11,'(2(a,i5))')                                                     &
           "There are ",neq,"  equations and the half-bandwidth is", nband
allocate(kv(neq*(nband+1)),mv(neq*(nband+1)),x0(0:neq),d1x0(0:neq),x1(0:neq),&
       d2x0(0:neq),loads(0:neq),d1x1(0:neq),d2x1(0:neq),f1(neq*(nband+1)))
  kv=.0; mv=.0; x0=.0; d1x0=.0; d2x0=.0
!--------------- element stiffness and mass integration and assembly-----------
elements_2: do iel = 1 , nels
            num = g_num( : , iel ); coord = transpose(g_coord(: , num ))
            g = g_g( : , iel )     ; km=0.0    ; area = .0 ; emm = .0
         gauss_points_1: do i = 1 , nip
              call shape_der (der,points,i) ; jac = matmul(der,coord)
              det = determinant(jac); call invert(jac)
              deriv = matmul(jac,der) ; call beemat (bee,deriv)
            km = km + matmul(matmul(transpose(bee),dee),bee) *det* weights(i)
              area = area + det*weights(i); call shape_fun(fun,points,i)
              if(consistent) then
               call ecmat(ecm,fun,ndof,nodof); ecm=ecm*det*weights(i)*rho
```

```
              emm = emm + ecm
            end if
        end do gauss_points_1
      if(.not.consistent) then
        do i=1,ndof; emm(i,i)=area*rho*.2 ; end do
        do i=1,13,4 ; emm(i,i)=emm(3,3)*.25; end do
        do i=2,14,4 ; emm(i,i)=emm(3,3)*.25 ; end do
      end if
   call formkv (kv,km,g,neq); call formkv(mv,emm,g,neq)
 end do elements_2
!----------------------------factorisation-----------------------------------
   c1=(1.-theta)*dtim; c2=beta-c1; c3=alpha+1./(theta*dtim); c4=beta+theta*dtim
   f1 = c3*mv + c4*kv  ;     call banred(f1,neq)
!----------------------- time stepping loop ----------------------------------
   tim = .0
   write(11,'(a)') "  Time t    cos(omega*t) Displacement"
  timesteps: do i = 1 , nstep
            tim = tim + dtim  ; loads = .0
            x1 = c3*x0 + d1x0/theta
            loads(neq)=theta*dtim*cos(omega*tim)+c1*cos(omega*(tim-dtim))
            call linmul(mv,x1,d1x1)  ; d1x1 = loads + d1x1
            loads = c2*x0; call linmul(kv,loads,x1); x1 = x1 + d1x1
            call bacsub(f1,x1)
            d1x1=(x1-x0)/(theta*dtim)-d1x0*(1.-theta)/theta
            d2x1=(d1x1-d1x0)/(theta*dtim)-d2x0*(1.-theta)/theta
            if(i/npri*npri==i)write(11,'(3e12.4)')tim,cos(omega*tim),x1(neq)
            x0 = x1; d1x0 = d1x1; d2x0 = d2x1
  end do timesteps
end program p111
```

where it is often called the "Crank–Nicolson" approach. In second-order problems it is also known as the "Newmark $\beta = 1/4$" method in which form it was used in Program 11.0.

The formulation was described in section 3.17.1 where it was shown that in principle the algorithm is the same as that used in Program 8.0. To step from one time instant to the next, a set of simultaneous equations has to be solved. Since the differential equations are often linearised, this is not as great a numerical task as might be supposed because the equation coefficients are constant and need be factorised only once before the time-stepping procedure commences (see equation (3.128)). Velocities and accelerations are computed by ancillary equations (3.129) and (3.130).

The problem layout and data are reproduced as Figure 11.4. The structure of the program is contained in Figure 11.5. Turning to the coding, the new variables are the simple variables:

ALPHA, BETA	Rayleigh damping parameters α and β (see 3.124)
AREA	element area

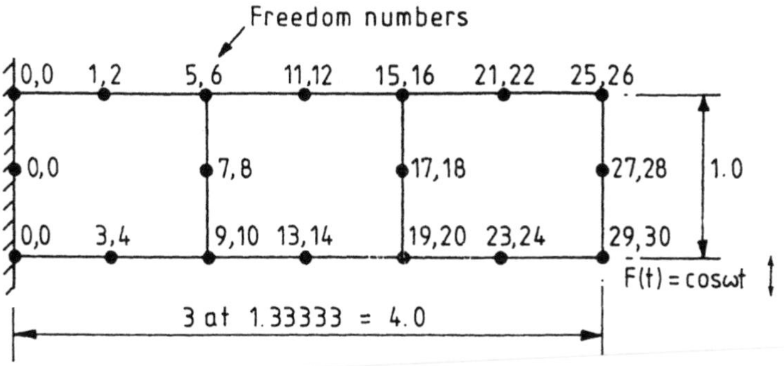

Structure and element data	NELS	NYE	NN	NIP			
	3	1	18	9			
	AA	BB	RHO	E	V	ALPHA	BETA
	1.33333	1.	1.	1.	.3	.005	.272
Time integration data	NSTEP	NPRI	THETA			OMEGA	
	20	1	.5 (Program 11.1) 1.4(Program 11.2)			.3	
Node freedom data	NR						
	3						
	K,NF(:,K),I=1,NR						
	1 0 0 2 0 0 3 0 0						

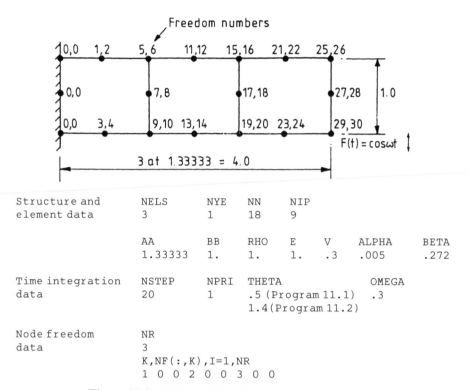

Figure 11.4 Mesh and data for Programs 11.1 and 11.2

and the arrays:

`F1`	$(\alpha + 1/\theta/\Delta t)\mathbf{MM} + (\beta + \theta\Delta t)\mathbf{KM}$(see 3.128)
`X0,D1X0,D2X0`	displacement, velocity, acceleration at time 0
`X1,D1X1,D2X1`	displacement, velocity, acceleration at time 1

The logical variable `CONSISTENT` is `.TRUE.` or `.FALSE.` as the case may be.

Up to the section headed "element stiffness and mass integration and assembly" the program's task is the familiar one of generating the global stiffness and mass matrices, in this case stored as the vectors `KV` and `MV` respectively. The matrix arising on the left-hand side of (3.128) is then created, called `F1` and factorised using `BANRED`.

In the time-stepping loop, the matrix-by-vector multiplications and vector additions specified on the right-hand side of (3.128) are carried out and equation solution is completed by `BACSUB`. It then remains only to compute the new velocities and accelerations using (3.129) and (3.130).

In this example, the lumped mass option has been chosen (`CONSISTENT = .FALSE.`) and, using (3.125), damping constants $\alpha = 0.005$ and $\beta = 0.272$

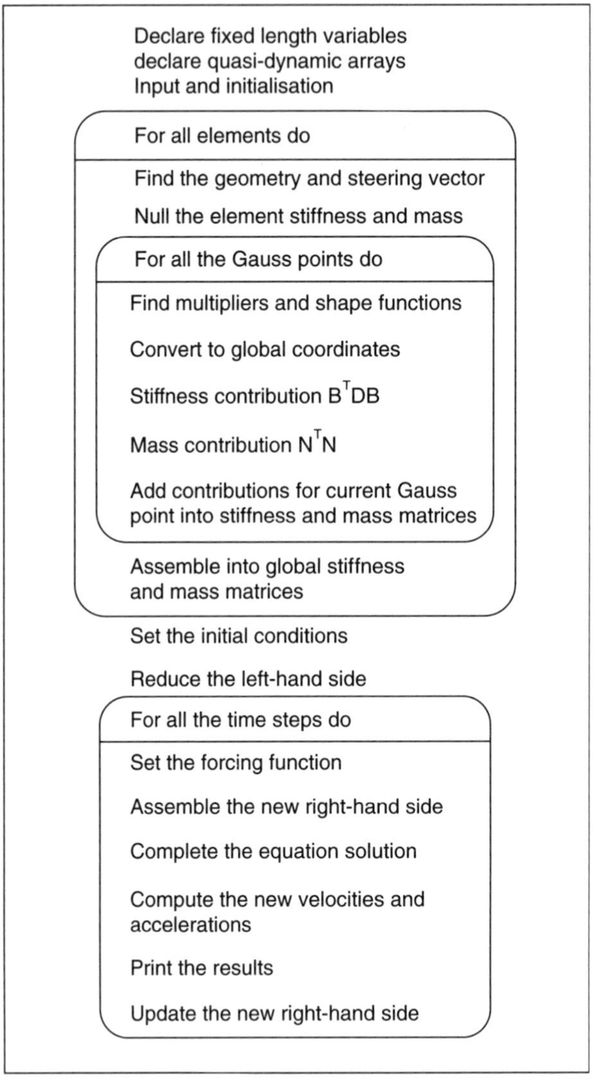

Figure 11.5 Structure chart for Program 11.1

yield damping ratios (conventionally called γ) close to 0.05 for the first three natural frequencies ω. The results are listed as Figure 11.6, where only the first complete cycle of forcing is shown. The displacements are considerably disturbed by the initial condition "transients", but once the response has settled down, say after 200 time steps, the average amplitude is 45.59 with a phase shift of about 90° relative to the loading function.

```
Global coordinates
Node    1         0.0000E+00  0.0000E+00
Node    2         0.0000E+00 -0.5000E00
Node    3         0.0000E+00 -0.1000E+01
Node    4         0.6667E00  0.0000E+00
Node    5         0.6667E00 -0.1000E+01
Node    6         0.1333E+01  0.0000E+00
Node    7         0.1333E+01 -0.5000E00
Node    8         0.1333E+01 -0.1000E+01
Node    9         0.2000E+01  0.0000E+00
Node   10         0.2000E+01 -0.1000E+01
Node   11         0.2667E+01  0.0000E+00
Node   12         0.2667E+01 -0.5000E00
Node   13         0.2667E+01 -0.1000E+01
Node   14         0.3333E+01  0.0000E+00
Node   15         0.3333E+01 -0.1000E+01
Node   16         0.4000E+01  0.0000E+00
Node   17         0.4000E+01 -0.5000E00
Node   18         0.4000E+01 -0.1000E+01
Global node numbers
Element     1            3    2    1    4    6    7    8    5
Element     2            8    7    6    9   11   12   13   10
Element     3           13   12   11   14   16   17   18   15
There are    30  equations and the half-bandwidth is   15
  Time t     cos(omega*t) Displacement
  0.1047E+01  0.9511E00  0.2291E+01
  0.2094E+01  0.8090E00  0.5473E+01
  0.3142E+01  0.5878E00  0.7771E+01
  0.4189E+01  0.3090E00  0.1108E+02
  0.5236E+01 -0.4371E-07  0.1470E+02
  0.6283E+01 -0.3090E00  0.1735E+02
  0.7330E+01 -0.5878E00  0.1866E+02
  0.8378E+01 -0.8090E00  0.1882E+02
  0.9425E+01 -0.9511E00  0.1768E+02
  0.1047E+02 -0.1000E+01  0.1520E+02
  0.1152E+02 -0.9511E00  0.1154E+02
  0.1257E+02 -0.8090E00  0.7068E+01
  0.1361E+02 -0.5878E00  0.2195E+01
  0.1466E+02 -0.3090E00 -0.2633E+01
  0.1571E+02 -0.4649E-06 -0.6880E+01
  0.1676E+02  0.3090E00 -0.9971E+01
  0.1780E+02  0.5878E00 -0.1147E+02
  0.1885E+02  0.8090E00 -0.1116E+02
  0.1990E+02  0.9511E00 -0.9027E+01
  0.2094E+02  0.1000E+01 -0.5314E+01
```

Figure 11.6 Results from Program 11.1

PROGRAM 11.2: FORCED VIBRATION OF A RECTANGULAR ELASTIC SOLID IN PLANE STRAIN USING EIGHT-NODE QUADRILATERALS — LUMPED OR CONSISTENT MASS, IMPLICIT INTEGRATION BY WILSON THETA METHOD

This algorithm is described in section 3.17.2. The essential step is that shown in (3.131), which is of exactly the same form as (3.128), and so this program can

```
program p112
!-----------------------------------------------------------------------
!      program 11.2 forced vibration of a rectangular elastic
!      solid in plane strain using uniform 8-node quadrilateral elements
!      numbered in the y direction - lumped or consistent mass
!      implicit integration by Wilson method
!-----------------------------------------------------------------------
use new_library;     use  geometry_lib   ;      implicit none
integer::nels,nye,neq,nband,nn,nr,nip,nodof=2,nod=8,nst=3,ndof,           &
        i,k,iel,ndim=2,nstep,npri
real   ::aa,bb,e,v,det,rho,alpha,beta,omega,theta,period,pi,dtim,area,   &
        c1,c2,c3,c4,c5,c6,c7,c8,time
logical:: consistent = .false.; character(len=15) :: element='quadrilateral'
!---------------------------- dynamic arrays--------------------------------
real    ,allocatable :: kv(:),mv(:),loads(:),points(:,:),dee(:,:),coord(:,:),&
                        fun(:),jac(:,:), der(:,:),deriv(:,:), weights(:),    &
                        bee(:,:),km(:,:),g_coord(:,:),                       &
                        emm(:,:),ecm(:,:),f1(:),x0(:),d1x0(:),d2x0(:),       &
                        x1(:),d1x1(:),d2x1(:)
integer, allocatable :: nf(:,:), g(:) , num(:)  , g_num(:,:) , g_g(:,:)
!------------------------input and initialisation---------------------------
 open (10,file='p112.dat',status=    'old',action='read')
 open (11,file='p112.res',status='replace',action='write')
 read (10,*) nels,nye,nn,nip,aa,bb,rho,e,v,                               &
             alpha,beta,nstep,npri,theta,omega
 ndof=nod*nodof
 allocate ( nf(nodof,nn), points(nip,ndim),g(ndof), g_coord(ndim,nn),     &
           dee(nst,nst),coord(nod,ndim),jac(ndim,ndim),weights(nip),      &
           der(ndim,nod), deriv(ndim,nod), bee(nst,ndof), km(ndof,ndof),  &
           num(nod),g_num(nod,nels),g_g(ndof,nels),                       &
           emm(ndof,ndof),ecm(ndof,ndof),fun(nod))
 nf=1; read(10,*) nr ; if(nr>0)read(10,*)(k,nf(:,k),i=1,nr)
 call formnf (nf);neq=maxval(nf)
 nband = 0      ; pi=acos(-1.)    ; period = 2.*pi/omega ; dtim =period/20.
 call deemat (dee,e,v); call sample(element,points,weights)
!----------------loop the elements to find bandwidth and neq-------------------
 elements_1: do iel = 1 , nels
             call geometry_8qy(iel,nye,aa,bb,coord,num)
             call num_to_g ( num , nf , g
             g_num(:,iel)=num; g_coord(:,num)=transpose(coord);g_g(:,iel) = g
             if(nband<bandwidth(g))nband=bandwidth(g)
            end do elements_1
   write(11,'(a)') "Global coordinates "
   do k=1,nn;write(11,'(a,i5,a,2e12.4)')"Node",k,"       ",g_coord(:,k);end do
   write(11,'(a)') "Global node numbers "
   do k = 1 , nels; write(11,'(a,i5,a,8i5)')                               &
                              "Element ",k,"        ",g_num(:,k); end do
   write(11,'(2(a,i5))')                                                    &
           "There are ",neq,"  equations and the half-bandwidth is", nband
allocate(kv(neq*(nband+1)),mv(neq*(nband+1)),x0(0:neq),d1x0(0:neq),x1(0:neq),&
      d2x0(0:neq),loads(0:neq),d1x1(0:neq),d2x1(0:neq),f1(neq*(nband+1)))
  kv=.0; mv=.0; x0=.0; d1x0=.0; d2x0=.0
!------------ element stiffness and mass integration and assembly--------------
elements_2: do iel = 1 , nels
            num = g_num( : , iel ); coord = transpose(g_coord(: , num ))
            g = g_g( : , iel )    ; km=0.0   ; area = .0 ; emm = .0
        gauss_points_1: do i = 1 , nip
              call shape_der (der,points,i) ; jac = matmul(der,coord)
              det = determinant(jac)  ; call invert(jac)
              deriv = matmul(jac,der) ; call beemat (bee,deriv)
            km = km + matmul(matmul(transpose(bee),dee),bee) *det* weights(i)
              area = area + det*weights(i); call shape_fun(fun,points,i)
              if(consistent) then
               call ecmat(ecm,fun,ndof,nodof); ecm=ecm*det*weights(i)*rho
```

```
              emm = emm + ecm
            end if
        end do gauss_points_1
     if(.not.consistent) then
       do i=1,ndof; emm(i,i)=area*rho*.2 ; end do
       do i=1,13,4 ; emm(i,i)=emm(3,3)*.25; end do
       do i=2,14,4 ; emm(i,i)=emm(3,3)*.25 ; end do
     end if
   call formkv (kv,km,g,neq); call formkv(mv,emm,g,neq)
 end do elements_2
!----------------------------factorisation-----------------------------------
   c1=6./(theta*dtim)**2; c2=6./(theta*dtim); c3=dtim**2/6.; c4 = 2.
   c5=3.*alpha/(theta*dtim); c6=3.*beta/(theta*dtim); c7=.5*alpha*theta*dtim
   c8=.5*beta*theta*dtim;     f1 = (c1+c5)*mv +(1.+c6)*kv  ; call banred(f1,neq)
!------------------------ time stepping loop ---------------------------------
   time = .0
   write(11,'(a)') " Time t     cos(omega*t) Displacement"
  timesteps: do i = 1 , nstep
              time = time + dtim  ; loads = .0
              x1 = (c1+c5)*x0 + (c2+2.*alpha)*d1x0 + (2. + c7)*d2x0
              loads(neq)=theta*cos(omega*time)+(1.-theta)*cos(omega*(time-dtim))
              call linmul(mv,x1,d1x1)  ; d1x1 = loads + d1x1
              loads = c6*x0 + 2.*beta*d1x0 + c8 * d2x0
              call linmul(kv,loads,x1); x1 = x1 + d1x1
              call bacsub(f1,x1)
              d2x1=(x1-x0)*c1-d1x0*c2-d2x0*c4 ;   d2x1=d2x0+(d2x1-d2x0)/theta
              d1x1=d1x0+.5*dtim*(d2x1+d2x0); x1=x0+dtim*d1x0+2.*c3*d2x0+d2x1*c3
              if(i/npri*npri==i)write(11,'(3e12.4)')time,cos(omega*time),x1(neq)
              x0 = x1; d1x0 = d1x1; d2x0 = d2x1
  end do timesteps
end program p112
```

be expected to resemble the previous one very closely. The structure chart of Figure 11.5 is again appropriate and the problem layout and data are again those of Figure 11.4. No new variables are involved, but the parameter θ has a stability limit of about 1.4 compared with 0.5 in the previous algorithm. All that need be said is that the `F1` matrix is now constructed as demanded by (3.131). The remaining steps of (3.132)–(3.135) are carried out within the section headed "time-stepping loop".

The results for one cycle are listed as Figure 11.7. Again these reflect mainly the influence of the start-up conditions and after 200 or so time steps the amplitude of vibration has settled down to 45.13, although with a slightly greater phase shift than was computed by the previous program.

```
Global coordinates
Node    1          0.0000E+00  0.0000E+00
Node    2          0.0000E+00 -0.5000E00
Node    3          0.0000E+00 -0.1000E+01
Node    4          0.6667E00   0.0000E+00
Node    5          0.6667E00  -0.1000E+01
Node    6          0.1333E+01  0.0000E+00
Node    7          0.1333E+01 -0.5000E00
Node    8          0.1333E+01 -0.1000E+01
Node    9          0.2000E+01  0.0000E+00
Node   10          0.2000E+01 -0.1000E+01
Node   11          0.2667E+01  0.0000E+00
Node   12          0.2667E+01 -0.5000E00
Node   13          0.2667E+01 -0.1000E+01
Node   14          0.3333E+01  0.0000E+00
Node   15          0.3333E+01 -0.1000E+01
Node   16          0.4000E+01  0.0000E+00
Node   17          0.4000E+01 -0.5000E00
Node   18          0.4000E+01 -0.1000E+01
Global node numbers
Element     1          3     2     1     4     6     7     8     5
Element     2          8     7     6     9    11    12    13    10
Element     3         13    12    11    14    16    17    18    15
There are    30  equations and the half-bandwidth is    15
  Time t    cos(omega*t) Displacement
  0.1047E+01  0.9511E00  0.4557E00
  0.2094E+01  0.8090E00  0.2699E+01
  0.3142E+01  0.5878E00  0.5166E+01
  0.4189E+01  0.3090E00  0.7263E+01
  0.5236E+01 -0.4371E-07  0.9409E+01
  0.6283E+01 -0.3090E00  0.1138E+02
  0.7330E+01 -0.5878E00  0.1244E+02
  0.8378E+01 -0.8090E00  0.1213E+02
  0.9425E+01 -0.9511E00  0.1041E+02
  0.1047E+02 -0.1000E+01  0.7466E+01
  0.1152E+02 -0.9511E00  0.3475E+01
  0.1257E+02 -0.8090E00 -0.1297E+01
  0.1361E+02 -0.5878E00 -0.6473E+01
  0.1466E+02 -0.3090E00 -0.1158E+02
  0.1571E+02 -0.4649E-06 -0.1612E+02
  0.1676E+02  0.3090E00 -0.1962E+02
  0.1780E+02  0.5878E00 -0.2171E+02
  0.1885E+02  0.8090E00 -0.2211E+02
  0.1990E+02  0.9511E00 -0.2076E+02
  0.2094E+02  0.1000E+01 -0.1780E+02
```

Figure 11.7 Results from Program 11.2

PROGRAM 11.3: FORCED VIBRATION OF AN ELASTIC SOLID IN PLANE STRAIN USING FOUR-NODE QUADRILATERALS — LUMPED OR CONSISTENT MASS, MIXED EXPLICIT/ IMPLICIT INTEGRATION

The next program in this chapter combines the methods of implicit and explicit (see Program 11.5) integration in a single program. Although not "mesh free" this procedure should allow economical bandwidths of the assembled matrices. The idea is (e.g. Key, 1980) that a mesh may contain only a few elements which have a very small explicit stability limit and are therefore best integrated implicitly. The remainder of the mesh can be successfully integrated explicitly at reasonable time steps.

```
program p113
!-------------------------------------------------------------------------
!      program 11.3 forced vibration of a rectangular elastic
!      solid in plane strain using uniform 4-node quadrilateral elements
!      numbered in the y direction - lumped or consistent mass
!      mixed explicit/implicit integration
!-------------------------------------------------------------------------
use new_library;    use  geometry_lib   ;      implicit none
integer::nels,nxe,nye,neq,nband,nn,nr,nip,nodof=2,nod=4,nst=3,ndof,        &
        i,k,iel,ndim=2,nstep
real   ::aa,bb,e,v,det,rho,beta,gamma,dtim,area,c1,c2,time
character (len=15) :: element = 'quadrilateral'
!---------------------------- dynamic arrays--------------------------------
real    ,allocatable :: kv(:),mv(:),loads(:),points(:,:),dee(:,:),coord(:,:),&
                        fun(:),jac(:,:), der(:,:),deriv(:,:), weights(:),    &
                        bee(:,:),km(:,:),g_coord(:,:),                      &
                        emm(:,:),ecm(:,:),x0(:),d1x0(:),d2x0(:),            &
                        x1(:),d1x1(:),d2x1(:)
integer, allocatable :: nf(:,:), g(:),num(:),g_num(:,:),g_g(:,:),kdiag(:)
logical, allocatable :: type(:)
!----------------------input and initialisation----------------------------
 open (10,file='p113.dat',status=    'old',action='read')
 open (11,file='p113.res',status='replace',action='write')
 read (10,*) nxe,nye,nn,nip,aa,bb,rho,e,v,                                &
            gamma,beta,nstep,dtim
 nels = nxe*nye ; ndof=nod*nodof ;c1=1./dtim/dtim/beta;c2=gamma/dtim/beta
 allocate ( nf(nodof,nn), points(nip,ndim),g(ndof), g_coord(ndim,nn),     &
          dee(nst,nst),coord(nod,ndim),jac(ndim,ndim),weights(nip),       &
          der(ndim,nod), deriv(ndim,nod), bee(nst,ndof), km(ndof,ndof),   &
          num(nod),g_num(nod,nels),g_g(ndof,nels),type(nels),             &
          emm(ndof,ndof),ecm(ndof,ndof),fun(nod))
 read(10,*) type
 nf=1; read(10,*) nr ; if(nr>0)read(10,*)(k,nf(:,k),i=1,nr)
 call formnf (nf);neq=maxval(nf) ; allocate(kdiag(neq))
 call deemat (dee,e,v);call sample(element,points,weights); kdiag= 0;nband = 0
!------------loop the elements to set globals and find nband ------------------
 elements_1: do iel = 1 , nels
             call geometry_4qy(iel,nye,aa,bb,coord,num)
             call num_to_g (num , nf , g )
             g_num(:,iel)=num; g_coord(:,num)=transpose(coord);g_g(:,iel) = g
             if(nband<bandwidth(g)) nband = bandwidth(g)
             if (.not.type(iel)) call fkdiag(kdiag,g)
 end do elements_1
     where(kdiag==0)kdiag=1   ; kdiag(1) = 1
     do i=2,neq; kdiag(i)=kdiag(i) + kdiag(i-1); end do
   write(11,'(a)') "Global coordinates "
   do k=1,nn;write(11,'(a,i5,a,2e12.4)')"Node",k,"       ",g_coord(:,k);end do
   write(11,'(a)') "Global node numbers "
   do k = 1 , nels; write(11,'(a,i5,a,4i5)')                               &
                          "Element ",k,"        ",g_num(:,k); end do
   write(11,'(3(a,i5))')                                                   &
          "There are ",neq," equations,nband=",nband," Sky store=",kdiag(neq)
 allocate(kv(kdiag(neq)),mv(neq*(nband+1)),x0(0:neq),d1x0(0:neq),x1(0:neq), &
       d2x0(0:neq),loads(0:neq),d1x1(0:neq),d2x1(0:neq))
       kv= .0 ; mv=.0
!-------------- element stiffness and mass integration and assembly------------
elements_2: do iel = 1 , nels
            num = g_num( : , iel ); coord = transpose(g_coord(: , num ))
            g = g_g( : , iel )    ; km=0.0   ; area = .0 ; emm = .0
            if(type(iel)) then ; do i=1,ndof; emm(i,i) = 1.; end do; end if
        gauss_points_1: do i = 1 , nip
              call shape_der (der,points,i) ; jac = matmul(der,coord)
              det = determinant(jac)  ; call invert(jac)
              deriv = matmul(jac,der) ; call beemat (bee,deriv)
```

```
            km = km + matmul(matmul(transpose(bee),dee),bee) *det* weights(i)
              area = area + det*weights(i); call shape_fun(fun,points,i)
              if(.not. type(iel)) then
               call ecmat(ecm,fun,ndof,nodof); ecm=ecm*det*weights(i)*rho*c1
               emm = emm + ecm
             end if
         end do gauss_points_1
        area = area/nod * rho
      if( type(iel) ) then
        do i=1,ndof; emm(i,i)=emm(i,i)*area*c1 ; end do
        call fsparv(kv,emm,g,kdiag)    ; km = - km
        call formkv(mv,km,g,neq); call formkv(mv,emm,g,neq)
      else
        call fsparv(kv,km,g,kdiag) ; call fsparv(kv,emm,g,kdiag)
        call formkv(mv,emm,g,neq)
      end if
 end do elements_2
!-----------------   initial conditions and factorisation of l.h.s. ----------
  x0 = .0 ;  d1x0 = 1. ;  d2x0 = .0;   call sparin(kv,kdiag)
!---------------------- time stepping loop ---------------------------------
   time = .0
   write(11,'(a)') "   Time  displacement velocity acceleration"
  timesteps: do i = 1 , nstep
            time = time + dtim  ; loads = .0
            d1x1=x0+d1x0*dtim+d2x0*.5*dtim*dtim*(1.-2.*beta)
            call linmul(mv,d1x1,x1); x1=loads+x1 ; call spabac(kv,x1,kdiag)
            d2x1=(x1-d1x1)/dtim/dtim/beta
            d1x1=d1x0+d2x0*dtim*(1.-gamma)+d2x1*dtim*gamma
            write(11,'(f8.5,3e12.4)')time,x1(neq),d1x1(neq),d2x1(neq)
            x0 = x1; d1x0 = d1x1; d2x0 = d2x1
  end do timesteps
end program p113
```

The recurrence relations (3.128)–(3.130) are cast in a slightly different form:

$$\left(\frac{1}{\Delta t^2\beta}\mathbf{MM}+\mathbf{KM}\right)\mathbf{r}_1=\mathbf{F}_1+\frac{1}{\Delta t^2\beta}\mathbf{MM}\bar{\mathbf{r}}_1 \tag{11.4}$$

for implicit elements and

$$\left(\frac{1}{\Delta t^2\beta}\mathbf{MM}\right)\mathbf{r}_1=\mathbf{F}_1+\left(\frac{1}{\Delta t^2\beta}\mathbf{MM}-\mathbf{KM}\right)\bar{\mathbf{r}}_1 \tag{11.5}$$

for explicit elements where

$$\bar{\mathbf{r}}_1=\mathbf{r}_0+\Delta t\dot{\mathbf{r}}_0+1/2\Delta t^2(1-2\beta)\ddot{\mathbf{r}}_0 \tag{11.6}$$

Accelerations and velocities are obtained from

$$\ddot{\mathbf{r}}_1=(\mathbf{r}_1-\bar{\mathbf{r}}_1)/(\Delta t^2\beta) \tag{11.7}$$

and

$$\dot{\mathbf{r}}_1=\dot{\mathbf{r}}_0+\Delta t(1-\gamma)\ddot{\mathbf{r}}_0+\Delta t\gamma\ddot{\mathbf{r}}_1 \tag{11.8}$$

The time integration parameters are the "Newmark" ones introduced in Program 11.0 and conventionally called $\beta=1/4$ and $\gamma=1/2$ corresponding to $\theta=1/2$ in Program 8.0.

When an explicit element is not coupled to an implicit one, the half-bandwidth of the assembled equation coefficient matrix will only be 1, whereas the full half-bandwidth will apply for implicit elements.

This is clearly a case where variable bandwidth storage is essential. The assembly routine for "skyline" storage FSPARV was used for example in Program 5.10 and is used again here (see Smith, 1984).

The problem chosen is illustrated in Figure 11.8. An elastic rod is constrained to vibrate in the axial direction only, and fixed at the right-hand end. Initial conditions are that a uniform unit velocity at $t = 0$ is applied to all freedoms in the mesh. The appropriate structure chart is Figure 11.5 for implicit integration.

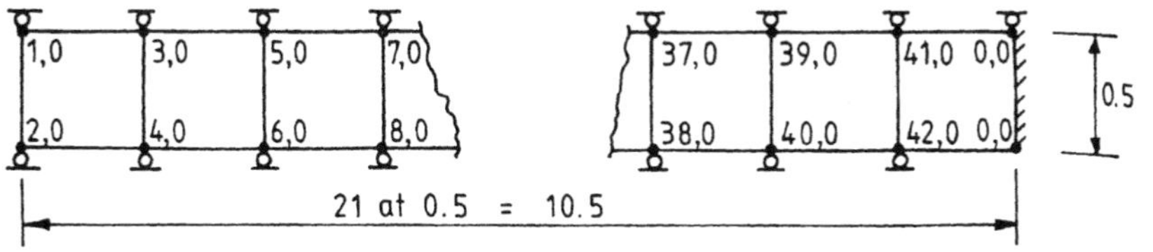

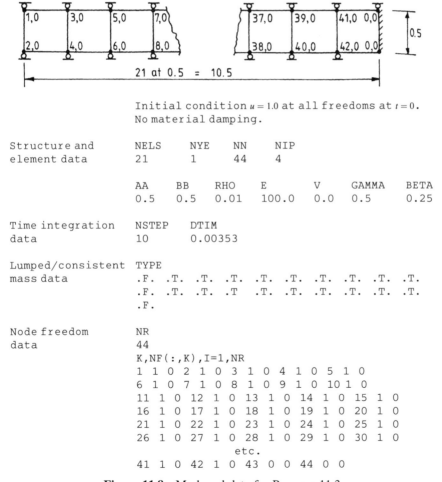

Initial condition $u = 1.0$ at all freedoms at $t = 0$.
No material damping.

```
Structure and       NELS    NYE    NN    NIP
element data        21      1      44    4

                    AA    BB    RHO    E      V     GAMMA   BETA
                    0.5   0.5   0.01   100.0  0.0   0.5     0.25

Time integration    NSTEP   DTIM
data                10      0.00353

Lumped/consistent   TYPE
mass data           .F.  .T.  .T.  .T.  .T.  .T.  .T.  .T.  .T.  .T.
                    .F.  .T.  .T.  .T   .T.  .T.  .T.  .T.  .T.  .T.
                    .F.

Node freedom        NR
data                44
                    K,NF(:,K),I=1,NR
                    1 1 0 2 1 0 3 1 0 4 1 0 5 1 0
                    6 1 0 7 1 0 8 1 0 9 1 0 10 1 0
                    11 1 0 12 1 0 13 1 0 14 1 0 15 1 0
                    16 1 0 17 1 0 18 1 0 19 1 0 20 1 0
                    21 1 0 22 1 0 23 1 0 24 1 0 25 1 0
                    26 1 0 27 1 0 28 1 0 29 1 0 30 1 0
                                  etc.
                    41 1 0 42 1 0 43 0 0 44 0 0
```

Figure 11.8 Mesh and data for Program 11.3

The only new parameters are GAMMA and BETA, used in the time integration (not to be confused with damping parameters), and constants $C1 = 1/\beta\Delta t^2$ and $C2 = \gamma/\Delta t\beta$. No damping is considered.

The logical array TYPE is used to distinguish between elements with lumped or consistent mass. In the present example, elements 1, 11 and 21 have consistent mass and the others have lumped mass. Also, the lumped elements are explicitly integrated while the consistent ones are implicitly integrated.

Early in the program, the variable bandwidth store is set up, using the knowledge that attached explicit elements have a half-bandwidth of 1.

```
Global coordinates
Node    1          0.0000E+00  0.0000E+00
Node    2          0.0000E+00 -0.5000E00
Node    3          0.5000E00   0.0000E+00
Node    4          0.5000E00  -0.5000E00
Node    5          0.1000E+01  0.0000E+00
Node    6          0.1000E+01 -0.5000E00
Node    7          0.1500E+01  0.0000E+00
Node    8          0.1500E+01 -0.5000E00
.............................................................

Node   40          0.9500E+01 -0.5000E00
Node   41          0.1000E+02  0.0000E+00
Node   42          0.1000E+02 -0.5000E00
Node   43          0.1050E+02  0.0000E+00
Node   44          0.1050E+02 -0.5000E00
Global node numbers
Element     1             2    1    3    4
Element     2             4    3    5    6
Element     3             6    5    7    8
Element     4             8    7    9   10
Element     5            10    9   11   12
Element     6            12   11   13   14
Element     7            14   13   15   16
Element     8            16   15   17   18
Element     9            18   17   19   20
Element    10            20   19   21   22
Element    11            22   21   23   24
Element    12            24   23   25   26
Element    13            26   25   27   28
Element    14            28   27   29   30
Element    15            30   29   31   32
Element    16            32   31   33   34
Element    17            34   33   35   36
Element    18            36   35   37   38
Element    19            38   37   39   40
Element    20            40   39   41   42
Element    21            42   41   43   44
There are     42 equations,nband=    3 Sky store=    55
   Time   displacement velocity acceleration
 0.00353  0.3071E-02  0.7398E00 -0.1474E+03
 0.00706  0.4783E-02  0.2301E00 -0.1414E+03
 0.01059  0.5072E-02 -0.6625E-01 -0.2649E+02
 0.01412  0.4883E-02 -0.4069E-01  0.4097E+02
 0.01765  0.4906E-02  0.5353E-01  0.1240E+02
 0.02118  0.5055E-02  0.3102E-01 -0.2515E+02
 0.02471  0.5054E-02 -0.3168E-01 -0.1037E+02
 0.02824  0.4961E-02 -0.2123E-01  0.1629E+02
 0.03177  0.4962E-02  0.2199E-01  0.8200E+01
 0.03530  0.5029E-02  0.1601E-01 -0.1159E+02
```

Figure 11.9 Results from Program 11.3

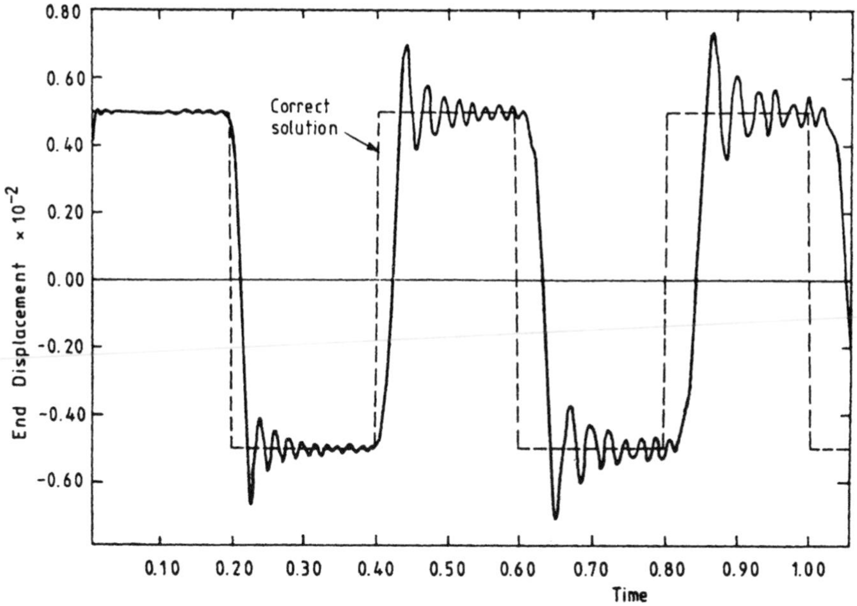

Figure 11.10 Displacement near support versus time from Program 11.3

Then the stiffness and mass are integrated as usual. When `TYPE` is `.T.` (explicit element) the element lumped mass is assembled into `KV` by subroutine `FDIAGV` while the element `MM` – `KM`(see equation (11.5)) is assembled into `MM` by subroutine `FORMKV`. Conversely, when the element is an implicit one, `MM` + `KM` are assembled into `KV` by subroutine `FSPARV` while the consistent element mass is assembled into `MM` again by subroutine `FORMKV`.

The initial conditions are then set, with the starting velocity, `D1X0`, equal to 1.0. The global matrix factorisation (sparse matrix) is done by `SPARIN` and the time-stepping loop is entered. Equations (11.4) and (11.5) require the usual matrix-by-vector multiplication on the right-hand side (by `LINMUL` in this case; if the right-hand side matrix is sparse, the appropriate routine is `LINMUL_SKY`). Equation solution is completed by `SPABAC` and it remains only to update velocities, accelerations, etc., for the next time step from (11.7) and (11.8).

The results are listed in Figure 11.9 for the first 10 steps. The displacements close to the support (freedom 42) are compared graphically with the exact solution in Figure 11.10. Despite some spurious oscillations the response is reasonably modelled.

11.1 MESH-FREE STRATEGIES

Typical mesh-free strategies can preserve the unconditional stability of the "implicit" procedures such as those used in Programs 11.1 and 11.2 by replacing the direct equation solution by an iterative approach such as p.c.g. Alternatively, as was done in Chapter 8, a purely explicit time integration procedure can be adopted, with its inherent stability limitations. Product EBE techniques are also available (Wong et al, 1989) but are not dealt with in this book.

Program 11.4 is adapted from Program 11.1 (implicit integration by the "theta" method) in the same way as Program 8.4 was adapted from Program 8.0. Equation solution is accomplished on every time step by preconditioned conjugate gradients using diagonal preconditioning of the left-hand-side matrix in equation (3.128).

The element stiffness matrices `KM` are stored as `STORE_KM` with the mass matrices `MM` as `STORE_MM`. The only new variables are `TOL`, the tolerance assigned in the p.c.g. iteration process and `LIMIT`, an iteration limiter in case convergence should be unacceptably slow. Otherwise the data listed in Figure 11.11 is the same as in Figure 11.4.

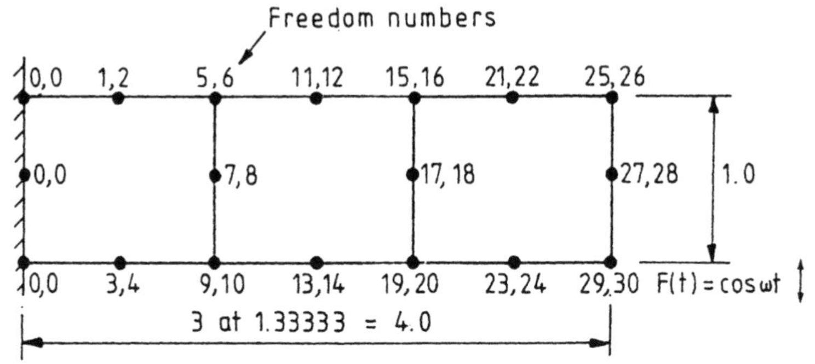

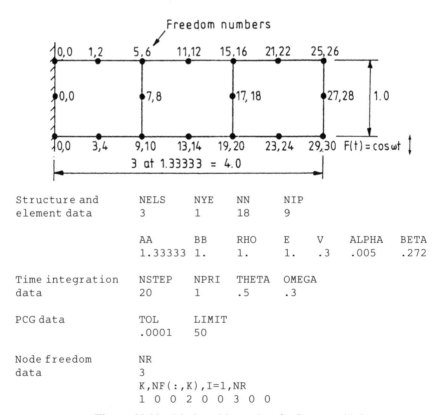

```
Structure and      NELS     NYE    NN     NIP
element data       3        1      18     9

                   AA       BB     RHO    E     V     ALPHA   BETA
                   1.33333  1.     1.     1.    .3    .005    .272

Time integration   NSTEP    NPRI   THETA  OMEGA
data               20       1      .5     .3

PCG data           TOL      LIMIT
                   .0001    50

Node freedom       NR
data               3
                   K,NF(:,K),I=1,NR
                   1 0 0 2 0 0 3 0 0
```

Figure 11.11 Mesh and input data for Program 11.4

PROGRAM 11.4: FORCED VIBRATION OF A RECTANGULAR ELASTIC SOLID IN PLANE STRAIN USING EIGHT-NODE QUADRILATERALS—LUMPED OR CONSISTENT MASS, IMPLICIT INTEGRATION BY THETA METHOD, MESH-FREE STRATEGY BY PCG

```
program p114
!-----------------------------------------------------------------------
!      program 11.4 forced vibration of a rectangular elastic
!      solid in plane strain using uniform 8-node quadrilateral elements
!      numbered in the y direction - lumped or consistent mass
!      implicit integration by theta method : pcg version
!-----------------------------------------------------------------------
use new_library;     use  geometry_lib   ;      implicit none
integer::nels,nye,neq,nn,nr,nip,nodof=2,nod=8,nst=3,ndof,                &
        i,k,iel,ndim=2,nstep,npri,iters,limit
real::aa,bb,e,v,det,rho,alpha,beta,omega,theta,period,pi,dtim,area,      &
      c1,c2,c3,c4,time,tol,big,up; character(len=15)::element='quadrilateral'
logical :: consistent = .false.  , converged
!----------------------------- dynamic arrays--------------------------------
real    ,allocatable :: loads(:),points(:,:),dee(:,:),coord(:,:),temp(:,:),  &
                        fun(:),jac(:,:), der(:,:),deriv(:,:), weights(:),    &
                        bee(:,:),km(:,:),g_coord(:,:),x1(:),d1x1(:),d2x1(:), &
                        emm(:,:),ecm(:,:),x0(:),d1x0(:),d2x0(:),             &
                        store_km(:,:,:),store_mm(:,:,:),u(:),p(:),d(:),      &
                        x(:),xnew(:),pmul(:),utemp(:),diag_precon(:)
integer, allocatable :: nf(:,:), g(:) , num(:)  , g_num(:,:) , g_g(:,:)
!-----------------------input and initialisation-------------------------------
 open (10,file='p114.dat',status=    'old',action='read')
 open (11,file='p114.res',status='replace',action='write')
 read (10,*) nels,nye,nn,nip,aa,bb,rho,e,v,                              &
             alpha,beta,nstep,npri,theta,omega,tol,limit
 ndof=nod*nodof
 allocate ( nf(nodof,nn), points(nip,ndim),g(ndof), g_coord(ndim,nn),    &
          dee(nst,nst),coord(nod,ndim),jac(ndim,ndim),weights(nip),      &
          der(ndim,nod), deriv(ndim,nod), bee(nst,ndof), km(ndof,ndof),  &
          num(nod),g_num(nod,nels),g_g(ndof,nels),emm(ndof,ndof),        &
          ecm(ndof,ndof),fun(nod),store_km(ndof,ndof,nels),utemp(ndof),  &
          pmul(ndof),store_mm(ndof,ndof,nels),temp(ndof,ndof))
 nf=1; read(10,*) nr ; if(nr>0)read(10,*)(k,nf(:,k),i=1,nr)
 call formnf (nf);neq=maxval(nf)
 pi=acos(-1.)   ; period = 2.*pi/omega ; dtim =period/20.
 c1=(1.-theta)*dtim; c2=beta-c1; c3=alpha+1./(theta*dtim); c4=beta+theta*dtim
 call deemat (dee,e,v); call sample(element,points,weights)
!-------------- loop the elements to find neq and store globals ---------------
 elements_1: do iel = 1 , nels
             call geometry_8qy(iel,nye,aa,bb,coord,num)
             call num_to_g ( num , nf , g ) ; g_num(:,iel)=num
             g_coord(:,num)=transpose(coord);g_g(:,iel) = g
 end do elements_1
   write(11,'(a)') "Global coordinates "
   do k=1,nn;write(11,'(a,i5,a,2e12.4)')"Node",k,"        ",g_coord(:,k);end do
   write(11,'(a)') "Global node numbers "
   do k = 1 , nels; write(11,'(a,i5,a,8i5)')                             &
                            "Element ",k,"        ",g_num(:,k); end do
   write(11,'(a,i5,a)') "There are ",neq," equations to be solved"
 allocate(x0(0:neq),d1x0(0:neq),x1(0:neq),diag_precon(0:neq),u(0:neq),   &
         d2x0(0:neq),loads(0:neq),d1x1(0:neq),d2x1(0:neq),               &
         d(0:neq),p(0:neq),x(0:neq),xnew(0:neq))
   xnew=.0; p=.0; diag_precon=.0  ;     store_km = .0; store_mm = .0
!------ element stiffness and mass integration ,storage and preconditioner ----
 elements_2: do iel = 1 , nels
            num = g_num( : , iel ); coord = transpose(g_coord(: , num ))
            g = g_g( : , iel )    ; km=0.0    ; area = .0 ; emm = .0
         gauss_points_1: do i = 1 , nip
             call shape_der (der,points,i) ; jac = matmul(der,coord)
             det = determinant(jac)  ; call invert(jac)
             deriv = matmul(jac,der) ; call beemat (bee,deriv)
           km = km + matmul(matmul(transpose(bee),dee),bee) *det* weights(i)
             area = area + det*weights(i); call shape_fun(fun,points,i)
```

```
              if(consistent) then
               call ecmat(ecm,fun,ndof,nodof); ecm=ecm*det*weights(i)*rho
               emm = emm + ecm
              end if
          end do gauss_points_1
       if(.not.consistent) then
         do i=1,ndof; emm(i,i)=area*rho*.2 ; end do
         do i=1,13,4 ; emm(i,i)=emm(3,3)*.25; end do
         do i=2,14,4 ; emm(i,i)=emm(3,3)*.25 ; end do
       end if
   store_km (: , : , iel ) = km ; store_mm( : , : , iel ) = emm
  do k=1,ndof;diag_precon(g(k))=diag_precon(g(k))+emm(k,k)*c3+km(k,k)*c4;end do
 end do elements_2
 diag_precon(1:neq) = 1. / diag_precon(1:neq); diag_precon(0) = .0
!-----------------------initial conditions -----------------------------------
           x0 = .0; d1x0 = .0; d2x0 = .0
!----------------------- time stepping loop ----------------------------------
   time = .0
   write(11,'(a)') "    Time t  cos(omega*t) Displacement Iterations"
  timesteps: do i = 1 , nstep
              time = time + dtim    ; loads = .0
    u = .0
    elements_3 : do iel = 1 , nels    ! gather for rhs multiply
                    g = g_g( : , iel ) ; km = store_km(:, :, iel)
                    emm = store_mm(:,:,iel) ; pmul = x0(g); temp=km*c2+emm*c3
                    utemp = matmul(temp , pmul)
                    u ( g ) = u ( g ) + utemp   ! scatter
                    pmul = d1x0(g) ! velocity bit
                    temp=emm/theta  ; utemp=matmul(temp,pmul);u(g)=u(g)+utemp
    end do elements_3
    loads(neq)=theta*dtim*cos(omega*time)+c1*cos(omega*(time-dtim))
    u(0) = .0; loads = u +   loads
!------------------ solve simultaneous equations by pcg ----------------------
      d = diag_precon*loads; p = d; x = .0
      iters = 0
     iterations  :      do
             iters = iters + 1     ;    u = 0.
      elements_4 : do iel = 1, nels
                     g = g_g( : , iel ) ; km = store_km(:,:,iel)
                     emm=store_mm(:,:,iel); temp=emm*c3+km*c4 ;  pmul = p(g)
                     utemp = matmul(temp,pmul); u(g) = u(g)+ utemp
      end do elements_4    ; u(0) = .0
!--------------------------pcg equation solution------------------------------
          up=dot_product(loads,d); alpha= up/ dot_product(p,u)
          xnew = x + p* alpha ; loads=loads - u*alpha;  d = diag_precon*loads
          beta=dot_product(loads,d)/up; p=d+p*beta
          big = .0; converged = .true.
          u = xnew ; where(u<.0) u = -u; big = maxval(u)
          u = (xnew - x)/big ;where(u<.0) u=-u ; big = maxval(u)
          if(big>tol) converged=.false.   ;    x=xnew
          if(converged .or. iters==limit) exit
     end do iterations
              x1=xnew
              d1x1=(x1-x0)/(theta*dtim)-d1x0*(1.-theta)/theta
              d2x1=(d1x1-d1x0)/(theta*dtim)-d2x0*(1.-theta)/theta
             if(i/npri*npri==i) then
               write(11,'(3e12.4,i10)')time,cos(omega*time),x1(neq),iters
             end if
              x0 = x1; d1x0 = d1x1; d2x0 = d2x1
  end do timesteps
end program p114
```

The results are listed as Figure 11.12.

Notice that this version of the p.c.g. solution is taking approximately NEQ/2 iterations to converge, a similar convergence rate as was usual in Chapters 5 and 6. In Chapter 8, the convergence rate was much quicker at approximately

```
Global coordinates
Node    1          0.0000E+00  0.0000E+00
Node    2          0.0000E+00 -0.5000E00
Node    3          0.0000E+00 -0.1000E+01
Node    4          0.6667E00   0.0000E+00
Node    5          0.6667E00  -0.1000E+01
Node    6          0.1333E+01  0.0000E+00
Node    7          0.1333E+01 -0.5000E00
Node    8          0.1333E+01 -0.1000E+01
Node    9          0.2000E+01  0.0000E+00
Node   10          0.2000E+01 -0.1000E+01
Node   11          0.2667E+01  0.0000E+00
Node   12          0.2667E+01 -0.5000E00
Node   13          0.2667E+01 -0.1000E+01
Node   14          0.3333E+01  0.0000E+00
Node   15          0.3333E+01 -0.1000E+01
Node   16          0.4000E+01  0.0000E+00
Node   17          0.4000E+01 -0.5000E00
Node   18          0.4000E+01 -0.1000E+01
Global node numbers
Element     1            3    2    1    4    6    7    8    5
Element     2            8    7    6    9   11   12   13   10
Element     3           13   12   11   14   16   17   18   15
There are    30  equations to be solved
    Time t  cos(omega*t) Displacement Iterations
  0.1047E+01  0.9511E00   0.2291E+01         14
  0.2094E+01  0.8090E00   0.5473E+01         14
  0.3142E+01  0.5878E00   0.7771E+01         14
  0.4189E+01  0.3090E00   0.1108E+02         14
  0.5236E+01 -0.4371E-07  0.1470E+02         14
  0.6283E+01 -0.3090E00   0.1734E+02         14
  0.7330E+01 -0.5878E00   0.1866E+02         13
  0.8378E+01 -0.8090E00   0.1882E+02         14
  0.9425E+01 -0.9511E00   0.1768E+02         14
  0.1047E+02 -0.1000E+01  0.1519E+02         15
  0.1152E+02 -0.9511E00   0.1154E+02         15
  0.1257E+02 -0.8090E00   0.7067E+01         15
  0.1361E+02 -0.5878E00   0.219 E+01         14
  0.1466E+02 -0.3090E00  -0.2629E+01         14
  0.1571E+02 -0.4649E-06-0.6870E+01          14
  0.1676E+02  0.3090E00  -0.9952E+01         15
  0.1780E+02  0.5878E00  -0.1144E+02         15
  0.1885E+02  0.8090E00  -0.1112E+02         15
  0.1990E+02  0.9511E00  -0.8975E+01         15
  0.2094E+02  0.1000E+01-0.5250E+01          16
```

Figure 11.12 Results from Program 11.4

NEQ/10. Fortunately, as problem sizes increase, the iteration count, as a proportion of number of equations to be solved, drops very rapidly.

PROGRAM 11.5: FORCED VIBRATION OF A RECTANGULAR ELASTO-PLASTIC SOLID IN PLANE STRAIN USING EIGHT-NODE QUADRILATERALS—LUMPED MASS, EXPLICIT INTEGRATION

In the same way as was done for first-order problems in Program 8.5, θ can be set to zero in second-order recurrence formulae such as (3.128). Then the only matrix remaining on the left-hand side of the equation is **MM**; if this is lumped

```
   program p115
!-----------------------------------------------------------------------
!      program 11.5 forced vibration of an elastic-plastic(Von Mises) solid
!      using 8-node quadrilateral elements; viscoplastic strain method
!      rectangular mesh : lumped mass , explicit integration
!-----------------------------------------------------------------------
 use new_library      ; use geometry_lib     ;      implicit none
 integer::nels,nxe,neq,nn,nr,nip,nodof=2,nod=8,nst=4,ndof,loaded_nodes,      &
          i,k,iel,ndim=2,jj,nstep ,npri
 real    ::aa,bb,rho,dtim,time,e,v,det,sbary,pload,sigm,f,fnew,fac,          &
          area,sbar,dsbar,lode_theta
 character (len = 15) :: element = 'quadrilateral'
!--------------------------- dynamic arrays-----------------------------------
 real    ,allocatable :: points(:,:),bdylds(:),x1(:),d1x1(:),stress(:),      &
                         pl(:,:),emm(:),d2x1(:),tensor(:,:,:),etensor(:,:,:),&
                         val(:,:),mm(:),dee(:,:),coord(:,:),jac(:,:),        &
                         weights(:), der(:,:),deriv(:,:),bee(:,:),eld(:),    &
                         eps(:),sigma(:),bload(:),eload(:),g_coord(:,:)
 integer, allocatable :: nf(:,:) , g(:), no(:) ,num(:), g_num(:,:) ,g_g(:,:)
!-----------------------input and initialisation------------------------------
  open (10,file='p115.dat',status=    'old',action='read')
  open (11,file='p115.res',status='replace',action='write')
  read (10,*) aa,bb,sbary,e,v,rho,pload,   &
              nels,nxe,nn,nip,loaded_nodes,dtim,nstep,npri
  ndof=nod*nodof
  allocate (nf(nodof,nn), points(nip,ndim),weights(nip),g_coord(ndim,nn),    &
            num(nod),dee(nst,nst),tensor(nst,nip,nels),no(loaded_nodes),     &
            coord(nod,ndim), pl(nst,nst), etensor(nst,nip,nels),             &
            jac(ndim,ndim),der(ndim,nod),deriv(ndim,nod),g_num(nod,nels),    &
            bee(nst,ndof),eld(ndof),eps(nst),sigma(nst),emm(ndof),           &
            bload(ndof),eload(ndof),g(ndof), stress(nst),                    &
            val(loaded_nodes,ndim),g_g(ndof,nels))
  nf=1; read(10,*) nr ; if(nr>0) read(10,*)(k,nf(:,k),i=1,nr)
  call formnf(nf); neq=maxval(nf)
    read(10,*)(no(i),val(i,:),i=1,loaded_nodes)
  ! loop the elements to set up global arrays
      elements_1:   do iel = 1 , nels
                      call geometry_8qx(iel,nxe,aa,bb,coord,num)
                      call num_to_g (num , nf , g); g_num(:,iel)=num
                      g_coord(:,num)=transpose(coord); g_g( : , iel ) = g
      end do elements_1
    write(11,'(a)') "Global coordinates "
    do k=1,nn;write(11,'(a,i5,a,2e12.4)')"Node",k,"       ",g_coord(:,k);end do
    write(11,'(a)') "Global node numbers "
    do k = 1 , nels; write(11,'(a,i5,a,8i5)')                                &
                               "Element ",k,"        ",g_num(:,k); end do
    write(11,'(a,i5,a)') "There are ",neq," equations to be solved"
  allocate(bdylds(0:neq),x1(0:neq),d1x1(0:neq),d2x1(0:neq),mm(0:neq))
  tensor = .0;  etensor = .0
  x1=0.0; d1x1=0.0; d2x1=0.0; mm=0.0  ;  call sample(element,points,weights)
!--------------------- explicit integration loop ----------------------------
write(11,'(a)') "   Time     Displacement  Velocity   Acceleration"
time = .0
write(11,'(4e12.4)')time,x1(neq),d1x1(neq),d2x1(neq)
time_steps : do jj = 1 , nstep
!------------------------ apply the load -----------------------------------
time = time + dtim
x1 = x1 +(d1x1+d2x1*dtim*.5)*dtim  ;  bdylds = .0
!-------------------- element stress-strain relationship ---------------------
elements_2: do iel = 1 , nels
               num = g_num(:,iel) ; coord = transpose(g_coord(: , num ))
                 g = g_g( : , iel ) ; area = 0.0 ; bload = .0 ; eld = x1 ( g )
            gauss_pts_1:  do i =1 , nip  ;  dee = .0 ; call deemat(dee,e,v)
              call shape_der (der,points,i);  jac = matmul(der,coord)
```

```
              det = determinant(jac)   ;    call invert(jac)
              deriv = matmul(jac,der) ;  call beemat (bee,deriv)
              area = area + det * weights(i)*rho; eps = matmul ( bee , eld )
              eps = eps - etensor( : , i , iel )
              sigma= matmul ( dee , eps ); stress = sigma+tensor (: , i, iel)
              call invar(stress,sigm,dsbar,lode_theta); fnew = dsbar - sbary
!---------------------- check whether yield is violated ----------------------
          if(fnew>=.0) then
              stress= tensor(:,i,iel); call invar(stress,sigm,sbar,lode_theta)
              f = sbar - sbary; fac = fnew/(fnew - f)
              stress = tensor ( : , i , iel )+(1.-fac) * sigma
              call vmpl(e,v,stress,pl); dee = dee - fac * pl
          end if
              sigma = matmul(dee ,eps) ;sigma = sigma + tensor( : , i , iel )
              eload=matmul(sigma,bee); bload= bload+ eload * det * weights(i)
!----------------------update Gauss point stresses and strains ---------------
              tensor( : , i , iel) = sigma
              etensor( : , i , iel ) = etensor( : , i , iel ) + eps
           end do gauss_pts_1
           bdylds ( g ) = bdylds ( g ) - bload  ; bdylds(0) = .0
           if( jj == 1) then
             emm = .2 * area; emm(1:13:4)=.05*area; emm(2:14:4)=.05*area
             mm ( g ) = mm ( g ) + emm
           end if
 end do elements_2
      do i=1,loaded_nodes
         bdylds(nf(:,no(i)))=bdylds(nf(:,no(i)))+val(i,:)*pload
      end do
      bdylds(1:neq) = bdylds(1:neq) / mm(1:neq)
      d1x1=d1x1+(d2x1+bdylds)*.5*dtim  ;  d2x1 = bdylds
      if(jj==jj/npri*npri)write(11,'(4e12.4)')time,x1(neq),d1x1(neq),d2x1(neq)
end do time_steps
end program p115
```

(diagonalised), the new solution $\mathbf{r}_1$ can be computed without solving simultaneous equations at all. Further, the right-hand-side products can again be completed using element-by-element summation and so no global matrices are involved. This procedure is particularly attractive in nonlinear problems where the stiffness **KM** is a function of, for example, strain. In the present program, nonlinearity is introduced in the form of elasto-plasticity, which was described in Chapter 6. The nomenclature used is therefore drawn from the earlier programs in this chapter and from those in Chapter 6, particularly Program 6.0 which dealt with von Mises solids.

The structure chart for the program is shown in Figure 11.13 and the problem layout and data in Figure 11.14. All the variables used, and their significance, are listed as follows:

Scalar integers and reals:

NELS	number of elements (6)
NXE	number of elements in x direction (1)
NYE	number of elements in y direction (6)
NEQ	total number of non-zero freedoms in mesh (50)
NN	total number of nodes in mesh (33)
NR	number of restrained nodes (15)

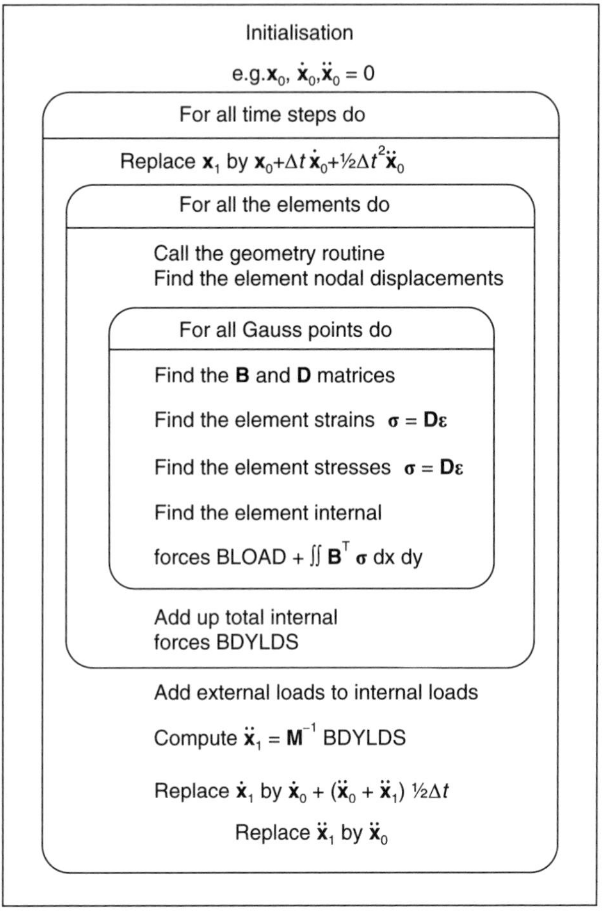

Figure 11.13 Structure chart for Program 11.5

`NIP`	number of "Gauss points" (4)
`AA`	size of elements in x direction (1.)
`BB`	size of elements in y direction (2.5)
`RHO`	mass of elements per unit volume (0.733×10^{-3})
`E`	Young's modulus (3.0×10^{7})
`SBARY`	von Mises yield stress (50 000)
`PLOAD`	load multiplier (180.0)
`DTIM`	time step (1.0×10^{-6})
`NSTEP`	number of time steps (300)
`NPRI`	printing control (every `NPRI` results printed) (50)
`NL`	number of loaded freedoms (13)
`TIME`	accumulated time

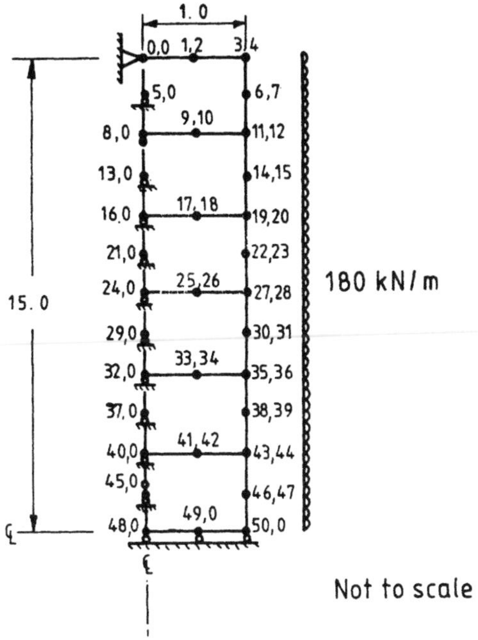

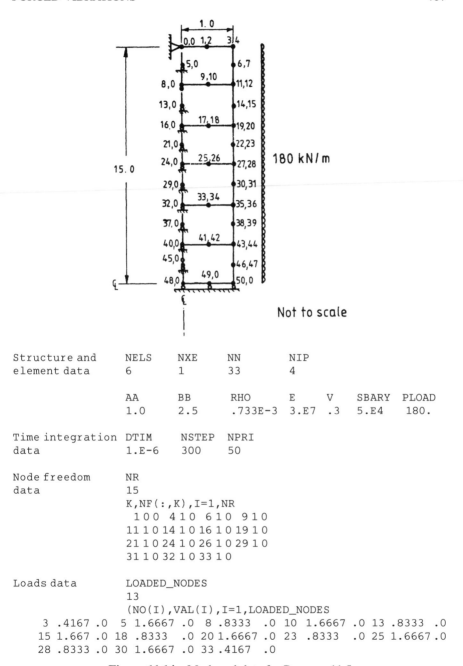

```
Structure and      NELS     NXE     NN         NIP
element data       6        1       33         4

                   AA       BB      RHO        E      V     SBARY   PLOAD
                   1.0      2.5     .733E-3    3.E7   .3    5.E4     180.

Time integration   DTIM     NSTEP   NPRI
data               1.E-6    300     50

Node freedom       NR
data               15
                   K,NF(:,K),I=1,NR
                    1 0 0  4 1 0  6 1 0  9 1 0
                   11 1 0 14 1 0 16 1 0 19 1 0
                   21 1 0 24 1 0 26 1 0 29 1 0
                   31 1 0 32 1 0 33 1 0

Loads data         LOADED_NODES
                   13
                   (NO(I),VAL(I),I=1,LOADED_NODES
    3 .4167 .0  5 1.6667 .0  8 .8333  .0 10 1.6667 .0 13 .8333  .0
   15 1.667 .0 18 .8333  .0 20 1.6667 .0 23 .8333  .0 25 1.6667 .0
   28 .8333 .0 30 1.6667 .0 33 .4167  .0
```

Figure 11.14 Mesh and data for Program 11.5

AREA	element area (includes multiple ρ)
DSBAR	deviatoric stress invariant $\bar{\sigma}$
FNEW	new value of yield function
LODE_THETA	Lode angle (see Chapter 6)
F	yield function
SBAR	deviatoric stress invariant $\bar{\sigma}$
FAC	factor (see Figure 6.7)

Allocatable arrays not usually changed:

DEE	stress–strain matrix
PL	plasticity matrix
SAMP	Gaussian abscissae and weights
COORD	element coordinates
JAC	Jacobian matrix
DER	derivatives of shape functions (local)
DERIV	derivatives of shape functions (global)
BEE	strain–displacement matrix
ELD	element nodal displacements
FUN	element shape functions
EMM	element mass matrix
STRESS	vector of current Gauss point stresses
EPS	vector of Gauss point strains
SIGMA	vector of elastic Gauss point element stresses
BT	transpose of BEE
BLOAD	element 'internal' nodal forces
G	steering vector

Allocatable arrays changed with problem size:

VAL	values of nodal freedom loads
X1,D1X1,D2X1	displacement, velocity, acceleration
MM	global mass matrix
BDYLDS	global "body-loads" vector (see Chapter 6)
SX,SY,TXY,SZ	accumulated Gauss point stresses
EX,EY,GXY,EZ	accumulated Gauss point strains
NF	node freedom array

Turning to the program code, after input and initialisation, the von Mises plastic stress–strain matrix PL is formed by subroutine VMPL. The remainder of the program is a large explicit integration time-stepping loop. The displacement X1 is updated and then, scanning all elements and Gauss points, new strains can be computed. The constitutive relation then determines the appropriate level of stress and hence the **D** matrix which should operate (whether or not the yield stress has been violated). The difference between the true stresses and the elastic ones is redistributed as "body-loads" BDYLDS, whence the new accelerations D2X1 can be found and integrated to find the new velocities, D1X1. Then the next cycle of displacements can be updated.

The results from the program are printed in Figure 11.15 in the form of elapsed time, displacement, velocity and acceleration at the upper surface centreline of the beam. Figure 11.16 gives the centreline displacement of the beam as a function of time computed over the first 10 000 time steps. The development of permanent, plastic deformation is clearly demonstrated.

```
Global coordinates
Node    1          0.0000E+00  0.0000E+00
Node    2          0.5000E00  0.0000E+00
Node    3          0.1000E+01  0.0000E+00
Node    4          0.0000E+00 -0.1250E+01
Node    5          0.1000E+01 -0.1250E+01
Node    6          0.0000E+00 -0.2500E+01
Node    7          0.5000E00 -0.2500E+01
Node    8          0.1000E+01 -0.2500E+01
Node    9          0.0000E+00 -0.3750E+01
Node   10          0.1000E+01 -0.3750E+01
Node   11          0.0000E+00 -0.5000E+01
Node   12          0.5000E00 -0.5000E+01
Node   13          0.1000E+01 -0.5000E+01
Node   14          0.0000E+00 -0.6250E+01
Node   15          0.1000E+01 -0.6250E+01
Node   16          0.0000E+00 -0.7500E+01
Node   17          0.5000E00 -0.7500E+01
Node   18          0.1000E+01 -0.7500E+01
Node   19          0.0000E+00 -0.8750E+01
Node   20          0.1000E+01 -0.8750E+01
Node   21          0.0000E+00 -0.1000E+02
Node   22          0.5000E00 -0.1000E+02
Node   23          0.1000E+01 -0.1000E+02
Node   24          0.0000E+00 -0.1125E+02
Node   25          0.1000E+01 -0.1125E+02
Node   26          0.0000E+00 -0.1250E+02
Node   27          0.5000E00 -0.1250E+02
Node   28          0.1000E+01 -0.1250E+02
Node   29          0.0000E+00 -0.1375E+02
Node   30          0.1000E+01 -0.1375E+02
Node   31          0.0000E+00 -0.1500E+02
Node   32          0.5000E00 -0.1500E+02
Node   33          0.1000E+01 -0.1500E+02
Global node numbers
Element     1            6    4    1    2    3    5    8    7
Element     2           11    9    6    7    8   10   13   12
Element     3           16   14   11   12   13   15   18   17
Element     4           21   19   16   17   18   20   23   22
Element     5           26   24   21   22   23   25   28   27
Element     6           31   29   26   27   28   30   33   32
There are     50 equations to be solved
   Time       Displacement  Velocity    Acceleration
 0.0000E+00  0.0000E+00  0.0000E+00  0.0000E+00
 0.5000E-04  0.2995E-03  0.1199E+02  0.2301E+06
 0.1000E-03  0.1214E-02  0.2381E+02  0.4809E+06
 0.1500E-03  0.2684E-02  0.3354E+02  0.4907E+06
 0.2000E-03  0.4867E-02  0.5286E+02 -0.8493E+06
 0.2500E-03  0.8084E-02  0.7435E+02  0.6141E+06
 0.3000E-03  0.1231E-01  0.9414E+02  0.6372E+04
```

Figure 11.15 Results from Program 11.5

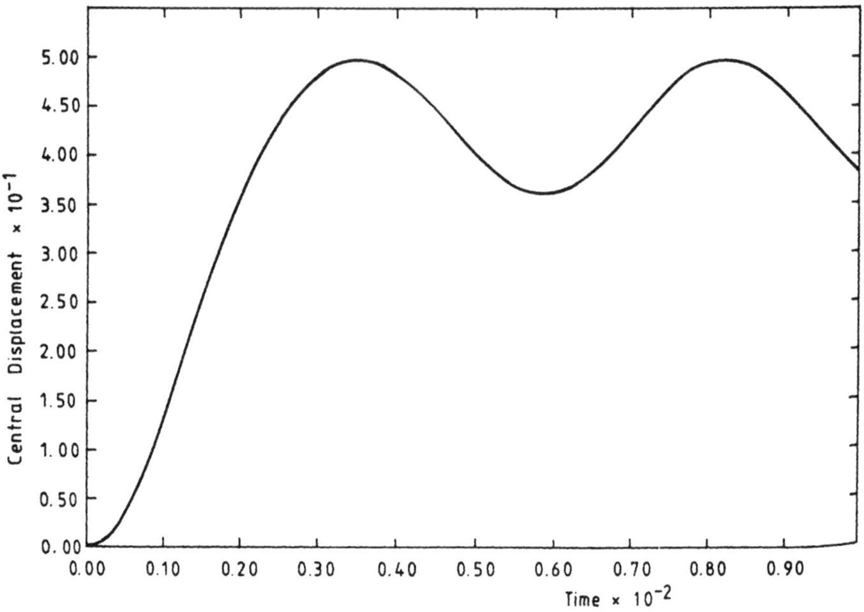

Figure 11.16 Plot of results from Program 11.4

11.2 EXAMPLES

1. The undamped beam shown in Figure 11.17 is initially at rest and subjected to a suddenly applied moment of one unit at its left support. Using a single finite element, a time step of 1 second and the constant acceleration method ($\beta = 1/4$, $\gamma = 1/2$), estimate the rotation at both ends of the beam after 2 seconds.
2. Repeat the previous question assuming 5% damping. Use Rayleigh damping by assuming the mass matrix damping parameter equals zero. The fundamental natural frequency of the beam is $\omega_1 = 0.48$.
3. The undamped cantilever shown in Figure 11.18 is initially at rest and subjeted to a suddenly applied load at its mid-span. Using two finite elements, a time step of 1 second and the linear acceleration method ($\beta = 1/6$, $\gamma = 1/2$), estimate the deflection under the load after 2 seconds.
4. The undamped cantilever shown in Figure 11.19 is initially at rest and subjected to a suddenly applied load and moment at its tip. Using one finite element, a time step of 1 second and the constant acceleration method ($\beta = 1/4$, $\gamma = 1/2$), estimate the deflection and rotation at the tip after 2 seconds.
5. A simply supported beam of length, $L = 1$ and properties $EI = 1.0$, $\rho A = 3.7572$ is subjected to a constant transverse force, $P = 48$, which moves

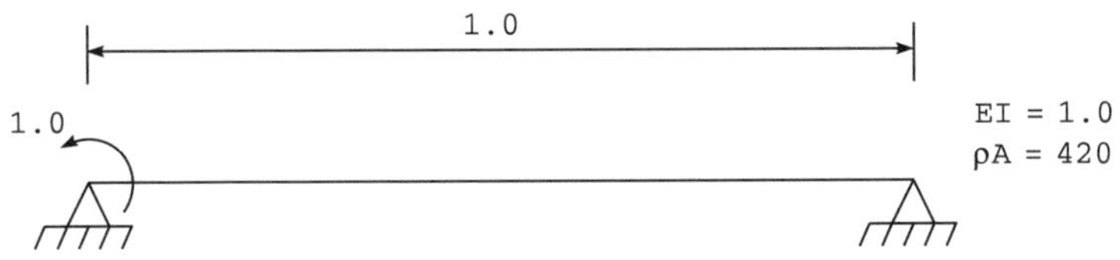

Figure 11.17

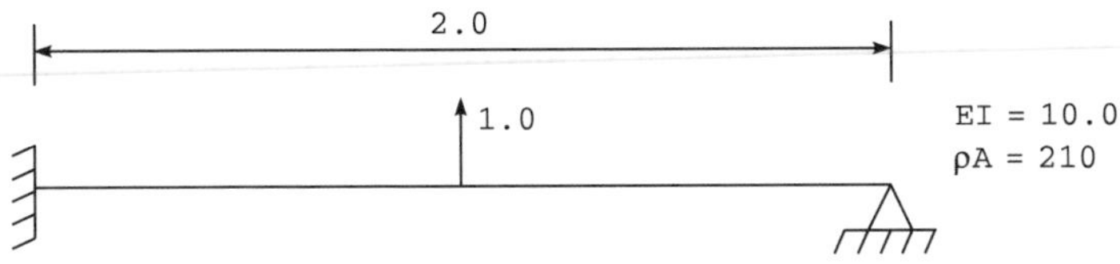

Figure 11.18

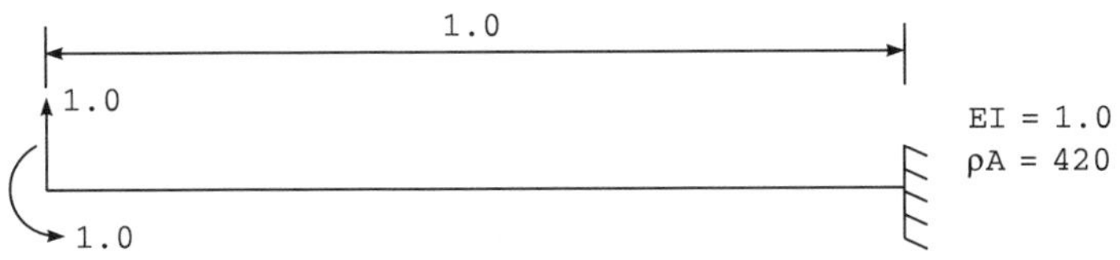

Figure 11.19

across the beam from the left to right support with a constant velocity, $U = 1$. By discretising the beam into a four elements compute the time dependent response of the centreline of the beam. Show that the centreline deflection reaches a maximum value that is approximately 1.743 times the deflection that would have been obtained if the load had been placed statically at the centre of the beam. Use equivalent fixed end moments and reactions to model the effect of the moving load at four locations within each element.

Compare your result with the analytical solution given by:

$$v(L/2, t) = \frac{2PL^3}{\pi^4 EI}\left[\frac{\sin c\,\omega_1 t - c\sin\omega_1 t}{1 - c^2}\right]$$

where

$$\omega_1 = \frac{\pi^2}{L^2}\sqrt{\frac{EI}{\rho A}} \quad \text{and} \quad c = \frac{\pi U}{\omega_1 L}$$

11.3 REFERENCES

Bathe, K. J. (1996) Numerical Methods in Finite Element Analysis, 3rd edn. Prentice-Hall, Englewood Cliffs, New Jersey.

Key, S. W. (1980) Transient response by time integration: a review of implicit and explicit operators. In *Advance in Structural Dynamics*, ed. J. Donea, pp. 71–95. Applied Science, London.

Smith, I. M. (1984) Adaptability of truly modular software. *Engineering Computations*, **1**(1), 25–35.

Warburton, G. B. (1964) *The Dynamical Behaviour of Structures*. Pergamon Press, Oxford.

Wong, S. W., Smith, I. M. and Gladwell, I. (1989) PCG methods for transient FE analysis. Part II: Second order problems, *IJNME*, 28(7), 1567–1576.

Appendix 1
Consistent Nodal Loads

PLANE ELEMENTS (2-D)

Width of loaded face = 1 unit.

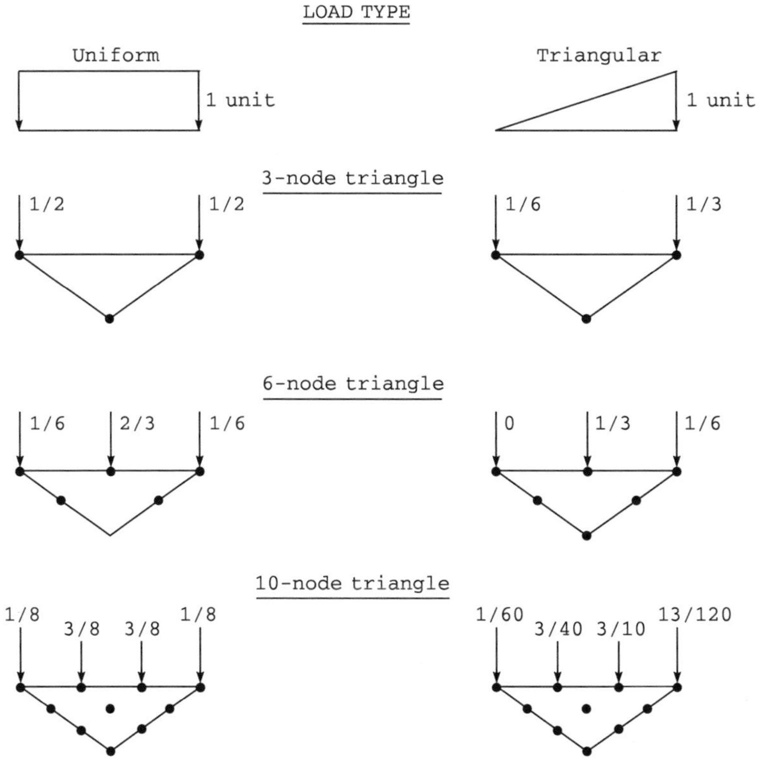

15-node triangle

4-node quadrilateral

8- and 9-node quadrilaterals

AXISYMMETRIC ELEMENTS (2-D)

Width of loaded face = 1 unit.
Loading over 1 radian.

LOAD TYPE

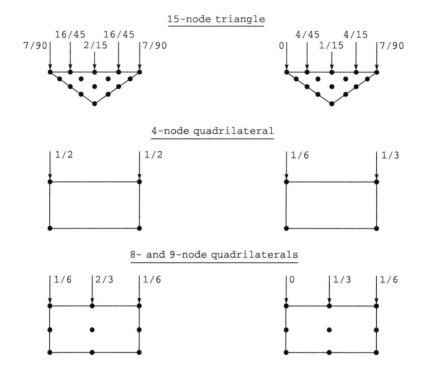

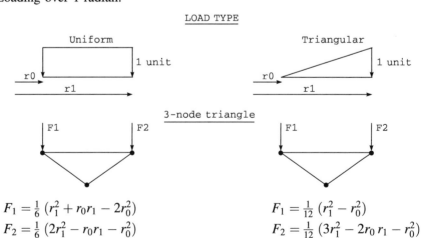

3-node triangle

$$F_1 = \tfrac{1}{6}\left(r_1^2 + r_0 r_1 - 2r_0^2\right)$$
$$F_2 = \tfrac{1}{6}\left(2r_1^2 - r_0 r_1 - r_0^2\right)$$

$$F_1 = \tfrac{1}{12}\left(r_1^2 - r_0^2\right)$$
$$F_2 = \tfrac{1}{12}\left(3r_1^2 - 2r_0 r_1 - r_0^2\right)$$

6-node triangle

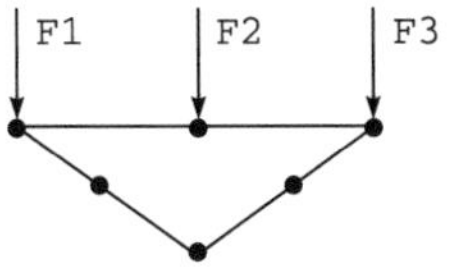

$$F_1 = \tfrac{1}{6}\,(r_0 r_1 - r_0^2)$$
$$F_2 = \tfrac{1}{3}\,(r_1^2 - r_0^2)$$
$$F_3 = \tfrac{1}{6}(r_1^2 - r_0 r_1)$$

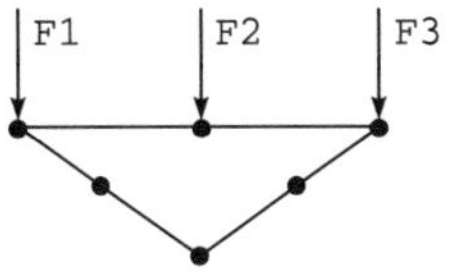

$$F_1 = -\tfrac{1}{60}\,(r_1^2 - 2r_0 r_1 + r_0^2)$$
$$F_2 = \tfrac{1}{15}\,(3r_1^2 - r_0 r_1 - 2r_0^2)$$
$$F_3 = \tfrac{1}{60}(9r_1^2 - 8r_0 r_1 - r_0^2)$$

10-node triangle

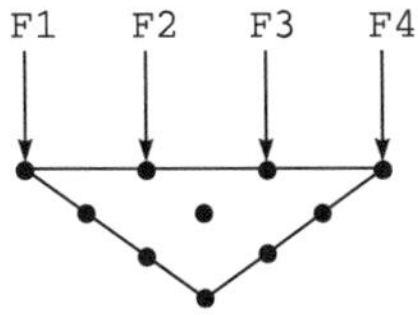

$$F_1 = \tfrac{1}{120}\,(2r_1^2 + 11r_0 r_1 - 13r_0^2)$$
$$F_2 = \tfrac{1}{40}\,(3r_1^2 + 9r_0 r_1 - 12r_0^2)$$
$$F_3 = \tfrac{1}{40}(12r_1^2 - 9r_0 r_1 - 3r_0^2)$$
$$F_4 = \tfrac{1}{120}\,(13r_1^2 - 11r_0 r_1 - 2r_0^2)$$

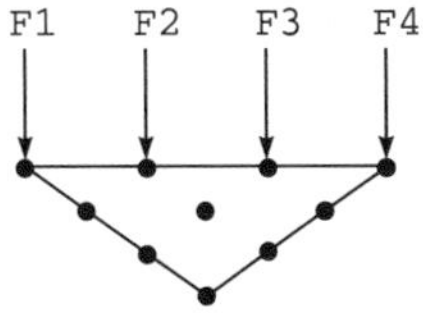

$$F_1 = \tfrac{1}{120}\,(r_1^2 - r_0^2)$$
$$F_2 = \tfrac{3}{40}\,(r_0 r_1 - r_0^2)$$
$$F_3 = \tfrac{3}{40}(3r_1^2 - 2r_0 r_1 - r_0^2)$$
$$F_4 = \tfrac{1}{120}\,(12r_1^2 - 11r_0 r_1 - r_0^2)$$

15-node triangle

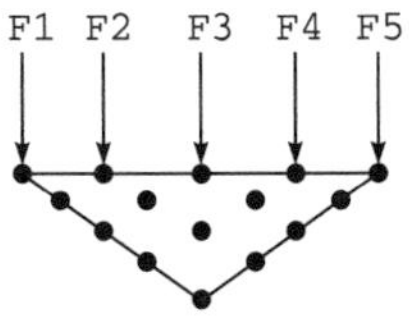

$$F_1 = \tfrac{7}{90}\,(r_0 r_1 - r_0^2)$$
$$F_2 = \tfrac{4}{45}\,(r_1^2 + 2r_0 r_1 - 3r_0^2)$$
$$F_3 = \tfrac{1}{15}(r_1^2 - r_0^2)$$
$$F_4 = \tfrac{4}{45}\,(3r_1^2 - 2r_0 r_1 - r_0^2)$$
$$F_5 = \tfrac{7}{90}\,(r_1^2 - r_0 r_1)$$

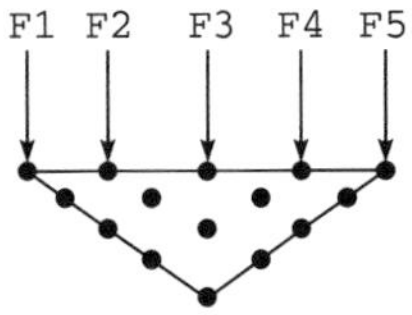

$$F_1 = -\tfrac{1}{252}\,(r_1^2 - 2r_0 r_1 + r_0^2)$$
$$F_2 = \tfrac{4}{315}\,(3r_1^2 + r_0 r_1 - 4r_0^2)$$
$$F_3 = \tfrac{1}{105}(r_1^2 + 5r_0 r_1 - 6r_0^2)$$
$$F_4 = \tfrac{4}{315}\,(17r_1^2 - 13r_0 r_1 - 4r_0^2)$$
$$F_5 = \tfrac{1}{1260}\,(93r_1^2 - 88r_0 r_1 - 5r_0^2)$$

<u>4-node quadrilateral</u>

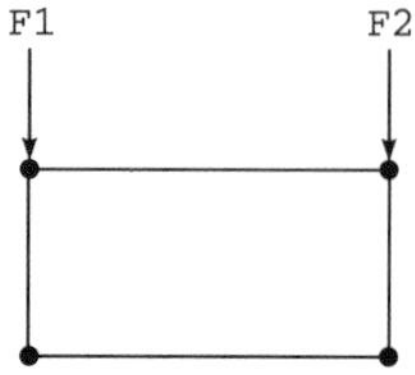

$$F_1 = \tfrac{1}{6}(r_1^2 + r_0 r_1 - 2r_0^2)$$
$$F_2 = \tfrac{1}{6}(2r_1^2 - r_0 r_1 - r_0^2)$$

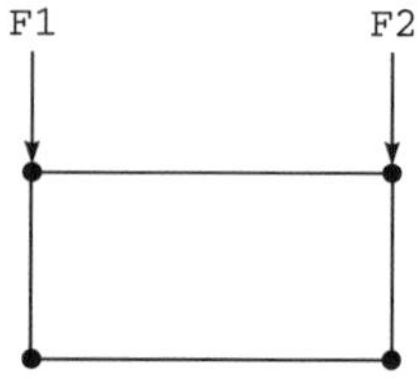

$$F_1 = \tfrac{1}{12}\,(r_1^2 - r_0^2)$$
$$F_2 = \tfrac{1}{12}\,(3r_1^2 - 2r_0 r_1 - r_0^2)$$

<u>8- and 9-node quadrilaterals</u>

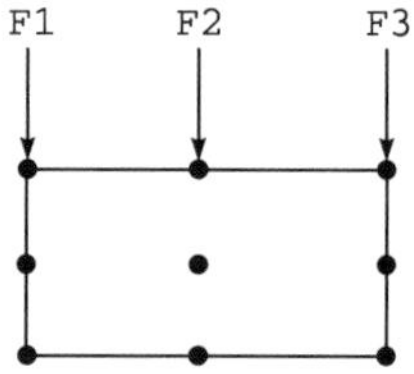

$$F_1 = \tfrac{1}{6}(r_0 r_1 - r_0^2)$$
$$F_2 = \tfrac{1}{3}(r_1^2 - r_0^2)$$
$$F_3 = \tfrac{1}{6}\,(r_1^2 - r_0 r_1)$$

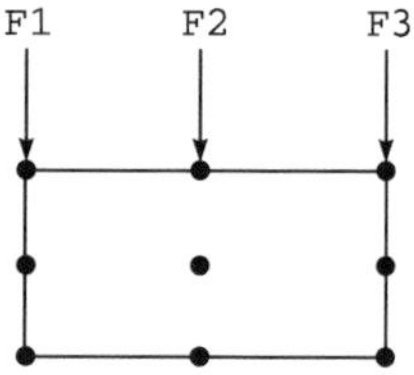

$$F_1 = -\tfrac{1}{60}\,(r_1^2 - 2r_0 r_1 + r_0^2)$$
$$F_2 = \tfrac{1}{15}\,(3r_1^2 - r_0 r_1 - 2r_0^2)$$
$$F_3 = \tfrac{1}{60}\,(9r_1^2 - 8r_0 r_1 - r_0^2)$$

THREE-DIMENSIONAL ELEMENTS (3-D)

Area of loaded face = 1 unit.
Unit stress applied.

<u>4-node tetrahedron</u>

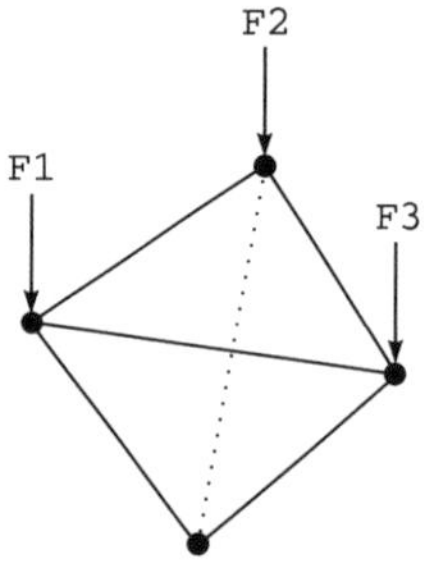

$$F_1 = F_2 = F_3 = \tfrac{1}{3}$$

<u>8-node hexahedron</u>

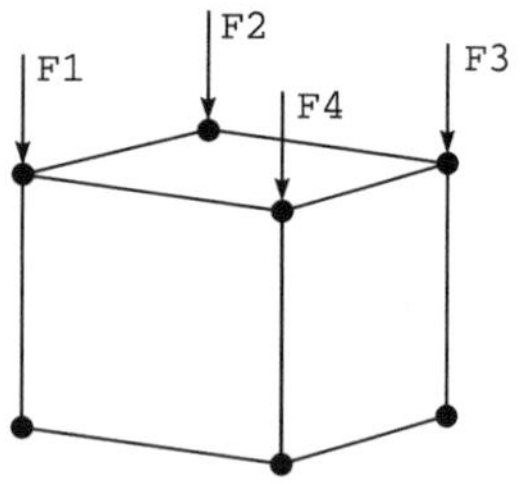

$$F_1 = F_2 = F_3 = F_4 = \tfrac{1}{4}$$

14-node hexahedron (type 5)

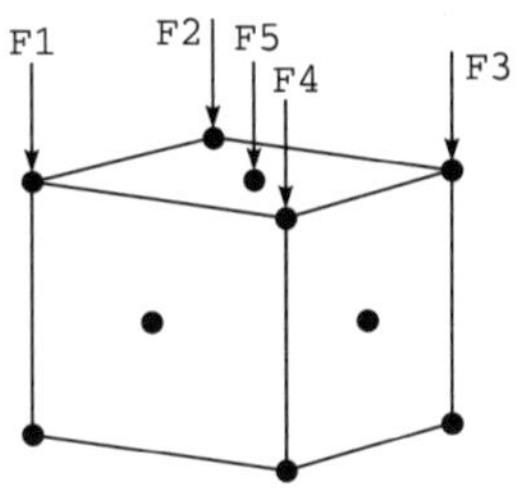

$F_1 = F_2 = F_3 = F_4 = \frac{1}{12}$
$F_5 = \frac{2}{3}$

20-node hexahedron (square)

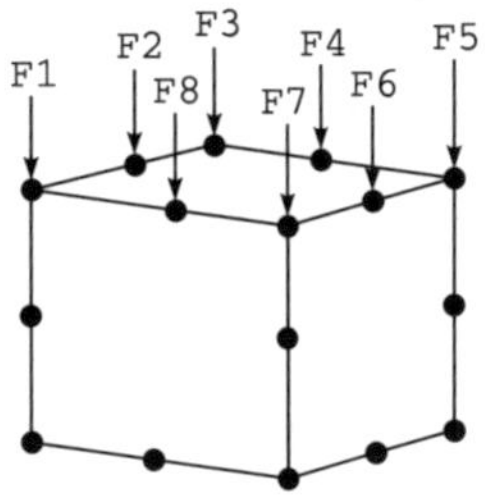

$F_1 = F_3 = F_5 = F_7 = -\frac{1}{12}$
$F_2 = F_4 = F_6 = F_8 = \frac{1}{3}$

Appendix 2
Plastic Stress–Strain Matrices and Plastic Potential Derivatives

A PLASTIC STRESS–STRAIN MATRICES

1 VON MISES

$$\mathbf{D}^{\mathbf{P}} = \frac{2G}{t^2}\begin{bmatrix} s_x^2 & s_x s_y & s_x \tau_{xy} & s_z s_x \\ & s_y^2 & s_y \tau_{xy} & s_y s_z \\ & & \tau_{xy}^2 & s_z \tau_{xy} \\ \text{symmetrical} & & & s_z^2 \end{bmatrix}$$

where

$G =$ shear modulus

$t =$ second deviatoric stress invariant (equation (6.3))

$s_x = (2\sigma_x - \sigma_y - \sigma_z)/3$, etc.

2 MOHR–COLOUMB

If not near a corner, that is $\sin\theta \leq 0.49$, then

$$\mathbf{D}^{\mathbf{P}} = \frac{E}{2(1+\nu)(1-2\nu)(1-2\nu+\sin\phi\,\sin\psi)}\mathbf{A}$$

where

$$\mathbf{A} = \begin{bmatrix} R_1C_1 & R_1C_2 & R_1C_3 & R_1C_4 \\ R_2C_1 & R_2C_2 & R_2C_3 & R_2C_4 \\ R_3C_1 & R_3C_2 & R_3C_3 & R_3C_4 \\ R_4C_1 & R_4C_2 & R_4C_3 & R_4C_4 \end{bmatrix}$$

and

$$C_1 = \sin\phi + k_1(1-2\nu)\sin\alpha$$
$$C_2 = \sin\phi - k_1(1-2\nu)\sin\alpha$$
$$C_3 = k_2(1-2\nu)\cos\alpha$$
$$C_4 = 2\nu\sin\phi$$
$$R_1 = \sin\psi + k_1(1-2\nu)\sin\alpha$$

$$
\begin{aligned}
R_2 &= \sin\psi - k_1(1-2\nu)\sin\alpha \\
R_3 &= k_2(1-2\nu)\cos\alpha \\
R_4 &= 2\nu\sin\psi \\
\alpha &= \arctan\left|\frac{\sigma_x - \sigma_y}{2\tau_{xy}}\right| \\
k_1 &= \begin{cases} 1 & \text{if } |\sigma_y| \geq |\sigma_x| \\ -1 & \text{if } |\sigma_x| > |\sigma_y| \end{cases} \\
k_2 &= \begin{cases} 1 & \text{if } \tau_{xy} \geq 0 \\ -1 & \text{if } \tau_{xy} < 0 \end{cases}
\end{aligned}
$$

If near a corner, that is $\sin\theta > 0.49$, then

$$\mathbf{D^P} = \frac{E}{(1+\nu)(1-2\nu)[K_\phi \sin\psi + C_\phi C_\psi t^2(1-2\nu)]}\mathbf{A}$$

where **A** is defined as before with

$$
\begin{aligned}
C_1 &= K_\phi + C_\phi\left[(1-\nu)s_x + \nu(s_y + s_z)\right] \\
C_2 &= K_\phi + C_\phi\left[(1-\nu)s_y + \nu(s_z + s_x)\right] \\
C_3 &= C_\phi(1-2\nu)\tau_{xy} \\
C_4 &= K_\phi + C_\phi\left[(1-\nu)s_z + \nu(s_x + s_y)\right] \\
R_1 &= K_\psi + C_\psi\left[(1-\nu)s_x + \nu(s_y + s_z)\right] \\
R_2 &= K_\psi + C_\psi\left[(1-\nu)s_y + \nu(s_z + s_x)\right] \\
R_3 &= C_\psi(1-2\nu)\tau_{xy} \\
R_4 &= K_\psi + C_\psi\left[(1-\nu)s_z + \nu(s_x + s_y)\right]
\end{aligned}
$$

where

$$
\begin{aligned}
K_\phi &= \frac{\sin\phi}{3}(1+\nu) \\
K_\psi &= \frac{\sin\psi}{3}(1+\nu) \\
C_\phi &= \frac{\sqrt{6}}{4t}\left(1 \pm \frac{\sin\phi}{3}\right) \\
C_\psi &= \frac{\sqrt{6}}{4t}\left(1 \pm \frac{\sin\psi}{3}\right)
\end{aligned}
$$

In the expressions for C_ϕ and C_ψ, the positive sign is valid if $\theta \simeq -30°$ and the negative sign is valid if $\theta \simeq 30°$.

B PLASTIC POTENTIAL DERIVATIVES

$$\frac{\partial Q}{\partial \boldsymbol{\sigma}} = (\mathrm{DQ1}\mathbf{M}^1 + \mathrm{DQ2}\mathbf{M}^2 + \mathrm{DQ3}\mathbf{M}^3)\boldsymbol{\sigma}$$

where

$$\mathbf{M}^1 = \frac{1}{3(\sigma_x + \sigma_y + \sigma_z)} \begin{bmatrix} 1 & 1 & 1 & 0 & 0 & 0 \\ & 1 & 1 & 0 & 0 & 0 \\ & & 1 & 0 & 0 & 0 \\ & & & 0 & 0 & 0 \\ & & & & 0 & 0 \\ \text{symmetrical} & & & & & 0 \end{bmatrix}$$

$$\mathbf{M}^2 = \frac{1}{3} \begin{bmatrix} 2 & -1 & -1 & 0 & 0 & 0 \\ & 2 & -1 & 0 & 0 & 0 \\ & & 2 & 0 & 0 & 0 \\ & & & 6 & 0 & 0 \\ & & & & 6 & 0 \\ \text{symmetrical} & & & & & 6 \end{bmatrix}$$

$$\mathbf{M}^3 = \frac{1}{3} \begin{bmatrix} s_x & s_z & s_y & \tau_{xy} & -2\tau_{yz} & \tau_{zx} \\ & s_y & s_x & \tau_{xy} & \tau_{yz} & -2\tau_{zx} \\ & & s_z & -2\tau_{xy} & \tau_{yz} & \tau_{zx} \\ & & & -3s_z & 3\tau_{zx} & 3\tau_{yz} \\ & & & & -3s_x & 3\tau_{xy} \\ \text{symmetrical} & & & & & -3s_y \end{bmatrix}$$

$$\boldsymbol{\sigma} = \begin{Bmatrix} \sigma_x \\ \sigma_y \\ \sigma_z \\ \tau_{xy} \\ \tau_{yz} \\ \tau_{zx} \end{Bmatrix}$$

1 VON MISES

$$\mathrm{DQ1} = 0$$

$$\mathrm{DQ2} = \sqrt{\frac{3}{2}}\frac{1}{t}$$

$$\mathrm{DQ3} = 0$$

2 MOHR–COLOUMB

$$\text{DQ1} = \sin \psi$$

$$\text{DQ2} = \frac{\cos \theta}{\sqrt{2}t} \left[1 + \tan \theta \tan 3\theta + \frac{\sin \psi}{\sqrt{3}} (\tan 3\theta - \tan \theta) \right]$$

$$\text{DQ3} = \frac{\sqrt{3} \sin \theta + \sin \psi \cos \theta}{t^2 \cos 3\theta}$$

where

t = second deviatoric stress invariant (equation (6.3))

θ = Lode angle (equation (6.3))

ψ = dilation angle

Appendix 3 Geometry Subroutines

Mesh generation constitutes a subject on its own and is not the focus of the present book. All mesh generation systems will provide as output nodal coordinates (G_COORD in the book terminology) and nodal connectivities (G_NUM). Several programs in the book, for example Program 5.9, illustrate how this information can be simply read as data by an analysis program.

As a supplement to mesh generation programs, this book uses some 20 small "geometry" subroutines for simple (usually rectangular or cuboidal) mesh arrangements. They are listed in full below so that users are encouraged to develop similar routines of their own. The routines all use a uniform nomenclature for parameters and have the essential job of delivering as output a real array called COORD, which holds the nodal coordinates of the element and an integer array NUM, which holds the node numbers.

The conversion from node numbers to "freedom numbers", held in integer array G, is carried out in nearly all programs by the subroutine NUM_TO_G.

The naming convention is that a geometry subroutine name contains the characters "GEOMETRY_". Following the underscore is the number of nodes and then "L" for "line", "T" for "triangle", "Q" for "quadrilateral" and "B" for "brick" or hexahedron. Then there is a coordinate, for example "X" denoting that node numbering is in the X direction. For bricks there have to be two coordinates, for example, "XZ", denoting the planes which are numbered first. If the mesh consists of elements of varying size, the character "V" is found.

Thus, for example, GEOMETRY_8QXV sets up a mesh of rectangular, eight-node, quadrilateral elements, numbering in the X direction. The elements can have varying sizes.

There are a few special purpose geometry subroutines concerned with analysis of slopes (Chapter 6), free-surface flow (Chapter 7) and construction processes (Chapter 6). These are not listed in the book but can of course be inspected on disc or the Internet.

```
module geometry_lib
contains
!----------------Node to freedom number conversion ---------------------------
 subroutine num_to_g(num,nf,g)
 !finds the g vector from num and nf
 implicit none
 integer,intent(in)::num(:),nf(:,:)  ; integer,intent(out):: g(:)
 integer::i,k,nod,nodof ; nod=ubound(num,1) ; nodof=ubound(nf,1)
  do i = 1 , nod
      k = i*nodof  ; g(k-nodof+1:k) = nf( : , num(i) )
  end do
 return
 end subroutine num_to_g
!------------------------------ Lines  -------------------------------------
subroutine geometry_2l(iel,ell,coord,num)
! node numbers, nodal coordinates and steering vectors for
! a line of (nonuniform) beam elements
 implicit none
  integer,intent(in)::iel; integer,intent(out)::num(:)
  real,intent(in)::ell; real,intent(out)::coord(:,:)
  num=(/iel,iel+1/)
  if(iel==1)then;coord(1,1)=.0; coord(2,1)=ell; else
    coord(1,1)=coord(2,1); coord(2,1)=coord(2,1)+ell; end if
 end subroutine geometry_2l
!---------------------------- Triangles -------------------------------------
subroutine geometry_3tx(iel,nxe,aa,bb,coord,num)
!      this subroutine forms the coordinates and node vector
!      for a rectangular mesh of uniform 3-node triangles
!      counting in the x-direction    ; local numbering clockwise
  implicit none
  real,intent(in):: aa,bb; integer,intent(in):: iel,nxe
  real,intent(out) :: coord(:,:); integer,intent(out):: num(:)
  integer::ip,iq,jel
      jel= (2*nxe+iel-1)/(2*nxe)
      if(iel/2*2==iel)then; iq=2*jel; else; iq=2*jel-1; end if
      ip= (iel-2*nxe*(jel-1)+1)/2
      if(mod(iq,2)/=0) then
       num(1)=(nxe+1)*(iq-1)/2+ip  ;   num(3)=(nxe+1)*(iq+1)/2+ip
       num(2)=num(1)+1             ;    coord(1,1)=(ip-1)*aa
       coord(1,2)=-(iq-1)/2*bb     ;    coord(3,1)=(ip-1)*aa
       coord(2,2)=coord(1,2)       ;    coord(2,1)=ip*aa
       coord(3,2)=-(iq+1)/2*bb
      else
       num(1)=(nxe+1)*iq/2+ip+1    ;    num(3)=(nxe+1)*(iq-2)/2+ip+1
       num(2)=num(1)-1              ;    coord(1,1)=ip*aa
       coord(1,2)=-iq/2*bb          ;    coord(3,1)=ip*aa
       coord(3,2)=-(iq-2)/2*bb      ;    coord(2,1)=(ip-1)*aa
       coord(2,2)=coord(1,2)
      end if
     return
   end subroutine geometry_3tx
subroutine geometry_6tx(iel,nxe,aa,bb,coord,num)
!      this subroutine forms the coordinates and nodal vector
!      for a rectangular mesh of uniform 6-node triangles
!      counting in the x-direction ; local numbering clockwise
  implicit none
  real ,intent(in):: aa,bb; integer,intent(in):: iel,nxe
  real,intent(out) :: coord(:,:); integer,intent(out)::  num(:)
  integer::ip,iq,jel,i
      jel= (2*nxe+iel-1)/(2*nxe)
      if(iel/2*2==iel)then; iq=2*jel; else; iq=2*jel-1; end if
      ip= (iel-2*nxe*(jel-1)+1)/2
      if(mod(iq,2)/=0) then
       num(1)=(iq-1)*(2*nxe+1)+2*ip-1     ;    num(2)=num(1)+1
```

```
        num(3)=num(1)+2                      ;    num(4)=num(2)+1
        num(6)=(iq-1)*(2*nxe+1)+2*nxe+2*ip ;    num(5)=(iq+1)*(2*nxe+1)+2*ip-1
        coord(1,1)=(ip-1)*aa                 ;    coord(1,2)=-(iq-1)/2*bb
        coord(5,1)=(ip-1)*aa                 ;    coord(5,2)=-(iq+1)/2*bb
        coord(3,1)=ip*aa                     ;    coord(3,2)=coord(1,2)
       else
        num(1)=iq*(2*nxe+1)+2*ip+1          ; num(6)=(iq-2)*(2*nxe+1)+2*nxe+2*ip+2
        num(5)=(iq-2)*(2*nxe+1)+2*ip+1       ; num(4)=num(2)-1
        num(3)=num(1)-2                      ; num(2)=num(1)-1
        coord(1,1)=ip*aa                     ; coord(1,2)=-iq/2*bb
        coord(5,1)=ip*aa                     ; coord(5,2)=-(iq-2)/2*bb
        coord(3,1)=(ip-1)*aa                 ; coord(3,2)=coord(1,2)
       end if
       do  i=1,2
       coord(2,i)=.5*(coord(1,i)+coord(3,i))
       coord(4,i)=.5*(coord(3,i)+coord(5,i))
       coord(6,i)=.5*(coord(5,i)+coord(1,i))
       end do
       return
    end subroutine geometry_6tx
   subroutine geometry_15tyv(iel,nye,width,depth,coord,num)
!       this subroutine forms the coordinates and node vector
!       for a rectangular mesh of nonuniform 15-node triangles
!       counting in the y-direction ; local numbering clockwise
 implicit none
 real,intent(in) :: width(:),depth(:) ; real,intent(out) :: coord(:,:)
 integer, intent(in) ::  iel,nye; integer,intent(out)::num(:)
 integer::ip,iq,jel,i , fac1,fac2
   jel = (iel - 1)/nye; ip= jel+1; iq=iel-nye*jel
    if(mod(iq,2)/=0) then
      fac1=4*(2*nye+1)*(ip-1)+2*iq-1 ; num(1)=fac1;  num(12)=fac1+1
      num(11)=fac1+2 ;  num(10)=fac1+3 ; num(9)=fac1+4 ; num(8)=fac1+2*nye+4
      num(7)=fac1+4*nye+4 ;  num(6)=fac1+6*nye+4 ;  num(5)=fac1+8*nye+4
      num(4)=fac1+6*nye+3 ;  num(3)=fac1+4*nye+2 ;  num(2)=fac1+2*nye+1
      num(13)=fac1+2*nye+2 ;  num(15)=fac1+2*nye+3  ;  num(14)=fac1+4*nye+3
      coord(1,1)=width(ip) ;  coord(1,2)=depth((iq+1)/2)
      coord(9,1)=width(ip)   ;  coord(9,2)=depth((iq+3)/2)
      coord(5,1)=width(ip+1)   ;  coord(5,2)=depth((iq+1)/2)
    else
      fac2=4*(2*nye+1)*(ip-1)+2*iq+8*nye+5 ;  num(1)=fac2 ;  num(12)=fac2-1
      num(11)=fac2-2 ;  num(10)=fac2-3 ;  num(9)=fac2-4; num(8)=fac2-2*nye-4
      num(7)=fac2-4*nye-4 ; num(6)=fac2-6*nye-4  ;  num(5)=fac2-8*nye-4
      num(4)=fac2-6*nye-3 ; num(3)=fac2-4*nye-2 ; num(2)=fac2-2*nye-1
      num(13)=fac2-2*nye-2  ; num(15)=fac2-2*nye-3 ; num(14)=fac2-4*nye-3
      coord(1,1)=width(ip+1) ;  coord(1,2)=depth((iq+2)/2)
      coord(9,1)=width(ip+1) ;  coord(9,2)=depth(iq/2)
      coord(5,1)=width(ip)   ;   coord(5,2)=depth((iq+2)/2)
    end if
    do  i=1,2
      coord(3,i)=.5*(coord(1,i)+coord(5,i))
      coord(7,i)=.5*(coord(5,i)+coord(9,i))
      coord(11,i)=.5*(coord(9,i)+coord(1,i))
      coord(2,i)=.5*(coord(1,i)+coord(3,i))
      coord(4,i)=.5*(coord(3,i)+coord(5,i))
      coord(6,i)=.5*(coord(5,i)+coord(7,i))
      coord(8,i)=.5*(coord(7,i)+coord(9,i))
      coord(10,i)=.5*(coord(9,i)+coord(11,i))
      coord(12,i)=.5*(coord(11,i)+coord(1,i))
      coord(15,i)=.5*(coord(7,i)+coord(11,i))
      coord(14,i)=.5*(coord(3,i)+coord(7,i))
      coord(13,i)=.5*(coord(2,i)+coord(15,i))
    end do
  return
 end subroutine geometry_15tyv
```

```
!---------------------- Quadrilaterals -----------------------------------
subroutine geometry_4qx(iel,nxe,aa,bb,coord,num)
! coordinates and nodal vectors for equal four node quad
! elements, numbering in x
implicit none
integer,intent(in)::iel,nxe; real,intent(in)::aa,bb
real,intent(out)::coord(:,:); integer,intent(out)::num(:)
integer :: ip,iq    ; iq=(iel-1)/nxe+1; ip=iel-(iq-1)*nxe
   num=(/iq*(nxe+1)+ip,(iq-1)*(nxe+1)+ip,   &
         (iq-1)*(nxe+1)+ip+1, iq*(nxe+1)+ip+1/)
   coord(1:2,1)=(ip-1)*aa; coord(3:4,1)=ip*aa
   coord(1:4:3,2)=-iq*bb; coord(2:3,2)=-(iq-1)*bb
 return
end subroutine geometry_4qx
   subroutine geometry_4qy(iel,nye,aa,bb,coord,num)
   ! rectangles of equal 4-node quads numbered in y
   implicit none
   integer,intent(in)::iel,nye; real,intent(in)::aa,bb
   real,intent(out)::coord(:,:);integer,intent(out)::num(:)
       num=(/iel+(iel-1)/nye+1,iel+(iel-1)/nye,iel+(iel-1)/nye+nye+1,  &
             iel+(iel-1)/nye+nye+2/)
       coord(1:2,1)= aa*((iel-1)/nye); coord(3:4,1)=aa*((iel-1)/nye+1)
       coord(1:4:3,2)=-(iel-((iel-1)/nye)*nye)*bb; coord(2:3,2)=coord(1,2)+bb
     return
   end subroutine geometry_4qy
subroutine geometry_4qyv(iel,nye,width,depth,coord,num)
!  coordinates and steering vector for a variable rectangular
!  mesh of 4-node quad elements numbering in the y-direction
  implicit none
  real,intent(in)::width(:),depth(:); integer,intent(in)::iel,nye
  real,intent(out)::coord(:,:); integer,intent(out):: num(:)
  integer:: ip,iq; ip=(iel-1)/nye+1; iq=iel-(ip-1)*nye
  num(1)=(ip-1)*(nye+1)+iq+1; num(2)=num(1)-1
  num(3)=ip*(nye+1)+iq;num(4)= num(3) + 1
  coord(1:2,1)=width(ip); coord(3:4,1)=width(ip+1)
  coord(1,2)=depth(iq+1); coord(2:3,2)=depth(iq); coord(4,2)=coord(1,2)
 return
end subroutine geometry_4qyv
subroutine geometry_8qx(iel,nxe,aa,bb,coord,num)
! coordinates and steering vector for a rectangular mesh of
! equal  8-node  elements  numbering in x
implicit none
 real,intent(out)::coord(:,:); integer,intent(out)::num(:)
 integer,intent(in)::iel,nxe; real,intent(in)::aa,bb
 integer:: ip,iq ; iq=(iel-1)/nxe+1; ip=iel-(iq-1)*nxe
 num(1)=iq*(3*nxe+2)+2*ip-1; num(2)=iq*(3*nxe+2)+ip-nxe-1
 num(3)=(iq-1)*(3*nxe+2)+2*ip-1; num(4)=num(3)+1
 num(5)=num(4)+1; num(6)=num(2)+1; num(7)=num(1)+2; num(8)=num(1)+1
 coord(1:3,1)=aa*(ip-1); coord(5:7,1)=aa*ip
 coord(4,1)=.5*(coord(3,1)+coord(5,1))
 coord(8,1)=.5*(coord(7,1)+coord(1,1))
 coord(1,2)=-bb*iq; coord(7:8,2)=-bb*iq
 coord(3:5,2)=-bb*(iq-1); coord(2,2)=.5*(coord(1,2)+coord(3,2))
 coord(6,2)=.5*(coord(5,2)+coord(7,2))
return
end subroutine geometry_8qx
subroutine geometry_8qy(iel,nye,aa,bb,coord,num)
!  coordinates and steering vector for a constant rectangular
!  mesh of 8-node quad elements numbering in the y-direction
  implicit none
  real,intent(in):: aa,bb ; integer,intent(in):: iel,nye
  real,intent(out)::coord(:,:); integer,intent(out):: num(:)
  integer:: ip,iq; ip=(iel-1)/nye+1; iq=iel-(ip-1)*nye
  num(1)=(ip-1)*(3*nye+2)+2*iq+1; num(2)=num(1)-1; num(3)=num(1)-2
```

```
   num(4)=(ip-1)*(3*nye+2)+2*nye+iq+1;num(5)=ip*(3*nye+2)+2*iq-1
   num(6)=num(5)+1; num(7)=num(5)+2; num(8)=num(4)+1
   coord(1:3,1)=(ip-1)*aa; coord(5:7,1)=ip*aa
   coord(4,1)=.5*(coord(3,1)+coord(5,1))
   coord(8,1)=.5*(coord(7,1)+coord(1,1))
   coord(1,2)=-iq*bb; coord(7:8,2)=-iq*bb; coord(3:5,2)=-(iq-1)*bb
   coord(2,2)=.5*(coord(1,2)+coord(3,2))
   coord(6,2)=.5*(coord(5,2)+coord(7,2))
  return
end subroutine geometry_8qy
subroutine geometry_8qxv(iel,nxe,width,depth,coord,num)
! nodal coordinates and node vector for a variable mesh of
! 8-node quadrilaterals numbering in the x-direction
implicit none
   integer,intent(in)::iel,nxe;real,intent(in)::width(:),depth(:)
   real,intent(out)::coord(:,:); integer,intent(out)::num(:)
   integer::ip,iq; iq=(iel-1)/nxe+1; ip=iel-(iq-1)*nxe
   num(1)=iq*(3*nxe+2)+2*ip-1; num(2)=iq*(3*nxe+2)+ip-nxe-1
   num(3)=(iq-1)*(3*nxe+2)+2*ip-1; num(4)=num(3)+1;num(5)=num(4)+1
   num(6)=num(2)+1; num(7)=num(1)+2; num(8)=num(1)+1
   coord(1:3,1)=width(ip); coord(5:7,1)=width(ip+1)
   coord(4,1)=.5*(coord(3,1)+coord(5,1));coord(8,1)=.5*(coord(7,1)+coord(1,1))
   coord(1,2)=depth(iq+1); coord(7:8,2)=depth(iq+1); coord(3:5,2)=depth(iq)
   coord(2,2)=.5*(coord(1,2)+coord(3,2));coord(6,2)=.5*(coord(5,2)+coord(7,2))
  return
end subroutine geometry_8qxv
subroutine geometry_8qyv(iel,nye,width,depth,coord,num)
!  coordinates and steering vector for a variable rectangular
!  mesh of 8-node quad elements numbering in the y-direction
  implicit none
  real,intent(in)::width(:),depth(:); integer,intent(in)::iel,nye
  real,intent(out)::coord(:,:); integer,intent(out)::num(:)
  integer::ip,iq; ip=(iel-1)/nye+1; iq=iel-(ip-1)*nye
  num(1)=(ip-1)*(3*nye+2)+2*iq+1; num(2)=num(1)-1; num(3)=num(1)-2
  num(4)=(ip-1)*(3*nye+2)+2*nye+iq+1;num(5)=ip*(3*nye+2)+2*iq-1
  num(6)=num(5)+1; num(7)=num(5)+2; num(8)=num(4)+1
  coord(1:3,1)=width(ip); coord(5:7,1)=width(ip+1)
  coord(4,1)=.5*(coord(3,1)+coord(5,1))
  coord(8,1)=.5*(coord(7,1)+coord(1,1))
  coord(1,2)=depth(iq+1); coord(7:8,2)=depth(iq+1); coord(3:5,2)=depth(iq)
  coord(2,2)=.5*(coord(1,2)+coord(3,2))
  coord(6,2)=.5*(coord(5,2)+coord(7,2))
 return
end subroutine geometry_8qyv
 subroutine geometry_9qx(iel,nxe,aa,bb,coord,num)
!      this subroutine forms the coordinates and steering vector
!      for equal 9-node Lagrangian quads counting in x-direction
 implicit none
 real,intent(out)::coord(:,:); integer,intent(out)::num(:)
 integer,intent(in)::iel,nxe; real,intent(in)::aa,bb
 integer:: ip,iq ;iq=(iel-1)/nxe+1;ip=iel-(iq-1)*nxe
   num(1)=iq*(4*nxe+2)+2*ip-1 ; num(2)=iq*(4*nxe+2)+2*ip-nxe-4
   num(3)= (iq-1)*(4*nxe+2)+2*ip-1 ;    num(4)=num(3)+1
   num(5)=num(4)+1; num(6)=num(2)+2 ;  num(7)=num(1)+2
   num(8)=num(1)+1        ;    num(9)=num(2)+1
   coord(1,1)=(ip-1)*aa  ;  coord(3,1)=(ip-1)*aa   ; coord(5,1)=ip*aa
   coord(7,1)=ip*aa ;    coord(1,2)=-iq*bb ;  coord(3,2)=-(iq-1)*bb
   coord(5,2)=-(iq-1)*bb     ;   coord(7,2)=-iq*bb
   coord(2,1)=.5*(coord(1,1)+coord(3,1)); coord(2,2)=.5*(coord(1,2)+coord(3,2))
   coord(4,1)=.5*(coord(3,1)+coord(5,1)); coord(4,2)=.5*(coord(3,2)+coord(5,2))
   coord(6,1)=.5*(coord(5,1)+coord(7,1)); coord(6,2)=.5*(coord(5,2)+coord(7,2))
   coord(8,1)=.5*(coord(1,1)+coord(7,1)); coord(8,2)=.5*(coord(1,2)+coord(7,2))
   coord(9,1)=.5*(coord(2,1)+coord(6,1)); coord(9,2)=.5*(coord(4,2)+coord(8,2))
  return
```

```
 end subroutine geometry_9qx
!-----------------------Hexahedra "Bricks" ------------------------------------
 subroutine geometry_8bxz(iel,nxe,nze,aa,bb,cc,coord,num)
!      this subroutine forms the coordinates and nodal vector
!      for boxes of 8-node brick elements counting x-z planes in y-direction
 implicit none
   integer,intent(in)::iel,nxe,nze;integer,intent(out)::num(:)
   real,intent(in)::aa,bb,cc; real,intent(out)::coord(:,:)
   integer::ip,iq,is,iplane
   iq=(iel-1)/(nxe*nze)+1 ; iplane = iel -(iq-1)*nxe*nze
   is=(iplane-1)/nxe+1; ip = iplane-(is-1)*nxe
   num(1)=(iq-1)*(nxe+1)*(nze+1)+is*(nxe+1)+ip ;  num(2)=num(1)-nxe-1
   num(3)=num(2)+1 ;    num(4)=num(1)+1 ; num(5)=num(1)+(nxe+1)*(nze+1)
   num(6)=num(5)-nxe-1 ;    num(7)=num(6)+1     ;    num(8)=num(5)+1
   coord(1,1)=(ip-1)*aa ; coord(2,1)=(ip-1)*aa ;  coord(5,1)=(ip-1)*aa
   coord(6,1)=(ip-1)*aa ; coord(3,1)=ip*aa ;   coord(4,1)=ip*aa
   coord(7,1)=ip*aa   ;   coord(8,1)=ip*aa
   coord(1,2)=(iq-1)*bb  ;   coord(2,2)=(iq-1)*bb ; coord(3,2)=(iq-1)*bb
   coord(4,2)=(iq-1)*bb  ;   coord(5,2)=iq*bb  ;     coord(6,2)=iq*bb
   coord(7,2)=iq*bb ;     coord(8,2)=iq*bb ;    coord(1,3)=-is*cc
   coord(4,3)=-is*cc  ;   coord(5,3)=-is*cc ;    coord(8,3)=-is*cc
   coord(2,3)=-(is-1)*cc ; coord(3,3)=-(is-1)*cc  ;  coord(6,3)=-(is-1)*cc
   coord(7,3)=-(is-1)*cc
  return
 end subroutine geometry_8bxz
subroutine geometry_20bxz(iel,nxe,nze,aa,bb,cc,coord,num)
! nodal vector and nodal coordinates for boxes of 20-node
! bricks counting x-z planes in the y-direction
implicit none
  integer,intent(in)::iel,nxe,nze; real,intent(in)::aa,bb,cc
  real,intent(out)::coord(:,:); integer,intent(out)::num(:)
  integer::fac1,fac2,ip,iq,is,iplane
  iq = (iel-1)/(nxe*nze)+1; iplane = iel-(iq-1)*nxe*nze
  is = (iplane-1)/nxe+1 ; ip = iplane-(is-1)*nxe
  fac1=((2*nxe+1)*(nze+1)+(2*nze+1)*(nxe+1))*(iq-1)
  fac2=((2*nxe+1)*(nze+1)+(2*nze+1)*(nxe+1))*iq
  num(1)=fac1+(3*nxe+2)*is+2*ip-1
  num(2)=fac1+(3*nxe+2)*is-nxe+ip-1; num(3)=num(1)-3*nxe-2
  num(4)=num(3)+1; num(5)=num(4)+1; num(6)=num(2)+1
  num(7)=num(1)+2; num(8)=num(1)+1
  num(9)=fac2-(nxe+1)*(nze+1)+(nxe+1)*is+ip
  num(10)=num(9)-nxe-1; num(11)=num(10)+1; num(12)=num(9)+1
  num(13)=fac2+(3*nxe+2)*is+2*ip-1
  num(14)=fac2+(3*nxe+2)*is-nxe+ip-1
  num(15)=num(13)-3*nxe-2; num(16)=num(15)+1; num(17)=num(16)+1
  num(18)=num(14)+1; num(19)=num(13)+2; num(20)=num(13)+1
  coord(1:3,1)=(ip-1)*aa; coord(9:10,1)=(ip-1)*aa; coord(13:15,1)=(ip-1)*aa
  coord(5:7,1)=ip*aa; coord(11:12,1)=ip*aa; coord(17:19,1)=ip*aa
  coord(4,1)=.5*(coord(3,1)+coord(5,1));coord(8,1)=.5*(coord(1,1)+coord(7,1))
  coord(16,1)=.5*(coord(15,1)+coord(17,1))
  coord(20,1)=.5*(coord(13,1)+coord(19,1))
  coord(1:8,2)=(iq-1)*bb; coord(13:20,2)=iq*bb
  coord(9,2)=.5*(coord(1,2)+coord(13,2))
  coord(10,2)=.5*(coord(3,2)+coord(15,2))
  coord(11,2)=.5*(coord(5,2)+coord(17,2))
  coord(12,2)=.5*(coord(7,2)+coord(19,2))
  coord(1,3)=-is*cc; coord(7:9,3)=-is*cc; coord(12:13,3)=-is*cc
  coord(19:20,3)=-is*cc; coord(3:5,3)=-(is-1)*cc
  coord(10:11,3)=-(is-1)*cc; coord(15:17,3)=-(is-1)*cc
  coord(2,3)=.5*(coord(1,3)+coord(3,3))
  coord(6,3)=.5*(coord(5,3)+coord(7,3))
  coord(14,3)=.5*(coord(13,3)+coord(15,3))
  coord(18,3)=.5*(coord(17,3)+coord(19,3))

end subroutine geometry_20bxz
end module geometry_lib
```

Appendix 4
Alphabetic List of Building Block Subroutines

All subroutines used in programs, with the exception of the "geometry" routines described in Appendix 3, are listed below. Subroutines marked with an asterisk are so-called "black box" routines, mostly concerned with standard linear algebra. Although the coding of these routines is available to users, it is not believed to be of direct interest, and so is not reproduced in this book. However, the remaining subroutines, not asterisked, are concerned with specific operations in finite element analysis—description of shape functions, etc.—and so the code is helpful to users in understanding the programs in the book and in creating new ones. Therefore, these routines are listed in full in Appendix 5. In the parameter lists, underlined parameters are those returned by the routine as output. The other parameters are input. Subroutines in parentheses are normally used in conjunction with the routines they follow.

Subroutine name	Parameters	Action of subroutine
ANALY4*	KMA, COORD,E,V	Forms four-node quadrilateral elastic stiffness matrix KMA for an element with nodal coordinates COORD, Young's modulus E and Poisson's ratio V.
BACSUB (BANRED)	BK, LOADS	Performs complete Gaussian forward and backward substitution on matrix BK factorised by BANRED. Solution LOADS overwrites RHS.
BANDRED* (BISECT)	A,D,E,E2	Transforms real symmetric band matrix A to tridiagonal form by Jacobi rotations. The transformed diagonal is in D and the off-diagonal in E. Vector E2 is working space.
BANDWIDTH	G	Function returning the maximum bandwidth for element steering vectors G.
BANMUL	KB,LOADS,ANS	Matrix by vector multiplication of symmetric band matrix KB by vector LOADS to yield result ANS.

BANMUL_K*	KB,LOADS,ANS	As BANMUL but start address of LOADS and ANS is 1.
BANRED (BACSUB)	BK,N	Gaussian factorisation of a real symmetric band matrix BK, stored as a lower rectangle in columns. The order of the matrix is N.
BANTMUL	KB,LOADS,ANS	Matrix by vector multiplication of unsymmetrical band matrix KB by vector LOADS to yield result ANS.
BEAM_KM	KM,EI,ELL	Forms stiffness matrix KM of a beam element of length ELL and stiffness EI. Bending terms only.
BEAM_KP	KP,COORD,PAX	Forms "stability" matrix KP of a beam element with coordinates COORD due to axial load PAX.
BEAM_MM	MM,FS,ELL	Forms "mass" matrix MM of a beam element of length ELL. Factor FS can be mass, foundation stiffness, etc.
BEE4	COORD,POINTS,I,DET, BEE	Analytical formation of BEE matrix of a four-node quadrilateral with nodal coordinates COORD. Jacobian determinant DET at Ith Gauss point sampled at POINTS.
BEEMAT	BEE,DERIV	Returns BEE matrix for shape function derivatives DERIV.
BISECT* (BANDRED)	D,E,ACHAPS,IFAIL	Returns eigenvalues of tridiagonal matrix whose leading diagonal is in D and off-diagonal in E. Tolerance ACHAPS, failure flag IFAIL.
BMATAXI	BEE,RADIUS,COORD, DERIV,FUN	Strain–displacement matrix BEE for axisymmetric strain. Element shape functions FUN and derivatives DERIV. Nodal coordinates COORD and radius of current Gauss point is RADIUS.
BMAT_NONAXI	BEE,RADIUS,COORD, DERIV,FUN,IFLAG,ITH	Strain–displacement matrix BEE for axisymmetric elements with non-axisymmetric load. Shape functions FUN and derivatives DERIV. Nodal coordinates COORD and radius of current Gauss point RADIUS. For Ith

		harmonic, IFLAG is +1 for symmetry, −1 for antisymmetry.
CHECON	LOADS,OLDLDS,TOL, CONVERGED	Logical variable CONVERGED set to .FALSE. if relative change in, not underscore vectors LOADS and OLDLDS is greater than TOL and updates OLDLDS to LOADS.
CHOBAC (CHOLIN)	KB, LOADS	Performs complete forward and backward substitution on matrix KB factorised by Cholin. Solution LOADS overwrites RHS.
CHOBK1* (CHOLIN)	KB,LOADS	Performs forward substitution on matrix KB factorised by CHOLIN. Solution LOADS overwrites RHS.
CHOBK2* (CHOLIN)	KB, LOADS	Performs backward substitution on matrix KB produced by CHOLIN and CHOBAC. Solution LOADS overwrites RHS.
CHOLIN	KB	Performs Choleski factorisation of a real symmetric band matrix KB, stored as a lower rectangle by rows.
DEEMAT	DEE,E,V	Returns elastic DEE matrix for Young's modulus E and Poisson's ratio V.
DETERMINANT	JAC	Function returning the determinant of the 1 × 1, 2 × 2 or 3 × 3 Jacobian matrix JAC.
ECMAT	ECM,FUN,NDOF,NODOF	Element consistent mass matrix ECM for element with NODOF degrees of freedom per node and NDOF total freedoms. Shape functions are FUN.
FKDIAG	KDIAG,G	For a skyline system, returns maximum bandwidth for each row for steering vectors G. Row lengths are in array KDIAG.
FKDIAG_K*	KDIAG,G,G_MIN	As FKDIAG but block option with minimum G value G_MIN.
FMACAT	VMFL,TEMP,ACAT	Forms ACAT matrix (see Chapter 6).
FMBEAM	DER2,FUN,POINTS,I,	Returns beam element shape

	ELL	functions FUN and second derivatives DER2 (local coordinates) for Ith sampling point in POINTS. Element length ELL.
FMKDKE	KM,KP,C,KE,KD, THETA	Builds coupled matrices KE and KD from elastic stiffness matrix KP. Coupling matrix is C and time integration scalar THETA.
FMRMAT	VMFL,DSBAR,DLAM,DEE, TEMP, RMAT	Forms *t*-matrix RMAT (see Chapter 6).
FORMAA	FLOW,RMAT,MODEE	Forms modification to DEE matrix (see Chapter 6).
FORMBAND_K*	KB,KM,G,G_MIN	Builds symmetrical element matrices KM into lower rectangular global band matrix KB for elements with steering vector G. Block option with minimum G value G_MIN.
FORMBLOCK_K*	BIGK,KM,G,G_MIN	Builds element matrices KM into square global matrix BIGK for elements with steering vector G. Block option with minimum G value G_MIN.
FORMKB	KB,KM,G	Builds symmetrical element matrices KM into lower rectangular global band matrix KB for elements with steering vector G.
FORMKU	KU,KM,G	Builds symmetrical element matrices KM into upper rectangular global band matrix KU for elements with steering vector G.
FORMKV	BK,KM,G,N	Builds symmetrical element matrices KM into lower rectangular global band matrix BK (stored as a vector by columns). Order of matrix BK is N and elements have steering vectors G.
FORMLUMP	DIAG,EMM,G	Forms global mass matrix as a diagonal DIAG from consistent element mass matrices EMM and steering vectors G.
FORMM	STRESS,M1,M2,M3	M1,M2,M3 matrices (see Appendix 2B) from stress components in array STRESS.

FORMNF	<u>NF</u>	Returns node freedom array NF from input array of 1s and 0s.
FORMTB	<u>KB</u>,KM,G	Assembles element matrices KM into an (unsymmetrical) band matrix KB. Element steering vectors are G.
FORMUPV	<u>KE</u>,C11,C12,C21, C23,C32	Builds unsymmetric element matrix KE for use in *u–v–p* version of Navier–Stokes equations. Constituent matrices C11,C12, C21,C23,C32.
FSPARV	<u>BK</u>,KM,G,KDIAG	Builds symmetrical element matrices KM into lower triangular global matrix BK using skyline storage controlled by vector KDIAG. Element steering vectors are G.
FSPARV_K*	<u>BK</u>,KM,G,G_MIN,KDIAG	As FSPARV, but block option with minimum G value G_MIN.
GAUSS_BAND* (SOLVE_BAND)	<u>PB</u>,WORK	Gaussian factorisation of an unsymmetric band matrix PB. Work array WORK.
GLOB_TO_ AXIAL	<u>AXIAL</u>,GLOBAL,COORD	Transforms global end reactions GLOBAL for an element with coordinates COORD, to scalar AXIAL force.
GLOB_TO_LOC	<u>LOCAL</u>,GLOBAL,GAMMA COORD	As LOC_TO_GLOB but in reverse.
HINGE	COORD,HOLDR, ACTION,<u>REACT</u>, PROP,IEL,ETYPE, GAMMA	End forces and moments REACT: (see Chapter 4).
INTERP	K,DT,RT,RL,<u>AL</u>,NTP	Forms load/time functions by interpolation. See Program 11.0.
INVAR	STRESS,<u>SIGM</u>,<u>DSBAR</u>, <u>THETA</u>	First invariant SIGM, second invariant DSBAR and Lode angle THETA for a stress tensor STRESS.
INVERT	<u>MATRIX</u>	Inverts a small square matrix MATRIX on to itself.
KVDET*	KV,N,IW,<u>DET</u>,<u>KSC</u>	Computes determinant DET of symmetrical band matrix KV stored (upper triangle) as a vector. The order of KV is N and the half bandwidth IW. Number of negative pivots KSC.

LANCZ1* (LANCZ2)	N,EL,ER,ACC,LEIG,LX, LALFA,LP,ITAPE, IFLAG,U,V,EIG,JEIG, NEIG,X,DEL,NU,ALFA, BETA	Lanczos method for eigenvalues—see Chapter 10.
LANCZ2* (LANCZ1)	N,LALFA,LP,ITOPE, EIG, JEIG,NEIG,ALFA, BETA, LZ,JFLAG,Y,W,Z	Lanczos method for eigenvectors—see Chapter 10.
LINMUL	BK,DISPS,LOADS	Matrix by vector multiplication for a symmetric lower rectangular band matrix BK, stored as a vector, by DISPS to yield LOADS.
LINMUL_SKY	BP,DISPS,LOADS,KDIAG	Matrix by vector multiplication for a symmetric lower band matrix BP, skyline storage controlled by vector KDIAG. Multiply by DISPS to yield LOADS.
LINMUL_SKY_K*	BP,DISPS,LOADS, KDIAG	As LINMUL_SKY but start address of DISPS and LOADS is 1.
LOC_TO_GLOB	LOCAL,GLOBAL,GAMMA, COORD	Transformation of local end reactions LOCAL to GLOBAL. Element coordinates COORD and geometry property GAMMA.
MOCOPL	PHI,PSI,E,V, STRESS,PL	Plastic stress–strain matrix PL from stresses STRESS, angle of friction PHI, dilation PSI and elastic stiffness E, V for a Mohr–Coulomb material.
MOCOUF	PHI,C,SIGM,DSBAR, THETA,F	Mohr–Coulomb yield function F from invariants SIGM and DSBAR. Lode angle THETA (radians) angle of friction PHI and cohesion C.
MOCOUQ	PSI,DSBAR,THETA, DQ1,DQ2,DQ3	Derivatives of Mohr–Coulomb function DQ1,DQ2,DQ3 (Appendix 2B) for angle of dilation PSI, invariant DSBAR and Lode angle THETA.
PIN_JOINTED	KM,EA,COORD	Stiffness matrix KM of a pin-jointed element with coordinates COORD and axial stiffness EA.
READBF	NF,NODOF,NRB	Returns node freedom array NF for NODOF degrees of freedom per node. There are NRB blocks of data to be read.

RIGID_ JOINTED	KM,PROP,GAMMA, ETYPE,IEL,COORD	Stiffness matrix KM of a general beam–column element. Properties PROP,GAMMA; element type ETYPE; element number IEL with coordinates COORD.
SAMPLE	ELEMENT,S,WT	Returns element sampling points S and weights WT for character variable ELEMENT
SEEP4*	COORD,PERM,KC	Forms analytically computed "conductivity" matrix KC for a four-node quadrilateral with nodal coordinates COORD and permeability tensor PERM.
SHAPE_DER	DER,POINTS,I	Returns element shape function derivatives DER (local coordinates) for Ith sampling point coordinate in POINTS.
SHAPE_FUN	FUN,POINTS,I	Returns element shape functions FUN (local coordinates) for Ith sampling point coordinate in POINTS.
SOLVE_BAND* (GAUSS_BAND)	PB,COPY,ANS	Gaussian back-substitution on factorised unsymmetric band matrix PB. Solution ANS overwrites RHS. Work array COPY.
SPABAC (SPARIN)	A,B,KDIAG	Gaussian forward and backward substitution on lower triangular variable bandwidth matrix A. Bandwidth control by array KDIAG. Solution B overwrites RHS.
SPABAC_GAUSS* (SPARIN_GAUSS)	KV,LOADS,KDIAG	As SPABAC but skyline array KV stored as a vector. Bandwidth control by KDIAG. Solution LOADS overwrites RHS.
SPARIN (SPABAC)	A,KDIAG	Gaussian factorisation of lower triangular variable bandwidth matrix A. Bandwidth control by array KDIAG.
SPARIN_GAUSS* (SPABAC_GAUSS)	KV,KDIAG	As SPARIN but skyline array KV stored as a vector. Variable bandwidth control by array KDIAG.

VMFLOW	STRESS,DSBAR,VMFL	Von Mises flow vector VMFL from stress tensor STRESS and second invariant DSBAR.
VMPL	E,V,STRESS,PL	Forms plastic matrix PL for a von Mises material from stress tensor STRESS and elastic moduli E and V.

Appendix 5
Listings of Special Purpose Routines

```
module new_library
contains
!------------------------- building global matrices ----------------------
subroutine formkb(kb,km,g)
! lower triangular global stiffness kb stored as kb(n,iw+1)
implicit none
real,intent(in)::km(:,:);real,intent(out)::kb(:,:)
integer,intent(in)::g(:);integer::iw,idof,i,j,icd
idof=size(km,1);  iw=size(kb,2)-1
   do i=1,idof
      if(g(i)>0) then
         do j=1,idof
            if(g(j)>0) then
               icd=g(j)-g(i)+iw+1
               if(icd-iw-1<=0) kb(g(i),icd)= kb(g(i),icd) +km(i,j)
            end if
         end do
      end if
   end do
 return
end subroutine formkb
 subroutine formku(ku,km,g)
 !      this subroutine assembles element matrices into symmetrical
 !      global matrix(stored as an upper rectangle)
      real,intent(in):: km(:,:) ; real,intent(out)::ku(:,:)
      integer,intent(in):: g(:) ;integer::i,j,icd,ndof; ndof=ubound(km,1)
      do  i=1,ndof
      if(g(i)/=0) then
        do  j=1,ndof
          if(g(j)/=0) then
           icd=g(j)-g(i)+1
           if(icd>=1) ku(g(i),icd)=ku(g(i),icd)+km(i,j)
          end if
        end do
      end if
     end do
   return
 end subroutine formku
subroutine formkv(bk,km,g,n)
!global stiffness matrix stored as a vector (upper triangle)
implicit none
real,intent(in)::km(:,:);real,intent(out)::bk(:)
integer,intent(in)::g(:),n
integer::idof,i,j,icd,ival
idof=size(km,1)
     do i=1,idof
        if(g(i)/=0) then
           do j=1,idof
              if(g(j)/=0) then
                 icd=g(j)-g(i)+1
                 if(icd-1>=0) then
                    ival=n*(icd-1)+g(i)
                    bk(ival)=bk(ival)+km(i,j)
                 end if
               end if
           end do
        end if
     end do
return
end subroutine formkv
 subroutine formlump(diag,emm,g)
!       this subroutine forms the global mass matrix as vector diag
  implicit none
   real,intent(in)::emm(:,:)  ; real,intent(out)::diag(0:)
```

```
   integer,intent(in)::g(:); integer::i,ndof; ndof=ubound(emm,1)
     do  i=1,ndof;   diag(g(i))=diag(g(i))+emm(i,i) ; end do
    return
 end subroutine formlump
subroutine fsparv(bk,km,g,kdiag)
! assembly of element matrices into skyline global matrix
implicit none
real,intent(in)::km(:,:); integer,intent(in)::g(:),kdiag(:)
real,intent(out)::bk(:) ;  integer::i,idof,k,j,iw,ival
 idof=ubound(g,1)
   do i=1,idof
      k=g(i)
      if(k/=0) then
         do j=1,idof
            if(g(j)/=0) then
               iw=k-g(j)
               if(iw>=0) then
                   ival=kdiag(k)-iw
                   bk(ival)=bk(ival)+km(i,j)
                end if
            end if
         end do
      end if
   end do
 return
end subroutine fsparv
subroutine formtb(kb,km,g)
! assembles unsymmetrical band matrix kb from constituent km
  implicit none
  real,intent(in)::km(:,:); integer,intent(in)::g(:)
  real,intent(out)::kb(:,:); integer::i,j,idof,icd,iw
  idof=size(km,1); iw=(size(kb,2)-1)/2
  do i=1,idof
     if(g(i)/=0) then
        do j=1,idof
           if(g(j)/=0) then
              icd=g(j)-g(i)+iw+1
              kb(g(i),icd)=kb(g(i),icd)+km(i,j)
           end if
        end do
     end if
  end do
 return
end subroutine formtb
!---------------------matrix by vector multiplication ------------------
subroutine banmul(kb,loads,ans)
implicit none
real,intent(in)::kb(:,:),loads(0:); real,intent(out):: ans(0:)
integer :: neq,nband,i,j ; real :: x
neq = ubound(kb,1) ; nband = ubound(kb,2) -1
  do i = 1 , neq
     x = .0
     do j = nband + 1 , 1 , -1
        if(i+j>nband+1) x = x +kb(i,j)*loads(i+j-nband-1)
     end do
     do j = nband , 1 , -1
        if(i-j<neq - nband) x = x + kb(i-j+nband+1,j)*loads(i-j+nband+1)
     end do
     ans(i) = x
  end do
 return
end subroutine banmul
subroutine bantmul(kb,loads,ans)
! multiplies unsymmetrical band kb by vector loads
```

```
! could be much improved for vector processors
! look out for chance for zero-sized arrays
implicit none
  real,intent(in)::kb(:,:),loads(0:); real,intent(out)::ans(0:)
  integer::i,j,k,l,m,n,iw; real::x
  n=size(kb,1); l=size(kb,2); iw=(l-1)/2
    do i=1,n
       x=.0; k=iw+2
       do j=1,l
          k=k-1; m=i-k+1
          if(m<=n.and.m>=1)x=x+kb(i,j)*loads(m)
       end do
       ans(i)=x
    end do
 return
end subroutine bantmul
subroutine linmul(bk,disps,loads)
! matrix-vector multiply for symmetric matrix bk
! stored in upper triangular form as a vector
implicit none
  real,intent(in)::bk(:),disps(0:); real,intent(out)::loads(0:)
   integer::i,j,n,iw; real::x; n=ubound(disps,1);iw=ubound(bk,1)/n-1
    do i=1,n
       x=0.0
       do j=1,iw+1
          if(i+j<=n+1)     x=x+bk(n*(j-1)+i)*disps(i+j-1)
       end do
       do j=2,iw+1
          if(i-j+1>=1)     x=x+bk((n-1)*(j-1)+i)*disps(i-j+1)
       end do
       loads(i)=x
    end do
  return
end subroutine linmul
subroutine linmul_sky(bp,disps,loads,kdiag)
! skyline product of symmetric matrix and a vector
implicit none
 real,intent(in)::bp(:),disps(0:);real,intent(out)::loads(0:)
 integer,intent(in)::kdiag(:); integer::n,i,j,low,lup,k; real::x
  n=ubound(disps,1)
  do i = 1 , n
     x = .0 ; lup=kdiag(i)
     if(i==1)low=lup; if(i/=1)low=kdiag(i-1)+1
     do j = low , lup
        x = x + bp(j) * disps(i + j - lup)
     end do
     loads(i) = x
     if(i == 1) cycle   ; lup = lup - 1
     do j = low , lup
        k = i + j -lup - 1
        loads(k) = loads(k) + bp(j)*disps(i)
     end do
  end do
 return
end subroutine linmul_sky
!------------------ element matrix formation --------------------------------
 subroutine beam_km(km,ei,ell)
!      this subroutine forms the stiffness matrix of a
!      line beam element(bending only)
 implicit none
   real,intent(in)::ei,ell; real,intent(out):: km(:,:)
   km(1,1)=12.*ei/(ell*ell*ell) ;km(3,3)=km(1,1)
   km(1,2)=6.*ei/(ell*ell) ;  km(2,1)=km(1,2) ; km(1,4)=km(1,2)
   km(4,1)=km(1,4) ;  km(1,3)=-km(1,1) ; km(3,1)=km(1,3) ; km(3,4)=-km(1,2)
```

```
    km(4,3)=km(3,4) ; km(2,3)=km(3,4) ;  km(3,2)=km(2,3); km(2,2)=4.*ei/ell
    km(4,4)=km(2,2) ;km(2,4)=2.*ei/ell ; km(4,2)=km(2,4)
   return
  end subroutine beam_km
 subroutine beam_kp(kp,coord,pax)
 !      this subroutine forms the terms of the beam stiffness
 !      matrix due to axial loading
 implicit none
 real,intent(in)::coord(:,:),pax; real,intent(out)::kp(:,:)
 real::x1,x2,y1,y2,ell,c,s,pre(6,4),k_1d(4,4)
 integer::ndim
  ndim=ubound(coord,2)
  select case(ndim)
  case(1)
  x1=coord(1,1); x2=coord(2,1)
  ell=x2-x1
  kp(1,1)=1.2/ell; kp(1,2)=0.1       ; kp(2,1)=0.1; kp(1,3)=-1.2/ell
  kp(3,1)=-1.2/ell; kp(1,4)=0.1      ; kp(4,1)=0.1; kp(2,2)=2.0*ell/15.0
  kp(2,3)=-0.1; kp(3,2)=-0.1         ; kp(2,4)=-ell/30.0; kp(4,2)=-ell/30.0
  kp(3,3)=1.2/ell; kp(3,4)=-0.1      ; kp(4,3)=-0.1; kp(4,4)=2.0*ell/15.0
  kp=kp*pax
  case(2)
  x1=coord(1,1); y1=coord(1,2)       ; x2=coord(2,1); y2=coord(2,2)
  ell=sqrt((y2-y1)**2+(x2-x1)**2)    ; c=(x2-x1)/ell; s=(y2-y1)/ell
  pre=0.0
  pre(1,1)=-s; pre(2,1)=c; pre(3,2)=1.0; pre(4,3)=-s; pre(5,3)=c; pre(6,4)=1.0
  k_1d(1,1)=1.2/ell;    k_1d(1,2)=0.1
  k_1d(2,1)=0.1;        k_1d(1,3)=-1.2/ell
  k_1d(3,1)=-1.2/ell;   k_1d(1,4)=0.1
  k_1d(4,1)=0.1;        k_1d(2,2)=2.0*ell/15.0
  k_1d(2,3)=-0.1;       k_1d(3,2)=-0.1
  k_1d(2,4)=-ell/30.0; k_1d(4,2)=-ell/30.0
  k_1d(3,3)=1.2/ell;    k_1d(3,4)=-0.1
  k_1d(4,3)=-0.1;       k_1d(4,4)=2.0*ell/15.0
  kp=matmul(matmul(pre,k_1d),transpose(pre))
  kp=kp*pax
  end select
  return
 end subroutine beam_kp
  subroutine beam_mm(mm,fs,ell)
  implicit none
  real,intent(in)::fs,ell
  real,intent(out)::mm(:,:)
  real::fac
 !     this subroutine forms the mass matrix of a beam element
  fac=(fs*ell)/420.
  mm(1,1)=156.*fac; mm(3,3)=mm(1,1); mm(1,2)=22.*ell*fac; mm(2,1)=mm(1,2)
  mm(3,4)=-mm(1,2); mm(4,3)=mm(3,4); mm(1,3)=54.*fac; mm(3,1)=mm(1,3)
  mm(1,4)=-13.*ell*fac; mm(4,1)=mm(1,4); mm(2,3)=-mm(1,4)
  mm(3,2)=mm(2,3); mm(2,2)=4.*(ell**2)*fac; mm(4,4)=mm(2,2)
  mm(2,4)=-3.*(ell**2)*fac; mm(4,2)=mm(2,4)
  return
  end subroutine beam_mm
 subroutine ecmat(ecm,fun,ndof,nodof)
 implicit none
  real,intent(in)::fun(:); real,intent(out)::ecm(:,:)
  integer,intent(in):: nodof,ndof
  integer:: nod,i,j; real::nt(ndof,nodof),tn(nodof,ndof)
  nod = ndof/nodof; nt = .0; tn = .0
   do i = 1 , nod ; do j = 1 , nodof
      nt((i-1)*nodof+j,j) = fun(i); tn(j,(i-1)*nodof+j) = fun(i)
   end do; end do
   ecm = matmul( nt , tn )
  return
```

```
end subroutine ecmat
  subroutine fmkdke(km,kp,c,ke,kd,theta)
  ! builds up 'coupled' stiffnesses ke and kd from 'elastic'
  ! stiffness km , fluid stiffness kp and coupling matrix c
  implicit none
  real,intent(in)::km(:,:),kp(:,:),c(:,:),theta
  real,intent(out)::ke(:,:),kd(:,:)
  integer::idof; idof=size(km,1)
  ke(1:idof,1:idof)=theta*km; ke(1:idof,idof+1:)=theta*c
  ke(idof+1:,1:idof)=theta*transpose(c); ke(idof+1:,idof+1:)=-theta**2*kp
  kd(1:idof,1:idof)=(theta-1.)*km; kd(1:idof,idof+1:)=(theta-1.)*c
  kd(idof+1:,1:idof)=ke(idof+1:,1:idof)
  kd(idof+1:,idof+1:)=theta*(1.-theta)*kp
  end subroutine fmkdke
 subroutine formke(km,kp,c,ke,theta)
 ! ke matrix for incremental Biot
 implicit none
    real,intent(in)::km(:,:),kp(:,:),c(:,:),theta;real,intent(out)::ke(:,:)
    integer::ndof; ndof = ubound(km,1)
    ke(:ndof, :ndof)=km ;    ke(:ndof,ndof+1:)=c
    ke(ndof+1:,:ndof)=transpose(c) ; ke(ndof+1:,ndof+1:) = -theta * kp
  return
 end subroutine formke
 subroutine formupv(ke,c11,c12,c21,c23,c32)
!      this subroutine forms the unsymmetrical stiffness matrix
!      for the u-v-p version of the Navier-Stokes equations
 implicit none
  real,intent(in):: c11(:,:),c21(:,:),c23(:,:),c32(:,:),c12(:,:)
  real,intent(out):: ke(:,:)  ; integer::nod,nodf,ntot
   nod = ubound(c11,1); nodf = ubound(c21,1) ;ntot=nod+nodf+nod
   ke(1:nod,1:nod)=c11  ;  ke(1:nod,nod+1:nod+nodf)=c12
   ke(nod+1:nod+nodf,1:nod)=c21  ; ke(nod+1:nod+nodf,nod+nodf+1:ntot)=c23
   ke(nod+nodf+1:ntot,nod+1:nod+nodf)=c32
   ke(nod+nodf+1:ntot,nod+nodf+1:ntot)=c11
  return
 end subroutine formupv
function formxi(fsoil,fmax,rf,rm,ro) result(xi)
! soil spring stiffness in coupled pile-soil analysis
implicit none     ; real :: xi
 real,intent(in)::fsoil,fmax,rf,rm,ro;  real::phi
 phi = fsoil*ro*rf/fmax
 xi = log((rm-phi)/(ro-phi))+phi*(rm-ro)/((rm-phi)*(ro-phi))
return
end function formxi
subroutine rod_km(km,ea,length)
  implicit none
  real,intent(in):: ea,length; real,intent(out)::km(:,:)
  km(1,1)=1. ; km(2,2)=1.; km(1,2)=-1. ; km(2,1)=-1.; km=km*ea/length
end subroutine rod_km
!------------------------ equation  solution ---------------------------------
subroutine bacsub(bk,loads)
! performs the complete gaussian backsubstitution
implicit none
real,intent(in)::bk(:);real,intent(in out)::loads(0:)
integer::nkb,k,i,jn,jj,i1,n,iw;real::sum
n = ubound(loads,1); iw = ubound(bk,1)/n - 1
loads(1)=loads(1)/bk(1)
   do i=2,n
      sum=loads(i);i1=i-1 ; nkb=i-iw
      if(nkb<=0)nkb=1
      do k=nkb,i1
         jn=(i-k)*n+k;sum=sum-bk(jn)*loads(k)
      end do
      loads(i)=sum/bk(i)
```

```
      end do
      do jj=2,n
         i=n-jj+1;sum=.0;i1=i+1;nkb=i+iw
         if(nkb-n>0)nkb=n
         do k=i1,nkb
              jn=(k-i)*n+i   ; sum=sum+bk(jn)*loads(k)
         end do
         loads(i)=loads(i)-sum/bk(i)
      end do
   return
   end subroutine bacsub
   subroutine banred(bk,n)
   ! gaussian reduction on a vector stored as an upper triangle
   implicit none
   real,intent(in out)::bk(:);integer,intent(in)::n
   integer::i,il1,kbl,j,ij,nkb,m,ni,nj,iw ; real::sum
    iw = ubound(bk,1)/n-1
          do i=2,n
             il1=i-1;kbl=il1+iw+1
             if(kbl-n>0)kbl=n
             do j=i,kbl
                ij=(j-i)*n+i;sum=bk(ij);nkb=j-iw
                if(nkb<=0)nkb=1
                if(nkb-il1<=0)then
                   do m=nkb,il1
                      ni=(i-m)*n+m ; nj=(j-m)*n+m
                      sum=sum-bk(ni)*bk(nj)/bk(m)
                   end do
                end if
                bk(ij)=sum
              end do
          end do
   return
   end subroutine banred
   subroutine chobac(kb,loads)
   !Choleski back-substitution
   implicit none
   real,intent(in)::kb(:,:);real,intent(in out)::loads(0:)
   integer::iw,n,i,j,k,l,m; real::x
   n=size(kb,1); iw=size(kb,2)-1
   loads(1)=loads(1)/kb(1,iw+1)
       do i=2,n
          x=.0;k=1
          if(i<=iw+1)k=iw-i+2
          do j=k,iw; x=x+kb(i,j)*loads(i+j-iw-1); end do
          loads(i)=(loads(i)-x)/kb(i,iw+1)
       end do
       loads(n)=loads(n)/kb(n,iw+1)
       do i=n-1,1,-1
          x=0.0; l=i+iw
          if(i>n-iw)l=n;  m=i+1
          do j=m,l; x=x+kb(j,iw+i-j+1)*loads(j); end do
          loads(i)=(loads(i)-x)/kb(i,iw+1)
       end do
     return
   end subroutine chobac
   subroutine cholin(kb)
   ! Choleski reduction on kb(l,iw+1) stored as a lower triangle
   implicit none
   real,intent(in out)::kb(:,:);integer::i,j,k,l,ia,ib,n,iw;real::x
   n=ubound(kb,1); iw=ubound(kb,2)-1
      do i=1,n
         x=.0
         do j=1,iw; x=x+kb(i,j)**2; end do
```

```
      kb(i,iw+1)=sqrt(kb(i,iw+1)-x)
      do k=1,iw
         x=.0
         if(i+k<=n) then
           if(k/=iw) then
             do l=iw-k,1,-1
                x=x+kb(i+k,l)*kb(i,l+k)
             end do
           end if
           ia=i+k; ib=iw-k+1
           kb(ia,ib)=(kb(ia,ib)-x)/kb(i,iw+1)
         end if
      end do
   end do
 return
end subroutine cholin
subroutine sparin(a,kdiag)
! Choleski factorisation of variable bandwidth matrix a
! stored as a vector and overwritten
implicit none
real,intent(in out)::a(:);integer,intent(in)::kdiag(:)
integer::n,i,ki,l,kj,j,ll,m,k; real::x
 n=ubound(kdiag,1)  ; a(1)=sqrt(a(1))
 do i=2,n
    ki=kdiag(i)-i;  l=kdiag(i-1)-ki+1
    do j=l,i
       x=a(ki+j);  kj=kdiag(j)-j
       if(j/=1) then
          ll=kdiag(j-1)-kj+1; ll=max0(l,ll)
          if(ll/=j) then
              m=j-1
              do k=ll,m ; x=x-a(ki+k)*a(kj+k) ; end do
          end if
       end if
       a(ki+j)=x/a(kj+j)
    end do
    a(ki+i)=sqrt(x)
 end do
 return
end subroutine sparin
subroutine spabac(a,b,kdiag)
! Choleski forward and backward substitution combined
! variable bandwidth factorised matrix a stored as a vector
implicit none
real,intent(in)::a(:);real,intent(in out)::b(0:);integer,intent(in)::kdiag(:)
integer::n,i,ki,l,m,j,it,k; real::x
n=ubound(kdiag,1)
 b(1)=b(1)/a(1)
  do i=2,n
     ki=kdiag(i)-i;  l=kdiag(i-1)-ki+1 ; x=b(i)
     if(l/=i) then
        m=i-1
        do j=l,m ; x=x-a(ki+j)*b(j); end do
     end if
     b(i)=x/a(ki+i)
  end do
  do it=2,n
     i=n+2-it; ki=kdiag(i)-i; x=b(i)/a(ki+i); b(i)=x; l=kdiag(i-1)-ki+1
     if(l/=i) then
       m=i-1
       do k=l,m; b(k)=b(k)-x*a(ki+k); end do
     end if
  end do
 b(1)=b(1)/a(1)
```

```
 return
end subroutine spabac
!---------------------- plasticity ----------------------------------------
subroutine vmflow(stress,dsbar,vmfl)
! Forms the von Mises flow vector
implicit none
real,intent(in)::stress(:),dsbar; real,intent(out)::vmfl(:)
real::sigm; sigm=(stress(1)+stress(2)+stress(4))/3.
vmfl(1)=stress(1)-sigm;vmfl(2)=stress(2)-sigm
vmfl(3)=stress(3)*2. ; vmfl(4)=stress(4)-sigm
vmfl = vmfl*1.5/dsbar
return
end subroutine vmflow
subroutine formaa(flow,rmat,modee)
! Modification to dee matrix
implicit none
real,intent(in)::flow(:),rmat(:,:); real,intent(out)::modee(:,:)
real::flowt(1,4),flowa(4,1)    ; flowt(1,:) = flow ; flowa(:,1) = flow
modee = matmul(matmul(matmul(rmat,flowa),flowt),rmat)
return
end subroutine formaa
subroutine fmrmat(vmfl,dsbar,dlam,dee,temp,rmat)
! Forms the r matrix
implicit none
real,intent(in)::vmfl(:),dsbar,dlam,dee(:,:),temp(:,:)
real,intent(out)::rmat(:,:); real::acat(4,4),acatc(4,4),qmat(4,4)
integer::i,j  ; real:: con
do i=1,4;do j=1,4;acat(i,j)=vmfl(i)*vmfl(j);end do;end do
acat = (temp-acat)/dsbar; acatc = matmul(dee,acat); qmat = acatc*dlam
do i=1,4; qmat(i,i)=qmat(i,i)+1.; end do
 do i=1,4
  con = qmat(i,i); qmat(i,i) = 1.
  qmat(i,:) = qmat(i,:)/con
  do j=1,4
   if(j/=i) then
    con = qmat(j,i); qmat(j,i)=0.0
    qmat(j,:) = qmat(j,:) - qmat(i,:)*con
   end if
  end do
 end do
 rmat = matmul(qmat,dee)
return
end subroutine fmrmat
subroutine fmacat(vmfl,temp,acat)
! Intermediate step
implicit none
real,intent(in)::vmfl(:),temp(:,:);real,intent(out)::acat(:,:);integer::i,j
do i=1,4; do j=1,4; acat(i,j)=vmfl(i)*vmfl(j); end do; end do
acat = temp - acat
return
end subroutine fmacat
subroutine vmpl(e,v,stress,pl)
! plastic matrix for a von-Mises material
implicit none
real,intent(in)::e,v,stress(:); real,intent(out)::pl(:,:)
real::sx,sy,txy,sz,dsbar,ee,term(4); integer::i,j,nst; nst = ubound(stress,1)
  sx=stress(1); sy=stress(2); txy=stress(3); sz=stress(4)
  dsbar=sqrt((sx-sy)**2+(sy-sz)**2+(sz-sx)**2+6.*txy**2)/sqrt(2.)
  ee=1.5*e/((1.+v)*dsbar*dsbar)
  term(1)=(2.*sx-sy-sz)/3.; term(2)=(2.*sy-sz-sx)/3.
  term(3)=txy              ; term(4)=(2.*sz-sx-sy)/3.
  do i=1,nst;do j=1,nst;pl(i,j)=term(i)*term(j)*ee;pl(j,i)=pl(i,j);end do;end do
return
end subroutine vmpl
```

```
 subroutine mocouf(phi,c,sigm,dsbar,theta,f)
!      this subroutine calculates the value of the yield function
!      for a Mohr-Coulomb material (phi in degrees)
 implicit none
 real,intent(in)::phi,c,sigm,dsbar,theta   ; real,intent(out)::f
 real::phir,snph,csph,csth,snth
 phir=phi*4.*atan(1.)/180.
 snph=sin(phir) ;  csph=cos(phir) ; csth=cos(theta);  snth=sin(theta)
 f=snph*sigm+dsbar*(csth/sqrt(3.)-snth*snph/3.)-c*csph
 return
 end  subroutine mocouf
 subroutine mocouq(psi,dsbar,theta,dq1,dq2,dq3)
!      this subroutine forms the derivatives of a Mohr-Coulomb
!      potential function with respect to the three invariants
!      psi in degrees
 implicit none
 real,intent(in)::psi,dsbar,theta; real,intent(out)::dq1,dq2,dq3
 real::psir,snth,snps,sq3,c1,csth,cs3th,tn3th,tnth
 psir=psi*4.*atan(1.)/180. ;  snth=sin(theta) ;   snps=sin(psir)
 sq3=sqrt(3.)   ; dq1=snps
     if(abs(snth).gt..49)then
       c1=1.
       if(snth.lt.0.)c1=-1.
       dq2=(sq3*.5-c1*snps*.5/sq3)*sq3*.5/dsbar ;   dq3=0.
     else
       csth=cos(theta); cs3th=cos(3.*theta);tn3th=tan(3.*theta); tnth=snth/csth
       dq2=sq3*csth/dsbar*((1.+tnth*tn3th)+snps*(tn3th-tnth)/sq3)*.5
       dq3=1.5*(sq3*snth+snps*csth)/(cs3th*dsbar*dsbar)
     end if
 return
 end subroutine mocouq
 subroutine mocopl(phi,psi,e,v,stress,pl)
!      this subroutine forms the plastic stress/strain matrix
!      for a mohr-coulomb material  (phi,psi in degrees)
 implicit none
  real,intent(in)::stress(:),phi,psi,e,v  ;real,intent(out)::pl(:,:)
  integer::i,j;  real::row(4),col(4),sx,sy,txy,sz,pi,phir,psir,&
                           dx,dy,dz,d2,d3,th,snth,sig,rph,rps,cps,snps,sq3,&
                           cc,cph ,alp,ca,sa,dd ,snph,ee,s1,s2
  sx=stress(1);  sy=stress(2);  txy=stress(3) ;  sz=stress(4)
  pi=4.*atan(1.) ; phir=phi*pi/180.; psir=psi*pi/180.;  snph=sin(phir)
  snps=sin(psir)      ;  sq3=sqrt(3.) ;   cc=1.-2.*v
  dx=(2.*sx-sy-sz)/3.    ;   dy=(2.*sy-sz-sx)/3. ;  dz=(2.*sz-sx-sy)/3.
  d2=sqrt(-dx*dy-dy*dz-dz*dx+txy*txy) ;  d3=dx*dy*dz-dz*txy*txy
  th=-3.*sq3*d3/(2.*d2**3)
  if(th.gt.1.)th=1.   ;  if(th.lt.-1.)th=-1.
  th=asin(th)/3.   ;    snth=sin(th)
  if(abs(snth).gt..49)then
     sig=-1.
     if(snth.lt.0.)sig=1.
     rph=snph*(1.+v)/3.;  rps=snps*(1.+v)/3. ; cps=.25*sq3/d2*(1.+sig*snps/3.)
     cph=.25*sq3/d2*(1.+sig*snph/3.)
     col(1)=rph+cph*((1.-v)*dx+v*(dy+dz));col(2)=rph+cph*((1.-v)*dy+v*(dz+dx))
     col(3)=cph*cc*txy    ; col(4)=rph+cph*((1.-v)*dz+v*(dx+dy))
     row(1)=rps+cps*((1.-v)*dx+v*(dy+dz));row(2)=rps+cps*((1.-v)*dy+v*(dz+dx))
     row(3)=cps*cc*txy ;   row(4)=rps+cps*((1.-v)*dz+v*(dx+dy))
     ee=e/((1.+v)*cc*(rph*snps+2.*cph*cps*d2*d2*cc))
 else
     alp=atan(abs((sx-sy)/(2.*txy))) ; ca=cos(alp);  sa=sin(alp)
     dd=cc*sa ;  s1=1.  ;  s2=1.
     if((sx-sy).lt..0)s1=-1.
     if(txy.lt..0)s2=-1.
     col(1)=snph+s1*dd ; col(2)=snph-s1*dd;col(3)=s2*cc*ca; col(4)=2.*v*snph
     row(1)=snps+s1*dd ;row(2)=snps-s1*dd; row(3)=s2*cc*ca ; row(4)=2.*v*snps
```

```
      ee=e/(2.*(1.+v)*cc*(snph*snps+cc))
  end if
  do  i=1,4; do j=1,4 ; pl(i,j)=ee*row(i)*col(j) ; end do ; end do
  return
 end subroutine mocopl
!------------------------- invariants ----------------------------------------
subroutine invar(stress,sigm,dsbar,theta)
! forms the stress invariants in 2-d or 3-d
implicit none
  real,intent(in)::stress(:);real,intent(out)::sigm,dsbar,theta
  real::sx,sy,sz,txy,dx,dy,dz,xj3,sine,s1,s2,s3,s4,s5,s6,ds1,ds2,ds3,d2,d3,sq3
  integer :: nst ; nst = ubound(stress,1)
 select case (nst)
 case(4)
  sx=stress(1); sy=stress(2); txy=stress(3); sz=stress(4)
  sigm=(sx+sy+sz)/3.
  dsbar=sqrt((sx-sy)**2+(sy-sz)**2+(sz-sx)**2+6.*txy**2)/sqrt(2.)
  if(dsbar<1.e-10) then
     theta=.0
  else
     dx=(2.*sx-sy-sz)/3.; dy=(2.*sy-sz-sx)/3.; dz=(2.*sz-sx-sy)/3.
     xj3=dx*dy*dz-dz*txy**2
     sine=-13.5*xj3/dsbar**3
     if(sine>1.) sine=1.
     if(sine<-1.) sine=-1.
     theta=asin(sine)/3.
  end if
 case(6)
  sq3=sqrt(3.);  s1=stress(1)  ;  s2=stress(2)
  s3=stress(3) ;  s4=stress(4);  s5=stress(5);  s6=stress(6)
  sigm=(s1+s2+s3)/3.
  d2=((s1-s2)**2+(s2-s3)**2+(s3-s1)**2)/6.+s4*s4+s5*s5+s6*s6
  ds1=s1-sigm ;  ds2=s2-sigm  ;  ds3=s3-sigm
  d3=ds1*ds2*ds3-ds1*s5*s5-ds2*s6*s6-ds3*s4*s4+2.*s4*s5*s6
  dsbar=sq3*sqrt(d2)
  if(dsbar==0.)then
      theta=0.
    else
      sine=-3.*sq3*d3/(2.*sqrt(d2)**3)
      if(sine>1.)sine=1. ;  if(sine<-1.)sine=-1. ; theta=asin(sine)/3.
  end if
 case default
  print*,"wrong size for nst in invar"
 end select
 return
end subroutine invar
subroutine formm(stress,m1,m2,m3)
! forms the derivatives of the invariants with respect to stress 2- or 3-d
 implicit none
  real,intent(in)::stress(:); real,intent(out)::m1(:,:),m2(:,:),m3(:,:)
  real::sx,sy,txy,tyz,tzx,sz,dx,dy,dz,sigm ; integer::nst , i , j
  nst=ubound(stress,1)
  select case (nst)
  case(4)
  sx=stress(1); sy=stress(2); txy=stress(3); sz=stress(4)
  dx=(2.*sx-sy-sz)/3.; dy=(2.*sy-sz-sx)/3.; dz=(2.*sz-sx-sy)/3.
  sigm=(sx+sy+sz)/3.
  m1=.0; m2=.0; m3=.0
  m1(1,1:2)=1.; m1(2,1:2)=1.; m1(4,1:2)=1.
  m1(1,4)=1.; m1(4,4)=1.; m1(2,4)=1.
  m1=m1/9./sigm
  m2(1,1)=.666666666666666; m2(2,2)=.666666666666666; m2(4,4)=.666666666666666
  m2(2,4)=-.333333333333333;m2(4,2)=-.333333333333333;m2(1,2)=-.333333333333333
  m2(2,1)=-.333333333333333;m2(1,4)=-.333333333333333;m2(4,1)=-.333333333333333
```

```
  m2(3,3)=2.; m3(3,3)=-dz
  m3(1:2,3)=txy/3.; m3(3,1:2)=txy/3.; m3(3,4)=-2.*txy/3.; m3(4,3)=-2.*txy/3.
  m3(1,1)=dx/3.; m3(2,4)=dx/3.; m3(4,2)=dx/3.
  m3(2,2)=dy/3.; m3(1,4)=dy/3.; m3(4,1)=dy/3.
  m3(4,4)=dz/3.; m3(1,2)=dz/3.; m3(2,1)=dz/3.
 case(6)
  sx=stress(1); sy=stress(2)    ;   sz=stress(3)
  txy=stress(4)   ;   tyz=stress(5) ;   tzx=stress(6)
  sigm=(sx+sy+sz)/3.
  dx=sx-sigm  ;   dy=sy-sigm ;  dz=sz-sigm
  m1 = .0; m2 = .0; m1(1:3,1:3) = 1./(3.*sigm)
  do  i=1,3 ; m2(i,i)=2. ;  m2(i+3,i+3)=6. ; end do
  m2(1,2)=-1.; m2(1,3)=-1. ; m2(2,3)=-1.; m3(1,1)=dx
  m3(1,2)=dz ; m3(1,3)=dy ; m3(1,4)=txy  ;  m3(1,5)=-2.*tyz
  m3(1,6)=tzx ; m3(2,2)=dy ; m3(2,3)=dx ; m3(2,4)=txy
  m3(2,5)=tyz ; m3(2,6)=-2.*tzx ;  m3(3,3)=dz
  m3(3,4)=-2.*txy; m3(3,5)=tyz ;  m3(3,6)=tzx
  m3(4,4)=-3.*dz ;  m3(4,5)=3.*tzx;  m3(4,6)=3.*tyz
  m3(5,5)=-3.*dx;  m3(5,6)=3.*txy ;  m3(6,6)=-3.*dy
  do  i=1,6 ;  do  j=i,6
      m1(i,j)=m1(i,j)/3.;  m1(j,i)=m1(i,j) ;  m2(i,j)=m2(i,j)/3.
      m2(j,i)=m2(i,j)   ;  m3(i,j)=m3(i,j)/3. ; m3(j,i)=m3(i,j)
  end do; end do
 case default
  print*,"wrong size for nst in formm"
 end select
 return
 end subroutine formm
!-------------------------- node freedom array --------------------------------
 subroutine formnf(nf)
   ! reform nf
   implicit none
   integer,intent(in out)::nf(:,:)
    integer:: i,j,m
    m=0
    do j=1,ubound(nf,2)
       do i=1,ubound(nf,1)
          if(nf(i,j)/=0) then
             m=m+1; nf(i,j)=m
          end if
       end do
    end do
    return
 end subroutine formnf
 subroutine readbf(nf,nodof,nrb)
!----- blocks of input of nf data  -----------------------
      implicit none
      integer,intent(out)::nf(:,:); integer,intent(in)::nodof,nrb
      integer::nfd(nodof),i,j,l,n,if,it,is,nn;nn=ubound(nf,2)
      nf=1
      do l=1,nrb
       read(10,*)if,it,is,nfd
       do i=if,min(nn,it),is
         nf(:,i) = nf(:,i) * nfd(:)
       end do
      end do
      n=0
      do j = 1,nn  ;    do i= 1,nodof
          if(nf(i,j) /= 0) then
           n=n+1     ;   nf(i,j)=n
          end if
      end do      ;    end do
     return
 end subroutine readbf
```

```
!------------------------- B matrix formulation ---------------------------------
 subroutine beemat(bee,deriv)
 ! bee matrix for 2-d elasticity or elastoplasticity (ih=3 or 4 respectively)
 ! or for 3-d (ih = 6)
 implicit none
 real,intent(in)::deriv(:,:);  real,intent(out)::bee(:,:)
 ! local variables
 integer::k,l,m,n , ih,nod; real::x,y,z
 bee=0. ; ih = ubound(bee,1); nod = ubound(deriv,2)
     select case (ih)
       case(3,4)
        do m=1,nod
           k=2*m; l=k-1; x=deriv(1,m); y=deriv(2,m)
           bee(1,l)=x; bee(3,k)=x; bee(2,k)=y; bee(3,l)=y
        end do
       case(6)
        do m=1,nod
           n=3*m;  k=n-1; l=k-1
           x=deriv(1,m); y=deriv(2,m); z=deriv(3,m)
           bee(1,l)=x; bee(4,k)=x; bee(6,n)=x
           bee(2,k)=y; bee(4,l)=y; bee(5,n)=y
           bee(3,n)=z; bee(5,k)=z; bee(6,l)=z
        end do
       case default
        print*,'wrong dimension for nst in bee matrix'
      end select
  return
 end subroutine beemat
 subroutine bmataxi(bee,radius,coord,deriv,fun)
 ! b matrix for axisymmetry
 real,intent(in)::deriv(:,:),fun(:),coord(:,:);real,intent(out)::bee(:,:),radius
 integer::nod ,k,l,m; real :: x,y
  radius = sum(fun * coord(:,1))  ; nod = ubound(deriv , 2) ; bee = .0
  do m = 1 , nod
   k=2*m; l = k-1 ; x = deriv(1,m); bee(1,l) = x; bee(3 , k) = x
   y = deriv(2,m); bee(2,k)=y; bee(3,l) = y; bee(4,l)=fun(m)/radius
  end do
  return
 end subroutine bmataxi
  subroutine bmat_nonaxi(bee,radius,coord,deriv,fun,iflag,lth)
!      this subroutine forms the strain-displacement matrix for
!      axisymmetric solids subjected to non-axisymmetric loading
 implicit none
  real,intent(in)::deriv(:,:),fun(:),coord(:,:)
  real,intent(out):: bee(:,:),radius; integer,intent(in)::iflag,lth
  integer::nod,k,l,m,n; nod = ubound(deriv , 2 ) ; bee = .0
  radius = sum(fun * coord(:,1))
    do  m=1,nod
      n=3*m      ;    k=n-1      ;     l=k-1
      bee(1,l)=deriv(1,m);   bee(2,k)=deriv(2,m);    bee(3,l)=fun(m)/radius
      bee(3,n)=iflag*lth*bee(3,l) ; bee(4,l)=deriv(2,m); bee(4,k)=deriv(1,m)
      bee(5,k)=-iflag*lth*fun(m)/radius ;   bee(5,n)=deriv(2,m)
      bee(6,l)=bee(5,k) ; bee(6,n)=deriv(1,m)-fun(m)/radius
    end do
   return
  end subroutine bmat_nonaxi
subroutine bee4(coord,points,i,det,bee)
!  analytical version of the bee matrix for a 4-node quad
 implicit none
 real,intent(in):: coord(:,:),points(:,:)
 real,intent(out):: det,bee(:,:)
 integer,intent(in):: i
 real:: x1,x2,x3,x4,y1,y2,y3,y4,xi,et,&
        x12,x13,x14,x23,x24,x34,y12,y13,y14,y23,y24,y34,&
```

```
        xy12,xy13,xy14,xy23,xy24,xy34,xifac,etfac,const,den,&
        dn1dx,dn2dx,dn3dx,dn4dx,dn1dy,dn2dy,dn3dy,dn4dy
 x1=coord(1,1); x2=coord(2,1); x3=coord(3,1); x4=coord(4,1)
 y1=coord(1,2); y2=coord(2,2); y3=coord(3,2); y4=coord(4,2)
 xi=points(i,1); et=points(i,2)
 x12=x1-x2; x13=x1-x3; x14=x1-x4; x23=x2-x3; x24=x2-x4; x34=x3-x4
 y12=y1-y2; y13=y1-y3; y14=y1-y4; y23=y2-y3; y24=y2-y4; y34=y3-y4
 xy12=x1*y2-y1*x2; xy13=x1*y3-y1*x3; xy14=x1*y4-y1*x4; xy23=x2*y3-y2*x3
 xy24=x2*y4-y2*x4; xy34=x3*y4-y3*x4
 xifac= xy12-xy13+xy24-xy34; etfac= xy13-xy14-xy23+xy24
 const=-xy12+xy14-xy23-xy34; den=xifac*xi+etfac*et+const; det=0.125*den
 dn1dx=( y23*xi+y34*et-y24)/den; dn2dx=(-y14*xi-y34*et+y13)/den
 dn3dx=( y14*xi-y12*et+y24)/den; dn4dx=(-y23*xi+y12*et-y13)/den
 dn1dy=(-x23*xi-x34*et+x24)/den; dn2dy=( x14*xi+x34*et-x13)/den
 dn3dy=(-x14*xi+x12*et-x24)/den; dn4dy=( x23*xi-x12*et+x13)/den
 bee(1,1)=dn1dx; bee(1,2)=0.; bee(1,3)=dn2dx; bee(1,4)=0.
 bee(1,5)=dn3dx; bee(1,6)=0.; bee(1,7)=dn4dx; bee(1,8)=0.
 bee(2,1)=0.; bee(2,2)=dn1dy; bee(2,3)=0.; bee(2,4)=dn2dy
 bee(2,5)=0.; bee(2,6)=dn3dy; bee(2,7)=0.; bee(2,8)=dn4dy
 bee(3,1)=dn1dy; bee(3,2)=dn1dx; bee(3,3)=dn2dy; bee(3,4)=dn2dx
 bee(3,5)=dn3dy; bee(3,6)=dn3dx; bee(3,7)=dn4dy; bee(3,8)=dn4dx
return
end subroutine bee4
!------------------------ bandwidth determination ---------------------
   function bandwidth(g) result(nband)
   ! finds the element bandwidth from g
   implicit none     ; integer :: nband
   integer,intent(in)::g(:)
      nband= maxval(g,1,g>0)-minval(g,1,g>0)
   end function bandwidth
subroutine fkdiag(kdiag,g)
! finds the maximum bandwidth for each freedom
implicit none
 integer,intent(in)::g(:); integer,intent(out)::kdiag(:)
 integer::idof,i,iwp1,j,im,k
  idof=size(g)
  do i = 1,idof
     iwp1=1
     if(g(i)/=0) then
        do j=1,idof
           if(g(j)/=0) then
              im=g(i)-g(j)+1
              if(im>iwp1) iwp1=im
           end if
        end do
        k=g(i);    if(iwp1>kdiag(k))kdiag(k)=iwp1
     end if
  end do
 return
end subroutine fkdiag
!------------------- determinants and inversion -----------------------
 function determinant (jac) result(det)
 ! returns the determinant of a 1x1 2x2 3x3 jacobian matrix
 implicit none    ; real :: det
 real,intent(in)::jac(:,:); integer:: it ; it = ubound(jac,1)
 select case (it)
   case (1)
    det=1.0
   case (2)
    det=jac(1,1)*jac(2,2) - jac(1,2) * jac(2,1)
   case (3)
    det= jac(1,1)*(jac(2,2) * jac(3,3) -jac(3,2) * jac(2,3))
    det= det-jac(1,2)*(jac(2,1)*jac(3,3)-jac(3,1)*jac(2,3))
    det= det+jac(1,3)*(jac(2,1)*jac(3,2)-jac(3,1)*jac(2,2))
```

```
   case default
    print*,' wrong dimension for jacobian matrix'
 end select
 return
 end function determinant
  subroutine invert(matrix)
  ! invert a small square matrix onto itself
  implicit none
  real,intent(in out)::matrix(:,:)
  integer::i,k,n; real::con  ; n= ubound(matrix,1)
  do k=1,n
     con=matrix(k,k); matrix(k,k)=1.
     matrix(k,:)=matrix(k,:)/con
     do i=1,n
        if(i/=k) then
           con=matrix(i,k); matrix(i,k)=0.0
           matrix(i,:)=matrix(i,:) - matrix(k,:)*con
        end if
     end do
  end do
  return
 end subroutine invert
!--------------------------- convergence testing -----------------------------
subroutine checon(loads,oldlds,tol,converged)
! sets converged to .false. if relative change in loads and
! oldlds is greater than tol and updates oldlds
implicit none
real,intent(in)::loads(0:),tol;real,intent(in out)::oldlds(0:)
logical,intent(out)::converged
  converged=.true.
  converged=(maxval(abs(loads-oldlds))/maxval(abs(loads))<=tol)
  oldlds=loads
 return
end subroutine checon
!-------------- load function interpolation ----------------------------------
 subroutine interp(k,dt,rt,rl,al,ntp)
!
!     this subroutine forms the load/time functions by interpolation
!     if dt is not an exact multiple it stops one short
!
 real,intent(in)::dt,rt(:),rl(:)
 integer,intent(in)::k,ntp; real,intent(out)::al(:,:)
 integer::np,i,j; real::t,val
 np=size(rt); al(1,k)=rl(1); t=rt(1)
 do j=2,ntp
   t=t+dt
   do i=2,np
     if(t.le.rt(i))then
       val=rl(i-1)+((t-rt(i-1))/(rt(i)-rt(i-1)))*(rl(i)-rl(i-1))
       exit
     end if
   end do
   al(j,k)=val
 end do
 return
 end subroutine interp
!-------------------------D matrix formulation -----------------------------
 subroutine deemat(dee,e,v)
 ! returns the elastic dee matrix for given ih
 ! ih=3,plane strain; =4,axisymmetry or plane strain elastoplasticity
 ! =6 , three dimensional
 implicit none
   real,intent(in)::e,v; real,intent(out)::dee(:,:)
 ! local variables
```

```
   real::v1,v2,c,vv; integer :: i,ih;  dee=0.0  ; ih = ubound(dee,1)
         v1 = 1. - v; c = e/((1.+v)*(1.-2.*v))
   select case (ih)
          case(3)
             dee(1,1)=v1*c; dee(2,2)=v1*c; dee(1,2)=v*c; dee(2,1)=v*c
             dee(3,3)=.5*c*(1.-2.*v)
          case(4)
             dee(1,1)=v1*c; dee(2,2)=v1*c; dee(4,4)=v1*c
             dee(3,3)=.5*c*(1.-2.*v) ; dee(1,2)=v*c; dee(2,1)=v*c
             dee(1,4)=v*c; dee(4,1)=v*c; dee(2,4)=v*c; dee(4,2)=v*c
          case(6)
             v2=v/(1.-v); vv=(1.-2.*v)/(1.-v)*.5
             do i=1,3; dee(i,i)=1.;end do; do i=4,6; dee(i,i)=vv; end do
             dee(1,2)=v2; dee(2,1)=v2; dee(1,3)=v2; dee(3,1)=v2
             dee(2,3)=v2; dee(3,2)=v2
             dee = dee*e/(2.*(1.+v)*vv)
          case default
             print*,'wrong size for dee matrix'
   end select
 return
 end subroutine deemat
!--------------------- numerical integration ------------------------------
 subroutine sample(element,s,wt)
 ! returns the local coordinates of the integrating points
 implicit none
  real,intent(out)::s(:,:),wt(:)  ; character(*),intent(in):: element
  integer::nip ;  real:: root3, r15 , w(3),v(9),b,c
  root3 = 1./sqrt(3.)   ;  r15 = .2*sqrt(15.)
  nip = ubound( s , 1 )
         w = (/5./9.,8./9.,5./9./); v=(/5./9.*w,8./9.*w,5./9.*w/)
     select case (element)
            case('line')
             select case(nip)
              case(1)
               s(1,1)=0.  ;  wt(1)=2.
              case(2)
               s(1,1)=root3  ; s(2,1)=-s(1,1)  ;  wt(1)=1.  ; wt(2)=1.
              case(3)
               s(1,1)=r15 ; s(2,1)=.0      ; s(3,1)=-s(1,1)
               wt = w
              case(4)
               s(1,1)=.861136311594053  ; s(2,1)=.339981043584856
               s(3,1)=-s(2,1)  ; s(4,1)=-s(1,1)
               wt(1)=.347854845137454 ; wt(2)=.652145154862546
               wt(3)=wt(2) ; wt(4)=wt(1)
              case(5)
               s(1,1)=.906179845938664 ; s(2,1)=.538469310105683
               s(3,1)=.0 ; s(4,1)=-s(2,1) ; s(5,1)=-s(1,1)
               wt(1)=.236926885056189 ; wt(2)=.478628670499366
               wt(3)=.568888888888889 ; wt(4)=wt(2) ; wt(5)=wt(1)
              case(6)
               s(1,1)=.932469514203152 ; s(2,1)=.661209386466265
               s(3,1)=.238619186083197
               s(4,1)=-s(3,1) ; s(5,1)=-s(2,1) ; s(6,1)=-s(1,1)
               wt(1)=.171324492379170 ; wt(2)=.360761573048139
               wt(3)=.467913934572691
               wt(4)=wt(3); wt(5)=wt(2) ; wt(6)=wt(1)
                     case default
                      print*,"wrong number of integrating points for a line"
             end select
            case('triangle')
             select case(nip)
              case(1)   ! for triangles weights multiplied by .5
                s(1,1)=1./3.  ; s(1,2)=1./3.  ;  wt(1)= .5
```

```
              case(3)
               s(1,1)=.5 ;  s(1,2)=.5 ;  s(2,1)=.5
               s(2,2)=0.;  s(3,1)=0.  ;  s(3,2)=.5
               wt(1)=1./3.  ;  wt(2)=wt(1) ; wt(3)=wt(1)   ; wt = .5*wt
              case(6)
s(1,1)=.816847572980459  ; s(1,2)=.091576213509771
s(2,1)=s(1,2);  s(2,2)=s(1,1) ;  s(3,1)=s(1,2); s(3,2)=s(1,2)
s(4,1)=.108103018168070 ;  s(4,2)=.445948490915965
s(5,1)=s(4,2) ;   s(5,2)=s(4,1) ;  s(6,1)=s(4,2)  ; s(6,2)=s(4,2)
wt(1)=.109951743655322 ;   wt(2)=wt(1)  ;   wt(3)=wt(1)
wt(4)=.223381589678011 ;   wt(5)=wt(4)  ;   wt(6)=wt(4)    ; wt = .5*wt
              case(7)
s(1,1)=1./3. ; s(1,2)=1./3.;s(2,1)=.797426985353087 ;s(2,2)=.101286507323456
s(3,1)=s(2,2) ;  s(3,2)=s(2,1) ; s(4,1)=s(2,2) ;  s(4,2)=s(2,2)
s(5,1)=.470142064105115 ;   s(5,2)=.059715871789770
s(6,1)=s(5,2) ; s(6,2)=s(5,1);  s(7,1)=s(5,1);  s(7,2)=s(5,1)
wt(1)=.225 ; wt(2)=.125939180544827 ;  wt(3)=wt(2);  wt(4)=wt(2)
wt(5)=.132394152788506;  wt(6)=wt(5)      ;  wt(7)=wt(5)     ;wt = .5*wt
              case(12)
s(1,1)=.873821971016996 ; s(1,2)=.063089014491502
s(2,1)=s(1,2) ;  s(2,2)=s(1,1);  s(3,1)=s(1,2) ;  s(3,2)=s(1,2)
s(4,1)=.501426509658179 ;  s(4,2)=.249286745170910
s(5,1)=s(4,2); s(5,2)=s(4,1)   ;  s(6,1)=s(4,2) ;  s(6,2)=s(4,2)
s(7,1)=.636502499121399 ;      s(7,2)=.310352451033785
s(8,1)=s(7,1) ;  s(8,2)=.053145049844816 ;  s(9,1)=s(7,2) ; s(9,2)=s(7,1)
s(10,1)=s(7,2) ; s(10,2)=s(8,2) ; s(11,1)=s(8,2);   s(11,2)=s(7,1)
s(12,1)=s(8,2) ;  s(12,2)=s(7,2)
wt(1)=.050844906370207 ; wt(2)=wt(1); wt(3)=wt(1)
wt(4)=.116786275726379 ; wt(5)=wt(4); wt(6)=wt(4)
wt(7)=.082851075618374 ; wt(8:12)=wt(7)            ; wt = .5*wt
              case(16)
s(1,1)=1./3. ;  s(1,2)=1./3.  ;  s(2,1)=.658861384496478
s(2,2)=.170569307751761 ; s(3,1)=s(2,2)   ;  s(3,2)=s(2,1)
s(4,1)=s(2,2)  ; s(4,2)=s(2,2)
s(5,1)=.898905543365938 ; s(5,2)=.050547228317031
s(6,1)=s(5,2);  s(6,2)=s(5,1) ; s(7,1)=s(5,2)  ;  s(7,2)=s(5,2)
s(8,1)=.081414823414554; s(8,2)=.459292588292723
s(9,1)=s(8,2)  ;  s(9,2)=s(8,1);  s(10,1)=s(8,2) ;  s(10,2)=s(8,2)
s(11,1)=.008394777409958; s(11,2)=.263112829634638
s(12,1)=s(11,1)   ;  s(12,2)=.728492392955404
s(13,1)=s(11,2) ;   s(13,2)=s(11,1)  ;  s(14,1)=s(11,2); s(14,2)=s(12,2)
s(15,1)=s(12,2) ;  s(15,2)=s(11,1) ;  s(16,1)=s(12,2) ;  s(16,2)=s(11,2)
wt(1)=.144315607677787 ; wt(2)=.103217370534718 ; wt(3)=wt(2); wt(4)=wt(2)
wt(5)=.032458497623198 ; wt(6)=wt(5)   ;  wt(7)=wt(5)
wt(8)=.095091634267284 ; wt(9)=wt(8)   ;  wt(10)=wt(8)
wt(11)=.027230314174435 ; wt(12:16) = wt(11)  ;     wt = .5*wt
              case default
                 print*,"wrong number of integrating points for a triangle"
             end select
            case ('quadrilateral')
             select case (nip)
              case(1)
                s(1,1) = .0 ; wt(1) = 4.
              case(4)
                s(1,1)=-root3; s(1,2)= root3
                s(2,1)= root3; s(2,2)= root3
                s(3,1)=-root3; s(3,2)=-root3
                s(4,1)= root3; s(4,2)=-root3
                wt = 1.0
              case(9)
                s(1:7:3,1) = -r15; s(2:8:3,1) = .0
                s(3:9:3,1) =  r15; s(1:3,2)   = r15
                s(4:6,2)   =  .0 ; s(7:9,2)   =-r15
                     wt= v
```

```
     case default
       print*,"wrong number of integrating points for a quadrilateral"
   end select
   case('tetrahedron')
    select case(nip)
     case(1)             ! for tetrahedra weights multiplied by 1/6
        s(1,1)=.25     ; s(1,2)=.25  ;  s(1,3)=.25   ; wt(1)=1./6.
     case(4)
      s(1,1)=.58541020 ; s(1,2)=.13819660  ;  s(1,3)=s(1,2)
      s(2,2)=s(1,1) ; s(2,3)=s(1,2)  ;  s(2,1)=s(1,2)
      s(3,3)=s(1,1) ; s(3,1)=s(1,2)  ;  s(3,2)=s(1,2)
      s(4,1)=s(1,2) ; s(4,2)=s(1,2)  ;  s(4,3)=s(1,2) ; wt(1:4)=.25/6.
     case(5)
       s(1,1)=.25  ;  s(1,2)=.25   ; s(1,3)=.25 ;  s(2,1)=.5
       s(2,2)=1./6. ;  s(2,3)=s(2,2);  s(3,2)=.5
       s(3,3)=1./6.  ;   s(3,1)=s(3,3)   ;    s(4,3)=.5
       s(4,1)=1./6. ;    s(4,2)=s(4,1);    s(5,1)=1./6.
       s(5,2)=s(5,1) ;  s(5,3)=s(5,1)
       wt(1)=-.8  ;  wt(2)=9./20. ;   wt(3:5)=wt(2)   ; wt =wt/6.
     case(6)
wt = 4./3.          ;  s(6,3) = 1.
s(1,1)=-1. ;s(2,1)=1. ; s(3,2)=-1. ; s(4,2)=1. ;  s(5,3)=-1.
     case default
      print*,"wrong number of integrating points for a tetrahedron"
   end select
   case('hexahedron')
    select case ( nip )
     case(1)
            s(1,1) = .0 ; wt(1) = 8.
     case(8)
            s(1,1)= root3;s(1,2)= root3;s(1,3)= root3
            s(2,1)= root3;s(2,2)= root3;s(2,3)=-root3
            s(3,1)= root3;s(3,2)=-root3;s(3,3)= root3
            s(4,1)= root3;s(4,2)=-root3;s(4,3)=-root3
            s(5,1)=-root3;s(5,2)= root3;s(5,3)= root3
            s(6,1)=-root3;s(6,2)=-root3;s(6,3)= root3
            s(7,1)=-root3;s(7,2)= root3;s(7,3)=-root3
            s(8,1)=-root3;s(8,2)=-root3;s(8,3)=-root3
            wt = 1.0
     case(14)
 b=0.795822426     ;      c=0.758786911
 wt(1:6)=0.886426593   ; wt(7:) =  0.335180055
 s(1,1)=-b ; s(2,1)=b  ;  s(3,2)=-b ;   s(4,2)=b
 s(5,3)=-b   ;     s(6,3)=b
 s(7:,:) = c
 s(7,1)=-c  ;  s(7,2)=-c  ; s(7,3)=-c ; s(8,2)=-c ;   s(8,3)=-c
 s(9,1)=-c  ;  s(9,3)=-c  ; s(10,3)=-c; s(11,1)=-c
 s(11,2)=-c ;  s(12,2)=-c ; s(13,1)=-c
     case(15)
 b=1.    ;      c=0.674199862
 wt(1)=1.564444444 ;  wt(2:7)=0.355555556  ; wt(8:15)=0.537777778
 s(2,1)=-b  ;    s(3,1)=b  ;    s(4,2)=-b  ;    s(5,2)=b
 s(6,3)=-b  ;    s(7,3)=b  ;    s(8:,:)=c  ;    s(8,1)=-c
 s(8,2)=-c  ;    s(8,3)=-c ;    s(9,2)=-c  ;    s(9,3)=-c
 s(10,1)=-c ;    s(10,3)=-c  ;  s(11,3)=-c ;    s(12,1)=-c
 s(12,2)=-c ;    s(13,2)=-c  ;  s(14,1)=-c
     case(27)
            wt = (/5./9.*v,8./9.*v,5./9.*v/)
            s(1:7:3,1) = -r15; s(2:8:3,1) = .0
            s(3:9:3,1) =  r15; s(1:3,3)   = r15
            s(4:6,3)   =  .0 ; s(7:9,3)   =-r15
            s(1:9,2)   = -r15
            s(10:16:3,1) = -r15; s(11:17:3,1) = .0
            s(12:18:3,1) =  r15; s(10:12,3)   = r15
```

```
                    s(13:15,3)   =  .0 ; s(16:18,3)   =-r15
                    s(10:18,2)   = .0
                    s(19:25:3,1) = -r15; s(20:26:3,1) = .0
                    s(21:27:3,1) =  r15; s(19:21,3)   = r15
                    s(22:24,3)   =  .0 ; s(25:27,3)   =-r15
                    s(19:27,2)   =  r15
               case default
                 print*,"wrong number of integrating points for a hexahedron"
             end select
           case default
             print*,"not a valid element type"
     end select
   return
 end subroutine sample
!---------------------- shape functions and derivatives ----------------------
 subroutine shape_der(der,points,i)
 implicit none
 integer,intent(in):: i; real,intent(in)::points(:,:)
 real,intent(out)::der(:,:)
  real::eta,xi,zeta,xi0,eta0,zeta0,etam,etap,xim,xip,c1,c2,c3 ! local variables
  real:: t1,t2,t3,t4,t5,t6,t7,t8,t9 ,x2p1,x2m1,e2p1,e2m1,zetam,zetap,x,y,z
  integer :: xii(20), etai(20), zetai(20) ,l,ndim , nod   ! local variables
  ndim = ubound(der , 1); nod = ubound(der , 2)
  select case (ndim)
   case(1) ! one dimensional case
         xi=points(i,1)
     select case (nod)
         case(2)
           der(1,1)=-0.5 ; der(1,2)=0.5
         case(3)
           t1=-1.-xi ; t2=-xi  ; t3=1.-xi
           der(1,1)=-(t3+t2)/2.  ; der(1,2)=(t3+t1)
           der(1,3)=-(t2+t1)/2.
         case(4)
           t1=-1.-xi ; t2=-1./3.-xi ; t3=1./3.-xi ; t4=1.-xi
           der(1,1)=-(t3*t4+t2*t4+t2*t3)*9./16.
           der(1,2)=(t3*t4+t1*t4+t1*t3)*27./16.
           der(1,3)=-(t2*t4+t1*t4+t1*t2)*27./16.
           der(1,4)=(t2*t3+t1*t3+t1*t2)*9./16.
         case(5)
           t1=-1.-xi ; t2=-0.5-xi ; t3=-xi ; t4=0.5-xi ; t5=1.-xi
           der(1,1)=-(t3*t4*t5+t2*t4*t5+t2*t3*t5+t2*t3*t4)*2./3.
           der(1,2)=(t3*t4*t5+t1*t4*t5+t1*t3*t5+t1*t3*t4)*8./3.
           der(1,3)=-(t2*t4*t5+t1*t4*t5+t1*t2*t5+t1*t2*t4)*4.
           der(1,4)=(t2*t3*t5+t1*t3*t5+t1*t2*t5+t1*t2*t3)*8./3.
           der(1,5)=-(t2*t3*t4+t1*t3*t4+t1*t2*t4+t1*t2*t3)*2./3.
     case default
       print*,"wrong number of nodes in shape_der"
     end select
   case(2)      ! two dimensional elements
       xi=points(i,1); eta=points(i,2) ; c1=xi ; c2=eta ; c3=1.-c1-c2
       etam=.25*(1.-eta); etap=.25*(1.+eta); xim=.25*(1.-xi); xip=.25*(1.+xi)
       x2p1=2.*xi+1. ;   x2m1=2.*xi-1. ;  e2p1=2.*eta+1. ;   e2m1=2.*eta-1.
     select case (nod)
      case(3)
       der(1,1)=1.;der(1,3)=0.;der(1,2)=-1.
       der(2,1)=0.;der(2,3)=1.;der(2,2)=-1.
      case(6)
       der(1,1)=4.*c1-1. ;  der(1,6)=4.*c2;  der(1,5)=0.  ; der(1,4)=-4.*c2
       der(1,3)=-(4.*c3-1.);  der(1,2)=4.*(c3-c1);   der(2,1)=0.
       der(2,6)=4.*c1 ; der(2,5)=4.*c2-1.; der(2,4)=4.*(c3-c2)
       der(2,3)=-(4.*c3-1.)  ; der(2,2)=-4.*c1
      case(15)
       t1=c1-.25  ;  t2=c1-.5 ;  t3=c1-.75   ;   t4=c2-.25
```

```
      t5=c2-.5   ;  t6=c2-.75 ;  t7=c3-.25  ;   t8=c3-.5 ;  t9=c3-.75
      der(1,1)=32./3.*(t2*t3*(t1+c1)+c1*t1*(t3+t2))
      der(1,12)=128./3.*c2*(t2*(t1+c1)+c1*t1) ;  der(1,11)=64.*c2*t4*(t1+c1)
      der(1,10)=128./3.*c2*t4*t5  ; der(1,9)=0. ; der(1,8)=-128./3.*c2*t4*t5
      der(1,7)=-64.*c2*t4*(t7+c3) ; der(1,6)=-128./3.*c2*(t8*(t7+c3)+c3*t7)
      der(1,5)=-32./3.*(t8*t9*(t7+c3)+c3*t7*(t8+t9))
      der(1,4)=128./3.*(c3*t7*t8-c1*(t8*(t7+c3)+c3*t7))
      der(1,3)=64.*(c3*t7*(t1+c1)-c1*t1*(t7+c3))
      der(1,2)=128./3.*(c3*(t2*(t1+c1)+c1*t1)-c1*t1*t2)
      der(1,13)=128.*c2*(c3*(t1+c1)-c1*t1) ;  der(1,15)=128.*c2*t4*(c3-c1)
      der(1,14)=128.*c2*(c3*t7-c1*(t7+c3))
      der(2,1)=0.0 ;  der(2,12)=128./3.*c1*t1*t2;  der(2,11)=64.*c1*t1*(t4+c2)
      der(2,10)=128./3.*c1*(t5*(t4+c2)+c2*t4)
      der(2,9)=32./3.*(t5*t6*(t4+c2)+c2*t4*(t6+t5))
      der(2,8)=128./3.*((c3*(t5*(t4+c2)+c2*t4))-c2*t4*t5)
      der(2,7)=64.*(c3*t7*(t4+c2)-c2*t4*(t7+c3))
      der(2,6)=128./3.*(c3*t7*t8-c2*(t8*(t7+c3)+c3*t7))
      der(2,5)=-32./3.*(t8*t9*(t7+c3)+c3*t7*(t8+t9))
      der(2,4)=-128./3.*c1*(t8*(t7+c3)+c3*t7)
      der(2,3)=-64.*c1*t1*(t7+c3)  ;  der(2,2)=-128./3.*c1*t1*t2
      der(2,13)=128.*c1*t1*(c3-c2)
      der(2,15)=128.*c1*(c3*(t4+c2)-c2*t4)
      der(2,14)=128.*c1*(c3*t7-c2*(c3+t7))
     case (4)
      der(1,1)=-etam; der(1,2)=-etap; der(1,3)=etap; der(1,4)=etam
      der(2,1)=-xim; der(2,2)=xim; der(2,3)=xip; der(2,4)=-xip
     case(8)
      der(1,1)=etam*(2.*xi+eta); der(1,2)=-8.*etam*etap
      der(1,3)=etap*(2.*xi-eta); der(1,4)=-4.*etap*xi
      der(1,5)=etap*(2.*xi+eta); der(1,6)=8.*etap*etam
      der(1,7)=etam*(2.*xi-eta); der(1,8)=-4.*etam*xi
      der(2,1)=xim*(xi+2.*eta); der(2,2)=-4.*xim*eta
      der(2,3)=xim*(2.*eta-xi); der(2,4)=8.*xim*xip
      der(2,5)=xip*(xi+2.*eta); der(2,6)=-4.*xip*eta
      der(2,7)=xip*(2.*eta-xi); der(2,8)=-8.*xim*xip
    case(9)
      etam = eta - 1.; etap = eta + 1.; xim = xi - 1.; xip = xi + 1.
      der(1,1)=.25*x2m1*eta*etam  ;   der(1,2)=-.5*x2m1*etap*etam
      der(1,3)=.25*x2m1*eta*etap  ;   der(1,4)=-xi*eta*etap
      der(1,5)=.25*x2p1*eta*etap  ;   der(1,6)=-.5*x2p1*etap*etam
      der(1,7)=.25*x2p1*eta*etam  ;   der(1,8)=-xi*eta*etam
      der(1,9)=2.*xi*etap*etam    ;   der(2,1)=.25*xi*xim*e2m1
      der(2,2)=-xi*xim*eta        ;   der(2,3)=.25*xi*xim*e2p1
      der(2,4)=-.5*xip*xim*e2p1   ;   der(2,5)=.25*xi*xip*e2p1
      der(2,6)=-xi*xip*eta        ;   der(2,7)=.25*xi*xip*e2m1
      der(2,8)=-.5*xip*xim*e2m1   ;   der(2,9)=2.*xip*xim*eta
    case default
      print*,"wrong number of nodes in shape_der"
    end select
  case(3)   ! three dimensional elements
      xi=points(i,1); eta=points(i,2); zeta=points(i,3)
      etam=1.-eta ; xim=1.-xi;  zetam=1.-zeta
      etap=eta+1. ; xip=xi+1. ;  zetap=zeta+1.
   select case (nod)
    case(4)
     der(1:3,1:4) = .0
     der(1,1)=1.;  der(2,2)=1.  ;  der(3,3)=1.
     der(1,4)=-1. ;  der(2,4)=-1. ;  der(3,4)=-1.
    case(8)
     der(1,1)=-.125*etam*zetam    ;   der(1,2)=-.125*etam*zetap
     der(1,3)=.125*etam*zetap     ;   der(1,4)=.125*etam*zetam
     der(1,5)=-.125*etap*zetam    ;   der(1,6)=-.125*etap*zetap
     der(1,7)=.125*etap*zetap     ;   der(1,8)=.125*etap*zetam
     der(2,1)=-.125*xim*zetam     ;   der(2,2)=-.125*xim*zetap
```

```
      der(2,3)=-.125*xip*zetap      ;   der(2,4)=-.125*xip*zetam
      der(2,5)=.125*xim*zetam       ;   der(2,6)=.125*xim*zetap
      der(2,7)=.125*xip*zetap       ;   der(2,8)=.125*xip*zetam
      der(3,1)=-.125*xim*etam       ;   der(3,2)=.125*xim*etam
      der(3,3)=.125*xip*etam        ;   der(3,4)=-.125*xip*etam
      der(3,5)=-.125*xim*etap       ;   der(3,6)=.125*xim*etap
      der(3,7)=.125*xip*etap        ;   der(3,8)=-.125*xip*etap
 case(14) ! type 6 element
  x= points(i,1)    ;   y= points(i,2)   ;    z= points(i,3)
  der(1,1)=((2.*x*y+2.*x*z+4.*x+y*z+y+z)*(y-1.)*(z-1.))/8.
  der(1,2)=((2.*x*y-2.*x*z-4.*x+y*z+y-z)*(y+1.)*(z-1.))/8.
  der(1,3)=((2.*x*y+2.*x*z+4.*x-y*z-y-z)*(y-1.)*(z-1.))/8.
  der(1,4)=((2.*x*y-2.*x*z-4.*x-y*z-y+z)*(y+1.)*(z-1.))/8.
  der(1,5)=-((2.*x*y-2.*x*z+4.*x-y*z+y-z)*(y-1.)*(z+1.))/8.
  der(1,6)=-((2.*x*y+2.*x*z-4.*x-y*z+y+z)*(y+1.)*(z+1.))/8.
  der(1,7)=-((2.*x*y-2.*x*z+4.*x+y*z-y+z)*(y-1.)*(z+1.))/8.
  der(1,8)=-((2.*x*y+2.*x*z-4.*x+y*z-y-z)*(y+1.)*(z+1.))/8.
  der(1,9)=-(y+1.)*(y-1.)*(z-1.)*x  ;   der(1,10)=(y+1.)*(y-1.)*(z+1.)*x
  der(1,11)=-(y-1.)*(z+1.)*(z-1.)*x ;   der(1,12)=(y+1.)*(z+1.)*(z-1.)*x
  der(1,13)=-((y+1.)*(y-1.)*(z+1.)*(z-1.))/2.
  der(1,14)=((y+1.)*(y-1.)*(z+1.)*(z-1.))/2.
  der(2,1)=((2.*x*y+x*z+x+2.*y*z+4.*y+z)*(x-1.)*(z-1.))/8.
  der(2,2)=((2.*x*y-x*z-x+2.*y*z+4.*y-z)*(x-1.)*(z-1.))/8.
  der(2,3)=((2.*x*y+x*z+x-2.*y*z-4.*y-z)*(x+1.)*(z-1.))/8.
  der(2,4)=((2.*x*y-x*z-x-2.*y*z-4.*y+z)*(x+1.)*(z-1.))/8.
  der(2,5)=-((2.*x*y-x*z+x-2.*y*z+4.*y-z)*(x-1.)*(z+1.))/8.
  der(2,6)=-((2.*x*y+x*z-x-2.*y*z+4.*y+z)*(x-1.)*(z+1.))/8.
  der(2,7)=-((2.*x*y-x*z+x+2.*y*z-4.*y+z)*(x+1.)*(z+1.))/8.
  der(2,8)=-((2.*x*y+x*z-x+2.*y*z-4.*y-z)*(x+1.)*(z+1.))/8.
  der(2,9)=-(x+1.)*(x-1.)*(z-1.)*y
  der(2,10)=(x+1.)*(x-1.)*(z+1.)*y
  der(2,11)=-((x+1.)*(x-1.)*(z+1.)*(z-1.))/2.
  der(2,12)=((x+1.)*(x-1.)*(z+1.)*(z-1.))/2.
  der(2,13)=-(x-1.)*(z+1.)*(z-1.)*y
  der(2,14)=(x+1.)*(z+1.)*(z-1.)*y
  der(3,1)=((x*y+2.*x*z+x+2.*y*z+y+4.*z)*(x-1.)*(y-1.))/8.
  der(3,2)=((x*y-2.*x*z-x+2.*y*z+y-4.*z)*(x-1.)*(y+1.))/8.
  der(3,3)=((x*y+2.*x*z+x-2.*y*z-y-4.*z)*(x+1.)*(y-1.))/8.
  der(3,4)=((x*y-2.*x*z-x-2.*y*z-y+4.*z)*(x+1.)*(y+1.))/8.
  der(3,5)=-((x*y-2.*x*z+x-2.*y*z+y-4.*z)*(x-1.)*(y-1.))/8.
  der(3,6)=-((x*y+2.*x*z-x-2.*y*z+y+4.*z)*(x-1.)*(y+1.))/8.
  der(3,7)=-((x*y-2.*x*z+x+2.*y*z-y+4.*z)*(x+1.)*(y-1.))/8.
  der(3,8)=-((x*y+2.*x*z-x+2.*y*z-y-4.*z)*(x+1.)*(y+1.))/8.
  der(3,9)=-((x+1.)*(x-1.)*(y+1.)*(y-1.))/2.
  der(3,10)=((x+1.)*(x-1.)*(y+1.)*(y-1.))/2.
  der(3,11)=-(x+1.)*(x-1.)*(y-1.)*z  ; der(3,12)=(x+1.)*(x-1.)*(y+1.)*z
  der(3,13)=-(x-1.)*(y+1.)*(y-1.)*z  ; der(3,14)=(x+1.)*(y+1.)*(y-1.)*z
   case(20)
     xii=(/-1,-1,-1,0,1,1,1,0,-1,-1,1,1,-1,-1,-1,0,1,1,1,0/)
     etai=(/-1,-1,-1,-1,-1,-1,-1,-1,0,0,0,0,1,1,1,1,1,1,1,1/)
     zetai=(/-1,0,1,1,1,0,-1,-1,-1,1,1,-1,-1,0,1,1,1,0,-1,-1/)
     do l=1,20
        xi0=xi*xii(l); eta0=eta*etai(l); zeta0=zeta*zetai(l)
        if(l==4.or.l==8.or.l==16.or.l==20) then
         der(1,l)=-.5*xi*(1.+eta0)*(1.+zeta0)
         der(2,l)=.25*etai(l)*(1.-xi*xi)*(1.+zeta0)
         der(3,l)=.25*zetai(l)*(1.-xi*xi)*(1.+eta0)
        else if(l>=9.and.l<=12)then
         der(1,l)=.25*xii(l)*(1.-eta*eta)*(1.+zeta0)
         der(2,l)=-.5*eta*(1.+xi0)*(1.+zeta0)
         der(3,l)=.25*zetai(l)*(1.+xi0)*(1.-eta*eta)
        else if(l==2.or.l==6.or.l==14.or.l==18) then
         der(1,l)=.25*xii(l)*(1.+eta0)*(1.-zeta*zeta)
         der(2,l)=.25*etai(l)*(1.+xi0)*(1.-zeta*zeta)
```

```
          der(3,1)=-.5*zeta*(1.+xi0)*(1.+eta0)
         else
          der(1,1)=.125*xii(1)*(1.+eta0)*(1.+zeta0)*(2.*xi0+eta0+zeta0-1.)
          der(2,1)=.125*etai(1)*(1.+xi0)*(1.+zeta0)*(xi0+2.*eta0+zeta0-1.)
          der(3,1)=.125*zetai(1)*(1.+xi0)*(1.+eta0)*(xi0+eta0+2.*zeta0-1.)
         end if
      end do
     case default
       print*,"wrong number of nodes in shape_der"
   end select
  case default
   print*,"wrong number of dimensions in shape_der"
  end select
 return
 end subroutine shape_der
 subroutine shape_fun(fun,points,i)
  implicit none
  integer,intent(in):: i; real,intent(in)::points(:,:)
  real,intent(out)::fun(:)
  real :: eta,xi,etam,etap,xim,xip,zetam,zetap,c1,c2,c3      !local variabl
  real :: t1,t2,t3,t4,t5,t6,t7,t8,t9,x,y,z
  real :: zeta,xi0,eta0,zeta0; integer::xii(20),etai(20),zetai(20),l,ndim,
        ndim = ubound(points , 2 ); nod = ubound(fun , 1 )
    select case (ndim)
      case(1) ! one dimensional cases
           xi=points(i,1)
        select case(nod)
         case(2)
           t1=-1.-xi ; t2=1.-xi
           fun(1)=t2/2. ; fun(2)=-t1/2.
         case(3)
           t1=-1.-xi ; t2=-xi ; t3=1.-xi
           fun(1)=t2*t3/2. ; fun(2)=-t1*t3 ; fun(3)=t1*t2/2.
         case(4)
           t1=-1.-xi ; t2=-1./3.-xi ; t3=1./3.-xi ; t4=1.-xi
           fun(1)=t2*t3*t4*9./16.  ; fun(2)=-t1*t3*t4*27./16.
           fun(3)=t1*t2*t4*27./16. ; fun(4)=-t1*t2*t3*9./16.
         case(5)
           t1=-1.-xi ; t2=-0.5-xi ; t3=-xi ; t4=0.5-xi ; t5=1.-xi
           fun(1)=t2*t3*t4*t5*2./3. ; fun(2)=-t1*t3*t4*t5*8./3.
           fun(3)=t1*t2*t4*t5*4. ; fun(4)=-t1*t2*t3*t5*8./3.
           fun(5)=t1*t2*t3*t4*2./3.
          case default
             print*,"wrong number of nodes in shape_fun"
        end select
      case(2) ! two dimensional cases
            c1=points(i,1); c2=points(i,2); c3=1.-c1-c2
            xi=points(i,1);  eta=points(i,2)
            etam=.25*(1.-eta); etap=.25*(1.+eta)
            xim=.25*(1.-xi); xip=.25*(1.+xi)
        select case(nod)
          case(3)
            fun = (/c1,c3,c2/)
          case(6)
            fun(1)=(2.*c1-1.)*c1 ;  fun(6)=4.*c1*c2 ;  fun(5)=(2.*c2-1.)*c
            fun(4)=4.*c2*c3      ;  fun(3)=(2.*c3-1.)*c3 ; fun(2)=4.*c3*c1
          case(15)
            t1=c1-.25  ;  t2=c1-.5 ;  t3=c1-.75   ;   t4=c2-.25
            t5=c2-.5   ;  t6=c2-.75 ;  t7=c3-.25  ;   t8=c3-.5 ;  t9=c3-.7
            fun(1)=32./3.*c1*t1*t2*t3    ;  fun(12)=128./3.*c1*c2*t1*t2
            fun(11)=64.*c1*c2*t1*t4      ;  fun(10)=128./3.*c1*c2*t4*t5
            fun(9)=32./3.*c2*t4*t5*t6    ;  fun(8)=128./3.*c2*c3*t4*t5
            fun(7)=64.*c2*c3*t4*t7       ;  fun(6)=128./3.*c2*c3*t7*t8
            fun(5)=32./3.*c3*t7*t8*t9    ;  fun(4)=128./3.*c3*c1*t7*t8
```

```
          fun(3)=64.*c3*c1*t1*t7      ;  fun(2)=128./3.*c3*c1*t1*t2
          fun(13)=128.*c1*c2*t1*c3    ;  fun(15)=128.*c1*c2*c3*t4
          fun(14)=128.*c1*c2*c3*t7
        case(4)
          fun=(/4.*xim*etam,4.*xim*etap,4.*xip*etap,4.*xip*etam/)
        case(8)
          fun=(/4.*etam*xim*(-xi-eta-1.),32.*etam*xim*etap,&
                4.*etap*xim*(-xi+eta-1.),32.*xim*xip*etap, &
                4.*etap*xip*(xi+eta-1.), 32.*etap*xip*etam,&
                4.*xip*etam*(xi-eta-1.), 32.*xim*xip*etam/)
        case(9)
          etam = eta - 1.; etap= eta + 1.; xim = xi - 1.; xip = xi + 1.
          fun=(/.25*xi*xim*eta*etam,-.5*xi*xim*etap*etam,&
                .25*xi*xim*eta*etap,-.5*xip*xim*eta*etap,&
                .25*xi*xip*eta*etap,-.5*xi*xip*etap*etam,&
                .25*xi*xip*eta*etam,-.5*xip*xim*eta*etam,xip*xim*etap*etam/)
        case default
           print*,"wrong number of nodes in shape_fun"
      end select
    case(3) ! three dimensional cases
     xi=points(i,1); eta=points(i,2); zeta=points(i,3)
     etam=1.-eta ;  xim=1.-xi  ;  zetam=1.-zeta
     etap=eta+1. ;  xip=xi+1.   ;  zetap=zeta+1.
     select case(nod)
      case(4)
       fun(1)=xi   ;   fun(2)= eta ;  fun(3)=zeta
       fun(4)=1.-fun(1)-fun(2)-fun(3)
      case(8)
       fun=(/.125*xim*etam*zetam,.125*xim*etam*zetap,.125*xip*etam*zetap,&
            .125*xip*etam*zetam,.125*xim*etap*zetam,.125*xim*etap*zetap,&
            .125*xip*etap*zetap,.125*xip*etap*zetam/)
      case(14) !type 6 element
  x = points(i,1);  y = points(i,2);  z = points(i,3)
fun(1)=((x*y+x*z+2.*x+y*z+2.*y+2.*z+2.)*(x-1.)*(y-1.)*(z-1.))/8.
fun(2)=((x*y-x*z-2.*x+y*z+2.*y-2.*z-2.)*(x-1.)*(y+1.)*(z-1.))/8.
fun(3)=((x*y+x*z+2.*x-y*z-2.*y-2.*z-2.)*(x+1.)*(y-1.)*(z-1.))/8.
fun(4)=((x*y-x*z-2.*x-y*z-2.*y+2.*z+2.)*(x+1.)*(y+1.)*(z-1.))/8.
fun(5)=-((x*y-x*z+2.*x-y*z+2.*y-2.*z+2.)*(x-1.)*(y-1.)*(z+1.))/8.
fun(6)=-((x*y+x*z-2.*x-y*z+2.*y+2.*z-2.)*(x-1.)*(y+1.)*(z+1.))/8.
fun(7)=-((x*y-x*z+2.*x+y*z-2.*y+2.*z-2.)*(x+1.)*(y-1.)*(z+1.))/8.
fun(8)=-((x*y+x*z-2.*x+y*z-2.*y-2.*z+2.)*(x+1.)*(y+1.)*(z+1.))/8.
fun(9)=-((x+1.)*(x-1.)*(y+1.)*(y-1.)*(z-1.))/2.
fun(10)=((x+1.)*(x-1.)*(y+1.)*(y-1.)*(z+1.))/2.
fun(11)=-((x+1.)*(x-1.)*(y-1.)*(z+1.)*(z-1.))/2.
fun(12)=((x+1.)*(x-1.)*(y+1.)*(z+1.)*(z-1.))/2.
fun(13)=-((x-1.)*(y+1.)*(y-1.)*(z+1.)*(z-1.))/2.
fun(14)=((x+1.)*(y+1.)*(y-1.)*(z+1.)*(z-1.))/2.
      case(20)
         xii=(/-1,-1,-1,0,1,1,1,0,-1,-1,1,1,-1,-1,-1,0,1,1,1,0/)
         etai=(/-1,-1,-1,-1,-1,-1,-1,-1,0,0,0,0,1,1,1,1,1,1,1,1/)
         zetai=(/-1,0,1,1,1,0,-1,-1,-1,1,1,-1,-1,0,1,1,1,0,-1,-1/)
         do l=1,20
          xi0=xi*xii(l); eta0=eta*etai(l); zeta0=zeta*zetai(l)
          if(l==4.or.l==8.or.l==16.or.l==20) then
            fun(l)=.25*(1.-xi*xi)*(1.+eta0)*(1.+zeta0)
          else if(l>=9.and.l<=12)then
            fun(l)=.25*(1.+xi0)*(1.-eta*eta)*(1.+zeta0)
          else if(l==2.or.l==6.or.l==14.or.l==18) then
            fun(l)=.25*(1.+xi0)*(1.+eta0)*(1.-zeta*zeta)
          else
            fun(l)=.125*(1.+xi0)*(1.+eta0)*(1.+zeta0)*(xi0+eta0+zeta0-2)
          end if
         end do
        case default
```

```
          print*,"wrong number of nodes in shape_fun"
        end select
      case default
        print*,"wrong number of dimensions in shape_fun"
    end select
  return
 end subroutine shape_fun
 subroutine fmbeam(der2,fun,points,i,ell)
!
! this subroutine forms the beam shape functions
! and their 2nd derivatives in local coordinates
!
 implicit none
 real,intent(in)::points(:,:),ell
 integer,intent(in)::i
 real,intent(out)::der2(:),fun(:)
 real::xi,xi2,xi3
 xi=points(i,1); xi2=xi*xi; xi3=xi2*xi
 fun(1)=.25*(xi3-3.*xi+2.); fun(2)=.125*ell*(xi3-xi2-xi+1.)
 fun(3)=.25*(-xi3+3.*xi+2.); fun(4)=.125*ell*(xi3+xi2-xi-1.)
 der2(1)=1.5*xi; der2(2)=.25*ell*(3.*xi-1.)
 der2(3)=-1.5*xi; der2(4)=.25*ell*(3.*xi+1.)
 end subroutine fmbeam
!----------------------local to global conversion ----------------------
subroutine loc_to_glob(local,global,gamma,coord)
!      this subroutine transforms the local end reactions and
!      moments into the global system (3-d)
implicit none
real,intent(in)::local(:),gamma,coord(:,:)
real,intent(out)::global(:)
real::t(12,12),r0(3,3),x1,x2,y1,y2,z1,z2,xl,yl,zl,&
      pi,gamrad,cg,sg,den,ell,x,sum
integer::i,j,k
     x1=coord(1,1); y1=coord(1,2); z1=coord(1,3)
     x2=coord(2,1); y2=coord(2,2); z2=coord(2,3)
     xl=x2-x1; yl=y2-y1; zl=z2-z1; ell=sqrt(xl*xl+yl*yl+zl*zl)
     t=0.0
     pi=acos(-1.); gamrad=gamma*pi/180.; cg=cos(gamrad); sg=sin(gamrad)
     den=ell*sqrt(xl*xl+zl*zl)
     if(den /= 0.0)then
      r0(1,1)=xl/ell; r0(2,1)=yl/ell; r0(3,1)=zl/ell
      r0(1,2)=(-xl*yl*cg-ell*zl*sg)/den; r0(2,2)=den*cg/(ell*ell)
      r0(3,2)=(-yl*zl*cg+ell*xl*sg)/den; r0(1,3)=(xl*yl*sg-ell*zl*cg)/den
      r0(2,3)=-den*sg/(ell*ell); r0(3,3)=(yl*zl*sg+ell*xl*cg)/den
     else
      r0(1,1)=0.; r0(3,1)=0.; r0(2,2)=0.; r0(2,3)=0.
      r0(2,1)=1.; r0(1,2)=-cg; r0(3,3)=cg; r0(3,2)=sg; r0(1,3)=sg
     end if
     do i=1,3; do j=1,3
      x=r0(i,j)
      do k=0,9,3; t(i+k,j+k)=x; end do
     end do; end do
     do i=1,12
      sum=0.
       do j=1,12
        sum=sum+t(i,j)*local(j)
       end do
      global(i)=sum
     end do
      return
end subroutine loc_to_glob
subroutine glob_to_loc(local,global,gamma,coord)
!
       this subroutine transforms the global end reactions and
```

```
!      moments into the local system (2-d, 3-d)
!
implicit none
real,intent(in)::global(:),gamma,coord(:,:)
real,intent(out)::local(:)
real::t(12,12),r0(3,3),x1,x2,y1,y2,z1,z2,xl,yl,zl,&
      pi,gamrad,cg,sg,den,ell,x,sum
integer::i,j,k,ndim
ndim=ubound(coord,2)
select case(ndim)
  case(2)
      x1=coord(1,1); y1=coord(1,2)
      x2=coord(2,1); y2=coord(2,2)
      ell=sqrt((x2-x1)**2+(y2-y1)**2)
      cg=(x2-x1)/ell
      sg=(y2-y1)/ell
      local(1)=cg*global(1)+sg*global(2)
      local(2)=cg*global(2)-sg*global(1)
      local(3)=global(3)
      local(4)=cg*global(4)+sg*global(5)
      local(5)=cg*global(5)-sg*global(4)
      local(6)=global(6)
  case(3)
     x1=coord(1,1); y1=coord(1,2); z1=coord(1,3)
     x2=coord(2,1); y2=coord(2,2); z2=coord(2,3)
     xl=x2-x1; yl=y2-y1; zl=z2-z1; ell=sqrt(xl*xl+yl*yl+zl*zl)
     t=0.0
     pi=acos(-1.); gamrad=gamma*pi/180.; cg=cos(gamrad); sg=sin(gamrad)
     den=ell*sqrt(xl*xl+zl*zl)
     if(den /= 0.0)then
      r0(1,1)=xl/ell; r0(1,2)=yl/ell; r0(1,3)=zl/ell
      r0(2,1)=(-xl*yl*cg-ell*zl*sg)/den; r0(2,2)=den*cg/(ell*ell)
      r0(2,3)=(-yl*zl*cg+ell*xl*sg)/den; r0(3,1)=(xl*yl*sg-ell*zl*cg)/den
      r0(3,2)=-den*sg/(ell*ell); r0(3,3)=(yl*zl*sg+ell*xl*cg)/den
     else
      r0(1,1)=0.; r0(1,3)=0.; r0(2,2)=0.; r0(3,2)=0.
      r0(1,2)=1.; r0(2,1)=-cg; r0(3,3)=cg; r0(2,3)=sg; r0(3,1)=sg
     end if
     do i=1,3; do j=1,3
      x=r0(i,j)
      do k=0,9,3; t(i+k,j+k)=x; end do
     end do; end do
     do i=1,12
      sum=0.
       do j=1,12
        sum=sum+t(i,j)*global(j)
       end do
       local(i)=sum
     end do
   end select
   return
end subroutine glob_to_loc
!---------------------------- line structures --------------------------
subroutine hinge(coord,holdr,action,react,prop,iel,etype,gamma)
!
!      this subroutine forms the end forces and moments to be
!      applied to a member if a joint has gone plastic
!
implicit none
real,intent(in):: holdr(:,:),coord(:,:),action(:),prop(:,:),gamma(:)
real,intent(out):: react(:)
integer,intent(in)::etype(:),iel
real::ell,x1,x2,y1,y2,z1,z2,csch,snch,bm1,bm2,bm3,bm4,bm5,&
      s1,s2,s3,s4,s5,mpy,mpz,mpx
```

```
real::global(12),local(12),total(12),gam
integer::ndim
ndim=ubound(coord,2)
bm1=0.; bm2=0.; bm3=0.; bm4=0.; bm5=0.
total(:)=holdr(:,iel)
select case(ndim)
  case(1)
    mpy=prop(2,etype(iel))
    ell=coord(2,1)-coord(1,1)
    s1=total(2)+action(2); s2=total(4)+action(4)
  if(abs(s1) > mpy)then
    if(s1 >  0.)bm1= mpy-s1
    if(s1 <= 0.)bm1=-mpy-s1
  end if
  if(abs(s2) > mpy)then
    if(s2 >  0.)bm2= mpy-s2
    if(s2 <= 0.)bm2=-mpy-s2
  end if
    react(1)= (bm1+bm2)/ell; react(2)=bm1
    react(3)=-react(1);      react(4)=bm2
  case(2)
    mpy=prop(3,etype(iel))
    x1=coord(1,1); y1=coord(1,2); x2=coord(2,1); y2=coord(2,2)
    ell=sqrt((y2-y1)**2+(x2-x1)**2)
    csch=(x2-x1)/ell; snch=(y2-y1)/ell
    s1=total(3)+action(3); s2=total(6)+action(6)
  if(abs(s1) > mpy)then
    if(s1 >  0.)bm1= mpy-s1
    if(s1 <= 0.)bm1=-mpy-s1
  end if
  if(abs(s2) > mpy)then
    if(s2 >  0.)bm2= mpy-s2
    if(s2 <= 0.)bm2=-mpy-s2
  end if
    react(1)=-(bm1+bm2)*snch/ell; react(2)= (bm1+bm2)*csch/ell; react(3)=bm1
    react(4)=-react(1); react(5)=-react(2); react(6)=bm2
  case(3)
    gam=gamma(iel)
    mpy=prop(5,etype(iel))
    mpz=prop(6,etype(iel))
    mpx=prop(7,etype(iel))
    x1=coord(1,1); y1=coord(1,2); z1=coord(1,3)
    x2=coord(2,1); y2=coord(2,2); z2=coord(2,3)
    ell=sqrt((z2-z1)**2+(y2-y1)**2+(x2-x1)**2)
    global=total+action
    call glob_to_loc(local,global,gam,coord)
    global=0.0
    s1=local(5); s2=local(11)
    if(abs(s1) > mpy)then
      if(s1 >  0.)bm1= mpy-s1
      if(s1 <= 0.)bm1=-mpy-s1
    end if
      if(abs(s2) > mpy)then
      if(s2 >  0.)bm2= mpy-s2
      if(s2 <= 0.)bm2=-mpy-s2
    end if
      local( 3)=-(bm1+bm2)/ell
      local( 9)=-local(3)
      local( 5)= bm1
      local(11)= bm2
    s3=local(6); s4=local(12)
    if(abs(s3) > mpz)then
      if(s3 >  0.)bm1= mpz-s3
      if(s3 <= 0.)bm1=-mpz-s3
```

```
    end if
      if(abs(s4) > mpy)then
      if(s4 >  0.)bm2= mpz-s4
      if(s4 <= 0.)bm2=-mpz-s4
    end if
      local( 2)=(bm3+bm4)/ell
      local( 8)=-local(2)
      local( 6)= bm3
      local(12)= bm4
      s5=local(4)
      if(abs(s5) > mpx)then
        if(s5 >  0.)global(4)= mpx-s5
        if(s5 <= 0.)global(4)=-mpx-s5
      end if
      local(10)=-local(4)
      call loc_to_glob(local,react,gam,coord)
  end select
return
end subroutine hinge
 subroutine rigid_jointed(km,prop,gamma,etype,iel,coord)
!
!      this subroutine forms the stiffness matrix of a
!      general beam/column element(1-d, 2-d or 3-d)
!
 implicit none
   real,intent(in)::gamma(:),coord(:,:),prop(:,:)
   integer,intent(in)::etype(:),iel
   real,intent(out):: km(:,:)
   integer::ndim,i,j,k
   real::ell,x1,x2,y1,y2,z1,z2,c,s,e1,e2,e3,e4,pi,xl,yl,zl,cg,sg,den
   real::ea,ei,eiy,eiz,gj
   real::a1,a2,a3,a4,a5,a6,a7,a8,sum,gamrad,x
   real::t(12,12),tt(12,12),cc(12,12),r0(3,3)
   ndim=ubound(coord,2)
   select case(ndim)
    case(1)
     ei=prop(1,etype(iel)); ell=coord(2,1)-coord(1,1)
     km(1,1)=12.*ei/(ell*ell*ell) ;km(3,3)=km(1,1)
     km(1,2)=6.*ei/(ell*ell) ;  km(2,1)=km(1,2) ; km(1,4)=km(1,2)
     km(4,1)=km(1,4) ;  km(1,3)=-km(1,1) ; km(3,1)=km(1,3) ; km(3,4)=-km(1,2)
     km(4,3)=km(3,4) ; km(2,3)=km(3,4) ;  km(3,2)=km(2,3); km(2,2)=4.*ei/ell
     km(4,4)=km(2,2) ;km(2,4)=2.*ei/ell ; km(4,2)=km(2,4)
    case(2)
     ea=prop(1,etype(iel)); ei=prop(2,etype(iel))
     x1=coord(1,1); y1=coord(1,2); x2=coord(2,1); y2=coord(2,2)
     ell=sqrt((y2-y1)**2+(x2-x1)**2)
     c=(x2-x1)/ell; s=(y2-y1)/ell
     e1=ea/ell; e2=12.*ei/(ell*ell*ell); e3=ei/ell; e4=6.*ei/(ell*ell)
     km(1,1)=c*c*e1+s*s*e2; km(4,4)=km(1,1); km(1,2)=s*c*(e1-e2)
     km(2,1)=km(1,2); km(4,5)=km(1,2); km(5,4)=km(4,5); km(1,3)=-s*e4
     km(3,1)=km(1,3); km(1,6)=km(1,3); km(6,1)=km(1,6); km(3,4)=s*e4
     km(4,3)=km(3,4); km(4,6)=km(3,4); km(6,4)=km(4,6); km(1,4)=-km(1,1)
     km(4,1)=km(1,4); km(1,5)=s*c*(-e1+e2); km(5,1)=km(1,5); km(2,4)=km(1,5)
     km(4,2)=km(2,4); km(2,2)=s*s*e1+c*c*e2; km(5,5)=km(2,2); km(2,5)=-km(2,2)
     km(5,2)=km(2,5); km(2,3)=c*e4; km(3,2)=km(2,3); km(2,6)=km(2,3)
     km(6,2)=km(2,6); km(3,3)=4.*e3; km(6,6)=km(3,3); km(3,5)=-c*e4
     km(5,3)=km(3,5); km(5,6)=km(3,5); km(6,5)=km(5,6); km(3,6)=2.*e3
     km(6,3)=km(3,6)
    case(3)
     ea =prop(1,etype(iel)); eiy=prop(2,etype(iel))
     eiz=prop(3,etype(iel)); gj =prop(4,etype(iel))
     x1=coord(1,1); y1=coord(1,2); z1=coord(1,3)
     x2=coord(2,1); y2=coord(2,2); z2=coord(2,3)
     xl=x2-x1; yl=y2-y1; zl=z2-z1; ell=sqrt(xl*xl+yl*yl+zl*zl)
```

```
      km=0.0; t=0.0; tt=0.0
      a1=ea/ell; a2=12.*eiz/(ell*ell*ell); a3=12.*eiy/(ell*ell*ell)
      a4=6.*eiz/(ell*ell); a5=6.*eiy/(ell*ell); a6=4.*eiz/ell
      a7=4.*eiy/ell; a8=gj/ell
      km(1,1)=a1; km(7,7)=a1; km(1,7)=-a1; km(7,1)=-a1; km(2,2)=a2; km(8,8)=a2
      km(2,8)=-a2; km(8,2)=-a2; km(3,3)=a3; km(9,9)=a3; km(3,9)=-a3; km(9,3)=-a3
      km(4,4)=a8; km(10,10)=a8; km(4,10)=-a8; km(10,4)=-a8; km(5,5)=a7
      km(11,11)=a7; km(5,11)=.5*a7; km(11,5)=.5*a7; km(6,6)=a6; km(12,12)=a6
      km(6,12)=.5*a6; km(12,6)=.5*a6; km(2,6)=a4; km(6,2)=a4; km(2,12)=a4
      km(12,2)=a4; km(6,8)=-a4; km(8,6)=-a4; km(8,12)=-a4; km(12,8)=-a4
      km(5,9)=a5; km(9,5)=a5; km(9,11)=a5; km(11,9)=a5; km(3,5)=-a5
      km(5,3)=-a5; km(3,11)=-a5; km(11,3)=-a5
      pi=acos(-1.); gamrad=gamma(iel)*pi/180.; cg=cos(gamrad); sg=sin(gamrad)
      den=ell*sqrt(xl*xl+zl*zl)
      if(den /= 0.0)then
       r0(1,1)=xl/ell; r0(1,2)=yl/ell; r0(1,3)=zl/ell
       r0(2,1)=(-xl*yl*cg-ell*zl*sg)/den; r0(2,2)=den*cg/(ell*ell)
       r0(2,3)=(-yl*zl*cg+ell*xl*sg)/den; r0(3,1)=(xl*yl*sg-ell*zl*cg)/den
       r0(3,2)=-den*sg/(ell*ell); r0(3,3)=(yl*zl*sg+ell*xl*cg)/den
      else
       r0(1,1)=0.; r0(1,3)=0.; r0(2,2)=0.; r0(3,2)=0.
       r0(1,2)=1.; r0(2,1)=-cg; r0(3,3)=cg; r0(2,3)=sg; r0(3,1)=sg
      end if
      do i=1,3; do j=1,3
       x=r0(i,j)
       do k=0,9,3; t(i+k,j+k)=x; tt(j+k,i+k)=x; end do
      end do; end do
      do i=1,12; do j=1,12
       sum=0.0
       do k=1,12; sum=sum+km(i,k)*t(k,j); end do
       cc(i,j)=sum
      end do; end do
      do i=1,12; do j=1,12
       sum=0.
       do k=1,12; sum=sum+tt(i,k)*cc(k,j); end do
       km(i,j)=sum
      end do; end do
  end select
 return
 end subroutine rigid_jointed
 subroutine pin_jointed(km,ea,coord)
       this subroutine forms the stiffness matrix of a
       general rod element(1-d, 2-d or 3-d)
 implicit none
   real,intent(in)::ea,coord(:,:); real,intent(out):: km(:,:)
   integer::ndim ,i,j
   real::eaol,ell,cs,sn,x1,x2,y1,y2,z1,z2,a,b,c,d,e,f,xl,yl,zl
   ndim=ubound(coord,2)
   select case(ndim)
    case(1)
      ell=coord(2,1)-coord(1,1)
      eaol=ea/ell
      km(1,1)=eaol;  km(1,2)=-eaol
      km(2,1)=-eaol; km(2,2)=eaol
    case(2)
      x1=coord(1,1); y1=coord(1,2)
      x2=coord(2,1); y2=coord(2,2)
      ell=sqrt((y2-y1)**2+(x2-x1)**2)
      cs=(x2-x1)/ell; sn=(y2-y1)/ell
      a=cs*cs; b=sn*sn; c=cs*sn
      km(1,1)=a; km(3,3)=a; km(1,3)=-a; km(3,1)=-a; km(2,2)=b; km(4,4)=b
      km(2,4)=-b; km(4,2)=-b; km(1,2)=c; km(2,1)=c; km(3,4)=c; km(4,3)=c
      km(1,4)=-c; km(4,1)=-c; km(2,3)=-c; km(3,2)=-c
      do i=1,4; do j=1,4; km(i,j)=km(i,j)*ea/ell; end do; end do
```

```
      case(3)
        x1=coord(1,1); y1=coord(1,2); z1=coord(1,3)
        x2=coord(2,1); y2=coord(2,2); z2=coord(2,3)
        xl=x2-x1; yl=y2-y1; zl=z2-z1
        ell=sqrt(xl*xl+yl*yl+zl*zl)
        xl=xl/ell; yl=yl/ell; zl=zl/ell
        a=xl*xl; b=yl*yl; c=zl*zl; d=xl*yl; e=yl*zl; f=zl*xl
        km(1,1)=a; km(4,4)=a; km(2,2)=b; km(5,5)=b; km(3,3)=c; km(6,6)=c
        km(1,2)=d; km(2,1)=d; km(4,5)=d; km(5,4)=d; km(2,3)=e; km(3,2)=e
        km(5,6)=e; km(6,5)=e; km(1,3)=f; km(3,1)=f; km(4,6)=f; km(6,4)=f
        km(1,4)=-a; km(4,1)=-a; km(2,5)=-b; km(5,2)=-b; km(3,6)=-c; km(6,3)=-c
        km(1,5)=-d; km(5,1)=-d; km(2,4)=-d; km(4,2)=-d; km(2,6)=-e; km(6,2)=-e
        km(3,5)=-e; km(5,3)=-e; km(1,6)=-f; km(6,1)=-f; km(3,4)=-f; km(4,3)=-f
        do i=1,6; do j=1,6; km(i,j)=km(i,j)*ea/ell; end do; end do
     end select
   return
   end subroutine pin_jointed
   subroutine glob_to_axial(axial,global,coord)
!        this subroutine transforms the global end reactions
!        into an axial force for rod elements (2-d or 3-d)
   implicit none
   real,intent(in)::global(:),coord(:,:)
   real,intent(out)::axial
   real::add,ell
   integer::ndim,i
   ndim=ubound(coord,2)
   add = 0.
   do i = 1 , ndim
     add = add + (coord(2,i)-coord(1,i))**2
   end do
   ell=sqrt(add)
   axial = 0.
   do i = 1 , ndim
     axial = axial + (coord(2,i)-coord(1,i))/ell*global(ndim+i)
   end do
   return
   end subroutine glob_to_axial

  end module new_library
```

Appendix 6
Solutions to Examples

CHAPTER 4

1. $$[k] = \frac{EA}{3L}\begin{bmatrix} 7 & -8 & 1 \\ & 16 & -8 \\ & & 7 \end{bmatrix}$$

   Less elements would be needed. Better predictions *between* the nodes.

2. $$[m] = 30\begin{bmatrix} 4 & 2 & -1 \\ & 16 & 2 \\ & & 4 \end{bmatrix}$$

3. $$\begin{bmatrix} (8 - P/15) & (-24 + P/10) \\ (-48 + P/5) & (192 - 4.8P) \end{bmatrix} \begin{Bmatrix} U_1 \\ U_2 \end{Bmatrix} = \begin{Bmatrix} 0 \\ 1 \end{Bmatrix}$$

4. 0.188

5. 0.0098, 0.0588

6. 1975

7. 0.0422, 0.0313, 0.0172

8. $u_{\text{tip}} = 0.075$, $\theta_{\text{tip}} = 0.125$, $\theta_{\text{supp}} = 0.025$

9. $u = -0.165$, $v = -0.1347$, $\theta = 0.03$

10. $u_1 = -0.0037$, $u_2 = 0.00095$, $u_3 = -0.001$, $u_4 = 0.00155$

11. $P < \dfrac{320EI}{L^3} + \dfrac{8kL}{3}$ simply supported

    $P < \dfrac{960EI}{L^3} + \dfrac{13kL}{7}$ clamped

12. $\theta = \dfrac{qL^2/12}{2EI/L + 7kL^3/420}$ clamped

13. 0.489

14. (a) 12, exact solution $=\pi^2$
    (b) 2.486, exact solution $=\pi^2/4; 0.638 : 1.000$
    (c) 40.0, exact solution $=4\pi^2$
    (d) 120.0, exact solution $=8.18\pi^2$

CHAPTER 5

4. $\delta_x \approx 0.2,\ \delta_y \approx -0.2$

5. $F_3 = 5L\sigma/24,\ F_4 = 8L\sigma/24,\ F_5 = -L\sigma/24$

6. One possibility : $1, x, y, z, xy, yz, zx, x^2y, xy^2, y^2z, yz^2, z^2x, zx^2, xyz$

7. $P = 80$

8. 0.55, 0.19

9. $\delta_h = 0.0192,\ \delta_v = 0.0087$

10. Using two-point Gaussian integration, $m_{11} = 0.0185, m_{12} = -0.0185$.
    Using three-point Gaussian integration, $m_{11} = 0.0333, m_{12} = -0.0333$.

11. $F_1 = F_3 = 1/12$

$$\mathbf{B} = \begin{bmatrix} -0.5 & 0 & -0.5 & 0 & 0.5 & 0 & 0.5 & 0 \\ 0 & -0.5 & 0 & 0.5 & 0 & 0.5 & 0 & -0.5 \\ -0.5 & -0.5 & 0.5 & -0.5 & 0.5 & 0.5 & -0.5 & 0.5 \\ 0.5 & 0 & 0.5 & 0 & 0.5 & 0 & 0.5 & 0 \end{bmatrix}$$

$\varepsilon_r = 0.01, \varepsilon_z = -0.0388, \varepsilon_\theta = 0.01, \gamma_{rz} = 0.0122$

12. $\delta_v \approx 0.045\,\mathrm{m}$

CHAPTER 7

1. $$\mathbf{k} = \frac{k}{6}\begin{bmatrix} 4 & -1 & -2 & -1 \\ & 4 & -1 & -2 \\ & & 4 & -1 \\ & & & 4 \end{bmatrix}$$

2. $T_A = 37.78, T_B = 10.00, T_C = T_D = 21.11$

3. 75.0

4. $$\mathbf{f} = \begin{bmatrix} (1-y)^2+(1-x)^2 & y(1-y)-(1-x)^2 & -y(1-y)-x(1-x) & -(1-y)^2+x(1-x) \\ & y^2+(1-x)^2 & -y^2+x(1-x) & -y(1-y)-x(1-x) \\ & & y^2+x^2 & y(1-y)-x^2 \\ & & & (1-y)^2+x^2 \end{bmatrix}$$

$k_{11} = \frac{2}{3}k$

5. 68.3

6. 27.27, 81.82, flow rate = 272.7

7. Using 0.5 m square elements throughout, $Q \approx 24\,\text{m}^3/(\text{day m})$

CHAPTER 8

1. Ans: numerical 0.40, analytical 0.41

2. Ans: analytical $u = e^{-21t/2}$

3. $(\mathbf{M} + \Delta t\theta\mathbf{K})\mathbf{U}_{i+1} = (\mathbf{M} - \Delta t(1-\theta)\mathbf{K})\mathbf{U}_i$

   where $\mathbf{K} = \begin{bmatrix} 2 & -2 & 0 \\ & 4 & -2 \\ & & 2 \end{bmatrix}$, $\mathbf{M} = \frac{1}{12}\begin{bmatrix} 2 & 1 & 0 \\ & 4 & 1 \\ & & 2 \end{bmatrix}$ and $\mathbf{U}_0 = \begin{Bmatrix} 100 \\ 0 \\ 0 \end{Bmatrix}$

4. Using consistent 'mass': $U_2 = 20.9$, $U_1 = 35.6$. Using lumped "mass": $U_2 = 20.0$, $U_1 = 38.6$.

5. Using consistent ''mass'', $U = 40\%$.

CHAPTER 10

1. Using eight beam elements of equal length, "exact" solutions in parentheses.
   (a) $\omega_1 = 223\,\text{s}^{-1}(225)$, $\omega_2 = 877\,\text{s}^{-1}(899)$
   Data file:
   8 1 3
   0.4 18.82e-4
   0.1 0.1 0.1 0.1 0.1 0.1 0.1 0.1
   2
   1 0 1 9 0 1

(b) $\omega_1 = 506\,\mathrm{s}^{-1}(510)$, $\omega_2 = 1364\,\mathrm{s}^{-1}(1405)$
Data file:
8 1 3
0.4 18.82e-4
0.1 0.1 0.1 0.1 0.1 0.1 0.1 0.1
2
1 0 0 9 0 0

(c) $\omega_1 = 349\,\mathrm{s}^{-1}(351)$, $\omega_2 = 1107\,\mathrm{s}^{-1}(1138)$
Data file:
8 1 3
0.4 18.82e-4
0.1 0.1 0.1 0.1 0.1 0.1 0.1 0.1
2
1 0 0 9 0 1

(d) $\omega_1 = 79\,\mathrm{s}^{-1}(80)$, $\omega_2 = 480\,\mathrm{s}^{-1}(502)$
Data file:
8 1 3
0.4 18.82e-4
0.1 0.1 0.1 0.1 0.1 0.1 0.1 0.1
1
1 0 0

2. Using a string of eight, eight-node elements of equal size, "exact" solutions in parentheses.

(a) $\omega_1 = 219\,\mathrm{s}^{-1}(225)$, $\omega_2 = 824\,\mathrm{s}^{-1}(899)$
Data file:
8 1 43
0.1 0.1 0.01882 4800. 0. 5
2
2 0 0 42 0 0

(b) $\omega_1 = 474\,\mathrm{s}^{-1}(510)$, $\omega_2 = 1196\,\mathrm{s}^{-1}(1405)$
Data file:
8 1 43
0.1 0.1 0.01882 4800. 0.5
6
1 0 0 2 0 0 3 0 0
41 0 0 42 0 0 43 0 0

(c) $\omega_1 = 335\,\mathrm{s}^{-1}(351)$, $\omega_2 = 1007\,\mathrm{s}^{-1}(1138)$
Data file:
8 1 43
0.1 0.1 0.01882 4800. 0. 5
4
1 0 0 2 0 0 3 0 0 42 0 0

(d) $\omega_1 = 80\,\mathrm{s}^{-1}(80)$, $\omega_2 = 462\,\mathrm{s}^{-1}(502)$
Data file:
8 1 43
0.1 0.1 0.01882 4800. 0. 5
3
1 0 0 2 0 0 3 0 0

3. Using five, four-node elements, $\omega_3 = 2.70 \times 10^4\,\mathrm{s}^{-1}\,(2.70 \times 10^4), \omega_6 = 7.84 \times 10^4\,\mathrm{s}^{-1}(8.10 \times 10^4)$ — "exact" solutions in parentheses.
4. Using five rod elements, $\omega_1 = 2.70 \times 10^4\,\mathrm{s}^{-1}\,(2.70 \times 10^4), \omega_2 = 7.84 \times 10^4\,\mathrm{s}^{-1}(8.10 \times 10^4)$ — "exact" solutions in parentheses.
5. Using four rod elements, $\omega_1 = 24615\,\mathrm{s}^{-1}, \delta_\mathrm{A} : \delta_\mathrm{B} = 0.82 : 1; \omega_2 = 59\,380\,\mathrm{s}^{-1}, \delta_\mathrm{A} : \delta_\mathrm{B} = -0.08 : -1.$

   Data file:
   4 2 2
   1.355e5 5.050e-3
   6.783e4 2.525e-3
   1 1 2 2
   0.127 0.127 0.0635 0.0635
   1
   1 0

6. Lumped : $\dfrac{2}{L}\sqrt{\dfrac{E}{\rho}}$

   Consistent : $\dfrac{2\sqrt{3}}{L}\sqrt{\dfrac{E}{\rho}}$

CHAPTER 11

1. $\theta_1(2) = 0.284, \theta_2(2) = 0.036$
2. If $fm = 0$, then $\xi_1 = \dfrac{\omega_1 fk}{2}$.

   $\xi = 0.05$ and $\omega_1 = 0.48$, hence $f\,k = 0.208$.

   $\theta_1(2) = 0.254, \theta_2(2) = 0.015$
3. $u(2) = 0.077$
4. $u(2) = 0.092, \theta(2) = 0.653$
5. Using $\beta = 1/4$ and $\gamma = 1/2$, together with a time step of 0.01 sec, the maximum deflection is computed as 1.728 after 0.76 sec.

Author Index

Italic page numbers indicate references in full

Subject Index